A DICTIONARY OF
SOUTH AFRICAN ENGLISH

A Dictionary of South African English

New Enlarged Edition

JEAN BRANFORD

1980

Oxford University Press

Cape Town

Oxford University Press

OXFORD LONDON GLASGOW
NEW YORK TORONTO MELBOURNE WELLINGTON
NAIROBI DAR ES SALAAM CAPE TOWN
KUALA LUMPUR SINGAPORE HONG KONG TOKYO
DELHI BOMBAY CALCUTTA MADRAS KARACHI

ISBN 0 19 570177 1

Set in 10pt on 10pt and 8pt on 8pt Times Roman

Printed in South Africa by Citadel Press, Lansdowne, Cape.
Published by Oxford University Press, Harrington House, Barrack Street,
Cape Town 8001, South Africa

Contents

Preface

This is not a conventional dictionary. It deals with an unconventional part of the English vocabulary, namely that peculiar to or originating in South Africa, and it treats that part in a manner which is not strictly orthodox. It has been designed with that dual purpose which Horace ascribes to poets in his *Ars Poetica:*

> *Aut prodesse . . . aut delectare*
> 'Either to instruct or to delight.'

In this case it is my earnest hope that it will do both. Like all dictionaries, it is intended to be useful; but unlike many, it is intended also to give pleasure and amusement.

In preparing the dictionary I have had the good fortune to be treating some of the most expressive language possible, and have had access to what must surely be a uniquely varied body of source material. From this have been chosen for their utility, validity, or pure pleasure, the quotations which are the spirit of this text. They have been chosen with love, care, and also laughter; some of them with sorrow and dismay. On the body of the text too, love, care and effort have been expended over a number of years, but like any other work of man, it is far from perfect.

In this respect I feel I cannot do better than quote my remarkable predecessor, Charles Pettman, who wrote in his *Africanderisms* in 1913:

> In all the author has aimed at accuracy; he would be
> foolish, however, to suppose that there are no mistakes,
> but trusts that they will not be so many as to detract
> from the usefulness of the book.

J.B.

PREFACE TO THE SECOND EDITION

On June 27 1842 – a hundred and thirty seven years ago – John Appleyard, the pioneer of the Xhosa Bible, wrote of his grammar of the Xhosa language. 'I intend to revise and improve it, if spared, next year when I hope to have a little more experience.' It has fallen to me both to be spared and to be given the opportunity to revise this book, just over a year since its first publication. Like Appleyard, I hope too that I have a little more experience, and am grateful indeed for this opportunity to try to improve it.

Our language, growing and changing around us daily, brings the experience of words and their behaviour often too thick and fast to process or assimilate adequately in the mind, let alone on paper.

Dr Johnson himself, a founding father of English lexicography, realized that language is not static but dynamic, and expressed in his *Preface to a Dictionary* (1755) the impossibility of trying to 'fix' a living language '. . . No dictionary of a living tongue can ever be perfect, since while it is hastening to publication, some words are budding and some falling away . . .' So I trust I shall be forgiven if some of the dead heads remain and if some of the newly-budded blooms have failed to appear in this text. I trust, too, that I may be forgiven if some of those which do appear are not exclusively South African. It becomes more and more evident that many items which seem to us as South African as we are ourselves nevertheless occur in the English of other parts of the world – even to the 'Are you coming with?' or 'John-and-them' of certain American dialects.

Thus in R. K. Tongue's little book on 'E.S.M.' (*The English of Singapore and Malaysia*) which was given me last year in Singapore, I find that there the regular short cut for 'are they?', 'did he?' or 'really is that so?' is *is it?* E.S.M. gave me several more surprises of the same kind, and though the frequency and the origin of some such phrases is very likely *here* to be due to the influence of Afrikaans, it is evident that quite often English is simplified, both here and in other parts of the world far removed from our own, by rather similar processes.

In many cases of informal usage it is extremely difficult to establish whether a given item is 'South African' or not. The standard dictionaries may or may not help. Thus *tiffy, tiffie* (artificer), common in South African service usage, is traced back in Partridge's *A Dictionary of Slang and Unconventional English* to 'nautical use' in the 1890's. It does not however, have the range of variants with which ingenious South African Servicemen have invested it, from the *soul tiffy* who officiates at church parades to the *pot-tiffy* in the mess. *Zol*, similarly, figures in Partridge, backdated to 1946 and identified as South African, though there are indications that it may relate to the argot of the drug traffic on the Mexican-USA border, (*American Speech*, May 1955). But for many other quite common items we find no documentation whatever. Thus for *longdrop* (pit privy) which I think may be South African, a search in eighteen different reference texts brought no information at all except the established eighteenth century cant usage 'the drop' for the gallows.

Another typically South African expression as it seems to me – 'to go farming', as others go hunting, shooting, fishing or swimming – I am similarly unable to trace. No dictionary I have available has this form for embracing a profession as opposed to the temporary pursuit of an amusement. I have also heard 'go nursing' and 'go teaching' and although a recent *Farmer's Weekly* remarks 'The dream of many city people is to go farming on their own piece of ground after retirement,' I am restrained, perhaps wrongly, by its very ordinariness, from including it.

In the same way the expression 'hold thumbs' suggested to me by Mr Sydney Kentridge S.C., regularly used by children, students and adults and even found in our press, has defied research, though it does exist with a diminutive, in the singular, in German.

My objective is still simply that of attempting a coverage of some of the many loan words and other terms characteristic of English in South Africa, which one can with reasonable accuracy call 'South Africanisms'. This text does not purport to cover any of the Standard English common to all English-speech communities whether of the Old World or the New. There are, however, terms which we have in common with some other English-speech communities but which are, I think, 'South African' enough to merit inclusion. Among these are some irresistible bits of Yiddish, like *chutzpah* and *schlep* and some military terms such as *chopper. tiffy* and *webbing*. There are too, some improbable-seeming terms like *banana-republic, coals* and *Christmas box* which seem to be acquiring meanings peculiar to themselves in the South African context. About the inclusion of all these I must again fall back on the precedent of that better workman in this field, the Reverend Charles Pettman, who wrote,

> It has been difficult sometimes to decide what to admit to the Glossary and what to exclude. A few words have been included that could not be termed 'Africanderisms' but no word has been admitted that had not some special interest for South Africans.

Apart from these problem terms of the dialectologist's 'half-world', new, indisputably South African items are emerging, or older ones coming to the surface. Some old words are being given new meanings. A new *meercat* is being driven about for underground work in the mines, while the *ratel*, the *eland* and the *buffel* have exchanged legs for wheels and are being put to work on the Border, and a *sprinkaan* with them, by our Defence Force. Even at the time of writing I find there is another *kudu*, described as a 'mine protected vehicle' in use in Rhodesia. Conversely, some remarkably modern-looking words are coming to light in surprisingly early settings. Sir James Alexander, whose *Expedition to the Interior* was published in 1838, remarked with disfavour upon the habits of 'rookers' of dagga, but was himself told to 'Loop!' by an angry old woman on a remote farm. It is to this amazing man, who was responsible for the placing of Cleopatra's Needle on the Thames Embankment, that I owe my earliest-traced instance of *voetsek* (*voortsek*) (1837) which appeared in the first edition.

It is self-evident that the function of a second edition is to provide material which will bring or try to bring that of the first up to date. Our language, however, moves too fast for there to be any hope of making good so comprehensive an intention. One can only hope to present some main areas of growth, to expand on others which were perhaps skimped formerly and, in the case of certain entries, to provide illustrative quotations which were lacking before. This new edition contains roughly 450 complete new entries and a further 300 improved upon or otherwise expanded, where new material or new meanings have come to light.

One major area of vocabulary growth in South African English is the language of National Servicemen. Some of this is reflected in the first edition but more and more of it is now moving into focus for the man or woman in the street – father, mother, wife or girl-friend of someone in the Forces. Our situation is now in some ways parallel to that of the British stay-at-homes during the Anglo-Boer War in 'the parlours of Brixton and the pubs of Highgate' described by Stuart Cloete. Suddenly the language of conflict in Africa has expanded into a language of explicit warfare and military experience. It is therefore with deliberate intent and at the Publisher's request that I have included a number of items that I have recently been able to collect from National Servicemen (NSM). As my informants in this field have hastened to assure me, many if not most, of the terms are unprintable, so those which appear here are perforce a selection. It is a selection however from a fast-growing area of our language,of which Brigadier J. H. Picard wrote in 1975 'One must bear in mind that military English in South Africa is not merely a 'taal' for 'takhare' or 'backvelders', neither is it a language for 'lang hare and dik brille' or 'Indoenas'. It is a practical means of communication, spontaneously accepted and used by thousands of National Servicemen. Judiciously selected, some of these 'South Africanisms' would certainly enrich South African English.' Like much of the usage of the very young, some of this language has a freshness of its own. The judicious selection will, however, be made not by me but by time. Only time will show which of these many lively terms will last long enough really to earn a place in this story.

As I pointed out in the Introduction to the first edition (p xii paragraph 6) it is probably more usual to consult a dictionary for the rare or unknown word rather than for the familiar. In 1747 in his *Plan of a Dictionary* Dr Johnson wrote of his reader encountering Milton's '. . . pining atrophy, *Marasmus*, and wide-wafting pestilence'.

> He will with equal expectation look into his dictionary for the word *marasmus*, as for *atrophy*, or *pestilence* and will have reason to complain if he does not find it.

Dr Johnson was writing of an ideal situation and an ideal dictionary which few if any could hope to achieve. This new edition, however, does accordingly contain some words which are far from the beaten track of the verbal 'bundu basher' but they are, or will be, I think and hope, of interest to others as they have been to me.

Among these are offshoots of two particular African streams, one ancient and one modern, both represented in the first text but which I have attempted to follow a little further here.

The modern stream is that of the jargon of the townships – the *staffriders*, the *spoilers*, the *tsotsis*, *kings*, *queens* and *pennylines*, some of whose colourful vocabulary is now well established in print in black literary journals and newspapers.

The ancient stream is one which was cluttered with enough conflicting traditions to occupy, if not bewilder, such a doughty Christian and anthropologist as Bishop Callaway, with years of research. This is that of the origin of man and the naming of the Deity. From the days of the earliest missionaries these have been a matter of conflict even among experts. Today when members of established Churches are pressing for the incorporation of traditional African rites and terminology into their church practices, this has again become a matter for discussion among clergy and laity alike; and a small collection of terms has joined the *Unkulunkulu* and *Tixo* of the first edition. These include *Qamata*, *Uhlanga* and *Umvelinqangi*, also *uvuko*, the ancient name for the 'reawakening' equated by some modern African Christians with the resurrection of the dead – those *abadala* 'old ones', the ancestors, or the *abaphansi*, those who are in the earth below.

In the enormous field of vocabulary covering African customs and beliefs those represented here are few indeed, and I am aware of it. I am also aware that over-enthusiasm for this fascinating material would result in the incorporation of a plethora of anthropological terms remote from daily experience, and upon which I am not qualified to expound.

Having no wish to be a 'bootlegger' or even a *bandbreker* I have not trespassed upon the fascinating ground of the South African element in the 'Clingo' of Citizen Band Radio. This has been handled elsewhere and many of the expressions well established in our usage appear here without CB labels and in some cases without their special CB meanings e.g. *spook* (a policeman), *teff* (food), *kneehalter* (husband) or *touleier* (initiator of a three way conversation). Others with the same or similar meanings include some 'straight' terms like *maat*, *goodies* or *ou*, some items of *fanakalo* (q.v.) like *buka*, *kaya*, *madala*, *maningi*, *ayikona*, *kahle*, *phumula* and *lala*, and some which are what one might call general 'words for the road' *tackies* (tyres), *gooi ankers* (brake), *draai* (turn) and *wapad* (highway). There are also the expressive Yiddish and Hebrew loan words *chutzpah*, *kugel*, *gabba* and *gevalt*. Many more will, I am sure, find their way into what has always been a hospitable form of English, and the language of 'CBoks' is therefore likely to prove another area of growth to watch.

While it has, in the limited time available, been possible to expand and augment this work, time and the exigencies of typesetting have not permitted the cutting and pruning which in many cases might have been thought desirable. This must therefore be regarded as a project for the future.

I would like to express my gratitude to my colleagues on the Dictionary Project of the Institute for the Study of English in Africa: to Mrs Margaret Britz for her daily support in more ways than I can enumerate and to Mrs Sybil Kopke who cheerfully undertook this most difficult typing assignment; to the Hon. Mr Justice J. P. G. Eksteen and to Mr Sydney Kentridge S.C.; also to colleagues at Rhodes University, Mr Andrew Lang, Mr Isadore Pinchuck, Miss Leonie Prozesky and Mr Andrew Tracey; my neighbours at St Paul's College – Mr Sabelo Sillie, Mr Klaus Kühne and Mr Thabo Letlala who took time off from their theological studies to help me; Mrs Estelle Brink of the Albany Museum herbarium; Miss Hilda Grobler of Natal University (Durban), Mr. Angus Rose of Pietermaritzburg, Miss Louise Sloman at Oxford University Press; Mr H. C. Davis of Knysna, Mr Joe Ellmore of the Grahamstown MOTH Shell-hole, Mrs Mimi Chan of the University of Hong Kong for some more instances of 'Chinglish', and Professor Charles Kreidler of Georgetown University, Washington D.C. for most valuable suggestions; those many reviewers whose constructive criticisms of my omissions I have found most helpful; and to Brian Mullins in particular, John Beachyhead, Stewart MacSimon, Martin Britz, my son Justin, and the many other young servicemen who have answered with endless patience and good nature the questions of what must have seemed a most eccentric old 'toppie'; finally to Pen, my daughter, who has handled the editing of this project, and William Branford, my husband, both of whom have handled me and it with more patience than I deserve.

Jean Branford
Pieterkoen
George
Cape Province
June 1979

Acknowledgements

I wish first to express my thanks to the Dictionary Committee and the Board of the Institute for the Study of English in Africa of Rhodes University for making it possible for me to undertake this work. My gratitude is due also to the Human Sciences Research Council, without whose financial support for the Rhodes University research programme for a Dictionary of South African English on Historical Principles from its inception in 1970 this present work could not have been contemplated; and also to the Department of National Education whose subvention of the Research Unit since 1975 has made its completion possible.

I am most deeply indebted to Professor Johan Smuts, my supervisor of studies, and to Professor André de Villiers, Director of the Institute, both of whom have given generously of their time and unstintingly of their specialized knowledge far beyond my deserts.

To my colleagues on the Dictionary staff, Margaret Britz, Assistant Research Officer, who prepared the appendix of first dates and whose endless patience and good humour have always supported me, and Dorothy Muggeridge, who has twice typed the entire manuscript, my love and thanks are due in no small measure

My special thanks are due to Mr C. J. Skead for the many hours of guidance he has given me in the complexities of scientific names of birds and beasts; and to the instructors of my younger days, Professor W. S. Mackie and Professor Dorothy Cavers. I am indebted too, to my former colleagues Penelope M. Silva and the late J. D. Walker upon some of whose wide reading I have been fortunate enough to draw.

My debts to friends known and unknown are more numerous than I can well compute but I would like to record my gratitude to many colleagues at Rhodes University: Dr M. V. Aldridge, Mr Michael Berning, Professor R. Beuthin, Professor André P. Brink, Professor Guy Butler, Mr J. S. Claughton, Miss A. C. Dick, Dr D. S. Henderson, Dr A. P. Hendrikse, Mr Peter Jackson, Dr Amy Jacot-Guillarmod, Mrs Margaret Smith, Mr C. Z. Gebeda now of Fort Hare and to Professor L. W. Lanham of Witwatersrand University;

To fellow Africana enthusiasts: Mrs Margaret Rainier, Dr and Mrs J. V. L. Rennie, Professor Gordon Richings and Mrs Rita Snyman;

To many fellow citizens of the Eastern Cape, including the Very Rev. Godfrey Ashby, Dr Bertram Brayshaw, Mrs Estelle Brink, Mr and Mrs Francis Bowker, Mr Brian Hobson, the Hon. Mr Justice D. D. V. Kannemeyer, Mr Norman Mka, Mr C. G. Mullins, Mrs Aurelia Nyathikazi, Mr and Mrs Geoff Palmer and Advocate Geo. Randell, and especially to my former neighbour Sue Allanson;

To friends from elsewhere in the Republic, Miss Yvonne Cloete and Mr J. S. B. Marais of Pretoria; Mr Patrick Kohler and Mrs Audrey Ryan of the S.A.B.C.; Mr Philip Birkinshaw, Dr Patricia McMagh, Mr Eric Rosenthal, Dr Pieter and Madeleine van Biljon and Gerda Pretorius of Cape Town; Mr Marius Le Roux of Stellenbosch; to John and Brenda Hartdegen of George and Hjalmar Thesen of Knysna; to Mrs Zuleika Mayat and Mrs Bibi Daoud Mall of Durban, to Mr R. D. Guy of Empangeni, Zululand and Terry Shean of the *Sunday Times* for photographs;

Also to those from further afield: Mr Robert Burchfield, editor of the Oxford English Dictionaries, Dr Alice Heim of the Psychological Laboratory, Cambridge; Mrs Emilia Andreyeva of Moscow; Mr and Mrs Hugh Lewis of Singapore; Mr and Mrs Dick Welch of Sydney, and Professor Y. Matsumura of the University of Himeji, Japan.

From all these, and many more, material, help and kindness have come to me.

I should like too, to pay tribute to other and better workmen, my predecessors in this field; the great and original spirit of Charles Pettman, and to Dr C. P. Swart and Dr M. D. W. Jeffreys who continued the work which he began.

To my husband William Branford, for more than I can adequately express here, both in his personal capacity and as designer of the Rhodes Dictionary project.

Finally, for his many gifts and his example of meticulous scholarship, to my father, Alfred Gordon-Brown of Cape Town, to whom I dedicate this work.

JEAN BRANFORD
Grahamstown

Note:
'South African English'

South African English – the English of South Africans of whatever race, colour or national group – is in every sense, culturally, lexically, grammatically and phonologically, a 'mixed bag'. It is not, as I see it, only that complex of forms spoken by what are sometimes called 'ESSAS' (English Speaking South Africans), White or Black, but a *lingua franca* among those to whom English is, and many to whom English is not, their mother tongue. It teems therefore with words, ideas, structures and concepts from many languages and many cultures, and if my sample collection of these serves to illustrate this fact, it will fulfil the function which I have envisaged for it.

Introduction

'To every man the domain of "common words" widens out in the direction of his own reading, research, business, provincial or foreign residence, and contracts in the direction with which he has no practical connection: no man's English is *all* English' – Sir James Murray, *Introduction* to *A New English Dictionary on Historical Principles* (1888).

1. 'South Africanisms' in our English

Sir James Murray's statement, made many years ago, that 'no man's English is *all* English', is particularly true for South Africans. As members of a multilingual society, we freely borrow from the languages around us, from Afrikaans to Zulu. Some of these borrowings are ephemeral; others, like *sopie*, have obtained a permanent footing in South African English but are unlikely to pass beyond it. Still others, like *trek* and *apartheid*, have become part of 'world English' and are likely to remain so.

Early in this century, as Stuart Cloete pointed out in *Rags of Glory* (1963), the Anglo-Boer War 'brought the African veld into the parlours of Brixton and the pubs of Highgate' with words like *donga*, *dorp*, *drift* and *kop* in news reports and soldiers' letters home. More recently the vocabulary of a different and yet deeper conflict – *apartheid*, *verkramp*, *amandhla*, *hippo*, *troopie* and *terr* – has begun to figure increasingly in the international press.

The South African origins or provenance of words like these are probably consciously felt by most overseas readers. In other cases, a word may have lost, for the user elsewhere, its specifically South African flavour. The South African origin of *trek* must have long been forgotten by big city commuters, by those who re-enacted the westward *trek* of pioneer wagons in the bi-centennial year of the U.S.A., and by British holiday makers *pony-trekking* in the hill country or, more adventurously, *Mini-trekking* abroad. The executive who *commandeers* a company car – (or who even decides to *Stellenbosch* an unsatisfactory colleague) may well be unaware of the fact that he is using a word of South African origin; and many of the specialized *commando* shock troops of World War II may not have known, or wanted to know, that their name echoes back to that of the militia of the early Cape frontiersmen.

But while few would dispute that *trek*, *commando* and *apartheid* are now well established in world English, there are many South African purists who react with horror to the notion of designating as 'English' – or even as 'South African English' – some of our more casual and informal loan-words such as *lekker*, *bakkie* or *braai*, even though many of these are terms which South Africans of all ages find an integral and almost indispensable part of their everyday usage. One of several reasons for preferring the term *South Africanisms* is to smooth the hackles or allay the alarms of the purists.

A further reason is that many English words – for instance, *land*, *location* and *camp* – have meanings in South Africa which differ from their usual established significations in British English. South Africa is, of course, not unique in this. The phenomenon is even more frequent in the United States (consider *gas* meaning 'petrol' and *depot* meaning 'railway station'). Indeed, during World War II, an illustrated word list was compiled by *Life* magazine to explain to American G.I.s what *biscuit*, *cookie*, *vest*, *bum* and other such items meant in British English.

A further group – including the useful *dingus*, which is the property of much of the English-speaking world – consists of terms which we share with some English-speaking communities but not with all. We share, for instance, with Australia the ubiquitous fencing *dropper*, or the *fossicker* of the past. Some of these are marked in the text as 'non-SAE', though this is simply a marker of convenience. South Africanisms they are, but they are 'other English' though not perhaps 'world English' as well.

I have also included a few words which might well be regarded as not South African at all – notably *rust, rinderpest* and *'flu* – all of which have local significance as historical rather than as clinical manifestations in South Africa. *Redwater*, similarly, has, in South Africa, a rather wider than usual application. With these are *Almanac, the Company, Court Calendar, District, Division, party* and *settler*, all of which have special signification or overtones surviving from the earlier years of Colonial administration at the Cape.

Lastly, these South Africanisms include, as I interpret them, a number of English phrases or usages, whose frequency in the English of South Africans – as opposed to their actual existence in English itself – can be attributed in part to the influence of Afrikaans, or possibly sometimes to that of other languages of this country. Cases in point are: *in place of* for 'instead of'; *just now* for 'in a little while'; *come there* for 'arrive'; *wait on* for 'wait for'; *come right* for 'resolve itself'; *rather* for 'instead'; *doesn't want to* for 'won't'; *busy (with)* for 'engaged in', or 'in the process of'; *is it?* for 'really?', and the prefix *there-* as a substitute for the pronoun 'it'.

2. The User

This text has been designed basically with three types of reader or user in mind. Firstly it has been written for South Africans of all racial groups, including the English speaker interested in the finer detail of the dialect with which he has grown up and the Afrikaans speaker with an interest in how much and in what way his language has permeated that of his English-speaking fellow-countryman. Secondly, it is for the 'stranger within our gates', tourist or immigrant, who may need a guide to the many unfamiliar terms which he will encounter in this country. These he will meet in shops, in the Press, in advertisements, in daily conversation, place names and even in menus, finding a new vocabulary spanning almost every aspect of experience. I hope that for him it will provide an interesting and helpful key to what has developed from the original polyglot 'Tavern of the Seas'. Thirdly, it has been planned as a handbook for the overseas student of South African literature, in the hope that in the copious illustrations from South African texts of all kinds, he will find background material to supplement those which he happens to be studying.

3. The Material

> 'It has been difficult sometimes to decide what to admit to the Glossary and what to exclude. A few words have been included that could not be termed "Africanderisms", but no word has been admitted that had not some special interest for South Africans.' Pettman, *Africanderisms* (1913)

Makers of dictionaries generally agree that their job is to provide an objective record of the actual practices of writers and speakers, rather than a reflection of personal prejudice or taste. But however objective the lexicographer may wish to be, the decision to include or exclude an item is finally his personal choice and responsibility. This is never easy. Even Dr Johnson in his *Plan of a Dictionary* (1747) remarked 'It was not easy to determine by what rule of distinction the words of this dictionary were to be chosen.' For the dialect lexicographer, who is in any case dealing with only a part of the English vocabulary, the choice is already, to a certain extent, made in that what he treats is that section of English in South Africa which might be called the *non-standard*. If it were standard there would be no need for such a text as this, as everything in it would appear in ordinary English dictionaries. Thus the standard vocabulary, unless it be adapted for some special purpose other than those it serves in 'world English', has no place here. Since lexicography is a continuing task, I shall at all times be grateful for material which might be included in any future revision of this text.

The items included were collected over a number of years and initially grouped into thirty-four categories chosen to cover the principal fields of South African English experience. These range from (1) *Address, modes of*, to (34) *Writing, Education and the Arts*. In between were such categories as *Dishes and Cookery; Church and State; Farming and Domestic Animals; Games, Dances and Diversions; Health, Moods, Medicine and Witchcraft; African World; Landscape and Places; Drinking and Smoking*, and so on. Under each of these headings items were assembled in a small notebook, and afterwards each on a card in one of thirty-four bundles. It was only when the original collection was more or less complete that the categories were broken up, and the whole alphabetized so that the drafting of entries could be done in order. Alphabetization was thus secondary to categorization in the design of the text as a whole.

The words that found their way into the categorized scheme came from many places and in many ways. There are two primary ways of observing dialect usage: firstly, listening to people speak, and secondly, reading. Very often usages first encountered in speech are reinforced by being afterwards found in print, in newspapers, books and magazines. While the second method is usually the more reliable, the first is very often more interesting. Items were noted down from the speech of people in many walks of life and over a wide range of situations, including shopping, farming, travelling and telephone calls. Material acquired from reading is of varying reliability as a reflex of the dialect of the country. This is because it ranges from instances of a highly conscious literary device adopted for imparting a specifically local atmosphere to that of the genuine bread-and-butter usage of newspaper reports and advertisements. It would, however, be unfair to call the literary use of loan words an artificial device. There are many instances where a loan word must be used, not for local colour or the creation of atmosphere, but for the simple reason that there is no English equivalent for it, or that the thing it designates does not exist elsewhere.

It is possible that a distinction should be drawn between literary and non-literary sources: between South African stories, plays, novels and poetry on the one hand, and more utilitarian newspapers, magazines, recipe books and histories on the other. This, however, would not accommodate two of the earliest and most valuable groups of source materials available to us. One of these is the array of narratives by early traveller-naturalists such as Burchell and (in translation) Thunberg, Kolb, Sparrman, Le Vaillant, Lichtenstein and Latrobe. The second consists of people with a somewhat longer commitment to South Africa, notably travellers such as Alexander and Thompson and the missionaries, settlers, administrators, military men and their wives. For all of these, the need to describe new experiences, new peoples, unfamiliar landscapes, food, flora and fauna forced them to use a new vocabulary, without which their accounts would not have been valid, nor probably as satisfying to the taste of nineteenth-century readers with an interest, half-fascinated, half-fearful, in Darkest Africa.

These early loan-words reflect, of course, an important general property of English, its assimilative capacity. This was demonstrated early in the general history of the language by extensive borrowings from Latin, French and later Greek, and by less extensive but still important assimilations of Scandinavian material. The English of South Africa has long been a receptacle for what is most vivid, viable and apposite from many tongues: notably of course from Dutch, its successor Afrikaans, and from the Bantu languages, but also from others ranging from the Far East to the heart of Europe. The etymologies, for which I must again stress my debt to Professor J. Smuts, reflect something of the 'banquet of languages', to borrow Otto Jespersen's phrase, from which South African English in its present manifestations has grown.

One special interest regularly shown by early writers was in the place names of South Africa, upon which they made frequent comment. Included in this text are a number of 'place-name formatives', e.g. *-berg, -bosch, -kloof* and *-pan* which regularly appear in place names (*Stellenbosch, Du Toitspan*) as prefixes or suffixes, as well as in many cases as independent words. Among these are features of the landscape, names of birds, beasts and plants, adjectives, verbs, participles, and occasional abstract nouns. While particular place names like *Villiersdorp* have no place in a dictionary, this collection of place name 'pieces' or building bricks, has been designed for the visitor or stranger who is unfamiliar with what is commonplace to most South Africans, to enable him to construct the meanings of many of the place names encountered in his travels or reading. It is also interesting in its own right as a regular source of new and recent coinages such as *Verwoerdburg*.

A full list of the word sources quoted appears on page 300. This does not include dictionaries and other reference or linguistic texts consulted.

Four words of particular complexity – *boer, kaffir, kraal* and *veld* with their compounds – have been divided up under headings according to their separate and different meanings or uses in South Africa. A note explaining the presentation of each of these four items precedes it.

In attempting a reasonably broad coverage of 'South Africanisms' I have included some grammatical material to reflect, for instance, such well-established colloquial usages as 'Are you coming with?' The text, as a result, contains some unusual headings: some of these are only parts of words such as *-ed* or *-ie*, and others include *omissions; prepositions, adjective with infinitive, negative, uses of, third person form of address* and *redundancies.* These give clues to certain grammatical idiosyncracies in South African English usage, particularly the short cuts taken, found at *omissions*, the extraneous matter, listed as *redundancies*, and South African uses of prepositions. All prepositions discussed are given as head-words in the text, but under *prepositions* they are simply listed together. Many of the unfamiliar usages of dialect grammar are direct translations or transliterations, usually from Afrikaans.

It should be borne in mind that all of the three thousand-odd items included are not likely to occur in the dialect of any single South African. The vocabulary reflects the usage of persons in many differing walks of life, differing backgrounds and widely differing working or recreational milieux. Thus to the farmer with a whole semi-specialist terminology of his own, that of the miner may be almost, or totally, unknown. Someone whose contacts with Africans have been minimal, or who has never had the opportunity or inclination to read the black press, may well be sceptical of the African-language borrowings or otherwise Africanized English terms included here. Nevertheless it is, I hope, a representative collection of some of the forms, features, adaptations and loan words characteristic of English in South Africa.

4. *Quotations*

The illustrative quotations, as indicated in the Preface, have been chosen with the two purposes of making the text both useful and readable. If a quotation appears overlong, this is frequently because it contains and illustrates items other than the one below which it appears.

In most quotations of recent date, the names of persons and institutions have been removed, particularly where their inclusion might harm or give offence, as in murder or other criminal cases. But names have in certain instances been retained where the identity of the speaker or person mentioned has been of importance to the sense, for example in the case of an item of specifically African English usage to show that the speaker quoted is indeed an African.

5. *Cross-references*

The cross-references among definitions are basically of two kinds. The first directs the user to equivalent items, e.g. '*miltsiekte* see *gifsiekte*', '*nagmaalhuis* see *kerkhuis*' (and vice versa) or to related items: '*rixdollar* see *schelling, skilling, stuiver*'. This type of cross-reference is also made by the sign (q.v.), e.g. '*boeremusiek*: rhythmic country-style dance music played by a *boereorkes* (q.v.)'.

The second kind refers the user to a quotation under another head-word in which the word may be seen in a different context, or even several, e.g. '*alles sal reg kom*: see also quots. at *toe maar* and *moenie worry nie, moenie panic nie,*'; or '*roman*: see also quot. at *allewêreld*' (where there is a description of a large varied haul of fish).

This method of cross-referencing ensures that attention is drawn to other and possibly useful material on the one hand, and on the other that the word in question can be seen in as many different illustrative quotations as the text can afford.

There is one further type of reference, not strictly a cross-reference, which will be found at certain entries. These references are to items or usages from other variants of English comparable in form or idea with the South African terms e.g. '*randlord* cf. Anglo-Indian *nabob*, Hong Kong *tai-pan*'. While this is not part of conventional lexicography, I think it is of great interest to match certain recurrent themes in English vocabularies across the world. Thus *brak* is paralleled by Anglo-Indian *pye-dog* and Australian *mong; sugar baron* by Canadian *sawdust nobility, lumber king* and British *merchant prince; verkrampte* by Australian *wowser,* Canadian *mossback; trek wagon* by United States *prairie schooner; Anglikaans* by Canadian *Franglish, Franglais*. While these cross-references are not as many as I could wish, they give a fair number of instances of striking parallels of experience for English speech communities far apart in space and time.

6. Limitations of Choice

There have been several fields in which the inclusion of items has been consciously restricted. Dialect dictionaries tend, I feel, to over-invest in three particular areas: flora, fauna and the more ephemeral colloquialisms. However, the categorization system has made it possible to control the proportion of items of different kinds to be included here.

It is not, I think, justifiable to attempt a detailed treatment of flora and fauna in a dictionary, since it is not designed as a biological glossary. Too many of these infrequently-used terms would result in an unwieldy volume, and one of dubious value, since standard textbooks deal with them with the competence and detail required. I have therefore attempted to include only those which are likely to be fairly often encountered. The longer *Dictionary of South African English on Historical Principles* will naturally include far more material of this sort. Similarly, the names of peoples and tribes, of which hundreds and perhaps thousands can be found in our source materials, have been kept relatively few.

Colloquialisms inevitably form a large and significant part of the usage of any dialect speech community, large or small. It has been difficult, in the case of some colourful but possibly ephemeral terms, to resolve to exclude them. Unlike Australia, South Africa has as yet no great exponent of the colloquial idiom like C. J. Dennis, whose work of 1914–18 in particular continues to delight Australians today, as recent new editions prove. Nor have we a modern text on South African English to match the range and scholarship of Sidney J. Baker's *The Australian Language*. At a more popular and anecdotal level, John O'Grady's *Aussie English* and *They're a Weird Mob* have run to nearly a million and a half copies. It is evident however from the success of Robin Malan's *Ah Big Yaws* (closely related to Afferbeck Lauder's *Let Stalk Strine*), and the following enjoyed by 'Blossom Broadbeam' (Jenny Hobbs) of *Darling*, that there is a widespread interest in colloquial South African English. It is from this latter witty satirist that many illustrative quotations for items common in speech, but seldom to be seen in print, have been taken. A few items of the highly sectional argot of motor cyclists have been illustrated from *Bike S.A.*

A further problem of inclusion is that of rare or obsolete items. Some terms here may be thought to be too esoteric for a general handbook of this sort; but it remains true that a dictionary is more liable to be consulted by a user wishing for information on the unusual word than on the everyday one. To the South African user in particular, it is not the items of everyday vocabulary that are likely to be of interest, but the rarer and unfamiliar ones. It is therefore without apology that some rare historical and legal terms are included, some African journalese, and odd usages heard among farmers. But rare birds, beasts and flowers, for reasons outlined earlier, have been generally excluded.

7. Status labels

Some guidance as to levels or fields of use is offered by means of status-labels.

As is suggested elsewhere, any usage peculiar to a dialect can be regarded as *non-standard* in relation to the internationally-recognized literary norm. For this reason the label *non-standard* is not used at all. The status label *substandard* does appear, but it is restricted to points of usage rather than of vocabulary. Thus it marks the use of *by* for standard *at* ('I was working by a chemist's down in Jeppe') and other structures translated, or transliterated, from Afrikaans, e.g. 'I'm busy having new sunglasses made'; or redundancies such as *little* in 'a small little tin' ('n klein blik*kie*). It is seldom used for lexical items, particularly nouns, as a criticism of their use or content, except where the term itself is a transliteration, e.g. *youth, the* (Afrikaans *die jeug*).

Much dialect usage is naturally informal; but for the lexicographer this poses a very real problem. Few would dare to attempt to lay down a hard and fast line through that hazy territory in which the colloquial shades off into slang. The words themselves, even, may be one or the other, depending on the nature of the context in which they are found. There has therefore been no attempt at a classification based on any rule of thumb. The label *colloq.* has been used on the whole for words or usages informal in varying degrees in speech *or* writing, and the label *slang* for terms so very informal that they are seldom found in print. No scheme could possibly hope to fit this shifting and unstable ground, and what is given here can only hope to be a guide. Where terms are genuinely obscene and objectionable this is stated in the definition itself. A label such as *taboo* 'not used', would be unrealistic for a term which *is*, and has therefore been avoided.

Furthermore, in a multiracial and multilingual country, there are many words which, while they are completely harmless in the view of a large number of people, will and do give hurt and offence to others. These include 'boy', 'girl', 'coon', 'coolie', 'hotnot', 'spider', 'rock' and many more. For these no suitable status label could be devised, so the definition usually includes the term 'offensive', or 'objectionable', as a description of how such words are regarded by many members of a mixed population. The inclusion of such items in this text is purely a record of their use and existence in the language, and should not be interpreted as a mark of approval or signal of their acceptability: nor should the presence of certain of the illustrative material be regarded as endorsement of the sentiments it contains.

The labels *rare* or *regional* are given on a basis of personal experience or opinion only. Certain terms such as *mali*, stated by Pettman to have been 'common among Colonists', are now possibly better known or more used among certain groups, possibly the older generation or people from the Eastern Cape. It is often observed that the usage of one age group is obscure to another and while it is obvious that the use of colloquial or other dialect terms must depend upon individual and the generation or area to which he belongs, this is not something which can be readily pinpointed or labelled by the lexicographer.

8. Spelling: Orthographic conventions

There are four aspects of spelling in this text which require brief comment.

Firstly, the set rules of the spelling of compound words in Afrikaans are frequently not followed when these compounds are used in English, so that usage as we have found it is often at variance with what the correct Afrikaans forms would be. Usage varies, though usually if one item is English and the other Afrikaans (as with *trek-ox* or *rooibos tea*) they are separated or hyphenated. Even this is not a reliable rule of thumb, however, as *Kaffirboom* shows; and the English tendency to maintain separation is not followed in *camelthorn* and *dryland*. These do of course appear as *camel('s) thorn* and *dry land* and the lack of consistency in our sources may well be reflected in the text in spite of conscious effort to maintain a uniform system.

Secondly, the spelling of African-language loan words is regularly found to differ from the correct forms of the 'new' orthographies. These reflect the aspirated consonants [kʰ] [tʰ] [pʰ] as *kh, th* and *ph* to indicate a contrast with the unaspirated forms [k] [t] [p]. A related contrast is indicated by the spellings *bh* and *b*. These contrasts are seldom reflected in our sources, though in the text, wherever possible, both forms are given, e.g. '*khehla* see *kehla*', '*thwasa* see *twasa*', '*bhuti* see *buti*', '*phuthu* see *putu*'.

In South African loan usages of African-language nouns the initial vowels of prefixes are frequently dropped or lost, so that *iNkosi* becomes *Nkosi; intombi, ntombi; imbongi, mbongi* (even *mbongo*) and *umnumzane, mnumzane*. As far as possible both forms are given in the text, usually with the prefixed vowels in parentheses. In plural nouns the prefixes are more frequently retained (as in *imishologu, izinyanya* or *izibongo*), possibly because of their relative rarity. In others, such as *abafazi* or *amadoda*, they are retained with or without the initial *a*, which does not prevent the relatively common addition of an anglicized -*s* plural, particularly for plural forms such as *Basotho, Bantu* or *Mashona*, which are not perceived as plurals by the majority of English speakers. The infinitive verb prefix *uku-* is used in the etymologies only, with the exception of *ukuthwasa*, which, as it is used by anthropologists as a noun, appears as a cross-reference item in the text.

A third area of orthographic difficulty is that of the scientific names of flora. Authorities differ in their treatment of such names as *Solanum burchellii* or *Gardenia thunbergii* which incorporate a personal name. Although modern botanists now largely use lower case letters in botanical names in preference to the more conservative capitals (thus matching the convention long established in the names of fauna), the early sources from which much of our material has been drawn do not. In the interests of standardization, however, the now prevailing lower-case convention has been followed in the text.

Lastly, the now obsolete Dutch spelling *sch-*, which in Afrikaans is usually *sk-*, appears in much of the earlier source material for this work, and has therefore been retained in such words as *schans* and *schepel*. When the commoner spelling appears from our data to be *sk-*, as in *skerm, skelm* and *skof(t)*, the *sch-* forms are given as cross-references only -'*schelm* see *skelm*' etc. – even where both spelling forms may be found in the illustrative material. The -*sch* spelling, now -*s* in Afrikaans as in *bos*, is retained usually only in such place names as *Stellenbosch* or *Rondebosch*, or in historical material, such as the quotation at *bosch*, or terms in which the pronunciation is now [ʃ] as in *schelling* or *schlenter*.

9. Pronunciation

The system of phonetic transcription has been evolved with some care in the effort to make it both simple and flexible enough to indicate a number of various sounds unfamiliar to the non-South African or non-speaker of Afrikaans. The key has been prepared to provide illustrations of the sounds of the non-English words heard in

South African English. For such a sound, an analogue is given from a European language (French, German or Italian) as a rough approximation. Thus the spelling *ui* is given as [œï] and described as being 'as in French coup d'*œil*' and its obverse *eu* as [ïœ] 'as in French mons*ieur*'. In the case of some other non-English diphthongs the nearest English approximation is given, e.g. for *ee* (as in h*ee*mraad) [ɪə] 'as in b*ee*r' is suggested, and for *oo* (as in b*oo*m) [ʊə] 'as in p*oo*r'. These, it must be stressed, are given for simplicity and convenience since detailed articulatory descriptions would be of little interest to non-specialist readers.

The pronunciation of loan words is always an awkward issue. Too-perfect French pronunciation of names, places, dishes and people, or for that matter unrecognizably Spanish renderings of the same type, can only jangle the English speech chain; and the same is true of too-perfectly enunciated Afrikaans or African loan words in the speech of South Africans. This is therefore a problem for the lexicographer who wishes to give an idea of how words unknown to the non-South African should and do sound. It makes it necessary to attempt a mean between the wholly or partially anglicized rendering on the one hand, and the authentic pronunciation in the original language on the other.

One device adopted is the placing of *r*, sounded by some speakers and left silent by others, within small parentheses (r) to indicate its optional status. Another is the alternative endings shown in words with diminutive suffixes *-tjie* or *-djie*, which have a number of differing manifestations.

Not shown in most cases are those plurals which for most Afrikaans-influenced speakers are [s] and for English speakers are [z], though it has been possible in a plural noun such as *sousboontjies* to indicate the Afrikaans form.

Also not normally shown is a phonetic transcription for English words, as the re-production of a South African accent in the pronunciation of English is hardly part of the lexicographer's brief. Only when the actual stress pattern deviates from the standard norm, as in such a word as *cooldrink*, is a transcription included.

The South African English speaker's rendering of the Afrikaans or African-language loan words in his own vocabulary must inevitably be directly related to the linguistic influences to which he is exposed. It is self-evident that the country-dweller in a largely Afrikaans community is more likely to roll r's and produce vowels more closely re-lated to those of the original Afrikaans pronunciation of the loan words he uses than is a city-dweller in a largely English-speaking environment. It is, however, interest-ing to note that some fluent speakers of African languages omit clicks in place names or loan words such as *mngqusho* or *Xhosa*, even though they can, and regularly do, produce them in speech. Variant pronunciations are given in some cases. They are, of course, too few, and can by no means hope to cover variations from speaker to speaker, or variations within the same speaker in different contexts or situations. *Quot homines tot sententiae* holds true of pronunciation also. Such variations are often of vowel length or quality. More significant variations occur in stress patterns, e.g. in the main stress in such words as *likkewaan*, *bobbejaan*, *Shangaan* and *kabeljou*.

Variation is especially difficult to reflect for non-English sounds (including at times the clicks) which may or may not be given their 'authentic' sound-values – for want of a better word. This is firstly and obviously according to the preference or habit of the speaker himself, and secondly and less obviously, the circumstances in which the speaker finds himself at the time, since he may or may not wish to accommodate his speech to that of his companions, which may differ significantly from his own.

Accommodation or non-accommodation can reflect either his wish to level out the differences, or maintain or exaggerate them; or simply a lack of articulatory adapta-tion. An employer of African staff, for example, may make, in their company, a conscious effort to Africanize his pronunciation of African loan words which he would not consider proper or desirable among White companions.

Among children, of course, speech adaptation and accommodation under social and environmental pressure is only too well known, and regularly observable in any playground. The New York slum child from a poor but well-spoken home, when asked why he didn't speak to 'the kids he played with on the street' as he did to his parents, succintly replied to the research worker: 'I couldn't live here if I did.'

It is impossible to attempt to lay down a hard and fast standard for the pronunciation of loan words, and what appears in this text can at best only hope to be a guide. The sound values suggested are usually approximations to a far greater variety of sounds than can be shown in a simple text.

10. Form of Entry

The normal form of entry is as follows:

putu ['putu] *n.*

Traditional African preparation of *mealie meal* (q.v.) cooked until it forms dry crumbs: equiv. of *krummelpap* (q.v.) eaten by Africans with meat and gravy etc. or with *calabash milk* (q.v.) or *maas* (q.v.).⫿ It is also a popular breakfast food among Whites, served instead of porridge; see second quot. [Ngu. *uphuthu* crumb porridge, anything crumbly, e.g. earth]

Soon one of Mazibe's wives . . . entered on her hands and knees. Permitted to kneel but not to stand, she adroitly balanced a baby strapped to her back in a cloth sling, while serving us roast chicken and uphuthu, a kind of hominy that has been a staple Zulu dish for centuries. *National Geographic Mag.* 6.12.71

. . . her first move is to the kitchen where Lukas her cook is preparing breakfast . . . Whatever else might be in the offing a large pot of putu – crumb mealie meal porridge – will be ready for eating. *Fair Lady* Jan. 1972

Each entry takes more or less the same form, namely:

1. The headword, *putu* in the above example, in bold type.
2. A *phonetic transcription*, usually only in the case of items of non-English origin, in square brackets.
3. The *grammatical designation*, *n.* for 'noun' in the case of *putu*, followed for 'countable' nouns, e.g. *voorkamer* or *land*, by the marker or markers of their plural forms. This is not given for *putu* which, like English *porridge*, normally acts as a 'noncount' noun: we seldom convert 'porridge' to a 'count' noun by pluralizing it as 'porridges'. In the case of names of game birds or animals, e.g. *eland*, which are regularly used, like *sheep*, with an unmarked plural, the sign ∅ (zero) is used.
4. The *definition*, with cross-references if any, e.g. from *putu* to '*mealie meal* (q.v.)'. This sometimes includes an extra note marked ⫿.
5. The etymology, again in square brackets: this for *putu* relates the word to the Nguni *uputhu* from which it derives.
6. The illustrative quotations.

Forms which occur as prefixes or suffixes are preceded and/or followed by hyphens to indicate that these are not normally 'free' forms. They occur quite frequently among the place name formatives mentioned above, and also in the names of flora and fauna. Many forms such as *veld* and *berg* appear both free and in compounds.

Latin names of flora and fauna have been taken from the newest available checklists and texts. These are given either at the beginning or in the body of a definition. Cross-referenced equivalent terms are marked (q.v.) – 'which see' – and will be found as head words elsewhere in their alphabetical sequence. Other cross-references given are explained in Section 5.

In the case of alternative spellings which are frequent in African loan words, these also appear as head words in the alphabetical sequence: '*phuthu* see *putu*'.

PRONUNCIATION KEY

TABLE OF PHONETIC SYMBOLS

Approximate sound values of symbols for South African English.

CONSONANTS

[p, b, t, d, g, k, f, v, s, z, h, m, n, l] have their usual agreed English speech value.
Non-English sounds are marked *.

r   like normal English initial 'r' as in *rat*. This *r* in initial position is rolled as in Scottish English by some S.A.E. speakers.

*(r)   the parentheses indicate that some speakers omit this sound altogether. This (r) when sounded is rolled as in Scottish English or Afrikaans by some S.A.E. speakers e.g. poo*r*t [pʊə(r)t].

ŋ   as in ri*ng*.

dʒ   as in *j*u*dg*e.

tʃ   as in *ch*ur*ch*. This is a frequent S.A.E. pronunciation of *[c], the palatal plosive (see below).

*c   this sound is rare for S.A.E. speakers who usually pronounce it as [tʃ] (see above) or [k]. In Afrikaans it is spelt *tj* or *dj* and sounds like a [k] produced forward on the hard palate, e.g. naar*tj*ie [narcĭ].

θ   as in *th*in, pi*th*.

ð   as in *th*en, ti*th*e.

ʃ   as in *sh*ine, fini*sh*.

ʒ   as in plea*s*ure.

j   as in *y*ellow. The Afrikaans spelling is *j* e.g. *j*a, yes.

*x   as in Scottish lo*ch*, German a*ch*. The Afrikaans spelling is *g* or *gg* e.g. *g*o*gg*a, insect.

*ɬ   as in Welsh *Ll*andudno: found in African language borrowings spelt *hl* e.g. *hl*onipa reverence, ka*hl*e well, also in all Zulu place names containing this combination e.g. *Hl*u*hl*uwe, Ma*hl*abatini; often erroneously pronounced [ʃl] as in German *sch*loss (see first quot. at *kehla*).

Note: In words borrowed from the African languages the *c*, *x* and *q* spellings represent clicks of three different basic types.
c represents the dental click formed behind the teeth on the teeth ridge, rather like the English 'dismay sound' variously spelt *tut-tut*, *tch tch*, *tsk tsk*.
x or *xh* represents the lateral click, formed at the side of the mouth, and *q* the palatal click formed at the hard palate. Approximations are not available for these.
In the interests of simplicity, therefore, they are all transcribed as [k] sounds, a quite usual way of rendering them.
The presence of a click in a word is indicated by [+] following the transcription thus:
Tixo [ˈtĭkɔ̃ +].

VOWELS AND DIPHTHONGS

Vowels or diphthongs which have no English equivalent are marked *.
The example and/or description following each is the nearest approximation to the sound.
Note: 'high', 'low', 'front', 'back' and 'central' refer to the position of the tongue in the mouth.

VOWELS

*y   as in French r*u*e, German ü*b*er. This is pronounced like the sound in p*ea* but with closely rounded lips. The Afrikaans spelling is *uu* e.g. s*uu*rveld   sour grass veld, occ. *u* e.g. s*u*ring   sorrel.

i   as in p*ea*.
Note: A form of this sound, very short and slightly lower, is used by most S.A.E. speakers for the pronunciation of final *y* as in cit*y*, unlike the [ɪ] of British English in the same position. Where this occurs in the text the symbol [ï] is used.

*ï   short as in German *i*ch, French r*i*z. The Afrikaans spelling is usually *ie* as in r*ie*m   thong, occ. *e* before *tj* e.g. ribb*e*tjie   rib chop.

ɪ   as in p*i*ck.
Note: Unless it is initial or in combination with [g], [k] or [ŋ], preceded by [h] or followed by [ʃ], when the symbol [ɪ] is used, this sound is pronounced in a lower form, further back, see [ɪ] below.

*ɪ   Similar to the sound in b*e*cause, d*e*gree pronounced stressed. This is only an approximation: see note on [ɪ] above. The [ɪ] symbol is used for *i* spellings other than the combinations described there, and for the S.A.E. rendering of Afrikaans *i* spellings, rendered [ə̄] by Afrikaans speakers and some speakers of S.A.E.

e   as in p*e*n.

*e: as in French m*e*re, German *äh*nlich, the [e] sound pronounced long. The Afrikaans spelling is either *ê* as in s*ê* say, or k*ê*rel fellow, *e* before *r* as in v*e*r far, occ. *è* as in n*è* not so? See also diphthong [ɛə].

æ   as in p*a*n.

*a   short as in German *a*ch, Italian *a*ltro, French *a* la mode. This sound occurs with many variations in both Afrikaans and the African languages which contribute to S.A.E., spelt *a*. This description is only an

approximation since the variants are between the extremes of the [ʌ] of h*u*t, on the one hand and the [ɒ] of h*o*t on the other.

ɑ   as in p*a*r, p*a*lm, used in this text for most Afrikaans 'aa' spellings and for some *a* spellings.

ɒ   as in *o*n, s*o*ck.

ɔ   as in c*o*rn, c*a*ll. This symbol is also used to transcribe long 'o' sounds in words borrowed from African languages e.g. lob*o*la   bride price.

*ɔ̆   Similar to [ɔ] but short and pronounced stressed, something like the *o* in German Gott. The quality and duration of the sound are equivalent to that of the first syllable of *au*thority, pronounced stressed. [This does not make it equivalent in quality or duration to the first syllable of *au*thor.] The Afrikaans spelling is *o* as in k*o*ppie   hillock.

ʊ   as in b*oo*k, p*u*ll.

u   as in b*oo*t, r*u*le.

*ŭ   as in German H*u*nd. Similar to the sound in b*oo*t, but pronounced short. Spellings are *oe* as in Afrikaans st*oe*p   open verandah, and *u* as in k*u*du (Afrikaans k*oe*doe) in both stressed and unstressed syllables.

ə   as in butt*er*, *a*bout. The unstressed central 'neutral' vowel.

*ə̄   the same vowel stressed, a sound said to be unique to S.A.E. It is standard in Afrikaans e.g. sin [sə̄n] sentence. Some S.A.E. speakers and most Afrikaans speakers use this sound for *i* spellings e.g. pit [pə̄t] Afrikaans 'stone', 'pip', for which the symbol [ɪ] is used in this text.

*œ   as in French b*œu*f (pronounced short). A short semi-low front vowel with slight lip rounding. The

Afrikaans spelling is *u* as in -r*u*s [-roes] rest.

ɜ as in p*er*t, b*ur*n, b*ir*d, *ear*th.

ʌ as in b*u*t.

DIPHTHONGS

Note: In the interests of simplicity and convenience the closest English diphthong has been taken as an example wherever possible.

ɪə similar to the sound in p*ier*. Afrikaans spellings *ee* as in g*ee*l yellow, and *e* as in br*e*die stew, are rendered as a diphthong by S.A.E. speakers, though not always before *f*.

ʊə similar to the sound in p*oo*r. Afrikaans spellings *oo* as in b*oo*m tree and *o* as in d*o*minee minister are rendered as a diphthong by S.A.E. speakers though not always before *f*.
Note: These two diphthongs could more accurately be given with their first elements [ĭ] and [ŭ] respectively. However, since [ɪə] as in b*eer*, f*ear*, p*ier* and [ʊə] as in p*oor*, s*ure*, are regular English diphthongs they are given as more convenient approximations to the S.A.E. sounds. These are pure vowels not diphthongs for many Afrikaans speakers.

ɛə as in p*air*, p*ear*. This is also a frequent pronunciation of the Afrikaans long *e* before *r* as in k*ê*rel, v*e*r, p*e*rske. See [e:].

eɪ as in p*ay*. The Afrikaans spelling is *y* as in vr*y* court (*vb.*) *ei* as *ei*na ouch! *ey* as in br*ey* curry (hides).

aɪ as in p*ie*, but of slightly longer duration. The Afrikaans spelling is *aai* as in br*aai* grill, or *ai* as in asseg*ai* spear.
Note: Before *-tj* or *-dj* *aa* and *a* spellings are pronounced [aɪ] e.g. v*aa*tje a small barrel (vat), rooib*aa*djie redcoat, l*a*tjie twig.

ɔɪ as in pl*oy*, c*oi*n. Like the [aɪ] sound this tends to be slightly lengthened.

əʊ as in *toe*, *coa*l. This is an approximation to the sound of the Afrikaans *ou* spellings as in juffr*ou* mistress (teacher).

aʊ as in pr*ow*.

*œĭ as in French coup d'*œil*. This sound is similar to that of English d*ay* pronounced with rounded lips. The Afrikaans spelling is *ui* e.g. m*ui*svoël mousebird (occasionally *uy* in proper names e.g. *Uy*s, or in Dutch borrowings). This is often pronounced [eɪ] as in p*ay* by S.A.E. speakers.

*ĭœ as in French mons*ieu*r: the obverse of the [œĭ] diphthong, pronounced like the sound of *ear* with lip-rounding. For Afrikaans speakers this is a pure vowel spelt *eu* as in d*eu*rmekaar muddled, vern*eu*k deceive.
Note: This is often pronounced [ɪə] as in f*ear* by S.A.E. speakers.

OTHER MARKERS

: placed after a vowel indicates that it is of exceptional length.

˘ above a symbol indicates that the sound is short.

( ) indicates that the sound within the parentheses is frequently dispensed with or dropped.

ˌ placed below a resonant consonant [m̩] [n̩] or [l̩] indicates that the consonant is syllabic i.e. it serves as vowel and consonant in one e.g. bʌtn̩ (button) lɪtl̩ (little).

ˈ placed before a syllable indicates that it carries the primary stress or emphasis in that word e.g. [ˈbɑskɪt] [krɪsˈænθəməm].

ˌ placed before a syllable indicates that it carries secondary stress in that word e.g. [ˈhændˌbæg] [ˈkæbɪdʒˌlif].

Where alternative pronunciations of single syllables occur the alternative is given for the syllable only.

For final syllables the alternative follows a comma and a hyphen thus [(ɪ)nˈdɑbə, -a]: for initial syllables it follows a comma and precedes a hyphen thus [ˌjəˈfrəʊ, ˌjœf-]: for medial syllables it follows a comma and is placed between two hyphens, thus [afrɪˈkăndə(r), -ˈkænd-, -ˈkand-].

ABBREVIATIONS AND LABELLING

A few items are more fully explained in the Grammatical Notes which follow this list. In the list below, all abbreviations are italicized throughout; others may be italicized or not according to their position in the entry.

abbr.	abbreviation/abbreviation of	*Du.*	Dutch, Nederlands
acc.	according to	*dub.*	dubious/doubtful
acronym	a word, often a name, formed from the initial letters and/or syllables of the words of a compound designation, e.g. *NGK* from *Nederduits Gereformeerde Kerk* or *Swakara* from *South West African Karakul:* see also 'portmanteau' word	*E.*	East/ern
		ed.	edition/edited by
		e.g.	*exempli gratia*, for example
		Eng.	English
		equiv.	equivalent/equivalent of/to
		erron.	erroneous/ly
		esp.	especially
		etc.	et cetera
		etym.	etymology
adj.	adjective	*fam.*	family
adv.	adverb/adverbial	*Farm.*	farming
adv. m.	adverb of manner	*figur.*	figurative/ly
adv. p.	adverb of place	*fr.*	from
adv. t.	adverb of time	*Fr.*	French
Advt.	advertisement	*freq.*	frequently
affix	prefix or suffix	*Ger.*	German
Afk.	Afrikaans	*Gk.*	Greek
Afr.	African	*Hebr.*	Hebrew
Afr.E.	African English, i.e. typical of the English spoken and written by Africans in S.A.	*hist.*	historical
		Hong K.	Hong Kong English (usually a Chinese loan-word)
agent. suffix	agentive suffix, e.g. *-er* as in *teacher*, marking the doer or performer of the action designated by a verb	*Hott.*	Hottentot
		ibid.	*ibidem*, the same, usually of a source of quotation
analg.	analogy/analogous to	*idiom*	idiom/atically
Anglo-Ind.	Anglo-Indian (English): see also *Ind.E.*	*i.e.*	*id est*, that is
		imp.	imperative
Arab.	Arabic	*Ind.E.*	Indian English, i.e. typical of the English used by Indians in S.A., mostly in Natal
Army	term used mainly in the Army		
		indef.	indefinite/non-definite: of articles or pronouns
art.	article (definite *the* or non-definite *a/an*)		
		inflect.	inflected (form)/inflection
attrib.	attributive/ly; of a noun modifier or of an adjective	*intensifier*	See Grammatical Notes
		interj.	interjection, exclamation
Austral.	Australian English	*intrns.*	intransitive, see under *vb.*
Bantu	Bantu language (this includes languages of the Nguni group)	*intrns,*	(intransitive verb) in Grammatical Notes
		Jam. Eng.	Jamaican English
Brit.	British English (used of dialect or 'standard')	*Jewish*	In use mostly among the Jewish group
C.	Century, e.g. *C.18* for 'eighteenth century'	*Journ.*	Journal
		Lat.	Latin
Canad.	Canadian English	*lit.*	literal/ly
cf.	*confer*, compare	*Malay*	Malay language/Malaysian
cit.	cited by	*Mining*	Terms characteristic of gold and diamond mining; see also *Sect.* below
Cape of G.H.	Cape of Good Hope		
cogn.	cognate with, used of words of common or related origin	*mistrans.*	mistranslation of
		modifier	a word or phrase (not an adjective) qualifying a noun; see *n. modifier* in Grammatical Notes
colloq.	colloquial: informal in speech *or* writing		
demon.	demonstrative		
deriv.	derivatively / derivation / derived	*MS.*	manuscript
		n.	noun
descr.	description	*N.*	North/ern
dial.	dialect/dialectal	*n. abstr.*	abstract noun
dimin.	diminutive/diminutive form		

Nama	Nama Hottentot
neg.	negative
Ngu.	Nguni (Zulu, Xhosa, Swazi)
n. modifier	noun used to modify (qualify) another noun: see Grammatical Notes
n. prop.	proper noun or proper name
N.Z.	New Zealand (usually Maori)
object.	object (of a verb)
obs.	obsolete, no longer in use
occ.	occasionally
O.E.	Old English (Anglo-Saxon)
O.E.D.	*Oxford English Dictionary.* This means the twelve-volume text. If a reference is to the *Shorter, Concise* or *Pocket Oxford Dictionary*, or to one of the *O.E.D. Supplements*, this is specified
O.I.	'Oral Informant/s' in the case of quotations from speech rather than a written source
onomat.	onomatopoeic
orig.	origin/originally
partic. also *vb. partic.*	participle or participial, of a verb or phrase
pass.	passive
past	past tense
personif.	personifying, usually of suffixes, see also *agent. suffix*
phr./s	phrase/s
pl.	plural
Port.	Portuguese
portmanteau word	a combination of two words such as *Australorp* Australian/Orpington: see also *acronym*
poss.	possibly
predic.	predicative; see Grammatical Notes
prefix	an affix, e.g. *agter-*, placed before a word to change or modify its meanings, as in *agterlaaier, agterskot*, etc.
prep.	preposition/al
pres.	present (tense)
presum.	presumably
prn.	pronoun
prob.	probably
pron.	pronunciation/pronounced
pub.	published by/in
qn.	question
quot.	quotation
(q.v.)	*quod vide*, which see
rare	seldom occurring (in our experience)
reg.	regional: this label can at present be only tentatively assigned for S.A.E.
rel.	related to/relative
remin.	reminiscences / reminiscences of
S.	South/ern
S.A.	South Africa
S.A.E.	South African English
S.Afr.	South African
Scottish	Scottish English
Sect.	Sectional jargon of certain occupational, professional, trade or other groups
sic	thus
sig.	signifying, meaning
sing.	singular
slang	slang usage, seldom found written
Sotho	Sesotho (Sesuto)
sp.	spelling (form/s)
Span.	Spanish
spp.	species (pl.)
subj.	subject (of a verb)
substandard	substandard usage: see Grammatical Notes
suffix	an affix placed at the end of a word to change or modify its function or meaning, e.g. *-heid* as in *apartheid.*
trans.	translation/translation of
translit.	transliteration of
trns.	transitive
U.S.	United States usage
usu.	usually
vb.	verb
vb. intrns.	intransitive verb: see Grammatical Notes
vbl.	verbal as in *vbl. n.* 'verbal noun'
vb. trns.	transitive verb: see Grammatical Notes
viz.	*videlicet*, namely
W.	West/ern
Xh.	Xhosa
Zu.	Zulu
Ø	zero marker, usually in 'unmarked plurals' e.g. *bontebok* and many other terms for game, wild animals, fish or birds: cf. *sheep*
~	swung dash used instead of a repetition of the headword
[ ]	single square brackets enclose phonetic transcriptions and etymologies
ℙ	information or comment not strictly part of a definition

Note: The names of seldom-used languages, Yiddish, Japanese, Persian, Telegu, Hindi, Chinese, etc., which do nevertheless appear in the text, are given in full and are not listed above.

GRAMMATICAL NOTES

Attributive:	An attributive, in our terminology, is a word that modifies a noun, e.g. an adjective such as *verligte* as in 'a verligte outlook' or a noun such as *Zulu* in 'a Zulu village'. See also *Predicative* and *Modifier*.
Intensifier:	Intensifiers, e.g. *very* in 'a very hot day' are a subclass of what traditional grammar calls 'adverbs of

degree'. Intensifiers, however, pattern with adjectives and adverbs but not, as do normal adverbs, with verbs. Quirk and Greenbaum (1973) point out that intensifiers 'are not limited to indicating an increase in intensity: they indicate a point in the intensity scale which may be high or low'. A high point for example would be indicated by *extremely*, a low one by *somewhat*.

Modifier: A modifier is any item (word or phrase) standing as adjunct to another. In 'the fat cat with white whiskers' the modifiers of *cat* are *with white whiskers* and *fat*. A one-word modifier normally precedes the item that it modifies: a multi-word modifier normally follows it, e.g. 'for Africa' in 'He has money for Africa', although usually it remains an 'attribute' of the noun. See *Attributive*.

N.modifier: A noun modifier is a noun that modifies another noun, e.g. *Zulu* in 'a Zulu village'. Nouns in these patterns, e.g. 'a Zulu village', 'a Voortrekker costume', are distinguishable from adjectives in similar patterns, e.g. 'a large village', 'a bright costume'. Unlike adjectives, noun modifiers cannot be marked as comparatives: *a brighter costume* is English but *a Voortrekker-er costume* is not. Noun modifiers, moreover, do not readily take predicative position: compare:
the costume was bright
? the costume was Voortrekker.

Predicative: The 'predicative' position of an adjective is that of *fat* in
The poodle is fat
i.e. after a verb like *be* or *seem*. Contrast the pattern
A fat poodle
in which the adjective is in 'attributive' position before the noun it modifies. Similar positions are taken by nouns: compare 'attributive' *Voortrekker* in
His Voortrekker ancestors
with 'predicative' *Voortrekker* in
His ancestors were Voortrekkers.

Substandard: This status label is applied to grammatical structures considered inappropriate in speech or texts representing formalized standard usage. These are of two kinds. The first are patterns 'borrowed' or carried over in translation from Afrikaans, e.g. 'The rinderpest was *busy decimating* their herds', or 'The tree is *capable to withstand* frost'. The second group are translated or transliterated forms: e.g. *by* (*by*, at), *a person* (*'n mens*, one), *forget* (*vergeet*, leave behind), *full of* (*vol*, covered in), *so long* (*so lank*, in the meantime), etc. *Substandard* is not used of a noun unless the noun is itself a transliterated form, e.g. *youth, the* (*die jeug*, young people).

Vb. trns: (transitive verb) marks a verb which normally takes an object, e.g. *brei* as in
Karel breied the hides.
Here *the hides* has the status of object, for which the standard test is that in the passive version of the sentence it takes subject position:
The hides were breied by Karel.

Vb. intrns: (intransitive verb) marks verbs which do not take an object, e.g. verbs of 'complete predication' like *skinder* in
Mary and Sannie are skindering in the laundry
and for non-transitive verbs which cannot take a passive but nevertheless require a complement, e.g. *have*. See the entry for *omissions*.

Note: In many cases, e.g. *trek* which may act as noun, noun modifier or verb, a word has more than one grammatical function. This has usually been shown in entries, e.g. by *n. and vb.* for 'noun and verb' or *n. and n. modifier*. Wherever possible, each function has been illustrated in a quotation.

REFERENCE

Quirk and Greenbaum (1973): *A University Grammar of English* Longman.

A

a *art.* See *articles.*

aand- [ɑ:nt] *n. prefix* See *avond-.*

aandag [ˈɑ:nˌdax] *interj.* Attention: used with double intention in advertising 'smalls', primarily in order to head the column alphabetically. [*Afk. fr. Du. aandacht cf. Ger. Achtung*]
Aandag! Massage and sauna. Call at the Executive Penthouse and meet Betty, Lucelle and Ortrud. *E.P. Herald Advt.* 20.11.74

ELECTRICAL   FURNITURE
APPLIANCES   AANDAG
AANDAG   A DEPOSIT NO PROBLEM
We repair . . .
                                    Argus 2.7.79

aandblom [ˈɑ:ntˌblɔm] *n. pl.* -me. Any of several flower species, including varieties of *Hesperantha* and *Gladiolus,* which emit their strongest scent towards evening. Also known as *aandpypie :* see quot. at (2) *Afrikaner.* [*Afk. fr. Du. avond* evening + *blo(e)m* flower *cogn.* bloom (*pypie cogn.* pipe + *dimin. suffix -ie*)]
. . . it being then nearly dusk, the delightful fragrance of the Avond-bloem (evening flower), a species of Ixia (Hesperanthera), began to fill the air, and led to the discovery of the plants. In the day time their flowers, which, though white within, are of a dusky colour on the outside, and, being then quite closed, do not readily catch the eye. Burchell *Travels I* 1822

aantree [ˈɑnˌtriə] *vb. Sect. Army.* Also *tree aan,* to fall in, form up: see also quot. at *wit* [*Afk. tree aan*]
Don't ring back – I've got to aantree in a moment – fall in. *O. I. Serviceman* 19.1.79

aap [ɑp] *n. dimin. apie pl.* -s. Monkey: affixed to the names of some primates or plants: *nagapie* (q.v.); *apiesdoring* (q.v.); *aapsekos* (food) *Rothmannia capensis; aapsnuif,* see *devil's snuffbox; aapstert steekgras,* see *steekgras:* also in place name *Apiesrivier.* [*Afk. fr. Du. aap* monkey *cogn.* ape]

Aapkas, the [ˌɑpˈkas] *n. prop. Sect. Army.* The jumping tower used for practice by paratroopers in training: see *Parabat.* [*Afk. aap cogn.* ape, *kas* box, *cogn.* case]
The Aapkas is the name given to the jumping tower. It is here that the trainee paratrooper experiences the first feeling of what it is like to jump from an aircraft. Marks *Our S.Afr. Army Today* 1977

aar [ɑ:(r)] *n.* An underground watercourse often indicated by more luxuriant surface growth above it: found in place name De Aar and in combination

~*bossie* generic name of plants preferring this habitat. [*Afk. fr. Du. ader* bloodvessel *cogn.* artery]
The aarbosje, or 'water-finder', *Selago leptostachya,* . . . is a useful forage plant for goats, being a stand-by in times of drought. It is found generally on gebroken veld. Wallace *Farming Industries* 1896

aardvark [ˈɑrtˌfa(r)k] *n. pl.* Ø, -s. The African ant-bear *Orycteropus afer,* a burrowing, insectivorous mammal of nocturnal habits. See also quot. at *rooikat.* [*Afk. fr. Du. erd/aard* earth + *vark* pig]
. . . the great ant-eater is found throughout every part of South Africa. This animal . . . is called by the Dutch colonists the *Aardvark* or earth pig. It burrows underground making large holes . . . It is destitute of teeth, and lives entirely upon ants. Thompson *Travels II* 1827

aardwolf [ˈɑrtˌvɔlf] *n. Proteles cristatus* the hyena dog, maned jackal, often called *maanhaar* (maned) *jakkals.* See quots at *wolf* and *strandwolf.* [*Afk. aard/ erd* earth + *wolf* hyena]
When outspanned for the night, our dogs started an Aard Wolf (Proteles Lalandii) from the long grass, pursued it and brought it to bay. Leyland *Adventures* 1866

aas [ɑs] *n.* Carrion: prefixed to names of plants ~ *blom* (q.v.) ; ~ *bossie;* ~ *kelk;* ~ *uintjie;* and bird *aasvoël* (q.v.) sig. bait: usu. in combination *rooi aas* (q.v.). [*Afk. fr. Du. aas* bait, carrion]
Fishermen from George who recently made the observation that good old red bait, or 'rooi aas' might possibly be an aphrodisiac, are perhaps on the threshold of some scientific discovery! *Evening Post* 3.3.73

aasblom [ˈɑsˌblɔm] *n. pl.* -me. Carrion flower: any of several species of *Stapelia* having a smell like putrid flesh which attracts flies and other insects. [*Afk. fr. Du. aas* bait, carrion + *blom* flower *cogn.* bloom]

aasvoël [ˈɑsˌfuəl] *n. pl.* -s. Vulture: any of several species of large scavenger, of the Gypætidae: *Gyps coprotheres* (*G. kolbii* etc.). [*Afk. aas* carrion + *voël fr. Du vogel* bird *cogn.* fowl]
I remember . . . the slow but startled flutter skywards of the *aasvoëls* disturbed from their carrion. Birkby *Thirstland Treks* 1936

abadala [ˌabaˈdala] *pl. n. Afr.E. lit.* 'Old ones': the spirits of the forefathers: see also *amadhlozi* and *amatongo.* ⦿ At memorials for the ~ or the *abaphansi* (q.v.) drink is poured on the ground as a libation to them. [African Informant, 70,

1979] [*Ngu. pl. prefix aba-, -dala* old, *abadala* ancestors]

Then too if a man has been very lucky . . . he calls his clansmen together to thank 'the old ones' (*abadala i.e. the shades*). Wilson & Mafeje *Langa* 1963

abafazi [ˌabaˈfazĭ] *pl. n.* Women: see quots at *fazi* and *umfazi*. ⚏ A sign commonly seen on African women's public lavatories: see quot. [*Ngu. pl. prefix aba + fazi* woman, wife]

The first thing that hits the eye are the signboards 'Blacks only' wherever you look. Here and there behind some backstreet bush there will be 'White Amadodas this way' and 'White Abafazis the other way'. *Drum* July 1971

abakhaya [ˌabaˈkaɪa] *pl. n. Urban Afr.E.* A group of migrant Africans from the same village, clan or country area; ⚏ Members of such a group tend to identify with and protect one another where possible: see also *homeboy*. [*Ngu. mkhaya* one from home]

For a migrant in town by far the most important group is that of his abakhaya or homeboys. This is the group within which he lives and eats and probably works . . . They address one another as *mkhaya* or *home-boy*. Wilson & Mafeje *Langa* 1963

abakwetha [ˌabaˈkweta] *pl. n. (occ. -s).* Recently circumcised Xhosa initiates to manhood, living apart from other people for a certain period and usu. daubed with white clay. *cf. Austral. kipper* aboriginal initiate. [*Xh. pl. prefix aba + kweta fr. umkwetha* a circumcised youth]

. . . this year more than 200 abakwetha were in the bush. Included . . . were many youths from institutions and colleges who took advantage of the long summer holidays to qualify for manhood in the bush. *E.P. Herald* 16.2.74

abaphansi [ˌabaˈpansĭ] *pl. n. Afr.E. lit.* 'Those below': the spirits of the forefathers: see also *abadala, amadhlozi* and *amatongo* [*Ngu. pl. prefix aba, phansi* under, below. ⚏ *phansi* is also a *Zu. hlonipa* (q.v.) word for *umhlaba* earth]

Umhlaba the earth, is a name given to the Amatongo, that is the Abapansi, or Subterraneans . . . on the Zambesi, Azimo or Bazimo is used for the good spirits of the departed. Callaway *Religious System of the Amazulu* 1884

He feels his children should follow the example of Abaphansi-badimo, our Zulu and Sotho forefathers. Venter *Soweto* 1977

abba [ˈaba] *vb. usu. trns.* also *n.* To carry (a child) on the back: as *n.* a pick-a-back [piggy-back] ride. see also *piggyback*

deriv. fr. ~*hart* (*Afk.*) and second quot. [*See quot. prob. fr. Nama* to carry, *also Bushman aba* carry]

The name given to this new industrial growth point is Babelegi a word which in Setswana is derived from 'abba': to carry a baby on one's back. A visit to Babelegi explains the choice . . . At Babelegi industrialists and workers really 'abba' each other. *Panorama* Apr. 1974

. . . the word abba . . . has an interesting rebirth in the recent use of the term 'abbahart' in Afrikaans for the piggy-back heart operation. *Argus* 19.4.78

abelungu [ˌabeˈluŋgŭ] *pl. n.* Whites: *pl. of (u)mlungu* (q.v.). [*Ngu. pl. prefix abe + lungu* white people]

ace, (all) on my, *adv. m. phr.* Also *Austral.* On my own, or by myself: see *eis, on my.*

ach *interj. usu. Ag* (q.v.).

'Ach! Don't be so frightened, man' . . . 'Don't be a blerry fool. Sit!' Gordon *Four People* 1964

actuarius [ˈaktjuˈɑrɪəs] *n.* An official of the Dutch Reformed Church Synod, also occ. *actuary.* [*Lat. actuarius* an amanuensis, a keeper of accounts or records]

The Synodical Commission is composed of the Moderator, the Scriba, the Actuarius, five acting ministers, and three elders or retired elders nominated by the Synod. *Cape Town Directory* 1866

Adamastor [*n. prop.*] The spirit believed by the Portuguese in the early days to inhabit the 'Cape of Storms'. ⚏ The title of a collection of verse by Roy Campbell (1930) named after this spirit in the *Lusiads* of Camoëns (1572). See second quot.

The low sun whitens on the flying squalls,
Against the cliffs the long grey surge is rolled
Where Adamastor from his marble halls
Threatens the sons of Lusus as of old.
Campbell *Rounding the Cape* (*Adamastor* 1930)

. . . Portuguese had regarded the Cape with feelings of suspicion and superstition . . . To them it was the Cape of Adamastor, the vengeful spirit of storms who in the Lusiad had appeared . . . to Vasco da Gama predicting the woes that would befall those who sailed on to India. De Kiewiet *Hist. of SA* 1941

adjective with infinitive *substandard. trans. fr. Afk.* in inappropriate English contexts, e.g. I am lazy to get up (Ek is lui om op te staan).

The leaves . . . soon drop to the ground leaving the tree rather stark and bare but quite capable to withstand even the severest frost. *Farmer's Weekly* 30.5.73

Administrator *n. pl. -s usu. capitalized.* The chief executive officer in each province 'in whose name all executive acts relating to provincial affairs therein shall be done' *S. Africa Act 1909.*

A Grahamstown resident . . . has appealed to the Administrator to request the mayor to call a meeting of ratepayers. . . . hoping for action from the Administrator in terms of section 259 of the Cape Municipal Ordinance. *E.P. Herald* 2.8.74

advocate *n. pl.* -s. *equiv. Brit.* Barrister: see quot. at *side-bar.* ~ is also the Scottish designation. ⫐ In S.A. the titular use is commonly found, see second quot.

'. . . it is correct that the South African Advocate is the equivalent of the English barrister. The official designation of a member of my profession is an advocate of the Supreme Court . . . the term is also used as a description of a particular function of a barrister's calling, namely the conduct of cases in court. Thus one might say of an English barrister that he is a very learned lawyer but he is not a good advocate.' *Letter* Mr. Sydney Kentridge S.C. 23.9.78

The . . . peculiarity of the South African usage of the word is its use as an appellation . . . It is very common in South Africa to hear a member of the bar described as 'Advocate Jones'. It is however an undesirable usage and . . . one to be avoided. *Ibid.*

advokaat [ˌadfʊˈkɑt -vŏ- -vŭ-] *n.* A thick sweet Dutch liqueur of egg yolks and brandy, see quot. [*Afk. sp. fr. Du. advocaat abbr. advocatenborrel* advocates + drink]

How can we make our own Advokaat liqueur?
ADVOKAAT – 6 egg yolks 1 cup sugar ¼ teaspoon vanilla essence 2 or more wineglasses brandy *Farmer's Weekly* 27.2.74

aeroplane See *ai-ai.*

Af [æf] *n. pl.* -s. African: used among White speakers, esp. Rhodesians. [*abbr. Af*rican].

Those dogs of yours bitten any Afs yet? O.I. 1976
You can't tell me a Jo'burg Af doesn't pick up a smattering of English and/or Afrikaans. *Sunday Times* 21.5.78

afdak [ˈafˌdak] *n. pl.* -s. A lean-to, shed or shelter; also loosely used of any plain, sloping roof without a pitch. *cf. Anglo-Ind.* and *Hong K. godown,* store, warehouse, and *Austral. skillion (skilling),* lean-to. [*Afk. af* down + *dak* roof]

Nice house, except for the afdak. O.I. 1970

afkak parade [ˈafˌkak paˈradə] *n. phr. Sect. Army. slang* [vulgar]. Pack drill 'at the double', see *kak off:* a form of punishment drill: see also *motivation P.T.*

Africa, for *attrib. phr. after n.* Used to indicate a vast, unspecified number or amount e.g. *money* ~ : see also second quot. at *ghoen.*

There are still bottles for Africa around but in the not too distant future glass of any variety will be a thing of the past. *Argus* 3.10.70

African *n. pl.* -s and *adj.* **1.** A Black, i.e. negroid, inhabitant of S.A. Formerly officially designated *Bantu* (q.v.) by the Department of Interior: see quots. at *Bantu* and *European.*

The terms African, Bantu and Native will be used as synonymous terms when discussing the Negroes of South Africa. All white people in South Africa, whether or not born in Europe, are referred to as Europeans. Longmore *Dispossessed* 1959

2. Of or pertaining to (1) ~s, their languages or cultures.

The Nationalist Party saw the main issue involved as the establishment of African languages as official languages alongside English and Afrikaans in various African territories. *Daily News* 4.3.71

3. *rare* Any South African, long *obs.* now revived.

I arrived at the house of farmer *Van der Spoei,* who was a widower, and an African born. *Trans.* Sparrman *Voyages I* 1786
. . . the Burger's assistant editor rapped FAIR LADY's knuckles for distinguishing Afrikaners and Africans. 'We are all Africans,' said he. 'Some of us are white and some of us are black.' The man who shouted from the Public Gallery in Parliament: 'Don't call us coloureds. We are South Africans' patently agreed with him. *Fair Lady* 13.10.76

In combination: *white* ~

After all, we are White Africans. Africa is our only home and we shall have to learn to get on with the rest of Africa, or perish. E. G. Malherbe *cit.* Spottiswoode *S.A. Road Ahead* 1960
If they can't see themselves as White Africans or simply as Africans, then their heart is just not going to be in separate and equal development, or in federalism – or anything else. *Sunday Tribune* 1.4.73

Africana [ˌafrɪˈkɑnə, æf-] *collective n. usu. capitalized.* Books, pictures, furniture and *objets d'art* of S. Afr. provenance or interest: see quot. at *Cape Triangular.* [*Lat. adj. Africanus -a, -um* of Africa, African]

Africana is the term used for all those items, large and small, of historical importance and interest to southern Africa, although not necessarily made or manufactured in this country. These rarities and antiquities include stamps, medals, books, musical instruments, implements, costumes, paintings, silver, glass, porcelain, china, carvings, furniture, clocks, watches and other memorabilia of our cultural heritage. *The 1820* Feb. 1974

African Court Calendar *hist. obs.* See *Court Calendar.*

Africander [afrɪ'kǎndə(r), æf-, -'kæn-] *n. pl.* -s, also *n. modifier.* **1.** Tall, indigenous, humped usu. *red* (q.v.) cattle with long horns, hardy and resistant to disease, used as a beef breed and for draught purposes ; see *trek ox* and quots. at *red, Zebu* and (3) *Zulu.*

Afrikander cattle, the only breed indigenous to South Africa, have always been the foundation of the country's beef industry. *Panorama* July 1971

2. Often attrib. ~ *sheep.* The indigenous *Cape* or *fat tailed sheep* (q.v.) also known as *ronderib* (q.v.).

Africanders. For Sale Ronderib blinkhaar Africander rams R30 and studs R75. Also young ewes at R25. *Farmer's Weekly Advt.* 27.2.74

3. *obs.* An *Afrikaner* (q.v.), see quot. at *Dutch.*

African Hotel *n. obs.* Travellers' or colonists' term for a rough-and-ready usu. country hostelry in S.A. *cf. Canad. stopping house, keepover.*

We put up at an African Hotel. Such places were to be found in all parts of the Colony, from 20 to 40 miles apart, the proprietors of which were invariably large cattle and sheep breeders. Buck Adams *Narrative* 1884

African print *n.* **1.** See *Kaffir print.*

2. See *Swazi print.*

African time *n. phr., modifier etc. colloq.* Late, behind schedule: jocular reference to unpunctuality used by both Blacks and Whites. *Informant* Mr. Clifford Sileya, Salisbury, 1977.

Two fifteen African time means four o'clock. Mackenzie *A Dragon to Kill* 1963
. . . Communion this morning . . . Real African time set-up here in . . . The service was at 7.30 and at eight o'clock the first person rocks up. *Letter, Schoolgirl* August 1974

Afrikaans [afrɪ'kɑns] *n.* **1.** *n. prop.* The language of the *Afrikaner* (q.v.) people, evolved as a distinct language from what was formerly known as *Cape Dutch* (q.v.); one of the two official languages of S.A. (see *bilingual*). [*fr. Du. adj. Afrikaans(ch)* of Africa]

Afrikaans. The language of the Afrikaner, a much simplified and beautiful version of the language of Holland, though it is held in contempt by some ignorant English-speaking South Africans, and indeed by some Hollanders. Paton *Cry, Beloved Country* 1948

During its brief life (100 years) Afrikaans, once a child of protest and dissent has triumphed over public contempt and private slight, to become the mother tongue of the majority of White and Brown men in this corner of the sub-continent. *Sunday Times* 17.8.75

In 1974 the Pretoria Government Education Department ruled that students in Soweto's schools…must take some subjects in Afrikaans, the Dutch-based language that is one of the two official languages for white South Africa. *Time* 28.6.76

In frequent combination ~ *-speaking.*
. . . many immigrant wives would like to get jobs, but they don't speak Afrikaans, and find it difficult to get jobs in predominantly Afrikaans-speaking areas in the country. *The 1820* Apr. 1971

2. *n. modifier* or *adj.* Of or pertaining to the language or its speakers.

Afrikaans words like trek, kop, donga, dorp and drift – hill, gully, village, and ford – used by the war correspondents, brought the African veld into the parlours of Brixton and the pubs of Highgate. Cloete *Rags of Glory* 1963

And we were very Afrikaans – but the old type of Afrikaner, not the kind one sees today. *Sunday Tribune* 1.4.73

Afrikaans farm *n. pl.* -s. A farm where only *Afrikaans* (q.v.) is spoken, open during school vacations to English-speaking children as paying guests, enabling them to combine language learning with a farm holiday.

Addresses of Afrikaans Farms are: . . . Grahamstown Headmistress's letter 1973

Afrikander see *Africander.*

Afrikaner[1] [ˌafrɪ'kɑnə(r)] *n., n. prefix* and *adj.* **1.** *pl.* -s. An *Afrikaans*-speaking S. African. See quot. at (1) *Dutch* and (2) *Afrikaans* also at (2) *Vaderland.* [*Afrikaan(s)* + *personif. suffix -er cf.* British-*er*].

2. *n. prefix* ~*dom* [~ˌdɔm] *n. abstr.* the ~ nation and/or its patriotic spirit. *cf.* Christendom. ~*skap*, the state of being an ~. [*suffix cogn. Eng.* -*ship*]

Addressing the Rapportryers, an organisation of hard-core Afrikaner Nationalists whose object in life is to uphold and promote Afrikanerdom, Mr Vorster hinted that all racial groups would have to share certain facilities. *E.P. Herald* 2.10.76

3. *attrib.* Of or pertaining to the ~ people or nation. See (2) *Afrikaans* and quot. above.

Afrikaner[2] *n. pl.* -s. Any of several indigenous species of *Gladiolus* including those of the *aandblom* (q.v.) having various colours, brown ~, pink ~, also the red ∼ or *sandveldlelie.*

G. grandis (brown Afrikander) . . . bears
flowers of a brownish hue which give off a sweet
scent in the evening. Van der Spuy *Gardening
in S.A.* 1953

Afrikanerism [afrɪˈkɑnərˌɪzm̩] *n. pl.* -s. A
linguistic usage transferred from Afrikaans to English *cf.* Americanism, Ca-
nadianism etc., formerly *Africanderism*
(*obs.*)

Afriks [ˈafˌrɪks, ˈɒf-] *n. colloq.* Also *Afrix:*
among English-speaking children: Afri-
kaans language and literature as a school
subject, see also (3) *Boet* and (3) *Dutch.*
[*abbr. Afrikaans*]

I got another result today, Afriks, and I got
57% – not bad considering . . . Letter Schoolgirl 9.7.74

Ek sê vir jou I nave improved my speech so
much that when I speak Afriks I'm recognized
as an Englishman. *Letter, Serviceman* 13.1.79

afslag [ˈafˌslax] *n. hist.* Downward bidding, an early Dutch method of auctioneering, see *opslag*[1], also *strykgeld.*
[*Afk. fr. Du. afslag* reducing auction]
. . . the property was sold by afslag, or downward bidding. The auctioneer then asked for a
buyer at a figure generally about double that of
the highest bid. When there was no response
he gradually lowered the figure until someone
shouted out 'Mine' and the deal was concluded.
Gordon-Brown *S.Afr. Heritage* 1965

afterclap *n. pl.* -s. The tailboard or the
canvas flap at the rear of the *tent* (q.v.) of
a covered wagon. [*trans. Afk. agter fr.
Du. achter* hind + *klap* flap]

He was mending the afterclap of the wagon,
stitching it up where it was torn. Cloete *Hill of
Doves* 1942

after-ox *n. pl.* -en. One of the hindmost
pair in a *span* (q.v.) of draught oxen:
equivalent of 'wheeler' in a team of
horses. *cf. Jam. Eng.* tongue cattle, *Austral. poler.* [*trans. Afk. agteros* (q.v.)]

I remember some of them now far better
than many of the men known then and since:
Achmoed and Bakir, the big after-oxen who
carried the disselboom contentedly through
the trek and were spared all other work to
save them for emergencies. FitzPatrick *Jock
of the Bushveld* 1907

afval [ˈaf(f)al] *n.* ∼ A traditional S. Afr.
dish made usu. on farms consisting of
sheep's head and trotters: see *kop-en-
pootjies,* and occ. tripe. [*Afk. afval cogn.*
offal ‖This differs from the *harslag* or
pluck]
. . . the superb food which will always be the
food of her childhood . . . waterblommetjie-

bredie, bobotie, afval, and . . . a myriad of
crayfish. *Sunday Times* 19.2.78

ag [ax] *interj.* Oh: a frequent exclamation
expressing various moods, often indicating irritation or exasperation if used
alone or in combination with *man* (q.v.)
or *sis* (*sies*) (q.v.) see also quot. at *ach.*
[*Afk. fr. Du. ach as in Ger.*]

'*Ag,* old Bushes,' . . . 'he won't fight all
three of us. And I am myself quite strong.'
Philip *Caravan Caravel* 1973

again *adv. particle substandard* freq. used
in S.A.E. redundantly, not signifying
repetition but as a type of intensifier
roughly equiv. to 'really' or 'indeed';
see also *something else.* [*trans. fr. Afk.
weer as in* 'That is something else again'
(*Dit is weer iets anders*)]

Cold in London is nothing new, a cold in the
head is something else again. *Fair Lady* 6.9.71

You can't buy glamour. Queen Elizabeth II
is one of the richest women in the world . . .
she isn't glamorous. Jackie Onassis is something
else again . . . moves in a glamorous world of
her own. *Darling* 26.11.75

agter- [ˈaxtə(r)] *adj. prefix* Hind, rear:
[*Afk. fr. Du. achter cogn.* after] **1.** Commonly prefixed to nouns, usu. with equiv.
form with *prefix voor-* (fore) (q.v.) *cf.*
fore deck, after deck, *e.g.* ∼ *karos* see
kaross; ∼ *kis* see *wakis;* ∼ *laaier* (q.v.);
∼ *ryer* (q.v.) rider ; ∼ *slag* (q.v.) ; ∼ *skot*
(q.v.).

2. Back, behind: found in place names.
Agtersneeuberg, Agtertang, Agterplaas.

agterkamer [ˈaxtə(r)ˌkamə(r)] See *voorkamer.*

agterlaaier [ˈaxtə(r)ˌlaɪə(r)] *n. pl.* -s. A
breech-loading gun: see also *voorlaaier.*
[*Afk. agter fr. Du. achter* hind, rear +
laai fr. Du. laden load + *agent. suffix* -er]
. . . only a few burghers possessed breech-
loading rifles, achterlaaiers, as we call them.
De Wet *Three Years War* 1903 *cit.* Swart
Africanderisms: Supp. 1934

agteros [ˈaxtərˌɔs] *n.* **1.** See quot. at
after-ox.

2. *figur. colloq.* A plodder; see also
sukkelaar cf. Austral. battler : also one
who lags behind as in *Afk.* proverb ' ∼
kom ook in die kraal' roughly equiv. to
'slow but sure': see quot. at *voorbok.*
[*Afk. fr. Du. achter* hind, rear *cogn.* after
+ *os cogn.* ox]

agterryer [ˈaxtə(r)ˌreɪə(r)] *n. pl.* -s. **1.** *obs.*
A mounted groom or other attendant.
[*Afk. fr. Du. achter* behind *cogn.* after +

rijder cogn. rider]
. . . two boors on hərseback, attended by two Hottentot *achter-ryders* . . . passing by, halted . . . These *achter-ryders* . . . correspond to many of our English grooms. Burchell *Travels II* 1824

2. *Prison slang:* a man in charge of a labour gang or *span* (q.v.). [*presum. fr. above*]
In a big prison in the Transvaal all the agter-ryers (boss-boys) in a big span working on a site on which a dam was being built were Big Five. *Sunday Times* 10.10.76

agterskot [ˈaxtə(r)ˌskɔt] *n.* Final payment see quot. at *voorskot.* [*Afk. agter, cogn.* after + *skot,* shot *sig.* payment]
. . . woolgrowers were losing interest on the 'agterskot' which would only be paid out at the end of the wool season. *Evening Post* 30.9.72

agterslag [ˈaxtə(r)ˌslax] *n. pl.* -s. The thong or lash of a whip: see quot. at *voorslag.* [*Afk. fr. Du. achter* hind + *slag* lash]

aia [ˈaɪə] *n. pl.* -s. Child's nurse, usu. coloured, also an old coloured woman: see quot. at *ayah:* also occ. mode of address with first name; feminine of *outa:* see quot. at *outa, cf. Anglo-Ind. ayah,* nursemaid, servant, *Hong K. amah,* female servant [*Port. aia* nursemaid *prob. fr. Hindi aya* female servant]

ai-ai [ˈaɪˌaɪ] *n.* Absolute alcohol used for lacing drinks illicitly and sometimes causing death; also known as *aeroplane.* [*etym. dub. poss. fr. Xh. hayi* no, *poss. fr.* A.A.]
. . . a couple collapsed and died in an Orlando West shebeen after drinking a scale of maiza doped with deadly ai-ai. *Post* 18.1.70

aid centre *n.* A centre to which persons arrested on pass offences are referred for investigation and assistance: see also second quot. at (1) *pass.*
. . . the identification documents . . . had not been re-submitted to the Aid Centre for further investigation. In terms of a Ministerial directive all Africans arrested for petty offences such as a failure to produce a reference book, should first be referred to an Aid Centre so that their cases are investigated. *E.P. Herald* 8.3.74

aikona [ˌaɪˈkɔnə] *interj. colloq.* An emphatic negative 'never!' or 'not on your life!' *cf. Austral. ba(a)l.* [*Ngu. hayi* no + *khona* be here, present]
Boesman: You mean that day you get a bloody good hiding.
Lena: Aikona! I'll go to the Police.
Fugard *Boesman & Lena* 1969

Idiomatic? Aikona – it's idiotic! . . . If that is the way to teach a boy to speak his home language then, aikona, I'd rather he didn't learn it. *E.P. Herald* 15.9.71

akkewani [akəˈvanĭ] *n.* A plant of the genus *Cymbopogon* with odorous roots used when dried to discourage moth. [*Malay akar* root + *wangi* fragrant]
The creeping and fibrous roots of this grass have a peculiar and rather ferulaceous smell. By the name of Akarwanie they are known to most colonists, and serve as a sure preventive against the destruction of wearing apparel etc. by moths and other noxious vermin. Van Pappe in *Cape of G.H. Almanac* 1856

alderman *n. pl.* -men. See *raadsheer.* ↗ Not equiv. of *Brit.* ~.

alfkoord [ˈalfˌkʊə(r)t] *n.* One of several corruptions of *albacore,* the edible fish *Seriola lalandi* also known as *halfkoord* (q.v.) and Cape *Yellowtail* (q.v.) [*fr. albacore prob. fr. Arab. albukr* young camel]

algemene handelaar [ˈalxəˌmɪənə ˈhandəlaːr] *n.* See *general dealer.* [*Afk. algemeen(e)* general + *attrib. suffix* + *handelaar* trader]
At one time the 'algemene handelaar' was as much a part of the South African countryside as the sunshine and the wide, open spaces. But today there are only a few left. *Panorama* Apr. 1974

alie [ˈalĭ] *n. pl.* -s. A marble; *cf. Eng.* and *U.S. ally, blood-ally, Canad. smoke alley.* See also *ghoen, ironie, yakkie.* [*fr. Du. albast, Lat. albastrum* marble]
. . . two boys were content to stand looking on. Both had sold all their ghoens and allies to swell the P.E. School feeding fund. *E.P. Herald* 3.4.72

alikreukel [ˌalĭˈkrǐækəl] *n. pl.* -s. An edible mollusc *Turbo sarmaticus* of the Turbaniidae also known as *arikreukel, ollycrock* (q.v.): see quot. [*Gk. (h)alikochlos* sea snail]
The menu will include Danish mussels, oysters, chokka, crayfish thermidor, nasi goreng and arikreukel, a delicacy much favoured by gourmets. *Het Suid Western* 17.5.71

allemagtig [ˌaləˈmaxtɪx] *interj.* See *magtig.*
The Boer expostulated *Alamachtigs* with considerable vigour, and spat contradictions. Cohen *Remin. Johannesburg* 1924

alles sal regkom [ˈaləs ˌsal ˈrexˌkɔm] *idiom. Afk.* saying 'Everything will resolve itself, turn out well': an optimistic or consolatory remark; see *come right,* and quots. at *moenie worry* and *toe maar.* [*Afk. alles* everything + *sal* will + *reg*

right + *kom* come]
'Moenie worry nie – alles sal regkom' is the mood. The rest of the world, Africa in particular, is in a mess. *Argus* 5.5.73

alles van die beste ['aləs ˌfan dï 'bestə] *idiom.* All the best: usu. a farewell greeting. [*Afk. lit.* all of the best]
   There are handshakes, warmly grasped elbows and deep, meaningful looks. Totsiens! Alles van die beste, oom! *Argus* 28.10.72

allewêreld [ˌaləˈveːrəlt] *interj.* An exclamation, usu. of surprise, *cf. allemagtig*; see *magtig.* [*Afk. alle* all + *wêreld* cogn. world]
   And what a catch! Alle Wêreld! Surely there had never been such a catch – kabeljou, geelbek, Roman, stompneus, Miss Lucy, Hottentot and steenbras. Meiring *Candle in Wind* 1959

alliteral concord Also *alliterative concord:* see *euphonic concord.*

almanac ['ɔlməˌnæk, ˌalmaˈnak] *n. pl.* -s. *hist.* The early Court Calendar, directory-cum-handbook produced at the Cape from 1801 onwards; see quot. at *Court Calendar.* [*Du. almanak* calendar]
   The South African Journal is also published every two months, besides two almanacks of an excellent character every year. Webster *Voyage I* 1834
In combination *Kaffir* ~ ; two flowers *Haemanthus katherinae* and *H. magnificus* the flowering period of which is said to be taken by some Africans to be the correct sowing time for crops.

aloe ['æˌləʊ] *n. pl.* -s. Generic name both for a large number of species of African aloe and for certain other superficially similar genera e.g. *Agave*, the 'American ~' see *garingboom*. ⫞ ~ *juice* is still produced commercially esp. fr. the sap of *A. ferox* for the manufacture of the drug *Aloes:* see quot. at *sea-cat.*

already *adv.t. substandard.* Used redundantly or in deviant word- or time-sequence in S.A.E. e.g. 'He came when I was asleep already.' [*trans. Afk. al* already, yet, before now, *a particle used for indicating perfective e.g. Ek het dit gedoen* I did it, *Ek het dit al gedoen* I have done it (already)]
   ... the old Baas had told her to pay for the sugar and coffee she had bought three weeks ago already! Meiring *Candle in Wind* 1959

amaas [aˈmaːs] *n.* see *maas.*

amabunu [amaˈbunŭ] *pl. n.* African name for the *Afrikaners (Boers)*, usu. derogatory. [*Zu. pl. prefix ama* + *bunu fr. Afk.*

boer]
   One chief told me he would like nothing better than to be allowed to lead an *impi* against the Amabuna (Boers). Mitford *Through Zulu Country.* 1883 cit. M. D. W. Jeffreys
   ... asked her whether she had not heard of white people hereabouts ... in the years before 'lo ma Bunu' (the Boer) and the Red Coats had fought. McMagh *Dinner of Herbs* 1968

amadaki [ˌamaˈdakï] *pl. n. colloq. Afr. E.* African people: by extension used of the brown African prints: see *German Print, bloudruk, Kaffir print, Duitse sis*, also *amagerimani.* [*prob. fr.* darkie (*pl. prefix ama-*)]
   AMADAKI
   R1,60 per metre
   AMAGERIMANI
   R1,50 per metre
   Available at
   Commercial Road, Phone ...
   Advt. Indaba 23.3.79

amadhlozi [ˌamaˈdlɔzï] *pl. n.* Ancestral spirits: also *amatongo* (Xh. *imishologu* and *izinyanya*): see quot. at *amatongo.* [*Ngu. pl. prefix ama* + (*i*)*dhlozi* ancestral spirit]
   To qualify as a top witchdoctor, he had to leave town life to call his amadlozi (spirits) of his dead ancestors near a spruit at Protea. *Post* 28.4.68
   Then Chief ... told his followers that it was only through the intercession of his madlozis that he was still standing there to address them. *Drum* 8.11.75

amadoda [ˌamaˈdɔda] *pl. n. occ. pl.* -s. Grown men. ⫞ Also a sign on African men's public lavatories: see quot. at *abafazi* also *madoda.* [*Ngu. pl. prefix ama* + (*in*)*doda* grown man]
   The rite of circumcision is performed on all the young men at the age of sixteen who are thus made men or amadodas. Bisset *Sport & War* 1875

amadumbi [ˌamaˈdŭmbï] *pl. n.* See *madumbi.*

amagerimani [ˌamaˈdʒerïmanï] *pl. n.* ~ *colloq. Afr. E.* Used of the blue *German prints* (q.v.) also *Duitse sis*, and *bloudruk:* see quot. at *amadaki.* [*fr.* German + *pl. prefix ama-*]

amagoduka [ˌamaɣɔˈduːga, -ka] *pl. n.* Migrant labourers: 'returning ones' who go home at regular intervals: see quots. at *goduka* and *u-clever.* [*Ngu. pl. prefix ama* + *-goduka* to return home]
   Hundreds of Amagoduka -- migrant labourers – went on the rampage in Langa on Saturday night hitting everyone in sight with knobkerries. *E.P. Herald* 26.3.73

amajoni [ˌamaˈdʒɔnĭ] *pl. n.* African term for White soldiers. [*Ngu. pl. prefix ama +(i)joni poss. fr. Johnny,* (q.v.) *analg.* 'Tommy', *or fr. vb. ukujoina fr. Eng.* join]
When the *amajoni* are mustered . . . the trumpet is blown. Mitford *Romance of Cape Frontier* 1891 *cit.* Pettman

amalaita [ˌamaˈlaɪta] *pl. n. occ. pl.* -s. Street thugs operating after dark, or stick fighters; see also *tsotsi.* [*Ngu. pl. prefix ama + laita fr. -layitha (Zu.),* a street desperado, hooligan, *acc. some fr. Eng. vb.* 'loiter', *more prob. rel. Eng.* 'light' *fr. use of flashlight on victims*]
. . . he was beaten up by the amelitas, and now he feels that the gay life is overrated. Cowin *Bushveld, Bananas Bounty* 1954

amandla [aˈmandla] *n. interj.* Power: usu. ~ *ngawetu;* see quot. [*Zu. pl. n. amandla* power]
'. . . giving the Black Power salute and shouting in reply . . . amandla (power) ngawetu (is ours).' *E.P. Herald* 2.11.73
. . . some students began picking up stones. Shouting '*Amandhla* (power)', they moved haltingly towards the police. *Time* 28.6.76

amanzi [aˈmanzĭ, -mɑn-] *n.* Water: found in *Zu.* place names e.g. Amanzimtoti (sweet), Amanzimnyama (black). [*Ngu. amanzi* water]

amapakathi [ˌamapaˈgatĭ, -ˈka-] *pl. n.* The inner circle of advisers, closest to an African Chief, *cf.* privy councillors: also *sing. umphakati.* [*Xhosa, Ngu. pl. prefix ama + phakathi* inside one]
Their curiosity . . . and . . . pilfering, became so troublesome that I was obliged to ask the chief to appoint one of his men as a sort of sentry or police officer. This was readily granted, and an old umpagate . . . with great good-will and a few hard strokes . . . soon dispersed the crowd. Shaw Nov. 1823 *cit.* Sadler 1967
Great chiefs (*incosee incoolo*) are assisted by *amapakati* or counsellors. These are experienced old men, or wise young ones. Alexander *Western Africa I* 1837

amaphepha [ˌamaˈpepa] *pl. n. (erron. pl.* -s). *colloq. Afr. E.* Paper money, notes of any denomination: see also *mali.* [*Ngu. fr.* paper]
This means these people must fork out a thousand maphephas for costs and the cops want their pound of flesh. *Post* 10.11.70

amaQaba [amaˈkaba+] *pl. n.* see *Qaba*

amasoka See quot. at *bachelor quarters*

amatongo [ˌamaˈtɔ̆ŋgɔ̆] *pl. n.* Also *amadhlozi,* ancestral spirits revered by tribal *Africans* (q.v.) as agents of protection, also of prophecy via the medium of the witchdoctor (q.v.) in *Xh.* known as *imishologu* or *izinyanya.* [*Ngu. pl. prefix ama + (i)thongo* ancestral spirit]
And those who died in the fight will now become Amatongo.
And those who escaped, whose national Amatongo looked on them and saved them say, 'We have been saved by the Amadhlozi of our people'.
When they come back from the army, they sacrifice cattle to the Amatongo. Callaway *Religious System of Amazulu* 1869

amatopi [amaˈtɔ̆pĭ] *pl. n. sing. itopi. Urban Afr. E.* Persons of a mature age group: adopted in the form 'toppies' 'old toppies' or 'old tops' *sig.* 'oldies': see *toppie;* also *uMac, uscuse me, uclever.* [*fr. topi, Persian* hat, see quot.]
. . . still respectable is the middle aged and elderly type known as amatopi from topi the pith helmet worn by an earlier generation of Europeans . . . the *amatopi* proper are over 45. Wilson & Mafeje *Langa* 1963

amatungulu [ˈamaˌtŭŋˈgulŭ] *pl. n. Carissa macrocarpa,* a spiny plant popular for hedges: the tart-flavoured scarlet fruit, also known as Natal plum and *noem-noem-bessie* or *num-num,* is edible and used for making jelly and jam. [*Ngu. plant name, Zu. form -thungulu*]
Carissa grandiflora . . . called Knum-Knum by the Boers, amatungulu by the Kaffirs a bush with sweet smelling flowers and red gooseberry-like berries with a pleasant flavour. Krauss *Travel Journal* 1838–40 trans. Spohr 1973

American aloe *n. Agave americana:* see quot. at *garingboom.*

anaboom, anatree See *apiesdoring.*

A.N.C.[1] [ˈeɪˌenˈsiː] *n.* The African National Congress now banned in S.A., having a 'thumbs up' sign, and song *Mayibuye i Afrika* (q.v.) [*acronym*]
. . . used code names . . . for instance . . . leader of the banned African National Congress, was known as 'the King'. The ANC itself was known as 'mama'. *Sunday Times* 9.11.75

A.N.C.[2] *n.* African National Council (Rhodesian): see quot. at *Zimbabwe.* [*acronym*]

A.N.C.[3] See *ante-nuptial contract.* [*acronym*]
. . . we were married by ANC . . . In the divorce summons you can claim . . . the R3 000 which he is bound to pay you in terms of the ANC. *Darling* 22.12.76

-and-them [ˈ-ən(d)-ðəm, -ðem] part of *n.*

phr. colloq. Used preceded by a name to refer to a group of people, not necessarily a family, previously mentioned or known to both speakers as in 'When are Bill-and-them coming?'; *cf. Jam. Eng. an-dem* either as *pl. the cow-an-dem* cows, or *John-an-dem* John and company; also *Southern U.S. you-all* as mode of address to more than one person. [*fr. Afk. hulle* them *e.g. (Jan)-hulle sig.* Jan and his family, Jan and company *etc.*]

... I'll have to keep this one quiet or my boet and them'll kill themselves larfing. *Darling* 29.1.75

angels' food *n.* Fruit salad: also *Jam. Eng. cf. U.S. angels' food (cake).*

... bowls of sliced paw-paw, oranges, grenadillas, bananas and pineapples. 'Angels' food,' said Nan. 'A Ball would not be complete without Angels' Food.' Westwood *Bright Wilderness* 1970

angler See *monk* (Lophiidae fam.).

Anglikaans [ˌæŋglɪˈkɑns] *n.* A mythical hybrid language compounded of English and *Afrikaans* (q.v.) *cf. Canad. Franglish, Franglais.* [Angl- *fr.* English (*Lat. angl-*) + *-kaans fr.* Afrikaans]

... If Anglikaans includes all these and all departures from traditional English and 'correct' Afrikaans in favour of South Africanisms of every variety, is there any likelihood that Anglikaans will eventually become the dominant South African vernacular? *Cape Times* 8.1.72

Anglo-Boer War See *Boer War.*

Angolsh goulash See *rat pack*

animal unit See *stock (unit).*

anker [ˈæŋkə, ˈaŋkə(r)] *n. hist.* An old *Du.* and *Ger.* liquid measure formerly in use at the Cape, of between 30 and 40 litres, or a quarter of one *aum* (q.v.), also a keg or cask containing this amount. [*Du. anker measure*]

LIQUID MEASURE. 16 Flasks, equal to 1 Anker. – 4 Ankers equal to 1 Aum. 4 Aums, equal to 1 Leaguer. *Greig's Almanac* 1833

antbear See *aardvark.*

ante-nuptial contract *n. pl.* -s. An agreement made by two persons intending to marry, regarding their property and usually also the legal capacity of the wife, the effect being that each retains full control over his or her separate property during the marriage. ⸿ Persons married in South Africa without an ~ are automatically married *in community of property* which results in the

assets of each of them becoming the common property of both, under the control of the husband. (In Natal until 1956 *postnuptial contracts* were also permitted.) See also *kinderbewys, boedelhouder* and *boedelscheiding, A.N.C.*[3].

ant-heap [ˈæntˌhip] *n. pl.* -s. Anthill: often hollowed out by early pioneers for use as bake-ovens. See also quot. at *mamba.* [*trans. Afk. miershoop, miers* ant + *hoop* heap]

At times we baked our own bread in ovens of antheaps. *Drum* Aug. 1971

anyswortel [ˌaneɪsˌvɔ(r)ˈtəl] *n. pl.* -s. Any of several species of *Annesorrhiza* having anise-flavoured tubers which were eaten in early times by indigenous tribes and by travellers of the period. [*Afk. anys fr. Du. anijs cogn.* anise + *wortel* root]

'The root of the Anise (anys-wortel) was eaten here roasted, and tasted well; it is either roasted in the embers, or boiled in milk, or else stewed with meat.' Thunberg *Travels I* trans. 1795

apache-snor [snɔr] *n. phr. pl.*-re *Sect. Army slang.* Used of a usually somewhat sparse or *yl* (q.v.) moustache as grown by *oumanne* (q.v.) as soon as they are permitted to do so, to show their status. *cf. Harley Davidson* (q.v.) [*Afk. snor,* moustache]

Apache-snor-moustache ... 'a patchie here a patchie there' ... *Informant* H. C. Davies *Letter* 2.2.79

apartheid [əˈpɑtˈheɪt, aˈpart(h)eɪt] *n.* 'Separateness': in use since the late 1940s sig. racial separation at various levels: *grand* ~ or *separate development* (q.v.) as a major political policy, as in the setting up of African *homelands* (q.v.) or *Bantustans* (q.v.) or *petty* ~ such as separate entrance doors, park benches and other amenities for White and non-White: see also quot. at (3) *Xhosa,* [*Afk. apart* separate + *n. forming suffix -heid equiv. Eng.* -ness. *cogn.* -hood *fr. O.E. had* state of being]

These people converted to Islam to escape from the oldest *apartheid* that is known to man – which is represented by the caste system of the Hindu society. *Commonwealth* June 1971

As the leader of a Bantustan, you have accepted separate development. Does this mean you accept apartheid? 'Stripped of the racial aspects that humiliate and hurt black men, apartheid – or separate development as the government prefers to call it – simply boils down to partition.' Gatsha Buthelezi *cit.*

Newsweek 27.11.72

For more than a generation the cornerstone of South Africa's internal policy has been *apartheid*, or 'separate development' for the country's 16 million blacks, 3 million whites, 2 million 'coloureds'. *Time* 15.10.73

Petty apartheid was probably the widest front on which a man's dignity was assaulted and ... almost the entire population was acutely aware of how it operated. *E.P. Herald* 9.11.73 *also figur*. In combinations freq. non-S.A.E. *age* ~, *cultural* ~, *ecclesiastical* ~, *industrial* ~, also *white* ~.

The Archbishop of Canterbury, Dr Fisher, drew a parallel yesterday between the political *apartheid* which he has seen in South Africa, separating the nation, and ecclesiastical *apartheid* which prevented unity among the churches. *Times* 1955 5 July 6/3 *cit.* OED Supp. '72

Conservative Leader Margaret Thatcher castigated the proposals as being 'not for industrial democracy but for industrial apartheid'. *Time* 14.2.77

. . . the . . . Nursery School. Thank heaven there is no 'white apartheid' there. Children are children whatever language they speak. *Het Suid Western* 19.4.73

apiesdoring [ˈɑpɪsˌdʊərɪŋ] *n. pl.* -s. Any of several species of *Acacia* having high leafy crowns affording cover to monkeys, including *A. albida* known as *anaboom/tree*. [*Afk. aap* monkey *cogn.* ape + *dimin. suffix* -ie + *doring fr.* Du. *doorn cogn.* thorn]

appelkoossiekte [ˈapelˌkuəsˈ(s)iktə] *n.* See *apricot sickness* [*Afk. appelkoos fr.* Du. *abrikoos* apricot + *siekte fr.* Du. *ziekte* disease]

Apple Express *n. prop.* The narrow gauge railway between Port Elizabeth and Avontuur, which traverses the fruit growing area of the Lang Kloof [opened 1905].

Exactly 70 years after the first tiny trains started running on Port Elizabeth's 'Apple Express' railway, powerful diesel locomotives have been introduced to modernise what is probably the busiest and most useful narrow-gauge line in the world. *The 1820* May 1974

apricot sickness *n.* Diarrhoea, occ. with vomiting, prevalent in the summer stoned-fruit season, and thought to be caused by eating too much, or unripe fruit.

'Apricot sickness' . . . gained its name because apricots were the first spring fruits which the early Cape farmers produced. They suffered from the familiar griping pains and diarrhoea, and rightly blamed the apricots. Green *Land of Afternoon* 1949

Whether you call it the 'trots', a 'runny tummy', 'Apricot sickness', 'gastric flu' . . . puts a stop to it. *Sunday Times Advt.* 22.12.74

Arab (merchant) *n. pl.* -s. See *Bombay Merchant*.

The white colonists, through ignorance and lack of interest, referred to all Indians as coolies. It was only when educated, intelligent Bombay Merchants and shipowners came upon the scene with cargoes of rice and condiments for sale, that the townsfolk learned to discriminate between what they then designated 'Arab merchants' and 'coolie shopkeepers'. Tait *Durban Story* 1961

arad See *borrie*, also *Indian terms*.

area *n. pl.* -s. *Sect. Army.* The ~ immediately surrounding each individual National Serviceman's bed, locker (see *kas*) etc. in his *bungalow* (q.v.). ⸿ Other men's ~s must usu. be crossed on '*taxis*' (q.v.): see quot. at *varkpan*.

Argentine ant *n. pl.* -s. *Iridomyrmex humilis:* a small blackish-brown ant introduced into *S.A.*, the *U.S.* and *Austral.* from S. America, a common household pest, also troublesome in orchards.

arikreukel [ˌɑrɪˈkrɪœkəl] *n.* see quots. at *alikreukel* and at *ollycrock*.

arm [ɑm, arəm] *n. pl.* -s. *Sect. Drug users* A measure of *dagga* (q.v.) [*unknown*].

AN INITIATE'S GUIDE TO DAGGA SMOKING
Pill, slow boat: Dagga cigarette
Katchie aksie parcel: Ten or five cent roll of dagga
Arem (arm): One rand parcel of dagga *Drum* 27.8.67

We smoked a whole arm of boom and drank out a big can of vlam. Muller *Whitey* 1977

. . . it contained four 'arms' of dagga. An 'arm' weighs about 2 kilograms. *E.P. Herald* 5.6.79

army worm *n. pl.* -s. The destructive larva of any of several moths, so named from the massed line in which they advance across the veld: also *U.S.* [see quot. also *masonga*]

ARMY WORM (Laphygma exempta) This pest has captured the imagination of the public as its appearance on farm lands, is often announced in the press . . . The larvae march together in the mass, eating only grass, maize and other members of the grass family. Eliovson *Gardening for S.A.* 1960

articles Certain non-standard uses of the definite and indefinite articles are characteristic of S.A.E. **1.** redundant *the* in addressing persons: see *third person address* and quot. at (2) *kaross*.

2. redundant *a* or *an:* see also quot. at (3).

They questioned several people living in the house for many hours and came within a half an hour of making an important arrest. *E.P.*

Herald 12.3.74
3. *pron. sp. a* for *an :* see also second
quot. at *môre is nog 'n dag* [*fr. Afk. 'n*
[ə] *indef. art.* a, an *fr. Du. een*].
    . . . decorated with a half a olive, bits of
gherkin. *Darling* 24.12.75
arvie, arvey *n. colloq.* Afternoon ; also
'sarvie: usu. among children *cf. Austral.*
arvo, afto. [*presum. fr. Eng.* after(noon)]
    I'm there at the municipal baths one arvey
trying to catch a tan. *Darling* 9.10.74
A.S.B. [ˈeɪˌesˈbiː, ˈɑˌes-bɪə] *n.* The *Afri-*
kaanse Studente Bond: an organization
of *Afk.* speaking university students:
see also *NUSAS.* [*acronym*]
    The Afrikaanse Studentebond (ASB) has
come a long way during the past ten years.
Having attended eight annual congresses in
this period I have no doubt that the ASB is
breathing a different spirit today. *Sunday Times*
14.7.74
ask *vb. trns. substandard.* Used as equiv.
of 'ask for'. [*trans. Afk. vra* ask for,
usu. with concrete object]
    'Roast fowl, roast . . . potatoes, fried steak . . .
vetkoekies . . . ' '. . . we wouldn't dare give you
any of those things . . . maar ask something
reasonable.' Stormberg *Mrs P. de Bruyn* 1920
    I must . . . ask boodle from the boss. *Drum*
22.9.75
askoek [ˈasˌkŭk] *n. pl.* -s, -ke. *occ. as-*
brood : dough cake baked in hot embers:
see also (2) *stormjager, maagbom cf.*
Austral. damper, devil-on-the-coals, Jam.
Eng. bammy. [*Afk. fr. Du. as cogn.* ash
+ *koek cogn.* cake]
    Another way of making bread is what is
called an *Ash cookie.* It is something akin to
an Australian 'damper'. Browning *Fighting*
& Farming in S.A. 1880 cit. Pettman
    On trek, of course, there is still *asbrood,*
dough baked in the ashes. Green *Karoo* 1955
~ **slaan** *n.* A lively dance formerly of
the *Hottentots* (q.v.) in which the click-
ing of heels was said to sound like
knocking ~*s* together, now, loosely,
to dance a reel. [*Afk. fr. Du. slaan* beat,
strike]
ASSA [ˈæsə] *n. pl.* -s *acronym* *A*frikaans
*S*peaking *S*outh *A*frican: see also *ESSA*
Assa plus Essa – a tower of Babel
    To be born with an ASSA heart and an ESSA
head, is, as they say in the suburbs, only complicated, hey ! I am a member of the Progressive
Federal Party, and my ASSA heart urges me
that . . . good intentions are no use unless . . .
translated into action. My ESSA head says,
well I vote for the party so if there's a meeting
. . . why should I go . . .? Whereas my Nationalist relatives go to meetings at any time even

if they have to go on crutches. *Sunday Times*
8.10.78
assegai [ˈæsəˌgaɪ] *n. pl.* -s, also *vb. trns.*
1. Spear, either short, for stabbing, as
introduced by Shaka for the Zulu armies,
or long, for throwing, usu. with an iron
blade: used from earliest times by Africans both in hunting and war. See
(*u*)*mkonto* [*orig. Berber zagayah* lance,
spear + *Arab. prefix al* the]
    A police task force . . . have made 53 arrests
and 83 battle axes, 11 assegais and 73 battle
sticks have been confiscated. *Daily Dispatch*
6.10.71
2. *vb. trns.* To stab with an assegai.
[*fr. n.* ~]
    The Zulu lives long and if you give a penny
to an ancient sitting on the kerb in Durban
today you may place it in the hand that assegai'd
that Prince: Reed *South of Suez* 1950
assegaibos [ˈæsəˌgaɪˈbɔs, ˌasəˈxaɪ-] *n.*
Grewia occidentalis, the strong, resistant
branches of which were formerly used
by Hottentots and Africans for *asse-*
gai (q.v.) shafts. Also known as *pylhout*
(q.v.): found in place names Assegaibos-
rivier, Assegaayenbosch. [*assegai* (q.v.)
spear, lance + *bos* bush]
assegai wood, assegaihout [ˈæsəˌgaɪˈwʊd,
ˌasəˈxaɪˌhəʊt] *n.* The timber of *Curtisia*
dentata or *C. faginae* used from early
times in wagon-making: also known as
assegaiboom, see *assegaibos.* [*Afk. -hout*
wood, timber + *-boom* tree]
    Rafters of keur-boom taken from the river
banks, and Papa has had made for her a set
of chairs from assegay-wood grown at Pigot
Park, this similar to plain mahogany. Kate
Pigot *Diary* 1826 cit. Fitzroy 1955
atjar [ˈaˌtʃa(r)] *n.* A hot pickle or relish of
fruit or vegetable usu. in oil or vinegar
prepared with chillis and other curry
spices: see quot. at *fish oil.* [*prob. fr.*
Persian a(*t*)*sjar* sour]
    Green Mango Atjar Cut the flesh from green
mangoes . . . add to taste some borrie and a few
chillies and about 2oz. of fenugreek . . .
(. . . obtainable in Indian shops under the name
of meti.) Gerber *Cape Cookery* 1950.
atshitshi [aˈtʃiˌtʃi] *n. Prob. Afr. Urban:*
Dagga (3) (q.v.) see also *zoll, makulu*
and *boom.*[2] [*unknown*]
    . . . a stranger to the tsotsi's dangerous world
could still save his throat if he has some knowledge of basic words . . . Atshitshi – marijuana,
also known as dagga . . . Venter *Soweto* 1977
    . . . light him an atshitshi fuse and the town-
ship blues will wail from his battered guitar.
Ibid.

aum [ɔːm] *n. pl.* -s. An old *Du.* or *Ger.* liquid measure usu. for liquor consisting of 4 *ankers* (q.v.) i.e. between 120 and 160 litres. See quot. at *anker*, also *half* ~, see quot. [*fr. Du. aam*]

WINE MERCHANTS, EXPORTERS, CAPTAINS AND OTHERS are informed, that they can be supplied at the Cape Distillery with good Brandies, at very moderate Prices, by the Half-Aum, and upwards. *Greig's Almanac Advt.* 1833

auntie [ˈɑnti] *n. pl.* -s. **1.** Mode of friendly but usu. respectful address or reference to an older woman, not necessarily a blood relation, with or without a Christian name, *esp.* among country-bred children; occ. with irony between adults: see first quot. *cf. Jam. Eng. aunt*, stepmother. [*transferred fr. earlier Afk. use; see also tannie*]

Sorry, Auntie. Better go to Veeplaas. Maybe you're there. Fugard *Boesman & Lena* 1969
'This legislation is not aimed at . . . if the poor old soul feels like demonstrating I will ask the police not to notice what the poor old Auntie is doing.' *Daily Dispatch* 12.5.73
'Oh auntie,' I say, 'we're lost . . . if auntie could tell us where to catch the bus' . . . 'No ma'am.' (I figure she's not so keen on the 'auntie' because she sort of blinks when I say it.) *Fair Lady* 13.4.77

2. *Afr.E.* A *shebeen queen* (q.v.) or other dealer in illicit liquor; see also quot. at *Ma-*. [*unknown*]

. . . each time I partake of this brew I like to do it in the privacy of some aunty's place. *Drum* 8.10.73
The shebeen queens are never married or young; township people seem to feel that, because death and the courts have robbed them of breadwinners, widows and divorcees deserve the right to this illicit trade. They are always addressed as auntie or sister, with a note of deference. Venter *Soweto* 1977

Australian bug *n. pl.* -s. A scale pest *Icerya purchasi*, which attacks a great variety of plants; see also *mealy bug*: *U.S.* and *Austral. cottony cushion scale*.

autumn fever See *blue tongue*.

auxiliaries *pl. n.* Rhodesian: see quot.: Auxiliary troops formed primarily of terrorist *'joiners'* (q.v.) see quot. at *National Scout.*

. . . his blue denim uniform marked him as a member of Rhodesia's newest and most controversial military force, the 'auxiliaries', built around former terrorists who have changed allegiance. Or, as the local slang has is, 'terrs who have come on-side'. *Sunday Times* 15.4.79

avie See *arvey, arvie*.

avond- [ˈɑˌvɔ̃nt] *n.* Evening; prefixed to nouns as in place name Avondrust(rest), or flower name *avondbloem* (see *aandblom*); in *Afk.* form *aand* in *aandbossie*, *aandganna, aandpypie* etc. Also place name Aandster(star). [*Du. avond, Afk. aand* evening]

avondbloem See *aandblom*.

ayah *n. pl.* -s. See *aia*.

. . . the Malays played an important part in life at the Cape during the 19th century. The women did the cooking and washing and acted as *ayahs* to the children. Du Plessis *Cape Malays* 1944

Azania [əˈzeɪnɪə, -ˈzɑn-] *n. prop.* A name used by many Blacks in referring to South Africa. [*prob. fr. Arab. Zanj, still common usage for Africans whose skins are dark, and found in prefix Zanzi*bar *and infix* Tan*zania*.] ℙ Azania was mentioned in the *Periplus of the Erythraean Sea circa* AD 60 as 'the continent of Azania'. Acc. Davidson the Azanian Civilization was of Iron Age peoples of East and South East Africa contemporary with that of Zimbabwe.

In the hinterland the Azanians were probably Bantu-speaking peoples as well; although this does not settle their racial type. This racial type may have been bushmanoid or negroid; or, as seems more likely, it may already have shown a mixture and mingling of various African stocks. The only certainty is that the Azanians were a purely African people. Davidson *Old Africa Rediscovered* 1959
Most Blacks preferred to call South Africa 'Azania' the Saso terrorism trial heard in the Supreme Court here yesterday . . . Blacks saw the name 'South Africa' as one which Whites had given the country. It was called Azania before White colonisation and Blacks saw it as the correct name. *E.P. Herald* 17.6.76
When our people are planning . . . a revolt in any country the first thing they do is to alter the country's name. 'So Rhodesia is known as Zimbabwe, South West Africa as Namibia and South Africa as Azania.' Credo Mutwa *cit. E.P. Herald* 23.9.76
A letter which arrived from Kenya this week bears a commemorative Steve Biko stamp . . . In the centre is a tombstone on which is written . . . Born 18.12.1946, Died 12.9.1977 One Azania, one nation. Chisholm *cit. Cape Times* 20.1.79

B

BAAB [bɑːb] *n. pl.* -s *Bantu Affairs Administration Board.* [*acronym*]

The final riot damage figure . . includes damage to the property of the Bantu Affairs Administration Board and township residents, mainly shop-keepers . . . The Higher Primary School . . . was the only BAAB property still covered by insurance. *E.P. Herald* 23.8.76

baadjie [ˈbaɪkĭ, -cĭ] *n. pl.* -s *colloq.* Jacket. In combination *bloubaadjie* (q.v.), *rooibaadjie* (q.v.). [*Afk. fr. Malay baju* jacket]

They wear a shirt, with sleeves left wide and open at the wrists and *baadjies*, or hip-jackets, in the pockets of which their hands are inserted in a very Frankish fashion. *Cape Monthly Magazine* Dec 1861 *cit.* Du Plessis & Lückoff *Malay Quarter* 1953

-baai [baɪ] *n.* Bay: found in place names *e.g.* Mosselbaai (mussel), Oesterbaai (oyster), Stilbaai; also *n. prop. die ~*, see also *Bay*, Port Elizabeth. [*Afk. fr. Du. baai cogn.* bay]

The Transvaal was populated largely by farming folk. They had to travel hundreds of kilometres to 'Die Baai' (Port Elizabeth) for essentials. *Panorama* Jan. 1975

According to our perceptive informant there should be no P.E. students at Rhodes because of the UPE drawcard – sorry to disappoint you . . . but the ous from the 'Baai' are here in droves. *Rhodeo* 29.4.76

baaken [ˈbɑkən] *n. pl.* -s. See *baken* also *beacon.*

baardman [ˈbɑ:(r)t₁man] *n. pl.* Ø. Any of the four spp. of the S.Afr. fish of the genus *Umbrina* (Sciaenidae fam.) having a fleshy filament hanging from the jaw esp. *Umbrina capensis* also known as *tasselfish* or belvis. [*Afk. baard cogn.* beard + *man*]

The dikkop and stompneus, of course, owe their names to their shape, while the baardman has a feeler under the jaw. Green *Grow Lovely* 1951

baas [bɑ:s] *n. pl.* -s. Master, Sir: mode of address usu. by non-Whites to the master or employer, often with definite article in the *third person* (q.v.) *occ. my ~* for emphasis or when making a request: see also *basie.* Also mode of reference to the master, usu. with definite *art.* Combinations: *groot*(big) *~ ; klein* (small) *~ ; makulu* (q.v.) *~ ; oubaas* (q.v.); *wit ~* (q.v.); see also quots. at *karros* and *ja; cf. Jam. Eng. backra* when used as equiv. of boss, master, *Anglo-Ind. sahib, Hong K. lo-ban.* [*Afk. fr. Du. baas* master, captain]

I therefore took leave of the *baas,* an appellation given to all the Christians here, particularly to bailiffs and farmers. *trans.* Sparrman *Voyage I* 1786

At length, with a show of confidence, he

pushed the money under the glass partition and said, 'Please, my baas, a third class return to Umtata.' Gordon *Four People* 1964

baasboets [ˈbɑ:s₁bŭts] *pl. n. colloq. slang.* Great friends, boon companions: see also *bokpal(s), pellie blue, bad friends* and *kwaaivriende.* [*Afk. fr. Du. baas* chief, master + *boet* brother + *pl.* -s]

They've been baasboets since they first went to school together. *O.I.* 1974

baasskap [ˈbɑ:s₁(s)kap] *n. abstr.* Dominion, mastery: usu. in political sense of White supremacy. [*Afk. fr. Du. baas* master + *suffix -skap* state of being, condition, *cogn. Eng.* -ship]

The 'either or brigade' who said there must be either complete integration or white baasskap. *Star* 25.10.72

babala(grass) [bəˈbɑlə] *n. Pennisetum americanum,* large grass cultivated largely by Africans for grain similar to *kaffircorn* (q.v.) and also for silage: known also as *kaffermanna* (q.v.) and *kaffir millet.*

babbala(a)s [ˈbabə₁las, -las] *n. slang.* A hangover: as *modifier* see quot. at *dronkie: cf. U.S. katzenjammer.* [*Afk. fr. Zu. i-babalazi* after-effects of a drinking bout]

. . . the price of a bout of bacchanalian altar is still the groggy head, unsteady hand, furred tongue, rubber knees and the floating stomach – the age-old 'babbelas'. *Cape Herald* 22.9.73

babalaas: A condition a wine-lover should rarely (if ever) experience: a hangover. The word, an Afrikaans version of the Zulu *babalazi* is also used as an adjective. De Jongh *Encyc. S.Afr. Wine* 1976

In combination *~dop,* a *regmaker* (q.v.).

. . . on their way to the kitchen to buy their morning wine; the old babalaasdop, the hair of the dog . . . Muller *Whitev* 1977

babbie-shop *n.* An offensive mode of reference to an Indian trader's store or business. [*unknown poss. rel. Hindi baboo* (Mr, father, gentleman) pejorative among Eng.speakers for an Indian clerk, or poss. rel. *Malay babi* hog]

Indian businessmen in Cape Town are up in arms over statements made last week by . . . In his interview . . . used the term 'babbieshop' which is regarded by Indian people in the same manner as Coloured people regard the term 'hotnot'. *Sunday Times* 21.11.76

. . . the life to be seen in the myriad little streets . . . teenagers standing on the corner invariably near a Babi-shop singing the latest songs of the hit parade or liedjies handed down through the generations and now remembered by only a few. *Voice* 4 – 10.3.79

babiana [ˌbabiˈɑnə] *n. pl.* -s. See *bobbe-jaantjies. [corruption of Du. baviaan* baboon]

baboon spider See *bobbejaan spider, bobbejaanspinnekop.*

babotie See *bobotie.*

baby *n.* A sifting machine for sorting diamondiferous gravel on alluvial diamond diggings, in the C19; see quot. *cf. Canad. grizzly* (for gold). [*fr. n. prop.* Babe]

> The earth and gravel from the claim was first thrown into a wooden frame with swing rockers . . . a native rocked the frame, which is called the 'baby' . . .
> I . . . thought the name of this contrivance referred to its likeness to a cradle, but this was not so. An American with the appropriate name of Babe arrived on the diggings in the eighteen seventies and introduced this machine, which was already in use on the Australian goldfields. They used to say that he was the only Babe who had ever rocked his own cradle. Morton *In Search of S.A.* 1948

bachelor(s') quarters *pl. n.* Barracks or other quarters in urban townships for usu. migrant labourers without, or unaccompanied by, wives and families: see *goduka* and *amasoka,* first quot. below.

> Enog . . . walked swiftly in the direction of the bachelors' quarters where lived the amasoka –the wifeless ones, who either had no wife, or had not the permission to bring a wife from the homelands to live in Cape Town. Louw *20 Days* 1963
> Thus black 'labour force units' are tolerated in urban 'bachelor quarters' while their 'superfluous appendages' are . . . 'repatriated' to 'resettlement camps'. *Cape Times* 5.6.70

> **Bachelor families move out.**
> The conversion of the long standing bachelor quarters in the Peninsula black townships has left many men, women and children without shelter. With the wholesale conversion of these brick blocks . . . occupied by bachelors since 1956 . . . 800 houses will be available to families. *Indaba* 16.2.79

backveld [ˈbækˌfelt] *n.* and *n. modifier.* Rural areas remote from city life: also *attrib.* ~ *farmers etc.*: see also *bundu, g(r)amadoelas: cf. Austral. outback, backblocks, back country, U.S. boondocks, sticks.* [*Eng.* back *trans. Afk.* agter + *veld* country]

> . . . the sheer weight of . . . prejudice that you find in nine out of ten backveld farmers . . . Pretoria might still be more English than the English, but the backveld was Dutch to the marrow . . . and seething with discontent. Brett Young *City of Gold* 1940

backvelder [ˈbækˌfeltə(r), -de(r)] *n. pl.* -s.

One from the *backveld* (q.v.); a rustic, primitive or unsophisticated person: see quot. at *gawie.* [*backveld* (q.v.) + *personif. suffix* -er]

> . . . most of them are backvelders, the old, old Dutch type of whose stubbornness we hear so much all through the pages of South African history. Goold-Adams *S.A. Today* 1936

B.A.D. [ˈbiˌeɪˈdiː] *n. prop. Bantu Administration and Development,* a ministerial portfolio. [*acronym*]

> **The end of BAD**
> The new official name for the Department of Bantu Administration and Development is the Department of Plural Relations and Development. And the first PRD Minister . . . made it clear last night the word 'bantu' is to disappear from Government vocabulary . . . said that the new name was positive and reflected the plural nature of the South African population without any racial connotations. He also stressed that all homelands should be turned into full fatherlands and developed as quickly as possible. *Daily Dispatch* 16.2.78

bad friends *n. phr. Usu.* in phr. ~ *with:* at enmity, not on speaking terms: as *n.* enemies, usu. temporarily: see also *baasboets* and quot. at *kwaaivriende.* [*trans. Afk.* kwaaivriende (q.v.) *kwaai* bad-tempered + *vriend(e) cogn.* friend(s)]

> I'm bad friends with her this term. Schoolgirl 12, 1970

-bad [bat] *n. suffix.* Mineral spring: found in names, *e.g.* Warmbad, of places where there are hot springs: *cf. Ger.* Wiesbaden *etc.* [*Afk. fr. Du.* bad *cogn.* bath]

bafta [ˈbaftə, ˈbæf-] *n. pl.* -s. *Obs.* Cotton material from India, similar to calico: see quot. at *voerschitz.* [*fr. Persian bafta* woven]

> . . . swapping baftas (calico), punjums (loose trousers), and voerschitz (cotton gown-pieces), pronounced 'foossy', against oxen and sheep. Duff Gordon *Letters* 1862

bag [bæg] *n. pl.* -s. Unit of measurement usu. of grain holding 90kg, see *muid;* potential yield is calculated as ~*s per morgen* (q.v.). In combination *school* ~ or *book* ~ (q.v.) a satchel or suitcase for books.

> . . . inspects the freshly-harvested wheat crop which yielded 45 to 58 bags a morgen under irrigation in spite of hail damage. *Farmer's Weekly* 27.2.74

bagger [ˈbægə] *n. pl.* -s. Also *bagre:* a marine fish of the Ariidae, *Tachysurus feliceps,* thought in the early days to be poisonous, perhaps because of its ugli-

ness. ⚑ In fact mucus on the dorsal and pectoral spines is toxic. [*unknown*]
... the bagre, a very bad species of fish, and supposed to be of a poisonous quality. Percival *Account of Cape of G.H.* 1804

baie ['baɪə] *intensifier. occ. prn.* Very: *usu.* with adjectives, e.g. ~ *bang* (q.v.) and ~ *mooi* (q.v.) see quot. below: in *Afk.* a *n./prn. sig.* much, many, esp. in combination ~ *dankie* see second quot.: ~ *dae*, see *min dae* and quot. at *varkpan*. [*Malay banyak* a lot]
At length I succeeded and began to operate on the phiz of my sable friend, who appeared to enjoy it very much, exclaiming repeatedly 'banya mooi' – very nice. Buck Adams *Narrative* 1884
... she spoke a few words in English and brought the house down with a Japanese version of 'baie dankie'. *Sunday Times* 9.12.73

bake *vb. trns. substandard.* To fry. [*translit. Afk. bak* fry (also bake)]
Heat a very little butter in a small pan ... Bake over a low flame until underside is light brown. Turn and bake on the other side. Roll up. Gerber *Cape Cookery* 1950

baken ['bɑkən] *n. pl.* -s. A landmark, natural or man-made: a surveyor's boundary mark, also beacon; in place name Baaken's River: see quot. at *over*. [*Afk. fr. Du. baaken cogn.* beacon]
If the farmer is supposed to have put his baaken, or stake, or landmark, a little too near to that of his neighbour, the Feld-wagt-meester or peace officer of the division is called in by the latter to pace the distance, for which he gets three dollars. Barrow *Travels* 1801 *cit.* Pettman

bakgat ['bak,xat] *interj.* or *predic. adj. occ. adv. m. slang*, splendid, 'posh', excellent. *cf. Austral. beaut; E. Afr. maridadi.* [*etym. dub. poss. fr. Eng.* buck, a dandy]
It's going to be cold tonight ... Doesman's all right. Two bottles and a pondokkie. Bakgat! Fugard *Boesman & Lena* 1969

bakkie ['bakɪ] *n. pl.* -s. **1.** A light truck or van with a cabin and open back for conveying goods, animals or persons. Commercially hired to farmers and others under trade name *Rent-a-* ~ . [*see at* (2) ~]
... thrives on tough, rugged conditions and feels as at home in the bundu as the bull leader of a rhino herd. But there the resemblance ends. On or off the road our ... Bakkie is sweet-tempered, comfortable. *E.P. Herald Advt.* 21.5.71
2. A basin or other container. [*Afk. bak* container (+ *dimin. suffix* -(k)ie)]
... Nico ... took to joining Klein Hannes

behind the fowl-hok at five to twelve, carrying a bakkie of laying mash as an excuse. *New S. Afr. Writing* 4 (no date)

bakkis ['bak(k)ɪs] *n. pl.* -te. A wooden trough on legs in which dough was made and kneaded. [*Afk. fr. Du. bakken cogn.* bake + *kis(t) cogn.* chest]
Yellow wood dough mixer or baking trough (bakkis). Fehr *Treasures at Castle* 1963
... this large box on legs in which dough was kneaded and the baked bread was stored ... The lid of the bakkis is detachable ... Bakkiste have neither handles nor hinges. Baraitser & Obholzer *Cape Country Furniture* 1971

bakkop(slang) ['bak,(k)ɔp (slaŋ)] *n. pl.* -s. The ringed cobra: see *rinkhals.* [*Afk. fr. Du. bak* basin + *kop* head *cogn. Ger. Kopf*]
For a bite from a bakkop or a puff-adder or a ringhals, a sharp knife and permanganate of potash crystals are nearly always efficacious. Bosman *Mafeking Road* 1947

bakoond ['bak,ʊənt] *n.* Brick or stone oven usu. with an iron door, built either into the side of a wide kitchen hearth or outside the house. *Du.* form in place name Bakoven. [*Afk. bak* bake + *oond* (*Du. oven*) oven]
Until recently bread was still baked in the spacious bakoond which Mr Duckitt showed us in the enormous chimney stack. *Farmer's Weekly* 25.4.73

bakore ['bak,ʊərə] *pl. n. colloq.* Protruding ears. [*Afk. bak* bowl + *o(o)r* ear + *pl.* -*e*]
My children say I have bakore so I keep my hair over them. O.I. Italian S. Afr. ex Florence 1972

balie ['bɑːlɪ] *n. pl.* -s. Small tub or vat: *usu.* in combination *sout* ~ (salt), *botter* ~ (butter), *pekel* ~ (pickle): articles in kitchen use in former times: see quot. at *brandewynketel.* [*Afk. balie poss. cogn. vb.* bail, *or* barrel]
... items such as a 'pekelbalie' (tub for pickling meat) and on the boekenhout table in the middle ... a 'teegoedbalie' used for washing dishes. *Panorama* Jan. 1975

baliestoel ['bɑːlɪ,stʊl] *n.* A 'tub chair' in which the back and arms, which are *usu.* caned or *riempie(d)* (q.v.), form a single curve, a design said to be peculiar to the Cape: see quot. at *tub chair.* [*Afk. balie* (q.v.) + *stoel* chair *cogn.* stool]

balk ['balk] *n. pl.* -e. Ceiling or floor beam, baulk of timber. [*Afk. fr. Du. balk* beam *cogn.* baulk]

balsak ['bal,sak] *n. pl.* -ke *Sect. Army.* Army kitbag.

When soldiers check up on their personal kit, they must account for their *mosdoppies* (plastic inner helmets), their *staaldakke* (steel helmets) and their *balsakke* (kitbags). Picard *Eng. Usage in S.A.* Vol. 6 No. 1 May 1975

Yes I know it's cold but my jersey's at the very bottom of my balsak! *Serviceman ex Oudtshoorn* Jan. 1977

bamboo fish *n. pl. ∅. Sarpa salpa* of the Sparidae: known as *strepie* (q.v.) *striped karanteen, mooinooi(en)tjie, stinkfish* and *bamvoosie* : see quot. at *moo.nooi(en)tjie.*

bambus, bamboos [ˌbamˈbŭs] *hist.* A cylindrical wooden vessel made by the Hottentots as a container for milk, also used upon occasions as a drum resonator: see *bombos, poss. fr. bamboos, bambus* [*Hott.*].

He provided himself with a clean wooden milk vessel, or bambus. Alexander *Expedition II* 1838

One . . . held before her a bambus, in which was a little water, and over the top of it was stretched a piece of sheep-skin . . . beaten with the fore-finger of the right-hand, whilst the pitch was regulated by the fore-finger and thumb of the left. *Ibid*

bamvoosie [ˌbamˈvuəsĭ] *n. pl.* -s. See *bamboofish.*

ban *vb. trns.* To prohibit an individual by means of a ~*ning* order from attending or addressing gatherings of whatsoever nature in terms of the Suppression of Communism Act of 1950: also usu. confining or restricting his activities to a defined area.

He was released in 1969 but banned and confined to the Kimberley district. *Daily Dispatch* 18.5.71

Johannesburg United Party city councillors yesterday called upon the Government immediately to revoke the banning orders on the eight student Nusas leaders. *E.P. Herald* 8.3.73

banana (boy) *n. pl.* -s. *colloq.* Nickname for one born or long resident in subtropical Natal: see also *blikoor, vaalpens, kaapenaar, cf. Austral. Bananalander,* a Queenslander.

Panana Boys Natalians should lose the title of Banana Boys to the new boys of the tropical fruit-growing districts. *Daily Dispatch* 25.9.71

banana republic *n. pl.* -s. Non-SAE but now poss. acquiring an extension of meaning or *metaph.* use as a pejorative term for a quasi-independent or impoverished African state: see quots. at *Zimbabwe-Rhodesia* and *RhodZim.*

bandiet [ˌbanˈdĭt] *n. pl.* -e. Convict: *pl.* form mistaken in more than one early text for 'banditti' (brigands). ❡Long term prisoners, ~*e*, are hired in gangs as farm labour for which they receive a small wage: see quot. at *katkop.* [*Afk. bandiet* convict *fr. Du. bandiet* robber, brigand, *cogn.* bandit]

Two redoubts were accordingly marked out: and with . . . a party of banditti, as the Dutch call convicts . . . the bush was soon cleared away for these works. Alexander *Western Africa II* 1837

The Boers not infrequently shut their doors in his face, telling him he is a 'bandit' or convict. Gray *Journal II* 1851

bandom [ˈbantɔm] *n. pl.* -s. *Sect. Mining* Also *bantom* and anglicization 'bantam': a banded pebble of a type indicative of the presence of diamonds. [*Afk. band cogn.* band + *om* (a)round]

And there was I at the sorting table listening to the old hands at the game. 'See these waterworn stones with hoops like a beer barrel round them – "bandoms" we call them,' said a digger. 'Well, when you get "bandoms" you get diamonds, and when you get "bandoms" that size you expect something good.' Green *Secret Hid Away* 1956

bang [ˈbaŋ] *adj. colloq.* Scared, afraid: ~*broek* coward, funk, see *papbroek;* ~*ie(s) slang:* persons who visit *shebeens* (q.v.) in groups for self protection. [*Afk. bang* afraid (*broek* trousers, *cogn.* breeches)]

Don't be bang . . . we are not murderers, not even of women and children. Cloete *Rags of Glory* 1963

A better type shebeen is where 'Bangies' hang out, as the tsotsis say, referring to the customers who go there in groups, for fear of clashing with teenage thugs. Becker *cit. The 1820* July 1973

bangalala [ˌbaŋaˈlala] *n.* An aphrodisiac prepared *usu.* from the powdered roots of certain species of *Rhyncosia.* ❡A decoction of the roots in milk is also used. [*Zu. bangalala* plant name *fr. vb. u(lu)bangalala* to rage furiously]

. . . sold at a price the potion that he claimed gave him his immortality. His virility, he claimed, came from the herb *ibangalala,* and you could buy that from him too. *Scope* 8.9.72

bank [baŋk] *n.* Bench, usu. in *dimin.* form, see *bankie,* or in combination *stoep* (q.v.) ~ ; *tuin(garden)* ~ ; *sit* ~ seat ; *rus* ~ (q.v.). [*Afk. fr. Du. bank cogn.* bench]

banket [ˈbænkət, ˌbaŋˈket] *n.* and *n. modifier.* Gold bearing conglomerate. [*Afk. fr. Du. banket* confectionery *cogn.* banquet]

These river pebbles of resistant quartz were to become embedded in a matrix . . . this formation is called banket from the Dutch word, 'banquet' – confectionery. . . . from the appearance that can be compared to almonds lying in a brown sugary base. *The 1820* Feb. 1974

From the beginning, reef mining on the Witwatersrand was predominantly a capitalistic venture. The 'banket' reef had to be mined, crushed in stamps and pulped in order that its gold content could be extracted. *Cambridge Hist. VIII* 1936 ed. Walker

bankie ['baŋkĭ] *n. pl.* -s. A low, usu. oblong stool, often with sectioned seat of cane or *riempie*. [*Afk.fr. Du. bank* bench + *dimin. suffix* -ie]

. . . large yellowwood table; a bunkie (*sic*) *Grocott's Mail Advt.* 30.8.74

Bantu[1] ['băntŭ, 'ban-, 'bæn-] *pl. n.* usu. capitalized *lit.* People: former official designation of black *Africans* (q.v.) by Government. See quots. at *African, BAD* and *classification.* ¶The term is disliked by African people partly on political and partly on linguistic grounds; esp. erron. forms ~*s* and *a* ~: see fourth quot. [*fr. Bantu pl. prefix aba* + -*ntu stem sig.* person (*sing. form um(u)ntu* one person) (*Afk. form. Bantoe pl.* -s)]

The principal Act defined a Bantu as a person who in fact is, or is generally accepted as, a member of any aboriginal race or tribe of Africa. *S.A.I.R.R. Survey* 1969

. . . you're one of the good Bantoes hey. I can see it. Fugard *Boesman & Lena* 1969

I can't understand why Whites call Africans Bantu for the word Bantu means people. *Post* 6.6.71

In an editorial headed 'Please call us Africans' The World newspaper says if there is a word which annoys Black people it is 'Bantu', a term which many non-Blacks like to apply to African people. Yet many in Government or other official circles as well as private people continue to insist on calling us Bantu and even forcing our own people against their will to call themselves Bantu or Bantus. You can imagine the embarrassment when our educated people like teachers, professional men and others have to grit their teeth and speak of themselves and their people as Bantu, whose use is neither African nor grammatical. 'We are Africans and want to be called so.' *Evening Post* 16.3.73

A Police lieutenant was told by Judge J. H. Snyman not to use the word 'Bantu' during . . . the Commission of Inquiry at the University of the North. *Friend* 21.11.74

Also *attrib.* and in combinations: ~ *Affairs and Development*, a Government Department, see *B.A.D* and *BAAB.* ~

beer see *tshwala* and quots at *pathapatha* and *scale;* ~ *Education,* ~*stan* (q.v.)

The Africans never asked for Bantu education and were never consulted. *Evening Post* 30.9.72.

The Regional Director of Bantu Education . . . has condemned the 'senseless' boycotting of African schools where attendance is voluntary . . . The pupils did not like the label 'Bantu', which to them meant inferior or different. If this was a matter about which Black people generally – not only the pupils – felt strongly, they could approach the Minister with a request to alter the name of the department. *E.P. Herald* 2.11.76.

Bantu[2] *n. prop.* (*pl.*). An extensive group of negroid peoples of southern and central Africa; of or pertaining to any of the languages spoken by these peoples; in freq. combination ~ *language(s)*, hence ~*ist* etc. ¶These include the *Nguni* (q.v.) group of languages.

Bantustan *n. pl.* -s *usu. capitalized.* Term used for an African *homeland* (q.v.) freq. derogatory or facetious: see quots. at -*stan, Pretoria* and *mlungu* [*Bantu* + *Hindi* -*stan* country of]

Not a single country, apart from South Africa, recognises Bantustan sovereignty and only a handful of relatively obscure politicians could be persuaded to attend the Umtata goings-on. *Sunday Times* 31.10.76

BaPedi [ba'pedĭ] *pl. n.* See *Pedi.*

Barberton ['babətən] *n. prop. usu. capitalized.* An illicitly concocted liquor sold in *shebeens* (q.v.) see also *shimiyaan cf. Austral., N.Z. Hokonui* (place name) illicit spirits. [*fr. name of town* Barberton, Transvaal]

The ingredients of *barberton* drink are uncertain but it must contain yeast and mealie-meal . . . A policeman's evidence that a concoction of which he produces a sample is *barberton* is *prima facie* proof of this fact. Sisson *S. Afr. Judicial Dict.* 1960

Before the liquor laws were relaxed to allow blacks to buy alcohol, the shebeen queens brewed concoctions in their backyards; dangerous liquids like methylated spirits were mixed with the fermenting juices of dead animals and rotting plants to create potential killers. Names like Barberton and Skokiaan were given to the poison and people told funny stories about it. But what it did to the human brain was no laughing matter. Over-indulgence could result in death. Venter *Soweto* 1977

Barberton daisy ['babətən-] *n. pl.* -ies. *Gerbera jamesoni*, a daisy-like flower *usu.* red or coral coloured abundant in the Barberton district: numerous hybrids of many colours have been developed.

Double Barberton Daisies . . . Nurseries can now supply a limited number of various high quality Barberton daisy plants. *Farmer's Weekly Advt.* 30.5.73

Barolong *pl. n.* anglicized *pl.* -s. *sing.* (*M*)*orolong.* A *Tswana* (q.v.) people, living mainly in the Western Transvaal and Botswana: a member of this people.

The Barolong take their name from their earliest recorded chief Morolong, under whom, according to tradition, they migrated from a country in the far north. . . . and settled their first permanent residence somewhere near Mafeking. *Native Tribes of the Transvaal* 1905

baruti [ˌbaˈrutĭ] *pl. n.* Priests: see quot. at *moruti.* [*Sotho pl. prefix -ba* + (*mo*)*ruti* a priest]

basboom (ˈbasˌbuəm] *n.* See *wattle.* [*Afk. bas* bark + *boom* tree *cogn. Ger. Baum*]

basela [ˌbaˈseːla] *n.* See *bonsella.*

basics *pl. n.* The initial period of training given to all National Servicemen in common, regardless of what branch of the Forces they may enter afterwards. [*fr. basic*]

. . . you soon learn to sleep any time, anywhere. During your basics you don't get your head down till about 10 p.m., and stand-to is at four in the morning. *Darling* 7.2.79

basie [ˈbaːsĭ] *n. pl.* -s. Affectionate *dimin.* form of *baas* (q.v.) *usu.* addressed to a child, or by an old servant to a younger member of his employer's family [*Afk. fr. Du. baas* master (*cogn.* boss) + *dimin. suffix* -ie]

Put your ear to the ground, my basie, and you will hear them coming. McMagh *Dinner of Herbs* 1968

Basotho [ˌbaˈsutŭ] *pl. n.* (*erron. pl.* -s.) The people of *Lesotho* (q.v.) see quots. at *Mosotho, Russians* and *Sesotho.* [*Bantu pl. prefix ba* + (*mo*)*Sotho, freq. among Whites sig. a single citizen of Lesotho (formerly Basutoland)* see quot.]

Ask anyone in Soweto, and if that person is honest he'll tell you that the situation between the Zulus and the Basothos [sic] is dynamite. On the mines a compound manager will think twice before he puts a Zulu and a Basotho together. Venter *Soweto* 1977

Basotho Qwaqwa *n. prop.* See *Qwaqwa.*

Basta(a)rd [ˈbastəd, ˈbastərt] *n. pl.* -s, *usu. capitalized.* **1.** Also *Baster:* a member of a Coloured tribe descended from the union of Whites with Hottentot women, *usu.* known as the *Rehoboth* ~*s* living in the Rehoboth area of S.W.A.

from 1868: see also second quot. at (2) ~. [*Du. bastaard Afk. baster* bastard]

In this territory the term '*baster*' when it is ascribed to a person's race is well known to refer to members of the Rehoboth Bastard Community. Sisson *S. Afr. Judicial Dict.* 1960

. . . Rehoboth, home of the intriguing race group Die Basters. . . . whereas in the Republic the people of mixed origin prefer the designation, coloured, the folk in Rehoboth with admitted mixed blood, are emphatic that they are not coloured. They insist on being addressed as Baster. At the same time their equivalent race group in South Africa view the title, Baster, as an insult. Jackie Heyns *cit. Drum* Mar. 1979

2. *hist.* used of any established half-caste or other mixed race, including the *Griquas* (q.v.) and other Basters. [*as for* (1) ~]

Oct. 1811 The existence of this little community of Hottentots, was well known to the colonists under the name of the Bastaards, because the whole of them were at that time, all of the Mixed Race. Burchell *Travels I* 1822

The Basters were the descendants of Dutch colonist and frontiersman fathers, and Namaqua and Cape Khoi Khoin mothers. Their culture was neither Khoi Khoin nor Dutch but is best described as a synthesis of the two traditions. Carstens *Coloured Reserve* 1966

baster [ˈbastər] *n. pl.* -s. **1.** *n.* see *Basta(a)rd*

2. *n. prefix* in numerous plant names esp. of timber trees *e.g.* ~ *swartstinkhout* (see *stinkwood*); ~ *geelhout* (see *yellowwood*), *usu. sig.* false, mock, non-genuine hence inferior, cf. *aap-*

3. One of the seven race *classifications* (q.v.) used on personal identity cards relating to S. Africans who are not 'Asian', 'Chinese', 'Bantu' or 'White': see quot. at *classification*

Basuto [ˌbaˈsutŭ] *pl. n.* Earlier sp. form of *Basotho* (q.v.).

Basuto blanket [bəˈsutŭ] *n. pl.* -s. A many coloured blanket with either geometric or animal designs forming part of the dress of *Basotho* (q.v.) tribesmen. Popular also for bedding in S.A.: see quots. at *homeboy* and *Russian.* [*earlier sp. of Basotho*]

Basuto pony [bəˌsutŭ] *n. pl.* -ies. A hardy, surefooted mountain pony orig. bred in *Lesotho* (q.v.).

Basuto ponies stand about 14 to 14½ hands at the withers, and are extremely hardy, active and sure-footed. They are reared in a hilly rugged country . . . freely using their feet and limbs

from the first. Wallace *Farming Industries* 1896
bataleur See *dassievanger*.
bath *n. pl.* -s. Freq. used for a galvanized
wash-tub: see quots. at *zinc*.
Used bath as boat : Boys drowned
Two young boys drowned when they went
boating on a dam in a zinc bath . . . They found
a big zinc bath in the water and pulled it out . . .
later pushed the bath back . . . and climbed in.
The bath drifted into the middle . . . tipped over
and the two fell in. *Cape Times* 20.1.79
bathing box *n.phr pl.* -es *prob. reg. Cape
Town* A peaked-roofed wooden hut on
low stilts, usu. privately owned, and used
as a changing room: found on certain
Cape beaches. ⫿ Unlike archaic 'bathing
machines' ~*es* are not mobile, but provide privacy inside and shade underneath. These are not *equiv.* of Victorian
~*es*, prob. so named by *analg.* with
'hunting box'. *Harper's magazine* 1883
cit. OED ". . . 'bathing boxes' (as the
seaside cottages are called) perched
about on the hillsides."
Muizenberg . . . There were wooden bathing
boxes built on short stilts, beneath which the
younger generation could discuss politics or
something during the evenings. *Capetonian*
May 1979
battle stick *n. pl.* -s. African weapon used
in the sport of stick fighting, also in
faction fights (q.v.) [*prob. trans. Ngu.
induku* stick for fighting or walking]
⫿*Battle axe* is also in current use; see
quot. at *knobkerrie*.
A police task force of about 60 from Umtata
have made 53 arrests and 83 battle sticks have
been confiscated. *Daily Dispatch* 6.10.71
BaVenda *pl. n.* See *Venda*.
baviaan [ˈbavĭˌan, ˌbavĭˈ ɑn] *n. prefix.*
Baboon: usu. prefixed with possessive
-s to *n.* In place names e.g. Baviaanskloof, Baviaansrivier. ~ *spider* see *bobbejaan spider, bobbejaanspinnekop;* ~
stouw see *bobbejaantou;* ~ *boud* see *snaphaan.* [*Du. baviaan fr. Fr. babouin* ba-
boon]
A large kind of monkey, with a long greenish-
brown fur (Cercopithecus ursinus), called
Baviaan by the colonists, inhabits this mountain. Burchell *Travels I* 1822
Bay, the *n. prop. colloq. usu. reg.* E.
Cape: Port Elizabeth: see also *baai: cf.
Berg, the.* [*abbr.* Algoa Bay]
From the Bay to Grahamstown, roasted
coffee had been issued to the men. McKay *Last
Kaffir War* 1871
bayete [ˌbaïjeːte, -de] *interj.* Zulu royal

greeting orig. only for the paramount
chief: latterly as compliment to a few
favoured persons: see also quot. at *imbongi* [*Zu. bayede* Hail (your Majesty)]
. . . the catering staff of the motel – about
20 Zulus – appeared and greeted their Paramount Chief with shouts of 'Bayete'. *Rand
Daily Mail* 28.7.71
The Zulus in the play Umabatha will give
Princess Margaret a full-throated greeting of
'Bayete' when she goes on stage. *Daily Dispatch* 7.4.72
beach thongs *pl. n.* (Poss. non S.A.E.):
Sandals for casual wear consisting of a
sole and two thongs from between the
big and second toes to the sides, formerly called Indian sandals: see *champals,* also *slip-slops.*
Beach thongs are very popular with the out-
crowd especially the type with the flowers on
top. *Personality* 5.6.69
beacon *n. pl.* -s. See *baken.*
The beacons of new farms, still unfenced,
were piling up on the veld. *Daily Dispatch*
29.7.72
beadwork, African *n.* Beads worked into
ornamental geometric designs in which
colour and pattern have definite significance. ⫿ Love messages and tokens
can be exchanged by means of ~ *articles :*
see also quot. at *Ndebele cf. Canad.
wampum beads/belt/reader.*
Fine beadwork in bright colours and complicated designs is not only an artistic decoration
but also has a symbolic meaning. *Panorama*
Sept. 1973
Could any bead workers send me specimens,
fully described, of the purposes, colours and
meaning of design . . . There should be plenty
of DRUM readers who understand the
language of beads. *Drum* 22.4.74
beast *n. pl.* -s. A single head of cattle:
ooo quots. at *biltong* and *-bok; Royal
~* an ox slaughtered by hand by warriors at a Zulu coronation ceremony:
see quot. below. [*Afk. bees fr. Du. beest*
head of cattle, animal, *prob. cogn. Lat.
bos* cow]
beest – As employed in South Africa this
word is restricted to bovine animals: a cow,
ox or bull. Pettman *Africanderisms* 1913
Warriors close in on the royal beast, grabbing
its tail and twisting its neck until it dies . . .
Other beasts swarm around. *Drum* 1.1.72
beaten out *partic. phr.* see *tramped out.*
bechu see *beshu.*
Bechuana [ˌbeˈtʃʊɑnə, -a] *n. pl.* -s. hist.
form. A member of a Bantu people
speaking *Tswana* (q.v.): see also *Bo-*

puthatswana, Botswana, Sechuana, Tswana, and first quot. at *European*.

The Bechuanas are universally much attached to children. Livingstone *Missionary Travels* 1857

Bechuanaland, *n. prop.* see *Botswana*.

becreep [bəˈkrĭp, bĭ-] *vb. trns.* To stalk, creep up upon: *~ing* cap (*obs.*): a camouflage covering for *hunters* (q.v.) made of animal skin: *cf. U.S. 'Davy Crocket' cap*, also *Canad. creep* to stalk. [*translit. Afk. bekruip. fr. Du. bekruipen* to stalk]

... my Gun in her hand readey to give it to me in cace aney Kaffer or Kaffers should trie to becreep us. Goldswain *Chronicle I* 1819–36

... jackets, and shoes, were cast off, and leather trousers were rolled up to prevent noise, the rhinoceros was *becrept*, the hunters sat down in the bushes ... Henrick immediately 'becrept' him ... Two rhinoceroses, an old dam and her weaned calf, were ... cautiously 'becrept'. Alexander *Expedition* II 1838

-bedacht [bəˈdaxt] *partic.* Mindful, prepared: in place name preceded by *adv. Welbedacht*. See also *-gedag, gedacht*. [*Du. partic. of vb. bedenken* to consider]

beefwood *n.* Red-coloured wood of the *Casuarina* used for panelling in Dutch furniture: also *Austral.* see quot. at *bluegum*. [*fr. colour*]

Some time ago ... examined one of the supports of this early jetty and found that it was made of beef-wood, the *Casuarina equisetifolia* which was frequently used for panels in old Cape furniture. Fairbridge *Anne Barnard at Cape* 1924

beerdrink *n. pl.* -s. A ceremonial or social occasion for the drinking of African beer; see (*u*)*tshwala*. [*prob. trans. Xh. intselo* beer drinking]

Two tribesmen were killed and three others injured when two groups clashed at a dance and beerdrink in the Flagstaff district. *Daily Dispatch* 20.4.71

beerhall *n. pl.* -s. African drinking establishment, municipally owned, for the sale of 'Bantu beer': see *maiza*, (*u*)*tshwala*, *kaffir beer*.

Closing beerhalls on Sundays would simply encourage shebeens, Bloemfontein's UBC was told this week. *Post* 6.6.71

By week's end blacks – angered by the mindless vandalism – turned on the rioters. Residents of one township beat up a gang that tried to wreck a beer hall. *Time* 28.6.76

begging-hand *n. Herschelia spathulata* and *H. charpenteriana:* see *moederkappie*.

begrafnisrys [bəˈxrafnɪsˌreɪs] *n. lit.* 'Fu-

neral rice' *yellow rice* (q.v.) or *geelrys* (q.v.) with raisins, still a favourite Sunday dish: also *begrafniskoek* (cake). [*Afk. begrafnis* funeral, burial *cogn.* grave + *rys cogn.* rice]

After a funeral there was always the special *begrafnisrys*, yellow rice with raisins. Green *When Journey's Over* 1972

bek [bek] *n.* Mouth. *suffix* as in *grootbek, dikbek, geelbek,* [*Afk. bek* mouth (*vulgar as mouth of person, standard as mouth of animal*) *cogn.* beak]

'Hou jou bek!' 'Kaffirboetie!' 'Liberalist!' Under the Cohen Code our boys would have to drag up some better guns than that. *Personality* 5.3.71

bekfluitjie See *mondfluitjie*.

bekslaner hek [ˈbekˌslɑnə(r) ˈhek] *n. pl.* -ke. *colloq.* A type of farm gate: see also *concertina gate* [*Afk. bek* (vulgar) mouth *cogn.* beak, *slaan* strike, hit, *agent. suffix* -er, *hek* gate]

There is even the gate that attacks you – They call it the bekslaner hek, which means 'smack-you-in-the-mouth gate'. No penny is ever better spent than the coin which is ... tossed ... to the little piccanins who are sometimes ... the guardian spirits of these varied obstructions. Morton *In Search of S.A.* 1948

Bel-en-ry-na [ˈbel enˈ reɪ ˈna] *modifier* see second quot. at *Ride-Safe* and quot. below.

The women, called the 'Bel en Ry Na' girls, formed a vital part of the Ride Safe scheme ... There were two Bel en Ry Na contacts in Port Elizabeth and three in Uitenhage. *E.P. Herald* 3.5.79

bell *vb. trns. colloq.* To ring someone up on the telephone, *U.S. to call up.* [*translit. Afk. bel* to telephone, ring up]

OK so we'll fix it for Saturday night then. I'll bell some of the chicks meanwhile and organise the graze. *Darling* 1.9.76

beneek(te) [bəˈnɪək(te)] *adj. slang.* Contrary, impossible, unreasonable, crazy. [*Afk. beneuk* as above + *attrib. suffix* -*te*]

Must I tell you why? Listen! ... We're whiteman's rubbish. That's why he's so beneeked with us. He can't get rid of his rubbish. He throws it away ... His rubbish is people. Fugard *Boesman & Lena* 1969

benoudheid [bəˈnəʊtˌheɪt] *n.* and *n. abstr.* Anxiety, nervousness, tightness in the chest: also in place names Benoudheid (anxiety, oppression), Benoudheidsfontein. [*Afk. fr. Du. benauwd* suffocating + *heid* -ness]

She smiled kindly ... quickly returning with

a bottle of red drops in her hand. 'They are very good for "benaawdheit"; my mother always drinks them,' she said. Schreiner *African Farm* 1883

In combination ~*druppels* one of the (*Old*) *Dutch Medicines* (q.v.) used as a sedative for nervous conditions, also for colic: see quot. above.

-berg- [bɜg, berx] *n.* Mountain: [*Afk. berg* mountain] **1.** *prefix* in numerous plant and animal names indicating a hilly or mountainous habitat: ~*adder, Bitis atropos;* ~*lelie* (q.v.). ~*veld,* mountainous farming land *usu.* fit only for grazing; ~ *wind* (q.v.).
2. *Berg, the, n. prop.* [*reg.* Natal] The Drakensberg.

Some years ago Nomkubulwana was said to have revealed herself to some natives over the Berg to whom she transmitted a sealed package. *Lantern* 1966
Veronica who spend a deadly fortnight in the Berg playing croquet with a plump bore . . . A year later . . . was heard to cluck dreamily about having had a terrific time in the mountains last vacation. *Darling* 10.11.76

3. Mountain: found in place names *e.g.* Bergvliet, Bergplaas, Bergville, Katberg, Sneeuwberg, Dwarsberg, Swartberg.

Berg Damara *n. pl.* -s. See *Damara.*

berghaan ['berx₁han, 'bɜg-] *n.* See *dassicvanger.* [*berg* mountain + *haan* cock]
The game birds mocked me from the thicket; a brace of white *berghaan* circled far up in the blue. Buchan *Prester John* 1910

berghaas ['berx₁(h)as] *n. pl.* -e. Mountain hare *Pedetes capensis* (*cafer*) also called *springhare, springhaas* (q.v.). [*Afk. berg* mountain + *haas cogn.* hare]

bergie ['bɜgĭ, 'berxĭ] *n. pl.* -s. *colloq. reg.* Cape: A vagrant living on the slopes of Table Mountain, Cape Town; *cf. Austral. bushy, bushie, bushranger.* [*Afk. berg* mountain + *personif. suffix -ie* (q.v.)]
Tamboerskloof residents – alarmed at the increasing 'bergie' population on the mountain slopes above their houses – called this week for firm action to remove the vagrants, who they say are a danger to their families and property . . . Cape Town's district commandant of police, said the 'bergies' were a 'burden of the police and a nuisance'. *Argus* 18.10.75

berglelie ['berx₁liəlĭ, 'bɜg-] *n. pl.* -s. Any of several bulbs with large showy lily-like flowers *usu.* of the Amaryllidaceae, including *Vallota speciosa* the George (Knysna) Lily, and species of *Nerine*

(q.v.). [*Afk. berg,* mountain + *lelie cogn.* lily]

bergsysie ['berx₁seɪsĭ] *n. pl.* -s. See (2) *dikbek:* The bully seed eater *Serinus sulphuratus,* also *S. alario* known as *dikbekkie* or *dikbeksysie.*

berg wind ['bɜg₁wɪnd, 'berx₁vɪnt] *n. pl.* -s. Hot, dry wind occurring even in winter. in certain parts esp. the S.W. Cape. *cf. Austral.* brickfielder, *Canad. Chinook wind* (warm dry wind in winter or spring) *deriv. rare adj.* bergish, bergy, of or pertaining to the weather. [*Afk. berg* mountain + wind]
Sweat it out in the berg wind yesterday? Well, at least the wind that blew in East London is not as bad as the Khamsin, a wind that blows frequently in the Western Desert and Egypt. *Daily Dispatch* 11.5.71

berry wax *n. obs.* High quality wax produced by boiling the berries of *Myrica cordifolia,* used for polishes and candles in the early days: see also quots. at *nooi* and (2) *mevrou.*
. . . the vegetable wax which the Vrow All°ng's slaves were stewing from berries, of which I saw a stock of green dull candles made. Barnard *Letters & Journals* 1797–1801

besembos ['bɪəsəm₁bɔs] *n.* Any of several varying species of shrubs of which brooms are made: the ~ of the Karoo including several species of *Rhus:* an '*invader plant*' see *plant migration.* [*Afk. fr. Du. cogn.* besom (broom) + *bosch cogn.* bush]

beshu ['be:ʃŭ] *n.* A hide covering worn over the buttocks by men: the rear part of the *umutsha;* see *moochi.* [*Ngu. Zu. ibeshu* skin covering for the buttocks]
Under the shirt-tails and in place of trousers was the traditional 'bechu' or fringed apron of oxhide as worn by generations of warrior ancestors. *Evening Post* 14.11.70

best *adj. substandard. usu.* among children: favourite, the one best-liked: *e.g. my* ~ *subject* not necessarily that in which highest marks are scored, but the most enjoyed: ~ *colour,* etc. [*poss. analg.* best *friend*]
An even funnier thing is happening to the titles of plays . . . We have the following titillating titles . . . 'You know I can't Hear You When the Water's running' . . . 'We Can't Pay. We Won't Pay' . . . But my best title is: 'Whose Life Is It Anyway?' Van Biljon *cit Sunday Times* 29.10.78

betoger [bə'tuəxə(r)] *n. pl.* -s. *colloq.*

A political demonstrator *cf. demo. betoging, n.* and *vb. partic.* Political demonstration or demonstrating. [*fr. Afk. betoog* demonstrate + *agent. suffix* -er]

I desperately needed some material . . . in order to crack a few jokes at your expense . . . love and kisses, all you blerrie betogers. *Rhodeo* 13.5.71

bewaarplaats [bə'va:(r)ˌplɑ(t)s] *n. pl.* -en. *obs.* Tailing site: dumping or storage site of a mine on which mining was not permitted. [*Du bewaar* preserve + *plaats* place *hence* depository]

. . . dumping places for debris or slimes from the mines. Originally the sites thus granted were not supposed to be auriferous . . . as the result however of improved methods of gold recovery not a few of these *bewaarplatsen* have now become very valuable. Licences for such *bewaartplatsen* ceased to be issued in 1902. Pettman *Africanderisms* 1913
I was no hand at this kind of business . . . *Mynpachts,* reefs and *bewaarplaatsen* were as Chinese to me. Cohen *Remin Johannesburg* 1924

bewertjie(s) ['bɪəvə(r)kǐs, -cǐs] *n. usu. pl.* Trembling grass: the graceful *Briza maxima* and *B. minor.* [*fr. Afk. bewe* shake, tremble + *agent. suffix* -(e)r + *dimin. suffix* -tjie + *pl.* -s]

. . . the little reservoir where 'beevertjies' hung their triangular little heads on delicate grass stalks under the firs. McMagh *Dinner of Herbs* 1968

bezitrecht [bə'sɪtrex(t)] *n.* Term in Roman Dutch law of land tenure: a title granted to a holder of land in the form of a certificate, giving him indisputable rights where legal evidence of transfer or title were lacking. [*Du. bezit* posession, *recht cogn.* right]

Later on the tenure was made more secure by the grant of what was called 'bezitrecht', which conferred an almost complete title upon the holder. Many years after, the holder was permitted to obtain a freehold title to Government stands on very easy terms, without any further payment of stand licences. Somerset Bell *Bygone Days* 1933

bhuti ['bu:tǐ] *n.* See *buti.*

bhajia ['baˌdʒɪə] *n. pl.* -s. Indian name for *chilliebite* (q.v.), a spiced savoury fritter: see *Indian terms.* [*prob. Urdu*]

While preparing batter for bhajias . . . always heat a tablespoonful of ghee and pour it with batter for the extra taste. *Leader* 24.7.70

bibi¹ ['biˌbǐ] *n. pl.* -s. *now rare.* An African woman: see also *Mary, fazi. cf. Austral. gin,* aboriginal woman. [*Ngu. Zu. (i)bibi* an inferior wife attached to

one of the chief huts in a *Zu.* kraal]

bibi² Mistress of an Indian household: as informal title 'wife of . . .' [*Hindi fr. Persian bibi* lady]

biesie(s)- ['bǐsǐ(s)] *n. pl.* -s Rush. [*Afk. fr. Du. biesje* rush] **1.** *prefix* to names of species of *Bobartia* e.g. *biesie(s)goed.* **2.** Rushes, reeds: found in place names e.g. Biesiesfontein, Biesiespoort, Biesiesvlei: see also *-riet-.*

biliary *n.* abbr. of *~fever,* freq. erron· 'billery': piroplasmosis carried in S.A· by infected ticks esp. to dogs and cats· [*fr. adj.* biliary, of or pertaining to bile, gall bladder *etc.*]

. . . field reports indicated the disease attacked the liver . . . The symptoms were similar to those of biliary. *Argus* 7.4.78

bilingual *adj.* In S.A., proficient in both official languages, English and Afrikaans (not in *any* two languages): see quots. at *wag-'n-bietjie* and *nogal.* [*Afk. tweetalig* (q.v.)]

FARM MANAGER required for dairy farm. Should be bilingual and must speak Xhosa. *Advt. E.P. Herald* 6.4.79

biltong ['bɪlˌtɒŋ, 'bɪlˌtɔ̃ŋ] *n.* Sun-dried salted strips of boneless meat cut usu. from the haunch of buck or beef: *game ~, beef ~,* also *ostrich ~* (q.v.): see also *toutjies, Cape biltong,* and quots at *stellasie* and *wildebeest. cf. U.S. jerky/charqui* (dried meat usu. beef), *Jam. Eng. jerked hog, jerk pork, Canad. dry meat* (not equiv. *pemmican,* ground *dry meat* mixed with fat). [*etym. dub. prob. Du. bil* buttock + *tong* tongue (here a strip or fillet) (*bil poss. cogn. piel* penis *fr. Lat. pilum* arrow)]

'. . . a good supply of Bel Tongue and bread was set before us. The old lady told us that her present husband was a Dutchman – I had guessed that the moment I saw the Bel Tong as this kind of meat is seldom found in the houses of Englishmen. It is made from beef cut from the very best beast procurable, cut into long strips and dried.' Buck Adams *Narrative* 1884
'Ah there, Piet! – be'ind' is stony kop. Witu 'is Boer bread an' biltong, an' 'is flask of awful Dop . . .' Kipling *Five Nations* 1903

binne ['bɪnə] *prep./adv.* Inside: found in street names e.g. Binnesingel: also in *kom ~* come in(side). [*Afk. binne* inside]

Adriaan knocked on the door of his father's study and opened it when he heard the voice calling: 'Kom binne – come in!' Louw *20 Days* 1963

biocafe [ˈbaɪ‚əʊˈkæfeɪ] *n. pl.* -s. Also *cafébio*: a cafe-cum-cinema offering continuous performances in which the patrons may watch a film while having light refreshments. [*'portmanteau word'* bioscope (q.v.) + *cafe*]

So it's farewell to Durban's last cafe-bio, closing because the building's to be demolished. Strange, I've always thought cafe-bios were demolished by the public . . . My favourite . . . ran a slide reading: 'Please do not throw lighted cigarette-ends at the screen or other patrons.' *Radio & TV* 21-28.11.76

bioscope [ˈbaɪə‚skəʊp] *n. pl.* -s. **1.** A cinema, picture house: sometimes also equiv. of 'the movies' as in (*substandard*) Let's go to ~ . [*fr. bioscope* cinema projector. *Gk. bios* life + *skopein* to look at]

. . . at the Zeerust bioscope they showed a film about an English lord. Bosman *Unto Dust* 1963

. . . that was where the new bioscope was going up that would have electric signs at night that you could see as far as Sephton's Nek. *Ibid. Bekkersdal Marathon* 1971

'It'll [T.V.] be like going to bioscope every night right here.' *Darling* 26.11.75

2. *figur. usu. Afr.E.* A spectacle, show, performance in the non-theatrical sense. [*fr.* (1) ~]

Those two that had the fight because somebody grabbed the wrong *broek*? The *ou* trying to catch his donkey. Or that other one . . . It was Bioscope man! And I watched it. Beginning to end. Fugard *Boesman & Lena* 1969

3. *substandard* A particular film.

. . . the chick . . . she was played by Corinne Clery in the bioscope . . . I'm going to tell my maat that this is the best bioscope about you [James Bond] that I've seen – what with the thrills and funnies and all. Monteath *cit. Sunday Times* 8.7.79

3. *figur. usu. Afr. E.* A spectacle, show, performance in the non-theatrical sense. [*fr.* (1) ~]

biriani [‚bɪrɪˈɑnĭ] *n.* An Indian dish often served on festive occasions: occ. vegetariän but usu. made with spiced marinated chicken or mutton cooked with rice, often with saffron, and *masoor* (brown lentils). ❡Commercially available in tins sp. breyani also, uncooked, in packets. [*prob. Urdu/Hindi*]

Biriani is the dish royal amongst all the exotic rice dishes of India. and remains 'the dish' to serve . . . to welcome house guests on their first day, or . . . the main course of the

menu in formal entertaining . . . the painstaking care which the housewife will take in the preparation of Biriani will commence when she selects her ingredients. *Indian Delights* ed. Mayat 1961

biskop [ˈbɪs‚kɔ̆p] *n. Cymatoceps nasutus* of the Sparidae: see *mussel- cracker* /crusher [*Afk. fr. Du. bisschop* cogn. bishop]

'Biskop' is the fishermen's corruption of the very old Dutch name 'beestkop', meaning animal or beast-head. 'Poenskop' – head of a hornless cow. Biden *Sea Angling Fishes of Cape* 1930

bite one's teeth *vb. phr. substandard.* To set one's teeth, be resolute (*usu.* in the face of misfortune). *cf. vasbyt* (q.v.) Brit. *bite on the bullet*, Austral. *to crack hardy/hearty*, endure patiently. [*trans. Afk. op jou tande byt*]

General Smuts appealed to the farmers to be calm and to bite their teeth. *Star* 16.6.34 *cit.* Swart *Africanderisms: Supp.* 1934

bitter(karoo)bos [ˈbɪtə(r)(ka‚rʊə)‚bɔ̆s] *n. Chrysocoma tenuifolia* a pernicious weed or 'plant pest' of the Karoid regions, which has spread over much veld owing to overstocking; it can be toxic to small stock and is the cause of *kaalsiekte* (q.v.) in lambs and kids transmitted through the milk of the ewes: see also *bloedpens.* [*Afk. bitter* + *bos* cogn. bush]

The most reliable method of preventing the disease is not to allow pregnant ewes and goats access to 'bitterbossie veld' for a period of at least 14 days before they are due to lamb, and for at least 14 days after lambing. *H'book for Farmers* 1937

bittere(i)nder [‚bɪtə(r)ˈeɪndə(r)] *n. pl.* -s. A die-hard of the Anglo-Boer War; one who will not accept defeat. *cf. handsupper* (q.v.). [*Afk. fr. Du. bitter* Du. *eind* end + *personif. suffix -er* as in Britisher]

Others who surrendered – 'hands-uppers' – thought the continued resistance madness and felt that the 'bitter-enders' would be responsible for the ruin of their country. Keppel-Jones *Hist. of S.A.* 1943

blaasop(pie) [ˈblɑs‚ɔ̆p(ĭ)] *n. pl.* -s. **1.** Any of several genera of fish of the Tetraodontidae fam., esp. *Amblyrhynchotes* and *Arothron* spp. which inflate their bodies when threatened: also known as puffers, pufferfish, *toby(fish)* (q.v.). [*Afk. blaas* blow + *op* up + *dimin. suffix* -(p)ie]

Certain members of the sunfish group – the blaasops or puffers are poisonous. Sunfish are non-poisonous, but tough and tasteless.

Green *S. Afr. Beachcomber* 1958
 Blame the sharptoothed doringhaai or some other similar little monster. Even blaasops have been known to chew through nylon with their parrot-like beaks. *E.P. Herald* 18.11.76
2. *Pneumora scutellaris* a grasshopper-like insect with an air-filled inflated body: also known as *gonya* (onomat). [*as above*]
3. A burrowing frog *Breviceps* spp. which has a habit of blowing itself up if alarmed. [*as above*]
 The Blaasop . . . often digs its way into a white-ant hill, and then comes out with the first rains. *Panorama* Sept. 1973

blaauw- [bləʊ] *n. prefix.* Blue: see *blou.*

Black *n. pl.* -s and *n. modifier.* **1.** Person of any dark-skinned racial group, *cf. U.S. colored:* now widely used in this sense by Non-White speakers and the Press. [*fr. black as in ∼ power, ∼ consciousness etc.*]
 It is situated about 37 miles from Stellenbosch, and inhabited by 105 whites and 280 blacks. *Greig's Almanac* 1833
 Students at the University of Natal Medical School, following the trend in other parts of the world, have decided to call themselves 'black' . . . In keeping with this decision they have decided to rename their section of the University of Natal (Black Section). It was previously known as the non-European Section. . . . the students no longer wished to be referred to as 'the negative of another group' . . . said that, in future, 'other students will be referred to as 'non-blacks'. *Daily News* 9.6.70
 Numbers of coloured and Indian students are now beginning to identify with Africans – actually calling themselves 'black' and declaring that 'black is beautiful'. *News/Check* 10.7.70
 For the Blacks, education, living conditions and transport facilities – all these vital things – are markedly inferior to those for the Whites. Oppenheimer, *Stock Exchange Chairman's Lecture* London 18.5.76
 Double shifts up in Black schools. In one year double shift classes in Coloured schools increased by 140. *Cape Herald* 6.7.76
2. Person of Bantu or other negroid origin: in contrast with *Coloured* (q.v.), *Indian*[2] (q.v.) and sometimes *Brown* (q.v.)
 Heavily armed riot police and Coloured and Black people clashed in running battles in many areas of the Cape Peninsula this week. *World* 12.9.76
 I feel nothing but sympathy for the many decent and law-abiding Blacks and Coloureds who must suffer as a result of the actions of the militants. *Argus* 20.9.76
Also in phr. *to go ∼* sig. to be proclaimed or *zoned* (q.v.) as an area exclusively for black occupation: see quot.

at *XDC.*
 PEDDIE DECLARED BLACK VILLAGE . . . It was business as usual yesterday in Peddie after an announcement by the Mayor of Peddie . . . that the village would go Black from September. *E.P. Herald* 2.8.74

blackjack[1] ['blæk,dʒæk] *n. pl.* -s. The spiky, adhesive seed of the weed *Bidens pilosa* which clings firmly to trousers and stockings often in large numbers. [*see first quot.*]
 Black jacks. – Bidens pilosa, L. The hooked seeds of this weed are so called because of their colour . . . The Kaffirs call them *Umhlaba- 'ngubo* 'the blanket stabbers'. Pettman *Africanderisms* 1913
 . . . the paths are full of blackjack and weeds . . . I spent quite 10 minutes pulling off blackjacks from my skirt and stockings. *Grocott's Mail* 19.4.74

blackjack[2] *n. pl.* -s. A member of the African Municipal Police, so called on account of their black uniforms: *flying ∼s* the flying squad of the same force.
 . . . a DRUM reporter who tried to enter the building was shooed away by a city council blackjack. *Drum* 1973

blackleg, blackquarter *n.* See *sponssiekte Brit. quarter evil.*

Black Sash *n. prop.* A women's political organization, founded in 1955, the members of which wear broad black diagonal sashes during picketing or other demonstrations. ⎡The *∼* also sponsors research into racial questions.
 Their first title, in full, was the Women's Defence of the Constitution League. Less formally, they were known in some circles as 'Weeping Winnies' – but one name which has stuck is the Black Sash. The name, in fact, is now official. *Cape Times* 22.5.70

black spot *n. pl.* -s. Black-inhabited area surrounded by White-occupied territory or areas zoned for White occupation.
 Since 1948 an estimated 175,788 Bantu people have been moved from black spots, small scheduled areas and outlying parts of other scheduled areas and resettled in the Bantu homelands. *Evening Post* 17.2.73

blacktail *n.* The S. Afr. marine fish *Diplodus sargus*, also known as *dassie* (q.v.), or *das*, having a black patch near the tail, hence the name *kolstert* [*Afk. kol* spot + *stert fr. Du. staart* tail]

blackwood ['blæk,wʊd] *n. pl.* -s. *Acacia melanoxylon:* a hardwood tree and its timber, *orig.* native of Australia also known erron. as *swarthout* (q.v.) and as *stinkboontjie* on account of the unpleasant smell of the seed when crushed. For instance, the blackwood is a tall, pyramidical tree, not unlike a beech in its compact foliage, but infinitely more graceful and feathery in its branchings. *Life at Cape* A Lady 1870

blanke ['blaŋkə] *n. pl.* -s. White person, *European* (q.v.) seen in combinations *slegs* ~*s*, ~*s alleen* (Whites only) in places restricted to Whites, and *nie-* ~*s* (non-Whites) where Whites are not permitted. [*Afk. blanke* white (*n.*) *cogn. Fr. blanc*]

...I still have my South African identity card with the words 'White-Blanke' on it. *Sunday Times* 8.10.72

...although Chief Minister ... he is denied some of the privileges enjoyed by some White hobos. Because of 'Blankes – Nie Blankes' signs there are places he just cannot enter. He is Black. *Drum* 8.11.73

blatjang ['blat₁jaŋ] *n.* Chutney or other spiced vinegar relish not equiv. of the Malay fish condiment of the same or similar names. [*Malay blachan various sp. forms*]

... here and there 'Blatjang' (a sort of chutney) was made from dried apricots, raisins, curry powder, cloves and other ingredients. Jackson *Trader on Veld* 1958

bleddy See *blerrie.*

blerrie, blerry *adj. slang.* Bloody; see also quots. at *betoger, plaasboer,* and *ach : cf. Brit. blurry, Austral. plurry.* [*corruption of* bloody]

The blerry women talked all the time and stirred up trouble. *Drum* 22.11.72

bles- ['bles] *n.* Blaze, prefixed to the names of several animals and birds which have a white mark on the head: as ~ *bok* (*buck*) (q.v.), ~ *mol* (*mole*) *Georychus capensis,* ~ *hoender* (q.v.). See quot. at *splinter new.* [*Afk. fr. Du. bles* blaze]

1 brown mare, with a white bles, two white legs. *Grahamstown Journ.* 13.12.1832

blesbok, blesbuck ['bles₁bɔk] *n. pl.* Ø. A S. Afr. antelope *Damaliscus dorcasphillipsi,* characterized by a large white blaze on the face. [*Afk. fr. Du. bles cogn.* blaze + *bok, cogn.* buck]

Because of the strong resemblance between the Bontebok and the Blesbuck (*Damaliscus dorcas phillipsi*) early writers often confused the two species and the large herds of bontebok described by Harris in 1838 were very probably blesbuck. *Farmer's Weekly* 12.3.71

bleshoender ['bleshŭn(d)ə(r)] *n. pl.* -s. Coot: *Fulica cristata* so named from the white 'blaze' on its head: also place name

Bleshoender. [*Afk. bles cogn.* blaze + *hoender* fowl]

There is also a shooting season of three months as these lakes are well known for their bleshoenders, duikers and certain types of wild duck. *Personality* 5.6.69

Blikkiesdorp ['blĭkĭs₁dɔ̃(r)p] *usu. n. prop.* Fictitious prototype of a dreary 'one-horse' town; also occ. a slum area: see *gopse, dorp; cf. U.S. podunk.* [*Afk. blik* tin + *dimin.* suffix + *pl.* -(k)ies + *dorp* town]

Are you bored stiff by that routine job somewhere in Blikkiesdorp? *Fair Lady* 6.9.71

Blikoor ['blĭk₁ʊə(r)] *n. pl.* -ore *usu.* capitalized. *colloq.* One born or long resident in the Orange Free State: *occ.* used of a Transvaler also, *cf. Austral. cornstalk* (New S. Wales). [*Afk. blik* tin + *ore* ears (*oor* + *pl.* -e)]

For the past 50 years and more Free Staters have been known as blikore (tin ears) Transvalers as Vaalpense and Cape Colonials as Woltone. *Star* 1.5.34 *cit.* Swart *Africanderisms: Supp.* 1934

bliksem ['blĭksəm, -sm̩] *n. and interj. slang.* An abusive mode of address or reference to someone, roughly equiv. of 'bastard', 'blackguard'. [*Afk. bliksem* lightning]

...became so enraged that he threw a heavy desk blotter ... and swore at him, calling him 'jou bliksem'. *Sunday Times* 2.9.72

blik, skop die *vb. phr.* See *skop die blik.*

blikskottel ['blĭk₁skɔ̃tl̩] *n. pl.* -s. *slang.* An abusive mode of address (or reference) to someone, occ. jocular as in 'blighter': *cf. bliksem,* a stronger term. Also (*rare*) a utensil: see *skottel.* [*Afk. blik* tin + *skottel* dish *cogn.* chattel]

... he came in late this morning man, covered in river mud and filth. I had ... to maar start the breakfast myself. I said listen here you blikskottel if you do this again you're out ... Roberts *Outside Life's Feast* 1975

blind *adj.* Of or pertaining to an estuary, lagoon or river mouth which lacks access to the sea, except in times of flood: as in Blind River (East London).

... we drove past the Gamtoos mouth without seeing it. It has become so badly silted up that a ridge of sand between the blind mouth and the beach completely cut off sight of the river. *E.P. Herald* 19.7.73

blink- [blĭŋk] *adj.* Shiny, gleaming: prefixed to certain nouns *e.g.* ~ *klip* (q.v.); ~ *haar* (q.v.); ~ *blaarboom Pterocarpus rotundifolius.* Also found in place

25

names Blinkwater, Cape; Blinkwater, Transvaal [*Afk. fr. Du. blinken* to shine, glitter]

blinkhaar ['blɪŋk₁(h)ɑ:r] *modifier.* Sheep breed: see *ronderib, Africander.* [*Afk. fr. Du. blinken* shine + *haar cogn.* hair]
   The Karakul ram is a treasure to be protected . . . against leopards and jackals; the rest of the flock of Afrikander, Persian or Blinkhaar ewes may cost no more than ten shillings each. Green *Where Men Dream* 1945

blinkklip[1] ['blɪŋk₁klɪp] *n. obs.* Powdered micreaceous ore, used as a cosmetic, valued for the native trade. [*Afk. blink* shining *fr. Du. blinken* to shine + *klip* stone]
   Hundred of pack-oxen were continually moving off to the westward, loaded with the most valuable effects of the inhabitants . . . red paint-stone, powder of the *blink-klip*, corn, carossses &c. &c. Thompson *Travels 1* 1827

blinkklip(pie)[2] ['blɪŋk₁klɪp(ĭ)] *n. pl.* -pe, (-s). A diamond. [*Afk. blink* shining + *klip* stone (+ *dimin. suffix* -ie)]
   There are Hottentots still living who remember Luderitz; they say he had two little boxes of 'blink klippies' (shining stones) and that he carried these diamonds with him when he sailed down the coast. Green *Where Men Dream* 1945

bloedpens ['blŭt₁pe:ns] *n.* Dysentery in lambs *usu.* caused by the ewes' eating *bitter bush/bos* (q.v.). [*Afk. bloed* blood + *pens* belly]

bloedsap ['blŭt₁sap] *n. pl.* -pe. See *Sap.*

-bloem- [-blŭm-] *n. prefix* and *suffix.* See *-blom:* flower: found in place names *e.g.* Bloemfontein, Bloemhof, Zonnebloem (Sunflower). [*Du. bloem* flower *cogn.* bloom]
   Our blood 'as truly mixed with yours – all down the Red Cross Train, We've bit the same thermometer in Bloeming-typhoidtein. Kipling *Five Nations* 1903

-blom [-blɔm] *n.* Flower: *suffix* to numerous flower names, *e.g. aand~, botter~, gou(d)s~ etc.* also *-bloem* as in *avond~* [*Afk. fr. Du. bloem* flower *cogn.* bloom]

-blommetjie ['blɔmĭkĭ, ¹-cĭ] *n. pl.* -s. Little flower: see *blom.* [*Afk. blom* flower + *dimin. suffix* -(m)etjie]

blood budgie *n. pl.* -s See (3) *Kaffir budgie* a mosquito.

bloodwood *n.* The tree *Pterocarpus angolensis* and its timber, which has a dark red colour: see *kiaat cf. Austral. bloodwood* any species of eucalyptus with red sap. [*trans. Afk. bloedhout*]

blou- [blǝʊ] *adj.* Blue: *prefix* to very numerous flower names: often sig. blue-purple, blue-green, mauve *etc.* ~ *tulp,* ~ *aalwee* (aloe), ~ *disa:* also to the names of animals and birds *e.g.* ~ *bok* (q.v.) (buck) *Cephalophus monticola;* ~ *valk* (falcon) *Elanus caeruleus;* ~ *kraan* (crane) *Anthropoides paridisea;* ~ *aap* (vervet monkey) *Cercopithecus aethiops, etc.* Also found in place names Bloukrans, Blouberg, Bloudrif, Blouhaak. [*Afk. fr. Du. blauw cogn.* blue]

bloubaadjie ['blǝʊ₁baɪkĭ, -cĭ] *n. pl.* -s. Blue-jacket [*Afk. fr. Du. blauw cogn.* blue + *baadjie* (q.v.) jacket (*rok* dress + *dimin. suffix* -(k)ie)]
   1. A provincial traffic policeman, so called from his blue uniform: a ~ usu. drives a blue provincial administration car.
   . . . there were only two deaths . . . over Easter . . . We believe that the main credit for this must go to the 'bloubaadjies' who were to be seen on active duty every few miles along the Garden Route. *Het Suid Western* 26.4.73
   2. *colloq.* A convict serving between nine and fifteen years, classified as a Habitual Criminal in terms of the Criminal Procedures Act of 1955: also *blourokkie.* [*fr. colour of prison uniform*]
   Oppas they don't get you. Blourokkie next time they catch you stealing. Fugard *Loesman & Lena* 1969

bloubok(kie) ['blǝʊ₁bŏk(ĭ)] *n. pl.* -ke, -s. *Cephalophus monticola,* see *blou-* (also *Hippotragus leucophaeus*) the smallest of the S. Afr. antelopes, bluish-grey· to brown in colour and about the size of a hare. *occ. colloq.* E. Cape for creme de menthe, see *green mamba*[2]
   Ruthless people have been shooting at . . . the tiny herd of rare little bloubokkies that live on the Mossel Bay golf course. There are only about 50 of this very rare species . . . left in South Africa. *Het Suid Western* 20.9.78

bloubossie ['blǝʊ₁bŏsĭ] *n. pl.* -s. Any of several species and genera esp. *Pteronia incana* a troublesome weed which encroaches on the (2) *veld* (q.v.). [*Afk. fr. Du. blauw cogn.* blue + *bos cogn.* bush + *dimin. suffix* -(s)ie]

bloudraad ['blǝʊ₁drɑt] *n. prob. Sect. Country:* Galvanized wire used for numerous purposes on farms and among country people. [*Afk. lit.* 'blue wire', *draad,* wire, *cogn.* thread]

If you can get a bit of bloudraad we can fix this trailer quite firmly. O.I. George, C.P. 1976

bloudruk [ˈblɔuˌdrœk] *n.* More often known as *German print* (q.v.), also *Kaffir print* (q.v.) [*Afk. fr. Du. blaauw* blue + *druk* print]

blougat [ˈblɔuxat] *n. colloq.* Also *blouie* (q.v.): a national serviceman who has completed his basic three months' training: see quot. also *ou man*. [*Afk. blou* blue + *gat* bottom, backside]
The new recruits are called *roofies*, a *roof* becomes a *blougat*, when he is halfway through his course, and when he has almost completed his training period he has *min dae* and is raised to the exalted ranks of the *oumanne*. J. H. Picard in *Eng. Usage in S.A.* Vol. 6 No. 1 May 1975

blouie [ˈblɔuĭ] *n. pl.* -s. See *blougat*.
Attention roofies, blouies and oumanne, you will read this. *Scope* 10.1.75

blourokkie [ˈblɔuˌrɔkĭ] *n. pl.* -s. See (2) *bloubaadjie*.

bloutrein [ˈblɔuˌtreɪn] *n. slang* Methylated spirit: also 'blue ocean', see *vlam*. [*Afk. blou cogn.* blue + *trein cogn.* train]

blueback *n. pl.* -s *hist.* Paper currency which was worth far less than its face value issued by the Transvaal (old S.A. Republic) Government in 1865: see *Zuid Afrikaanse Republiek cf. U.S.* greenback dollar (note), *Austral.* blueback formerly a five pound note.
The Republican paper currency was worth as little as its 'goodfors', pound 'Blue-back' notes were fetching a shilling apiece. Brett Young *City of Gold* 1940

bluebuck *n. pl.* Ø. See *bloubok(kie)*

blue ground *n.* Dark greyish blue diamondiferous clay of volcanic origin occurring in pipe-like formations usually below surface layers of *yellow ground* (q.v.) also known as *blue earth*, 'the blue' *Kimberlite* (q.v.) 'maiden blue', '(blue) stuff' see quots. at *cocopan* and *grease table*.
'Then it is clear that you have never seen the diamond diggings at Kimberley. . . . this is Solomon's Diamond Mine; look there,' I said, pointing to the strata of stiff blue clay. . . . the formation is the same. I'll be bound that if we went down there we should find "pipes" of soapy brecciated rock.' Haggard *Solomon's Mines* 1886.
. . . but the greatest proportion of diamonds were found in the unique geological formation of pipes of blue ground running deep into the earth. De Kiewiet *Hist. of S.A.* 1941.

bluegum *n. pl.* -s. Also *Austral.* Used loosely of any of several species of *Eucalyptus;* found as a place name Bluegums.
The day was a blazer and we were glad that the big meeting of the day was to be held within the shade of the bluegum and beefwood trees. Stormberg *Mrs P. de Bruyn* 1920

blue soap *n.* Soap made of animal fat and *orig.* vegetable lye; see *ganna* and *brakbos*, still in regular use all over S.A.: see also *boer soap*.
❡ . . . a species of soap made from the fat of beef and sheep, with the ashes of some particular plants; it resembles in appearance a bluish mottled marble. Percival *Account of Cape of G.H.* 1804
What most women had . . . were . . . mealie meal, samp, flour and sugar . . . sugar beans . . . yeast, long bars of blue soap and washing powder. *E.P. Herald* 5.7.75
I was carrying a leg of lamb and memories of women who can still make a cake of blue soap, smooth as marble, delicately mottled and veined with blue. Van Biljon *cit. Sunday Times* 16.5.76

blue tongue *n.* **1.** A disease of horses, also known as *autumn fever*, in which the head and/or tongue become swollen: see (2) *dikkop*. [*trans. Afk. bloutong fr. blauwtong*]
2. A serious virus disease of sheep which can also affect cattle. [*as above*]
Death, like a hailstorm in the wheat, rinderpest among the cattle, blue tongue among the sheep, could come at any moment. Krige *Dream & Desert* 1953

Blue Train *n. prop.* S.A.'s premier luxury train which travels between Pretoria and Cape Town: see also quot. at *Orange Express*.
A luxury hotel that can travel at 130km/h left Johannesburg today on its first official trip to Cape Town. It is South Africa's new Blue Train – one of the finest in the world. *S.Afr. Outlook* Mar. 1973
I've only passed through on the Blue Train – I mention this to give a touch of class . . . Van Biljon *cit. Sunday Times* 12.11.78

blushing bride *n. pl.* -s. Also *Pride of Franschoek: Serruria florida*, a member of the Protea family with drooping pale pink heads, hence the name; *cf. Jam. Eng. shamelady* (plant). [*poss. fr. Afk. skaamblom* 'bashful flower']
. . . the 'Blushing Bride' has reappeared in the Kloof fairly recently . . . The delicate shrub has pink flowers, the colour of a blush. According to Franschoek custom, a man takes off his hat when he encounters the 'Blushing Bride.' Green *Land of Afternoon* 1949

bo- [bu:, bʊə] *prep. and adj. prefix.* Upper, top: found in place names as *adj.* with *n. e.g.* Boplaas: also 'above': found in place names as *prep. usu.* hyphenated *e.g.* Bo-Wadrif, Bo-Kouga: see also *boven* and *onder* [*Afk. fr. Du. boven* above]

boardman *n. pl.* -men. A member of a *location* (q.v.) management board, or an Urban Bantu Council: see *U.B.C.*
Veteran Mamelodi Advisory Boardman . . . is going to ask the Board to press for women to be allowed to vote for hubbies in Board elections. *Post* 30.5.71

bobbejaan [ˌbɔ̃bəˈjɑ:n] *n.* Baboon: prefixed to numerous plant names often sig. bogus, inferior, e.g. ~*tou* (q.v.), ~*appel* (apple): found in place names Bobbejaanstert (tail), Bobbejaanskloof: also *Du.* form Baviaanskloof, Baviaansrivier [*Afk. fr. Middle Du. babiaen*]

bobbejaankruip, bobbejaanloop [ˌbɔ̃bəˈjɑn-krœɪp, -lʊəp] *vb. and n. Sect. Army.* To stalk in a particular manner, or the method of stalking itself: see *becreep* [*Afk.* baboon + kruip *cogn.* creep. *loop cogn.* lope]
Bobbejaan loop is stalking on all fours but on knuckles because there's a nerve in the palm that can mess up your hand if it's damaged. You put your knee exactly where your hand was so you don't break any sticks as you go. *O.I. Serviceman* 10.3.79

bobbejaan spanner [ˌbɔ̃bəˈjɑn-] *n. pl.* -s monkey wrench.
Ouma was always prepared for any eventuality. In the black bag which never left her side she kept bandages and medicine, needle and thread, and even a bobbejaan spanner. *New S. Afr. Writing* 3 (no date)

bobbejaan spinnekop [ˈbɔ̃bəˌjɑ:nˈspɪnə-ˌkɔp] *n. pl.* -s. Baboon spider: any of several large hairy spiders [also *Du. baviaan spinnekop*]. [*Afk. fr. Middle Du. babiaen* baboon + *spinnekop* spider]
Where the outspan place is are many spiders. The babiaan spinnekops with long brown hair on them. Vaughan *Diary circa* 1902

bobbejaantjie [ˌbɔ̃bəˈjaɪŋkĭ, ˈcĭ] *n. pl.* -s. Any species of *Babiana*, the bulbs of which are said to be eaten and enjoyed by baboons. [*Afk. bobbejaan fr. Middle Du. babiaen* baboon + *dimin. suffix* -*tjie*]

bobbejaantou [ˌbɔ̃bəˈjɑ:nˌtəʊ] *n.* Any of several liane-forming plants including

katdoring, Scutia myrtina also called *monkeyrope(s)*; and *baviaanstouw*, which can form an impenetrable undergrowth and do considerable harm to forest trees. [*Afk. fr. Middle Du. babiaen* baboon, *here poss. sig. spurious or inferior* + *touw* rope]
Among other parasitical plants, the baviaan's-tow (baboon's rope) protruded itself in all directions. Pringle *African Sketches* 1834

bobotie [bəˈbutĭ, -ˈbʊə-] *n.* A traditional Cape dish of curried minced meat, sometimes including dried apricots and/or almonds, covered with a savoury custard preparation: see quots. at *konfyt, borrie, frikkadel. cf. Canad. rubaboo, U.S. hobotee.* [*etym. dub. prob. fr. Malay boeboe, bubu fr. Javanese boemboe* spices, or *poss. burbur* pulp, soup]
Traditional South African dishes, including bobotie, were served. *Daily Dispatch* 4.4.72

bode [bʊədə] *n. pl.* -s. See *bo(o)de.*

boeboes [ˈbŭˌbŭs] *n. R(h)enosterbos* (q.v.) is so called in the E. Cape. [*fr. Xh.* name *ibhubhusi*]

boedelhouder [ˈbŭdəlˌhəʊdə(r)] *n. pl.* -s. Term in Roman-Dutch Law: see quot., also *boedelscheiding.* [*Du. boedel* estate *houder cogn.* holder *fr. vb.* houden to hold]
A boedelhouder is the survivor of persons married in community of property, whom the first-dying has by last will appointed executor, guardian and administrator of the joint estate during the minority of the children. In this manner community of property continues between the survivor and the children until the majority of the children. Sisson *S. Afr. Legal Dict.* 1960

Boedelkamer [ˈbŭdəlkɑmə(r)] *n. prop. hist.* The court having the handling of insolvent estates; *cf. Brit.* and *U.S. slang boodle, caboodle.* [*Du. boedel* property, estate + *kamer cogn.* chamber *fr. Lat. camera*]
A court called the Boodle Kaamer, comprised of president and two members, has the regulating of all insolvent estates. Ewart *Journal* 1811–1814

boedelscheiding [ˈbŭdəlˌskeɪdɪŋ] *n.* See quot. also *ante-nuptial contract.* [*Du. boedel* estate, goods, *scheiding fr. vb. scheiden* to separate, divide].
. . . should your husband . . . maladminister the joint estate, there is an old remedy in Roman Dutch law . . . This involves an application to the Supreme Court for *boedelscheiding* – separation of goods – which, if granted, results

in the division of the assets of the joint estate between the parties ... However, this remedy is rather outdated and is not often used in practice. The courts have in the past, without actually granting a *boedelscheiding* restricted the husband's marital power by prohibiting him from doing specific acts. *Darling* 18.12.76

boegoe [ˈbŭxŭ] *n.* See *buchu:* found in place name Boegoeberg.

boekenhout [ˈbŭkən₁həʊt] *n.* The tree *Rapanea melanophloeos* and its timber, also *Faurea saligna*, both having some similarity to the European beechwood: in place name Boekenhout; *cf. Austral.* *beech* for similar trees. [*fr. Du. beukeboom* beech]

Forest. – Felling of Timber: – Keurboom and Beukenhout this month. Oak in March or April, the other Cape timber all the year round. *Greig's Almanac* 1833

Furniture was mainly of wood indigenous to the Transvaal such as tambotie, boekenhout (South African beech), hardekool (leadwood), knoppiesdoring (knobthorn) and red and white syringa. *Panorama* Jan. 1975

boeke vat [ˈbŭkə ₁fat] *vb. phr.* as *n.* Family prayers usu. in the evening: see quot. [*Afk. boeke* books + *vat* to take]

When I was a small boy on the farm in the Free State my father observed the old Boer custom of 'boeke vat' performing religious devotions after the evening meal. The books of course, were the Bible and the hymn book. *E.P. Herald* 29.8.74

bo-en-onderdeur [ˈbʊə n ˈɔn(d)ə(r)₁dĭœ(r)] *n.* lit. 'Above-and-below-door'; see *stable door* and quot. at *onderdeur*. [*Afk. fr. Du. boven* above, upper + *en* and + *onder cogn.* under + *deur cogn.* door]

The kitchen invariably had its own outer door: always a 'top-and-bottom' door (bo-en-onderdeur) opening either to the side or the back of the house. Cook *Cape Kitchen* 1973

boep [bŭp]. *slang.* Paunch, pot belly: a *beer* ~ beer drinker's belly, see second quot.: *on* ~ , *slang usu. children*, pregnant: *cf. Austral.* bing(e)y, *Jam. Eng.* bangbelly, bang-gut. [*fr. Afk. boepens* paunch]

... for heaven's sake, if you're fat don't wear your trousers tight. Everywhere we've got to look at three layers of boep sticking out now. Peter Soldatos *cit. Darling* 9.6.76

Two bricks and a tickey high but with a boep bigger than Ouma's on account of how he works by the breweries and gets a staff discount. *Ibid.* 1.9.76

We ... believe in Father Christmas. We don't see him as a red faced old pagan with an enormous boep, a cotton wool beard and reindeer. *Het Suid Western* 22.12.76

boer Note: This term has several basic meanings and uses in S.A.E., *viz.* (1) an Afrikaner; (2) an early Dutch Colonist at the Cape; (3) *hist.* or in combination, a farmer; (4) a Republican fighter in either of the Boer Wars; (5) *slang:* a prison warder or policeman. As a prefix it occurs in numerous combinations, in the forms *Boer, boer-* or *boere-*. The first of these has a national or political significance, the two latter forms have numerous and not easily separable meanings such as 'Afrikaner style', 'home-made', 'folk', 'country style', 'of or pertaining to farming folk', 'rustic', 'indigenous' etc. For convenience therefore all compounds are listed alphabetically, those with the usual form *boere-* inserted between *boerbul* and *boer goat* in their alphabetical sequence, and those with national or political significance distinguished by a capital letter. [*Afk. fr. Du. boer* agriculturist]

boer [bʊə, bu:r] *n. pl.* -s, -e. **1.** An *Afrikaner* (q.v.) often in a political sense, see *Boerenasie*.

A Coloured ... member of the liaison committee ... told the Cillie Commission how his house had been stoned during the recent unrest. He said he had been told his home was stoned because 'I work with the Boere'. *E.P. Herald* 19.11.76

2. *hist.* often in combination *Dutch* ~ : an early Dutch inhabitant of the Cape.

In a large house ... inhabited by a Dutch settler of the name of ... we took up our abode for the night, where we experienced that hospitality so proverbial in the Cape Boors. Ewart *Journal* 1811–1814

The Boers will not allow their children to attend schools where natives or coloured children are taught. McKay *Last Kaffir War* 1871

3. A farmer: usu. in combination e.g. *trek* ~ (q.v.), *vee* ~ (q.v.).

Another afternoon I rode out ... to visit a respectable Dutch *böer* (farmer) Mynheer Botha, of Buffels Fontein. Alexander *Western Africa I* 1837

4. A fighter on the Republican side in either of the *Anglo-* ~ *Wars*.

Ubique means the dancing plain that changes rocks to Boers. Kipling *Five Nations* 1903

World War II – even the Great War – also proved that 'Boer and Brit' can work together harmoniously. *Evening Post* 29.4.72

5. ~ *e; Prison slang:* warders or the police.

Warders generally are *boere* ... you're a

canary if you *talk* . . . to the *boere*. *S.A. Gaol Argot, Eng. Usage in S.A.* Vol. 5 No. 1 May 1974

'The coast is clear, ou Whitey. The boere have gone the other way' . . . 'The boere were looking for me, all day . . .' Muller *Whitey* 1977

6. *Black Border Usage.* A member of the S. Afr. forces.

Swapo's secret strategy . . . Tasks 1 . . . 2 . . . 3 . . . 4 . . . 5. To ensure that the seizure of a boer prisoner of war is made a 'practice'. *Sunday Times* 12.3.78

We're called Boers – Vorster and his Boers – by the blacks on the other side. They yell 'kom Boers!' across the Border.

We found scrawled in paint 'Boere, julle –e, gaan huis-toe – julle is te min'. *O.I.s Ex Servicemen* Mar. 1979

7. boer *Urban Afr.* An African term of abuse for someone of mixed blood: also ~ *tjie.*

Zwelenzima woke up and cried . . . she gave me a dirty look . . . 'nyaa nyaa all night bloody boer' . . . I don't say he's white I don't say is a Xhosa, . . . but I do feel that he is . . . when other people say that he's white. Dike *First S. African* 1979.

'You are nothing but a boertjie wena' . . . he called me boertjie . . . tata . . . I say if that boy calls me a white man I'll beat him up.' *Ibid.*

boer- [bʊə, bu:r] *n.* See also *boere-* and compounds: *prefix* as in ~ *brandy,* ~ *tobacco* (q.v.); ~ *soap* (q.v.); ~ *wine/wyn* rough new wine; ~ *pumpkin,* a flat white variety; ~ *goat* (q.v.); ~ *perd* (q.v.); ~ *meal/meel* (q.v.); also in several plant names ~ *klawer,* ~ *lucerne,* (grasses); ~ *boon(boom)* (q.v.); ~ *turksvy* see *kaalblad;* etc.

Boer bank note(s) *n. usu. pl.* Currency printed on the field in the *Boer War:* ⸙These are now collectors' pieces of considerable value: see also *veldponde.*

A South African collector paid about R400 for three Boer banknotes issued from the field in 1902 . . . The notes, Zuid Afrikaansche Republiek £1, £5 and £10 denominations, are described as 'very rare indeed'. *E.P. Herald* 10.2.73

boerboon(boom) [ˈbʊə(r)ˌbʊən, (bʊəm), bu:r-] *n. pl.* -s, -bome. Any of several species of *Schotia* esp. *S. afra,* an evergreen flowering tree of which the timber is used. ⸙The roasted seeds were formerly eaten by country people hence the name. [*Afk. boer* farmer + *boon cogn.* bean (+ *boom* tree *cogn. Ger. Baum*)]

After 18 dormant years, the Schotia afra tree, commonly known as the pink boerboon,

is in full bloom in Settlers Park. *E.P. Herald* 23.1.74

boerbul [ˈbʊə(r)ˌbʊl, ˈbu:rˌbŭl] *n. pl.* -s. Boer mastiff dog: originally a crossbreed. [*fr. Afk. boerboel, boel* large dog, mastiff]

Beautiful Boerbull Puppies . . . Parents excellent watchdogs. *Advt. Farmer's Weekly* 3.1.68

Boerboel × Mastiff pups: excellent farm and watch dogs. *Advt. Grocott's Mail* 18.6.76

boere- [ˈbu:rə] *adj.* See also *boer-:* a frequent *prefix* to nouns ~ *beskuit;* ~ *haat;* ~ *liedjies;* ~ *musiek;* ~ *nasie;* ~ *orkes;* ~ *seun;* ~ *sport;* ~ *vrou;* ~ *wors* all (q.v.) sig. either *Afrikaner* (*adj.* or *modifier*) (q.v.) or 'folk', 'country-style' etc., also 'rustic' see quot.

One of the first offerings from the South African film industry is a boere 'Trinity', entitled 'My Naam is Dingetjie'. *E.P. Herald* 24.5.75

boerebeskuit [ˈbu:rəbəˌskœït] *n. pl.* ∅. Country-style bun rusks, sweetened or unsweetened, commercially available in various flavours, often made of *mosbolletjies* (q.v.) see quot. at *boeretroos* [*Afk. boere* farmers', country + *beskuit* rusk *cogn. Fr. biscuit*]

Boerebloed [ˈbu:rəˌblŭd] *n.* Inexpensive red wine: see also *Tassies.* [*Afk. boere* (q.v.) + *bloed cogn.* blood]

Pour me some wine . . . No, no . . . that's the good stuff. Costs a fortune. Pour me some 'Boerebloed' [. . . *pours her a glass of* . . .] *Paradise is Closing Down* Pieter-Dirk Uys 1978

boerebrood [ˈbu:rəˌbrʊət] *n.* Home-made, usually brown or whole-wheat bread. [*Afk. boere* country style + *brood cogn.* bread]

Boerebrood loaf tins 37cm × 14cm × 10cm – large 89c each. *E.P. Herald Advt.* 24.3.76

Boerechurch [ˈbu:rəˌtʃɜtʃ] *n. prop. Afr. E.* African usage for *Dutch Reformed Church* (q.v.) [*Afk. boere sig. Afrikaner* (*modifier*)]

The two Black ministers of the Black congregation Ds Molopi and Ds Meje . . . said they had to weather criticism from their people about the 'Boerechurch', but now they were able to hold their heads high. *E.P. Herald* 27.8.74

Boerehaat [ˈbu:rəˌhɑt] *n. usu.* capitalized. Political term coined in the early seventies sig. hatred of or a hate campaign against the *Afrikaner* (q.v.). See also (1) *Boer* and quot. at (2) *khaki.* [*Afk. boere* Afrikaners + *haat cogn.*

hate]

'Boerehaat' made its first appearance at Caledon a week ago when . . . Nationalist MP. claimed that a deliberate attempt was being made to discredit and destroy the Afrikaner. *Sunday Times* 22.10.72

deriv. ~ *er* one who hates the Afrikaner.

It certainly needs courage to speak up for the rights of the English section. Proof of this was the roar of 'boerehater' which greated Mr . . . Life presumably becomes very simple when a person who stands up for the elementary rights of the English section can be easily dismissed as a 'boerehater.' *Sunday Times* 1.9.74

boerekos [ˈbuːrəˌkɔs] *n.* Country food, farm fare [*Afk. boere* country-style + *kos fr. Du. kost* food, victuals]

. . . renowned for its handwritten menus and superb *boerekos* . . . Our traditional dishes . . . developed by a pioneering nation who put in a hard day's work . . . now seem a little rich or heavy for our modern tastes . . . But oh, for real boerekos every now and then. *Darling* 16.3.77

boereliedjie(s) [ˈbuːrəˌlĭkĭ, -cĭ] *n. usu. pl.* Afrikaans folk songs usu. fairly simple in words and structure: see also *boeremusiek.* [*Afk. boere* folk, country + *lied* song + *dimin. suffix -tjie* + *pl.* -s *cogn. Ger. lieder*]

boeremeisie [ˈbuːrəˌmeɪsĭ] *n. pl.* -s. A country girl: see quot. at *ware.* [*Afk. boere* country + *meisie* girl]

The story of how a young, sweet *boeremeisie* came out of the wilds of Witbank, went to London and became Miss World. *Darling* 27.10.76

boeremusiek [ˈbuːrəˌmœˈsĭk -mə-] *n.* Rhythmic country-style dance music played *usu.* by a *boereorkes* (q.v.) *cf.* U.S. *blue-grass.* [*Afk. boere* country, folk + *musiek cogn.* music]

Dominee says boeremusiek sessions will encourage evil.

. . . a dance hall where liquor is banned and young people dance to boeremusiek on Saturday nights – has started a bitter feud in the town. *Sunday Times* 10.9.72

Boerenasie (the) [ˈbuːrəˌnɑsĭ] *n. prop.* The Afrikaner people as a whole usu. esp. supporters of the *Nationalist* (q.v.) Government: see *Afrikanerdom:* also the name of a political organization. [*Afk. Boere* Afrikaners + *nasie cogn.* nation]

boereorkes [ˈbuːrə ɔrˌkes] *n.* A band, consisting usu. of a concertina or piano accordion, mouth organ, fiddle and/or guitar, sometimes a piano,

playing *boeremusiek* (q.v.) or *boereliedjies* (q.v.) usu. for dancing. [*Afk. boere* country people('s) + *orkes* orchestra, band]

. . . in time to the Boere-orkes music – and you simply can't keep your feet still when it's Boere-orkes music – and she is partnered by a young 'man also in Voortrekker costume. Bosman *Bekkersdal Marathon* 1971

boereraat [ˈbuːrəˌrɑt] *n. pl.* -rate. A country remedy or 'simple': see also *veld remedy.* [*Afk. boere* folk + *raat* specific, remedy]

. . . his cigarette caught her face. There was a big flap until a stage-hand came up with an old *boereraat* which covered the burn and cured it at the same time. *Darling* 9.4.75

boeresaal [ˈbuːrəˌsɑːl] *n. pl.* -s. A hall in a country town: poss. property of a Farmers' Co-op. [*Afk. boere* farmers' + *saal* hall]

The venue is a *boeresaal* at . . . in the sticks. *Cape Times* 3.7.71

boereseun [ˈbuːrəˌsĭœn] *n. pl.* -s. An unsophisticated country boy: not derogatory, not equiv. *hayseed, hick* etc. [*Afk. boere* country (people) + *seun* boy *cogn.* son]

boeresport [ˈbuːrəˌspɔ̆(r)t] *n.* **1.** Traditional Afrikaner games; see *jukskei, volkspele,* also *kussingslaan* (q.v.). [*Afk. boere* of or pertaining to the (1) *Boers* (q.v.) + *sport* games, amusements]

There will be boeresport for children and adults, a gymkhana, jukskei and a watermelon feast in the afternoon and a braaivleis at night. *Evening Post* 5.2.72

2. *slang.* Flirtation, cuddling.

Oom Skraal Piet said that sex . . . was 'the oldest boeresport in South Africa' – out dating jukskei and kennetjie by at least four generations. *Sunday Times* 28.3.77

boeretabak [ˈbuːrətəˌbak] *n.* See *boer tobacco.*

boeretroos [ˈbuːrəˌtruəs] *n.* Coffee: see quot. at *sakkoffie* [*Afk. boere* Afrikaner + *troos* comfort, consolation]

By the time boeretroos and boerbeskuit (coffee and rusks) are served guests have a . . . warm appreciation of Afrikaans tradition and lifestyle. *Panorama* Aug. 1975

boer(e)vrou [ˈbuːr(ə)ˌfrɐʊ] *n.* A countrywoman, understanding the domestic arts, soapmaking, sausage-making and other occupations befitting a farmer's wife: see quot. at *vrou:* also occ. used of a narrowminded usu. Afrikaner country woman of few ideas and limited taste. [*Afk. boere* farming folk + *vrou* woman]

boerewors [ˈbuːrəˌvɔ̆(r)s] *n*. Orig. home made country sausage, now commercially available everywhere, much in favour for *braais* (q.v.) consisting usu. of a mixture of beef and pork: see also *wors* and second quot. at *perlé*. [*Afk. boere* country-style + *wors* sausage, *cogn*, *Ger. Wurst*]

> She also . . . makes her own boerewors. 'I practically have to seduce the butcher to give me the skins for this' she said. *Daily Dispatch* 24.7.71
> Boerewors. 10lb beef, 6lb pork (the Voortrekkers used lean and fat game instead). Gerber *Cape Cookery* 1950

In combination: ~ *Western cf. Spaghetti Western.* See also at *sosatie* and first quot. at *boere-*

> She is soon to start filming a 'boerewors' Western 'My Naam is Dingetjie', which is a send-up of the Trinity films. *Evening Post* 21.1.75

boer goat [ˈbʊə ɡəʊt, buːr-] *n. pl.* -s. The hardy, indigenous S. Afr. goat. [*Afk. boer* 'country' *sig.* indigenous + *goat*]

> . . . The Boer goat . . . is a strong, coarse, hardy energetic animal, strongly resembling the English goat. Wallace *Farming Industries* 1896
> R6000 for boergoat ram. *E.P. Herald* 18.11.76

boermeel, boeremeal [ˈbʊə(r)ˌmɪəl, buːr-] *n*. Brown wheat flour containing a proportion of both bran and wheat germ, hence ~*bread*, scones etc: see quot. at *roti*. [*Afk. boer* country style + *meel* flour *cogn.* mill *vb.* grind]

boerperd [ˈbuːrˌpeːrt] *n. pl.* -e and *n. modifier*. A gaited horse described as 'almost indigenous' dating back to the time of Lord Charles Somerset and descended from English thoroughbreds: formerly known as the Cape horse. ¶There are only two breeds of gaited horses in S.A., the American saddle horse and the ~. In combination ~*championships*, ~ *shows* etc. [*Afk. boer* Afrikaner, indigenous + *perd fr. Du. paard* horse]

> Her research included an in-depth study of part-bred horses which included the Old Cape Horse from which the Boerperd originates. *Farmer's Weekly* 9.5.73
> The best child's five gaited riding horse . . . a boerperd gelding. *Ibid.* 24.4.74

boer soap *n*. See also *blue soap*. Home-boiled soap still made on farms, consisting of animal fat and lye formerly made of *ganna* (q.v.) bushes. [*Afk. boer*, country style]

> . . . home made bread and even riems . . . She also had for sale fine Boer-soap should one wish to indulge in the luxury of washing. McMagh *Dinner of Herbs* 1968

boer tobacco *n*. Home-cured tobacco for chewing or smoking: also *boeretabak;* see *pruimpie* and quot. at *swartwitpens* [*Afk. boer*, country style]

> The smell of that delightful shop remains . . . compounded as it was with the aroma of brown sugar, leather, soap, great twists of Boer tobacco and spices. McMagh *Dinner of Herbs* 1968
> . . . character which Indian traders gave . . . of an older, easier-going, more harmonious South Africa where the smell of boeretabak hung happily in the same air as incense from the East. *Post* 21.7.76

boerverneuker [ˈbuːrfə(r)ˌnĭœkə(r)] *n. pl.* -s. Also *boere-*, a usu. itinerant merchant trading on the ignorance and credulity of unsophisticated rustics: see *verneuk.* [*Afk. boer* countryman + *verneuk* cheat, deceive + *agent. suffix* -er]

> Dutchmen in those days had not learnt the value of money, but the sight of a heap of sovereigns had a magical effect on them: consequently they fell an easy prey to the experienced . . . Boer-Verneuker, who cheated them right and left. Cohen *Remin. Johannesburg* 1924

Boer War *n. pl.* -s. The South African War(s) of 1888 and 1899–1902 between the British and the (4) *Boers* (q.v.) also known as *Anglo-* ~ and (*Afk.*) *Vryheidsoorlog* (q.v.).

> Upon sections of South Africa's population the years immediately before and after the Boer War saw descend a dull apathy, even an indifference to the aspirations of civilized men. De Kiewiet *Hist. of S.A.* 1941

boesman [ˈbŭsˌman] *n. pl.* -s. *Afr.E.* A derogatory mode of address or reference to a Coloured person i.e. one of mixed blood, by an *African* (q.v.). [*Afk.* for Bushman fr. *Du. Boschjesman*]

> I'd like to ask Mr Mpanza why do most Africans refer to Indians and Coloureds as 'Boesmans' and 'm'kula's'?* These are everyday terms of abuse used by a majority of people. *Post* 16.5.71 [*Zu. fr. coolie*]

boesman(s) [ˈbŭsˌman(s)] *n. prefix*. Found in numerous plant names e.g. ~*s amandel* (almond), ~*sgras*, ~*stoontjies* (little toes) etc. [*Afk. Bushman* (q.v.) *fr. Du. Boschjesman*]

boet(ie) ['bŭt(ĭ)] *n. pl.* -s. Brother. [*Afk.
boet* brother (+*dimin. suffix -ie*)] **1.**
colloq. Mode of address or reference to
a friend, not necessarily a brother,
see quot. at *kussingslaan :* ~ *ie* to or of
a younger person or a child: also *ou* ~
(q.v.) as term of affection between
friends (see also *swaer*), or to an elder
brother (see also *ousus*): *klein* ~ to or of
a younger brother: see quot. at *ouboet*
cf. *buti* (q.v.) and *Austral. bing*(*h*)*i*,
brother.
  . . . it's a drag sharing a flat . . . but it's better
than sharing a bathroom . . . with Ouma and
my boet . . . when he takes off he's rugby
socks even the dog runs. *Darling* 24.12.75
  Charles, beloved father of . . . and . . . and
darling boet of . . . and . . . passed away peace-
fully. *E.P. Herald* 20.3.79
2. *boetie* erron. for *buti* (q.v.)
3. *Boet n. prop. colloq.* among children:
Afrikaans (q.v.) language and literature
as a school subject, e.g. 'I came bottom
in Boet'. See also *Afriks;* (3) *Dutch.*

boetebos(sie) ['bŭtə₁bɔs(ĭ)] *n. pl.* -ies, -se,
Ø. A plant pest *Xanthium spinosum*, a
type of burr weed the fruits of which do
serious damage to fleeces. ¶Its eradication was made compulsory on pain
of a fine by legislation in 1860: see
*proclaimed weed. cf. Austral. Bathurst
burr* (*Xanthium.*). [*Afk. boete* fine +
bos cogn. bush (+ *dimin. suffix -ie*)]
  Ϝoetebossi . . . Xanthium spinosum, so
called from the fact that a fine is the penalty
for failing to keep one's land free from this
pest. Pettman *Africanderisms* 1913

boetie [bŭtĭ] *n. pl.* -s. See (1) *boet* and (2)
boet, also *b*(*h*)*uti*, and quot. at *person.*

boetie-boetie ['bŭtĭ₁bŭtĭ] *modifier usu.
predic. colloq.* Over-friendly, flattering
often with an ulterior motive, also hand-
in-glove. [*Afk. boetie-boetie speel* (play)
to 'butter up' or flatter]
  . . . found that he and the other were boetie-
boetie again, confidential. Jacobson *Dance
in Sun* 1956

bof [bɔf] *n. pl.* -s. *slang.* usu. among children: An effeminate person or homo-
sexual, hence *adj.* ~ *ish: vbl n. boffing :*
see also *moffie; cf. Austral.* poofter,
Jam. Eng. bef, *Austral.* boof, a fool.
[*etym. dub. poss. fr. Brit. slang* boffing
masturbation *or Du.* boef rogue, vil-
lain]
  The acting in . . . was brilliant – one almost
forgot that they were just some old bofs. I

really enjoyed it. Letter Schoolgirl, 16 9.7.74

bohaai [₁bŭ'haɪ] *n. colloq.* A disturbance,
fuss, to-do, racket: see quot. at *volksleier. cf. Austral.* to kick up Bobsy-die *fr.
Brit. slang* Bobs-a-dying, *Anglo-Ind.*
bobbery. [*Afk. bohaai* fuss, noise, also
hoehaai cf. Brit. hoo-ha, *prob. rel. Fr.*
brouhaha]
  In all this controversy . . . one cannot but
suspect that much of the bohaai comes from
people hurt because their clothes have been
stolen while they were bathing. *Argus* 19.12.70

-bok [-bɔk] *n.* Buck: suffixed to names of
most varieties of S.A. antelopes *e.g.
spring* ~ , *bles* ~ , *bonte* ~ , *blou* ~ , all
(q.v.). [*Afk. bok cogn.* buck]
  . . . they called . . . horned cattle *beasts*, the
whole family of antelopes *boks* . . . trans.
Lichtenstein *Travels I* 1812

bok (bɔk) *n.* **1.** *slang.* Lover, see *bokkie.*
[*Afk. bok* flame, beau]
2. *slang.* One who is eager, a 'keen
starter': also as *adj. equiv.* of 'keen'.
[*unknown: poss. as at* (1) ~]
  Who? Me? I'm a bok for having fun and all,
but I'll have to keep this one quiet. *Darling*
29.1.75

'bok [bɔk] *n. pl.* -s. *abbr. Springbok:* a
member of a S. Afr. national or international sports team, *freq.* capitalized *Bok.*
  . . . on five overseas tours, the Boks' captain
has been a back and on five tours a forward.
Evening Post 29.4.72
  Italian ban may hit rugby Boks. *E.P. Herald*
22.8.74

bokaal [₁bŭ'kɑ:l] *n. pl.* -s. A large, lidded,
glass goblet, *occ.* used as a loving cup
and *freq.* engraved. [*Du. bokaal* beaker
fr. Gk. baukalis, vessel for liquids, *cogn.
Ger. Pokal* drinking cup]
  At the dinner which followed a Cape wedding
a large bokaal . . . was passed round the table
so that each guest might drink the health of the
family. Gordon-Brown *S. Afr. Heritage* 1965

bokbaaivygie ['bɔk₁baɪ'feɪxĭ] See *vygie.*

bokbaard ['bɔk₁bart] *n. colloq.* Goatee
beard: also a plant name, *Festuca
caprina*, a grass. [*Afk. fr. Du. bok*
goat + *baard cogn.* beard]
  Mr . . . has grey hair and a grey bokbaard.
He is about 5ft 6in tall. Col . . . is tall and clean
shaven. *Het Suid Western* 3.10.74

bok-bok ['bɔk₁bɔk] *n.* A boys' game of
great antiquity, similar to leap-frog:
see quot. *Brit.* High (*Hey*) Cockalorum.
¶*Abbr.* formula quoted below sig. ~
stand rigid, how many fingers on your
body? In Roman times: *Bucca Bucca quot*

33

sunt hic? – How many are here? It is illustrated in Breughel's painting of children's games. [*etym. dub. poss. cogn. buck*]

Bok-Bok, that most popular game among South African boys with its quaint formula – Bok, Bok, staan styf. Hoeveel vingers op jou lyf? followed by the collapse of the mound of players – has its foundation somewhere in the Middle Ages. De Kock *Fun They Had* 1955

bokkem(s) ['bɔ̆kəm(s)] *n. freq. pl.* Also *bokkoms :* salted, sun-dried fish, often *harders* (q.v.), also known as *Cape biltong; cf. Canad. Digby chips* salted dried herring fillets. [*Du. bokkem/ bokking* salt herring, *prob. rel. Italian baccala, Port. bacalhau* dried salt cod]

Some people call it 'Cape biltong', but as a rule it is better known as bokkem – delicious dried fish which is peculiar to the Western Cape Province. *Panorama* June 1970

bokkie[1] ['bɔ̆kĭ] *n. pl.* -s *colloq.* Usu. a term of endearment as address or reference to the loved one, boy- or girlfriend, lover: see *kêrel:* occ. vulgar, a coarse familiarity, see quot. *cf. Jam. Eng. bucky*, a man. [*Afk. bok* buck + *dimin. suffix* -(*k*)*ie*]

Someone bumped into her. 'Sorry, Bokkie,' he said, leering at her. 'Where are you going, Bokkie?' he asked. Meiring *Candle in Wind* 1959

bokkie[2] *Sect. Army.* An infantryman: also derogatory form *bokkop:* see second quot. at *tankjokkie.* [*fr.* springbok badge worn on green beret]

During exercises there is a wealth of typically Afrikaans-inspired English military jargon in evidence; the infantry are called *bokkies* as a term of endearment, whilst the 'scorn term' is *bokkoppe* toting *ketties.* Picard in *Eng Usage in S.A.* Vol. 6 No. 1 May 1975

bokmakierie ['bɔ̆kma₁kĭrĭ] *n. pl.* -s. *Malaconotus zeylonus*, a large green and yellow shrike with a loud distinctive call also known as *kokkewiet* (q.v.) from the cry of the male 'bacbakiri' and that of the female 'couit couit'. ⫽Also *Jan pierewiet, Jan frederik.* [*onomat.*]

In the English country side the peacock's cry announces changing weather; in the Cape the call to heed is that of the bokmakierie. Green *Land of Afternoon* 1949

bok pals ['bɔ̆k₁pæl(z)] *n. pl. colloq.* Great friend(s): see *baasboets, pellie blue.* [*Afk. bok cogn.* buck + *Eng.* pal(s)]

Boland ['buə₁lant] *n. prop. usu. capitalized* The Southern part of the Western Cape

Province: often used as *modifier e.g.* ~ *farmers,* ~ *climate;* also a peach variety. *deriv.* ~*ers*, the people of the area. [*Afk. bo fr. Du. boven* upper, top + *'land* region]

... delicious dried fish peculiar to the Western Cape Province. There it is eaten by discerning 'Bolanders'. Many of the inhabitants of this region have this delicacy strung up on their back verandah. *Panorama* June 1970

bollemakiesie [₁bɔ̆ləma'kĭsĭ] *n. pl.* -s. Somersault, head over heels, *cf. Austral. head over turkey.* [*Afk. bollemakiesie* somersault *etym. dub. various Du. forms*]

That poor girl did a bollemakiesie over her partner and fell flat on her face. O.I. 1970

bolo ['bəʊ₁ləʊ] *n.* A forequarter cut of beef. [*unknown*]

... usually you'll pay a little less for them than for hindquarter cuts. Bolo is a forequarter cut. It tastes like topside, and is cooked the same. *Sunday Times* 6.2.72

Bombay Merchant *n. pl.* -s. *obs.* An Indian merchant, also known as an *Arab merchant* in the early days in Natal; usu. a Muslim and of the *passenger* (q.v.) class: see quot. at *Arab.*

bombella (train) [₁bɔ̆m'belə] *n. pl.* -s. A third class train used by Africans: see *mbombela.* [*Afk. bombella*, railway truck for Africans, *thought to be fr. Xh. word meaning packed or stuffed in (not traceable)*]

All those who were grown up found it more profitable to go to a recruiting office and join the Bombella train, that took them to EGoli, Johannesburg, the city of gold. Westwood *Bright Wilderness* 1970

bombos ['bɔ̆m₁bɔ̆s] *n. colloq. poss. reg.* E. Cape: A container for liquid, recently usu. petrol. ⫽Poss. rel. Hottentot *bamboos*, a cylindrical wooden vessel for liquids, *obs.* [*unknown*]

bon chop *n. pl.* -s *colloq.* A haircut. [*poss. fr.* bonce, the head, *acc.* Partridge *Dict. of Slang.* used by schoolboys since circa 1870]

... if you ... laid in new make up and got a proper bon-chop at Clippers instead of a shampoo and set from Salon Dorene's – you'd look hang of a lot more with it – I zap them in ... for bon-chops and blow-dries ... Then we send it into town. *Darling* 1.2.78

bond *n. pl.* -s. Usual S. Afr. term for a mortgage bond on property, normally subject to set rates of interest: also *kusting(brief).*

Bond [bɔnt] *n. prop. abbr.* for either of two political organizations, the *Afrikaner* ~ and the *Broeder* ~, occ. for other organizations with *-bond* suffix when mentioned in context. [*Afk. bond* league cogn. Ger. *Bund*]

. . . head of the Broederbond, says that the Bond, like any other organisation, has a right to make representations on policy to the Government. *Sunday Times* 24.9.72

bones *pl. n.* The bones of elephants, baboons and monkeys representing the ancestors, parents and children respectively, used by witchdoctors for purposes of divination. [*prob. fr. Ngu. inhlola* diviners' bones, *or amathambo* bones]

. . . when asked how bones are read, stated that the secret . . . lay in the position in which the bones lay after being 'thrown' on a sacred mat. The 'patient' breathes his spirit on to the bones as they are placed in the cupped hands of the diviner. The diviner does the same. *Bona* Mar. 1974

bones, throw the *vb. phr.* See quots. at *bones, throw the bones,* (2) *dolos and gumboot dance. cf. Austral. to point the bone, Canad. throw medicine.*

bonsella [ˌbɒnˈselə] *n. pl.* -s. *colloq.* A present, gratuity: also *basela; cf. Canad. potlatch* (orig. gift: money or goods). *Anglo-Ind. baksheesh, buckshee.* [*Zu. ibhanselo* a gift, (*ukubansela vb.*) *Xh. ukubasele* to give a present, token of gratitude]

South Africa's brandy distillers are to receive a bonsella worth between R350 000 and R400 000 a year from taxpayers. *Sunday Times* 5.9.71

Bonsmara [ˌbɒnsˈmɑːrə] *n.* A S. Afr. bred cattle strain, a beef breed: see quot. [*fr. n. prop.* J. C. Bonsma]

Delegates to the conference look at Bonsmara bulls . . . The Bonsmara is an indigenous South African breed which combines three parts Shorthorn and five parts Afrikaner blood. *Panorama* June 1974

Bonus Bond *n. pl.* -s. Defence bonds, so called because of the possibility of the holder's number's being drawn for a prize.

Bonus Bond Winners . . . and unclaimed past prize winners are listed Monthly in the Bonus Bond Gazette. *E.P. Herald* 5.2.79

bont [bɔnt] *adj.* Pied, variegated, multicoloured, sometimes gaudy: in combination ~ *ebok* (q.v.), ~ *daeraad* (see *daeraad*), ~ *paling* (see *paling*), ~ *rok* (q.v.) ~ *tick* (q.v.), ~ *(e)veld* (q.v.): also found in place names e.g. Bontrand, Bonteheuwel [*Afk. bont* pied, multicoloured (+ *attrib. inflect. -e*)]

bontebok [ˈbɔntəˌbɔk] *n. pl.* -s. The rare S. Afr. antelope *Damaliscus dorcas dorcas* with a coat of reddish chocolate colour, white rump and a long white blaze down the face: now stringently protected. [*Afk. bonte attrib. adj.* pied + *bok* cogn. buck]

Among the buck . . . are four bontebok . . . The bontebok nearly met the same fate as the quagga and other interesting animals. It was estimated not so long ago that there were hardly more than two dozen bontebok left in South Africa – and that meant in the whole world. Green *Tavern of Seas* 1947

bont tick [bɔnt-, bɒnt-] *n. pl.* -s. The parasitic S. Afr. tick *Amblyomma hebraeum* which transmits *heartwater* (q.v.) a disease of sheep, goats and cattle.

The kudu is a carrier of the bont tick . . . which tick spreads the deadly heartwater disease. *Grocott's Mail* 14.4.71

bont(e) veld [ˈbɔnt(ə)ˌfelt] *n.* (1) *Veld* (q.v.) in which acacia trees abound, also called *elandveld* or *thorn veld.*

bontrok [ˈbɔntˌrɔk] *n. pl.* Ø. **1.** One of the names of the marine fish *Diplodus cervinus* usu. called *zebra* (q.v.), also used of the sand steenbras.

2. A species of stone-chat, *Saxicola torquata.* [*Afk. bont* colourful, pied + *rok* dress, *fr. Du. rok* coat (+ *dimin. suffix -ie*)]

bo(o)de [ˈbuədə] *n. hist. obs.* The messenger of the court. [*Du. and Afk. bode* messenger]

. . . one of my Hottentots was sent with a packet of letters to a farm-house . . . where the Tulbagh Boode who was to forward them to Cape Town, was expected to arrive on his way from Hantam. Burchell *Travels I* 1822

The officers of the Court consisted of a Fiscal . . . and some minor officials such as the *Bode* (Sheriff) etc. Meurant *Sixty Yrs. Ago* 1885

book *n. pl.* -s. *substandard.* Often used sig. magazine or other bound newspaper: also *comic* ~, a comic or strip cartoon paper: also *substandard Brit.*

. . . a carrier bag containing the dagga the police had found. Some of the drug was placed in a tin and some mixed with cigarette tobacco and placed in a comic book. *Daily Dispatch* 11.3.72

bookbag *n. pl.* -s. A satchel or other case

for school books. [*trans. Afk. boeksak* satchel]

I took the headmistress the bookbag we all used at . . . as a pattern. O.I. 1974

Book of Life *n.* A comprehensive personal identity document for White South Africans, initiated in 1971, including driving and gun licences and details of marital status.

The six foolscap-page form which people in South Africa will have to complete for the 'Book of Life' – the Government's new super-identity-card – will pry deeply into personal affairs. *Sunday Times* 14.11.71

book pass *n. pl.* -es. *Afk. E.* See *pass, reference book.*

boom [buəm] *n.* Tree [*Afk. fr. Du. boom* tree *cogn. Ger.* Baum] **1.** *suffixed* to the names of numerous S. Afr. trees e.g. *kaffir ~* (q.v.), *denne* (q.v.) *~*, *bas ~* (see wattle) etc: also found in place names e.g. Boomplaats, Maroelaboom, Wonderboom.

2. *slang.* Dagga, *Cannabis sativa*, Indian hemp, marijuana, bhang: *cf. Brit.* and *U.S. grass, pot* etc.: also *~ tee/tea.* See quot. at *arm.* [*etym. dub. boom* tree *cogn. Ger.* Baum, see *etym.* at *muti*]

. . . she and a boy friend went to a party . . . They went into a barn a few yards from the farmhouse and drank 'boom tea' (tea made from dagga). Some of them were 'feeling a bit funny'. *Cape Times* 20.7.71

Most widely used names in South Africa are pot, grass, weed, tea or boom. *Daily Dispatch* 4.9.71

boomdassie [ˈbuəmˌdasĭ] *n. pl.* -s. The tree hyrax: *Dendrohyrax arboreus*, see *dassie* [*Afk. boom* tree *cogn. Ger.* Baum + *dassie* hyrax *fr.* Du. *das* badger + *dimin. suffix -ie*]

boomslang [ˈbuəmˌslaŋ] *n. pl.* -s, -e. The greenish tree snake *Dispholidus typus* which preys upon birds: its bite is poisonous but said to be rare in human beings on account of the rear placing of the fangs: it varies from 2 to 3 metres in length. [*Afk. boom* tree *cogn. Ger.* Baum + *slang* snake]

Skaapstekers and boomslange are venomous, but they are backfanged snakes and not aggressive, so we do not worry about them. *The 1820* Apr. 1971

BophuthaTswana [bɔ̃ˌpŭtatˈtswɑna] *n. prop.* The *homeland* (q.v.) of the *Tswana* (q.v.) people consisting of territory in the central and W. Trans-

vaal, N.W. Cape and the Thaba 'Nchu area of the Orange Free State.

In terms of Proclamation 130 of 26 May 1972 Bophutha Tswana became a self-governing territory within the Republic with effect from 1 June of that year. Horrell *Afr. Homelands of S.A* 1973

Border, the *n. prop. usu. capitalized, colloq.* General term for the *Operational Area* (q.v.) freq. combinations *Border duty, on ~, going to ~, up on ~* etc.

The circumstances pertaining to the Operational Area are by their very nature abstractions from the normal routines and comforts of civilian life. The men on the border are isolated from their families for periods of up to three months or more and are separated from home by vast distances. *Paratus* Jan. 1979

border area *n. pl.* -s. See quot.

A border area is 'one where development takes place in a European area situated so closely to the Bantu areas, that families of Bantu employees engaged in that development, can be established in the Bantu areas in such a way that the employees can lead a full family life. *Tomlinson Report* 1946

border industry *n. usu. pl.* -ies. White-owned and White-managed enterprises at the edges of African *homelands* (q.v.) i.e. in a *border area* (q.v.).

The city is also the hub of many new border industries . . . In recent years the potential of the growing border industries has been investigated and found to be so promising that East London may well develop into one of the Republic's most important growth points. *Panorama* May 1973

borehole *n. pl.* -s. In S.A. a well drilled to tap an underground water source usu. operated by means of a windmill-type pump. *cf. Austral. bore* a waterhole for cattle, artesian well. Also in combination *~ water* (and windmill water).

. . . all good vlei lands with abundant stock water from permanent fountains and borehole. *Daily Dispatch Advt.* 11.3.72

Water Supply: Strong permanent boreholes Centrifugal pump with Lister engine. Windmills with catchment dams. *Farmer's Weekly Advt.* 27.2.74

borrie [ˈbɔ̆rĭ, ˈbɒrĭ] *n.* **1.** Turmeric: yellow coloured spice used in *geelrys* (q.v.) and other S. Afr. dishes: see quots. at *fish oil* and *atjar*, called by the Indians in S.A. *arad:* see *Indian terms Anglo-Ind. purree* (yellow colouring). [*Afk. fr. Malay boreh* turmeric]

. . . I'm going to make bobotie for lunch . . . with yellow rice, that is, rice cooked with sugar, salt, a knob of butter, a handful of raisins and a teaspoon of borrie – that's what you call turmeric. Westwood *Ross* 1975

2. *n. prefix* to plants, fruits etc. of deep yellow colouring: see quots. at *patat* and *pocket:* ~ *sweet potato;* ~ *quince;* ~ *hout* (wood); and ~ *vink* (finch).

borrow *vb. trns. Afr. E.* Used by Africans sig. lend, *cf.* S.A.E. lend = borrow as in 'Please borrow me five rand.' Also *slang*, see quot.: see also at *lend.* [*Ngu. (uku)boleka* borrow/lend]

> Myrt's main claim to fame is this floor length fun fur what she'll always borrow you for special occashuns if you ask nice. *Darling* 3.9.75

borsdruppels [ˈbɔrsˌdrœpəls, ˈbɔsˌdrəplz] *pl. n.* A *Dutch medicine* (q.v.) for the relief of asthma and other tightness of the chest: see *benoudheid.* [*Afk. fr. Du. borst,* chest *cogn.* breast + *druppels cogn.* drops]

> Why should Oom Piet, who had for years bought his borsdruppels just as regularly and from the same place as he ordered his tobacco now be prevented from doing so? *Sunday Times* 24.6.34 cit. Swart *Africanderisms Supp.* 1934

bos [bɔs] *n.* Bush. [*Afk. bos fr. Du. bosch* wood, forest, *cogn.* bush]

1. *prefix* in names of flora and fauna sig. forest e.g. ~ *lelie* (lily); ~ *buchu;* ~ *duif* (q.v.); ~ *bok* (q.v.); ~ *luis* (q.v.); ~ *vark* (pig) (q.v.) *Potomochoerus porcus koiropotamus.*

2. *suffix.* In numerous plant names: see *taaibos, bitterbos* etc.

> ...in none...are any of the elegant tribes of heaths ever seen under cultivation; and it is a curious fact that, among the colonists, these have not even a name, but, when spoken of, are indiscriminately called bosjes (bushes). Burchell *Travels I* 1822

3. Wood, forest: in place names e.g. Boshoek, Boskuil, Bosspruit, Blouleliesbos.

4. see *bossies.*

bosbok [ˈbɔsˌbɔk] *n. pl.* -ke. *Tragelaphus scriptus sylvaticus:* see *bushbuck:* also found in place name Bosbokrand. [*Afk. bos fr. Du. bosch* forest + *bok cogn.* buck]

bosbouers [ˈbɔsˌbəʊə(r)s, -z] *n. usu. pl.* Foresters: used in a somewhat derogatory sense of the primitive, uneducated people of the forest areas of the S.W. Cape: see *woodcutter.* [*Afk. bosbou* forestry + *personif. suffix* -er]

> There was reported to be much fear and

bitterness among the Knysna 'bosbouers'. Some had received white identity cards, but the majority had been classified as Coloured. *S.A.I.R.R. Survey* 1969

-bosch- [-bɒ ʃ-, ˈbɔ ʃ-, -bɔs-] *n. prefix and suffix.* Wood, forest; found in place names *e.g.* Boschheuwel, Rondebosch, Stellenbosch: see also *-bos-* [*Du. bosch* forest *cogn.* bush]

> We fairly tracked him down into a large *bosch,* or straggling thicket of brushwood and evergreens. Thompson *Travels II* 1827

bosdrag [ˈbɔsˌdrax] *n.* See *browns* [*Afk. bos cogn.* bush, *drag* dress, clothing]

bosduif [ˈbɔsˌdœïf] *n. pl.* -we [-və] The speckled Cape rock pigeon *Columba guinea,* troublesome to grain farmers. [*Afk. fr. Du. bosch* wood + *duif cogn.* dove, pigeon]

Boskop [ˈbɔsˌkɔp] *n. prop.* Place name: site of the discovery of the skull of ~ *man* of the late Pleistocene period orig. described as *Homo capensis,* a separate species, now regarded as a strain of *Homo sapiens* fr. which the Bushmen and Hottentots are prob. descended, hence ~ *race: deriv.* ~ *oid, adj.* characterized by a skull of the type of that of ~ *man:* see also *Kromdraai-man, Swartkrans-man* and (2) *Stellenbosch.* [*fr.* Boskop, Transvaal]

> *Boskop strain persists* . . . Prof Raymond Dart . . . said that Boskop man found in different parts of southern Africa was the ancestor of the Bush – Hottentot type and that there existed a Boskop strain in the African population. Boskop man was alive and well and living in southern Africa more than 20 000 years ago. *E.P. Herald* 21.8.74

boslemmer [ˈbɔsˌlemə(r)] *n. pl.* -s. See *Hernhutter.* [*Afk. fr. Du. bosch cogn.* bush + *lem* blade, cutter]

> For a fine fat sheep Mr S . . . was satisfied to take two Gnadenthal knives, called here boschlemmers, the goodness of which has long recommended them to the inhabitants of the Colony. Latrobe *Journal* (1815–16) 1821

bosluis [ˈbɔsˌlœïs] *n. pl.* -e. Tick; occ. *erron.* 'louse'. ~ *voël Ardeola ibis,* see *tickbird.* [*Afk. fr. Du. bosch cogn.* bush + *luis* tick *cogn.* louse]

> A noxious little insect annoying both to man and beast, the *acarus sanguisugus,* which Dr Clarke supposes to be the kind of louse which of old plagued the Egyptians. The Dutch colonists of the Cape call it the *bosch luis,* or wood-louse. Arbousset's 'Narrative', 1846 *cit.* Pettman

B.O.S.S. [bɒs] *n. prop.* Bureau of State

Security: also *attrib.* ~ *Bill.* [*acronym*]

The Prime Minister . . . is to be asked in Parliament to clear up allegations that Boss has been secretly funding Shaka's Spear, the Zulu opposition party. *E.P. Herald* 15.5.74

boss boy *n. pl.* -s. A man, usu. African, in charge of a gang of mine- or other workers: see also *induna.*

Kibuko had been a boss-boy of authority and strength that had won the respect and admiration of the people of Pomelo. Griffiths *Man of River* 1968

bossiedokter [ˈbɔsĭˌdɔktə(r)] *n. pl.* -s. A herbalist or naturopath: usu. derogatory or derisive. ⫿ A herbalist proper is *Afk. kruiedokter* (*kruie,* herbs): see also *herbalist.* [*Afk. bossie* little bush, *dokter cogn.* doctor]

Last year . . . referred to Mr . . . as a 'bossie-dokter' because . . . claims to have a diploma in naturopathy . . . *Sunday Times* 17.9.77

After eight right years of being called a bossie dokter . . . got fed up . . . He wanted a formal ruling on what he should be called. If he was to be Dr . . . as he had been in the past then he wouldn't stand for the 'bossie' bit. *Ibid.* 9.10.77

bossies, (bos) [ˈbɔsĭs] *adj. Sect. Army slang.* Of or pertaining to someone who has been serving too long in the bush: 'bush mad': cf. *Austral. Army,* '*troppo*' mad, referring generally to someone whose nerves are affected by the tropics, the heat or the war: not equiv. *Austral.* '*bushed*' *lit. or figur.* lost, or in a quandary. ⫿ Also in form *bos sig.* slightly mad, and other versions (obscene). [*fr. Afk. bos cogn.* bush]

When they come out a bit funny in the head – a bit like shell-shock really – we say 'he's bossies'. *O.I. Ex-Serviceman* 7.3.79

Here's Richard, this bossies ou I was telling you about – thinks he's still in the army. *Informant Ex Border* 6.4.79

bosvark See at *bos* (1). *Sect. Army.* Forces' term for either a light reconnaissance aircraft or an anti-mine vehicle. *O.I.s Servicemen* March-April 1979

Botha('s) Babe(s) [ˈbuəta] *n. usu. pl. colloq.* Members of the women's army at the S. Afr. Army Women's College at George C.P. so named after the Minister of Defence at the time of its inception: see also *soldoedie.* [*fr. n. prop.* P. W. Botha]

The Civil Defence College for Girls in George made history again this week by creating the first corporals among the 'Botha Babes'. *Daily Dispatch* 18.8.71

Botswana [ˌbɒtswənə, ˈbɔtsw-, -ana] *n. prop.* The country formerly known as

Bechuanaland: see also *Lesotho.*

botter [ˈbɔtə(r)] *n.* Butter [*Afk. botter cogn.* butter] **1.** Prefixed to names of articles for or containing butter, e.g. ~ *bak,* ~ *balie,* ~ *vaatjie :* see *bak, balie, vaatjie.*

With the botterbak goes the botterspaan or wooden butter-hand, consisting of a wide, spade-like paddle with raised edges and a thick handle. Baraitser & Obholzer *Cape Country Furniture* 1971

2. Prefixed to plant names sig. usu. yellow colour or buttery smell, or taste, e.g. ~ *blom* (any of many species, buttercup, Cape daisy etc.), ~ *boom* (tree) *Cotyledon paniculata. cf. Austral. butter-, Brit. buttercup* etc.

botterblom [ˈbɔtə(r)ˌblɔm] *n. pl.* -me, -s. *Gazania uniflora :* see *gousblom,* also (2) *botter.*

bottle brush *n. pl.* -es. Any of several species of *Greyia* the scarlet flowers and filaments of which resemble the Australian bottlebrush, a species of *Callistemon :* also one of the Proteaceae, *Nimetes argentea,* the silverleaved ~.

Another indigenous shrub with red flowers is the *Greyia Sutherlandii* or Natal bottlebrush. It has an interesting, rather gnarled growth habit and bears red flowers in early spring. *Evening Post* 11.3.72

bottle store *n. pl.* -s. A shop, usu. retail, in which bottled liquor is sold: see also *off-sales,* and quots. at *madolo, plural* and *international cf. Brit. wine merchant.*

. . . on the day before Christmas I had met the Archbishop in the bottle store and had said to him, *Your Grace, I am surprised to see you here.* Paton *Kontakion* 1969

-boven [ˈbuəvən] *prep.* Above: found in S. Afr. place names as *suffix* to nouns usu. with hyphen e.g. Waterval-boven, Welgemoed-boven: cf. *bo-* [*Du. boven* above]

boy *n. pl.* -s. Objectionable among certain groups: used alone to refer to a grown man, usu. African, who is a servant, even by older Africans. ⫿Now regarded as a discourtesy by many S. Africans, but in combination seldom intended as offensive: *garden* ~, *house* ~, *flat* ~, *delivery* ~ etc.: see also *girl.* [note inverted commas in first quot.] [*fr. old style Colonial* (*esp. Anglo-Ind.*) *term* '*boy*' *for a native servant prob. fr. boyi :* a personal servant (*English* 1609) *fr.*

Telegu or Malayalam for a palanquin bearer, from the name of the caste.]

Having set some of the 'boys' to cut off the best of the giraffe meat, we went to work to build a 'scherm' near one of the pools. Haggard *Solomon's Mines* 1886

I am sure ill-feeling between Blacks and Whites can soon end if the Blacks are prepared to complain when addressed by nasty names such as boy, girl. *Drum* 8.10.72

. . . the Zulu watchman, Mhlopi, encountered the flat-boy who worked in the block of flats lower down the road. Bosman *Willemsdorp* 1977

Fortunately, there was an Indian shop-keeper who needed a delivery boy. He applied for the work – not because he liked Indians, but he wanted to make money. Venter *Soweto* 1977

boy(s) *n. usu. pl. occ -∅* **1.** *Sect. Army.* Border usage: terrorists.

There were times when we tracked boy for two days through the bush. *Ex-Serviceman* Mar. 1979

2. *Afr. E.* African guerilla fighters.

People in the rural areas say that they have been instructed by the 'boys' (guerillas) to refrain from voting or risk certain death. *Voice* 16.12.78

~ *in the bush n. phr.* see quot.

Blacks have a favourite term of affection for the guerillas: the boys in the bush. *Time* 30.4.79

boykie, boytjie [ˈbɔɪkĭ, -ĭ] *colloq.* Little chap, fellow etc. also found as a pet name or nickname e.g. 'Boykie wasn't my ideal of a son-in-law, until . . .' (*Whisky advt.*)

. . . this little newspaper boytjie I've known for years . . . I said 'How are you?' 'No,' he said, 'I'm alright. My Mother died.' 'Oh I'm sorry,' I said. 'No,' he said, 'we don't mind.' Pieter-Dirk Uys cit. *Darling* 11.7.79

bra [bra] *n. pl.* -s. *Afr. E. Urban slang* Brother, freq. with given name e.g. 'Bra Victor at the bottle store'. *Drum* 8.3.74: see also quot. at *manne.* [*abbr.* brother]

You'll like my lanie bras.
You'll see . . .
Right, see you Ebrahim.
Right. Check you, bra.
cit. *Staffrider* May/June 1978

braai¹ [braɪ] *vb. trns. and intr.* To grill meat, poultry or fish usu. in the open air over the coals of a wood or charcoal fire. See quot. at *vleis.* cf. *U.S.* barbecue (*vb.*). [*Afk. vb. braai fr. Du.* braden roast, grill, *cogn.* broil]

. . . Murder and Robbery squad is urgently seeking members of a group of . . . coloured people who were braaing . . . on Sunday while two white men were assaulting a . . . white woman near by. *Cape Times* 14.1.75

. . . saying the cops had braaied boerewors whereas . . . said it was meat. *Drum* June 1976

braai² *n.* **1.** Anglicization of *vb.* as *n.* A gathering at which the main feature of the meal is *braaied* meat: see also (1) *braaivleis; cf. U.S.* barbecue and *Austral.* (*orig. U.S.*) cook-out. [*abbr. of* (1) *braaivleis* (q.v.)]

What do you need for a great braai? Chops, boerewors and . . . mieliepap. Crumbly or 'stywe' pap, you can't call your braai South African without it. *Sunday Times Advt.* 14.11.71

2. *n. pl.* -s. Any of several gadgets or pieces of equipment from built-in brick or stone fireplaces to small portable Hibachi (charcoal stoves) on which meat may be grilled: see quot. at *faggot* and as modifier, see quot. at *skinder.* [*prob. fr. vb.*]

Laze in the sun around the beautiful pool . . . or entertain your friends on the patio with its built-in braai. *Argus Advt.* 3.10.70

braaivleis [ˈbraɪˌfleɪs] *n. pl.* -es. **1.** Used with *art.* a,the:an informal outdoor social gathering or picnic at which the meat is cooked on a (2) *braai* or fire. [*fr. Afk.* braaivleisaand braai grill + vleis meat *cogn.* flesh + aand evening]

Under *braaivleis The Cape Times* has the second earliest dated entry, with a reference to a braai in aid of war funds in 1942. (The dreaded braaivleis – South Africa's secret military weapon?). *Cape Times* 2.7.73

2. *rare.* Used without *art.* to *sig.* the *braai*'d meat itself: see also (*braai*) *ribbetjie.* [*Afk.* braaivleis grilled meat *cogn.* flesh *Ger.* Fleisch]

. . . at the evening outspan the 18 members of the party will have a staple diet of braaivleis. *Evening Post* 3.4.71

braak [brɑːk] *vb.* **1.** To break up virgin or fallow land: as *n.* in place name Die Braak (fallow land); also Suurbraak. Combination ~ *land:* see quot. at *ou-land:* as *vbl. n.* ~ *ing: cf. Austral. to* break in (virgin land). [*Afk. fr. Du.* braak breaking]

Ploughing of virgin soil or braaking is done in September after rains. Wallace's *Farming Industries* 1896

Lands braaked during the winter absorb much of the rain falling after braaking, and a number of braakland plant-weeds can develop and thrive . . . the plant-growth on braaklands is sparse and not all . . . is edible. *H'book for Farmers* 1937

2. *slang* Vomit. [*Afk. fr. Du.* braken to vomit]

I asked politely did you have a thrash last night, because somebody's been braaking in the dustbin. *O.I.* 22.12.78

braambos ['brɑːmˌbɔs] *n*. Bramble, black-berry: any of several species of *Rubus* esp. the *R. pinnatus*: also place name Braambosspruit. [*Afk. fr. Du. braam cogn.* bramble + *bos cogn.* bush]

The fruit of the *Bramble* or *Blackberry* bush (Braambosch), ripens in the month of January It is equal in flavour and taste to that of Europe. The *roots* are astringent, and used in the form of a decoction against chronic diarrhoea etc. *Cape of G.H. Almanac* 1856

brak [bræk, brak] *adj*. Brackish: of or pertaining to water or soil containing alkaline salts. [*Afk. brak* saline, alka-line]

. . . we procured each of us a draught of very brack water; which, bad as it was, some-what relieved our thirst. Thompson *Travels I* 1827

Soils are generally termed 'brak' whenever they contain an injurious excess of various salts. Sometimes also they are dangerously alkaline. *H'book for Farmers* 1937

. . . he wondered what those lands were like that the holy woman's followers had sold. maybe it was just *brak* soil, and with *ganna* bushes. Bosman *Jurie Steyn's P.O.* 1971

Also found in S. Afr. place names usu. pertaining to water e.g. Brakpan, Klein Brakrivier, Brakspruit, Brakwal.

brak(kie) ['brak(ĭ)] *n. pl.* -s, -ke. Nonde-script mongrel dog: often used deroga-tively, equiv. of cur. *cf. U.S. pooch, Austral. mong, Brit. dial. tyke, Anglo-Ind. pye-dog*. Also ~ *kie* a friendly mode of addressing an unknown dog. [*Du. brak* setter *cogn.* archaic *Eng.* brach, *Fr.* braque hound]

I had a dog in Korsten. Just a *brak*. Once when we were sitting somewhere counting our bottles and eating he came and looked at us. Must have been a *kaffer hond*. He didn't bark. Fugard *Boesman & Lena* 1969

brakbos(sie), **brakbush** ['brakˌbɔs(ĭ)] *n. pl.* ∅, -s, -es. Generic name for many species, some of which grow and flour-ish in *brak* (q.v.) soils, others esp. spp. *Salsola* which produce vegetable lye used in the early days for soap making. See *ganna* also quot. at (1) *brak*. [*Afk. brak* (q.v.) + *bos fr. Du. bosch cogn.* bush]

. . . was covered principally with such shrubs and plants as afford alkali: these were the Kanna-bush and another whose name of *Brak-boschjes* (brackish bushes) indicates that their nature has been well observed by the inhabitants. Burchell *Travels II* 1824 *cit.* Pettman

. . . grazing consists of healthy Karoo

bushes, brak, ganna. *Farmer's Weekly Advt.* 2.2.74

brand- [brant] *n. prefix*. Fire, burning: found in place names e.g. Brandberg, Branddraai, Brandkop, Brandrivier: also in names of plants and flowers sig. either burning quality of sap etc. or a habitat of burnt ground e.g. ~ *lelie* (q.v.) ~ *gras* (grass) ~ *neutel* (nettle), ~ *bessie* (berry) etc. [*Afk.* blaze, fire *cogn. Eng.* (fire) brand]

brandewyn ['brandəˌveın] *n. Afk.* brandy: see quot. at *tiger's milk:* plant name ~ *bos Grewia flava* and its fruits ~ *bes-sie* (berry) from which an inferior 'brandy' was distilled: see also *brandy-wine*. [*Du. brandewijn fr. brantwyn* burnt wine]

brandewynketel ['brandəˌveınˌkɪətəl] *n. pl.* -s. A copper still with a long spout: now a sought-after collector's piece. [*Afk. brandewyn* (q.v.) + *ketel cogn.* kettle]

On lonely farms the grapes are still pressed in the *balies* (large vats) with bare feet . . . later the liquid *mos* is passed into the old *brandewyn-ketel* or still, made to very much the same pattern as those used by Tennessee '*moon-shiners*'. Green *Karoo* 1955

brandhoutboom ['brantˌhəutˌbuəm] See *kreupelboom*.

brandlelie ['brantˌlıəlĭ] *n. pl.* -s. Any of several species of *Cyrtanthus*, esp. *C. angustifolius*, also *rooipypie*, the typical habitat of which is burnt veld. [*Afk. brand* fire + *lelie cogn.* lily]

brandsiekte ['brantˌsĭktə] *n*. Scab, a highly contagious skin disease affecting sheep in S.A. Failure to report the disease renders the farmer liable to penalty. [*Afk. fr. Du. brand* burning, fire + *ziekte* disease]

They are subject also to a cutaneous disease that works great havoc . . . It is called by the farmers *brandt-siekte* or burning disease. Barrow *Travels I* 1801 *cit.* Pettman

brandsolder ['brantˌsɔldə(r)] *n*. A layer of bricks or clay laid over the ceiling of upper-storey rooms to catch burning thatch in the event of fire. ❡Early in-surance companies at the Cape quoted lower premiums for houses built with ~ s. [*Afk. fr. Du. brand* fire + *zolder* loft]

De Protecteur Fire and Life Assurance Company . . . Allowances made, as formerly, for substantial Brandzolders. *Cape of G.H. Almanac* for 1845

Fire was an ever present threat to town and city. To prevent it destroying his . . . furniture the farmer had a 'brandsolder' built into the roof. This was a layer of clay or thin bricks at ceiling level to catch the burning thatch. *Sunday Times* 6.8.72

brandywine *n. obs.* Brandy. [*fr. Du. brandewijn* brandy; see *brandewyn*]

From the refuse of the wine press, a strong spirit is distilled called brandy wine, in general use by the boors and the farmers. Ewart *Journal* 1811–1814

bray *n. and vb.* A speech mannerism: erron. sp. form: see *bry*, and (2) *brei*.

Breakfast Run, the *n. Sect. Motor Cyclists* The early morning ride to Hartebeespoort Dam outside Johannesburg, in which several hundreds of motor cyclists take part every Sunday. See second quot. at *kraak*. ⁋This is the best known of several Sunday ~s which take place regularly in various centres. e.g. Cape Town, Kimberley etc.

. . . crawl from their beds at the crack of dawn, battle with the inevitable hangover . . . and gather at Fourways to go on the now traditional Breakfast Run to Hartebeespoort Dam . . . Fanatics . . . have been on the Breakfast Run every Sunday for the past nine years . . . 'What else can you do at this time on a Sunday morning? Even the churches are closed!' *Darling* 17.6.79

Breakwater, the *n. prop. hist.* A convict station established at Cape Town in 1860 for prisoners sentenced to hard labour, i.e. building the Breakwater for Table Bay harbour. ⁋Time on the ~ was a common penalty for *IDB*: see quot. at (1) *I.D.B.*

Breakwater Convict Station: SHORT, John Superintendent Convict Department, Breakwater. *Cape Town Directory* 1866

It was not until 1923 that the Breakwater Prison was finally evacuated. Then it became a native location, and a government research laboratory was built in the old punishment yard. Green *Tavern of Seas* 1947

bredie [ˈbrɪədĭ, ˈbridĭ] *n. pl.* -s. A ragout or stew of meat, usu. mutton, and vegetable(s) named after the vegetable used as in *tomato* ~ below, see *tamatiebredie hotnotskool* (q.v.) ~, *spinach* ~, etc.; see also quot. at *wateruintjie cf. Canad. burgoo.* [*Afk. bredie prob. fr. Port. bredo* ragout]

The Cape bredie is a stew in which the vegetable has been reduced to a fairly thick consistency . . . tomato is the most popular among these bredies. *Farmer's Weekly* 25.4.73

breed(e) [brɪəd(ə)] *adj.* Wide, broad: found in place names e.g. Breede river. [*Afk. breed cogn.* broad]

brei, brey [breɪ] *vb. trns.* **1.** To curry hides by scraping, twisting and working until pliable by hand. [*Afk. brei* curry (*vb*), *fr. Du bereiden* to prepare]

The Hottentots can't 'bray' the skins as the Kafirs do. Duff Gordon *Letters* 1861–62

to ~ *riems.* To twist hide thongs, see *riem*, on a wheel (see *breipaal*) under pressure to render them soft and fit for tying and other uses.

Mrs K. was braing of reims, and making a kid kraal. Shone *Diary* 18.7.1862

2. *freq. erron.* form: see *bry* (in speech).

But I've heard some strange Afrikaans in my time, not excluding the 'Malmesbury brei' of the late Field-Marshal Jannie Smuts. *Daily Dispatch* 11.11.71

. . . an obscene midnight phone call . . . from a man who spoke with a brei. *Het Suid Western* 26.5.76

breipaal [ˈbreɪˌpɑːl] *n.* A device for *breying* riems, see *brei*, something like a gallows with a wheel for twisting the thongs, also a place name Breipaal: see third quot. at (1) *riem* [*Afk. brei* (q.v.) + *paal cogn.* pole]

breker [ˈbrɪəkə(r)] *n. pl.* -s. A 'tough,' or would-be tough, now freq. also used *colloq.* of a 'wide boy' or other exhibitionist esp. on the road, affecting leather-jacket and motor cyclists' style gear and behaviour. [*Afk. lit* breaker *fr. breek* to break]

A breeker [sic] is South African slang for a tough guy . . . a champion street fighter . . . an 'ou who can put the head and boot in better than the next 'ou'. *Sunday Times* 7.9.75

There's this breker in the smart embroidered jeans and black lummie, see . . . cruising slowly past on his iron . . . And I only smaak irons in a big way, man. Hobbs *Blossom* 1978

When the Joburg boys cribbed our Parow ou's and put oranges on their aerials – the Cape Town brekers went one better and put green fur on their dashboards. Lorraine cit. *Capetonian* May 1979

Bremer Bread *n. World War II Term:* Bread made of *Government flour* (q.v.) when white bread was not permitted. [*fr. n. prop.* Dr. Karl Bremer, then Minister of Health]

breyani [ˌbreɪˈɑnĭ] *n.* An Indian dish: see *biriani.*

Let's use our breyani and curry spices, our

fruits, game and fish and create brand new indigenous dishes. *Darling* 16.3.77

Curry – as South African as braaivleis rugby and sunny skies
... ideally suited as a way of preparing meat ... today vegetable curries and breyanis ... are as much in demand as the chicken, beef or mutton variety. *Fiat Lux* Oct. 1978

bright light *n.* pl. -s *colloq.* Rhodesian: a police reservist. [*unknown:* for poss. origin see second quot. below]
We have a security fence and 'bright lights' sleeping on our veranda ... The raid ... lasted about ten minutes. The bright lights were there and they fired back. *Fair Lady* 16.3.77
The men who leave office jobs in Salisbury to guard remote farm houses are called 'bright-lights', after the lights of the city they have left behind. *Sunday Times London Magazine* 9.10.77

bringal [ˈbrɪnˌdʒɔl] *n.* pl. -s. Also *brinjal:* *Solanum melongena* aubergine or egg fruit, long pear-shaped or globular purple vegetable used fried or as an ingredient of curries and similar dishes. See quot. at *kalya.* Also *Anglo-Ind., Jam. Eng. brown-jolly.* [*fr. Port. bringella* egg-fruit]
Many of the most common plants grown belong to this family – *Solanaceae* – including potatoes, tomatoes, brinjals and green peppers. *Evening Post* 30.6.73

Broederbond [ˈbrŭdə(r)ˌbɔnt] *n. prop.* A largely secret *Afrikaans* (q.v.) organization with limited membership, also known as the *Bond* (q.v.).
The Broederbond, a secret society composed of Afrikaners holding key jobs in all walks of life, has been subject to the same tensions and divisions as the Nationalist Party. *Sunday Times* 5.3.72
~ *er, n.* pl. -s. A member of the ~, also *Broeder.* [*Afk. fr. Du. broeder cogn.* brother + *bond* league, fellowship *cogn. Ger. Bund*]
During World War II the Smuts Government made disclosures about the Broeders and banned public servants from belonging to it (the Broederbond) on the grounds that it was a subversive body. *Sunday Times* 24.9.72

Mason and Broeder in 'idol' row.
The long-standing row between the Broederbond and the Freemasons erupted again last week when a prominent Freemason demanded R5 000 for alleged damages ... said the dominee had accused him and his fellow Masons of worshipping Jabuhlon, which he claimed was a Masonic idol. *Sunday Times* 22.10.78

Broederkring [ˈbrŭdə(r)ˌkrɪŋ] *n. prop.* The Ministers' Fraternal consisting of representatives of the *Nederduitse Re-*formeerde Kerk* (q.v.) (*D.R.C., N.G.K.*), the Gereformeerde Kerk (*Doppers* (q.v.)) and the *Hervormde Kerk* (q.v.) [*Afk. broeder cogn.* brother + *kring* circle]
In a statement, the Broederkring, a group of leading ministers of the three churches, said it fully supported the Dutch church's move. *S.A. Digest* 14.4.78

broedertwis [ˈbrŭdə(r)ˌtwɪs] *n.* Political term 'brothers' quarrel' usu. referring to a split in the ranks of the Afrikaners. [*Afk. fr. Du. broeder cogn.* brother + *twis(t)* (q.v.) quarrel]
The broedertwis bites deep between Afrikaner and Afrikaner as former Nationalists flock to the Republican Party. *Sunday Times* 2.4.78
That the National Party is in a state of ... confusion ... is no secret. Hardly a week passes without evidence of internal division – of the most savage kind of *broedertwis. Ibid.* 4.3.79

broek [brŭk] *n.* pl. -s, ∅. Trousers, pants, knickers for either sex; in S.A.E. usu. with pl. marker -s. Also *dimin.* ~ *ies* (q.v.) and in combination *klap* ~ (q.v.) *figur.slang: bang* ~ coward, 'scaredy cat'; see *bang. pap* ~ coward, poltroon; see *pap.* See also quots. at *loop* and (2) *bioscope.* [*Afk. broek* a pair of trousers, *cogn.* breeches, breeks]
... a clear memory of two little outjies in Khaki broeks. Fugard *Blood Knot* 1968

broekies [ˈbrŭkɪz] *pl. n.* Knickers usu. for children: in combination *children's* ~, dresses with matching ~ (also *now obs. shu-shu* ~, 'Hot pants' see quot. at *shu-shu*). [*Afk. broek* trousers + *dimin. suffix -ie* + pl. -s]
There were cries of goodbye and the little girls blew kisses to us; they had on new print dresses and broekies that Estelle had made for them. Rooke *Lover for Estelle* 1961

brom [brɔm] *vb. intr. colloq.* To complain, make a fuss: also *onomat.* see quot. at *gooi.* [*Afk. brom* growl, mutter]
K. didn't bring me back my car so I had to walk to the hairdresser's, bromming all the way I can tell you. *O.I.* 30.5.76

brommer [ˈbrɔmə(r)] *n.* pl. -s. Large noisy (bluebottle) fly or other buzzing insect [*in O.E.D. Supplement A-G 1972* as *brummer fly*] found in place name Brommersvlei. [*Afk. brom* mutter, growl + *agent. suffix -er*]

bromvoël [ˈbrɔmˌfʊəl] *n.* pl. -s. Usu. the turkey buzzard or ground hornbill *Bucorvus leadbeateri,* so named on account of its loud, booming call: also any of several other noisy birds including the *mahem* (q.v.) see also *rainbird*. [*Afk. brom* growl, mutter + *voël fr. Du. vogel cogn.* fowl]

... and if a person kill by accident a *mayhem*, (or Balearic crane) or one of those birds which the Colonists call *brom-vogel*, he is obliged to sacrifice a calf or young ox in atonement. Thompson *Travels II* 1827

-bron- [brɔn] *n. prefix and suffix.* Source, spring: found in place names, e.g. Heilbron, Suurbron, Brondal. [*Afk. bron* spring source (*heil* salvation, welfare)]

broodboom, kaffer [ˌkafə(r)ˈbrʊət ˌbʊəm] *n. pl.* -bome. Any of several species of *Encephalartos* esp. *E. caffer* and *E. cycadifolius*, often loosely called 'cycads', with large female cones of which part is edible. [*Afk. brood* bread + *boom cogn. Ger. Baum* tree]

brown *adj.* Of or pertaining to, usu. *Coloured* (q.v.) S. Africans: see quot. at *voorloper* and *not so.* [*prob. fr. Afk. bruinmens*, lit. brown person, *a Coloured*]

For Blacks and to a very high degree for Brown people as well, Afrikaans simply is the language of the oppressor . . . The Afrikaner has never admitted that the Coloured in fact is a Brown Afrikaner. Dreyer Kruger *cit. Sunday Times* 1.8.76

browns *pl. n. Sect. Army.* Everyday uniform, or occ. combat uniform, *usu. bosdrag* (q.v.) as opposed to *mooimoois* (q.v.) or '*step-out*' (*uitstap*) uniform.

The tunic's OK. I can manage that, but browns are hell to iron. You've got to have sharp lines straight up the sleeves and overall pants – leg pockets and all. *O.I. Serviceman* 14.4.79

bruidskis [ˈbrœɪts ˌkɪs] *n. pl.* -te. A dowel chest. see *kist cf. Austral. glory box; U.S. hope chest; Brit. bottom drawer; Afr. E. wedding box.* [*Afk. bruid* bride + possessive suffix -s + *kis fr. Du. kist cogn.* chest]

bry [breɪ] *vb. intrns. and n.* Also erron. forms *bray, brey* and *brei* (q.v.): a burr in speech with a rolling guttural 'r', a mannerism common in S.A. esp. in the Malmesbury district hence *Malmesbury* ~. ⸿The articulation of the 'r' is similar to that of French [*Afk. bry fr. Du. brijen fr. brouwen* to speak thickly]

... he shook his head, . . . 'Ag, there's *mos* plenty of guinea fowl here in the Cape, lady.

You don't have to get Trrrransvaal guinea fowl . . . (There was the bry again!) Kavanagh *Merry Peasants* 1963

buchu [bʉxʉ] *n.* Any of several species of Rutaceae the leaves of which have been used medicinally for stomach complaints since C17, also used cosmetically mixed with sheep's fat by the Hottentots for anointing their bodies: the name being extended to a number of other species. ⸿ ~ *leaves* are obtainable from the range of (*Old*) *Dutch medicines* (q.v.) species of *Agathosma, Barosma* and *Diosma* used medicinally freq. as diuretics; ~ *brandy*, an infusion of ~ *leaves* in brandy used mainly internally but also as a specific (sometimes in vinegar not brandy) for bruises and sprains. [*fr. Hott. buku, plant name*]

For cuts and bruises they use the leaves of the buku, and one or two other plants, with good effect. Thompson *Travels II* 1827

South Africans have for generations attached a particular value to kukumakranka brandy, buchu brandy, clove brandy etc., and these bottles occupied an honoured place in the medicine chest. Opperman *Spirit of Vine* 1968

buchu karoo bush [ˈbʉxʉ ˌkaˈruː, -rʊə] *n. pl.* -es. Any of several aromatic shrubs of the Karoid area some of which constitute a problem of plant migration and depletion of or encroachment on the (2) *veld* (q.v.).

buck¹ *n. pl.* -s. *obs.* 1. Used by the Settlers *sig.* 'goat' [*fr. Du. bok* goat]

Cut 50 lambs again today & about 150 Buck kids, purchased yesterday 97 Buck and kids Buck kids & lambs 255 Old Bucks . . . total in Buck krall. Collett *Diary* 13.10.1838

In combination ~*fat obs.* Goat's lard formerly much favoured for a variety of culinary and medicinal purposes.

We suffered exceedingly when we had buck-fat plasters put on our chests and some meddlesome old crone suggested that we be rubbed with honey back and front. McMagh *Dinner of Herbs* 1968

buck² *n. pl.* -s. *colloq. slang.* A *rand* (4) (q.v.). [*presum. fr. US buck* a dollar]

The local guardians of the law stung Miles fifteen bucks . . .a . . . magistrate reducing the fine to eight bucks pointed out that Traffic Officers could better spend their time preventing jay walking. *Rhodeo* Sept. 1978

bucket *n. pl.* -s **1.** *prob. obs.* Measure for potatoes (by retail) approximately 5kg.

Got 2 buckets potatos [*sic*] from Wilgemoed's. 21.5.1850 J. D. Lewins *MS Diary*

2. Measure used on farms for rations, usu. *mealie meal* (q.v.) or *mealies* (q.v.) *cf. bag, pocket, potwan(a),* also *gogogo.* [*lit. a bucket*]
In the early days of the Colony the bucket was often found to be a convenient measure when bartering. Pettman *Africanderisms* 1913

bucksail *n. pl.* -s. The canvas covering or 'sail' of a *buckwagon* (q.v.) [*fr. Afk. bokseil* bucksail, *seil* canvas, tarpaulin]
. . . Yankee Moore, whose store – a bucksail tent furnished with a counter of empty Rynbende gin-cases was . . . the busiest commercial establishment at de Kaap. Brett Young *City of Gold* 1940
We outspanned a mile or so farther on, drawing the five wagons up close together and getting what shelter we could by spreading buck-sails. Bosman *Mafeking Road* 1947

buckwagon *n. pl.* -s. A large clumsy transport wagon, for heavy loads, the 'buck' being the overlapping side rails or frame projecting beyond the wheels. [*fr. Afr. bokwa, bok* buck + *wa fr. Du. wagen cogn.* wagon]
. . . wait until help came along in the shape of some Boer's springless buck-wagon. Travelling on the buck-wagon was extremely dangerous. This type of farm wagon has its brake control situated at the back, in the form of a long wooden bar stretching across the two back wheels. Klein *Stagecoach Dust* 1937

buffalo grass *n.* A large-leafed fodder grass; any of several species of *Panicum* and others, including *Eragrostis,* see *oulandsgras,* said to have been favoured originally by the buffalo: also *buffelsgras.*
I looked . . . and about two hundred yards away saw a stembuck standing in the shade of a mimosa bush feeding briskly on the buffalo grass. FitzPatrick *Jock of Bushveld* 1907

buffel(s)-¹ [ˈbœfəl(s), ˈbəfəl(z)] *n. prefix.* Buffalo: in plant names e.g. ~ *gras* (see *buffalo grass*); ~ *sdoring* (thorn); ~ *horing* (horn). Also ~ *svoël* (bird), and in place names Buffeljagsrivier, Buffels baai, Buffelsvlei. [*Afk. fr. Du. buffel* buffalo]

buffel² [ˈbœfəl] *n. pl.* -s. *Sect. Army.* A mine-proofed armoured vehicle: ⫶ Information inadequate: informants do not agree upon the type of vehicle.

buite- [ˈbœitə] *adj. prefix.* Outside, outer: found in street names e.g. Buitekant Straat (*kant* side), Buite*singel* (q.v.), (also Buitencingle) [*Afk. fr. Du. buiten* outside]

buka [ˈbuga] *vb. colloq. prob. reg. Natal.* To look, take a look. [*Zu. ukubuka* to look]
The Zulu house servants and the Mosotho looked at each other . . . 'Buka!' said the Zulus. 'A Mosotho wearing a fez.' Lanham/Mopeli-Paulus *Blanket Boy's Moon* 1953
'. . . buka lo fuzz' (watch for the police). Molloy *S. Afr. C.B. Dict.* 1979

bully *n. pl.* ∅ -ies. See (2) *dikkop;* used also of two fish of the Clinidae fam. *C.* (*c*) *cottoides* and *C.* (*c*) *taurus.*

bult [bœlt] *n. pl.* -e, -s. A low ridge of ground or hillock usu. sandy rather than rocky: found in place names e.g. Droëbult, Bultfontein. [*Afk. fr. Du. bult* hump, hunch]
'. . . on the other side of the bult . . . Can you see through a bult – a bult about fifty paces high and half a mile over it?' – then the surveyor had to admit, of course, that no man could see through a bult. Bosman *Eekkersdal Marathon* 1971

bundu [ˈbundŭ] *n.* Wild, open country remote from civilization and cities: see also *backveld* and quot. at *mopane worm. cf. Austral. back of Bourke, boo-ay outback, N.Z. woop-woop, U.S. boondocks, Canad. the sticks.* [*prob. fr. Shona bundo* grasslands]
. . . he can work in the big city for the wife and children he left in the bundu. *Drum* 22.1.73
Game Ranger: 'I drove . . . over a thousand square kilometres of the most rugged country in the Transvaal . . . dried-up river beds and trackless bundu where you'd expect only a rhino to be at home. *E.P. Herald Advt.* 21.5.74
colloq. ~ *bash* usu. *vbl n.* ~ *bashing:* travel or travelling over very rough or difficult country. cf. *Austral. bush walking.* Also ~ *basher,* one who enjoys ~ *bashing* and rough or spartan travelling conditions. *cf. Austral. bushwalker,* (also *bushbashing/basher, bushwhacking/ whacker.*)
'Bundu-bashing' in pursuit of live game to sell . . . a four-wheel drive vehicle is completely worn out after two seasons of this work on his game ranch. *Star* 22.6.72
AG. 100 ('agbike') The most rugged bundu basher in the world . . . real elephant power designed for country where tar is veld and potholes are dongas. *Farmer's Weekly Advt.* 4.12.74

Bunga [ˈbŭŋga, ˈbungə] *n. prop.* The chief Council of the Transkeian Government, loosely used, largely by whites sig. the building housing the administrative offices: after Independence, the

44

Parliament of the Republic of Transkei. [*fr. Xh. ibhunga* council meeting]

Local councils and the Bunga catered to a limited extent for the tribesman and the reserve dweller, but those participating in the modern economy had access only to the dimly lit advisory ante-chambers of power. *Daliy Dispatch* 20.5.71

bungalow *n. pl.* -s *Sect. Army.* Barrack room ╠ Quarters in S. Afr. military camps are known as ~s.

I'm about to go to church with the other English 'ouens' . . . we're a dying race I'm afraid but there are a few of us in this bungalow – I'm in bungalow 49 if that's any help in the address. *Letter Serviceman* 13.1.79

All around the bungalow there are tall stories floating about . . . *Letter Serviceman* 21.1.79

bunny chow *n. Ind. E. prob. reg.* Natal. Vegetarian curry of beans and *dhal* see *Indian Terms*, sold usu. as 'take-away' food in a hollowed half-loaf of bread. ╠~ appears on the menu of the Royal Pantry [take-aways] in Smith Street Central Durban. [*fr. bhannia (various sp.) shopkeeper caste (vegetarian)* + *chow* food (*prob. fr. Peking Chinese chiao*)]

Bunny Chow 20 cents. Notice: Indian shop window, Durban Nov. 1972

Take-aways

UPPER CLASS	MIDDLE CLASS	LOWER CLASS
What? (on Thurs-days we dine out)	Chinese pizza, Kentucky Fried	Bunny chow, hot chips

Darling 4.4.79

burfee ['bɜ(r)ˌfi] *n.* Indian sweetmeat: milk fudge freq. spiced with *elachi*, see *Indian Terms*.

-burg [-bɜg, bœrx] *n. suffix.* City: found in place names e.g. Johannesburg, Middelburg, Lydenburg, Steynsburg. *cf.* Edinburgh; see also *-stad, -dorp.* [*Du. O.E. burg, burh* city *cogn.* borough, *Ger. Burg*]

burgher ['bɜgə] *n. pl.* -s. *hist.* A citizen of the Cape Colony, who was not a servant of the Dutch East India Company: also a citizen of one of the Transvaal and Free State republics. ╠ ~s were liable to ~ *duty,* military service, hence ~ *commando* (q.v.). The ~ *Senate,* dissolved in 1827, consisted of seven members, was responsible for civic affairs and was housed in the *Stadhouse/huis* (q.v.). A *free* ~ was one who was free from the control of the Dutch East India Company in the C17 and C18. ~ *ship,* the rights and privileges of being a ~. See quots. at *veldcornet* and *wardmaster.* [*Du. burgher* citizen *cogn.* burgess]

At the Searches of Shops, to be effected as aforesaid from time to time, by two Commissioners of the Burgher Senate, or by the Fiscal and two Commissionaires as aforesaid, at times not fixed and without previous notice. *Afr. Court Calendar* 1819

bush *n.* Bush [*Afk. bos fr. Du.* bosch *cogn.* bush] **1.** Used of both bushy and wooded country: *cf. Austral. bush.*

2. -bush *suffix* to names of any of numerous varieties of low indigenous shrub which intersperse the grass of the *veld* e.g. *karoo* ~, many of which are edible and many inedible or 'plant pests' or 'invader plants' e.g. *bitter* ~ (q.v.) and some varieties of blue ~.

3. As in *Austral. prefix* to other *n.* ~ *baby* (q.v.) ~ *buck* (q.v.) ~ *cart* (q.v.) ~ *mechanic* (q.v.) ~ *pig* ~ *soil* (q.v.) ~ *tea* (q.v.) ~ *veld* (q.v.)

bush baby See *nagapie.*

bushbuck *n. pl.* ∅. Usu. *Tragelaphus scriptus sylvaticus :* also used of any of several S. Afr. antelopes of medium size preferring for their habitat dense bush and woodland, and which can be very fierce if disturbed; see quot. [*fr. Afk. bos (Du. bosch) cogn.* bush + *bok cogn.* buck]

. . . manager of the . . . Game Reserve, was rushed to hospital . . . after being attacked by an enraged bushbuck yesterday afternoon. *E.P. Herald* 27.3.73

bush cart *n. pl.* -s. A sturdy mule-drawn conveyance bought in large numbers for the *U.D.F.* (q.v.) see quot.

The last seven of the bush carts ordered for the army by Mr O. Pirow, K.C. when he was Minister of Defence, will be sold at a War Stores Disposal Board auction here on Monday. More than 200 have already been sold and others are still being used by the U.D.F. for the carting of garbage. *E.P. Herald* 24.1.48

Bush leave *n. Sect. Army.* Extra leave granted to men who serve long periods in the *Red Area* (q.v.) ╠ *acc.* one ex-Border informant one day for every two weeks, see quot: see also *bush pay* and *bos(sies).*

'Bush leave' may be granted to men who serve extended periods of time in the operational area on the basis of one day for a month of time served, the spokesman said. *Sunday Times* 24.9.78

Bushman *n. pl.* -men. **1.** A primaeval

indigenous race of nomadic hunters of Southern Africa now largely living in the Kalahari desert: see quot. at *Boskop*. [*Du. Boschjesman, Afk. Boesman*]

> We saw in prison these Bushmen . . . this poor hunted race is now become almost extinct in this quarter. In color they resemble the Hottentot, but are more diminutive, and are easily distinguished by their greater activity, quicker eye and sprightliness of countenance. Philipps *Albany & Cafferland* 1827

In combination: ~ *paintings* lively primitive scenes of hunting etc. still found in caves all over S.A. ~ *rice*, termites.

> There are two species of ants which they chiefly feed upon – one of a black, and the other of a white colour. The latter is considered by them very palatable food, and is, from its appearance, called by the boors 'bushman's rice'. This rice has an acid, and not very unpleasant taste, but it must require a great quantity to satisfy a hungry man. Thompson *Travels I* 1827

2. The Khoisan language of the ~ characterised by numerous clicks.

> When enunciated with the appropriate clicks this word amounts to an emphatic negative in the Bushman language. *Sunday Times* 27.5.73

bush mechanic *n. pl.* -s. *colloq.* A very rough and ready workman *cf. Austral. bush carpenter.*

Bush pay *n. Sect. Army.* Danger money paid to men serving in the *Operational Area* (q.v.): see also *Bush leave* and *bos-(sies)*.

bush soil *n.* Soil rich in natural compost from wooded or bushy areas, favoured for potting and gardening purposes.

> Best clean bush and black garden soil . . . *E.P. Herald Advt.* 15.8.74

bush tea *n.* The dried leaves, or an infusion of them, of any of several shrubs having medicinal or stimulating properties esp. *rooibostee/tea* (q.v.) [*prob. fr. Afk. bossiestee*]

> In most of these Colonial stores *bush tea* can be bought. It costs sixpence a pound, looks like the clippings of a privet hedge, including the twigs, and is said to be a tonic. *Life in Cape Colony* by X.C. 1902 *cit.* Pettman

bushveld *n. Veld* (q.v.) composed largely of bush, often of a thorny or scrubby character. *cf. Canad. bushland.* [*Afk. bosveld*]

> Behind the Boers and the Britons, as they faced the fevers of the bushveld or the hardships of life in the mushroom camps of Kim-

berley and Johannesburg, lay the gracious Mother City in the Cape, the happy towns of its western or eastern districts with Dutch or English names. Mockford *Here Are S. Africans* 1944

> The veld consists of sweet grass and bushveld and the property is divided into 23 grazing and land camps. *Grocott's Mail* 17.11.72

In combination ~ *Ben n. prop.* Formerly a comic strip character in the *Farmer's Weekly*: also a hat of the style worn by him: see also *Van der Merwe.*

busy *adj.* **1.** *substandard.* Used redundantly in S.A.E. with an *-ing* form of the verb to indicate progressive aspect, sig. 'in the process of' or equiv. See quot. at *turf, black.* [*trans. besig (om te) which replaces in Afk. vb phrs. the Du. pres. partic. suffix -end(e) (equiv. Eng. -ing)* ¶It is retained only as *modifier e.g. lopende water*, running water.]

> I rushed in and found the two infants busy having convulsions – as though there were not enough troubles that day. McMagh *Dinner of Herbs* 1968

> This dreadful cattle disease [rinderpest] had swept down Africa and was now busy decimating their herds. *Personality* 21.8.71

2. 'Engaged' e.g. 'His line's busy, will you hold on please?' [*trans. Afk. besig cogn.* busy]

busy with *adj. phr.* 'Engaged upon', 'occupied with': as in 'He's *busy with* another call/patient etc. at the moment' see also *busy.* [*trans. Afk. besig* busy, engaged, occupied + *met* with]

> He said most members of the administration staff were busy with the investigation. *E.P. Herald* 28.2.74

butchery *n. pl.* -ies. A butcher's shop, where meat is sold, not an abattoir where slaughtering takes place. [*prob. fr. Afk. slagtery* butcher's shop]

> The Paarl butcher, . . . who was found dead on the floor of his butchery on Sunday morning, appears to have been strangled. *E.P. Herald* 6.11.73

buti [ˈbutĭ] *n. pl.* -s. **1.** *Afr. E.* Also *bhuti*: a frequent mode of address to African men by other Africans; see also *tata, sisi: cf. Austral.* (aboriginal) *bing(h)i, Jam. Eng. baada*, brother [*fr. Afk. boetie* brother]

> 'Have a double brandy.' He knocked that off without wincing . . . 'But Bhuti, I am struggling here.' Matshikiza *Chocolates for Wife* 1961

> Apparently an eight-year-old girl called out: 'Bhuti, Bhuti' and he went back to help her.

Both were drowned. *Indaba* 29.10.76

2. Also in the form *boetie* (also *sissie*, see *sisi*) a courteous mode of address or reference to servants at schools or universities (esp. E. Cape) by children, staff or students, poss. to avoid the term *boy* (q.v.) *cf. former U.S. use* of 'brother' for 'nigger'. 'He's gone where the good brothers go' (*Uncle Ned*). [fr. *bhuti*]

Warning: Only 'Boeties' in . . . overalls should be allowed to carry luggage into the Houses at the beginning of term. Grahamstown Headmistress's letter 1973

butter, household, table *n. Table* ~ 1st–2nd grade (not choice grade), *household* ~ 3rd grade (cooking butter).

Almost all accumulated stocks of table butter in South Africa had been sold out and only household butter and butter being turned out by the creameries was available. *E.P. Herald* 24.4.74

button spider *n. pl.* -s. Either of two poisonous S. Afr. spiders *Latrodectus indistinctus* or *L. geometricus* characterised by a red mark on the underside of the abdomen and closely rel. to the American black widow spider. [*prob. trans. Afk. knoop* button + *dimin. suffix -ie* + *spinnekop* spider]

But let's disregard snakes. Let's say that miraculously you escape all hurt from the snake world. That still leaves you in grave danger of meeting a button spider. *Argus* 14.2.73

butter-bread *n. Ind.E.* Indian term for bread-and-butter comparable with Sanskrit *dvanva* structure, a 'man-tiger', meaning a man and a tiger.

butter chilli *n. pl.* -s. *Ind.E.* S. Afr. Indian term for capsicum, green or sweet pepper, as opposed to the hot red chilli used in curries. [*cf. Jam. Eng. Indian pepper* capsicum]

y *prep. substandard.* Used sig. either 'beside' (alongside) or 'with' or 'at': see second quot. at *boep. cf. Austral.* locative *on* a place, equiv. 'at'. (*now poss. obs.*) [*transference of Afk. by* beside, with, at]

Daniels. What time did he miss the man – Was he at the shooting match by your place did he say he was then on his journey. *Hancock Notebook* 1822

Because the water drips directly by the plant, only the root area is irrigated and no water is wasted on 'dead' areas. *Farmer's Weekly* 1974 (date mislaid).

y-and-by *n. Afr.E. obs.* Cannon, field gun: thought to be onamatopoeically named: various explanations have been suggested – slow firing, reluctance to explain now, but by and by etc. [*Ngu. mbayi-mbayi* cannon]

By and By – The name by which cannon are known to the natives of Natal. It is said that inquiring in the early days what these cannon were, they were informed that they would learn *by and by*, hence the name, which seems to the native to represent the noise of the explosion – a primitive striving after meaning. Pettman *Africanderisms* 1913

bywoner [ˈbeɪˌvuənə(r)] *n. pl.* -s. A subfarmer, authorised squatter or sharecropper working part of another man's land, giving either a share in his profits or labour or both in exchange: see also *poor white*: also occ. *figur.* sig. second class citizen or dispossessed person. [*Afk. by* with, at + *woner* dweller *fr. Afk. woon* live + *agent. suffix -er*]

. . . the great body of landless bywoners who eked out an existence by the grace of the landowners . . . Native labour prevented the great body of bywoners and poor white from becoming an established and recognised class of labourers and wage-earners. De Kiewiet *Hist of S.A.* 1941

. . . bywoner existence not envisaged. *Farmer's Weekly Advt.* Jan. 1975

C

café *n. pl.* -s. Also known as a *Greek shop* (q.v.) or *tea room* (q.v.): in S.A. a ~ is seldom ever a tea or coffee-shop where refreshments are to be obtained. ℙ The ~ is now more technically called 'a convenience store': see quots. at *Greek shop* and below, though this term seems unlikely to gain spoken currency.

Tea-room troubles

SA's 12 000 café owners must move with the times or face extinction. Twelve years ago Athens born George . . . opened a small, typically South African café in Goodwood . . . Business thrived . . . he now owns five outlets . . . – 's Superettes are no longer poky cafés with fish lying next to ice cream in the freezer. They are known as convenience stores rather than cafés. *Financial Mail* 28.7.78

Three major local dairies will deliver fresh milk to both houses and cafés on Christmas day. *Cape Times* 16.12.78

café-bio See *bio-cafe*: also *tearoom cinema*

calabash [ˈkæləˌbæʃ] *n. pl.* -es. The bottle gourd much used by Africans for receptacles etc: also a vegetable see quot.

at *garum masala*. [*prob. fr. Arab.* *qa'rah yabisah* dry gourd]
. . . gave me some calabashes, and a slave taught me how to clean and prepare them. Nothing keeps water, wine, fruit, butter, so cool in summer as these receptacles, which are sometimes very large, and cannot receive injury as glass or stoneware would by being tossed about. Barnard *Letters & Journals* 1797–1801

In combination ~ *milk*, milk soured in a scooped out ~, see *maas*.
She then gave him a basin of coffee, some calabash milk and some flour porridge sprinkled with sugar. Metrowich *Frontier I lames* 1968

~ *piano* also *kaffir piano* (q.v.) and *mbira* (q.v.) see quots. at *mbira*.

~ *pipe* A tobacco pipe made of a small dried ~.
No farmer in my young days ever puffed away at anything except a pipe – preferably a calabash grown in the Western Province. Jackson *Trader on Veld* 1958

call *vb. trns. Afr. E. and Ind. E.* To invite. [*trans. Ngu. ukubiza* to call, *also* invite]
Why will people go to shebeens to buy liquor at inflated prices for cash even when bottle stores are open? Why not buy it and call friends if they need company and drink at their homes. *Daily Dispatch* 21.10.71

camelthorn(tree) *n. pl.* -s. See *kameeldoring* (*boom*).

came there *vb. phr. usu. past tense. substandard.* Arrived: past of 'come there': see quot. at *stick fast*. [*prob. trans. Afk. toe ons daar kom:* when we arrived there. (This also occurs in the present tense with future reference, e.g. As soon as we come there . . .)]

camp *n. pl.* -s. **1.** A fenced enclosure for grazing, equiv. of paddock; *land~* a fenced field for cultivation (see also *land*), *grazing ~, veld ~,* see (2) *veld cf. Austral. camp* mustering or resting place for stock. [*Afk. kamp* large paddock or run *fr. Lat. campus cogn. Fr. champs* field]
Six land camps and four veld camps jackal proof fencing in very good condition with abundant stock water. *Daily Dispatch* 4.3.72
2. *vb. trns. often pass.* Also ~ *off*, to divide (land) and enclose with fences. [*presum. fr. n.* camp (q.v.)]
The rest of the farm will be camped off in due course and eventually there will be 16 camps to accommodate the system. *Farmer's Weekly* 3.1.58

can [kæn] *n. pl.* -s. A glass jar usu. of 2 litre capacity, a container for wine. [*translit. Afk. kan* jar]
. . . sat in the kitchen drinking a can of wine Mrs B. had bought . . . The three accused left the house shortly afterwards and went on a spending spree . . . buying chips and several cans and bottles of wine. *Het Suid Western* 10.3.76
. . . an elderly woman was decanting wine from a six bottle can into a row of . . . pint bottles . . . Several full cans of wine stood on the floor. Muller *Whitey* 1977

canaribyter [kə'nɑrĭₗbeɪtə(r)] *n. pl.* -s. See *Jan Fiskaal.*
Fiscal and Canary-byter were the appellations given to a black and white bird (lanius collaris). Trans. Thunberg *Travels I* 1795

cancerbush *n. pl.* -s. Also *kankerbos(sie)*: *Sutherlandia frutescens:* a shrub the leaves of which were believed by early colonists to cure cancer; used also by the Hottentots for wounds and internally for fevers. ⟡Not to be confused with (1) *kankerbos, Euphorbia ingens,* see *naboom.* [*Afk. kanker cogn.* cancer + *bosch cogn.* bush (+ *dimin. suffix* -(s)ie]
Then there is the kankerbos, which has failed to provide a cure for cancer . . . early Afrikaans cookery book . . . 1898 advised the silvery kankerbos leaves for ordinary stomach troubles. Long before that Thunberg the botanist records that the leaves, dried and powdered, were applied to sore eyes. Green *Land of Afternoon* 1949

cane spirit *n. abbr. Cane:* also known *colloq.* as '*Natal Whisky*', a colourless spirit distilled from sugar cane and marketed under numerous trade names: see second quot. at *sluk* and quot. at *Transvaler.*
BREAK AWAY WITH –
Splash . . . with soda or cool it with cola. Enjoy it long with lemonade or tall with tonic and wedges of wild lime. And Break away today. With . . . The number one cane spirit. *Advt. Fair Lady* 25.10.78

canopy *n. pl.* -ies. A cover of fibreglass or metal for the rear section of an open *bakkie* (q.v.) or pick-up, converting it into a non-saloon vehicle which looks nevertheless not unlike a station-wagon. ⟡Not equiv. of old style 'canopy' or hood for an open tourer or 'convertible'. [*prob. fr. earlier usage*]
Motor Sundries Canopies Best value . . . stylish and strong, the only canopy which carries SABS glass-fibre laminate mark of quality, see our new models. *Cape Herald Advt.* 14.9.74
Bakkie Canopies. All models. *E.P. Herald Advt.* 3.10.74

canteen *n. pl.* -s. *now rare.* Formerly any drinking shop or public house, prob. fr. military usage. *cf. Canad. saloon.* [*Fr. cantine* canteen (military); *Italian cantina* cellar; *Span. cantina* small bar-room]

You'll find him up in the canteen if he's still alive . . . They walked round the scattered tents . . . At the top of the rise stood one larger than the rest which a notice-board, crudely inscribed, announced as the Square-face Canteen. Brett Young *City of Gold* 1940

~ **keeper** the proprietor of a ~, a publican.

Johnstone, Mrs dealer, high-street Jolly, John Canteenkeeper, east-barracks. *Cape of G.H. Almanac* 1843 (Albany Directory)

~ **wine** cheap wine supplied in bulk to ~*s*.

Cape *n.* Freq. prefixed to a variety of items, flora, fauna, household equipment etc. sig. of or pertaining to the Cape Colony e.g. ~ *biltong*, ~ *boy*, ~ *cobra*, ~ *Coloured*, ~ *Corps*, ~*doctor*, ~ *Dutch*, ~ *foot*, ~ *gooseberry*, ~ *honeysuckle*, ~ *salmon*, ~ *sheep*, ~ *smoke*, ~ *triangular*, all (q.v.): also sig. a promontory found in place names, (*Afk. form* Kaap) Cape St. Francis, Cape Agulhas, Kaap de Goede Hoop.

Cape biltong [ˈbɪlˌtɒŋ] *n.* See quot. at *bokkem(s)*.

Cape boy *n. pl.* -s. **1.** *prob. obs.* A Coloured (q.v.) person of mixed descent: see also (2) *Bastaard.*

The 'Cape-boys' form a mixed so-called 'bastard' class, descended from a variety of races, including Bushman, Hottentots, Mosambiques (with short wooly hair) and Malabaries (with long smooth hair), the latter having been brought to the country as slaves in the early days. Wallace *Farming Industries* 1896 **2.** *obs.* A St. Helena/Cape half breed: poss. erron.

. . . the island of St. Helena has added a bastard black element whose descendants are known as Cape Boys. Du Val *With a Show I* 1882

Cape Brandy *n.* Brandy distilled at the Cape from the earliest days after the establishment of vines: see *Cape smoke* also *dop.*

. . . the simple fare, which, served up twice a day, forms, with tea-water and the *soopie* or dram of Cape Brandy, the amount of their luxuries. Thompson *Travels I* 1827

Cape cart *n. pl.* -s. A two-wheeled hooded horse-drawn carriage. [*mistrans. of Afk. kapkar, kap* hood + *kar* cart]

A Cape cart is quite a colonial institution. It is a highly decorated dog-cart with seats capable of being reversed, as in a mail phaeton, and covered with a painted canvas hood, sunblinds in front and rear, and supplied with side curtains like a Hampton Court van. These carts are very light and very strong. A Lady *Life at Cape* 1870

Cape cobra *Naja nivea :* see *mfezi.*

Cape Coloured See *Coloured.*

Cape Corps *n.* Any of several Coloured or Hottentot regiments which fought in the Kaffir (Frontier) wars, World Wars I and II or subsequent military action: now a coloured regiment stationed in Cape Town, see quot. below.

A selection board for the Southern African Cape Corps Service Battalion will visit Grahamstown next Friday. Any young coloured men interested in applying should be at the Grahamstown Police Station at 9 a.m. next Friday. *Grocott's Mail* 14.8.73

Cape Doctor *n. prop.* The *South easter* (q.v.) so called at the Cape from early times as it was believed that in suddenly cooling the air it dispersed or blew away disease and germs during the hot summer months. *cf. Jam. Eng. Doctor* (wind from the sea) *Undertaker* (wind from the land). *Austral. Albany Doctor, Fremantle Doctor,* cool winds after heat.

Without the south-easter (or 'Cape Doctor') they must have fevers, etc.; and, though too rough a practitioner for me, he benefits the general health. Duff Gordon *Letters* 1861–62

The 'Cape-Doctor' is the name of Cape Town's private and personal monsoon, the South-Easter. . . . called the "Cape Doctor' because it always blows seaward and is said to carry away all the germs with it. Morton *In Search of S A* 1948

Cape Dutch *n.* **1.** An early term for the Afrikaans language, known as ~ when still regarded as being a dialect of Dutch: see also (2) *Dutch* and *Taal.*

She sat on a chair . . . with her feet on a wooden stove, and wiped her flat face with the corner of her apron, and drank coffee, and in Cape Dutch swore that the beloved weather was damned. Schreiner *African Farm* 1883 **2.** The early Dutch colonists at the Cape, also *attrib.* of or pertaining to the same.

. . . he would find it useful to avail himself, in all ordinary affairs, of the experience of the Cape Dutch colonists in his vicinity – a class of men not deficient in shrewdness, and who, if civilly treated, will be found generally useful and friendly neighbours. Thompson *Travels II* 1827

Cape Dutch *adj.* Of or pertaining to the whitewashed gabled style of architecture popular at the Cape in the C18, extended also to traditional Cape furniture styles.

Soon after 1750 their fortunes began to rise and spacious, gracious homes in what is now known as the Cape Dutch Style began to be built. Gordon-Brown *S. Afr. Heritage* 1965

Cape foot *n. pl.* feet. **1.** (*hist.*) A unit of land measurement now *obs.* still found in pre-*metrication* (q.v.) deeds, equiv. to 1.033 British Imperial feet.

The ratio of the Cape land-measure foot to the British Imperial foot was investigated by the Land-measure Commissioners, appointed by His Excellency the Governor on 19th June, 1858. They ascertained that 1,000 Cape feet are equal to 1,033 British Imperial feet. *Cape Town Directory* 1866.

2. A tapering foot usu. on a turned leg of table or chair typical of Cape Dutch furniture, having a broader ring or 'bracelet' above the tapered part of the foot.

The circular fluted leg terminating in a 'Cape' foot was used in a slender form by Sheraton in about 1800 . . . two of the earlier Cape chairs with arms where the rocker has been drilled to take the bottom of a 'Cape foot. Atmore *Cape Furniture* 1965

Cape gooseberry *n. pl.* -ies. The small gold fruit of *Physalis peruviana* with a pleasant fragrant tartness, used for jam and pies and eaten raw: also the plant, on which the fruits are individually enclosed in lantern-shaped papery casing, see quot. at *klapper*[1]. *U.S. ground cherry*, *Brit. golden berry*.

The gooseberries are not like ours at home. Each in a light, dry pod, and smooth, and yellow as gold. Mama believes they would make but an indifferent gooseberry fool, but they eat deliciously when fresh.
I mean to bring home an apronful of Cape Gooseberries, they grow there wild. Fitzroy *Dark Bright Land* 1955 *Ibid.*

Cape honeysuckle *n.* Also called *Tecoma* and *kaffir honeysuckle*: *Tecomaria capensis*: a dense woody shrub much used for garden hedges, 'tecoma', with clusters of deep orange tubular flowers frequented by *sunbirds* (q.v.) for nectar. [*fr. honeysuckle-shaped flowers*]

Cape lady *n. pl.* -ies. **1.** See *moonfish*.
2. See *frans madame*
3. See *pampelmoes*[2]

Caper tea *n.* A somewhat inferior blend of China tea prepared for marketing in the Cape Colony.

For beverage, the housewife refreshed her family with the rather coarse 'Caper' tea from China mixed specially for the Cape market. Tait *Durban Story* 1961

Cape salmon *n. pl.* ∅. **1.** *Geelbek* (q.v.) a firm-fleshed edible marine fish *Atractoscion aequidens* known freq. in Natal as 'salmon' *cf. Austral.* and *N. Z.* use of 'salmon' for species unrelated to the N. hemisphere salmon, and *Jam. Eng.* 'Jamaican salmon' (*Calipeva*).
2. *Elops machnata* or 'tenpounder' known as *skipjack, wildevis* and *springer*, all (q.v.).

Cape sheep *n. pl.* ∅. The indigenous sheep of S.A. which has hair not wool and a large fat tail: see *Africander* (*sheep*), *fat-tailed sheep, ronderib* and *blinkhaar*.

Everyone has heard of the immense tails of the Cape sheep, but the formation of them is not so well known. They consist of a mass of very nice sweet fat, which is exceedingly useful for domestic purposes, and consequently is much prized by the Dutch. Webster *Voyage I* 1834

Cape smoke *n. obs.* The earliest, rough *Cape Brandy* (q.v.) see quot. at *witblits*. [*prob. fr. Kaap + smaak* taste, savour (*Thomas Pringle*)]

In the skipper's cabin, over a glass of 'Cape Smoke' – the fiercest brandy that ever came from a still – I heard yarns about the risky trade of the sealers. Birkby *Thirstland Treks* 1936

Cape Triangular *n. pl.* -s. Any of several of the triangular stamps bearing the figure of Hope, issued at the Cape, many of which are valuable collector's pieces.

In the country of rich tradition, the Cape Triangular woodblock provisional issue of 1861 is precious Africana. Rare. Genuine. *E.P. Herald Advt.* 10.10.73

Cape wine *n. pl.* -s. *hist.* S. Afr. wines were formerly known as ~: usu. in earliest times sig. *Constantia* (q.v.) later a variety of wines made at the Cape.

ℙIn 1825 Brillat Savarin in '*The Philosopher in the Kitchen*' refers to the provenance of certain food and drink . . . 'some from Africa, . . . such as the Cape Wines,' and to a restaurant where 'the fortunate gastronome can wash his meal down with at least thirty kinds of wine, from Burgundy to Cape Wine and Tokay.'

Very unfavourable accounts from England of the price of, and demand for Cape Wines, – caused partly by a glut of the article, and partly by the disgraceful trash prepared and vended by some pretended Wine Merchants in this Colony. Entry for 6th Nov. 1830 *Greig's Almanac* 1831

Capey, Capie [ˈkeɪpĭ] *n. pl.* -s. *colloq.*
1. A Cape *Coloured* (q.v.) [*Cape* + personif. *suffix* -ie, -ey]
Just a little bit black. And a little bit white. He's a Capie through and through. Fugard *The Blood Knot* 1968
2. A *Kaapenaar* (q.v.) see quot. at *Transvaler*.
In consultation with the Editor . . . and Capeys of all races and provenances whose hearts are in the right place . . .I hereby announce our intention of declaring UDI. Steenkamp *cit. Capetonian* May 1969

captain *n. pl.* -s. **1.** *obs.* Mode of reference to a *Hottentot* (q.v.) chief or any other chief of an indigenous people *cf. Canad. captain* (Indian chief) [*fr. Afk. kafferkaptein* chief]
A chief or captain presides over each clan or kraal, being usually the person of greatest property; but his authority is extremely limited, and only obeyed so far as it meets the general approbation. Thompson *Travels II* 1827
2. Obsolescent official designation, partially revived as a substitute form for 'chief'. ⫿ It would appear that the Afrikaans term *kaptein* from which the use of 'captain' arose has been or is being superseded by *hoofman* for the English term 'chief': See also *Chief Minister.*
HOOFMAN GATSHA BUTHELEZI
HOOFMINISTER VAN KWAZULU *Buurman* Sept. 1978
. . . the chairman of the Zulu Territorial Authority and members of the Executive council. Third from the right is Captain Gatsha Buthelezi. *Panorama* Sept. 1970
. . . chiefly at the request of Captain Paulus Mopedi, to start its own mission at Witzieshoek . . . later inhabited by the Bakwena tribe . . .
. . . the captain was Paulus Mopedi brother of the Basotho paramount chief Moshesh. *Ibid* Jan. 1975

carbonaadjie [ˌkă(r)bɔ̃ˈnaɪkĭ, -cĭ] *n. pl.* -s. See *karbonaadjie.*
Coffee was immediately prepared and an hottentot cook having soon broiled the mutton carbonaadjie (chops or steaks) it was not long before supper was finished. Burchell *Travels I* 1822

carpenter See *doppie*[3].

casevac [ˈkæzəˌvæk] *n. pl.* -s. *Sect. Army.* Casualty evacuation: military rescue operation usu. by helicopter: see quot. at *chopper.* ['portmanteau' word]
Casualty evacuations are carried out at any time. You may find a helicopter crew has flown all day and then will do a night casevac as well. *Paratus* Jan. 1979

castor oil bush, castor oil plant, castor oil tree *n. pl.* -es. Also *kasterolieboom: Ricinus communis* a toxic tree-like shrub of up to 5 metres with prickly fruits containing seeds which yield castor oil.
For the Queen and princesses I was preparing a collection of flower-roots, and seeds of the castor-oil tree, so resembling beads that it was impossible, when strung into necklaces and mixed with gold ones, to suppose them anything else. Barnard *Letters & Journals* 1797–1801

catfish *n. pl.* ∅. Also *sea-cat:* common term for any of several species of octopus, at the Cape esp. *Octopus vulgaris:* also *Jam. Eng.* ⫿Also a name for various species of eel and sea-catfishes, *Photosus* and *Tachysurus* spp. see *bagger.* [*prob. fr. Afk. seekat* octopus]

cattie, catty [ˈkætĭ] *n. pl.* -s. *colloq.* A catapult: see also *mik.* [*abbr. catapult*]
You're threatened, laughed at, hunted, trapped. Children throw stones at you and shoot at you with catties; old ladies keep you in cages. Brink & Hewitt *Birds* 1973
I hear my boet screeching 'Footsack' . . . charging off the back stoep . . . catty at full stretch. *Darling* 24.12.75

certified *vb. part.* Applied to a wine bearing the *WO* seal of the S. Afr. Wine and Spirit Board: see *wine-of-origin* [*Afk.* Gesertifiseerd]

10. Colombar (Certified)	....	(Price)	
11. Riesling (Certified)	....	(„)	
	. . .		
12. Claret	....	(„)	
13. Cinsaut	....	(„)	
14. Pinotage (Certified)	....	(„)	
15. Cabernet Sauvignon (Certified)	....	(„)	

(. . . Wine Co-op Order form and price list 1979)

Ceylon rose See *Selonsroos.*

Chalifa(h) See *Khalifa.*

champals [ˈtʃampəlz] *pl. n.* Indian sandals: see also *beach thongs.*
With Compliments from – Sarie Distributors Consult us for . . . Kashmir Silk Saries – Indian Champals – Imitation Jewellery, *Advt.* (Calendar for 1974)

chana flour See *Indian terms.*

charra [ˈtʃara] *n. pl.* -s *slang* An offensive mode of address or reference to an Indian: *cf. coolie, koelie*[2].
'Churra! You wait!' Moosa could not bring himself to – he fled. McClure *Steam Pig* 1971

cheeky *adj.* Insubordinate: see quots.: also *parmantig, white* and *wit.*

In traditional African society people knew exactly how to behave. Every circumstance and every person required prescribed forms of behaviour and every African was brought up to know this. This gave self-assurance and dignity . . . All that has gone now – the knowledge and the dignity. Europeans call the modern African 'cheeky'. Don't they realize that behind this apparent cheekiness is the agonizing uncertainty . . .? Brandel-Syrier *Black Woman* 1962
The motorist said South Africa had given . . . 'the best part of South Africa'. And added: 'You know this . . . is getting cheeky now', apparently referring to the severing of diplomatic ties with Pretoria this year. *Voice* 11 – 17.10.78

cheers *interj. colloq.* An informal farewell: as in 'Cheers, see you!' etc. See *tjeers. cf. Brit. Cheerio, Austral. and N.Z. Hurroo, hurray,* sig. goodbye.

cheesa boy See *chisa.*

chicken parade *n. pl.* -s *Sect. Army.* Refuse collection in camp: so called from similarity between troops picking up leaves, papers, *stompies* (q.v.) etc., to fowls pecking up grain.
Ag pleez Major won't you take us to the Border . . .
Chorus
P.T., chase parades, inspections and sand bags, Tyre-straf, —ing beat and chicken parades – Ag Major how we'll miss Owambo at the end of this, Gyppo guts and —ing in the lilies all day.
Army version of song 'Ag Pleez Deddy' Serviceman Ex Border 27.4.79

Chicken Run, the *n. prop. colloq.* (Rhodesian). The exodus of Rhodesian Whites. See also yellow route and quots. at *take the gap* and *Owl Run.*
Two years ago I believed in Rhodesia. A year ago I could still hope . . . But now I'm a candidate for the Chicken Run. *Time* 28.3.77

chief *n. pl.* -s. *colloq.* Mode usu. of jocular address, or reference to an African, sometimes objectionable.
I always feel uncomfortable if a White calls me 'chief'. O.I. African University Lecturer Dec. 1973
Better move your car or the chief here will feel uneasy. Where should we park chief? O.I. 16.8.74
. . . an angry Zulu complaining to the white lady at the desk. The lady eventually told . . . that she would put him in the 'chief's room'. *Star* 29.8.78

Chief Minister *n. pl.* -s. The Chief Executive of an African *homeland* (q.v.). See also *captain.* [*trans. Afk. Hoofminister cf. Eersteminister* Prime Minister]
The Chief Minister of Transkei, (who is also

Minister of Finance), is elected by secret ballot by the members of the Assembly from amongst their number at the first session after a general election. Horrell *Afr. Homelands of S.A.* 1973
On December 6, 1963, Paramount Chief Matanzima . . . was elected . . . as Transkei's first Chief Minister. *Panorama* July 1976

Chilapalapa [ˌtʃïˈlapəˌlapə] *n. prop.* Rhodesian equiv. of *Fanagalo* (q.v.)
I flash my lights and it stops. Language problems. Finally in pidgin Portuguese, fractured English and badly molested Chilapalapa I get it through that I want a tow. *E.P. Herald* 4.3.74

chilliebite *n. pl.* -s. Indian savoury or snack, see also *bhajia:* a fritter of *chana flour* (q.v.) or *gram* (q.v.) containing chillies, onion or other vegetable: ~ *mix* a commercially available base for ~ batter.
On the corner of Diagonal Street . . . Chilibites are being offered for sale still warm from the pan. *Sunday Times* 1.5.77

Chimurenga See qůot. [*Shona*]
The insurgency, which the guerillas call *chimurenga* (liberation war in Shona, the principle Bantu language in Rhodesia) is now in its fifth year. It is spreading . . . 'It's worse this year than last' an off duty 'troopie' . . . declared. *Time* 28.3.77

chink *n. pl.* -s. *abbr. chinkerinchee* (q.v.)
'Chinks' grow only in the Western Province . . . New York flower shops sell 'chinks' at the equivalent of eightpence each. They call it 'Africa Star of Bethlehem'. Green *Land of Afternoon* 1949

chinkerinchee [ˈtʃɪŋkərɪŋˌtʃiː] *n. pl.* -s. Also *chincherinchee* and *colloq. 'chink'* (q.v.) various species of white flowered *Ornithogalum* esp. *O. thyrsoides,* poisonous to stock, which instinctively avoid them. ⁋Picked in bud these are exported overseas in the Northern winter where they last several weeks in water, see quots. at *chink* and (1) *kalkoentjie.* [*etym. dub. prob. fr. Du. tjienker (onomat.)* + *uintjie* bulb, *cf. Japanese chirin-chirin* tinkle]
As far back as 1794 Thunberg, the Swedish botanist, described them. 'Tinterinties' he said 'is a name given to a species of ornithogalum with a white flower, from the sound it produces when two stalks are rubbed together.' I have also seen the name spelt 'chickering ching'. Green *Land of Afternoon* 1949
. . . a bunch of green chinkerinchees for herself . . . with their green buds that climbed a long stalk and would open later into long-lasting white flowers. Whitney *Blue Fire* 1973

chisa boy [ˈtʃisa, -sə] *n. pl.* -s. *Sect.*

Mining. The man who lights the dynamite fuse for blasting on the mines; also *chisa stick* part of the fuse: usu. erron. sp. *cheesa.* [*Ngu. chisa* hot]

choc *n. pl.* -s. An offensive mode of reference to a non-white, usu. an African. [*abbr.* chocolate]

chokka [ˈtʃɒkə] *n.* Also *tjokka :* squid, cuttlefish, *Loligo* spp., often used for bait; ⫽The name is said to be onomatopoeic fr. sound it makes when caught: see also quots. at *rooiaas, alikreukel,* and *ollycrock.* [*poss. Port. see second quot.*]
    We trace where they are on the echo-sounder . . . Then we bait up with chokka, get up-wind and allow the boat to drift. *S.W. Herald* 2.7.74
    ⫽My steward tried to steer me away from another fish course called chocas, saying that it was suitable only for Portuguese passengers. . . . however, I had my squid, cooked in its own ink. Green *Sky Like Flame* 1954

choli [ˈtʃɒlĭ] *n. pl.* -s. The low-necked fitting bodice with tight elbow sleeves and freq. a bare midriff, worn with a sari: see *Indian Terms* [*prob. Hindi coli* bodice]

chommie [ˈtʃɔmĭ] *n. pl.* -s. *colloq.* Also *tjommie :* friend, pal etc. *cf. Austral. mate, cobber.* [*prob. fr. Eng.* chum]
    I whip up this file and turn round grinning . . . 'So watch it chommie.' *Darling* 19.1.75

chopper *n. pl.* -s. *Sect. Army.* Also *U.S.:* Helicopter: border term for S.A. Air Force helicopter: *cf. Austral.* chopper, a Tomahawk aircraft. [*acc. Partridge* 'Fleet Air Arm since *ca.* 1955']
    Within a short period of time, the crews of the helicopter had located the casualty and the 'chopper' was speeding towards the wounded man on its mission of mercy. *Paratus* Jan 1979

chorb [tʃɔːb] *n. pl.* -s. *slang* esp. among schoolchildren: pimple, pustule; in *pl.* ~s can sig. acne. [*unknown*]
    I *would* get a massive chorb on my nose just in time for the dance on Saturday, wouldn't I? *Schoolgirl* June 1974
    My boet is a skinny okie of fifteen with sticking-out ears and chorbs . . . He never stops eating. Hobbs *Blossom* 1978

chor-chor [ˈtʃɔrɪtʃɔr] See *grunter.*

chorrie [ˈtʃɔrĭ] *n. pl.* -s. *colloq.* Also *tjorrie, tjor :* motor car equiv. *U.S. jalopy, Austral.* bomb. In combination *knorchor* [*onomat.*], go-kart. [*Afk. tjor presum. onomat.*]
    Get into the old chorrie and drive off to school. Maclennan *Dawn Wind* 1970

chow tools *pl. n. Sect. Army.* A pocket 'knife-fork-spoon' set of cutlery designed to clip together and slip into a narrow case: also known as '*chow-spanners*' *Informant* H. C. Davies *Letter* 2.2.79 [*chow* food (*fr.* Chinese)]

Christenmens(ch) [ˈkrɪstənˌmens] *n. pl.* -e. *obs.* Also *Christian man :* A 'white man' both *lit.* and *figur :* see quot. at *settlaar.* [*Du. Christen cogn.* Christian + *mens(ch)* person *cogn.* man]
    Amongst other peculiarities ascribed to the lion, is his supposed propensity to prey on black men in preference to white, when he has the choice; or, as the Cape boers explain it, his discretion in refraining from the flesh of 'Christen-mensch' when '*Hottentot volk*' are to be come at. Thompson *Travels II* 1827
    He (the Dutchman) departs highly satisfied with you and calls you a 'nice man' and even 'a Christian man'. Webster *Voyage I* 1834

Christmas bee, Christmas beetle *n. pl.* -s. Any of several species of Cicada which are esp. shrill and noisy in the summer at Christmas time in S.A.
    The Christmas bee or cicada is another familiar insect of which only the male has the power of 'song'; and it must be confessed he is exceedingly persistent in the exhibition of his accomplishment, for a noisier insect it would be hard to find. *East London Dispatch* 16.2.1912 *cit.* Pettman
    . . . it was Christmas and . . . At night the Christmas bees made the longest-winded chorus I've ever heard. Stormberg *Mrs P. de Bruyn* 1920

Christmas box *n. pl.* -es colloq. Christmas present, in freq. use in SA esp. referring to yearly handouts exacted from householders by regular hawkers and vagrants as well as delivery personnel. ⫽ ~ is cited by Partridge *A Dictionary of Slang and Unconventional English* as 'low colloquial'. It is also given in the *Concise OED.* In S.A. it is usu. *abbr.* to 'Christmas' in *Afr. E.:* see *omissions* (5).
    . . . it was supplemented by a Christmas box of £5 from my chief at the end of the first year; another of £10 at the end of the second. Somerset Bell *Bygone Days* 1933

Christmas flower *n. pl.* -s. The hydrangea which is always in bloom at Christmas in S.A. and freq. in use in churches for Christmas decorations. *cf. Jam. Eng. Christmas Bush* (poinsettia), *Austral. Christmas bush, tree,* etc.
    Hydrangea . . . is known in many parts of Southern Africa as the 'Christmas flower' for it is in full bloom at this season of the year.

Van der Spuy *Gardening in S.A.* 1953

Church *n. prop. Afr. E.* The Anglican Church is so referred to by many Africans esp. *Zu.* or *Xh.* speaking. [*Ngu. iSheshi,* (Anglican) church *fr. Eng. church*]

Do you and Betty belong to the same Manyano? 'No, she's not a Methodist. She goes to Church.' Zulu Informant 1973

Mr Mfenyana said Xhosas first began to coin words such as 'urhulumente' for Government and 'itshetshi' for the Anglican Church, in the 19th Century. *E.P. Herald* 1.5.78

Church of the Province of South Africa *n. prop.* The Anglican Church (Episcopal) in Southern Africa and the Islands of St. Helena, Ascension and Tristan da Cunha (as distinct from the 'Church of England in South Africa') which has existed as a Province from the arrival of Bishop Robert Gray in 1848: also known as (*abbr.*) *Church of the Province C.P.S.A.* and by Africans as *Church* (q.v.). [*fr. ecclesiastical term* Province]

The Church of the Province of South Africa is that part of the Anglican communion within the Republic of South Africa . . . (etc.) . . . Its members are of African, European, Coloured and Asiatic descent, and its Prayer Book has to be translated into at least nine different languages. Peter Hinchliff *Anglican Church in S.A.* 1963

chutzpah [ˈxʊtspa] *n. slang formerly Sect. Jewish. Also U.S. prob. rare Brit.* Push, verve, 'go', 'cheek'. See quot. below. ⫫ ~ is freq. derogatory, see second quot., esp. in *U.S. slang sig.* 'brazen audacity, shameless impudence, nerve, gall', Barnhart *Dict. of New English* 1973. [*Yiddish fr. Hebrew* (+*Slav personif. suffix -nik*)]

Alison, their mother, is a devastating combination of French chic, British bounce and chutzpah. Jane Mullins *cit. Fair Lady* 14.3.79

We've lived through Charisma, Radical Chic and Happiness is a . . . We're now deep into Radical Cheek. It's no good being pretty or talented . . . You've got to have chutzpah if you want to succeed. Chutzpah . . . like most Yiddish words has a lot more moxie than its Anglo-Saxon equivalent. People with chutzpah are chutzpadiks* and South Africa is crawling with them. *Darling* 4.4.79

*prob. erron. Leo Rosten *The Joys of Yiddish* gives ~*nik.*

circumcision hut See *circumcision school, initiation school, inkhankhatha* and *ngcibi.*

There are still a few people who obey the old rules. Their adolescent sons are taken to cir-cumcision huts outside the townships. They stay there for five months . . . convinced that when a youth finally emerges from the circumcision hut, he has become a hardened man, . . . Simon . . . rejects the tribal ways. Educated as a Christian, he has never been inside a circumcision hut and scoffs at initiation rites. Venter *Soweto* 1977

circumcision school *n. pl. -s.* An *initiation school* (q.v.) in which *abakwetha* (q.v.) are circumcised according to tribal custom and admitted to the rights of manhood. Also *circumcision dance,* one of the ceremonies of the ~. *cf. Austral. bora* aboriginal initiation ceremony.

I passed my standard six in 1966 and my final aggregate was second highest. Now I've just come from my circumcision school . . . will they admit me back into the high school? *Drum* 8.4.72

The case is a sequel to the death of . . . 15, who was hacked to death with an axe . . . at Mpeko Location outside Umtata at a circumcision dance. *Daily Dispatch* 8.5.73

Ciskei [ˈsɪsˌkaɪ] *n. prop.* The *homeland* (q.v.) of the *Xhosa* (q.v.) people within the Republic of South Africa, as distinct from *Transkei* (q.v.), situated in the 'Border' area of the E. Cape Province. [*Lat. cis* this side + *Kei* name of river]

The Ciskei Constitution Proclamation, No. 187/1972, was gazetted on 28 July 1972. The Ciskei was declared a self-governing territory within the Republic. Horrell *Afr. Homelands in S.A.* 1973

citizenship certificate *n. pl. -s.* A certificate of citizenship issued to citizens of a *homeland* (q.v.) entitling them to vote in its elections: see quot. at *homeland.*

. . . falsely accused me of delaying the elections . . . I proposed to the Assembly that those who have not got certificates should use reference books. The Assembly refused so I bowed down to the will of the people that elections be held after a sufficient number of Zulus have their citizenship certificates. *Drum* 8.7.74

clapped *partic. adj.* Worn out, exhausted, e.g. I feel absolutely ~ : also ~ *out;* see quot. at *pellie. cf. Austral. stonkerd, Canad. bushed,* exhausted: [*poss. rel. Afk. 'dat dit klap' sig.* 'all out']

class *n. pl. -es.* Loosely used sig. *standard* (q.v.) or form in school, e.g. 'What class is he in?' sig. what standard or form has the pupil reached? [*prob. fr. Afk. klas*]

classification *n.* The official registration

of an individual as a member of one or other race group defined by the Population Registration Act. See also *re* ~, *classify*. [*fr. classify*[1] (q.v.)]

There are seven race classification categories used on personal identity cards relating to South Africans who are not Asian, Chinese, Bantu or White. These are 'Cape Malay', 'Griqua', 'Baster', 'Cape Coloured', 'Coloured', 'other Coloured', and 'Mixed'. *Rand Daily Mail* 29.1.71

... wrote on his behalf to the Minister of the Interior asking for his reclassification as 'Malay'. *Daily News* 24.4.71

classify[1] *vb. trns. freq. pass.* To assign an individual to a particular racial group in terms of the Population Registration Act: see also *re* ~ : to be ~ *ed*, to be so assigned by the Department of the Interior: see *classification*.

The Population Registration Act was originally conceived and implemented in 1950 soon after the coming to power of Dr Malan's newly elected Afrikaner Nationalist Government. Its purpose was to classify every citizen into one of three main racial categories – White, African and Coloured (with Indian being classified as one of the several Coloured sub-groups). *Star* 16.6.73

In correspondence dated September 27, 1972, he was classified as Black for the purpose of the Population Registration Act. Application has been made for his race reclassification. *E.P. Herald* 12.6.73

classify[2] *vb. trns. freq. pass.* To grade or assign a Government rating in terms of stars, maximum five, and other symbols, to hotels for the guidance of those wishing to stay in them, and to ensure that the standard of service be maintained: see *rotel*.

clevah, clever *n. pl.* -s. See *u-clever* and quot. at *la(a)itie*.

He says they're clevers and they're dangerous. Dike *First S. African* 1979

clingo See final quot. at *fanakolo*.

clinker See *klinker*.

closed hatches *adv. phr. n. Sect. Army. SSB* (q.v.) term *sig.* the condition of an armoured vehicle ready for battle using episcopes for observation usu. with both tower and driver's hatch shut down: *to drive* ~ .

We drove closed hatches in the Operation Zone. *Ex-Serviceman* 15.3.79

C.M.R. beetle [ˈsiˌemˈɑ] *n. pl.* -s. A garden pest, *Myaloris oculata*. [*acronym* **C**ape **M**ounted **R**ifles]

These large beetles are strongly marked with yellow bands across their backs. Their common name refers to the uniform of the Cape Mounted Rifles. . . . they should be dusted with D.D.T. dust or Malathion dust. Eliovson *Gardening for S.A.* 1960

coalman *n. pl.* -men. A black stoker on the S. Afr. Railways who does a fireman's duties for shunting in railway marshalling yards but not on journeys between centres.

The difference between a railways stoker and coalman has been cleared up. One is White and the other is Black. But the White stoker is also a learner driver while the Black coalman has got nothing actually to do with running the engine. The riddle of the footplate was cleared up . . . yesterday, following questions . . . on whether or not Blacks were being trained as firemen. . . . raised howls of laughter from Opposition benches as he said: 'No, but Blacks are being trained as coalmen.' *E.P. Herald* 13.3.75

coals *pl. n. Also U.S.:* The glowing or white-ashed embers of a wood or charcoal fire used for *braai*ing (q.v.) also quot. at *roosterkoek*. ⫽ In S.A. ~ does not usu. *sig.* the embers of a coal fire, which cannot safely be used to broil food. The *Brit. pl.* ~ as in 'a ton of ~ ' is not usu. heard in S.A. [both *Brit.* and *U.S.* but here *prob. fr. Afk. kole* coals, embers]

The men made the fires that would die down to give the 'coals' on which to grill chops . . . home-made sausages and skewered sosaaties lying ready in . . . delectable curry sauce. McMagh *Dinner of Herbs* 1968

Come For a Braai! . . . baste the chicken on the coals regularly . . . they should not dry out over the coals . . . Bake them directly on the hot coals . . . If you have an old pot and enough extra coals . . . Put the fish on the braai over medium coals . . . etc. *Darling* 1.2.78

cocopan [ˈkəʊˌkəʊˌpæn] *n. pl.* -s. A small tip truck on rails used on the mines for the transportation of *blue ground* (q.v.), gold-bearing ore etc: also known as *golovan* among Africans on the mines. [*etym. dub. poss. fr. nqukumbana/e*, Scotch cart (q.v.) *poss. fr. Afr. form koekepan*]

Various duties were assigned to the gangs; some loaded the pulverized blue ground into trucks for transport to the washing machines. Other tipped 'cocopans' (little half-ton trucks) containing the 'maiden blue' just out of the mines. Klein *Stagecoach Dust* 1937

Colenso Church *n. pl.* -es. One of the churches formerly of the 'Church of England in Natal' which resulted from the adherence of Bishop J. W. Colenso to the Church of England as by Law Established and his refusal to accept the authority of the *Church of the Province of South Africa* (q.v.). ❡The last of the ~es came back to the Province in 1958.

Within a short time of Baines' consecration all but one of the Colenso churches had come into the Province . . . The last remaining 'Colenso church' became a part of the province under the special terms a few years ago and the Colenso schism is at an end. Various similar troubles have arisen in other parts of the Province but these have no organic connection with the Colenso Church in Natal. Hinchliff *Anglican Church in S.A.* 1963

Coloured *n. pl.* -s and *adj./modifier.*
1. A S. African of mixed descent speaking either *Afrikaans* (q.v.) or English as his mother tongue. Also combination *Cape* ~ those ~*s* resident in the Cape Peninsula or surrounding area.

It may well serve as a model chapel. Its cost, exclusive of the purchase of the land was £1,300. The old chapel is now occupied by the coloured people for whose benefit regular services are held in the Dutch language. Shaw 1860 cit. Sadler 1967

Reports have reached us of gross exploitation of Cape Coloureds by White-owned business concerns. *Drum* Feb. 1970

2. A racial group in terms of the Population Registration Act; see *classification* and second quot. at (1) *Basta(a)rd*

. . . the Coloured group, as defined in our Population Registration Act, is a pretty variegated community. In fact it includes everybody who is neither White nor African. *Dispatch* 10.12.76

3. *modifier.* Of or pertaining to those of mixed descent: see (1) *Coloured.*

The Coloured people had been made to feel that they were God's stepchildren in the land of their birth. *Daily Dispatch* 19.5.71

4. *modifier.* Of or pertaining to possessions etc. of the ~ people e.g. ~ area, that zoned for occupation by ~ people; ~ school, a school exclusively for ~ children; ~ housing etc.

The five Coloured houses in the centre of the fields were almost submerged at the time of the flood. *Evening Post* 27.5.72

5. *obs.* Of or pertaining to any 'persons of colour' whether of mixed descent or not: *cf. U.S. colored.*

Before I conclude, I must say with regard to the Coloured tribes . . . that I look upon them with concerned pity. *Grahamstown Journ.* 30.10.1833

come down, to *vb. phr.* To be in spate or in flood: used of a river. *cf. Austral. run a banker, running bankers.* [*Afk. die rivier kom af* (*lit.* 'comes down'), the river is in spate]

. . . the river he comes down, my nonna . . . we must fly . . . His young mistress . . . looked around her wildly. How could the river possibly be coming down? There was not a cloud in the sky . . . she could hear the booming of the oncoming waters . . . the wall of grey water, pointed like an arrow, moving, it almost seemed without haste. McMagh *Dinner of Herbs* 1968

As they neared the river Kaspar heard a roaring sound. 'The river is coming down!' he shouted. 'Turn right.' . . . There must have been rain in the mountains. There was one chance – to get past the flood. . . . It was coming in a slowly moving wall . . . a wall that nothing would stop . . . 'We are safe . . . I have heard tell of the river coming down like that, but have never seen it.' Cloete *Watch for Dawn* 1939

come there See *came there* and quot. below.

. . . he was at the Willemsdorp Club, unhappy, wondering what on earth he had come there for. Bosman *Willemsdorp* 1977

come to hand *vb. phr.* Become available, be received, arrive (of inanimate subjects) [*translit. Afk. fr. Du. ter hand kom* come to hand]

Larger quantities of potatoes came to hand and values declined somewhat . . . *Farmer's Weekly* (no date) cit. *Webster's Third International Dict.* 1972 ed.

come out *vb. phr. intr. substandard.* Finish, come to an end used usu. of school, cinema shows etc. not of the people or crowd attending them, e.g. school/the bioscope/the show/the play comes out at . . . o'clock. [*metaphorical poss. trans. of similar use in Afk.*]

come right *vb. phr.* Equiv. of 'turn out well', 'be all right', 'resolve itself': see also *alles sal regkom* [*trans. Afk. regkom reg* right + *kom* *cogn.* come]

comma *n.* **1.** Decimal point, taught in schools in written and spoken use 'Thirteen ~ four' instead of 'point' as in pre-*metrication* (q.v.) usage.

2. The official replacement of the decimal point in the press, also commercial documents etc. being allegedly less easy to tamper with or deface than the full stop, full point or period.

. . . it is not 'ten comma seven' even though we have to use a comma instead of a point . . . In the old days we did not say 'ten full stop seven,' so now why must we say 'ten comma seven'? It always was a POINT, still is, and always will be, even though we write a point with a tail to it. *Sunday Times* 9.12.73

commandant [ˌkɔmənˈdant, ˈkɔm-, ˈkɔ̆m-] *n. pl.* -s One in command. [*fr. Fr. pres. partic. of commander* to command *Du. kommandant*] **1.** Substitute rank for that of Lieutenant Colonel, see quot. at (2) ~, in the S. Afr. Army, formerly used of any commanding officer.

Commandant in South African military history designates the office and rank of the commanding officer of the former Citizen Force in a given district. It also signified the officer in command of the former Transvaal State Artillery and of the Transvaal Volunteer and the Natal Militia organisations. *Std. Encyc. of S.A. III* 1971

2. Officer commanding a *commando* (q.v.) under the commando system of Military training.

In the South African Defence Force the rank of commandant was introduced in 1950 in substitution for that of Lieutenant-Colonel, and since 1969 the commandants under the present commando system have been organized in groups under chief commandants in order to facilitate central control. *Ibid.*

3. *hist.* (S.A.) Any officer-in-charge including a police officer. Also in combinations *field* ~, *veld* ~.

The landrost Mr Rivers (whom Lord Charles placed here in the room of our dear Major Jones, and he is not liked) is salaried and that for doing nothing in this matter of defence, there being a Commandant already, whereas the men must travel fifteen miles and more to drills and field days for which they are not paid, and can ill spare the time from their farms. Kate Pigot *Diary* 1822 *cit.* Fitzroy 1955

4. *hist.* In combination ~ *general.*

Official designation of the permanent head of the Department of Defence and of the South African Defence Force, carrying the rank of general. The term is believed to have been first used in the Cape Colony in 1802 when P. R. Botha acted as 'Commandant-General' over a number of commandants in connection with military operations, i.e. as commander-in-chief. *Std. Encyc. of S.A. III* 1971.

5. See quot. at (2) *Voortrekker.*

commandeer [ˌkɔmənˈdɪə] *vb. trns.* To press men into military service, or more usu. to seize goods, horses or vehicles for military use: now, loosely, to take arbitrary possession of: in *Brit.* and *U.S.* also. [*fr. Fr. commander* command, *here prob. cogn, mand*atory]

When supplies gave out he was empowered to forage round the farms in the vicinity and commandeer whatever he wanted: oxen, sheep,

cows, poultry . . . As payment for the commandeered stock he gave the farmer a voucher, equivalent to the value taken, which was redeemable in cash at the Landdrost's office. Klein *Stagecoach Dust* 1937

commando [kəˈmandəʊ] *n. pl.* -s **1.** *hist.* Burgher units of militia esp. in the Kaffir (Frontier) wars; see *Burgher duty :* See also quots at *veld-kommandant* and *veld-cornet.*

(16.2.1852) Mr Bradshaw came to our Laarger, to see how many people he could get to go on commando. Shone *Diary*

hence *on* ~ *:* taking part in a ~ expedition of reprisal or aggression often for the recapturing of cattle.

While at this place. I heard that a Commando (or expedition of armed boors) had been recently out against the Bushmen of the mountains, where they had shot thirty of these poor creatures. Thompson *Travels I* 1827

In current use also:

Replying to a motion urging the Defence Force to grant temporary exemption . . . to farmers engaged in sowing or reaping, he said that farmers who were mostly members of commandos could obtain exemption under certain circumstances. *E.P. Herald* 10.6.76

~ *tax* levied at the Cape fr. 1812 for the maintenance of a defence corps for the frontier.

2. A unit of the Boer Army in the S. Afr. Wars of 1888, 1889–1902.

3. A member of a unit, or the unit itself undergoing ~ training: see (2) *commandant*

South African servicemen sent for commando training learn their battle technique at the Danie Theron Combat school near Kimberley. *Sunday Times* 3.12.72

4. Found in place names usu. spelt Kommando, e.g. Kommando, Kommandokraal, also Commando Drift.

5. See quot. at (2) *Voortrekker.*

commandovoël [kŭˈmandŭˌfʊəl] *n. pl.* -s The stone plover *Burhinus capensis :* see (1) *dikkop.*

Committees [ˌkɔmɪˈtɪəs, -tiːz] *n. prob. pl.* Found in place names Committees Drift, Committees Flats: see quot. also *kommetje* [*uncertain*]

. . . the derivation of the word Committees, which is a corruption of the Dutch word kommetje, meaning a saucerlike hollow. *E.P. Herald* 9.4.74

community of property, in *adv. phr.* See *ante-nuptial contract, boedelhouder, boedelscheiding, kinderbewys,* and quots. at *divorce* and *A.N.C.*[3].

Compagnie, Jan *n. prop.* See *Company.*

Company, the *n. prop. hist.* The *Dutch East India Company* (q.v.) also *V.O.C.* (q.v.)
. . . the local regulations of the Company which stipulated that no one in the Colony might sell the produce of their labour on their own terms – that neither could they buy nor dispose of anything except at the Company's store and at the Company's price! Meurant *Sixty Yrs Ago* 1885
The Company felt that the establishment of an independent farming community close to the settlement would save the V.O.C. a considerable sum of money as the Free Burghers could supply the Company with fresh vegetables at a fixed price, equivalent to what would have been obtained with hired labour. *Panorama* May 1971

John or Jan ~ a humorous appellation *personif.* the Dutch East India Company. ℙ Also used of the British East India Company and in *Canad.* of the Hudson's Bay Company.
It appears that he was sent by Jan Compagnie to Castle Delmina . . . to preach and teach the natives there . . . remembering too, the value of impressing upon simple minds the power and prestige of the Dutch East India Company, he solved the problem of translating 'I am the Lord thy God . . .' by substituting 'I am John Compagnie . . .' De Kock *Those in Bondage* 1950

complain, can't *vb. phr. substandard* Response to qns. such as How goes it? How are you? sig. 'All right'. [*trans. Afk. kannie kla nie,* cannot complain]

compound [ˈkɒmˌpaʊnd] *n. pl.* -s. An enclosure in which orig. diamond mine-workers were housed and confined on the ~ system of serving articles for a fixed period: now loosely mine workers' or other African labourers' living quarters. [*by folk etym. prob. influenced by 'compound' fr. Malay kompong,* a village, cluster of buildings]
The notorious Katutura compound where thousand of Ovambo workers lived in over-crowded discomfort will be demolished within a few years. *Daily Dispatch* 12.3.73

concentration camp *n. pl.* -s. In S.A.E. those camps in which Boer women and children were detained in the S. Afr. Wars of 1888 and 1889–1902. See quot. at *Protected Village.*
Mr Botha gave it as his opinion that the death rate in the concentration camps in the first place was due to 'an entire want of proper accommodation' and 'want of proper food.' Subsequently the position improved and the authorities did everything they could. *E.P.*

Herald 25.10.1911
My wife came out of the concentration camp and we went together to look at our old farm. My wife had gone into the concentration camp with our two children, but she came out alone. Bosman *Mafeking Road* 1947
Will the Blacks remember with the same hatred the breaking up of their families . . . that the Afrikaner felt for the Englishman for fifty years after the Englishman confined Afrikaner women in concentration camps? *Daily Dispatch* 7.10.71

concertina gate *n. pl.* -s. A type of farm gate made like a section of fence, of usu. barbed wire strands with a pole at each end: *Austral. Mallee gate* 'little more than droppers and wires' See also *bekslaner hek.* [*fr. concertina,* accordion]
The name given to certain farm gates, made of barbed wire strands and parallel wooden poles, the frame of which resembles a concertina. Like this instrument they can also expand and contract. Also known as stomach-hitter. Swart *Africanderisms: Supp.* 1934
The noise of the car carried far in the still mountain air. They were approaching a 'concertina' wire gate when a silvery-haired man hastened jerkily along a path, waving a stick. Lighton *Out of the Strong* 1958

conditional exemption *n. pl.* -s. *Matriculation* (q.v.) exemption granted to a student for University entrance conditional upon his fulfilling certain requirements.

connection *Sect. Army.* Buddy, mate, as in 'Sight you, my connection' as a form of greeting.
I've got a connection in the airways – he'll fix my booking. *O.I. Serviceman* 1.4.79

Constantia [ˌkənˈstænʃə, -ʃ(ɪ)ə] *n. prop.* Historically renowned dessert wine made at ~ in the Cape and known in Europe from the C18: see also *Cape Wine* and quot. at (2) *hanepoot.* ℙIt was a costly luxury and seldom in plentiful supply, a fact which gave rise to the remark of the C18 wit, when a second bottle was not produced, 'Since we cannot double the Cape let us return to Madeira.' Also in combinations· *steen* ~, *Frontignac* ~ fr. names of grape varieties, also *red* and *white* ~ and ~ *wine.* ℙ The first quot. below pre-dates ~ : see Appendix of earliest dates. [*place name Constantia, Cape*]
They gave me some stuff under the name of Constantia, which to my palate was more like treacle-and-water than a rich and generous

wine. Hickey *Memoirs*: entry for 19.7.1777
A sweet luscious wine, well known in England by the name of Constantia, the produce of two farms lying close under the mountains. *Afr. Court Calendar* 1807
My dear . . . I have just recollected that I have some of the finest old Constantia wine in the house, that ever was tasted . . . My poor husband! how fond he was of it! Whenever he had a touch of his old cholicky gout, he said it did him more good than anything else in the world. Jane Austen *Sense and Sensibility* 1811

contact [ˈkɒntækt] *n. pl.* -s. *Sect. Army.* An encounter with opposing forces: see quot. at *floppy*. *O.I.s various Servicemen* 1979. ⟊ This sense of ~ does not appear in the major English or American dictionaries.
A reasonably safe area. It was only lightly infested by . . . forces until the auxiliaries . . . drove them out in January. It took only a couple of 'contacts' to get rid of them. *Sunday Times* 15.4.79

cooldrink [ˈkulˌdrɪŋk] *n. pl.* -s. Any soft drink, mineral water, squash or even fruit juice: *cf. Jam. Eng.* cool-drink (fermented or unfermented). [*translit. Afk.* koeldrank lit. cool drink]
Get me a cake. The best. Cooldrinks, peanuts and raisins. Fugard *People Living There* 1969
In an . . . ancient Buddhist rite, food, drink and flowers are left for the questing, spirit of the dead . . . where the seaman died . . . a small bowl of rice, a cup of soup, choice fruit, cooldrink, beer, cigarettes, matches, cakes and . . . flowers. *E.P. Herald* 6.4.74

coolie *n. pl.* -s. **1.** *hist. obs.* in S.A. A porter or bearer, listed in Cape directories, also in combinations *wharf* ~, *fish* ~, *market* ~: regulations for ~ *hire* were officially promulgated: see quot. [*prob. fr. Tamil, Urdu, Telegu etc.* kuli labourer, *also Chinese*]
Provisional Regulations respecting the COOLIES directed by the President and members of the Burgher Senate are authorised by his Excellency the Governor. Each coolie shall work during the summer from 6 o'clock in the morning till 8, from 9 till 12 and in the afternoon from 1 till 5 o'clock receiving per day . . . Done at a Meeting of the Burgher Senate at the Cape of Good Hope. 9th November 1815 *S.Afr. Almanac* 1827
2. Objectionable in S.A.E. now an offensive mode of address or reference to an Indian (Afk. *koelie* (q.v.)): see quots. at *Arab* (*merchant*), *boesman* and *karkoer*. *cf. Jam. Eng.* coolie (offensive) an E. Indian.
Also offensive in combinations ~ *pink*

shocking pink see *totty pink*; ~ *shop*; ~ *pan* plumbers' *slang*, an oriental floor level water-closet. See also quot. at *Dutchman*.
To say that a coolie is a 'South African Indian' is an insult to the Indian and the greatest travesty of words. *Rand Daily Mail* 26.2.71
Indian Council . . . man to man or sahib to coolie? *Sunday Tribune* 3.10.71
'Coolie' is a perfectly respectable word meaning people who earn their living honestly by toil and sweat and I do not share the common hypersensitivity to its use. Pat Poovalingam *cit. Optima* Vol 28 No 2 12.4.79

coon *n. pl.* -s also *n. modif.* **1.** Offensive mode of reference to an African esp. in combination ~ *boy*, ~ *girl*, cf. *U.S. coon*, negro. [*fr. raccoon U.S.*]
Salisbury – . . . a former MP, referred in public to Africans as 'coons' yesterday. Eton-educated . . . Questioned by a man in the audience about using the words 'coons' and 'kaffirs,' . . . said he had been quoting other people – an explanation the audience did not accept. *Daily Dispatch* 9.10.71
2. *Coloured* (q.v.) and *Malay* (q.v.) participators in the colourful ~ *carnival* at New Year in Cape Town: so named from the black and white raccoon-style make up similar to that of negro Christie minstrels.
Malay choirs . . . must not be confused with the coons who parade the streets on New Year's Day and hold their carnival at the Cycle Track and Rosebank. Some Malay singers join the Coons in these celebrations, but as individuals in Coon dress. Du Plessis *Cape Malays* 1944

Co-operation and Development, Department of See *B.A.D., P.R.D.* and *plural.*

copper *n. erron.* Used in S.A. for brass. [*fr. Afk.* (geel)koper brass lit. 'yellow copper']
So the small shrivelled body was laid out in the large, heavy coffin with its glistening copper fittings. Krige *Dream & Desert* 1953

cossie [ˈkɒzɪ] *n. pl.* -s. *substandard* Also *cozzie*: bathing suit: also *Austral.* [*abbr. costume*]
'. . . you won't be needing yore cozzie . . .'
'. . . aren't we going to goef?' 'Shore we gonna goef. I jis said no cozzie' . . . 'I get my goef alright. I also get sunburn all over.' *Darling* 12.2.75

costume *n. pl.* -s. See *cossie*

cotch *vb. intrns. slang* (vulgar) Puke, vomit 'throw up'. *cf. Austral. perk.* [*translit. Afk. vb.* kots puke]
. . . every three or so days Barbara sticks

her finger down her throat after meals and cotches so she doesn't have to digest the food and get fat. Letter Schoolgirl, 16, 9.7.74

[Cotyledon] *n. pl.* -s. A genus of plants of which several species are used in S.A. for the treatment of corns, plantar warts [verrucas] and abcesses. See *plakkie* and quot. below.

People were getting disability grants for their planters' [sic] warts farmers complained . . . 'And we through our taxes . . . are paying them.' . . . He said planters' warts were cured in a matter of weeks by the application of a cotyledon leaf. The matter is being taken up with the district surgeon. *E.P. Herald* 16.2.79

court *n. pl.* -s. *Afr.E.* African equiv. Magisterial District.

The address is Stanger, but her Court is Mapumulo. African Informant July 1974

Court Calendar, African *n. hist.* See *Almanac.*

The African Court Calendar is one of the famous Cape Almanacs, so important for research purposes, which commenced in 1801 with a manuscript copy and continued in varying forms to the present day with a break of only one year (1864). Gordon-Brown *S.Afr. Heritage* 1965

Cousin Jack *n. pl.* -s. *hist.* In S.A. one of the Cornish miners who immigrated to S.A. first to work the Namaqualand copper mines and afterwards the Kimberley Diamond fields, the E. Transvaal goldfields and finally the gold mines of the Witwatersrand and where the name ~ was much in use including '~*s' Corner*' in Johannesburg [Pritchard and Von Brandis streets]. ⫽ *acc. Websters Third International Dict.* '~ Cornishman, *esp:* a Cornish miner.' therefore non S.A.E.

Springbokfontein . . . where the old chimney of the smelting furnace bears witness to Cornish masonry skill. Built about 1890, local legend has it that the Cousin Jack who built it stood . . . on one leg and uncorked a bottle of whisky on top of the completed stack. Dickason *Cornish Immigrants to S.A.* 1978

cozzie See *cossie.*

crackers *pl. n. obs.* Leather trousers made in the early days, of partially tanned hide and worn by Settlers and military alike, so called from their habit of making a noise at every movement of the wearer: see allusion at *becreep* [1838]; also called *vel(d)broeks. cf. Canad. mitashes,* leather leggings.

I equipped myself in the proper style for a wild campaign: in broad-brimmed Spanish hat, blue jacket, brown buck-skin trousers, elegantly

termed 'crackers' in South Africa. Alexander *Western Africa I* 1837

crackling See *perlé* and quot. below. [*trans. Fr. pétillant* crackling] (Trade Name)
CRACKLING
PERLÉWYN — PERLÉ WINE
. . .
Winery
(Bottle Label) 6.4.79

crane flower *n. pl.* -s. *Strelitzia reginae,* also known as *piesangblom* (q.v.) and *geel piesang,* so called from its appearance

The rather striking resemblance of the flower to a crane's head, particularly of an enraged crane. Smith *S. Afr. Plants* 1966

Thirty dozen Strelitzias – the exotic crane flower which occurs in KwaZulu . . . decorated the marble hall and exhibition venue. *Panorama* Jan. 1975

crayfish *n. pl.* ∅ See *kreef.* Erron. form for crawfish (*Fr. langouste):* spiny rock lobster *Jasus lalandii* see also *kreef* ⫽As ~ is an important export product ~ *fishing* is strictly controlled. Removal of more than the licensed number, or of *undersized* ~, and ~ *smuggling* with or without a licence, are offences punishable by law, *cf. Canad. lobster cop.* (also *Jam. Eng. claw-fish by folk etym.*) [*presum. fr. Fr. ecrevisse* fresh water crustacean 'crayfish']

Crayfish are small river crustaceans, and the monsters from the sea that go with our mayonnaise are crawfish. This crawfish may not be enjoyed legally until it has grown to three and half inches in length. Green *Tavern of Seas* 1947

C.R.C. [ˈsiˌɑˈsiː] *n. prop.* The Coloured Representative Council: a body consisting of a majority of elected members and a smaller number of appointed members of the *Coloured* (q.v.) community, concerned with the political affairs of the Coloured people: see quot. at *tot system.* [*acronym*]

The Coloured Representative Council is expected to receive a full report on Wednesday about tomorrow's CRC talks with the Prime Minister, Mr B. J. Vorster, on the present unrest and detentions. *Argus* 20.9.76

cream of tartar (tree) *n. pl.* -s. *Adansonia digitata,* the baobab tree, see *kremetartboom:* also known as *lemonade tree.* Also *Austral. A. gregorii* (baobab).

crunchie *n. pl.* -s. *slang.* Offensive mode

of reference to an *Afrikaner* (q.v.). [*prob. fr. mealie cruncher* Afrikaner, *slang poss. fr. kransie* (q.v.)]
... referred to the use of these words 'crunch', 'hairyback' and 'rock spider' in the latest issue of *Wits Student*. He told the prosecuting officer . . . that these words were derogatory for Afrikaners. *Cape Times* 13.4.73

C.S.I.R. [ˈsiˌesˌaɪˈɑ] *n.* The Government's Council for *Sc*ientific and *I*ndustrial *R*esearch in Pretoria: see quot. at *monkey orange.* [*acronym*]
C.S.I.R. Chief on Fish Farms for S.A. . . . There is no reason why South Africa should not 'farm' fish off the coast for the market as Japan does . . . president of the Council for Scientific and Industrial Research, said here. *Argus* 3.10.70

cuca shop [ˈkukə] *n. phr. pl.* -s. *Sect. Army.* A trading store on the Angola border selling Cuca beer (Angolan product).
Ag pleez major won't you take us to a Cuca Shop, That's the right place if you want a dop
. . . *Servicemen's version of 'Ag Pleez Deddy':* Song 1977
Anything from a grass hut you crawl into on your belly to a four storey department store is a cuca shop on the border provided it has a Coca Cola sign up or a light on it. *Serviceman Ex Border* 2.4.79
. . . a guerilla gang had opened fire . . . from near a Cuca shop – an Owambo trading store . . . information was received that insurgents were hiding at a Cuca shop complex. *Dispatch* 21.9.79

culpable homicide *n. Also U.S.:* term used in S. Afr. law for *Brit.* 'manslaughter'. Informant Hon. Mr Justice D. D. V. Kannemeyer. [*Afk. and Du. manslag* homicide, manslaughter through negligence. *Afk. strafbare manslag* culpable homicide. (*strafbare* punishable)]
Culpable homicide differs from murder in that the unlawful killing is not accompanied by intention to kill, some lesser but nevertheless blameworthy state of mind being sufficient. Hahlo & Ellison Kahn *Union of S.A.* 1960
Culpable homicide consists in the unlawful killing of another person either negligently or intentionally but in circumstances of partial excuse. Hunt *S.Afr. Criminal Law & Procedure* 1970
The trial of a . . . doctor on a charge of culpable homicide yesterday focussed on instructions he gave to a nurse on injecting a patient . . . with an allegedly lethal dose of the drug . . . *E.P. Herald* 9.3.79

cumec [ˈkjuˌmek] *n. pl.* -s. A measure of water used in dam building and regulation: one cubic metre per second. [*pre-sum. acronym c*ubic *m*etre per second *analg. cusec* (cubic foot) per second]
. . . 1 800 cumecs had been gushing into the river for three days in an effort to lower the level of the huge Verwoerd Dam before the rainy season in March . . . To give an idea of the amount of water pouring from the dam round the clock he said two cubic metres of water (a cumec is one cubic metre a second) would fill a road petrol tanker. *E.P. Herald* 28.2.74

cushion *n. pl.* -s. *substandard.* Pillow (for a bed) See also *kussingslaan* [*prob. translit. Afk. kussing* pillow *cogn.* cushion]

cut *vb. trns.* To castrate an animal: see *kapater, tollie,* and quot. at *hamel* [*trans. Afk. sny* cut, *colloq.* castrate]
Cut 50 lambs again today and about 150 buck kids. Collett *M.S. Diary* 13.10.1838

D

dabbie(boom) [ˈdabī] *n. pl.* -s, (bome). The tamarisk, *Tamarix austro-africana.* ☞ There are numerous sp. forms of this, the vernacular Hottentot name. [*Hott.*]
The dubbee boom, or tamarisk tree, apparently the type of this part of Africa, and which I had constantly seen from the Kousie to the Kuisip, was now covered with white bloom. Alexander *Expedition Vol. II* 1838

dacha [ˈdaxa] *n.* Old sp. form of *dagga* (q.v.)

daeraad [ˈdaəˌrat, -rad] See *dageraad.*

dag [dax] *n.* **1.** Good day: a greeting. [*abbr. Afk. goeie dag fr. Du. goeden dag cogn.* good day]
They entered the hut, when my men having pointed out to them their master lying asleep, I was awakened by the sound of *Dag,* which they repeated till I uncovered my head and returned their salutation. Durchell *Travels II* 1824
. . . the grandfather . . . jogging silently on, Dutch-like, after a gruff 'Daag, mynheer'. Alexander *Western Africa II* 1837
2. Day: in place name Dagbreek. [*Afk. dag cogn.* day]

dageraad [ˈda(x)əˌrad, -rat] *n. pl.* ∅. Also *daeraad, daggerhead* (q.v.): of the marine fish *Chrysoblephus cristiceps:* of the Sparidae characterised by beautiful red and pink colouring hence by folk *etym. da(g)eraad* dawn. *cf.* Canad. *dore* (*fr. poisson dore* (gilded)) [*Afk. fr. Du. dorade, fr.* Port. *and* Span. *dorado* golden]
I used to think that the gorgeous dageraad

was named because it reminded someone of a lovely dawn at sea. Recently I heard a more reasonable but less romantic explanation. Dageraad is a corruption of the Portuguese word dorado, the fish that shines like gold. Cape fishermen do not use the word dageraad for dawn. They say dagbreek or daglumier. Green *S.Afr. Beachcomber* 1958

dagga [daxa, -ə] *n.* [*prob. fr. Hott.* [daxa-b] *Leonotis*] **1.** Usu. used of *Cannabis sativa* Indian hemp equiv. of hashish (bhang) introduced by the Dutch from the East, also called *mak* (*tame*) ~.

Besides smoking the leaves various native tribes employ the leaves young tops etc. as a remedy for various endemic fevers, anthrax dysentery etc. The cultivation of dagga is prohibited by law in South Africa yet it is commonly found as a weed round native kraals in Natal and is known as insangu.   Smith *S.Afr. Plants* 1966

2. Also any of several species of *Leonotis:* chewed and used medicinally by the Hottentots. *cf. Austral.* aboriginal *pituri*, a narcotic plant.

The original name for L. Leonurus was rendered as 'dacha' by the early colonists and subsequently as 'dagga', the present form. The name is preserved in the place names Commadagga Mountain and the town Commadagga no doubt from the abundance of L. leonurus or L. leonotis. *Ibid.*

3. The leaves and seeds of *Cannabis sativa* sold and smoked illegally in S.A., equiv. *U.S.* or *Brit. pot, grass* etc. *Jam. Eng. ganja, Anglo-Ind. bhang* also in S.A. *boom* (q.v.). In combinations ~ *offences,* ~ *running,* ~ *smoking,* ~ *zols* (q.v.), ~ *pips* etc.

The cannabis stativa [*sic*] or common hemp, called by the natives in the colony dakka, has a remarkable inebriating effect, insomuch that it was forbidden to be used among the native troops. Medical men might turn their attention towards attaining the particular virtue of hemp, as it might prove a useful auxiliary in the *materia medica*. I tried a tincture of it in a few cases; but the very limited practice of a small ship affords little or no opportunity for ascertaining the virtue of any new specific. Webster *Voyage I* 1834

Dagga is dangerous, though users try to convince others that measures taken against dagga (pot, marijuana, hemp, grass, hashish) are hysterical, that the habit is purely social. *Fair Lady* 24.11.71

4. Found in place names Daggaboersnek, Daggafontein.

5. sp. form of *dagha*, (q.v.)

6. Non S.A.E.: A marijuana cigarette: Term in use in narcotics argot on the Mexican border.

daggerhead *n. pl.* Ø, -s. Also *daggerheart;* prob. corruption of *dageraad* (q.v.) (usu. in Natal)

... about the sex life of red fish ... I have heard ... that the small red fish like daggerhead and roman, are all of one sex and that their sex changes with growth. *E.P. Herald* 14.10.76

dagha ['dɑɡa] *n. colloq.* (*Sect. Building*) *freq. sp dagga:* cement, mortar: also any of several mixtures for building and/or smearing floors, mud and clay, cowdung with mud, blood, earth etc. ⫽Pettman gives *vb* '*to dagher*', to smear with ~. *cf. Canad. to mud, mud up.* [*Ngu. uDaka* mud, clay]

I hired a hut constructed of *dagga* – that is, ant-heap mixed with cowdung. Clark *Old Drifter* 1936

A house made out of flattened and rusted paraffin tins, beaten flat and laid like this over pole-and-dagga walls. *E.P. Herald* 25.5.74

dakkamer ['dak₁(k)amə(r)] *n. pl.* -s. An extra room built usu. on the flat roof of a *Cape Dutch* (q.v.) town house, incorporated into the facade by means of a pseudo-gable: also occ. an attic; see *solder.* [*Afk. fr. Du. dak* roof + *kamer cogn.* chamber, *Lat. camera*]

We walked up the steep stairs that lead to the 'dakkamer', a small room built with the window facing the harbour. Here the inhabitants of the house watched for the arrival of the merchantmen. *Fair Lady* 8.8.73

-dakkies ['dakiz] *pl. n. suffix colloq. lit.* 'Little roofs': used with colour *adj.* e.g. *rooi-*(red) or *groen-*(green) sig. lunatic asylum. [*Afk. dak* roof + *dimin. suffix-*(*k*)*ie* + *pl. -s*]

My husband says I'll drive him into the Rooidakkies O.I. Grahamstown 1972

dakriet ['dak₁rĭt] *n. erron.* See *dekriet* (thatch) [*prob. fr. Afk. dak* roof + *riet* reed]

-dal- [dal] *n. prefix and suffix.* Vale, valley: found in S. Afr. place names equiv. *Brit. dale* and *vale* e.g. Hazendal (hares *Du.*), Genadendal (grace), Daljosophat (*n. prop.*), Rhenendal (*n. prop.*) etc. [*Afk. dal* vale, valley *cogn.* dale]

dam (dæm, dam] *n. pl.* -s. **1.** A reservoir in which water is retained by a man-made wall. See also quot. at *erf, water, cf. Anglo-Ind. tank,* pool, lake, artificial reservoir. ⫽As in *Canad.* 'the retaining wall of the dam and the waters within it' (*Dictionary of Canadianisms* 1969). In S.A. the reservoir itself, not as in *Brit.*

or *U.S.* the retaining wall only. ~ *scoop*, device for excavating for a ~.
Right next to the house was a large dam, fed from a watershed miles away from our farm. Jackson *Trader on Veld* 1958
2. Major reservoirs or irrigation schemes have ~ as *suffix* usu. to *n.* or *n. prop.* e.g. Le Roux Dam, Hendrik Verwoerd Dam, Haartebeestpoort Dam, Vaal Dam, Steenbras Dam.

Damara (ˈdæmərə] *n. prop. pl.* -s. Usu. in full *Berg, hill* or *mountain* ~ : One of a negroid people of South West Africa, speaking the language of the *Nama* Hottentots: see *Nama(qua)*. For *Cattle* ~ see *Herero.*
Leaders of the 65 000 Damara people – second largest Black nation in South West Africa – want a referendum to discover if the majority favour the Government's separate development policy. *E.P. Herald* 28.9.72

damba [ˈdæmbə, damba] *n. pl.* -s, ∅. *reg.* E. Cape and Transkei name for the marine fish elsewhere called *galjoen* (q.v.). [*etym. dub. poss. fr. Xh. iDamba*]
I read an article recently on the confusion that exists over the popular names of our common fish. The galjoen, for example, is known as a high-water or a damba. *Daily Dispatch* 20.6.72

dankie [ˈdaŋkĭ] *interj.* Thanks: also *baie* ~, thanks very much: see quot. at *baie.* [*Afk. fr. Du. dankje cogn.* 'thank ye']
We thought that with this council we could be better treated . . . Instead we are always required to say '*dankie baas*'. *Cape Times* 25.8.72

danne- [ˈdanə] *n. prefix.* Pine: see *denneboll/boom*

darem [ˈdarm, -əm] *adv. colloq.* Used often redundantly in S.A.E. as a rough equiv. of 'after all', 'at least', 'really', 'though' etc.: see quot. at *mos.* [*Afk. darem* all the same, though, however, *fr. daarom*]
Gavie, that was daarem not so bad, but (sarcastically) don't you think it a shame to sack the pantry so? Stormberg *Mrs P. de Bruyn* 1920
. . . something about bald guys, they darem love a bit of a cuddle. *Darling* 24.12.75

das [das] *n. pl.* -en. Earlier form for *dassie* (q.v.): also in combination *klip* ~ etc.: found in place names, Daskop, Dassen Island. [*Du. das* badger]
Our boat disturbed many of the *das* or coney that run along the rocks. This little creature . . .

seems to be of a species between the rabbit and the guineapig; it is esteemed good eating, and the flesh is white, but from its paws resembling those of a cat, it is not inviting. Philipps *Albany & Cafferland* 1827

dassie [ˈdasĭ] *n. pl.* -s. **1.** Also *das* (q.v.): the rock hyrax of the Procaviidae usu. *Procavia capensis* also in combination *boom* ~, *Dendrohyrax arboreus; klipdas, Procavia capensis.* [*Afk. fr. Du. dasje, das* badger + *dimin. suffix -je*]
12.8.1811. Here I procured for the first time, the Das or Dasje. This is of a brown colour, and has much the appearance of a rabbit; it is found in rocky places, where it takes shelter in the crevices. Its flesh is eatable; but the animal is exceedingly wary and difficult to get. Burchell *Travels I* 1822
2. The marine fish *Diplodus sargus* also called *blacktail* (q.v.) and *kolstert* ⫶*acc.* J. L. B. Smith 'about the best fighter of our inshore angling fishes'. [*prob. fr. colour*]
Dassie, wildeperd and galjoen were in mixed bags . . . in the gullies at Harmony over three days of fishing. *Argus* 14.5.71

dassiepis [ˈdasĭⸯpɪs] *n. Hyraceum:* the solidified urine of the *dassie* (q.v.) used medicinally by country people, and by certain tribes; also *klipsweet.* ⫶*Dassie buchu/boegoe, Pelargonium ramosissimum,* has leaves which smell like ~, a tincture of these is used for neuralgia. [*Afk. dassie* hyrax + *pis* urine]
And this is dassiepis – she held up a small lump of something that looked like gum or bitumen . . . It is good for hysterics, epilepsy, and convulsions of all kinds. Cloete *Hill of Doves* 1942
Early travellers in the Cape, men with scientific training, were puzzled by a peculiar substance they found in some caves, black masses like pitch. Thunberg thought it was bitumen, but the Bushmen and Hottentots knew better, and so did many farmers. They call the substance *klipsweet* or *dassipis.* In the Cape Pharmacopoeia it is listed more delicately as *hyracium.* Hyracium has been used as a medicine for centuries. Green *Karoo* 1955

dassievanger [ˈdasĭⸯfaŋə(r)] *n. pl.* -s. Also *berghaan:* either the *bataleur, Terathopius ecaudatus,* or Verraux's eagle, *Aquila verrauxii,* a large black eagle preying largely upon *dassies* (q.v.) [*Afk. dassie* hyrax + *vang* to catch + *agent. suffix -er*]
Most difficult of all birds to shoot . . . was the black eagle or dassievanger . . . they build their nests of sticks on the ledges of precipices. These powerful jet-black eagles live mainly on

dassies, though they take small antelopes and lambs. Green *Last Frontier* 1952

de [də] *art.* The: *definite art.*, see also *het*, found in place names e.g. De *Aar* (q.v.), De *Doorns* (q.v.) De *Hoek* (q.v.). Also *Afk.* form *die* in Die *Bos*, Die *Oog* (q.v.): see also *eye*. [*Du. de* the]

dear John(ny) *n. phr. pl.* -ies. A letter of dismissal formerly from a girl-friend, now acc. informants also from a boy to his girl-friend.
She even had the cheek to send me a Dear John, when as far as I was concerned there'd never been anything in it anyway. *O.I.* 3.3.79
His name is tottie – got a dear Johnny this morning, poor –. cit. *O.I. ex-Serviceman* 4.4.79

debris *n. prob. obs. in S.A.E.* See *tailings*, and quot. at *bewaarplaats*.

Deepavali [ˈdi(pə)ˌwalĭ, -valĭ] *n.* Also *Diwali Dewali Divali*: the Hindu Festival of Lights, the major religious holiday of the year. [*Sanskrit dīpāvali*, row of lights, *fr. dīpa* lamp + *āvalī* row, Hindi *dīvāli fr. Sanskrit*]
DEEPAVALI MESSAGE
. . . I wish to extend a message of goodwill to the Hindu commuity on the occasion of the Deepavali celebrations. The traditional lighting of the lamp, which symbolizes the triumph of light over darkness and good over evil has a special significance in these troubled times. *Fiat Lux* Oct. 1978

defamation *n. Sect. Legal.* In S. Afr. law ~ subsumes both libel and slander: an action for either wrong being one for ~, the law not distinguishing between the two as in Britain. Informant Hon. Mr Justice J. P. G. Eksteen.
Defamation is the unlawful publication by one person of anything which tends to injure the reputation of another person . . . 'Publication' means the communication or making known of the defamatory matter to at least one person other than the person defamed. Publication may take place by means of speech, writing, print, pictures, caricatures, effigies or even by insulting conduct. The distinction made in English law between 'libel' or written defamation, and 'slander' or oral defamation . . . is not known in South Africa. Wille *Principles of S. Afr. Law* 1945 ed.

dekriet [ˈdekˌrĭt] *n.* Also (erron.) *dakriet*: any of several species esp. of *Restio* used for thatching, also for making brooms (*besemriet*). [*Afk. dek*, to roof, cover + *riet cogn.* reed]
Most of the Cape Peninsula's thatch is the pale yellow dek-riet from the Riversdale and Albertinia dunes. Once looked upon as a dan-

gerous weed, the dek-riet is now carefully preserved. Green *Grow Lovely* 1951
. . . trees are scarce. There are occasional patches of 'dekriet', the reed used for thatching *Evening Post* 26.3.77

delict *n. pl.* -s. *Sect. Legal usu.* S. Afr. legal term for any form of injuria *Brit. tort.* 'Law of Delict' is equiv. 'Law of Tort'. Informant Hon. Mr Justice J. P. G. Eksteen
TORT (SEE DELICT)
A delict, *injuria*, in the wide sense, is an act committed or omitted by one person unlawfully which infringes the legal rights of another person to life, person, property, liberty or reputation, and which entitles the latter to claim redress . . . Wille *Principles of S.Afr. Law* 1945 ed.
. . . a senior lecturer in the law of delict and the author of a paper on civil actions against the police . . . *Sunday Times* 12.5.79

denne- [ˈdenə] *n. prefix.* Pine: ~*bol*, *n. pl.* -s. pine-cone: much favoured as kindling in S.A. Also *dannebol, donnyball.* ~ *pit/pip n. pl.* -s. The edible seeds, pine nuts (Italian *pignoli*) from the cones of *Pinus pinea*; see *stone pine* and quot. at *tammeletjie*. [*Afk. fr. Du: denne- cogn.* Ger. *Tannebaum* (+ *bol* cone *cogn.* ball) (~ *pit* pip)]
'Dennebol' (pine cone) is included, and an entry from the Cape Times of March 31, 1953, says that 50 dennebol pits could be bought for sixpence. The current price of dennepits, when available, has risen higher than pine trees. *Cape Times* 13.10.73
. . . cover with mayonnaise . . . to which you could add some finely chopped pistachio nuts (dannepits will do). Gerber *Cape Cookery* 1950

deproclaim *vb. trns. freq. pass.* See *proclaim.* [*official term*]
The Department of Planning has proposed that areas already proclaimed for Indian occupation in the town be deproclaimed and reproclaimed for White ownership and occupation. *Rand Daily Mail* 16.2.71

derm(s) [ˈderm, -rəm] *n. usu. pl.* **1.** *slang.* Guts, *lit.* not *figur.* as in 'my ~*s* is clapping/flapping together', vulgarism for 'hungry'. [*fr. Afk. my derms klap inmekaar Afr. fr. Du. darm* intestine]
'Sacrifice', he reckons, and . . . I get this sudden creepy feeling in my derms and a strong vibe I should . . . zap off. Hobbs *Blossom* 1978
2. Sheep's intestines used as sausagecasing, also *dik*(*thick*)~, the large intestine used fried or braai'd (q.v.) usu. among country people: *cf.* Brit. dial. *chitterlings.* [*Afk. fr. Du. darm*]
. . . the family lived on 'psalmpensies' and 'nersderms'*, the once despised offal of offal.

Burgess *A Life to Live* 1973 *rectum

dertiger [ˈdeː(r)tɪxə(r)] *n. pl.* -s. One of a group of Afrikaans poets of the nine-teen-thirties: see also *sestiger* and *sewentiger*. [*Afk. dertig* thirty + *personif. suffix -er*]

. . . there was the soaring renewal in Afrikaans literature, particularly the poetry of the 'Dertigers' (the generation of the 'thirties), notably N.P. van Wyk Louw, W.E.G. Louw, Uys Krige and others. Rosenberg *Sunflower to Sun* 1976

deurmekaar [ˈdiœ(r)məˌkɑːr] *adj. colloq.* Confused, muddled, disorganized: used of persons, places and situations. [*Afk. deurmekaar* muddled, confused]

He would not have minded, he said, if they had put his name to the colour brochure . . . but this pamphlet was too deurmekaar and smudgy. *Cape Times* 8.11.72

devil's snuffbox *n.* Also *aapsnuif, duiwelsnuif (dosie), monkey bomb, oumeidsnuif (doos), slangkop : Lycoperdon hyemale :* a puffball fungus forming round growths filled with dark brown snuff-like powder: these explode if kicked by children hence *colloq. monkey bomb.*

devil's thorn *n.* See *dubbeltjie.*

dhai *n.* See *Indian terms*

dhal, dholl *n.* See *Indian terms* and quot. at *masoor.*

Breads . . . were accompanied by chicken, samoosas, crisps, dal, pies and rolls. *Sunday Times* 8.4.79

dhunia *n.* See *Indian terms*, also called *Indian parsley.*

diaken [ˌdiˈakən] *n. pl.* -s. A deacon of the *Dutch Reformed Church* (q.v.) distinguishable on Sundays by a black frock coat ; see *manel, white tie* and *ouderling.* [*Afk. fr. Du. diaken cogn.* deacon *fr. Lat. diaconus, Gk. diakonos* server]

A newly elected diaken of the church came on his visit the other evening and got talking about his job in a different banking concern known for its solid image. *Star* 25.8.73

diamond 1. *n. (no pl.).* Used in the same way as 'gold', 'silver' or the name of any other mineral: see *blue ground, yellow ground, kimberlite.*

This time the geologists found abundant evidence of ilmenites and garnets of Kimberlitic quality – the two chief indicators of diamond. *Daily Dispatch* 14.2.72

2. *n. pl.* -s. In combinations *blue white ~,* the highest quality; *off colour* (q.v.) *~ ; industrial ~* not gem quality, used for industrial purposes only: see *melée, grit, stone(s), fancy.*

Diamond Trade Act, *n. prop.* Also *I.D.B. Act :* see *I.D.B.*

die [dɪ] *art.* See *de.*

died(e)rik(kie) [ˈdɪˌdrɪk(ɪ)] *n. pl.* -s. The bronze cuckoo, *Chrysococcyx caprius* so named from its call: see quot. [*presum. fr. Le Vaillant's onomat. 'didric'* (+ *dimin. suffix -(k)ie) sp. fr. Afk. name Diederik*]

The green golden cuckoo of the Cape . . . perched on the top of large trees, it continually repeats and with a varied modulation, these syllables di-di-didric, as distinctly as I have written them, for this reason I have named it the *didric.* Le Vaillant *Travels* 1796 *cit.* Pettman

dienaar [ˈdinɑːr] *n. pl.* -s. *lit.* One who serves, an official, see also *diender.* [*Du. dienen* to serve + *agent. suffix aar* servant, official]

. . . supplied by the Landdrost, Mr Ryneveld, with fresh horses and a *dienaar* (police man) to accompany me, I arrived in a few hours at Rondebosch. Thompson *Travels I* 1827

diender [ˈdində(r)] *n. pl.* -s. A police constable. [*Afk. fr. Du. diender* a policeman]

The Grand Bailiff had ordered the *dienders* (that is the police of the city . . . to open a way for their Highnesses . . . The officer in command came . . . and ordered these silly *dienders* driven away. The Grand Bailiff was violently enraged by this. *Boswell in Holland* 10 February 1764

He explained that the 'kaffers' had been to the charge office to ask the 'dienders' to arrest them for not carrying their passes, and that the police had closed all the doors because they did not have room. Louw *20 Days* 1963

dik¹ [dɪk] *adj. slang.* Stupid, dense; 'a ~ ou' (q.v.) a stupid fellow: see also *thick, toe.* [*Afk. dik* dense *cogn.* thick; *figur. in S.A.E.*]

dik² [dɪk] *adj.* Full to satiety, replete. [*Afk. dik (slang)* full]

. . . reported finding me – a huge sickle of watermelon in my grubby paws – . . . She offered me a slice of cake. 'Uh-uh, Dik.' Butler *Karoo Morning* 1977

dikbek [ˈdɪkˌbek] *n. adj. or adv. m.* **1.** *colloq.* One given to the sulks: as *adj.* sulky, pouting and as *adv. m.* sulkily, with an ill grace: also *diklip.* [*Afk. dik* thick, fat + *bek* (q.v.) *cogn.* beak, mouth]

. . . calls her dikbek because she never smiles. O.I. Schoolgirl, 1974

. . . like a morg for the next few days, everybody dik-lip with faces down to yere and the mutters. *Darling* 19.1.77

2. Also *bergsysie,* any of several species of the Fringillidae esp. *Serinus*

sulphuratus, the bully seed eater, or *geelbergsysie*. [*fr.* (1) ~]
3. See *panga²*. [*fr.* (1) ~]

dik dik [ˈdɪkˌdɪk] *n. pl.* -s. *Madoqua kirki damarensis*, a very small buck found mostly in S.W. Africa. [*onomat. fr. call*]

The dik-dik is one of the smallest antelope in existence, standing no more than 16 inches high at the back. *Life* 19.1.70

... the Damara dik dik is a much finer dish. Dik Dik occurs in and around the Kaokoveld ... So you have to travel far to taste this tiny and tender antelope. Green *Last Frontier* 1952

dikkop [ˈdɪ(k)ˌkɔp] *n. pl.* -s. Thick head. [*Afk. dik cogn.* thick + *kop* head *cogn. Ger. Kopf*] **1.** Also *commandovoël*: the stone plover (*Burhinus* spp. usu. *B. capensis*). ⫽The ~ lays its eggs on the ground and protects its nest by a 'display' with its wings and an aggressive jarring sound, unlike its usual call. *cf. Austral. thick head* (shrike).

You may not have thought of the dikkop as a shore bird but it has always been found off Robben Island and many of them nest in the dunes along Blaauwberg beach. Few see them, for they are elusive nightbirds and they know how to take cover. Dikkop is good eating, rather like the korhaan. Green *S. Afr. Beachcomber* 1958

2. Any of numerous species of fish of the Gobiidae fam., having joined pectoral fins: esp. *Caffergobius* and *Bathygobius* spp. also known as *bully*, goby. [*as at* (1) ~]

3. *no pl.* A form of horse-sickness also known as *bluetongue* (q.v.). characterised by swelling of the head and tongue. [*as at* (1) ~]

4. *no pl.* A disease in sheep, *Tribulosis*, caused by eating *dubbeltjies* (q.v.) also combination *geel* (*yellow*) ~. [as at(1) ~]

Dikkop disease is reaching near-epidemic proportions ... 150 sheep have been killed by the sickness, caused by eating dubbeltjie thorns and other plants which affect the liver and eyes of the sheep. There is no cure. *Daily Dispatch* 8.1.71

5. *figur.* Fathead, numskull: a mode of address or reference: see also *dik, thick, domkop, houtkop: cf. Austral. dilly, drongo, galah.* [*as at* (1) ~]

dikoog [ˈdɪkˌʊəx] *n.* Also *peuloog*, 'bulging eye': terms for the marine fish *jacopever* (q.v.) and *fransmadam* (q.v.) [*Afk. dik cogn.* thick, fat, *peul* bulge + *oog* eye]

Dingaan's apricot *n. pl.* -s. *Dovyalis caffra* the *Kei Apple* (q.v.) is so named in Natal.

The Kei apple or Dingaan's apricot, invaluable for forming thorny fences and yielding a pleasant fruit. Chapman *Travels II* 1868 *cit.* Pettman

dinges [ˈdɪŋəs] *n. pl.* -es. Also *dingus*. Not only S.A.E. ⫽Current in many parts of the Eng.-speaking world esp. where there has been Dutch influence. Common in S.A.E. as equiv. 'thingummyjig', 'whatsit', 'whatsisname' etc.; used where the name or term is unknown or cannot for the moment be recalled; see also *goodie. cf. Brit.* etc. *dingbat, Austral. thingo.* [*Afk. fr. Du. dinges cogn.* thing *poss. fr. Ger. Dings* (*posessive form*)]

In an emergency absolutely anything can be quickly described as a 'dingus': 'Pass me that "dingus" nurse' or 'It's that "dingus" in the differential again.' *Cape Times* 2.7.73

It's a thigamybob ... a dinges ... an OO-Jar! For sugar and spice. *Advt. Circular* 6.8.74

dinki [ˈdɪŋkɪ] *n. pl.* -s A 250 ml. bottle of wine; formerly a trade name for a single particular range, now loosely used of a bottle of this size containing wine (not spirits).

'I didn't know that the – range was being made in dinkis.' 'Yes, they're just new on the market.' *O.I. Bottle store, Durban* 29.4.78

dip *n. prop. colloq.* Jeyes fluid. [*fr. Afk. dip*, household term for Jeyes fluid]

dirt (road) *n. pl.* -s. A non-macadamised gravel road. Also known as 'the dirt' e.g. 'It's six miles on the tar and then ten on the dirt.'

... this patchwork of thatched buildings is reached ... on a dirt road through sheep country. *Panorama* Sept. 1975

disa [ˈdaɪsə] *n. pl.* -s. usu. *Disa uniflora*, red ~ or 'Pride of Table Mountain', best known of numerous species of Orchidaceae, esp. *Herscheliae* including *H. spathulata, H. lugens, H. graminifolia:* also various species vernacularly named *moederkappie* (q.v.).

There are 20 species of disa but the best known is the red disa or 'The Pride of Table Mountain' as it is popularly called. It is the rarest and richest of all. *Personality* 24.9.70

disselboom [ˈdɪsəlˌbuəm] *n. pl.* -s. The shaft or pole of an ox wagon to which the two *after-oxen* (q.v.) are yoked and the *trektou* (q.v.) stapled to carry the yokes for the rest of the team: also the single shaft of a horse-drawn cart or other vehicle. *cf. Jam. Eng. tongue* as in *tongue cattle.* [*Du. dissel* shaft + *boom cogn.* beam]

The wagon's front wheels pivoted freely as the single, short shaft, the disselboom, swung to the left or right with the turning span of plodding oxen. Mockford *Here are S. Africans* 1944

district *n. prop. pl.* -s. *hist.* One of the districts or divisions of the Cape Colony: also in combination ~ *Six* (q.v.).

This settlement is at the most Southern extremity of Africa, extending about 550 miles in length, from West to East, and 315 in breadth from South to North ... and is divided ... into six districts: the Cape district; that of Stellenbosch and Drakensteen; that of Zwellendam; that of Graaff Reinet; that of Uitenhage, and that of Tulbagh. *Afr. Court Calendar* 1807

The Colony was many years ago divided into four districts viz. the Cape district, district of Stellenbosch, of Swellendam and of Graaff Reinet. Ewart *Journal* 1811–1814

district road *n. pl.* -s. A *dirt* (q.v.) or tarred road, often numbered, maintained by the *Divisional Council* (q.v.) not by the central government: *cf. National Road* (q.v.).

District Six *n. prop.* Place name: (*hist.*) Formerly one of the six electoral districts of Cape Town. Latterly traditional *Cape Coloured* (q.v.) quarter now taken over under the *Group Areas* (q.v.) zoning to be an area for whites.

DISTRICT SIX THE FINAL ACT

Remember District Six? That colourful tangle of streets and memories where we used to live. Can you hear the banjos playing? The children laughing? The chattering of housewives? Do you remember? *Cape Herald* 27.4.74

diving goat *n. pl.* -s. *obs.* Early travellers' term for (1) *duiker* (q.v.).

The duiker-bock or diving-goat derives its name from its plunging and springing amongst the bushes when closely pursued. Percival *Account of Cape of G.H.* 1804

division *n. pl.* -s. *hist.* One of the historical districts into which the Cape Colony was officially divided: now in use only in *Divisional Council* (q.v.).

... erect the same with the money to be borrowed upon the security of the road rates of the said division to be levied under the Act No. 9. *Cape of G.H. Act* 1867

Divisional Council *n. pl.* -s. A Council whose area of jurisdiction is larger than that of municipalities, covering divisions of the country for the maintenance

of roads etc. ‖One division abuts upon another thus divisional boundaries are crossed while travelling.

It shall be lawful for the divisional council of any division, and it is hereby required, to employ, through the instrumentality of the field-cornets or otherwise, labourers to eradicate and burn the said weed wherever it may be found growing on public roads or crown lands, or on public outspan-places within such division. *Cape of G.H. Act* No. 27 of 1864

The days of the voortrekkers are gone, and Divisional councils no longer need outspans, members of the Albany Divisional Council decided at their monthly meeting. *Grocott's Mail* 27.7.71

divorce *vb. intrns.* Used intransitively in S.A.E. see quots. [*prob. fr. Afk. skei vb intrns. and trns. fr. Du. scheiden* separate, *also* divorce]

... socialites are shocked by the news that multimillionaire –'s daughter and son-in-law are to divorce. *Post* 14.3.71

As you are married in community of property, the money you inherited becomes part of your joint assets. Should you divorce, your husband could claim half of it. *Fair Lady* 14.6.72

Diwali See *Deepavali.*

doedie ['dŭdĭ] *n. pl.* -s. *slang.* Popsy, doll: see *soldoedie. cf. Austral. brush, sheila* (girl), *also Brit. and U.S. broad.* [*Afk. doedie (slang) equiv.* 'floozie', popsy, *etc.*]

... the thing I liked best was all those doedies. Makes a man wish he was still single. Ja, I want to go again next year for sure. *Personality Advt.* 18.10.74

doek [dŭk] *n. pl.* -s. Head-scarf or cloth tied about the head in various ways. Also *kop* ~ : occ. also worn by elderly African men under hats or by way of a night cap. *cf. Jam Eng.* tie-head (scarf): also *bandoo.* [*Afk.* (*kop* head) + *doek* cloth, scarf]

It is these women, the full-bosomed, aproned matrons of our townships with their *doeks* tightly framing their strong-boned faces, who come together every Thursday in their *Manyanos.* Brandel-Syrier *Black Woman* 1962

As another mark of respect for her father-in-law, the childless wife must wear the *doek* she sports as head-dress so low over her eyes that she cannot look him in the face. *Financial Mail* 22.10.76

doekpoeding ['dŭk₁pŭdɪŋ] *n. pl.* -s. Traditional style boiled pudding steamed in a cloth. [*Afk. doek* cloth + *poeding cogn.* pudding]

This was followed by our cook Ai Nettie Pekeur's stupendous 'doekpoeding' made with grated carrots – I wish I had the recipe today – and Ma's trifle. Van Biljon *cit. Sunday Times* 24.12.78

doepa [ˈdŭpa] *n.* **1.** Love potion, or amulet having magical properties for power e.g. over animals, or for love. *cf. Jam. Eng. tempting powder* (love philtre), *compelling powder.* [*Afk. fr. Malay dupa* incense]

His chief object, however, at present was to get us to make him dupas, or pastilles. Snyman having told him the tales current among the Malays at the Cape respecting their efficacy as love charms or other surgical properties. Baines *Explorations in S.W. Africa* 1864 cit. M.D.W. Jeffries

2. ~, ~*olie.* Two of the *Dutch Medicines* (q.v.) ~ for colic and flatulence, ~*olie* (oil) for coughs and bronchitis. [*presum. fr.* (1) ~ *Afk.* ~*n.* 'dope', charm: *as vb.* to drug, dope or bewitch]

doer [du:r] *adv./prep. slang.* freq. among children. Over there: equiv. of 'very far' or in *phr.* ~ *and gone,* e.g. 'He lives ~ and gone somewhere in the Karroo.' *cf.* to hell and gone. [*Afk. doer* yonder, far away]

And, of course, daar doer in die Kaap, Sir de Villiers Graaff and Mr Vorster . . . decide finally whether there is a fusion in their future. *Sunday Times* 17.8.75

doesn't want to *vb. phr. colloq.* See *want.*

dog See *hondjie.*

doiby [ˈdɔɪbĭ] *n. pl.* -s. *Sect. Army:* The inner plastic lining of the *staaldak* (q.v.) also *mosdoppie* (q.v.)

We wear plastic, or some similar compound, helmets called 'Doibies' which keep the sun out to a great extent. We have got all our kit . . . 13.1.79

The doiby I told you about is the inner for our main helmet – the STAALDAK – very strange names but quite apt. 21.1.79 *Letters Serviceman*

dolly *n. pl.* -ies. *prob. obs.* An early, primitive stamping machine used on S. Afr. gold fields: *[acc.* Pettman *orig. Austral.* [*unknown: also given as vb.* to dolly by *Webster's Third International Dict.*]

Crude dollies were constructed to act as mills. They consisted of an iron-shod wooden pole weighing about 100 pounds . . . suspended by riems from the end of a horizontal leverage pole about eighteen feet long and called a spring. Bulpin *Lost Trails* 1951

dolos [ˈdɔlɔ̃s] *n. pl.* -se. **1.** *usu. pl.* Children's playthings: formerly certain animal bones, used as oxen in games: also used of oxen modelled in clay: see *kleios.* [*etym. dub. poss. fr. Cape Du. dollen* to play + *os cogn.* ox, *poss. fr.* (2) ~]

These lichens give a red dye and are employed by children for dyeing their 'dolosse' the colour of Afrikaner cattle. Smith *S. Afr. Plants* 1966

The name is derived from '*dollen os*', which is what Voortrekker children called the small ankle bones of sheep and goats. *Star* 10.4.73

2. *usu. pl.* Bones used by a *witchdoctor* (q.v.) for divination; see *throw the bones* and quot. at *bones.* [*poss. fr.* (1) ~ *poss. fr.* Tswana or N. Sotho *in Dawula* diviners' bones + *pl.* -s *corrupted to* ~*os, poss. S.* Sotho *taola* (*sing*) a diviner's bone *or poss. fr* Xh. *indawule* acc. Kropf, *Fingo,* 'bones of different animals similar to dice used by witchdoctors']

The most familiar 'properties' of the witch-doctors are the 'knuckle-bones' known to the natives as 'daula' and to the Boers as 'dol os'. H.M.S.O. *Native Tribes Transvaal* 1905

It so happened that a cyclone hit the farm that afternoon and that established . . . as a medicine man . . . He started practising medicine with herbs and lion's fat and throwing dolosse like a typical withdoctor. *Daily Dispatch* 4.8.72

3. *usu. pl.* Formerly known as Merrifield blocks: interlocking concrete blocks called by the inventor ~*se* because of their shape, used all over the world to preserve beaches, shore lines and harbours from being washed away. [*fr. shape*]

They look like concrete jigsaw puzzles dumped on the beach. They are called dolosse and are the brainchild of Mr. E. M. Merrifield, East London's harbour engineer. *Star* 10.4.73

. . . rescued by civilians and helicopters when their ship was driven onto the dolosse protecting the new sea wall at Paarden Eiland. *E.P. Herald* 28.4.74

dom [dɔm] *adj. colloq.* Stupid, slow to comprehend: see also *dik, thick, toe.* [*Afk. dom cogn.* dumb] In combination ~*boek,* ~*kop,* ~*pass.* all (q.v.)

domboek [ˈdɔ̃mˌbŭk, -bʊk] *n. pl.* -s, -e. *colloq.* Among Africans *reference book* (q.v.) see also *dompas.* [*Afk. dom* stupid + *boek cogn.* book]

. . . simple hard-hitting English dialogue, 'to look at our real life you are too far away . . .', a man with no domboek is not a man. *E.P. Herald* 6.8.73

dominee [ˈdʊəmɪˌnĭ] *n. pl.* -s. Also *predikant* (q.v.) A minister of the *Dutch Reformed Church* (q.v.): a mode of

address, and in titular use with a surname: Dominee X; written *abbr. Ds.*
(*fr. dominus*) see quot. at *Boerechurch.*
cf. Canad. dominie schoolmaster, also
any minister esp. a Presbyterian. [*Du.
fr. Lat. dominus* master *etc. cogn.
dominee* teacher]

Many Afrikaans people here oppose the
view of a . . . dominee of the Dutch Reformed
Church, Ds. . . . that there should be no dancing at weddings. *Evening Post* 4.3.72

'I was later called to appear before the church
council to answer charges that I had called the
dominee a liar'. . . . The church wrote . . . telling
him that he had been censured and that unless
he showed regret for calling the dominee a liar
he would remain censured for a period of six
months. *Daily Dispatch* 3.8.74

domkop [ˈdɔmˌkɔp] *n. pl.* -s. *slang.* Fool:
mode of address or reference: *cf. Austral.
dilly, drongo* (fool). [*Afk. dom* stupid
+ *kop* head *cogn. Ger. Kopf*]

He had written 'Bantu'. The official must
have reached the end of his tether, because he
screamed 'domkop', scratched it all out and
then scribbled 'South African.' *Drum* 22.7.74

dompas(s) [ˈdɔmˌpas, -pas] *n. pl.* -es.
colloq. among Africans for *reference
book* (q.v.), see also *domboek.* [*Afk.
dom* (see *domboek*) + *pas cogn.* pass]

There are instances where Zulu-speaking
Africans get their 'dom-passes' stamped under
the influx control regulations and are told 'Go
to Gatsha, he must give you work.' *E.P. Herald*
15.10.73

'London . . . now I'm back to Soweto . . .
It's also back to my dompas which my wife has
kept safely for me during my absence,' Oswald
grinned. *Drum* 22.9.73

domsiekte [ˈdɔmˌsĭktə] *n.* A disease of
gravid ewes characterised by a stupid
appearance. [*Afk. dom* stupid + *siekte,
fr. Du. ziekte* disease]

Domsiekte in sheep. This disease has only
recently (1924) been reported . . . The cause is
as yet unknown, but conditions of drought and
vegetation play a role. It has only been observed in ewes in lamb during the last few
weeks of gestation, generally a few days before
parturition. *H'book for Farmers* 1937

donder¹ [ˈdɔnə(r), ˈdɔn(d)ə] *slang.* To
beat up, thrash, 'clobber': see quot. at
meid. See also *skiet and donder. cf. Austral. stoush,* to thrash, *U.S. dong.*
[*Afk. donder* to thrash *fr. Du. donderen*
to fulminate, swear, also bully]

. . . Le Roux said he would 'donder' Bidi
if he kicked the dogs. He came at Bidi and
struck him with the cricket bat. *Post* 22.8.71
Also figur.
The time has come to 'donner the H.N.P.'

and everything else that stands for stagnation
and reaction . . . if the only weapon to hand is
the N.P. . . . the U.P. will simply have to grin
and use it. *Sunday Times* 1.8.76

Professor attacks 'ek sal jou donder' principle . . . 'Security has come to mean 'donder-
ing' of any opposition that offers a serious challenge to the established party or policy. Prof.
S. A. Matthews *cit. E.P. Herald* 26.10.76

donder² *n. pl.* -s. *slang,* Equiv. of 'swine',
'bastard' etc. abusive mode of address
or reference to person or thing: see also
bliksem. [*Afk. donder* scoundrel]

Boesman [*the bulldozer*] slowly it comes . . .
slowly . . . big yellow *donner* with its jawbone
on the ground. One bite and there's a hole in
the earth! Fugard *Boesman & Lena* 1969

donder in See also *hell in, the,* and quot.
below.

Success surely breeds success and we get a
theatre in Pretoria for a transfer after our Blue
Fox run. Mother and Father Grundy were
however alive and well and Die Donner in!
Speak Vol I No. 2 Mar/Apr. 1978

donga [ˈdɒŋgə, ˈdɔŋga] *n. pl.* -s. A usu.
dry, eroded watercourse running only
in times of heavy rain. *cf. Anglo-Ind.
nullah, Austral. gully, breakaway.* In
combination *erosion ~s* found in areas
where the *veld* (q.v.) has been *tramped
out* (q.v.): see quot. at *bundu-basher.*
[*Ngu. -donga* washed out ravine, gully]

Thousands of miserable cattle and goats
roamed everywhere, making tracks that would
some day form cracks which successive rains
would open into gullies and dongas. Soil
erosion is one of the tragedies of the reserves.
Morton *In Search of S.A.* 1948

An uncalculable amount of soil has been
lost, wasted, and the land rendered down to
dusty dongas. Cowin *Bushveld, Bananas Bounty*
1954

donga *vb. Non S.A.E. Austral.* Army
World War II See S.A.E. *gyppo* and
quot. below.

donga 'a ravine or watercourse with steep
sides', a term from South Africa . . . *donga* any
gulley or depression in which men could settle
themselves to loaf: *dongarer,* a loafer; *to
donga* to loaf or *bludge;* and *dongaring* loafing
or bludging. Also used both in Tobruk and
New Guinea for a makeshift shelter. As late as
1965, *donga* was being used similarly by Australian army personnel (a member of Melbourne
University Regiment wrote that a donga was
'anywhere one sleeps – a tent, hut, etc.') Not
long after the end of World War II, the word
acquired use among New Guinea's white
population for a house. Sidney J. Baker *The
Australian Language* 1966 *ed.*

donker [ˈdɔŋkə(r)] *adj.* Dark: found in

S. Afr. place names prefixed to *n.* e.g. Donkergat, Donkerpoort. [*Afk. donker* dark]

donkermannetjie [ˌdɔ̃ŋkə(r)ˈmanĭkĭ, -cĭ] *n.* A children's game, similar to blind man's buff, played in a darkened room. [*Afk. donker* dark + *man* + *dimin. suffix -(ne)tjie* little man]

Donkey Church *n. prop.* The Bantu Methodist Church, formed as an offshoot of the Methodist Church in 1932–33, which includes among its ceremonies the re-enactment of the Palm Sunday procession with a donkey.

⟊Two African eyewitnesses state that the riderless donkey, accompanied by singing crowds, wore a wreath of *isundu* (palm) round its neck and a red shawl or blanket on its back. *Palm Sunday Grahamstown* 23.3.75

The Bantu Methodist Church . . . broke away . . . after a dispute in the Methodist Conference over church dues . . . People felt that they were being exploited to provide their superiors with cars, and yet Jesus Christ himself used a donkey . . . 'the Donkey Church', as it is popularly called, 'is becoming more and more respectable and one wonders if its leaders are still strong believers in donkey riding'. Wilson & Mafeje *Langa* 1963

doodgooi [ˈduət̩xɔɪ] *adj. and n. colloq.* Of or pertaining to cake or bread so heavy as to be *figur.* a lethal weapon *cf. Brit. sad, Jam. Eng. dough-dough,* used of a dumpling or insufficiently-baked bread, cake etc.: *occ.* as *n.* e.g. I've baked three ~*s* in a row. See also *throw dead.* [*Afk. doodgooier* a dumpling *fr. dood* dead + *gooi* throw + *agent. suffix -er*]

Doodgooi: A jocular name for a dumpling. It has been taken over by the Kaffirs in the form iDodroyi, the *r* being gutteral. Pettman *Africanderisms* 1913

. . . she did not take the trouble to mix yeast into her batch of dough the evening before, with the result that her baking was sad – real doodgooi. Butler *Karoo Morning* 1977

doofpot [ˈduəf̩pɔt] *n.* A cover or damper, sometimes bell-shaped, under which in former times remaining embers of a fire were preserved overnight for re-kindling in the morning. [(*Cape*) *Du. doofpot* an extinguisher for coals *fr. Du. doven* to extinguish, put out + *pot* vessel]

doorn [duə(r)n] *n.* Thorn: found in place names e.g. De Doorns, Doornfontein; see also *doring.* [*Du. doorn cogn.* thorn]

dop[1] [dɔp] *n. colloq.* Brandy; in combinations ~ *en/and dam* brandy and water, and ~ *brandy:* see also *Cape smoke, witblits* and quots, at *mampoer, biltong* and *horries.* [*Afk. dop* brandy *etym. dub. poss. fr.* belief brandy is distilled from *doppe* husks of grapes]

. . . the detectives made seventy arrests, destroying many gallons of kaffir beer and seized numerous bottles of dop. *E.P. Herald* 13.8.1916

dop[2] *n. pl.* -pe. *colloq.* A tot, little drink usu. of spirits: see also *sopie:* occ. *dimin.* ~ *pie* (q.v.), See quot. at *bab(b)alaas. cf. Canad. hooker, snort, smash: U.S. slug, Austral. nobbler.* [*Afk. steek 'n dop* have a tot, *poss. fr. dop* cup]

Let's have a dop first. I'm feeling the cold. Fugard *Boesman & Lena* 1969

dop[3] *n. slang.* Head, 'noddle'. *cf.* 'brain-pan'. [*Afk. dop* shell, empty vessel]

dop[4] *vb. intrns. slang.* To fail an examination. *cf.* plug, plough. [*Afk. vb. dop* fail, be ploughed]

dop[5] *n. pl.* -pe. *colloq.* A motor cyclist's crash helmet: see also *doiby* and *mos-doppie.*

Dumbo . . . scorns to wear a crash helmet (what he calls a dop) . . . he reckons nobody's gonna catch him looking like a blerry pampoen. Hobbs *Blossom* 1978

Dopper [ˈdɔ̃pə(r)] *n. prop. pl.* -s. A member of the Gereformeerde Kerk in S.A. or ~ *kerk, church* a Calvinistic church who were described by Bishop Merriman in 1853 as 'a sort of Dutch Church Puritans'. ⟊ ~*s* were formerly distinguishable by their style of dress, a short jacket instead of the frock coat or *manel* (q.v.) of the D.R.C. members whom they called *gatjaponners* (q.v.), and acc. some, the *klapbroek* (q.v.). See also quot. at *Hollander.* [*etym. dub. poss. fr. Du. doppen* to husk or shell, hence to strip away outer coverings, *poss. fr. Cape Du. doppen,* to gauge hence measure or judge. ⟊Various theories exist: *doper* (baptist), *dompen* (to suppress hence *domper*), *dop* (pudding basin) *fr.* hair cut, *dorper:* townsman, or *acc.* one ~ informant, from the bowler (*U.S.* derby) hat worn (*Afk. dophoed,* bowler, hard hat)]

These men (Doppers) were easily recognisable, for they clung to the old ways, not only in theology, but also in their dress; and, scorning belts or braces, kept up their trousers, which

were of the klapbroek kind with a flap in front, by means of draw-strings. Cloete *Turning Wheels* 1937
The 'Doppers', a strict calvinistic sect, were recognizable by their clothes, for they wore their jackets shorter than usual. Telford *Yesterday's Dress* 1972

doppie[1] ['dɔpǐ] *n. pl.* -s. *colloq.* An empty cartridge case: sometimes used of other empty articles such as eggshells or grape skins: see also dop[3]. [*Afk. dop* percussion cap, cartridge case]
... two children being christened ... one was wearing a camouflage cap and the other had a packet of doppies (cartridge cases) ... playing with them. This seemed to typify Rhodesia today. *Fair Lady* 16.3.77

loppie[2] ['dɔpǐ] *n. pl.* -s. *colloq.* A little drink, tot: see dop[2]. [dop[2] (q.v.) + *dimin.* suffix- (p)ie]
... one Saturday night Klaas, now 26, felt like celebrating and had a few extra doppies of his own. *Drum* Mar. 1971

loppie[3] ['dɔpǐ] *n. pl.* -s. *Argyrozona argyrozona* a small fish common in Cape waters also known as *Kaapenaar, silver fish, carpenter.* [*unknown*]
Doppies were cheaper than harders. Grain farmers gave their labourers fish for breakfast at harvest time, and doppies or hottentot and silver fish (sold in bunches of ten) served the purpose very well. The farmer could hang the strings of doppies in his loft until the fish became as dry as biltong. Green *S. Afr. Beachcomber* 1958

loring ['dʊərɪŋ] *n. pl.* -s. Thorn. [*Afk. doring fr. Du. doorn cogn.* thorn] **1.** Found in place names usu. prefixed as in Doringbaai, Doringbult, Doringkloof; also as suffix in Haakdoring, Ondersterdorings, Steekdorings; see also *doorn*
2. Suffixed to the names of various thorn trees or bushes often species of *Acacia* e.g. *kameel ~* (q.v.), *haak ~* (q.v.) etc.
3. Other compounds *~draad,* barbed wire; *~boom/bos* thorn tree/bush, *~ haai,* shark, see quot. at *blaasop.*
Our road now abounded in trees of the Cape Acacia, the colonial name of which is Doornboom (Thorn tree). Burchell *Travels I* 1822

Dormer ['dɔ:mə] *n. pl.* -s. *usu. capitalized.* A dual purpose mutton and wool breed, a Dorset horn-Merino cross: see also *Dorper.* [*'portmanteau word' Dorset + Merino*]
He has great faith in the *Dormer* breed, and says although essentially a mutton breed, they also give enough wool to substantially contri-

bute towards their keep. *Farmer's Weekly* 4 7.73

dorp [dɔp, dɔrp] *n. pl.* -s. Country town or village, sometimes *colloq.* in a derogatory sense sig. a backward or unprogressive place equiv. of 'dump' see *Blikkiesdorp occ. dimin. ~ ie. cf. U.S. podunk.* [*Afk. dorp* village *cogn.* thorp]
They visited all the dorps from the Cape – where they had started from – to Zeerust in the Transvaal, where Hannekie Roodt left the Company. Bosman *Unto Dust* 1963
Grahamstown's not a metropolis, but it's something more than a dorp. *Grocott's Mail* 13.7.71
Also found as *suffix* in numerous place names e.g. Villiersdorp, Machadodorp, Humansdorp, Klerksdorp. *cf. Brit. U.S.* place name *suffixes* -town, -ton, -ville.

Dorper ['dɔ:pə] *n. pl.* -s. *usu. capitalised.* Sheep: a mutton breed: cross between Dorset horn and Persian, producing almost no wool, *cf. Dormer.* [*'portmanteau word', Dorset + Persian*]
Dorpers – the wonder mutton sheep flourish under all conditions and in all parts of the country. *Farmer's Weekly* 21.4.72

Dorsland ['dɔ(r)s,lant] *n. prop. hist.* The desert country from the Transvaal to Angola crossed with great losses and privations by the *~ trekkers* in 1874: see also at *Thirstland* [*Afk. fr. Du. dorst cogn.* thirst + *land*]
'The Great Dorsland Trek', Koos Steyn shouted ... 'Anyway, we won't fare as badly as the Dorsland Trekkers ... because we've got less to lose. And joking that we are only five families, not more than about a dozen of us will die of thirst.' I thought it was bad luck ... to make jokes like that about the Dorsland Trek. Bosman *Mafeking Road* 1947

D.P. [,di'pi] *n. prop. abbr. Durban Poison* (q.v.)

-dra [dra] *vb.* or *n. suffix. Sect. Army suffix equiv.* 'Lift': used referring to various modes of carrying another person known as *maatjie-dra* 'buddy-carrying', part of the'buddy-aid' or 'makkerhulp' training: *skaap~* [sheep] carrying as a shepherd does a sheep, *baba~* cradled in the arms. [*Afk. fr. Du. dragen* to bear, carry] *O.I.s Servicemen* 1976, 1979

draai [draɪ] *n. pl.* -s, -e. Turn, bend, corner, also *figur.* twist. *Kaapse (Cape) ~* (q.v.) sharp corner or bend in the road encountered in travelling. Also found in place names as *n.* e.g. Kromdraai and *vb* e.g. Draaiomfontein. [*Afk. draai* turn, bend *n.* and *vb fr. Du. draaien* to turn]

drag *vb. slang* (usu. children). To bear a grievance, sulk, be out of temper, *cf. Austral. to chew the rag*, sulk. [*unknown*]
'What the hell are you dragging for now?'
'She always drags if she's not the centre of attention'. O.I. Children.

Drakensberger [ˈdrăkənsˌbɜgə] *n. pl.* -s. A beef cattle breed. [*presum. fr. n. prop. Drakensberg*]
. . . Drakensberg bulls from 1½ to 3 years old. Born and raised on severe Redwater and Gallsickness veld . . . Drakensberger for adaptability, fertility, weight-for-age and economical beef production. *Farmer's Weekly* 30.5.73

Draks, the [drǎks] *n. prop. abbr.* The Drakensberg: See *Berg, the.*
There was nothing for it but to dash to the Draks for the weekend. *O.I.* Nov. 1977

D.R.C. [ˈdiˌɑˈsi:] *n. and n. modif.* Dutch *R*eformed *C*hurch (q.v.) [*acronym*]
. . . dissociate himself from those DRC ministers who refused to pray for those men who were fighting for their country in the Second World War. *Daily Dispatch* 20.5.71

dre(c)k [drek] *n. slang* (vulgar) Excrement: see also *duiwelsdrek cf. Austral. Army drack*, bad, worthless, also a criminal, *fr. Ger. Dreck.* ❡ Also Yiddish, various meanings. [*Afk. drek* excrement, dung]
He offered some to a friend and solicited his opinion . . . he said 'Spatz, this wine is drek.' *Family Radio & T.V.* 6.11.77
Amid all the Disco dreck, good taste occasionally came through. *Darling* 20.12.78

driedoring [ˈdriˌduərɪŋ] *n. pl.* -s. *Rhigozum trichotomum*, a thorny shrub: see quot. at *soutbos*. [*Afk. drie cogn.* three + *doring fr. Du. doorn cogn.* thorn]
. . . parts comparatively unsuitable for the merino, i.e. areas where steekgras, driedoring and bush are prevalent. *H'book for ι armers* 1937

drift [drɪft] *n. pl.* -s. Ford [*Afk. fr. Du. drift fr.* earlier *doordrift* ford] **1.** A shallow fordable point in a river, now often a man-made, sometimes paved, ford where a river crosses a road: see quot. at (1) *khaki.*
When swollen by rains in the interior this river is a magnificent object . . . with considerable depth, and a full and rapid current. At other seasons it is easily forded, the water at the usual drift being shallow, and the banks of the river presenting an easy slope to the water's edge. *Cape of G.H. Almanac* 1843

Ubique means the tearin' drift where, breech-blocks jammed with mud,
The khaki muzzles duck an' lift across the khaki flood.
Kipling *Five Nations* 1903
Further up Renosterhoek, drifts are completely covered with silt and mud. Gardener's Drift is covered. *Grocott's Mail* 27.8.71
2. *drif(t)* Ford: found in place names e.g. Velddrif, Matjiesdrift, Committees Drift.

drill hall *n. pl.* -s. Used in S.A.E. to refer both to a gymnasium, esp. of a school, and a military ~: see *Turnhalle* and first quot. at *tackies.*

drink *vb. trns. substandard.* Used as equiv. of 'take' or 'swallow' (medicine or tablets): see quot. at *benoudheid.* [*poss. fr. Afk. pille/medisyne/drink*]
It's such a nuisance to leave my office to drink a pill every four hours. O.I. 1971

drink out *vb. phr. substandard* To empty, drain: see second quot. at *arm.* [*translit. Afk. uitdrink* finish, drain etc.]

droë [ˈdruə] See *droog.*

drogie(s) [ˈdruəxĭs] *n. usu. pl.* The astringent berries of *Scutia myrtina* (*katdoring*): see *droog-my-keel.* [*Afk. droog* dry + *dimin. suffix -ie*]

drol [drɔl] *n. pl.* -s, -le. *slang: lit.* 'Turd': an abusive means of address or reference to a person (vulgar): also animal droppings, *dimin.* ~*letjies* [*Afk.* dropping(s)]
Boesman (Shaking his head) You off your mind tonight. (to the old man) You're an expensive ou drol. Fugard *Boesman & Lena* 1969

dronkie [ˈdrɔŋkĭ] *n. pl.* -s. *slang.* 'Drunkie', boozer, 'alkie', *cf. Austral. bot, metho* etc. [*Afk. dronk* drunk + *personif. suffix -ie* (q.v.)]
Like worms, Babalas as the day you were born. That piece of ground was rotten with dronkies . . . falling over each other. Fugard *Boesman & Lena* 1969

dronklap [ˈdrɔŋkˌlap] *n. pl.* -pe. *slang* A drunkard. [*Afk. dronk cogn.* drunk, *lap* a clout]
'So you're a dronklap from East London,' he said sourly. 'Perhaps you'd better go back . . . We have enough hobos down here.' Muller *Whitey* 1977

droog [ˈdruəx] *predic. adj.* Dry: found in place name Droogas: more freq. in *attrib. inflect. droë* (q.v.) e.g. Droëbult, Droërivier, Droëvlakte. [*Afk. fr. Du. droog cogn.* dry]

droog-my-keel [ˈdruəxˌmeɪˈkɪəl] *n.* Any of

several astringent species including *Rhoicissus tridentata, Cyphostemma cirrhossum* and *Scutia myrtina* or their fruits, (see *drogies*) used by early colonists in throat infections, and *R. tridentata* in the treatment of diarrhoea and dysentery. [*Afk. droog cogn.* dry + *my* + *keel* throat]

. . . the little valleys were heavy with mimosa and Kei apple . . . and droog-mij-keel bushes. Scholefield *Wild Dog Running* 1972

dropper ['drɒpə] *n. pl.* -s. *Also Austral.* A slender pole of wood or metal used in fencing: unlike a post it is not planted: several ~*s* are placed between each post. In combination *fencing* ~. ▮'A *dropper* is a batten stapled to fence wires to keep them apart.' Sidney J. Baker *The Australian Language* 1966: see at *concertina gate.*

. . . What kind of fence . . . Is it the steel posts with anchoring wires that you cut? Or will it have standards that you pull out and bend the fence down by the droppers for the cattle to walk over. Bosman *Fekkersdal Marathon* 1971
   Resale of farming requirements . . . standards, droppers, . . . Creosoted non-bending wattle droppers. Standard length about 5cm thick. *Farmer's Weekly* 21.4.72
   . . . her father had a row with the German farmer . . . hitting the farmer with a dropper from the fence he was putting up at the time. Stander *Flight from Hunter* 1977

drostdy [ˌdrɔ̃s(t)'deɪ, 'drɒstĭ, 'drɒzdĭ] *n. pl.* -e, -ies. **1.** (*hist.*). The area of the *landdrost's* (q.v.) jurisdiction, the magisterial district. [*Du. drostdy fr. drost* a bailiff]

District of George. This district, situated on the south eastern coast of the Colony of the Cape of Good Hope, was separated from the district of Swellendam in the year 1811, and erected into a Drostdy, under Lord Caledon's government. *Greig's Almanac* 1831

2. The headquarters, office and home of the *landdrost* (q.v.), the residency, also ~ *house* [*fr.* (1) ~]

At the distance of half an hour's walk northwards from the village, is the Drostdy, or official residence of the landdrost, a modern erection, surrounded by the dwellings of the secretary and subordinate civil officers. Burchell *Travels I* 1822
   . . . has a neat church which will contain about 400 persons, the drostdy-house and several other public buildings. *Cape of G.H. Almanac* 1843

3. Drostdy: found in names of places in the neighbourhood or on the site of an extant or former ~ e.g. Drostdy-

straat, Drostdy Museum, Drostdy Arch. [*fr.* (2) ~]

druk [drœk, drək] *vb. usu. pass. slang* To stab: *to get* ~ *ed* to be stabbed in a fight.

I always said there were only three ways for . . . to get into trouble. The first was to get drukked (stabbed) in a bar, the second was a car prang. *Sunday Tribune* 16.7.72

dry diggings *pl. n.* Diggings producing non-alluvial diamonds: the dry process involving not washing as in river diggings, but exposure of the diamondiferous matrix to the weather. *cf. Canad. dry diggings* (gold).

The dry diggings are thirty miles to the southeast of Pniel, they are so called because the gems are not found in river-wash, but in dry tufa, which has apparently never been in contact with water. Lacy *Pictures of Travel* 1899 *cit.* Pettman

dryland *n. pl.* lands. That arable part of an *irrigation farm* (q.v.) not under irrigation; also, as modifier in ~ *farm/ing*, land or cultivation without irrigation. *cf. Canad.* and *Austral. dry farming.* See also second quot. at *rolbos.* [dry + land *fr. Afk. land* (q.v.) cultivated field, arable area]

Wheat plays an important role . . . and this year there is 51ha under irrigation and 171ha dryland. Zambesi No. 2 and Inia 66 are the irrigation varieties with Scheepers 69 on the dryland. *Farmer's Weekly* 4.7.73

ds. See *dominee* and quot. at *Boerechurch* [*abbr. dominus*]

DTA ['diˌti'eɪ] *acronym.* Democratic Turnhalle Alliance, South West Africa: See *Turnhalle.*

Tough DTA motion on report expected.

WINDHOEK – The Democratic Turnhalle Alliance is expected to table a tough motion in the South West African constituent assembly body rejecting key aspects of the report by the United Nations Secretary. *E.P. Herald* 5.3.79

dubbeltj:e ['dœbəlkĭ, -cĭ, də-] *n. pl.* -s
1. *hist.* The British twopenny piece introduced with the British occupation of the Cape in 1795. See also *rixdollar, skilling, schelling, stuiwer.* ▮The Zu. word for penny *indiblishi* (Xhosa for farthing) is *deriv. fr.* ~ [*Du. dubbel cogn.* double + *dimin. suffix -tjie*]

With the first British occupation of the Cape in 1795 two coins were introduced to help relieve the shortage of small money. One was the 'Dubbeltjie' – the British twopenny piece, also known as the 'Cartwheel'. It got its name because it was declared current at two stuiwers.

73

Daily Dispatch 5.4.71

2. Also *devil's thorn, duiweltjie, duwweltjie:* name applied to several species, more particularly to their sharp thorny fruits: *Emex Australis (Rumex spinosus),* and varieties of *Tribulus* esp. *T. terrestris* which causes Tribulosis, *dikkop* in sheep which browse upon it (see (4) *dikkop).* [*etym. dub. prob. fr. the two-four construction of the thorns; dubbel,* double: or *corruption of duiweltjie,* little devil]

Rumex spinosus . . . Dubbeltjies, very painful to the slave who has no shoes. Mackrill MS. *Diary* 1809

The soil was also sprinkled with the seed of a plant covered with prickles, making it very unpleasant to sit or lie down. These seeds are jocularly called by the colonists dubbeltjies (two-penny pieces). Thompson *Travels I* 1827

duh [dʒ] *n.* and *adj. slang* esp. among children: Stupid, a stupid person, derogatory. [*unknown poss. onomat.*]

. . . Bok-Bok may come over a bit sleepy, even duh sometimes . . . but when he gets to thinking, you'd be surprised what he comes up with. *Darling* 15.9.76.

duiker ['dœïkə(r), 'daɪk-] *n. pl.* Ø, -s.
1. *Sylvicapra grimmia,* or any of several other small species of S. Afr. antelope of the genus *Cephalophus* sometimes called *'diving goat'* (q.v.) [*Afk. duik fr. Du. duiken* to dive + *agent. suffix -er*]

One afternoon I went out to shoot duikerbok, (a brown deer with small upright horns, and its height two feet, named duiker from its diving under the bushes). Alexander *Western Africa I* 1837

2. Several types of cormorant (Phalacrocoracidae fam.) are so named because they feed by diving for fish. [*as at* (1) ~]

When the snoek were in, experienced old fishermen knew just where to find them, even without the aid of the gulls and duikers which share the small fish with the snoek. *Cape Times* 18.3.36

duin [dœïn] *n. pl.* -e. Sand dune: found in S. Afr. place names e.g. Duinefontein, De Duins: also a specific coastal area, see quot. at *sour fig* [*Afk. fr. Du. duin cogn.* dune]

duineveld ['dœïnə₁felt] *n.* Sandy dune lands usu. not far from the sea. *cf. Strandveld:* see first quot. at *sour fig.* [*Afk. duin cogn.* dune + *veld* country *cogn.* field]

Duitse sis ['dœïtsə 'sɪs] *n.* See *German*

print, voersis, and quot. at *sis*[2]. ⟨Most commonly known as *kaffir print* (q.v. [*Afk. Duits(e)* German + *sis fr. Du. sits cogn.* chintz]

duiwel(s) ['dœïvəl] *n.* Devil('s): found in S. Afr. place names e.g. Duiwelskloof, Duiwelspiek also in combinations: see below. [*Afk. duiwel cogn.* devil]

duiwelssnuifdoos ['dœïvəl'snœïf₁dʊəs] *n.* See *devil's snuff box.* [*Afk. snuif cogn.* snuff + *doos* box, container]

duiweltjie ['dœïvəlkï, -cï] See *dubbeltjie*

duiwelsdrek ['dœïvəlz₁drek] *n. Asafoetida,* one of the (*Old*) *Dutch Medicines* (q.v.) a plant resin, so called on account of its unpleasant smell: also known as *Satan's Dung* (q.v.) used for various medicinal purposes. ⟨Also used by S. Afr. Indians among whom it is known as *hing.* [*Afk. duiwels cogn.* devil's + *drek* excrement]

. . . he never regained consciousness after being fed the 'pill' made of curry powder, tumeric, garlic, pea powder, epsom salts and 'duiwelsdrek' during a ceremony at the . . . men's hostel on August 2, 1974. *E.P. Herald* 12.2.1976

dung floor *n. pl.* -s. See *misvloer,* also quot. at *pepper tree* [*trans. Afk. misvloer*]

Yesterday . . . finishing touches were being done to the two new dung-floor classrooms. *E.P. Herald* 22.4.74

dunkrimpsiekte ['dœn₁krɪmp₁sïktə] See *krimpsiekte*

dunsiekte ['dœn₁sïktə] *n. Seneciosis* in horses: a stock disease also variously known as Molteno disease, straining disease of cattle, stomach staggers, enzootic liver cirrhosis etc. according as it manifests in stock. ⟨It is caused by acute or sub-acute poisoning from eating any of several toxic species of *Senecio* esp. *S. retrorsus* known as ~ *bossie.* [*Afk. fr. Du. dun* thin + *ziekte* disease]

Chronic Senecio poisoning in horses is known as 'dunsiekte' or 'enzootic liver cirrhosis.' *H'Book for Farmers* 1937

dur see *duh*

Durban Beach *n. phr. Sect. Army: D.B.* coded acronym for Detention Barracks: similarly *Meat Pie, Military Police.*

If he gets rowdy and the Meat Pie get him, he's for Durban Beach no kidding. *O.I. Serviceman* Feb. 1979

Durban Poison *n.* Also *D.P.:* a particularly potent type of *dagga* (q.v.) cultivated in sub-tropical Natal. [*fr. place name* Durban]

Ask someone from overseas why South Africa is famous . . . a few . . . will grin and say: 'Durban Poison, Man.' Why? Because D.P. dagga, is just about the best high in the world, fit to rank alongside California Gold, Acapulco Gold, Zanzibar Green and Malawi Red in the dope smoker's list of all-time greats. *Darling* 8.11.78

Durbs [dɜbz] *n. prop. colloq.* Popular *abbr.* for Durban: see quot. at *ek sê.*

'What's it like up there?' Like Durbs at Christmas time without the sea' . . . 'Or the girls . . . just sand, thorn-bushes and *sun.*' *Darling* 7.2.79

dust devil *n. pl.* -s. A minor whirlwind visible crossing the veld like a column of dust. *cf. Austral. Darling shower*; *Cobar shower* etc. dust storm: see also quot. at *klompie.* [*unknown*]

Here we have only Karoo bushes . . . and the dustdevils making long curly chimneys in the sky when it is hot. Vaughan *Diary* circa 1902

A dust-devil went pirouetting in front of me, a whirling little dancing-dervish of a thing made up of scraps of dead grass and sand and other of nature's unconsidered trifles, caught up and whirled along by the wind. Reed *South of Suez* 1950

Dutch *pl. n.* **1.** The first colonists under the *Dutch East India Company* and their descendants.

Whenever mention is made of the Dutch in a more general sense, that part of the population of Cape Town, or of the colony, not English, is intended: since by far the greater proportion belongs to that nation; and all those who are born in the colony speak that language, and call themselves Africaanders, whether of Dutch, German, or French origin. Burchell *Travels I* 1822

2. *pl. n.* and *n. modifier.* The *Afrikaners* (q.v.)

Then Rhodes was anxious to link himself in the most obvious way with the Dutch . . . For the Dutch support, however, he had to work. There was hardly anything Rhodes did in Parliament which had not as its object the favour of the Dutch. He wanted Union. Millin *Rhodes* 1923

3. *n. prop.* The language of the (1) ∼ : also the dialect of Dutch from which *Afrikaans* (q.v.) is evolved. ❦The spoken language of the Cape-born ∼ differed from the ∼ of Holland. Also used to refer to *Afrikaans* sometimes in a derogatory sense: see third quot., *Taal* and *Afriks.*

Dutch was to be the official language, and no attempt would be made to suppress English, which would be allowed to be used in the law

courts and would be taught in the Schools, if so desired. O'Connor *Afrikander Rebellion* 1915

The belief that Colonials, English and Dutch speaking alike, abused their power over the natives was held almost as an article of faith by most people in England who took any interest in South African affairs. Walker *Lord de Villiers* 1925

In 1902 Genl. Hertzog stood in the Bloemfontein Post Office when someone walked in to send off a telegram. The Postmaster looked at the telegram and handed it back saying 'I do not dispatch telegrams in dirty Dutch.' *Daily Dispatch* 11.5.71

In combination *Cape* ∼ (q.v.) *kitchen* ∼ (q.v.) and *simplified* ∼ a form of the Dutch language with a simplified spelling system invented or developed for S. Afr. use with the intention of ousting Afrikaans (the Taal): see quot.

. . . a new language, which he called '*simplified Dutch*' was invented for the purpose (the Taal is not capable of being used as a medium of instruction beyond very elementary stages and the Dutch of Holland is NEITHER UNDERSTOOD NOR WANTED in South Africa). *E.P Herald* 19.5.1910

4. *adj.* Of or pertaining to the ∼ people or their language views etc. freq. in combination ∼ *boer:* see also quot. at *backveld.*

There was only three left in the hospital hut, namely myself, a blind man and a Dutch Burgher who had lost one arm. Buck Adams *Narrative* 1884

I fear that the proposed Constitution will not give satisfaction either to the English or to the Dutch party. *Letter* Lord de Villiers 12.12.1879 *cit.* Walker 1925

Dutch, kitchen *n. prop.* **1.** *colloq.* The form of *Afrikaans* (q.v.) spoken by servants or uneducated persons *cf. Canad. monkey French.* See also *kitchen kaffir, kitchen-* (*prefix*).

In the early days Afrikaans, as far as the written word was concerned, was in embryo. The coloured people, who spoke their version of it which we referred to as 'Kitchen Dutch', spoke English to us in our homes. Henshilwood *Cape Childhood* 1972

2. An offensive misnomer for *Cape Dutch* (q.v.) or *Afrikaans* (q.v.).

Dutch East India Company *n. prop.* The powerful company under the auspices and control of which the first settlement was made at the Cape (1652) as halfway house for victualling their ships: see quots. at *Company* and *Slave lodge.* ❦Referred to as *The Company* (q.v.) and jocularly personified as *John* or *Jan*

Company/Compagnie: see *Company,* also *V.O.C.* and *Here (Heeren) Seventien*

Dutchman *n. pl.* -men. *colloq.* Mode of reference to an *Afrikaner* (q.v.) sometimes derogatory.

Everyone knows that 'Kaffir' is regarded as an insult in South Africa. It is in the same category as 'hairy back', 'dutchmen' or 'coolies'. . . MP for Pinetown said. *E.P. Herald* 11.9.75

Dutch medicines *pl. n.* A series of household remedies much used and favoured in the country districts under this name: see also *huisapteek, borsdruppels, boegoe* (see *buchu*), *doepa, doepa olie, duiwelsdrek,* and quots. at *Old Dutch medicines* and *benoudheid.*

Juritz, Carel Friedrich, apothecary, chemist, and druggist depot of the patent Dutch medicines from the orphan house of Halle, 29 Loop Street. *Cape of G.H. Almanac* 1843
For 100 years Lennon's Dutch medicines have helped people to feel better. *Drum Advt.* Nov 1964

Dutch Reformed Church *n. prop.* Also *D.R.C.* (q.v.) and the *Kerk* (q.v.). The Calvinist Church to which the majority of *Afrikaans* (q.v.) speaking South Africans belong. [*trans. Nederduits Gereformeerde Kerk (N.G.K.)* (q.v.)]

Mr. Murray . . . is of the Church of Scotland, which in doctrine and discipline corresponds almost entirely with the Dutch Reformed communion. Thompson *Travels I* 1827
The whole district are members of the Dutch Reformed Church. There are three places of Worship in the district. *Greig's Almanac* 1831

duwweltjie ['dəvəlki, -ci] *n. pl.* -s. See (2) *dubbeltjie.* [*Afk. variant of dubbeltjie or corruption of duiweltjie,* little devil]

. . . a secondary plant growth (opslag) amongst which the few grass species and especially the well-known and widely spread duwweltjie (Tribulus sp.) can be included. *H'Eook for Farmers* 1937

dwaal [dwɑːl] *n.* and *vb.* **1.** *colloq. n.* A daze: to be in a ~, disoriented, not 'registering' hence *deriv. dwalie,* a nickname for a slow-witted person, or as *predic. adj.* 'she is terribly *dwalie* these days'. *cf. Jam. Eng. stray-minded.* [*fr.* ~ *vb.*]

Not lost? What way takes you to Berry's Corner twice, then back to where you started from? . . . The roads are crooked enough without you also being in a dwaal. Fugard *Eoesman & Lena* 1969
Tony's father . . . said: Tony phoned me at work . . . I was in such a 'dwaal' I thought it was my other son. *E.P Herald* 16.7.75

2. *vb.* To wander aimlessly, ~ around, to be lost: found in place names Dwaal and Dwaalboom; *ver*~ to stray: found in place name Verdwaalkloof. [*Afk. dwaal vb.* wander, roam]

dwaas [dwɑs] *n. pl.* (-e). *slang.* Fool, clot, idiot: see quot. at *gek.* [*Afk.fr. Du.* fool, silly fellow *cogn. O.E. dwæs* foolish]

Of course I recognise you; I'm not such a dwaas as that – even if you have cut your hair! *O.I.* 12.2.79
derivatively ~*heid* foolishness, stupidity.

Violence from the black man would not only be stupidity (dwaasheid) but ingratitude. *Sunday Times* 15.4.79

dwars [dwa(r)s] *adj. prefix.* Crosswise, diagonal: found in S. Afr. place names e.g. Dwarskloof, Dwarsberg, Dwarskersbos. [*Afk. dwars* transverse *cogn.* (a)thwart]

dwarswal ['dwa(r)s₁val] *n.* A transverse embankment, on the side of a road for drainage to divert stormwater, or in the veld to prevent erosion. ⊓Formerly a low bank or hump across a gravel road to divert storm water before the use of underground drainage pipes. [*Afk. dwars* diagonal + *wal* bank]

E

eaten out *modifier.* Of or pertaining to a *camp* (q.v.) or other area of (2) *veld* (q.v.) which has been over-grazed or overstocked: see also *tramped out, beaten out.* [*poss. trans. Afk. uitvreet* eat away, corrode]

The inspection party found it [the outspan] was badly eaten out, and councillors decided . . . to ask the Soil Conservation Department what it would cost to have the outspan planned. *Grocott's Mail* 27.7.71

eating house *n. pl.* -es. **1.** Formerly a place of refreshment, restaurant. [*presum. trans. Du. eethuis* restaurant]

Wardmasters shall take care, that no taps or Eating houses exist in their wards, without having proper and distinct signs affixed . . that there are no gambling nor any other houses, inconsistent with morality and good order. *Afr. Court Calendar* 1819
. . . getting the train across the bridge took well over three-quarters of an hour. There was an eating house on one side that did very well out of the wooden bridge. *E.P. Herald* 5.4.74
The mice who nest in the Piccadilly Line of the underground rail system come from their

hideouts to banquet in restaurants and cafes in the area. They are too cheeky and numerous for the army of cats kept specially by the eating houses. *E.P. Herald* 11.10.76.
2. Latterly but obsolescent, a place for non-whites to obtain meals e.g. *kaffir ~;* see (6) *kaffir.* [*prob. fr.* (1) *~*]
    Kaffir labourers appeared . . . on the charge of assaulting Wynne Guberu, an eating-house keeper. *E.P. Herald* 21.10.1911
    Regulations governing eating houses in the Coloured area of the Municipality of Bloemfontein, are published in the 'Official Gazette' of August 22. *The Friend* 25.8.30

eat-up *vb. trns. prob. obs.* To take revenge upon an enemy by taking possession of all his property and cattle. [*trans. Ngu. ukuDla* to eat]
    The Chief's cattle-folds are replenished from time to time by fines and occasional 'eating up' of delinquents, by which is meant the confiscation of the whole of their property, for alleged witchcraft, treason, or other great political crimes. Shaw *My Mission* 1860
    . . . no longer shall the witch-finder hunt you out so that ye shall be slain without a cause. No man shall die save he who offends against the laws. The 'eating up' of your kraals shall cease. Haggard *Solomon's Mines* 1886.

-ed, omission of ╟Dropping of the alveolar suffix *-ed* is not confined to S.A.E., but is very prevalent in S.A., e.g. *barb wire, three bedroom house, pickle onion, old fashion home, long sleeve shirt* etc.
    . . . half a cabbage with bits of cheese and red and green pickle onions stuck all over. *Darling* 24.12.75

eendrag maak mag [ˈɪənˌdrax ˌmɑk ˈmax] *idiom.* 'Union is strength' motto on the coat-of-arms of S.A. ╟The motto of the Republic of Transkei *Imbumba Yamanyama* is similar in meaning [*Afk. eendrag* union, unity + *maak* makes + *mag* strength, force *cogn. Ger. Mach*]
    The ideal of 'Eendracht maakt macht' which was the inspiration of the architects of Union in 1910 has . . . become obscured during recent years. *cit. S.A. Road Ahead ed.* Spottiswoode 1960

Egoli [eˈgɔlĭ] *n. prop. Zu.* name for Johannesburg –_ the City of Gold: see also quot. at *bombella train.* [*Zu. eGoli fr. Eng.* gold *e-* locative prefix]
    . . . Egoli is as green as Natal and twice as exciting . . . For a girl used to the deafening silence of the country and bed at nine with a good book, Egoli is a heady champagne. *Sunday Times* 9.12.73
    When he was 18, the call of Egoli gripped young Tshabalala, so he came to Johannesburg.

Drum Aug. 1971
egte [ˈextə] *attrib. adj. colloq.* Genuine: often in combination an *~ boer;* see also *ware. cf. Austral. dinkum, dinki-di, Anglo-Ind. pukka.* [*Afk. eg fr. Du. echt* real + *attrib. inflect. -te*]
Eid see *Labarang*
eiken [ˈeɪkən] *pl. n.* Oaks: found in S. Afr. place names prefixed to other *n.* e.g. Eikendal, Eikenhof. [*Du. eik* oak + *pl. suffix -en*]
eina [ˈeɪna] *interj.* An exclamation of pain: occ. used as *n.* See first quot. and second quot. at *sis¹* [*Hott. é + ná* exclamation of pain, or surprise]
    First aid for cuts – without the 'eina'. Sometimes treating your child causes more fuss than the actual wound. Sting! Burn! Ouch! Eina! *Personality* 24.9.71
    Eina, that stung! *Sunday Times* 7.10.73
eis, on my, your, etc. [eɪs] *adv. m. or predic. phr.* See *ace, on my, your* etc. (*Austral.*) [*translit. ace or fr. Afk. eie* own]
    . . . groovy okies . . . you can connect with if you rock along on your eis. Not that I ever get on my eis much . . . ou Blossom isn't short of what it takes. *Darling* 9.4.75
ek sê [ˌek ˈse:] *interj.* lit. 'I say': used to give emphasis to a statement, also as an exclamation of surprise: see also *telling you, I'm.* [*abbr. Afk. ek sê vir jou,* I'm telling you, *or as above*]
    . . . a great deal of South African. Lekker hey! And not only the girls . . . The biggest agency in London, the BIGGEST ek sê – is run by a cat from 'Durbs. *Sunday Times* 24.4.77
elachi See *Indian terms.*
eland¹ [ˈilənd, ˈɪəˌlant] *n. pl.* ∅. Elk. [*Afk. fr. Du. eland fr. Ger. Elentier* elk, moose]
1. *Taurotragus oryx,* the largest of the African antelopes believed to have been called by the Hottentots *kanna* (q.v.).
    . . . the Regulation which gives protection to the Hippopotamus and Bontebuck (that is, the 5th article of the proclamation of the 21st of March, of the year 1822), shall be henceforward also made applicable to the beautiful and scarce species of Deer called Eland, found in the George district. *Cape of G.H. Statutes* 1823
2. Found in place names e.g. Elandsputte, Elandsbaai, Elandskloof, Elandsdrif, Elands Height.
3. Prefixed to several plant names e.g. *~sboontjie* (bean), *~sdoring* (thorn), *~sertjie* (pea) etc. and in form *~veld* see *bontveld.*
eland² *Sect. Army.* An armoured vehicle carrying either a 60 or 90 mm gun:

~ 90, ~ 60:

Eland 90 armoured cars with turret mounted medium gun and light machine-gun . . . crew consists of a crew commander, a gunner and a a driver. Common to all armoured units is a tremendous sense of pride . . . for not only their vehicles but also their colleagues . . . they wear their black berets with pride. Marks *Our S Afr. Army Today* 1977

At the other end of the operational area the local population and the S.A. soldiers are guarded by members of 2 SSB with their Eland 90 armoured vehicles. . . . The commander of an Eland-type armoured car keeps a constant vigil from behind the menacing barrel of a 90 mm gun. *Paratus* Jan 1979

elephants' food *n. Portulacaria afra :* see *spekboom.*

A peculiar tree grows here in abundance, called 'speckboom' by the Dutch, known . . . as 'elephant's food'. The leaves are small, thick and juicy, and very sour. The elephants are fond of it as a food. Larson *Talbots (et al)* 1943

elephant's foot *n.* A plant with a large tortoise-shaped woody tuber above ground, *Dioscorea elephantipes* and *D. sylvatica,* also *olifantsvoet* and *Hottentot bread.* [*prob. trans. olifantsvoet*]

I found many curious plants among which was one called Elephant's foot. Patterson *Narrative* 1789 cit. Pettman

Patterson who first recorded the plant states that its large solid bulb (sic!) was eaten by the Hottentots and that they regarded it as very salubrious, whence called Hottentotsbrood. Smith *S. Afr. Plants* 1966

elephant's trunk *n.* See *halfmens.*

elf(t) [elf(t)] *n. pl.* -s. The edible marine fish *Pomatomus saltatrix* or shad, also known as *skipjack,* caught in large numbers in S. Afr. waters sometimes by means of *trek nets* (q.v.) [*Afk. fr. Du. elft* shad]

Another annual visitor that comes in shoals is the elf. It is rather less delicate than the harder but still makes an admirable contribution to the table. *Farmer's Weekly* 18.4.73

endorse out *vb. phr. usu. pass. To be ~ d out,* to be officially ordered to leave an urban area on account of lacking correct endorsements in a *reference book* (q.v.).

East London is the dumping ground for many thousands of blacks endorsed out of the Western Cape, yet there are not even enough jobs for those who were here in the first place. *Daily Dispatch* 12.3.73

(e)nkosi [(e)n'kɔsĭ] *interj.* Thank you, thanks: usu. *Afr. E.* but see first quot. [*Ngu. -nkosi* lord]

The word has, however, in the form *enkosi* come to be regarded, and is often used by colonists, as being the equivalent of the English 'thank you'. Pettman *Africanderisms* 1913

E, nkosi! or the simple vocative *nkosi!* is used . . . as the English 'thanks', to express gratitude to a giver by saying *uyinkosi,* you are a Lord. Kropf *Kaffir-English Dictionary* 1899 (1915)

erf [ɜf, erf, ɛəf] *pl.* -ven. An urban building lot, or *stand* (q.v.) of indefinite size, not a *farm* (q.v.) [*Afk. fr. Du. erf* inheritance, piece of land *cogn. Ger. Erbe*]

The erven or building lots, of which the village of Swellendam is composed, were first granted in 1750. *Greig's Almanac* 1831

The sale of the first two erven will commence on Erf No. 3344. That is the erf with all the stones. From there the Auctioneer will proceed to the top end of Amatola Row where another five erven will be sold. *Daily Dispatch Advt.* 5.8.71

🖙Certain towns which have public irrigation *furrows* (q.v.) distinguish between water *erven* and dry *erven.*

There are big houses, with gardens they are calling water erfs becos they only ones getting water in a furrow from the big dam at top [*sic*] of the street. Vaughan *Diary circa* 1902

A piece of ground marked off in a village or town for garden or building purposes. These erven may be either water- or dry-erven, as they carry the right to water for irrigation or not. Pettman *Africanderisms* 1913

erf *vb.* To inherit, succeed to (rare): see *erf.*

Spineless cactus can be a useful adjunct to grazing. . . . usually the farmer gets a patch going with leaves 'erfed' from a neighbour and when the stand has grown, the cows are invited in. *Farmer's Weekly* 21.3.79

erfpacht ['erf₁paxt] *n.* Roman Dutch legal term for *quitrent* (q.v.). [*Du. erf* inheritance, piece of land *pacht cogn.* pack]

It [the Company] doubled the rents of the loan farms and offered small farms adjoining them on *erfpacht* – this is a lease for fifteen years at a low rental with the promise of compensation for improvements on resumption. Walker *History of S.A.* 1928

esel ['ɪəsəl] *n.* Ass, donkey (poss. wild) found in S. Afr. place names, Eseljagpoort, Eseljagrivier, Eseljacht; also prefixed to several plant names *~bossie* (bush), *~oor* (ear), *~kos* (food). [*Afk. fr. Du. ezel cogn.* ass]

ESSA ['esə] *n. pl.* -s. *acronym* English *S*peaking *S*outh *A*frican of any race: see also quot. at *ASSA* and note p.000.

Professor Guy Butler . . . caused waves in the

pool of white South Africa when he said English-speaking South Africans (ESSAs) were becoming disenchanted with Afrikaans-speaking South Africans (ASSAs). What he ignored were the products of mixed marriages, And that my gabbas, is where there's the rub. *Sunday Times* 8.10.78
There is a fourth story which should . . . be included lest some should leave . . . feeling that wide though the definition of ESSA's is, it appears to include only whites. Francis Wilson *cit.* De Villiers (ed) *English Speaking S.A. Today* 1976

Ethiopia, Order of *n. prop.* An African separatist episcopal Church, founded in 1892, the clergy of which are ordained by Anglican Bishops but which is otherwise autonomous. See new quots.*
. . . a church service led by the Rev. E. M. Hopa, of the Order of Ethiopia, who blessed the leopard skin, assegaai and kierie of the new chief. *E.P. Herald* 16.12.74
. . . the new Bishop of Grahamstown performed his first ordination . . . on Sunday . . . The Rev. – was ordained to the priesthood and is to be placed in a parish by the Order of Ethiopia. *Grocott's Mail* 20.12.74
*Delegates discuss election of Order of Ethiopia Bishop.
Top Clergymen and delegates of the Church of the Province of South Africa and Order of Ethiopia have been unable to elect the first bishop of the Order of Ethiopia . . . The representatives included Canon M. Hopa, Provincial of the Order of Ethiopia. *Indaba* 9.2.79
*. . . a walk out at the synod last week . . . followed a heated debate over . . . the consecration of a bishop for the black-dominated order of Ethiopia, which falls under the authority of the church of the Province. Fears were expressed that the order might break away once it had its own bishop. *E.P. Herald* 18.6.79

Ethiopian *adj.* Of or pertaining to those African independent churches of the group called 'Ethiopians' as opposed to the *Zionists* (q.v.): see also *Ethiopia, Order of.*
These churches are without exception under African control with an all-African membership, and they can be divided into two categories: the 'Ethiopians' and the 'Zionists'. The Ethiopian churches are those that have remained very close in form and belief to the mission churches from which they seceded. The Zionist group is very different. This comprises mainly very small churches of a revivalist, pentecostal nature which combine African and Christian beliefs. West & Morris *Abantu* 1976

euphonic concord *n. phr.* The rule discovered, and so designated, by the Wesleyan missionary William Binnington Boyce, which governs the structure not only of Xhosa (then called (2) *Kaffir* (q.v.)) but the family of Bantu languages in general. ⫽ This was expounded in Boyce's *Kaffir Grammar* printed in Grahamstown in 1834 on the Wesleyan mission press.
. . . what Boyce discovered was that in the Kafir sentences the noun was the governing element and that all the other parts of speech were thrown immediately into an alliterative or euphonic concord with the subject noun.' The foundation of all future study of the Bantu family of languages was well and truly laid. Simple illustrations are; . . .
Baza babendula bonke abantu bati: (Then answered all the people and said,) *cit.* Mears Introd. (1971) to W. B. Boyce *Notes on South African Affairs* 1838

European *n. prop. pl.* -s. A white-skinned person i.e. one descended from the Caucasian races, is so called in S.A. without regard for any actual ties with Europe: now largely replaced by *White* (q.v.): *non-* ~ any person of colour whether of African, Asian or mixed ancestry: see also *Coloured, Black.*
The Bechuanas are described as a very interesting people, mild in their manners, naturally good humoured, with cheerful dispositions readily conforming to European habits, and very willingly exchanging their skins for English dress. Philipps *Albany & Cafferland* 1827
We sit in Africa and we are not Africans. We go to Holland, to France, and we suddenly realise that they lied to us. We are not Europeans. We go to England and we find out that we are Boers who try to live like the English here under the Southern Cross. *Evening Post* 17.2.73
. . . my wife . . . walked through the door marked 'non-European' at Jan Smuts Airport. 'But you're White!' I hissed. 'I'm not European!' she hissed back. 'I'm American.' *Natal Mercury* 8.4.78

everlasting(s) *n. usu. pl.* The papery, highly coloured blooms of several species of *Helichrysum* and *Helipterum :* so called on account of their property of lasting many years, seven or more, from which they gain the name *sewejaartjies* (q.v.).
The everlasting.flower . . . derives its name from appearing as fresh and in as high preservation after being seven years pulled as the day when it was first torn from the stalk . . . When first plucked it feels like an artificial flower of painted paper; indeed it is much more like an artificial than a natural one. Percival *Account of Cape of G.H.* 1804

excuse me *vb. phr. substandard.* Equiv. of 'I beg your pardon' freq. sig. 'I didn't

79

hear' or 'I didn't understand': see quot. at *niggie*. [*translit. Afk. ekskuus* I beg your pardon]

eye *n. pl.* -s. Fountainhead, source of a spring or river: see also *bron*. Place name in trans. Die Oog (the ~): see *oog*. [*trans. Afk. oog* fountainhead, source]
... the largest single 'eye' in the Kaokoveld a magnificent spring with a huge pool below. Green *Last Frontier* 1952
The springs deliver three-quarters of a million gallons of effervescent mineral water a day at 94 deg. F. There are five open-air pools and one totally enclosed pool over the eye of the main spring. *Daily Dispatch* 31.8.71

exemption, conditional *n. pl.* -s. See *conditional exemption* also *matriculation*.

exit permit *n. pl.* -s. A permit to leave the country without the right to return: granted to certain persons who, for political reasons, do not possess a valid passport.
Many of them were forced to leave on *exit permits*, thus being deprived of the right to come back to the land of their birth and enjoy the kinship of their loved ones. *Drum* Sept. 1968

ezel See *esel*.

F

faction fight *n. pl.* -s. A fight between rival African groups or factions usu. in tribal conditions or areas, but also in pursuit of location or clan vendettas among mineworkers or other urban Africans: see quot. at *hut*.
Yesterday afternoon a faction fight took place between two large bodies of Kafirs outside the town. *E.P. Herald* 10.4.1906
Seven killed in Transkei faction fight. Seven tribesmen were killed, one injured and 159 huts burned down in a faction fight in the Libode district. *Daily Dispatch* 11.5.72

faggot *n. pl.* -s. Small-sized extra-hard brick: see *klinker, klompie*. [*unknown*]
... Also faggots (stone walling) for fireplaces and braais. *Daily Dispatch* 14.7.71

fah-fee ['fa¹fi:] *n.* Also fa-fi: an illegal Chinese gambling game: a modified form of roulette with thirty-six named numbers, much played by urban Blacks. ⟟The banker who '*pulls* ~' is usually a Chinese, and the bets are collected from punters by ~ runners. [*unknown: prob. Chinese, poss. Zu. fr. fa* death, to die + *fi* dead person.]

I have heard some Zulu-speaking Fa-fi gamblers analysing the word as if it is of Zulu origin ... A Fah fee runner may travel from one place to another, from one home to another the whole day. Although Fah-Fee is illegal the Chinese bankers manage to get into the Townships ... each day ... They leave their cars parked some distance from the 'pulling house' in order to avoid arrest. Longmore cit. *S. Afr. Journ. of Science* Vol. 52 No. 12 July 1956
One of the many interesting assignments ... was one against a tough Indian thug named China who ran the biggest fah fee racket in Durban. *Drum* 8.10.72
How often have you heard that the fahfee man has 'pulled' number so-and-so because this is the Year of the Tiger? *Voice* 4 – 10.3.79

fair *n. pl.* -s. *hist.* A Government-sanctioned occasion for trade on the frontiers to permit occasional traffic, otherwise forbidden, between Colonists and the native tribes: see also *kaffir fair*.
A periodical 'fair' was established at Fort Willshire, where the colonial traders by scores, and the kaffirs by hundreds or thousands, met to exchange wares. Dugmore *Remin. Albany Settler* 1871

fall pregnant *vb. phr.* To become pregnant, conceive. ⟟*substandard* but in regular use in the press. [*prob. analg. fall ill or fr. archaic to fall (with child)*]
Since it is women who fall pregnant and give birth, it is time that two contentious issues of contraception and abortion should be decided by women for women. *Sunday Times* 27.5.73

family *pl. n.* Relations in general as opposed to one's immediate family. e.g. 'They have lots of family in the Free State.' ⟟'People' is also occ. used. [*presum. translit. Afk. familie* relations]
Now Britain and Rhodesia aren't even 'family' since the republic was declared, the chances are they never again will be. *Post* 15.3.70
Make sure your ... wine comes from some ... noble estate – preferably one run by your uncle. If you haven't got family in the trade try ... or ... *Darling* 1.10.75

Famo ['fam ͻ] *n. Prob. Reg.* Soweto S. Sotho term for a wild drinking party culminating in a sexual orgy: see quot. at *Russian* (2) [*presum. Sotho*]
There is the Famo, the South Sotho's regular orgy of sex and alcohol ... A Famo usually ends in a chaos of coupling ... probably the most sensual and abandoned of all Soweto parties. Venter *Soweto* 1977

Fanagalo, Fanakalo ['fanaga₁lͻ] *n. prop.* A pidgin language, a mixture of Zulu, *Afrikaans* (q.v.) and English, used and sometimes taught to facilitate communication between different ethnic and racial groups, esp. on the mines: formerly called *mine kaffir;* see also *kitchen kaffir.* cf. *Rhodesian Chilapalapa* (q.v.) *Canad. Chinook Jargon,* simplified trade language used between Indians and Whites. [*Ngu. fana* be like + *ka* possessive 'of' + *lo* this]

A more or less regulaized version of this ('Kitchen Kaffir') called Fanakolo, is used as the lingua franca on the gold-mines. Leo Marquard 1950 *cit. O.E.D. Supplement* 1972

But what is one to do if there are. . . . thousands of Black men working together, and speaking a variety of tongues, and unable to grasp one's instructions, let alone converse among themselves? 'Learn them Fanakalo!' came the reply one day when I put the question to a meeting of workmen . . . Then came the lessons in Fanakalo, and slowly like a stroke of magic the barriers lift. Becker *cit. Star* 17.3.72

As though to rub salt into the wound, the Evangelist Rev. . . . conducted the service in fanagalo, a despised form of pidgin English, which is bad enough anytime but bordered on the profane on such a solemn occasion . . . the memorial service in pidgin Zulu-fanagalo. *Drum* July 1978

Fanakolo . . . has taken to the air in a strangely altered form, heavily salted with English and Afrikaans in a simple non-grammatical 'clingo' (CB lingo) which bids fair to become a cult language that demonstrates the non-racial, all South African quality of the bond among CBers. Molloy *S.Afr. CB Dict.* 1979

fancy *n. pl.* -ies. Diamond trade term not exclusively S.A.E. Diamond of gemstone quality of a colour other than pure or blue white: not '*off colour*' (q.v.): see also quot. at *off-colour.*

'Fancies' are coloured diamonds, produced by eccentric conditions of chemical content or firing in the volcanic pipes. There is a wide range of hues – amber, pink, green, mauve, Fancies are not favoured by the public, and are therefore not greatly sought after by the trade. White *Land God Made* 1969

far *modifier.* Used as modifier of *n.* cf. Biblical 'a ~ country' or phr. the Far East: for place names see *ver-.* [*prob. fr. Afk. adj. ver*]

This isn't a far place. O.I. Child 1969

11 June Uncle Jesse came through in evening he is ploughing in far lands.*Emslie *Diary* 1901
*see *land*

Watch out for highway hypnosis when you have to travel a far distance. *Daily Dispatch* Dec. 1970

farm *n. pl.* -s. In freq. use in S.A.E. sig. a tract of land, worked or unworked, not necessarily having a homestead on it, but under its own title deed. A *grazing* ~, land to which stock (q.v.) may be moved for winter grazing etc. ~, *the, colloq.* sig. usu. the country, or a rustic background as in 'He comes from ~ ', 'She learned to drive on ~ ' etc.: see quot. at *boeke vat,* also *sies tog.*

The farm Platrug was the most unpromising of all for, unlike the other six farms, it had no house on it. Worse still it had no water – no spring, no fontein. McMagh *Dinner of Herbs* 1968

Off to the farm Thousands of jobless in Dar-es-Salaam are being rounded up and sent to rural areas for resettlement on farms. *E.P. Herald* 22.8.74

farm prison *n. pl.* -s. A prison in a farming area, providing convict labour to surrounding farms: see also *prison farm.*

farm school *n. pl.* -s. **1.** A European school for white children situated on a farm but financed by a Provincial Department, providing primary education for the children of neighbouring farms.

Anna was born . . . into a wealthy country family. . . . A private tutor took care of their preliminary schooling. . . . Anna went to a farm school when she was 11. *Fair Lady* 16.2.77
2. An African or Coloured primary school for the children of, usu. resident, farm labourers, of which the building is provided by the farmer and the teacher's salary by Government.

The buildings remain the property of the farmers concerned. The Department runs the schools and appoints the teachers, but the farmers have some say in the appointment of teachers and retain the right to visit the schools periodically. *S.A.I.R.R. Report* 1948–9

Like all those involved in African education . . . is concerned about money, specially . . . textbooks. These, the implements of learning, are what he needs most, even though desks and benches are also short in supply, as in most farm schools. *E.P Herald* 9.11.73

farm with *vb. phr. substandard.* Equiv. of Eng. *vb.* 'to farm' see also quot. at *Woltone.* [*trans. Afk. vb. phr. boer met, lit.* 'farm with]

. . . who farmed with wine grapes in the Cape for six years, believes that the future for wine production under irrigation . . . is very good. *Farmer's Weekly* 4.7.73

The idea of farming with various species of buck is becoming very popular . . . One of the main advantages of farming with buck in preference to livestock is that Springbuck are the only buck affected by ticks. *E.P. Herald* 25.5.74

fatcake *n. pl.* -s. Also *Austral.* See *vetkoek.* [*translit. Afk. vetkoek*]

father *n. pl.* -s. *Afr. E.* Actual father or his brothers, see quot., *big* ~ an elder brother, *small* ~ a younger, similarly *big mother, small mother.*

. . . twice in three years he has asked leave to attend his father's funeral. 'But you have already buried your father . . . How many fathers have you?' 'Six' replies the servant, . . . Bantu, rural and urban irrespective of social strata, usually have more than one father and more than one mother. For in Bantu society, my father together with all his brothers would be my fathers, and my mother together with all her sisters, would be my mothers. So it is quite feasible for the servant to attend the funeral of two of his fathers, or six for that matter within three years. Becker *cit. Lantern* 1966

Fatherland (cattle) *n.* and *n. modifier. obs.* Cattle imported from Holland, and their progeny hence *Bastard* ~, crossbreeds. [*trans. Vaderland(s)*]

. . . bulls have been imported . . . from Holland; these are called as well as their produce, Vaderland or Fatherland, and are certainly the best formed. Philipps *Albany & Cafferland* 1827

Hardy Bastard Fatherland Cattle are bred here; and Plettenberg Bay is celebrated for the excellence of its butter. *Greig's Almanac* 1833

fat lamb *n. pl.* -s. Lamb raised for slaughter, specially fed for quick maturity, hence ~ *production. cf. Austral. 'fats'.*

Here is a breed which has proved itself to be highly suitable for fat lamb or mutton production while at the same time producing a good fleece of wool . . . success of large scale fat lamb production is dependent on quick maturity which can only be profitably achieved by good quality and abundant feed. *Farmer's Weekly* 21.4.72

fat tackies *pl. n. colloq.* figur. Broad radial ply motor tyres: see also *tackie(s).*

He is . . . not above lying. He will announce that he fits Fat Tackies Radials (several times . . .), although everyone can see his car is actually fitted with tatty old cross-plys like everybody else. *Darling* 29.9.76

fat-tailed sheep *n. pl.* Ø. Also *Afric(k)ander, ronderib, blinkhaar, Cape sheep* all (q.v.). The indigenous S. Afr. sheep having hair not wool and a tail of solid fat much prized for culinary purposes.

¶Early travellers marvelled at the breed and the *tail trucks* (see below) on which their fat tails were carried. This breed cannot be crossed with woolled sheep since this practice introduces hair into the wool.

Farmers looked forward to his arrival and the exchange of his rolls of £5 for their attrac-

tive fat-tailed Afrikander sheep. Jackson *Trader on Veld* 1958

. . . we had been told by old Solders and Salers that had been to the Cape of Good Hope that they had seen the sheep climen up they hills with thear tailes maid fast to a little truck with two weals. Goldswain *Chronicle* I 1819–36

I had often laughed when I heard of sheep having trucks behind them to support the tail; some hundreds might be seen here with tail trucks. Buck Adams *Narrative* 1884

fazi [ˈfazĭ, ˈfa-] *n. pl.* aba-, -s. Loosely, an African woman, accurately a married woman only: see *abafazi Afr. E.* ~ *land* (jocular) marriage, the married state. [*Ngu. umFazi* woman, wife]

. . . the local natives not merely Swazies, but Shangaan and Zingeles, talk openly of the coming war . . . They declare the white men to be a lot of old women (umfazies), for they have no guns, and they say the Dutch will not help them. *E.P. Herald* 7.4.1906

Feast of Orange Leaves *n.* See *rampi sny.*

fence creeper *n. pl.* -s *colloq.* An animal usu. a bull which refuses to be confined, and knocks down and walks over fences.

¶ Also used of sheep, goats and pigs, which if ~s, wear triangular yokes made usu. of sticks to prevent their pushing their way through fences.

The municipal bull turned out to be an incurable fence creeper and had to be got rid of. *Grocott's Mail* 1973 (date unknown)

field, the *n. obs.* Travellers' term for the (1) *veld* (q.v.) and (2) *veld* (q.v.).

A man on the death of his wife is considered unclean, and must separate himself from society for two weeks, and fast for some days. He is not allowed to enter any kraal or dwelling, but must remain in the field, where food is brought to him, until the period of separation is expired. Thompson *Travels II* 1827

The field may be burnt but it is late . . . *Greig's Almanac* 1831 (*trans.* v.d. Stel's *Calendar for Farmers*)

field cornet, veld cornet *n. pl.* -s. *hist.* An office like that of *District Officer* in Brit. Colonial Service: a minor magistrate answerable to the *landdrost* (q.v.) similar to a Justice of Peace and having authority in his ~*cy*, a subdivision or 'ward' of a *district* (q.v.).

Each district is subdivided into a number of veld-cornetcies, in which the duty of the Veld-Cornet, (or Field-Cornet), is to put in execution all orders from the landdrost, to whom he is more immediately accountable. Burchell *Travels I* 1822

It shall be lawful for any justice of the peace, officer of police, field cornet, constable, or any owner or occupier of land to demand of any such native foreigner the production of his pass. *Cape of G.H. Act* 1867

A field-cornetcy was supposed to contain 150–200 men. Reitz *Commando* 1929

Fields, the *n. prop. hist.* The Diamond Fields of Kimberley: occ. *rare* used of the Goldfields [*abbr. the Diamond Fields*]

... outspanned here are several parties that have come from the Fields. Dugmore *Diary* 1871

"Baas, we are going after diamonds."

"Diamonds! why, then, you are steering in the wrong direction; you should head for the Fields". Haggard *Solomon's Mines* 1886.

Two of Hilda's sisters and their husbands were to travel to the Fields and live there. *Duckitt's Recipes* ed. Kuttel 1966 (Introduction)

fiemies ['fimĭs] *pl. n. slang*. Whims, fads: *to be full of* ~ to be fussy, capricious. [*Afk. vol fiemies wees* to be full of whims and fancies]

... could not get a Predikant because they were all vol fiemies. The one won't allow the wedding march – the other won't allow music afterwards. Letter Schoolgirl, S.W. Africa 11.4.75

Fingo ['fɪŋgəʊ] *n. prop. pl.* -s. A member of a tribe living in the E. Cape Province descended from refugees of tribes driven from Natal: see quot. ~ *village*, a grant of freehold land in return for loyal service in the Frontier Wars made in the reign of Queen Victoria. [*fr. Xh. amaMfengu-* pl. prefix *ama* + *mfengu* those looking for work *fr. vb. fenguza* to seek service]

The name Fingo, comes from ... amaMfingu, 'those who wander in search of service'. They were the remnants of tribes from Natal, broken up and scattered by Chaka's impis, who ... were treated as a subject people until brought across the Keiskama River tree by Sli Benjamin D'Urban in 1837. Later they were declared British subjects, and have remained loyal ever since, Fingo Levies serving in the various Kaffir Wars. Gordon-Brown 1941: footnote to Buck Adams *Narrative* 1884

finish(ed) and klaar [ˌfɪnɪʃ ən 'klɑː(r)] *idiom*. Emphatic expression sig. 'that's that', 'that's the end of it', etc. [*Afk. klaar* finished, done]

... 'Multiracial sport anywhere else in the world is okay. ... when in Rome do what the Romans tell you. Here ... I won't do it because I'm a White South African. Finish and klaar.' *Argus* 12.7.76

fink [fɪŋk, fɪŋk] *n. pl.* -s. Ploceidae fam.: Used of numerous species: see *vink*. [*anglicization Afk. fr. Du. vink cogn.* finch]

Barley or wheat where sown is very pro-

ductive ... Like the rice it has a great enemy in a small bird called the fink which perch upon it in flocks of several thousands. Ewart *Journal* 1811–14

firebucket *n. pl.* -s. *Sect. Army*. A large metal drinking cup kidney-shaped in section with a folding handle, into which a water bottle fits.

When we're on beat we use our firebuckets to make our coffee – otherwise we have a white Melamine cup. *O.I. Serviceman* 16.4.79

Fiscal *n. usu. capitalized. hist.* An official having numerous powers under the regime of the *Dutch East India Company* (q.v.), later *His Majesty's* ~ under the British. ⫼His office concerned both law enforcement and revenue, the ~ being chief of police, public prosecutor and collector of fines and taxes. [*Afk. fiskaal fr. Du. fiscaal* chief of the Treasury *Lat. fiscalis fr. fiscus* treasury]

Should the Wardmasters be neglectful hereof they will be fined 60 Rds, to be divided, three quarters for the Town Treasury, and one quarter for His Majesty's Fiscal. *Afr. Court Calendar* 1819

The Fiscal, being the head of the police, and the sitting magistrate, a great variety of business is daily transacted at his office. Burchell *Travels I* 1822

fiscal shrike *n. pl.* -s. *Lanius collaris*: see *Jan Fiskaal*.

fishcart, fishhorn *n. pl.* -s. Formerly the one-horse cart used in Cape Town by *Malay* (q.v.) fish vendors to ply their trade, accompanied by piercing blasts on a tin bugle, the *fish horn*, the traditional method of crying their wares.

Fish carts were driven on to the beach, and with a great whipping up of horses and elaborate use of the fish-horn, that 'pest of the Cape Town streets,' the vendors would be sent on their way. *Cape Times* 18.3.36

... the noise from a well kept garage is not a nuisance, nor is the blowing of fish horns by itinerant vendors of fish. Wille *Principles S. Afr Law* 1945 ed.

... hear again the clatter of the high wheeled fish carts ... with the 'strident bleat of the Fish horn riding triumphantly on the South Easter'. cit. *Star* 22.5.72

Coloured fishermen ... wearing high pointed straw hats, as can be seen in the paintings of Thomas Bowler ... blew high-pitched notes on fish-horns fashioned out of dried sea-kelp. *Panorama* June 1973

fish oil *n. substandard*. Frying or salad oil. [*prob. fr. Afk. visolie* cooking oil, ⫼Oil extracted from fish, whale etc., is *Afk. traan*]

Onion Atjar . . . Boil up 1 pint fish oil with a little pounded garlic, borrie, a few chillies and some curry powder. Gerber *Cape Cookery* 1950 EID SHOPPERS – Specials . . . Fish Oil . . . 750ml 67½c *Cape Herald Advt.* 14.9.76

fitter and turner *n. phr. pl.* -s. *Sect. Army.* Derogatory term for an army cook: see *tiffy, pot.*

'Any ou who can fit something into a pot and turn it into – .' *O.I. Serviceman* 15.2.79

flaaitaal See *fly taal* and quot. below.

. . . Tsotsis have their own secret language known as *flaaitaal* or *tsotsi* slang, which is continually being improved on and added to. These *flaaitaals* differ from city to city and even in different parts of the cities themselves. Longmore *The Dispossessed* 1959

flap *Euplectes progne :* see *sakabula.*

flat crown *n. pl.* -s. Familiar name for *Albizia adianthifolia* (also known as *A. fastigiata* and *A. gummifera*) esp. common in Natal. A tree of up to 18m high and 25m across, the canopy having a flat, spreading crown with clusters of red flowers 'like miniature shaving brushes' Smith *S. Afr. Plants* 1966.

The 'flat-crown' (should be flat-roof) – half a dozen naked branches, full of elbows, slant upwards like artificial supports, and fling a roof of delicate foliage out in a horizontal platform as flat as a floor. Mark Twain *More Tramps Abroad* cit. Pettman 1913

flat dog *n. phr. pl.* -s. *Border and Rhodesian slang.* A crocodile, esp. in phr. '(nearly) grazed by a ~ '.

Flat dogs Crocodiles, another natural menace. Also: 'walking handbags'. *Time* 30.4.79

flat(s) *n. usu. pl.* Plain, flat area: found in S. Afr. place names following other *n.* e.g. Cape Flats, Manley Flats, Bontebok Flats, Quagga Flats. [*trans. Afk. vlakte* (q.v.) plain(s)]

flatty *n. pl.* -ies. Also (2) *stompie :* the marine fish *Rhabdosargus holubi* also called *blinkvis*, silver bream, silvie: the Cape stumpnose.

. . . tells me some big flatties have been caught, up to 0,680kg (1½lb). (Another name for the 'flatty' is stumpnose or silver bream). *Grocott's Mail* 3.9.71

fleck *vb. trns* See *vlek.*

floppy *n. pl.* -ies. *Rhodesian* A terrorist: see also quot. at *terr.* [*unknown*]

'How did it go?' 'Positive. Small contact single floppy.' Dibb *Spotted Soldiers* cit. Fair *Lady* 25.10.78

flossie See *vlossie.*

'flu *n. Afr. E.* The influenza outbreak of 1917–18, used occ. by older Africans as a means of dating their birth or other events. *cf. Inkwenkwezi* ('star' sig. the appearance of Halley's Comet) and *brug* ('bridge' sig. the train disaster at Bloukrans E. Cape) also used to date events.

'How old is Idah?' . . . 'Flu'. Elderly African O.I. 1968

In the old days, time in South Africa was reckoned on being before or after the Drought, the 'Flu or the Rinderpest. Here's what the Rinderpest looked like. *E.P. Herald* 2.4.79

'fly' *n. colloq.* The *tsetse fly* (q.v.).

. . . the proximity or otherwise of 'Fly', should you be in a district where the dreaded Tsetse are found. Hemans *Log* 1935

fly taal ['flaɪˌtɑl] *n. pl.* -s. *Urban Afr. E.* The language used by younger town-bred Africans, varying from city to city and incomprehensible to people from rural districts, and acc. some to the older generation: see quots. also *flaaitaal.* [*Afk. taal* language]

Rhodes Man researches tsotsi talk for thesis

'In my studies . . . I would like to focus . . . on the modern language of the tsotsis . . . and 'fly taal' the language of 'hip' city slickers.' . . . 'Their jive talk is very difficult for adults to understand . . . if we could talk their language there wouldn't be so many problems.' *E.P. Herald* 1.5.78

I believe that the not so young people do use the 'fly taal'. The younger people tend to use new terms in 'their' language . . . the older group . . . keeping in touch with the younger group . . . add more to their 'fly taal' vocabulary . . . It should be understood that the use of the words depends on the city or the town. It is really difficult even for the 'fly taal' users to keep up with these words. *Informant* Mr Sabelo Sillie *Letter* 12.4.79

F.N. ['efˌen] *acronym pl.* -s. *Rhodesian and S.Afr.* A rifle: [*acronym trade mark* Fabrique Nationale]

Claver nodded, and releasing the catch on his FN, he stole down the path. Dibb *Spotted Soldiers* cit. Fair *Lady* 25.10.78

foefie slide [fʊfi, fʊ̃fi] *n. pl.* -s. Also *fuffie/(foofy) slide :* see quot. [*etym. dub. poss. fr. Afk. foefie* stunt, trick]

Of all the traditional South African games, one is unsung: the foefie slide. Now the foefie slide is . . . a thick wire or a rope secure to the highest point in the vicinity, usually a tree, and then slung down, sometimes over water, to an arm-high point where it's once again secured. A piece of iron pipe is threaded on to this rope or wire and you are supposed to launch yourself into space hanging from the pipe. *Fair*

Lady 31.10.73

foei tog [ˈfŭï ˌtɔx] *interj.* An exclamation of pity or sympathy, also as *'shame'* (q.v.) used to exclaim at something small, sweet or helpless. *cf. siestog* (q.v.) [*Afk. foeitog* for shame (*foei cogn. Eng.* fie)]
. . . admitted the cost of living had also increased, but he asked defensively didn't . . . know that the increase was part of a world wide phenomenon. 'Ag, foeitog!' he said. *Cape Times* 10.11.72
On the open road we tended to put up the tattered black cowl and push it protectively, which earned us fond smiles and foeitogs. *E.P. Herald* 9.5.74

fontein [ˌfɔnˈteïn] *n. pl.* -s. Spring. [*Afk. fr. Du. fontein* spring, fountain *cf. Fr. fontaine* spring, source] **1.** Natural spring or water-source: see *fountain* and quot. at *farm.*
. . . the farms exhibit signs of wealth, and here and there, where there is a 'fontein', there are patches of arable land covered with luxuriant crops. Gray *Journal I* 1849
2. Spring: found in numerous place names e.g. Witfontein, Grootfontein, Olifantsfontein, Rietfontein, Klipfontein etc.: see also *fountain* and quot. at *groot.*
. . . in dry countries, any circumstance relating to water, is of sufficient importance to distinguish that place. Thus it is that the Dutch word Fontein is made such liberal use of in every part of the Colony; the Hottentot word Kamma (water) is not less frequently found in the composition of the aboriginal names. Burchell *Travels I* 1822
But now, discharged, I fall away
To do with little things again . . .
Gawd, 'oo knows all I cannot say,
Look after me in Thamesfontein!
Kipling *Five Nations* 1903

footsack [ˈfʊtˌsæk] *interj. slang.* See *voe(r)tsek,* also quot. at *cattie/y.* [*anglicization translit. Afk. voertsek* (q.v.)]
. . . for prep. Used sig. 'of' esp. by children: I'm so scared for these exams. [*translit. Afk. vir* for, of, *esp. in phr. bang vir* scared of]
She's really skrikking for the dance. Schoolgirl 17, 1975

for Africa *colloq.* See *Africa, for.*

forget *vb. trns. substandard.* To leave behind (some concrete object), book, jersey etc.). [*translit. Afk. vergeet* leave behind]
But a friend borrowed them and forgot them outside on the lawn. They were stolen, of course, and the child's anguish knew no bounds. *Panorama* Dec. 1973

fort *n.* Found in place names as in British

sig. the site of a former military fortification e.g. Fort Brown, Fort Cox, Fort Mistake: see also *Pos(t).*

forty days *n. phr.* The start of *mindae* (q.v.). ⫟ When any *intake* (q.v.) of National Servicemen has ~ left to serve there is a party, usu. a *braai* (q.v) to celebrate this, after which a marked lack of military discipline is apparent: *mindae, minday signs; mindae, minday salutes* etc. *O.I.s Servicemen* 1979.

fortypercenters *pl. n.* A non-party organisation of English South Africans dedicated to achieving a fair share of power for the English in S.A. ⫟ The ~ was started in Nov. 1978 by Mr Douglas Alexander. [see first quot.]
The 'Fortypercenters' is not anti-Afrikaans . . . At the moment English speakers hold only five per cent of top jobs in the public service, but they constitute nearly forty per cent of the white population. *Sunday Times* 8.4.79
Fortypercenters are in no way a Broederbond
. . . True it was launched in response to arrogant Afrikaner Broederbond taunts that the English were free to form their own Broederbond 'if they so wished'. In taking up the challenge . . . It was made perfectly clear from the onset that the Fortypercenters would operate quite openly as watchdog and pressure group to push for a bigger say for the English in running South Africa. Douglas Alexander cit. *Ibid* 15.4.79

fossick [ˈfɒˌsɪk] *vb. intrns. colloq.* Term ex *Austral.* formerly to prospect for gold, hence ~*er* a prospector; now more loosely ~ *about, around* etc. sig. rummage, search or hunt.
An English ivory hunter . . . had his throat cut, and two prospectors, Charles Muller and George Anderson, fossicking in the northern bushveld . . . in June 1871, were attacked. Bulpin *Lost Trails* 1951

fountain *n. pl.* -s. A natural spring or water source: see also *fontein, eye, bron* and quot. at *trekbok :* also found in S. Afr. place names e.g. Seven Fountains Hope Fountain. [*translit. Afk. fontein* spring *cf. Fr. fontaine* spring]
. . . the arable land, . . . is not irrigated from the river, but from various fountains which issue from the steep woody kloofs of the Boschberg. Thompson *Travels II* 1827
The farm is divided into 12 grazing camps which are stockproof. The boundary fencing is jackal-proof. There is water in all camps which is provided by five water-holes in the river, five fountains and two windmills. *Grocott's Mail Advt.* 16.5.72

frans madame [ˈfrans məˈdɑm] *n. pl.* ∅. Also *french madam* (q.v.) *Boopsoidea inornata* a marine fish with bulging eyes and large red mouth usu. *jacopever* (q.v.), also *Cape lady, dikoog, peuloog, grootogie* [*Afk. Frans* French + *madame*]

Free Burgher *hist.* See *burgher* and quot. at *Company*.

Free State *n. prefix.* Used facetiously or satirically of something crude, backward or primitive by those living in other parts of the country. *cf. Austral. prefix bush-* in ~ *lawyer,* ~ *baptist* etc. and similar tendency in World War 1 term *Anzac button*, a nail to fasten trousers, S. Afr. *Rhodesian spanner*, a bottle opener. [*abbr. Orange Free State, S.A.*] In combinations: ~ *coal*, see also *Karroo coal*, dried cattle dung (see *mis*) used for fuel where wood is scarce; ~ *bolt*, a piece of wire used for fastening a door; ~ *nails*, stones to keep a roof or roof-patch in place; ~ *micrometer screw*, a *bobbejaan spanner* (q.v.); ~ *sandwich*, etc.

french madam *n. pl.* Ø. See *frans madame*
I caught the surprise of my fishing career – a french madam. It took my little spinner right on the surface . . . would never have believed a french madam would take a spinner. *E.P. Herald* 9.5.74

Friesland *n. pl.* -s and *modifier.* Holstein-Frisian cattle so called in S.A. also occ. *Fries* (modifier). [*fr. Afk. Friesbeeste.* Frisian cattle, and *Friesland*, the name of the province of the Netherlands where *Frisian* (*Afk. Fries*) is spoken]
18 Fries Tollies, 1 Fries cow with calf, 1 slaughter cow. *Advt. Farmers Weekly* 3.1.68
The large Friesland dairy herds which supply the Cape Town area with fresh milk may soon be making a . . . contribution towards relieving South Africa's growing shortage of red meat. *Argus* 19.4.73

frikkadel [ˌfrɪkəˈdel] *n. pl.* -s. Fried mince-meat balls equiv. of rissoles. See also quot. at *pastei*. [*Afk. fr. Fr. fricadelle* a meat ball]
. . . lumps of mutton, fried in the tallowy fat of the sheep's tail, or else – their only change of diet . . . the tasteless fricadel, kneaded balls of meat and onions, likewise swimming in grease. Boyle *Cape for Diamonds* 1873
Weary of Bangers and mash, 1001 ways with mince – babotie, spaghetti bols, frikkadels, meat loaf . . . *Cape Times* 11.7.73

fuffie slide See *foefie slide*.

full mouth *phr. modifier.* Also *Austral.* See *volbek.*

full of *predic. phr. modifier.* Covered with/in. [*mistrans. Afk. vol* covered with]
. . . a dead cat lay on the pavement full of flies. *Daily Dispatch* 13.3.72
Nursing cut elbows and a back full of red welts, she said . . . *Ibid* 2.6.72

fundi [ˈfʊndɪ] *n. pl.* -s. An expert: occ. as *adj.* equiv. of 'expert in' '*He's fundi at maths*' (*rare*). *cf. Anglo-Ind. pundit;* see quot. at *must*, also *fundis* (below) [*fr. Ngu. umfundisi* (q.v.) a teacher]
So there we have two problems for the language fundis. It is now in your court. Can you answer their prayers? *Cape Times* 3.7.73

fundis [ˌfʊnˈdis] *n. abbr. umfundisi* (q.v.). see *etym.* at *fundi.*
. . . Katongo urged the propriety of my borrowing a musket for him to carry. On my saying that the Caffres knew me to be a 'Fundis' (teacher) and why should a minister carry arms? Katongo said . . . he was not certain that if we fell in with any elephants they would recognise me as a 'Fundis'. Shaw 23.2.1825 *cit.* Sadler 1967

furrow *n. pl.* -s. A man-made water course usu. for irrigation purposes, urban or agricultural: see *sloot* and quot. at *trekbok*, also *lead water.* ▶Towns with *water erven:* see quot. at *erf*, have irrigation ~ *s* along the streets, usu. under control of a *water fiscal/fiskaal* (q.v.).
All furrows on the farm are cement. The property is situated between and riparian to the Damara and Letsitele Rivers. *Farmer's Weekly* 30.5.73
. . . this rural town . . . steeped in the history of the Transvaal when it was still the Zuid Afrikaansche Republiek. In those days, wide furrows ran down both sides of the streets. *Panorama* Apr. 1974

fuse *n. pl.* -s. *Urban Afr. E. slang* A cigarette: see quot. at *atshitshi*, also *tsotsi language, tsotsi taal* and *fly taal.*

fynbos [ˈfeɪnˌbɔs] *n.* General term for numerous species of narrow-leaved indigenous shrubs common to winter rainfall areas: elsewhere known as *macchie* or *maquis:* also used for *Euryops* spp., see *harpuisbos.* [*Afk. fr. Du. fijn* delicate + *bosch cogn.* bush]
. . . overgrown with the 'fynbos' rich in proteas or the coastal macchie . . . Krauss *Travel Journal* 1838–40 *trans.* Spohr 1973
The fynbos is fire-resistant and regenerates from perennial rootstocks. It is also extremely susceptible to invasion by 'exotic' plants . . .

areas where the pines and black wattles are . . . suppressing the indigenous vegetation chiefly the fynbos and sourveld types. Seagrief *Reading Signs* 24.5.76

The proposed Palmiet River dam . . . 'would be a sad blow' to the area's unique fynbos. *Argus* 9.10.76

G

G5 *n. pl.* -s. A S.Afr. artillery piece: see quot. below.

PM unveils G5, produced in SA in record time
Two brand-new South African weapons, the G5 artillery piece and the R4 service rifle were taken out of wraps in Cape Town yesterday. . . . Few new facts were disclosed about the G5 gun . . . it was not self-propelled, although it had a small motor for limited movement in the field and for elevating the gun. It fired the standard 155 mm Nato-type artillery shell, and could be used in an anti-tank role. *Cape Times* 26.4.79

ga [xa] *interj.* An exclamation of strong distaste or disgust: see also *sis*. [*Afk. ga* faugh]
Like the rest of your article, all that it boiled down to was a meaningless piece of invective, equivalent to saying 'Farmers, gha!' *Daily Dispatch* 21.7.71

gaatjeponner [ˌxaɪkĭˈpɔ̆nə(r), -cĭ-] See *gatjaponner*.

gabba [ˈxaba] *n. pl.* -s. *colloq.* Friend, pal etc. [*prob. fr. Hebrew chaver* comrade]
What he ignored were the products of mixed [Afrikaans-English] marriages, And that my gabbas, is where there's the rub. *Sunday Times* 8.10.78
In combination gat~. Army, Border usage: A serviceman's closest friend: *acc.* informant C. Cox 22.4.79, in the operational area ~s escort each other to relieve themselves, one remaining on guard, and clean their rifles together, only one weapon being unloaded at a time. ⫿Elsewhere derogatory, see *gat*.
Murray and I were gat gabbas on the Border so now we're at University we still share digs. *Ibid.*

galjoen [ˌxalˈjŭn] *n. pl.* ∅. Also *damba* (q.v.), *highwater*, blackfish, black bream the edible strongly-flavoured marine fish *Coracinus capensis*: also *streep* ~, *banded* ~ or *bastard* ~ *C. multifasciatus* (Coracinidae fam.) ⫿The name is said to be from their resemblance to a galleon in shape. [*Afk. fr. Du. galjoen fr. Lat. galea* galleon]
For the galjoen there can be no lukewarm

sentiments. You like it or dislike it intensely. It has a strong and distinctive flavour which most people at the Cape consider a special delicacy. *Farmer's Weekly* 18.4.73

galla [ˈxala] *vb. colloq.* To desire greatly: used of possessions or food etc. also by children meaning to tantalise. [*fr. Xh. ukuRhala, Zu. -Hala* to be greedy for]
Tom's galla'd for silver college tumblers for years so I gave him a pair for his twenty first. O.I. 10.6.75

gallsickness *n.* Also *galsiekte:* designation for most liver disorders in stock: ~ *veld*, areas in which such diseases are endemic: see also *biliary*. [*trans. Afk. galsiekte fr. Du. gal* bile *cogn.* gall + *ziekte* disease]
One of the diseases which had been under notice for a long time and which he had only been able to trace to ticks about a year ago, was also commonly known as gallsickness. *E.P. Herald* 10.10.1911
The animals are used to virulent heartwater, redwater and gallsickness veld. All animals have been injected against Bushveld diseases. *Farmer's Weekly Advt.* 21.4.72

galsiekte [ˈxalˌsĭktə] *n.* See *gallsickness:* ~ *bossie*, *Chenopodium ambrosioides* an aromatic plant thought to be a cure for ~. [*Afk. fr. Du. gal cogn.* gall + *ziekte* disease]

gamadoelas [ˌx(r)amaˈdʊləz, -as] See *gramadoelas*.

Gammat [xaˈmat, ˈxamakĭ, -tʃĭ -cĭ] *n. prop.* Comic Malay folk figure, subject of ~ *tjie jokes* in ~ *taal* (language): an extreme *Cape Coloured* (q.v.) dialect. [*corruption of Achmet fr. Mohammed*]
Laugh with Gammatjie – Send your jokes to: POST's Gammatjie Jokes Department. *Post* 10.5.70
You think you got troubles, Gammat – they got me for doing 51km/h, for storing 11 litres of petrol, for drunken driving, my wife's left me, and I got Meraai into trouble. *Drum* 22.3.74
. . . one request . . . is that they should publish Gammatjie's jokes. Then Cape Herald is the paper for grandparents, parents and children. *Cape Herald* 14.9.76
One of the interesting things about this country is the extraordinary mixes of language . . . when Pieter Dirk Uys uses *gammattaal*, he's legitimately using the language form which expresses the nature of the people he's writing about. Janet Suzman cit. *Darling* 16.3.77
. . . from his speech and mannerisms, and also, to some extent, from his name, Josias, you would be more inclined to classify him as a Cape Gamat. Bosman *Willemsdorp* 1977

game ranch/ing *n.* An alternative to *stock*

(q.v.) *farming.* ⟦Increasingly favoured, since indigenous buck require no special feeding and are more resistant to disease than domestic animals: see quot. at *farm with.*
> ... located in the heart of the Zululand game country . . . a highly developed 1133ha game ranch. Fully stocked with over 4 000 head of game. Comprising 15 different species, including White Rhino, Giraffe, Nyala, Kudu and Eland . . . Benefit by way of sale of game, meat and concession paid. *Farmer's Weekly* 27.2.74

ganna [ˈxana] *n.* Any of several species of *Salsola,* the ashes of which form white caustic alkali used in the early days for soap-making. ⟦Also called *kannabos* *prob. fr.* confusion of pron. of [x] in channa, see *kanna*[2] or *poss. fr. Hott.* name for *eland* (q.v.) which were said to feed on it: see *kanna*[1] and quot. at *brak.* [*prob. fr. kanna fr. Hott.*]
> ... The young shoots of the Channa bushes. The ashes of these saline plants produce a strong ley, and of this, mixed with the fat of the sheep, collected during the year, the women make an excellent soap. Lichtenstein *Travels* 1812

> Veld grazing consists of healthy Karoo bushes, brak, ganna, vleis, grass. *Farmer's Weekly* 27.2.74

gans [xăns, xans] *n. prefix.* Goose: found in S. Afr. place names e.g. Ganskraal, Gansbaai. [*Afk. fr. Du. gans* goose]

gap, to take the See *take the gap, to.*

garam/garum (ghurum) masala *n.* See *masala* also *Indian terms.*
> Cook until calabash is soft then garnish with a sprinkling of garum masala and chopped coriander. *Fair Lady* 21.7.76

garingboom [ˈxɑ:rɪŋˌbʊəm] *n. pl.* ∅. Also 'century plant': *Agave americana* introduced from central America: the despined leaves are chopped and used for fodder in time of drought. [*Afk. garing fr. Du. garen* thread. *cogn.* yarn + *boom* tree *cogn. Ger. Baum*]
> Proportionately, it probably has more American aloe (garingboom) and salt bush than any other farm in the country. And for very good reason. Throughout its long history . . . has never resorted to buying supplementary feed during times of drought. *Farmer's Weekly* 11.7.73

garrick See *leerfish/vis.*

gat [xat] *interj.* or *n.* Hole [*Afk. gat* hole, *hence also* vent, anus] **1.** Anus, vent, as an expletive (vulgar): see also *kaal* ~ and *se voet. cf. Jam. Eng. you bain.* **2.** *n. prefix* or *suffix.* Hole, opening,

pit etc. found in place names e.g. Moddergat, Makapansgat, Seekoegat, Gatberg. *cf. Canad. trou,* hole, in place names.

gat gabba See *gabba.*

gatjaponner(s) [ˈxatjaˌpɔ̃nə(r)] *n. usu. pl. rare.* Derogatory name given by the *Doppers* (q.v.) to the members of the *D.R.C.* (q.v.) on account of their customary wearing of the *manel* (q.v.) or frock coat. [*Afk. gat see* (1) gat + *japon* gown, robe + *-er personif. suffix*]
> ... a party of Voortrekkers, some Doppers ... others Gatjaponners ... formed a *laerplek* on the present town site. Religious arguments arose, and the Doppers moved off and founded a settlement . . . They called it Reddersburg because they had been 'saved' from those who remained behind. Not to be outdone, the Gatjaponners named their home Edenburg, an earthly paradise after the departure of the Doppers. Green *These Wonders* 1959

gattes, die [dĭ ˈxatəs] *pl. n. phr.* Urban *Afr. slang:* the police; see also *ore* and *boer(e)* (5). [*presum. fr. Afk. gat* vent, anus]
> ... a stranger to the tsotsi's dangerous world could still save his throat if he has some knowledge of basic words and phrases . . . Die gattes – the police . . . Venter *Soweto* 1977

gatvol [ˈxatfɔl] *adj. slang (vulgar)* Fed-up, disgusted, had enough: a cruder version of *keelvol* (q.v.) [see *gat* + *vol cogn.* full]

gaukum [ˈgəʊkəm] See *ghokum*

gavin(i) [gaˈvin(ĭ)] *n.* Also *govini:* a powerful spirit distilled from sugar cane: see quot. at *shimiyaan.* [*Zu. gavini* an intoxicant as above]

gawie [ˈxavĭ] *n. pl.* -s. *colloq.* A crude, backward or uneducated person, usu. a rustic: see also *backvelder, jaap, japie* and *takhaar* [*Afk. gawie* bumpkin, lout]
> These backvelders have today largely disappeared . . . naive, primitive, uneducated and in some cases even idiots. . . . travellers, social workers and school inspectors . . . referred to the backvelders as duine-molle, japies, takhare, gawies and so on. Meintjes *Manor House* 1974

Gazankulu [ˌgazaŋˈkulŭ] *n. prop.* The *homeland* (q.v.) in the Northern Transvaal of the *Shangaan* (q.v.) and *Tsonga* (q.v.) speaking peoples.
> Gazankulu was declared a self-governing territory within the Republic in terms of Proclamation R14 of 26 January 1973. Its seat of Government is Giyani, and Tsonga was recognized as an additional official language. Horrell

Afr. Homelands of S.A. 1973
gebroken veld [xə'brʊəkən₁felt] *n. prob. obs.* Mixed veld consisting of grass and bush: see (2) *veld* and quot. at *aarbossie*. [*Du.* gebroken *cogn.* broken *past partic.* of *vb.* breken to break]

The pasturage of these plains, with the exception of what the farmers term the 'Looge veld' which is covered with a sour wiry grass is principally what is termed in the Colony 'Gebroken veld', or a mixture of sour and sweet grass. *Cape of G.H. Almanac* 1856

-gedacht [xə'dax(t)] *vb. partic.* Thought (of): found in place names esp. of farms usu. with *adv. prefix* e.g. Nooitgedacht, Goedgedacht, also Afk. form *gedag:* see also *-bedacht* [*Afk.* gedag *fr. Du.* gedacht *past partic. of vb.* denken *cogn.* think]

gedoente [xə'dŭntə] *n. pl.* -s. *slang* Fuss, to do, 'carry on', doings etc. *cf. U.S. hassle, Jam. Eng. rig-jug* [*Afk.* bustle, to do, goings on etc]

Now that was a big gedoente loading the cow onto the little lorry. O.I. 1970

geel- [xɪəl] *adj.* Yellow, prefixed to numerous names in S.A. esp. of plants, also fauna generally, e.g. ~ *aandblom*, ~ *boegoekaroo*, ~ *boekenhout*, ~*bek* (q.v.), ~*stert* (q.v.), ~ *meerkat* etc. [*Afk. fr. Du.* geel yellow]

geelbek ['xɪəl₁bek] *n. pl.* Ø **1.** Also *Cape Salmon:* the edible marine fish *Atractoscion aequidens* which has bright yellow edges to the jaws and gill covers. ⟦Being firm fleshed it is favoured for the traditional dish *ingelegde vis* (q.v.): see also quot. at *allewêreld*. [*Afk. fr. Du.* geel yellow + *bek* mouth, bill *cogn.* beak]

... if the electronic aids of the trawlers track down all the anchovy, all the pilchards, all the bait fish round the Peninsula, ... no snoek, no game fish, no kob, no geelbek will come to these waters. *Argus* 4.6.71

2. A yellow-billed wild duck *Anas undulata* [*as above*]

This is a light-brown duck, and is easily identified by the bright yellow bill from which it takes its Boer name – geelbec. Bryden *Gun and Camera* 1893 *cit.* Pettman

3. *rare in S.A.E.* An offensive mode of reference to a coloured person [*see quot.*]

Coloured persons are so called in the Western Province on account of their yellowish colour. Swart *Africanderisms: Supp.* 1934

geelhond ['xɪəl₁hɔ̃nt] *n. prob. reg. Graaff-Reinet lit.* 'yellow dog', with*hond* (q.v.)

coloured and flavoured with liquorice. [*Afk.* geel yellow + *hond cogn.* hound]

For the ladies there was geelhond, the same distillation delicately treated with liquorice and served in dainty portions as a liqueur. Van Biljon *cit. Sunday Times* 6.11.77

geelhout ['xɪəl₁(h)əʊt] *n.* See *yellowwood.* [*Afk. fr. Du.* geel yellow + *hout* wood]
1. See *yellowwood*

Of the species peculiar to this country I have observed the geelhout: it grows to a very large size ... the wood is of a bright yellow colour: and much used for furniture. Percival *Account of Cape of G.H.* 1804
2. Yellowwood (tree) in place name Geelhoutboomberg.
3. See *oubos* (*ouhout*) (*Leucosidea sericea*).

geelrys ['xɪəl₁reɪs] *n.* See *yellow rice, begrafnisrys,* also *borrie* and quot. at *pastei.* [*Afk. fr. Du.* geel yellow + *rijs cogn.* rice]

... the meat dishes – chicken, mutton, 'geelrys' (yellow rice flavoured with saffron and condiments and coloured with turmeric). *Sunday Times* 18.2.73

geelstert ['xɪəl₁ste(r)t, -stɛət] *n.* See *yellowtail.*

geelvis ['xɪəl₁fɪs] *n.* See *yellowfish.*

-(ge)gund [(xə)'xœnt] *vb. partic.* Favoured: found in place names e.g. Welgegund, Goedgegund, Ongegund (also Misgund, begrudged, envied). [*Du. past partic. of vb.* gunnen to grant]

geilsiekte ['xeɪl₁sĭktə] *n.* Prussic acid poisoning, bloat, caused by eating luxuriant grass which has withered in sudden heat after rain. [*Afk.* geil lush, rank, *poss. rel. Du.* gijl ferment *n.* + *siekte fr. Du.* ziekte disease]

The term 'geilsiekte' as used by farmers does not signify a definite disease, but is a collective name for different ailments most of which are caused by plant poisoning. By far the most cases ... are due to hydrocyanic acid (prussic acid) poisoning caused through the ingestion of wilted green grass. *H'book for Farmers* 1937

geitjie ['xeɪkĭ-cĭ] *n. pl.* -s. Common name for any of several species of gecko. [*Afk. fr. Du.* geitje kid, *cogn.* goat]

... a small lizard, with blunt toes, and which is said to be so poisonous that its bite occasions death within an hour. ... Inhabits Namaqua land, where it is called Geitjie ... the specimens now being in the collection of the British Museum, as is also the Geitjie. Alexander *Expedition II* 1838

gek [xek] *n.* and *adj. slang* A fool, foolish, half-witted: see also *dwaas*. [*Afk. fr. Du.*

gek mad, besotted, stupid]
I left with wisdom just as full
As gekke tanta Mietjie. Bain *Kaatjie Kekkelbek
circa* 1838
   The word 'gek' is expressive enough. The
Hollanders have a couplet:
'Gekken en dwazen
Schryven op muren en glazen'*
Devitt *People & Places* 1945
(*Fools and idiots scrawl
Upon windows and wall.)
   'Jou gek,' I said. 'She's still asleep. Come
back later.' Maclennan *cit. Contrast* 45 Vol 12
July 1978

geleë [xə¹liə] See *gelegen*.

gelegen [xə¹liəxən] *vb. partic.* Situated:
found in place names e.g. Vergelegen,
Welgelegen, also Ongelegen. [*Du. gelegen past partic. of vb. liggen* to lie,
be situated. *Afk. geleë*]

gemeente [xə¹miəntə] *n. pl.* -s. Parish: the
congregation of a *D.R.C.* (q.v.) church.
[*Afk. gemeente* parish, community,
congregation]
   First in this pleasant gemeente was the Rev.
J. J. Beck. He never served any other community and he remained at his post for 53 years.
Green *Land of Afternoon* 1949
   . . . today the saga of the Kameeldrif gemeente is yet another frightening sign of the Government's Canute-like stand . . . The Ned. Geref.
church council sought Government permission
to erect a chapel for Blacks in the church
grounds . . . No, black churchgoers must attend
church . . . in their own areas. *Sunday Times*
1.8.76

gem (squash) *n. pl.* -s. Small green globular vegetable marrow, golden and hard-
skinned when mature: *colloq.* 'gems'.
   GEMS, well supplied, demand moderate.
Farmer's Weekly 21.4.72

gemors [xə¹mɔ(r)s] *n. slang.* A mess-up,
'hash', muddle [*Afk. n. gemors* mess
fr. vb. mors (*Du. morsen*) to make a
mess, spill, etc.]
   The walls are so tremendously thick, being
made of large rough boulders, lime and sand,
that the damage was mostly internal, said Mr
Duckitt, some walls were a real gemors. *Farmer's Weekly* 25.4.73

gemsbok [¹xems₁bɔk] *n. pl.* Ø (-s). One of
the S. Afr. antelopes, *Oryx gazella*, said
to be courageous enough to tackle a
lion. [*Du. gems* chamois + *bok cogn.*
buck]
   As we rode along, I observed several gems-
boks. This is a beautiful and noble-looking
antelope. His long, straight, sharp horns in-
cline a little backward, and it is said the animal
can use them with formidable effect in self
defence. Thompson *Travels I* 1827

general dealer *n. pl.* -s. Also *algemene
handelaar* (q.v.): a merchant, often in a
country trading store, carrying a widely
varied stock to meet the general needs
of a community. *cf. Canad. country
store, general store.* ▯A ~s licence
restricts the sale of certain classes of
merchandise, e.g. groceries after 6 p.m.:
only dairy products, confectionery and
cigarettes may then be sold: see *tearoom*.
[*trans. Afk. algemene* general + *handelaar* trader, dealer]
   J. C. TRUTER, General Dealer, Meade
Street, Has always on hand, a well assorted
stock of Staple and Fancy Goods. *George
Advertiser* 14.4.1870
   A roll of chewing tobacco, one pound of
flour, three yards of dress material and about
sixpence worth of 'London Mixture' sweets
. . . An order like this was commonplace a few
decades ago in a country general dealer's.
Panorama Apr. 1974

~ 's store
   The general dealer's store was the next step
in the chain of development following on the
itinerant pedlar who peddled his wares in the
wake of our pioneer forebears. Once the pioneers settled in a place the pedlars opened
stores to supply the local inhabitants with all
they needed. *Ibid.*

gentoo [¹dʒen₁tu] *n. pl.* -s *prob. reg.* Cape
Town. A prostitute: see quot: see also
jentoe. [*fr. n. prop.* Gentoo, *U.S. ship
wrecked Struys Bay,* 1846 (*poss. fr.* Gentoo, Hindi *sig.* a Telegu-speaking Hindu)]
   . . . the American sailing ship Gentoo . . .
Among the survivors were a number of young
servant girls who had been engaged by wealthy
people . . . The girls soon drifted out of respectable employment into New Street . . . and
Keerom Street, which had bad reputations.
They set up places which became known as
'Gentoo houses', with Malay orchestras to
provide dance music. To this day a loose girl
is called a 'gentoo' by the Cape Malays. Green
S. Afr. Beachcomber 1958
   He could hear the voice of Boats and his
gentoo . . . smacks of wet kisses, chuckles . . .
his gentoo on his lap, Her arms . . . around his
neck and her mouth wide with vulgar laughter.
Muller *Whitey* 1977
   In the kitchen the seamen and their girls were
gathered in a group, the gentoos clacking away
like harpies, carmine talons fluttering at powdered throats, rapacious eyes gleaming . . . *Ibid.*

German print *n.* Also *Duitse sis, bloudruk
Kaffir print* (q.v.) see *amagerimani* and
last quot. at *sisi.*
   . . . a miscellany of stuff . . . and there were

bolts of german print, the sprigged navy-blue cotton stuff from which the dresses and aprons of the house servants were made. McMagh *Dinner of Herbs* 1968
　　Some Xhosa women . . . dress in a material commonly called German print. Radio S.A. '*Woman's World*' 11.9.72

gerook [xə'rʊək] *adj. slang* 'High' *usu.* on *dagga* (q.v.): see *roker.* [*Afk. gerook* smoked, cured]
　　In any case if the police haven't got him he'll be so gerook by this time that he probably won't even recognise you. Muller *Whitey* 1977

gerry *pl.* -ies. *Sect. Army.* See quot. at *vlossie.* ⫫ This is *prob.* a *pron. sp.* for *gharry* (q.v.).

gesig [xə'sɪx] *n. suffix.* View, sight, also *gezicht :* found in place names, e.g. Schoongesig/gezicht: see also *sig.*

gesiggie [xə'sɪxĭ] *n. pl.* -s. Pansy, heart's-ease or viola. [*Afk. gesig* face, countenance + *dimin. suffix* -(*g*)*ie* 'little face']
　　Pansies have a charm all of their own, the Afrikaans name gesiggies, describes them so well. *Daily Dispatch* 11.3.72

gesondheid [xə'sɔnt₁(h)eɪt] *interj.* Good health, a toast equiv. *Fr. a votre santé :* see quot. at *Mayibuye i Afrika cf. N.Z. kia ora* (Maori) good health. [*Afk. gesondheid* health *cogn. Ger. Gesundheit cogn. Eng. adj.* sound]
　　The first night, Van der Merwe raised his glass and said 'Gesondheid'. When Broeder Van der Merwe says 'Gesondheid' he is not saying his name. He is saying 'Good health'. *Sunday Times* 3.12.72

gestoofde [xə'stʊəfdə] *vb. part. lit.* Stewed: used to denote any of numerous stews or braised dishes; ~ *patats,* ~ *vleis,* ~ *skaappootjies* (sheeps trotters): see *abbr.* in quot. at *patat.* [*Afk. fr. Du. stoven* to stew]

Gestoofde Patats
　　1 kg sweet potatoes peeled and sliced, . . . brown sugar . . . salt . . . stick cinnamon . . . water . . . *Fair Lady* 8.6.77

gevalt [gə'valt xəvɔlt] *n. colloq.* A row, noisy quarrel, 'rort'. [*Yiddish* a row ⫫ This now has an 'unofficial' *Afk.* form *gewald fr. Yiddish*]
　　'It wasn't a bust-up; it wasn't a rumpus; and it wasn't a gevalt – it was an altercation. And you didn't pull the cord roughly and with jabs, as if it were a bust-up, a rumpus, or a gevalt – but gently and firmly.' Sachs *Bosman as I knew Him* 1971

gevonde [xə'fɔndə] *vb. partic.* Found:

found in place names e.g. Welgevonde. [*past. partic. Du. vb. vinden cogn.* find]

-gezicht [xə'sɪx, -'zĭxt] *n.* See *gesig* [*Du. gezicht* view]

ghaap [ɡɑp] *n.* Or *guaap:* applied to several species of *Stapelia* (Asclepiadaceae) with edible succulent stems and foetid-smelling flowers. [*prob. Hott. name guaap, u' gaap*]
　　. . . a short fleshy plant, well known to the Hottentots by the name of Guaap . . . It has an insipid, yet cool and watery taste, and is much used by them for the purpose of quenching thirst, for which purpose, it would seem, nature has designed it, by placing it only in hot and arid tracts of country. Burchell *Travels I* 1822

gharretjie ['ɡarĭkĭ, -cĭ] *n. pl.* -s. From *graatjie meerkat* see *meerkat.* [*Afk. fr. Nama poss. fr. Bushman Xara-* yellow meerkat]

gharry ['ɡærĭ] *n. pl.* -es. *Sect. Army.* A landrover or jeep: see also *gerry* and quot. at *vlossie.* [*Hindi gāri* a horse-drawn vehicle]
　　We've got two long wheelbase gharries in the maintenance workshops and one ordinary wheelbase gharry for our troop. *O.I. Serviceman* 14.4.79

ghaukum see *ghokum*

ghee See *Indian terms* and quot. at *bhajia.*

ghieliemientjie [ˌɡĭlĭ'mĭŋkĭ] *n. pl.* -s, ∅. Any of numerous small species of *Barbus* of the Cyprinidae, e.g. *rooikol* (*red spot*) ~ ; *Barbus tangandensis, rooi oog* (*red eye*) ~ ; *B. rubellus, dwerg* (*dwarf*) ~ ; *B. puellus, skraal* ~ etc. [*poss. fr. Xh. ngcilimintye, poss. rel. Scottish gillie*]

ghoen [ɡʊn, ɡŭn] *n. pl.* -s. Either the stone used in hopscotch see third quot., or most freq. the marble, often of a slightly larger size, used for shooting or throwing at other marbles: see *ironie, alie,* (1) *glassie,* (2) *queen, yakkie.* [*etym. dub. poss. fr. Malay gundu,* a metal filled object for throwing at nuts, *or Hott.*]
　　. . . Both had sold all their ghoens and allies. *E.P. Herald* 3.4.74
　　. . . Marbles for Africa. 49c per bag (100 + 1 goen) Marble season has started. *Ibid. Advt.*
　　. . . the many varieties of hopscotch . . . the illustrations . . . are most artistic. One, showing a little girl hopping on to a square after her 'ghoen', has her hair tossing realistically. *Cape Times* 7.2.75

ghokum ['ɡəʊkəm] *n. pl.* -s. The fruit of several species of *Carpobrotus* esp. *C.*

edulis, used for jam. ⟦The leaves are used medicinally for a gargle: see *Hottentot fig* and *sour fig*. [*Hott. name (t')* *gaukum*]

Crystallised ghurkums [*sic*] and naartjies are novel delicacies which she expects to put on the market. *Argus* 1.2.24

ghomma ['gɔma, -mə] *n*. A Malay hand-drum used for the accompaniment of ~ *liedjies* (q.v.). [*prob. fr. Javanese gamelan(g) or ghom* gong *prob. rel. Ur-Bantu ngoma* (q.v.) drum]

The ghomma-liedjies, on the other hand, were all accompanied by the ghomma, a single headed drum made from a small cask with a skin nailed over one of the two open ends. Kirby 1939 *cit.* Du Plessis *Cape Malays* 1944

ghommaliedjie ['gɔmə₁lïkï, -cï] *n. pl.* -s. See quot. [*ghomma* (q.v.) + *Afk.* lied *song* + *dimin. suffix -jie cogn. Ger. Lied(er)*]

. . . and the ghommaliedjie, which is a picnic song. At a Malay picnic the participants form a circle, join hands, and walk slowly, singing a verse of a Dutch folk song. This is followed by a ghommaliedjie, led by a ghomma-player seated in the centre of the circle. Du Plessis *Cape Malays* 1944

ghurum masala See *garam masala*, also *Indian terms*.

ghwarrie See *guarri*.

giant protea *n*. See *king protea*.

gif [xɪf] *n*. Poison: prefixed to the names of numerous toxic plants, e.g. ~*bol* (q.v.), ~*boom* (q.v.), ~*bossie* etc. also ~*siekte* (q.v.). [*Afk. fr. Du. gift* poison]

gifbol ['xɪf₁bɔl] *n*. Any of several species of Amaryllidaceae with highly toxic bulbs and leaves, said by early naturalists to be the source of the poison used by the Bushman for arrows. [*Afk. gif fr. Du. gift* poison + *bol* bulb]

This plant is well known to the Bushman, on account of the virulent poison contained in its bulb. It is also known to the Colonists and Hottentots, by the name of Gift-bol (Poison-bulb). Burchell *Travels I* 1822

gifboom ['xɪf₁buəm] *n*. Name applied to several toxic species harmful to people and to stock: ⟦The seeds of one, *Hyaenanche globosa* were said to be crushed and used by Bushmen and Hottentots for tipping arrows: another is an 'invader plant' see quot., also *plant migration*. [*Afk. fr. Du. gift* poison + *boom* tree *cogn. Ger. Baum*]

In the dry sub-tropical route was found migration of the gifboom, melkbos (or spurge),

honey locust, blackthorn and skilpadkos (or vygie). *E.P. Herald* 28.2.73

gifsiekte ['xɪf₁sïktə] See *miltsiekte*.

girdle of famine *n. pl.* girdles. *obs* Traveller's term: see *hunger belt*.

The pangs of hunger pressed sore upon us and our only relief was to draw our 'girdles of famine' still tighter round our bodies Thompson *Travels II* 1827

Round his loins is a double thong, also thickly set with brass rings. This is regarded as a great ornament in South Africa; and it also serves as a girdle of famine to confine the stomach, if on a journey food runs short Alexander *Western Africa I* 1837

girl *n. pl.* -s. A female servant: regarded by many as offensive, though freq. used by black S. Africans also. ⟦Poss. orig. to avoid 'maid' for a married woman or the term 'servant', or by *analg. boy* (q.v.)

In combination *wash*~, *ironing* ~ etc

. . . I asked them 'Does your mother want a girl? Go ask your mother if she wants a girl. I would have gone Boesman. Fugard *Boesman & Lena* 1969

. . . when mom decided to do without a girl . . . she was costing too much . . . pa fixed up the girl's room in the backyard for him *Darling* 26.11.75

gits [xɪts] *interj. colloq.* An exclamation of dismay or surprise [*Afk. gits*, one of numerous euphemisms for *God*]

Critics . . . will inevitably find some of their favourite South Africanisms missing. What of 'gits' as an expression of surprise or self disgust . . .? Gordon Jackson *cit. To the Point* 14.4.78

give *vb. trns. substandard*. Teach: e.g. 'What subjects do you give? I give English and Geography.' [*trans. Afk. gee prob. abbr. onderwys gee* give instruction]

glasogie ['xlas₁uəxï] *n. pl.* -s. See *witogie* cf. *Austral. wax-eye*. [*Afk. glas cogn.* glass + *oog* eye + *dimin. suffix -ie*]

glassie *n. pl.* -s. **1.** *colloq.* among children A clear glass marble sometimes called *queen*, formerly *sodie* (*fr.* soda water bottle): see also *alie, ghoen, ironie, yak, kie*, and quot. at *it*.

They carry their marbles in bags sometime 100s big glassies and small glassies bloc alleys and commons. You shoot the marble out of the ring with a big glassy. Vaughan *Diary* circa 1902

2. Diamond trade term: see quot.

In their natural eight-sided shape, diamonds are known in the industry as 'glassies'. If, as often happens, they are twin stones or triangles they are called 'maccles'. White *Land God Made* 1969

gnu [nu:, nju:] *n. pl.* Ø, -s, -e. Also *wildebeest* (q.v.): any of several large ungainly S. Afr. antelopes of the *Connochaetes* with ox-like head and other characteristics, but with a mane and tail not unlike a horse. [*fr. Bushman nqu*]
. . . the gnu uniting the antelope, the horse and the ox. Rose *Four Yrs. in S.A.* 1829
Of all quadrupeds the gnoo is probably the most awkward and grotesque. Harris *Wild Sports* 1839

go-away bird *n. pl.* -s. Kwevoël, *Corythaixoides concolor*: see *loerie*, also quot. at *spooring*. [*onomat.*]
Among the less co-operative birds, or perhaps just one that understands the human being is Nkwenyane, the Grey Lourie with his harshly unmistakable cry of 'go away'. Bulpin *Lost Trails* 1951

goduka See *amagoduka* and quot. at *u-clever*.
There was the urgency of someone who had a long way to travel in his gait. It was doubtless a 'goduka' (migrant labourer) on his way home to his family after many months of work in the city. It might even have been years. Matshoba *cit. Staffrider* Vol. I No. 3 1978

goed[1] [xŭt] *n. suffix.* Material, substance: e.g. *matjies* ~ (q.v.), *hotnotskou* ~ (q.v.), *hotnotskooi* ~ (bedding). [*Afk. goed* substance]

goed(e)[2] [xŭt, -də] *adj. or adv. prefix.* Good, found in place names as *adj.* in plain form or with *attrib. inflect.* -e e.g. Goedgeloof (faith), Goedewil, Kaap de Goede Hoop, also sig. well: found in place names e.g. Goedverwag, Goedgegun. [*Afk. goed cogn.* good *adj.*, well *adv.*]

goef [gʊf, gŭf] *vb. intrns., also n. slang.* Swim, bathe, see quot. at *cossie. cf. Austral. bogey.* [*unknown*]
Yesterday arvey we went for a goef in the (censored) and Dumbo nearly stood on a crocodile . . . *Darling* 14.4.76

goëlery [ˌxʊələˈreɪ] *n.* Malay magic, occult happenings, conjuring see *slamaaier. cf. Canad. jonglerie* sorcery, trickery. [*Afk. goëlery fr. Du. goochelen cogn.* juggle, *Fr. jongleur, Lat. joculator*]
'Malay trickery' and 'goëlery' are now widely accepted and firmly believed terms to many of the Cape's white and non-whites. The traditional old Cape Malay khalifa is mainly responsible for the perpetuation of the popular Malay magic myth. *Cape Times* 6.6.70

goeters [ˈxŭtə(r)s] *pl. n. slang.* Goods and chattels, belongings, bits and pieces: see also *goodies cf. Canad. iktas,* goods, belongings. [*Afk. colloq. goeters prob. rel. goedere* property, goods *etc.*]
I've counted all the cutlery
So don't you drive away
With the goeters from the drive-in caff.
Pip Freedman (song) *Radio S.A.* 28.3.72

goffel [ˈgɔfəl] *n. pl.* -s. *slang* (prob. coloured usage) An ageing prostitute: also known as *afkop* ('no head') *fr.* habit of concealing face and head under bedclothes. *Coloured Informant ex Cape Town* June 1979 ¶ Loosely (*prob. erron.*) among whites, a coloured person.
'Who are these people?' 'Outies, goffels, rookers. Old ones who have no place. The bulldozer broke their place . . . They have no people. This one here's a goffel' . . . jerked the blanket to reveal a dishevelled, tattered female. Muller *Whitey* 1977

go garshly [ˈgɑʃlɪ] *imp. vb. phr. colloq.* Go slowly, carefully, proceed with caution: see also *go well, hamba kahle,* and sp. form in first quot. at *kehla.* [*anglicized corruption of Ngu. hamba kahle* go well, a leave-taking greeting: *freq. pron. of kahle* (q.v.) *among whites as above*]
Kahle . . . corrupted in kitchen kaffir to 'gaashly'. FitzPatrick *Jock of Bushveld* 1907

gogga [ˈxɔxɔ, -xa] *n. pl.* -s., *dimin.* ~ *tjie pl.* -s. *colloq.* Insect [*Afk. fr. Nama Xo Xo* insect or other small creature]
1. General term for any insect: see also *nunu* and second quot. at *perlé.*
The yellow clouds of sulphur billow out like a firework display as men and machines move into the orchards getting in before the destructive goggas. *Cape Times* 21.8.73
2. Term of endearment to a child or person of small stature.
Well, children, it's time to say goodbye to all you little goggatjies. *Radio S.A.* 30.5.72
3. In a political sense, a bogey, jinx.
I must congratulate . . . on coming up with a brand new political gogga – 'international liberalism'. *Sunday Times* 9.5.71
4. Angler's term for a non-edible fish.
. . . a small shark caught by Dave . . . won the prize for the heaviest 'gogga' (non-edible fish). *Grocott's Mail* 11.5.73

gogog(o) [ˈgɔˌgɔg(ɔ)] *n. pl.* -s. A paraffin tin usu. carried full of water on the head, also used as a measure of roughly two *buckets* (q.v.). *cf. Jam. Eng. tuku-tuku tukka-tukka,* gourd for carrying water. *Also poss. reg. E. Cape* used of a *can* (q.v.) or jar of wine. [*Ngu. igogogo* paraffin tin *rel. Zu. vb. -gogoza* (*onomat.*) to rumble, bang, as of empty tins]

When a woman carries water in a gog-gog on her head she puts leaves in the bucket to stop the water splashing when she walks. Griffiths *Man of River* 1968

gologo [₁gɔ̃ˈlɔ̃gɔ̃] *n. Afr. E.* African term for spirituous liquors. [*Zu. ugologo* European-type spirits *fr. Eng. grog*]

Have a double brandy. The liquor here is much weaker than what we drink back home . . . We call it gologo from grog. Matshikiza *Chocolates for Wife* 1961

golovan See *cocopan.*

gom [xɔm] *n. pl.* -s. *slang.* Also ~ *ie,* ~ *tor,* lout: a coarse or common person. [*fr. Afk. gomtor* lout]

That green pork-pie of his looks a proper gom's hat. O.I. 2.10.73

And so what if this ou . . . has got a bit of a gommy accent . . .? *Darling* 17.8.77

gom-gom *n. pl.* -s. (*obs.*). A Hottentot musical instrument, similar to the *gorah* (q.v.); ⫽Described by Kolb as being similar to a Jew's Harp. [*pron. and etym. unknown*]

When three or four of those Gom Goms are played upon in Concert by skilful Hands, I must confess I think the Harmony extremely agreeable. Kolb *Descr. Cape of G.H.* trans. Medley 1731

goniva(h) [ˈgɔnɪvə, -ĭva] *n. pl.* -s. *colloq. prob. obs.* A stolen or otherwise illicitly acquired diamond. [*Hebr. genavah* a stolen thing *fr. gannav,* a thief]

. . . then ship him to Europe like crooked specie or a special goniva – goniva, my poor innocents, is the polite term used among the learned professors of the game to denote a stolen diamond. Cohen *Remin. Johannesburg* 1924

gonya [ˈgɔ̃nja, xɔ̃n-] *n. pl.* -s. Species of *Pneumora :* see (2) *blaasop.* [*onomat.*]

goodfor [ˈgʊdˌfɔ̃] *n. pl.* -s. *hist.* An I.O.U.: see also quots. at *papbroek* and *blueback.* [*presum. abbr. good for the sum enumerated*]

No 'good for', 'I.O.U.', or other acknowledgement of debt, not being a promissory note, shall require to be stamped, so long as it shall be retained by the creditor to whom it was first delivered, and it may be paid by the debtor to such creditor without being stamped. *Cape Town Directory* 1866

Good Hope, Order of *n. prop.* S. Afr. Order of merit instituted by the *State President* (q.v.) of the Republic of S.A., Grand Master of the Order, in March 1973. ⫽The ~ can be awarded in five classes to foreign officials and persons of rank and standing. The ribbon is green and gold, and the badge, an eight pointed star, carries the legend *Spes Bona.* [*presum. fr. Cape of Good Hope*]

The State President . . . decorated President Stroessner with South Africa's 'Order of Good Hope – Special Class'. *Panorama* June 1974

goodie *n. pl.* -s. *colloq.* Any unspecified article, (occ. even person) of which the name may not be known: see also *dinges* (*dingus*), *goeters.* ⫽Not usu. sig. sweets or food in S.A.E. [*prob. fr. Afk. goed(ere)* goods, property]

Very delicately I would like to mention that a lot of the goodies, now reposing in elegant living rooms in Pretoria or Johannesburg, were found in abandoned henhouses, old lofts, dirty garages or sommer standing around in the rain. Van Biljon *cit. Star* 16.6.73

goodself, your *prn. phr. Ind. E.* Mode of written address, an item of 'commercialese', (see quot.) in use in Indian English, poss. to avoid the direct form 'you'. See note at *third person address.*

For instance, a letter written in typical commercialese might begin '*Your esteemed favour of even date to hand and we beg to thank your goodself for same*'. Fortunately things have changed for the better since Fowler wrote that – but it is by no means dead yet, is it? *E.P. Herald* 28.2.74

goodwill [ˈgʊdˌwɪl] *n. Ind. E.* Euphemistic term among S. Afr. Indians for key money charged by estate agents where accommodation is at a premium, occ. for 'protection money' extorted from business people by gangsters.

gooi [xɔɪ] *vb. slang.* Throw, fling: as *n.* a sexual fling (*vulgar*). [*Afk. fr. Du. gooien* to fling]

'Some other time' he reckons, and goois a wink at my boet and the ole man, who take the hint and scram. *Darling* 15.9.76

She loved the champers . . . Once the glass was flattened she sommer gooied it on the floor. The carpet prevented any breakage . . . *Daily Despatch* 28.9.78

Also in combination ~ *tackie* (motor cyclists' *slang*) accelerate.

Gooi tackie when the lights tune favour. O.I. 1976

and ~ *ankers,* to brake suddenly.

If yous bromm thru a speed trap do yous . . . Gooi ankers an' slaat the 'Gatso' to pieces? *Bike S.A.* Oct. 1976

gook *n. pl.* -s. *Rhodesian* A terrorist: see

quot. at *terr.* [*prob. fr. U.S.* gook]

goolab jambo See *Indian terms.*

gopse [xɔ̆psə] *n.* A low-grade or slummy area occ. *gops* a similar person. [*Afk. gops* slum area]

gorah [ˈgɔra, -ə] *n. pl.* -s. *obs.* A Hottentot musical instrument presum. similar to the *gom gom* (q.v.) [*etym. dub. poss. fr.* Bushman *ko//ha, poss. Nama*]
The Gorah, as to its appearance and form, may be more aptly compared to the bow of a violin, than to any other thing; but, in its principle and use, (it is quite different; being, in fact, that of a stringed, and wind instrument combined; and thus it agrees with the Æolian harp. Burchell *Travels I* 1822

gorrel [ˈxɔrəl] *n. pl.* -s. *slang.* Throat: windpipe. [*Afk. gorrel (pyp)* throat, windpipe]
He doesn't put fat in the pet's mince, just the gorrels and lungs and things, that's why it keeps so well. O.I. 1975

goudsboom [ˈxəʊtsˌbʊəm] *n.* See *kreupelbos.*

gousblom [ˈxəʊsˌblɔ̃m] *n. pl.* -s, -me. See also *botterblom.* [*Afk. fr. Du. goudsbloem* golden flower]
'A name applied to a number of species of *Compositae* . . . with large yellow heads and which are chiefly found in South West. The vernacular name was first applied to species of Arctotis and Gazania.' Smith *S. Afr. Plants* 1966
'Big orange stars. I have seen them in gardens in England –' 'Gousblom – gazanias, I mean –' Fairbridge *The Torch Bearer* 1915

government flour *n. obs.* The unsifted brown flour from which all bread in S.A. was made during World War II when white bread was not permitted by law. See *Bremer Bread.*

government sugar *n.* The least expensive type of brown sugar. [*poss. orig. fr.* '*Government subsidized*']
There's a world shortage of sugar, you can have only 5lbs of white but there's plenty of the brown government. O.I. Supermarket 1974

govini [gəˈvin(i)] See quot. at *shimiyaan.*

go well *imp. vb. phr.* Farewell greeting to those leaving: see also *hamba kahle, mooi loop, stay well. cf. Jam. Eng. walk good, drive good. E. Afr. kwa-heri,* (go/ be) at peace. [*trans. Ngu. hamba(ni) kahle* (q.v.)]
'Go well, umfundisi.' 'Stay well, inkosi'. Paton *Cry, Beloved Country* 1948

gown *n. pl.* -s. Dressing gown: in *Afr. E.* and *Ind. E.* called *morning* ~ (q.v.)

[*abbr. dressing gown*]
Winter gowns in cosy brushed terylene: two styles . . . sizes S.M.L. R12.99 and R13.99 *E.P. Herald Advt.* 4.3.74

gracht [xrax(t)] *n. pl.* -s. Canal, moat: found usu. in street names in Cape Town e.g. Heerengracht, Buitengracht. *cf. furrow.* [*Du. gracht Afk. grag* moat, canal, ditch]
. . . the streets lay almost due north and south, east and west, at as regular intervals as though scored with a ruler; the Heerengracht, Keisersgracht, Tuin Straat, . . . with their comfortable roomy dwellings and hospitable high stoeps. Fairbridge *Which Hath Been* 1913
'The Cape' had now become 'Cape Town'; and a pleasant town it was with oak-lined streets and rivulets – 'grachts' – running along them. Gordon-Brown *S. Afr. Heritage* 1965

Grahamstadter [ˈxrɑmˌstad(t)ər] *n. pl.* -s. Voortrekker name for a wagon made in Grahamstown, esp. for the stout eighteen-foot buckwagon half-tented and with a screw brake used extensively for *transport riding* (q.v.) throughout the interior. ℙThought to have been built only by wagonwrights named Brookshaw. [*Du. Grahamstadt* Grahamstown + *-er personif. suffix*]

gram(flour) *n.* See *Indian terms.*

gramadoelas [ˌxramaˈdŭlas, -əz] *pl. n. colloq.* Back of beyond, the wilds: see also *bundu cf. Austral. Back of Bourke, nevernever land/country* etc.: see *back-veld* [*poss. fr. amaduli pl. of Ngu. -duli* hill]
. . . she was a little girl staying in the lonely African wilds. *Gramadoelas* was the word that Aunt Susan used . . . It was loose talk about wilds and gramadoelas and tropics that gave the Marico a bad name, we said. Bosman *Jurie Steyn's P.O.* 1971
Single Girl's Medical kit: Snake-bite outfit. A must if you're in the *gramadoelas. Darling* 29.10.75

grandad(s) army, the *n. prop. Rhodesian* The call-up group of older men conscripted into the police field reserve for the emergency two-week election period beginning 12.4.79. ℙ acc. *Time* 30.4.79 also known as *Wombles.*
Like a rag-tag band of old warriors Rhodesia's grand-dads are marching into action. . . Terrorists beware. The grand-dads are going to war. They might be a bit long in the tooth and grey in the beard, but the determination is there [Above] Two of Rhodesia's grand-dad army. *Sunday Times* 8.4.79

grease table *n. pl.* -s. A sorting device used on S. Afr. diamond mines, see second quot.

⦿ . . . 'if those are not tables once used to wash the "stuff" I'm a Dutchman.' Haggard *Solomon's Mines* 1886.

. . . then gravitate down to the grease *tables*. These have sloping three-stepped decks which are vibrated . . . The water washes away the gravel while the diamonds stick to the grease and are periodically scraped off by hand. Hocking *Diamonds* 1973

Great *adj. freq. capitalized.* Of or pertaining to an African Chief as in ~ *place* (q.v.) ~ *hut/house* (q.v.), or first in position as being closest in relation to the chief, (not usu. in actual seniority) see quots. as in ~ *wife*, ~ *widow*, ~ *son.* [*trans. Bantu -kulu,* great]

. . . to decide as to the investment of two of his wives with the respective dignities of 'the great one' (omkulu) and 'the one of the right hand' (owasekunene). The mother of him who is to be the 'great son' may thus be the last wife the chief has taken. Dugmore *Diary* 1846–47

. . . when he was getting on in years he was expected to contract a political union with the daughter of some other powerful potentate and in this way enhance his own power and prestige. His newest acquisition became his great wife and their first male child became the heir to the throne. Metrowich *Frontier Flames* 1968

Great hut/house *n.* The chief's hut, the royal house. [*trans. Ngu.* (*i)ndlunkulu* chief hut, royal house]

INDHLUNKULU, a native term signifying the great house. Sisson *S. Afr. Judicial Dict.* 1960

Great Place *n. usu. capitalized.* The official residence or (6) *kraal* (q.v.) of the chief of an African tribe. See last quot. at *Robben Island.* [*trans. Ngu. -komkhulu* the great place, seat of the Chief]

. . . the loyal Middelburg and Lydenburg commandos, had sufficed to burn Sekukuni's Great Place and drive him northwards. Brett Young *City of Gold* 1940

Also found in place names: see quots.

Basutos to visit Rarabe Great Place. *Daily Dispatch* 8.8.72

Thousands of people of all races gathered at the Nyandeni Great Place . . . at the weekend to pay their last respects to Paramount Chief Victor Poto. *E.P. Herald* 22.4.74

Greek shop *n. pl. -s.* A café or *tearoom* (q.v.) often owned by a Greek or other person of Mediterranean or Middle E. origin. *cf. Jam. Eng. Chiney/China shop* (grocer).

. . . the ubiquitous corner cafe . . . Not a cafe at all, but a neighbourhood convenience store, this type of business is so dominated by the . . .

Greek community that it is also known as the 'Greek shop'. *Sunday Times* 24.7.77

greenie ['griniǐ] *n. pl. -s. slang.* Usu. among children. A ten *rand* (q.v.) note. *cf. U.S.* greenback, and earlier S. Afr. *blueback* (q.v.) also *Jam. Eng. plaintain leaf,* one pound note, and former *Austral. blueback* or *bluey,* a five pound note. [*fr. colour of note*]

green mamba *n. pl. -s.* **1.** *Dendroaspis angusticeps,* a poisonous snake: see *mamba.*

2. *figur.* Also *groen~* (green) Creme de menthe (peppermint liqueur). *cf. Brit.* (*ex Navy*) starboard light.

Words denoting liquor: Chwala, Groenmamba (Peppermint Liqueur), Snorts (tot), Tkierie (Kaffir beer). Opperman *Spirit of Vine* 1968

3. A *Putco* (q.v.) bus.

Having lived all their lives so far in Johannesburg in a tiny house . . . on a busy street down which incessant traffic and 'green mambas' screamed, it was understandable that they should feel vulnerable and exposed on the open veld. Roberts *Outside Life's Feast* 1975

gril [xrɪl] *n. pl. -le, -s. slang.* Shiver, shudder, 'the creeps'. [*Afk. gril* shudder]

This wall chart, attractive (if you don't mind the *grrrils* up and down your back) . . . is to be distributed to all hospitals, the army, the police force . . . which could benefit and help save victims' lives. *Cape Times* 21.6.73

Griqua ['grɪkwə, -wa] *n. pl. -s.* A member of a people of mixed descent, *Hottentot, White* and *Bushmen,* all (q.v.): see also (2) *Bastaard* and quot. at *classification.* ~*land,* the former home of this people. [*etym. dub. prob. fr. Hott. name*]

Coloured land call by Griquas Descendants of the Griqualand leader Adam Kok are demanding 1 000 square miles of fertile farmland near Kokstad to form the nucleus . . . of the country's first Colouredstan. . . . White farmers in the district are up in arms over the proposed 'Griquastan'. *E.P. Herald* 9.8.74

grit *n. Sect. Mining.* Diamond particles used for industrial purposes.

Industry uses mainly small diamonds or 'grit' which is set in a matrix to form a cutting, grinding or drilling surface. *Panorama* Dec. 1972

groen [xrŭn] *adj. prefix* Green, found in place names e.g. Groenvlei, Groenkloof. [*Afk. fr. Du. groen cogn.* green]

groenmamba See *green mamba* (1), (2).

Grondwet ['xrɔ̃ntₗvet] *n. prop.* The Constitution of the South African Republic, adopted in 1858, also used of the Con-

stitution of the Republic of Natal (Oct. 1838). [*Du. grondwet* constitution *fr. grond* foundation + *wet* law]
The first true fundamental law of the Trekkers is the *Grondwet* of Natal (October 1838). The *Grondwet* of Natal defined with some care the structure and functions of the legislature and of the judiciary, but said little about the executive. *Cambridge Hist. VIII* 1936 *ed.* Walker
The Rustenburg *Grondwet* of the South African Republic was a long and unwieldy document containing much miscellaneous material alien to a fundamental law. In general structure it resembled the Free State Constitution . . . but there were certain characteristics that reveal the traces of twenty years of Trek History. *Ibid.*

groot- [xrʊət-] *adj. prefix.* Big, great: found in place names Grootfontein, Grootbrakrivier. [*Afk. fr. Du. groot* big *cogn.* great]
Ubique means 'Entrain at once for Grootdefeatfontein'. Kipling *Five Nations* 1903

grootbek [ˈxrʊətˌbek] *n. pl.* -s. *slang.* A talkative usu. boastful person, a 'loudmouth'. *cf. Austral. ear-basher*, an inordinate talker, *Jam. Eng. moutimouti* (*fr. mouth*) [*Afk. groot* big *cogn.* great + *bek* (q.v.) mouth *cogn.* beak]
. . . he recognised the face on the photograph. It was that of Mussolini, heavy-jowled, at his most confident and aggressive. 'Look at him . . .' Mostert muttered, his voice rising. 'Just look at him, *grootbek* Caesar!' Krige *Dream & Desert* 1953

grootpraat [ˈxrʊətˌprat] *n. slang.* Boastful or indiscreet talk. *cf. Austral. spruiking, Jam. Eng. laba-laba, Anglo-Ind. buck,* bragging talk. [*Afk. groot* big + *praat* talk *cogn.* prate]
My fault with all my grootpraat. I asked them to dinner and we had a ghastly evening. O.I. 1972

Group Areas Act *n. prop.* Act 41 of 1950 which provided for the establishment in each urban area of a separate *Group area* (q.v.) for each race group, and prohibited occupation or ownership by members of any other racial group.
According to a proclamation under the Group Areas Act in 1968, Schornville has been zoned as part of the 'Whites only' Municipal area . . . This means that when alternative accommodation has been provided in Breidbach, Schornville's people will have to leave their homes and move to the village, seven kilometres from here. *Daily Dispatch* 17.5.71
hence *group area, n. pl.* -s. Part of an urban area which, under the *Group*

Areas Act (q.v.) is reserved for a particular racial group of the population, in which rights of occupation and/or ownership are restricted to members of that group.
. . . will be breaking the Group Areas Act if he lives with his elderly and ailing mother because, while she is classified as Malay and will shortly be moving to a Malay group area, he is classified as an Indian. *Daily News* 24.4.71
LENASIA, the Indian group area Johannesburg never wanted, has become a massive financial burden to the City Council. *Argus* 19.4.73

grunter [ˈɡrʌntə] *n. pl.* ∅. Also *U.S.* Any of several fishes which make a grunting noise when taken from the water: These include two of the Sparidae, *Lithognathus lithognathus*, the *pignose* ~ or *white steenbras* (q.v.), *Pagellus natalensis* the *red* ~, *red/rooi chor chor* and numerous species of *Pomadasys* and *Rhoncissus* known as *rock* ~, *silver* ~, *bull* ~, *spotted* ~ (see *tiger*), *purple* ~ etc.: see also *kno(o)rhaan cf. Austral. pigfish, trumpeter* and *Jam. Eng. grunt* (also with numerous compounds).
. . . the 'grunter king' who is renowned for his catches of outsize spotted grunter (tiger); . . . *Grocott's Mail* 14.1.72

grysbok [ˈxreɪsˌbɔk] *n. pl.* ∅. A small, greyish brown S. Afr. antelope *Raphicerus melanotis;* also *R. sharpei* (Sharp's ~). [*Afk. grys fr. Du. grijs* grey + *bok cogn.* buck]
The gries-bok is also of the size of a common deer, but bears a considerable resemblance to a goat; its colour is greyish and the hair loose and frizzled. This species . . . does a great deal of mischief to the gardens and vineyards in the night time. Percival *Account of Cape of G.H.* 1804
Humansdorp: Steenbok declared ordinary game; season June 1 to July 31. Grysbok declared ordinary game; season May 31 to July 31. *E.P. Herald* 19.3.73

GST [ˈdʒiˌesˈtiː] *acronym.* General Sales Tax (4%) introduced on all purchases over 12c at the end of June 1978.
The expected slump in retail sales after the June pre-gst spending spree has not happened, say retailers . . . is also 'extremely happy' with post-gst sales. His supermarket sales were not down . . . 'Gst has been a non-event in terms of July sales.' *Financial Mail* 28.7.78

guarriboom, guarribos [ˈɡwarɪˌbʊəm, -ˌbɔs] *n.* Any of several species of *Euclea* esp. *E. undulata*, or of the allied genus *Royena*, a shrub of up to 2,5m in height much browsed upon by stock, having succulent, slightly astringent fruit. ⫽~ *bessies* (berries) were used, fermented, by the Hottentots to make a type of vinegar: the roots of some species are used medicinally as in an infusion of the leaves (~ *tee/tea*): ~ *honey* from bees feeding on the nectar of the flowers of either *Euclea undulata* or *Royena*. [*prob. Hott. gwarri plant name for Euclea, Zu. umgwali*]

Guarri Bush (Euclea undulata) has a sweet taste and is eaten by the Natives . . . bruised and fermented they make vinegar. Mackrill *Diary* 1809
'Gwarrie' protection plea.
A University botanist and a lay authority on indigenous trees . . . fear that a demand for gwarrie wood will result from the news that Eastern Province braaivleis champion . . . attributed his success to . . . this particular indigenous wood. *E.P. Herald* 3.3.77

guava, to come on your, slip on your *vb. phr. slang, lit.* and *figur*. To slip up, make a fool of oneself etc. [*fr. Afk. vulgarism koejawel* guava]

I should also add that the tasting showed up how easy it is to slip on your guava when it comes to judging and making pronouncements about wine. Chisholm *cit. Cape Times* 14.4.78

gubu [ˈgubŭ] *n*. A *Zu*. musical instrument: see quot. [*Zu. u(lu)gubu Ngu. onomat.* a hollow sounding thing]

Their chief instrument, called a *gubu*, which is something like a one-string banjo, with an empty gourd for a drum. Clairmonte *The Africander* 1906 *cit*. Pettman
⫽Musical bow with single string fitted with calabash resonator attached to the stave near its lower end, the string (of sinew, hair or now brass wire) is struck by a stalk of tambootie grass. This is generally played by women to accompany the voice. Doke & Vilakazi *Zulu Dict*. 1948

gu(i)lder [ˈxœldər, ɡɪldə] *n. pl.* -s. *hist*. Unit of currency in use at the Cape, not equiv. to the guilder of Holland.

A guilder is now a nominal coin, one third of a rixdollar, which at the present exchange, is about 1s.6d. sterling. Bird *State of Cape of G.H.* 1823

gumboot dance *n. pl.* -es. A lively African dance performed in rubber boots ('Wellingtons') which are given resounding rhythmical slaps: see also *ngoma dancing*

What banishment means
Is never to see again,
Your inyanga's Fly whisk.
Nor the bones that expound our fate.
Oh! exile means

Never to see a Baca gumboot dance, never!
Poem: Paul Vilakazi *U.N.I.S.A. Eng. Studies Poetry* 1974
How about looking in at Crown Mines or City Deep and showing us the mine dancers, specially the gum boot-dance? I'm sure all races would enjoy that. *T.V. Times* 14.11.76

gwaai [gwaɪ] *n*. **1.** *colloq*. Tobacco, snuff. *cf. Canad. stemmo*. [*Zu. ugwayi* tobacco, snuff]

However, his handling of the 'gwaai' was as nothing compared with the old man's technique. *Evening Post* 14.11.70
2. *slang. n. pl.* -s. *poss. reg*. Natal: Cigarettes, or 'smokes' generally. [*fr*. (1) ~]

gwarrie See *guarri*

gyppo [ˈdʒɪpəʊ] *vb. n*. and *n. modifier. Sect. Army vb*. To dodge or wangle one's way out of doing something *cf. Austral. to donga* (q.v.) or *bludge*. e.g. To ~ out of doing P.T., going on a route march, etc.

Either the trick or dodge employed, or the person who employs it. *cf. Austral. dongarer, dongaring* etc. (q.v.) ⫽ Not equiv. World War II *abbr*. for *Egyptian*.

It's a lekker gyppo if you're good at sport – gets you out of doing all sorts of things and often you travel. *O.I Serviceman* 15.2.79
. . . he had shouted to Gunners . . . and . . . to 'see to him' as he was a 'gyppo' (lazy person). *Sunday Times* 24.9.78
If a soldier was 'gyppoing' – shirking his exercises – he should be given the benefit of the doubt. *E.P. Herald* 17.5.79
as modifier a ~ *gat* (vulgar), One who always ~ es his way out of work or distasteful tasks; *a* ~ *bed*, a bed 'squared off' etc., army-style at the top ready for inspection, the lower half only being used overnight by its occupant.

H

haakdoring [ˈhɑːkˌdʊərɪŋ] *n. pl.* -s. Any of several species of *Acacia* esp. *A. detinens* (*A. mellifera*) a deciduous tree of up to 7,5m with large curved thorns in pairs: foliage and seed pods are eaten by stock. [*Afk. fr. Du. haak* hook + *doring fr. Du. doorn cogn*. thorn]

The country through which we passed was much encumbered with the accursed *Haakdoorn* or *Wagt een beetje* (*Acacia detinens*), from which I had formerly suffered so severely Thompson *Travels I* 1827

haak-en-steek (**doring**) [ˌhɑkənˈstɪək] *n*

A large flat-topped deciduous thorn tree *Acacia tortilis* with long white straight spines and blackish curved spines. [*Afk.* *haak-en-steek* hook and stab *fr.* curved and straight thorns]

If you'll wear your Nagmaal jacket next time . . . I'll be glad to show you all over my farm where I'm not going to plant potatoes . . . That is, among the haak-en-steek thorns. Bosman *Bekkersdal Marathon* 1971

haarder See *harder.*

haas [hɑs] *n.* Hare, rabbit: [*Afk. haas,* rabbit, hare] **1.** Suffixed to names of species, e.g. *spring* ~, *berg* ~. **2.** Generic name for any type of hare or rabbit: also *dimin.* form *hasie.* **3.** Prefixed to certain plant names, e.g. ~ *oor* (ear), ~ *gras* etc. *cf. Brit. harebell.*

haasbek [ˈhɑsˌbek] *adj. colloq.* Toothless. usu. in form *go* ~. ⫽Not equiv. *hare-lip* (see quot.) nor of *Jam. Eng. buck mouth,* protruding teeth. [*Afk. haas* rabbit, hare + *bek* (q.v.) mouth *cogn.* beak]

Walk around toothless for a few months. Iu South Africa this is called 'going haasbek'. *Personality* 5.6.69

hadeda(h) [ˈhɑdĭˌdɑ̆] *n. pl.* -s. Various sp. forms. *Bostrychia hagedash:* also ~ *ibis,* a large ibis, greyish brown to olive coloured, with a harshly strident call. [*presum. onomat. fr. bird's call poss. influenced by name hagedash*]

Overhead a ragged formation of Hadedas tacked against the wind and hurled volleys of oaths at those below in high-pitched, raucous voices. Fulton *Swear to Apollo* 1970

hairy(back) *n. pl.* -s. *slang.* An offensive mode of reference to an *Afrikaner* (q.v.) also *hairy pl.* -ies. See also *rock(spider),* *crunchie, mealie cruncher* and quot. at *Dutchman.* [*unknown: poss. fr.* 'hairy at the heels' *sig.* crude *etc.*]

There should be a match . . . between the Springboks and an Invitation side from the hairiest hairy-back elements of the Universities of Cape Town and Stellenbosch. *Cape Times* 16.5.70

He said he did not consider words like 'rock spider' or 'hairy back' insulting to Afrikaners as a group, but he agreed that these and similar terms for Jews or Africans could harm race relations. *E.P. Herald* 17.4.73

Haj [hădʒ, hadʒ] *n. prop.* Not exclusively S.A.E. The pilgrimage to Mecca made by Indian Muslims and by Cape Malays. [*Arab. hajj*]

The Muslim community is in an uproar over the barring of well-wishers of pilgrims from the

. . . airport terminal building. Cape Herald received complaints from several irate Muslims who were refused permission to enter the terminal building to bid farewell to their friends and relatives who left for Haj last week. *Cape Herald* 14.9.74

deriv. ~ *ee,* ~ *i* mode of address or reference to one who has made the pilgrimage, also titular with surname.

Hadji Abduraghiem Johnson who was born in 1852 remembered seeing the ship and hearing Malays singing the song as they composed it. Bradlow *The Alabama* 1958

ha-ja [ˈhɑˌdʒɑ̆] *n. Afr. E. slang presum. half-jack* (q.v.) [*prob. abbr. half jack* (q.v.)]

I prefer this brew myself when I find myself not in the financial position to patronise Aunt Peggy for a ha-ja of mahoga. *Drum* 8.10.73

hakea [ˈhækɪə, hak-] *n.* One of the Proteaceae: a *proclaimed weed* (q.v.) in S.A.: see also *boetebos(sie). cf. Austral. hakea,* needlebush. [*fr.* C. L. von Hake *Ger. botanist* (died 1818)]

Plan Will Rid Kloof of Hakea. Payment will be refunded every six months on condition that the lessee has cleared hakea and other noxious weeds to the value of the six-monthly lease payment. The farm has an area of about 560ha and infestation of hakea . . . is bad. *Grocott's Mail* 25.6.71

halaal [haˈlɑːl] *adj.* Muslim equivalent of Jewish 'kosher': applied to food, particularly meat (and meat products) ritually slaughtered and which therefore may be eaten by Muslims. [*Arab. halăl-kar,* make lawful, *sig. slaughter in the ritually prescribed manner*]

Attention Muslim Public. Be advised that . . . Chickens have been declared non halaal . . . (Signed) Muslim Judicial Council. *E.P. Herald* 13.12.74

half-jack *n. pl.* -s. *Afr. E. colloq.* A half-bottle 375ml. also *slang, ha-ja* (q.v.) *cf. Austral. jackshay,* tin quart pot, *Canad. mickey* 12 fluid ounce bottle. [*presum. fr. half bottle*]

One took me outside and she gave him six beers and a half jack after he had threatened to go to the police. *Post* 25.7.71

halfkoord [ˈhalfˌkʊə(r)t] See *alfkoord.* [*prob. corruption of albacore*]

It is possible, however, that such mysterious names as bafaro, sancord, halfcord, and kartonkel are really Malay names. Green *Grow Lovely* 1951

halfmens [ˈhalfˌmeːns] *n. Pachypodium namaquanum:* also called *elephant's trunk,* a thick stemmed, usu. branchless spiny succulent of up to 2,5m high with a tuft of leaves at the top: ⲣThe name perpetuates a Hottentot belief that each is a transformed human being, therefore only 'half' a man. [*Afk. fr. Du. half* + mens *fr. Du. mensch* person *cogn.* man]

. . . three-metre-tall Pachypodium namaquanum – known to the Bushmen as a 'Half mens' . . . is a curiosity to travellers in the rocky country along the lower reaches of the Orange River. *E.P. Herald* 21.8.76
PACYPODIUM [sic.] NAMAQUANUM (halfmens). Very rare. 3-Year old plants at R20 each, includes GST. *Advt. Farmer's Weekly* 21.3.79

halim, haleem [ha'lim] *n.* Nourishing spiced soup containing wheat and barley and usu. chicken: distributed by Muslim charities: see also *Zakaat* and *naan* (q.v.)

Life is like a bubble – your charity keeps it afloat. Enclosed please find . . . Zakaat R——
Halim Distribution R——
Soofi Musjid Service Committee Appeal Nov. 1970
During Ramadaan Naan and Halim is distributed. *Appeal Anjuman Islam* Nov. 1970

hamba ['hamba] *interj. imp. vb. usu. offensive : equiv.* 'Push off!', 'Go away!': see also *voetsek, loop. cf. Canad. mush on.* [*imp. form Ngu. ukuhamba* to go]

Excepting in some words, the Caffers understood him very well . . . they say *hamba* for *get you gone* he says *kambu.* Thompson *Travels I* 1827
Hamba! Go! Do you think I want to see you? Jacobson *Long Way fr. London* 1953

hamba kahle [ˌhamba'gaɬe, ga-] *vb. phr.* African form of farewell lit. 'go well,' to which the reply is *sala kahle* 'stay well' (*pl. salani* to more than one person): see also *garshly* (go), *go well, stay well, mooi loop* and *kahle.* [*imp. sing. Ngu. ukuhamba* to go (*pl. hambani*) + *kahle* well]

HAMBA KAHLE! Mr Sebe off to Britain for a visit – and he gets a great farewell at East London airport. *Daily Dispatch* 14.9.73

hamel ['haməl *n. pl.* -s. A castrated ram, wether, in combination *wissel* ~ (q.v.) used to indicate the age or development in slaughter stock: see also *cut.* [*Afk. fr. Du. hamel* castrated sheep, *cf. vb. hamble* to be lame, limp, *cogn. O.E. hamelian* to maim]

An ox, a gelding, a hamel, a cut thing. Cloete *Rags of Glory* 1963
MERINO HAMELS WITH WOOL FOR

SALE 1640 MERINO HAMELS – unwisselled to 6 tooth – well grown and in good condition. *Grocott's Mail* 28.3.69

ham(m)erkop ['hamə(r)ˌkɔp] *n. pl.* -s. See (1) *hammerhead* [*Afk. hamer cogn.* hammer + *kop* head *cogn. Ger. Kopf*]

hammerhead *n. pl.* -s. **1.** *Scopus umbretta*, also *hamerkop*. A brown bird, with a large hammer-shaped head, frequenting marshy places. [*fr. shape of head*]

The 'hammerkop', or hammer-head, is another bird of a dull brown colour . . . which builds an immense nest, in which, like a magpie, it stores lost articles which it picks up . . . generally frequenting places where water is found in search of frogs and small fish. Wallace *Farming Industries* 1896
2. Also *hamerkop(haai)*: any of three species of *Sphyrna* (Sphyrnidae fam.) *S. lewini, S. mokarran* and *S. zygæna*: a shark with a hammer-shaped head, '. . . large specimens, fortunately rare in our waters, are among the most dreaded of marine creatures.' Smith *Sea Fishes of S.A.* 1961 ed. [*fr. shape of head* (+ *Afk. fr. Du. haai* shark)]

Send out your swimmer . . . and wait. You can pick up anything from a giant hammerhead to a fighting barracouta. *Farmer's Weekly* 12.5.71

hande skud ['handəˌskœt] *vb. phr. Sect. Army lit.* 'Shake hands' *e.g.* 'Do you want to ~ (or *skud hande*) with us?' *sig.* to have nothing to do with, or to dissociate oneself from [*Afk. hande* hands, *skud* to shake] ⲣ The phrase is used in numerous contexts *acc.* informants *esp.* if the person to whom it is addressed has been slacking, trying to *gyppo* (q.v.) out of duties or wangle extra leave. *O.I.s Various Servicemen* Jan.-Apr. 1979

handlanger ['hantˌlaŋə(r)] *n. pl.* -s. An unskilled assistant to a qualified *tradesman* (q.v.) usu. a *mason* (q.v.) or plumber, who hands tools, bricks etc. to his superior: not usu. an apprentice. [*Afk. fr. Du. handlanger* assistant, helper]

. . . building himself, brick by brick, with his wife . . . as his only handlanger. 'Robert is the brains . . . and I am the handlanger' . . . 'our concrete mixer . . . we reckon is worth two handlangers.' *Het Suid Western* (Property Supplement June 1976)

hands- [hans] *n. prefix* See *hans*.
hands-up *vb.* To surrender: see *handsupper* [*prob. translit. Afk. hen(d)sop fr. Eng.*

hands up]

. . . the sound of the firing decided Stefanus. He jumped on his horse . . . 'I am turning back,' he said 'I am going to hands-up to the English.' Bosman *Mafeking Road* 1947

handsupper [ˌhæn(d)zᶦʌpə] *n. pl.* -s. Used of a member of the Boer forces who surrendered to the British in the S. Afr. Wars: see *Boer War*, also quot. at *bitter-e(i)nder*. [*fr.* 'hands up', *gesture of surrender* + *agent.* or *personif. suffix* -er. see *hands-up vb.*]

. . . by his breeches and leggings he must be a khaki-Englander; and when he remonstrated in Afrikaans she denounced him as a renegade and 'hands-upper' which is even worse. Prance *Tante Rebella's Saga* 1937

They said that Koos was a hands upper and a traitor to his country . . . intimate with a man who had helped to bring about the downfall of the Afrikaner Nation. Bosman *Mafeking Road* 1947

hanepoot [ˈhɑnəˌpʊət] *n. pl.* -s. **1.** A sweet white grape, *Muscat d'Alexandrie*, used both as dessert fruit and for wine-making, said by some to be so named from the spread claw formation of the stems, (see first quot.): often corrupted to 'honey-pot' by association of ideas. [*see first quot.*]

Before we proceed to the vintage, we shall first describe the best types of grapes . . . It is the 'Haanen-Kloote' (Cock's testicle), but called Haanen-poote by the ladies of Africa. trans. Mentzel *Descr. of Cape* (Van Riebeeck Society 25, 1944)

Irrigation Farm . . . Vineyards include mainly established Sultanas, while French, Hanepoot, Muscadel and Hermitage varieties more than meet the requirements of an existing 251-leaguer K.W.V. quota. *Farmer's Weekly* 21.4.72

2. Sweet wine made from ~ grapes, unfortified for table use, fortified as a Muscatel dessert wine; bottled under the name ~ as *n. prop.*

There are two or three kinds of sweet wines made . . . The Hanepod made from a large white grape is very rich, but scarce and dear, and only used by the ladies at their parties in the same manner as the Constantia. The grapes from which this wine is made are chiefly dried and preserved for raisins to eat at desserts. Percival *Account of Cape of G.H.* 1804

hans- [hans] *modifier prefix.* Also *hands-*: Orphaned: equiv. hand-reared *cf. poddy* in *Austral. poddy-calf*, and *Jam. E. cossie fr.* cosset. In combination ~ *lam* (q.v.), ~ *piggie/varkie*, ~ *ossie* (*ox-calf*) see also *hansie.* [*Afk. fr. Du.* hands by hand, hand-reared]

hansie [ˈhansï] *n. pl.* -s. A young, hand-reared domestic animal, usu. a lamb but also a calf or pig: *cf. Canad. and U.S. dogie* motherless calf in a range herd, *Austral. sook* hand-reared calf [*fr. Afk.* hans orphan animal + *dimin. suffix* -ie]

. . . the farmer, or his wife, has to bottle feed the neglected hansies or lose them. *Cape Times* 6.6.73

Also a name freq. given to such an animal.

hanslam(metjie) [ˈhansˌlamïkï, -cï] *n. pl.* -s. A motherless, hand-reared lamb, often kept in or about the farmhouse as a pet: see *hans-, hansie. cf. Austral. poddy lamb. Jam. Eng. cossie-.* [*Afk. fr. Cape Du.* hans orphan + *lam cogn.* lamb + *dimin. suffix* -etjie]

When ewes died after lambing, the offspring were taken into the homestead and brought up as the 'hanslam'. The hanslam ran all over the place, a great nuisance to everybody, dropping its 'pralinees' indiscriminately. Jackson *Trader on Veld* 1958

. . . whatever the mysterious link between no rains and big lamb crops, one tangible result is the tremendous number of hanslammertjies being hand-reared this year. *Cape Times* 6.6.73

hap [hap] *usu. n. occ. vb. colloq.* Bite, mouthful: as in 'Give me a hap of your apple.' also dimin. form ~ *pie :* see *sout.* [*Afk.* hap bite *n. and vb.*]

Dog . . . goes to church . . . at the altar. I was afraid she might hap some of the bread out of the Dean's hand. *Het Suid Western* 2.3.77

. . . the sosaties sensational. (Above) A happie for the . . . hostess. *Darling* 20.7.77

hardegat [ˈha(r)dəˌxat] *adj. n. usu. Army.* Stubborn, an obstinate person, not necessarily equivalent *Brit.* or *U.S.* 'hard-arsed' [*hard* + *attrib. suffix* -e + *gat*, anus, vent]

I'm getting hardegat now. I'm not going to let his family hurt me any more. *O.I.* (*teenage girl*) 1974

The expression hardegat refers to a most stubborn person – this is of recent coinage. J. H. Picard in *Eng. Usage in S.A.* Vol. 6 No. 1 May 1975

hardekool [ˈha(r)dəˌkʊəl] *n. Combretum imberbe*, also known as *leadwood* on account of its exceptionally heavy timber. ℙ Ignited it is said to produce heat great enough to melt iron, and is favoured as *braai*wood, see quot., also quot. at *boekenhout.* [*Afk.* harde hard + *attrib. suffix* -e + kool ember]

. . . give a farmer . . . meat. . . grilled over burning maize cobs (Orange Free State and the

Transvaal Highveld), wingerd stompies (vine cuttings – the Cape Province), cow dung, hardekool wood or pine cones (the Transvaal bushveld) or just plain charcoal. *Panorama* Feb. 1976

harder [ˈhɑ(r)də(r)] *n. pl.* -s. Any of several S. Afr. species of mullet (Mugilidae) esp. *Mugil cephalus, Myxus capensis, Liza richardsoni* also known as *springer* (q.v.) similar to the herring: abundant in S. Cape waters and freq. used for *bokkems* (q.v.) [*Afk. harder fr. Du. harder, herder* mullet]

. . . the harder, somewhat of the flavour and appearance of our herring, but thicker. Percival *Account of Cape of G.H.* 1804

Most prized among the smaller fish is the harder, which is of the mullet family. But so fine is its flavour that the real harder-lover will always eat it simply grilled and served with lemon. *Farmer's Weekly* 18.4.73

hard pear, hardpeer [ˈhartˌpɪə(r)] *n. pl.* -s. Any of several species of *Olinia* including *O. capensis* and *O. cymosa* also called *rooibessie* (q.v.), all medium-sized forest trees, or their timber, a heavy hard wood used from the early days in wagon making. ¶Also used of *Strychnos henningsii* a tall forest tree of the E. Province, Natal and Transvaal: called *baster saffraan* (q.v.) [*Afk. fr. Du. peer cogn.* pear]

A typical wagon of the Great Trek period would have had wheelspokes made of either assegai or kershout, yellowwood wheel naves or hubs, yellowwood bed planks for the platform, wheel falloes of hard pear or saffraan. *E.P. Herald* 28.5.73

hard up [ˈhɑd ˌʌp] *adv. p. or modifier.* Right against, flush with. [*unknown poss. fr. printer's term*]

'You want this door hardup against the corner?' O.I. Builder 1969

This filter must be hardup against the side of the crate. I can't open it. O.I. Student 1970

hardveld [ˈhartˌfelt] *n.* An area of South West Africa inhabited by the *Herero* (q.v.) [*hard* + *veld* (q.v.)]

Despite its aridity, the plateau *hardveld* offered excellent grazing although enormous areas are needed to sustain a large herd, so the Herero settled and prospered on the empty savannah *Family of Man* Vol. 3 part 42 1975

Harley Davidson *n. pl.* -s. *Sect. Army.* A handlebar moustache; prerogative of sergeant majors and other N.C.O.s: see also *apache-snor. cf. Austral. Army brooms bass*, a big moustache [*fr.* trade name of motor cycle]

Harris buck See *swartwitpens.*

harpuisbos [ˌha(r)ˈpœĭsˌbɔ̆s] *n.* A name for all species of *Euryops*, resin-secreting shrubs, some of which are classified as 'undesirable' or 'invader' plants, troublesome particularly in *plant migration* (q.v.) in the Karroo. [*Afk. fr. Du. harpuis* resin + *bos cogn.* bush]

The harpuis (resin-pimple) plant is a species of *Euryops*, a showy composite with gay yellow blooms and a foliage not unlike that of a soft, rapidly grown, bushy young pine-tree . . . It should be remembered that its only important use is the supply of brushwood. Wallace *Farming Industries* 1896

Harry's Angels *pl. n.* Also *Harry's Flying Angels*, doctors participating in an air shuttle service operating between Johannesburg and centres in Swaziland, now also Transkei (see quot.) conveying medical specialists serving the area voluntarily. *cf. Austral. flying doctor.* [*see quots.*]

The hospital at Mbabane and the other at Hlatikulu . . . and a pathology laboratory at Manzini constitute the treatment centres where Harry's Angels operate. *Panorama* Oct. 1971

Harry's Angels for Transkei Mr Harry Oppenheimer had placed a company aircraft at the disposal of the team. He also gave financial assistance to the scheme – hence the term Harry's Angels. *Daily Dispatch* 17.7.74

harslag see *afval.*

hartebeest [ˈhɑtəˌbɪəs(t), ˈhart-] *n. pl.* ∅, -s. One of the largest S. Afr. antelopes *Alcelaphus bucelaphus* (*caama*), with handsome curving horns. [*Afk. fr. Du. hert/hart* stag + *beest Afk. bees* (q.v.) *poss. fr. Lat. bos sig.* Bovidae fam.]

We saw several harte-beasts, one of the largest species of deer, and the pride of the plain the spring buck. Philipps *Albany & Cafferland* 1827

Also found in place names e.g. Hartebeestpoort, Hartbeeskop, Hartebeesfontein, Hartebeesthoek.

hartbeeshuis, hartebeeshut, hartebeeshouse *n. pl.* -e, -s, -es. A temporary dwelling built of wattle-and-daub divided inside by rush or reed-mat screens made by *settlers* (q.v.) and *trekkers* (q.v.) [*etym. various prob. not fr. hartebees* (q.v.) *but hard* hard + *bies(sie)* reed + *huis cogn.* house, *also poss. fr. Hott. haru-b* reed mat + *huis*]

As the lack of tools . . . precluded the im-

mediate erection of permanent structures, most of the settlers contented themselves for the time being with thatched shelters more or less after the native fashion . . . then known on the frontier as 'hartebeest huts', 'hartbees-huise' or 'hardbieshuisies' (or, correctly, 'hardebiesieshuise'). Lewcock C19 *Architecture* 1963
. . . people living in . . . the Transvaal in the old days . . . in their hartbees houses and all . . . that fine spirit that the old Transvalers had inside themselves, in their wattle-and-daub and reed-and-daub houses. Bosman *Bekkersdal Marathon* 1971

hasie-hasie [ˈhɑsĭˈhɑsĭ] A children's game similar to that variously known as 'Red Rover' and 'Open Gates' in which one player tries to break through a line of other players. ⟡ Informants vary in their descriptions of the game, one of the many forms of 'catch'. [*Afk. haas* hare, *dimin-suffix -ie*]
Johnny's reminiscences in monologues, in duo with his wife Baby or in trio when his doddering hasie-hasie-playing ex-church-minister father joins in, have the lulling quality of someone else's reminiscences. *Speak* Vol. 1 No. 5 Oct-Nov 1978

hau [hɑʊ] *interj.* An exclamation usu. expressive of surprise or dismay among Zulus. [*Zu. hawu!* exclamation]
I couldn't translate R600 into Zulu so I told him 30 cattle. 'Hau!' he exclaimed. 'I can buy ten wives with that.' *The 1820* June 1971

have to *vb. substandard.* Used redundantly often in a 'cumulative effect' type of narrative e.g. 'First he had jaundice then he had three operations and then he got pneumonia – then after all that he *had to have* a coronary.' see quot. at *uitloop* and *ikey.* [*poss. trans. Afk. moet* must, have to (*without the sense of obligation unless emphasised intonationally*) see *must*]

hawker *n. pl.* -s. In S.A.E. sig. usu. a door-to-door seller of fruit and vegetables who requires a ~*s* licence to ply his trade: freq. in the Cape a *Malay* (q.v.) in Natal *Sammy* (q.v.) *Mary* (q.v.) see also *smous.* ⟡*acc. Bell's S. Afr. Legal Dict.* 'a person without a fixed place of business who goes about the streets with a stock of goods which he cries for sale.' [*fr. Eng. usage sig. usu.* pedlar]
As far as the city was concerned, the hawkers had ceased to exist – except for the hawker patrol. But even if Cape Town remains bliss-

fully unaware of the resentment and frustration this has caused among hawkers. *Cape Times* 15.5.71

hay *n.* Applied in S.A. as in the U.S. to a large number of dried fodder crops or stover, not exclusively grass, as in combinations *kidneybean* ~, *cowpea* ~, *bean* ~ as well as *redgrass (rooigras* (q.v.)) ~, *oat* ~ etc. all commercially supplied, see also quot. at *veld hay* [*fr. Afk. hooi* hay, forage]
First Grade Winter Fodder . . . Lucerne Hay (Cape) R21,00 Red Grass Hay R11,25 . . . Lucerne and Monkey Nut Hay and Pellets. *Farmer's Weekly* 21.4.72

headman [ˈhedˌmən, -ˌmæn] *n. pl.* -men. The head of a (5) *kraal* (q.v.) or petty chief of a section of a tribe: see (1) *induna.* [*fr.* head *sig.* 'chief' *adj.*]
Addressing the Chiefs and headmen, the Prince of Wales thanked them for coming great distances to see him. *E.P. Herald* 6.7.1925
Her application was supported by affidavits from her kraal headman and various old residents of the area. *Evening Post* 19.8.72

headring *n. pl.* -s A ring of wax usu. worked into the hair worn by older African men: a mark of status: see *kehla.* [*presum. trans. Zu. isicoco* headring]
One day . . . there came to us a grizzled worn-looking old kaffir, whose headring of polished black wax attested to his dignity as a kehla. FitzPatrick *Jock of Bushveld* 1907
They were village heads, prominent tribesmen who had earned the privilege of wearing the distinctive headring. Becker *Sandy Tracks* 1956

heartwater *n.* A disease of sheep, goats and cattle transmitted by the *bont tick* (q.v.) characterised by high fever and fluid in the thoracic cavity and pericardium: ~ *veld* grazing infested with the same tick. ⟡Animals raised on such *veld* (q.v.) acquire an immunity to the disease. *cf. gallsickness* (q.v.): see also *salted.* [*see first quot.*]
When they were cut open their heart cavities were filled with straw-coloured liquid. Having no name for it, the Boers called this new sickness heart-water. Cloete *Turning Wheels* 1937
All these cattle are in superb veld condition and are running on severe Heartwater, Redwater and Gallsickness veld. *Farmer's Weekly Advt.* 27.2.74

heemraad [ˈhɪəmˌrɑt, -ˌrɑdən] *n. pl.* -en. **1.** *hist.* A district council assisting the landdrost in the local court consisting of himself and three ~*en* in the administration of rural districts before the

establishment of British rule. [*Du. heem* local, *cogn*, *Ger. Heim, Eng.* home + *raad* council, councillor]
In country districts, the inferior courts consisted of the landdros, primarily an administrative officer, presiding over an appointed board of burghers known as heemraad. Hellmann & Abrahams *Race Relations H'book* 1949 **2.** *hist.* A member of the (1) ~, a petty magistrate under the *landdrost* (q.v.) similar to a British J.P. see first quot. at *Volksraad*. [*see* (1) ~]
My host, a jolly consequential-looking person, was, I found, a Mynheer Van Heerden, a *heemraad* and *kerkraad* of the district (i.e. a member of the district-court and churchwarden). Thompson *Travels II* 1827
. . . it was to have an unpaid magistracy by appointing special heemraden like English J. P. s Fitzroy *Dark Bright Land* 1955

heer [hɪə(r)] *n. pl.* -e, -en. *hist.* Gentleman: see *mynheer, meneer* (*O*) ~ ! *Afk.* mode in prayer: see also *here* (*interj.*) [*Du. heer* lord, hence gentleman]
. . . a sea-cow came out of the river, rushing upon us, with a hideous cry, as swift as an arrow out of a bow; at the same time, I heard the farmer call out, 'Heer Jesus!' . . . Sparrman *Voyages II* 1786
At my declining her offer of a bed-room, the good lady expressed surprise that the Heer should think his waggon better than the house. Burchell *Travels I* 1822
I met a Hottentot, who, asking me if I was not de engelsche heer (the English gentleman) presented me a letter. *Ibid.*

-heid [heɪt] *suffix.* Equiv. of *-ness n. forming suffix to adj.* as in *apart ~ , verkrampt ~ , verligt ~ , kragdadig ~* , all (q.v.): place names, e.g. Verlatenheid, Eensaamheid. [*Afk.* -*heid cogn.* Ger. -*heit*, *Eng.* -hood *fr. O.E.* -had state of being]

hek- *n. prefix, suffix.* Gate(way) found in place names Hekpoort, Hekplaas, Hekstroom, Niekerk's Hek. [*Afk. hek* gate]

helder- [ˈheldə(r)] *adj. prefix.* Clear, serene, found in place names e.g. Helderberg, Helderfontein, Helderzicht. [*Afk. fr. Du. helder* clear, bright]

hell in, the *predic. modifier. slang.* Furiously angry: also *moer* (q.v.) *in, the.* [*prob. analg. Afk. die josie in* furious, (*josie* the devil)]
Where did I find him . . . looking at the mud, the hell-in because we had lost all our things again. Just our clothes, and each other . . . Now you're really the hell-in. Nothing to laugh at. Fugard *Boesman & Lena* 1969
When your bony is . . . dirty, do you . . . Get

the moer in? *Eike S.A.* Oct. 1976

hell of a *intensifier, colloq.* See *helluva.*

hellout *intensifier, slang.* Used similarly to *helluva.* [*translit. Afk. slang, helluit* extremely, like hell]
I'm in a hellout hurry to get back . . . before my serial comes over the radio. *Darling* 23.4.75

helluva *intensifier colloq.* Also *hell of a:* in S.A.E. equiv. 'very' or 'extremely' with *adj.* or *adv.* e.g. 'a ~ good guy': ⫽Also as *Brit.* 'she gave him a ~ life' etc. See also *hellout.*
They were going to start some sort of a home for sick people. 'A sanatorium,' he said, remembering the word. 'This dorp is hell of a good for T.B. they say.' Jacobson *Dance in Sun* 1956
'It's up to you to make or break it. You must be prepared to be friendly and willing to meet people, otherwise you don't stand a chance. Life here can be helluva lonely, so you've just got to make a good start.' *Darling* 10.1.79

herald snake *Crotaphopeltis hotamboeia;* see *rooilip.*

herbalist *n. pl.* -s. An African medicine man dealing in herbs and other *materia medica*: see also *(i)nyanga, muti man, uhlaka. cf. witchdoctor, sangoma, isanusi, twasa.*
Initially it is necessary to distinguish two types of 'doctor' viz. the herbalist (*inyanga* . . .), and the diviner (*isangoma* . . .) Both are practitioners in the art of healing and both manipulate medicines to that end. . . . The herbalist is a specialist in medicines . . . unlike the diviner the herbalist does not commune with the shades nor can he divine – although in other respects their functions overlap. Hammond-Tooke *Bhaca Society* 1962
. . . it was accepted that witchcraft was an extenuating circumstance in murder. Herbalism was another manifestation of the belief in the supernatural. Some herbalists were charlatans, but some used what they believed were effective remedies. Judge *cit. Grocott's Mail* 9.5.72

herdboy *n. pl.* -s. A young boy who pursues the tribal occupation of herding livestock in rural areas.
The children who helped me . . . are the small shepherds or 'herdboys'. In the Transkei it is the men of the family who tend the cattle. The boys of ten to sixteen years herd the cattle and milk the cows, while their younger brothers of five to ten years herd the sheep, goats and calves. Broster *Red Blanket Valley* 1967
Mr Justice . . . started life as a herdboy looking after his father's cattle . . . Even then he dreamed of making law his career. *Drum* 8.3.74

Here [ˈhɪərə, ˈjɪərə, ˈjɪr(ə)] *interj. colloq.* Equiv. Oh Lord, Lawks, etc.: various pron. and sp. forms. [*Afk. Here fr. Du. Heer* lord]

'Oooo! *Yirra! Allemagtig!*' said Margaret, hopping from one foot to another. Always at this stage it seemed impossible that one should ever get in to water of this temperature – about 58 degrees. Mackenzie *Dragon to Kill* 1963
Big joke? Because I cried? No. 'Here' Boesman! Fugard *Boesman & Lena* 1969

Herero [ˌheˈreːrəʊ] *n. pl.* -s. A negroid people of South West Africa and their Bantu language: also a single member of this nation: see quot. at *hardveld.* ⸾The ~ s are also called Cattle *Damaras* (q.v.) and are particularly known for the elaborate Victorian-style dress of their women, adopted from that of the German missionaries' wives of the previous century. [*alleged onomat. fr. sound of a spear in flight*]

The Herero nation inhabited the northern plateau *hardveld* of what is today the disputed terrirory of Namibia. *Family of Man* Vol. 3, 42. 1975
Chief Clemens Kapuuo, head of the Herero delegation to the South West Africa constitutional talks and the man strongly tipped to become the territory's first head of state, has never felt more confident about the future. *Sunday Tribune* 29.8.76

Here Sewentien, Die *hist.* See quot., also at *V.O.C.* [*Du. Heeren* gentlemen + *zeventien cogn.* seventeen]

The States General of the Netherlands had succeeded in establishing a mighty organization by uniting the small trading associations into one great company, with the Here XVII as directors. . . . the letters VOC . . . (Vereenigde Oost Indische Compagnie). This mark had been decided on by the Here XVII or Council of Seventeen. *Panorama* Oct. 1973

Hernhutter [ˈhernˌhœtə(r)] *n. prop. pl.* -s. Also *boslemmer* (q.v.) a knife made under the direction ·of Moravian missionaries at Genadendal.

. . . the manufacture of the celebrated Moravian knives known as boslemmers or Hernhutters. Green *When Journey's Over* 1972
Herrnhutter knives made by Kühnel and four Hottentot apprentices . . . The knives in those days were known throughout the Colony and even overseas and for a long time were the main source of income for the mission station. *Panorama* Sept. 1974

erstigte [ˌherˈstɪxtə] *n. pl.* -s *or adj.* Reconstituted: as in ~ *Nasionale Party :* see *H.N.P.* ~ s the ultra-conservative splinter group of the nationalist party: ~ one of its members: as *adj.* of or pertaining to the party, its policies or

its members. [*Afk. her* re- + *stig* constitute, establish + *-te n. or adj. forming suffix*]

It was announced at a mass rally held in Pretoria on 24 October that a new political party, to be called the Herstigte Nasionale Party (reconstituted National Party) was to be formed. *S.A.I.R.R. Survey* 1969
The balance of the votes they lost undoubtedly went mostly to the United Party . . . and to the Herstigte Nationale Party . . . No one really expected this to happen, except the Herstigtes themselves. *Daily Dispatch* 25.2.72

Hervormde Kerk [ˌhe(r)ˈfɔ(r)mdə ˈkerk, ˈkɜk] *n. prop.* Smallest of the three Calvinist Churches to which most Afrikaners belong: ⸾The *Nederduits* ~ was the official church of the first S. Afr. Republic by the Constitution of 1858, see *Grondwet,* also quots. [*Afk. her* re- (again) + *vorm* constitute, shape + *-de partic. suffix*]

This [Hervormde] church comprising about 3 per cent of all Europeans in the Union was made the official Church of the Transvaal Republic by the Constitution of 1858. Swart *Africanderisms : Supp.* 1934
. . . the three federated Dutch Reformed Churches threatened to compile their own version of the metrical psalms independent of the 'Hervormde' and 'Gereformeerde' churches. *Friend* 16.2.34 cit. Swart *Ibid.*
The Hervormde N.G. Kerk smallest of the four Dutch Dutch Reformed Churches is taking the lead by eliminating racial discrimination in its internal structure. *E.P. Herald* 15.6.79.

het [het] *art.* The: definite art. see also *de :* found in place and other names, Het Kruis (Cross), Het Suid Western (newspaper, S.W. Cape) [*Du. definite art. (neuter) het*]

heuning- [ˈhɪœnɪŋ] *n.* Honey: found in place names e.g. Heuningneskloof (nest), Heuningspruit, Heuningvlei: also in names of plants etc. [*Afk. heuning cogn.* honey]

heuningbier [ˈhɪœnɪŋˌbiːr] *n.* See *honeybeer, karree.* [*Afk. heuning* honey + *bier* beer]

heuningblom [ˈhɪœnɪŋˌblɔm] *n. pl.* -me. *Melianthus comosus* and *M. major :* see *kruidjie-roer-my-nie.*

heuwel [ˈhɪœvəl] *n.* Hill: found in place names e.g. Klipheuwel, Bonteheuwel, Paleisheuwel. [*Afk. heuwel* hill]

hey [heɪ] *interj. colloq.* General but esp. prevalent in S.A.E.: used as an inter-

rogative equiv. of **1.** 'isn't/wasn't it?'
It was funny hey Boesman! Fugard *Boesman
& Lena* 1969
2. as a means of attracting attention or
response.
What do you think of that – hey? Jacobson
Dance in Sun 1956
3. redundantly as an utterance initiator
(also alone as equiv. of 'What?' or 'I
beg your pardon?')
Hey, how is it, man? I hear you came running
out? Matshikiza *Chocolates for Wife* 1961
hierjy ['hĭr‚jeɪ] *n. pl.* -'s. *slang* A lout,
gom (q.v.) [*Afk. hierjy lit.* here you! a
lout]
. . . Great variety of words and images . . .
such contemptuous ones as 'twee hierjy's' for
two nonentities. Van Heyningen & Berthoud
Uys Krige 1966
highveld ['haɪ‚felt] *n. prop.* and *n. modifier.*
The mountainous regions of the Trans-
vaal are so designated: also as modifier
~ *climate* etc.
Kleinberg was not very far from Willemsdorp in terms of miles. But Willemsdorp was
highveld. It stood on the edge of the great
plateau. Bosman *Willemdorp* 1977
him, that's *vb. phr. colloq.* Used as an
affirmative or agreement equiv. of 'Yes'
or 'That's it', when the object referred
to is non-animate. [*trans. Afk. dis hy
that's* it/him, *or poss. translit. ditsem
that's* it]
'Isn't that the . . . 's brand of whisky?' . . .
'That's him.' O.I. Bottle Store keeper 1970
Hindoo *n. prop. pl.* -s. and *modifier. hist.
obs.* A term used at the Cape for British
visitors from India often servants of the
East India Company, listed in Cape
Directories as '*Indian Visitors*': see
Indian¹: formerly *Brit. Anglo-Indian:*
see quots. at *Kaapenaar* and *koelie.*
[*poss. fr. ability to speak Hindustani*]
. . . three mistakes . . . First that the Society
of Cape Town is not fit for them; and secondly
that the Kapenaars have no wish that the
Hindoos should associate with them. The third
is that every Cape *spinster* (to use the elegant
Indian term) is looking out for a Hindoo
husband! Polson *Subaltern's Sick Leave* 1837
. . . Anglo Indians. These visitors were, by
the way, always known to Kapenaars as 'Hindoos'. Laidler *Tavern of Ocean* 1926
hing See *Indian terms,* also *duiwelsdrek.*
hippo ['hɪpəʊ] *n. pl.* -s. An armoured
police vehicle. [*presum. fr. appearance*]
Cops, helicopters, vans, formidable looking
hippos of the Terrorist Squad formed a road
block down the valley. *Drum* July 1976

At New Canada Railway Station, hard by
the giant yellow waste heaps of the gold mines,
the crowd ran up against another roadblock,
this one heavily manned and guarded by anti-
riot squads reinforced with a fleet of 'Hippo'
armored personnel carriers. *Time* 16.8.76
hlonipa [‚ɬɔˈnipa] *n.* and *n. modifier.* The
system of reverence and taboos ob-
served by the Nguni woman towards
her male relatives-in-law, involving
various gestures of respect and a whole
substitute vocabulary of ~ *words* enabling her to avoid speaking the radical
syllable of any one of their names; occ.
vb. trns. to ~ : see also *makoti* [Loosely,
a taboo, see second quot. [*Ngu. vb.
(uku-)hlonipha* to respect, show honour,
reverence]
The custom called hlonipa requires that
certain relatives by marriage shall never look
on each other's face, . . . more especially a
daughter-in-law and all her husband's male
relations . . . She is not allowed to pronounce
their names, even mentally; and, whenever the
emphatic syllable of either of their names occurs in any other word, she must avoid it, by
either substituting an entirely new word, or at
least another syllable in its place. This custom
has given rise to an almost distinct language
among the women. Shaw *My Mission* 1860
Behold, I make a decree, and it shall be
published from the mountains to the mountains:
your names, Incubu, Macumazahn, and Boug-
wan, shall be '*hlonipa*' even as the names of
dead kings, and he who speaks them shall die.
So shall your memory be preserved in the land
forever. Haggard *Solomon's Mines* 1886
H.N.P. ['eɪtʃ‚enˈpiː, ‚hɑ-] *n. prop.* The
Herenigde Nasionale Party. see *Nationalist Party* or the *Herstigte Nasionale
Party:* see also *Herstigte.* [*acronym*]
The cliff hanger 1948 victory . . . was the
result of an uneasy election pact between the
Herenigde Nasionale Party (H.N.P.) of D
Malan and the small badly organised Afrikane
Party. *Sunday Times* 27.5.73
. . . the protest meetings were attended mainly
by members of the Herstigte Nasionale Party
while a few U.P. members attended as 'spec
tators'. It did seem however, that the U.P
were 'as insistent as the H.N.P.' that the statue
be removed. *Rand Daily Mail* 8.3.71
hoek [hŭk, hʊk] *n.* Corner, angle. [*Afk.
fr. Du. hoek* corner *cogn.* hook] **1.** A
valley or indentation in or between
mountains.
. . . I got upon the ridge which divides the
hoek from another winding glen called Ganna
hoek. Thompson *Travels I* 1827
2. A corner, sometimes *dimin.* ~ *ie*
in compounds such as ~ *kas* corne

cupboard; *vry* ~ (*ie*), *colloq.* a nook indoors or out suitable for courting couples.
3. Angle, corner, found in place names e.g. Zwagershoek (also Swaers-), Olifantshoek, Driehoeksrivier, also *Hou*(*w*) (q.v.) *Hoek. cf. Hook of Holland.*

hoepelbeen [ˈhŭpəlˌbɪən] *n.* and *adj.* Bow legs, bandy-legged: found esp. in folk song, see quot. [*Afk. fr. Du. hoep*(*el*) hoop + *been* leg]

As I entered, the musicians, who were all smoking like chimneys, were playing and singing a favourite Dutch song, entitled . . . 'Janny met de hoopel been.' Cohen *Remin. of Kimberley* 1911

-hof [hɔf] *n. usu. suffix.* Court, garden, also law court: found in place names e.g. Eikenhof, Leeuwenhof, Bloemhof: also Hof Straat. [*Du. and Afk. hof* court]

hok [hɔk, hɒk] *n. pl.* -s. A cage or enclosure for animals or birds: pen, sty, run etc.; in combination *calf* ~, *fowl* ~, *ostrich* ~ : see also (1) *kraal* and quot. at (2) *bakkie.* [*Afk. hok* pen *etc.*]

Just look about how careful you had to be where you put your feet down on Chris Welman's front stoep. Half the time you didn't know if it was a front stoep or a fowl hok. Bosman *Bekkersdal Marathon* 1971

hokaai [ˈhŭˌkaɪ, hɔ-] *interj. colloq.* Exclamation usu. to halt animals equiv. of whoa! Also occ. to halt or check a driver of a motor vehicle. [*etym. dub. poss. fr. S. Sotho hoka* stop, *poss. extension Du. Ho* whoa (to a horse)]

It was not long before the loud 'hook haai! hook haai! (the order for stopping the oxen) were heard. Jackson *Trader on Veld* 1958

hol [hɔl] *adj.* Hollow: found in place names e.g. Hollaagte, Holbank. [*Afk. cogn.* hollow]

holbol [ˈhɔlˌbɔl] *adj.* Concavo-convex: used of moulding designs on (3) *Cape Dutch* (q.v.) gables. [*Afk. hol* hollow, concave + *bol* rounded, convex]

The cellar gable, which according to its date of 1804 should have been classical, is holbol, but it is not only grammar rules that have their exceptions! Opperman *Spirit of the Vine* 1968
. . . a fine thatched building with a *holbol* gable of alternate convex and concave curves and mouldings. Green *When Journey's Over* 1972

hold thumbs *vb. phr. colloq. equiv.* 'Cross fingers': a symbolic gesture wishing someone luck, hoping for the best etc. [*prob. fr. Afk. duime vashou,* hold thumbs, *rel. Ger. Halt Daumschen*]

'To hold thumbs' frequently occurs, even in print as a variant of 'to keep one's fingers crossed'. *English, South African, Std. Encyc. S.A.* Vol. 4, 1971
. . . When I went to Oxford people didn't know what I was talking about when I spoke of 'holding thumbs'. *cit. Letter* Guy Butler 6.10.78
'Don't worry Bloss . . . we'll all be holding thumbs for you extra hard this time.' Hobbs *Blossom* 1978
The potential is there . . . but the time has come for theory to be put into practice. Let's keep holding thumbs. *E.P. Herald* 9.3.79

Holland *modifier.* Used with *n.* instead of 'Dutch', e.g. 'Beetroot Salad – Holland style' (label). [*poss. to avoid confusion with Duits* German, *poss. translit. Afk. Hollands* Dutch]

VARIETY – Hamburgers, Hollandse Beef Stuk, Minute Steaks, Sosaties, Chicken Livers, Beef Olives. *Daily Dispatch Advt.* 17.5.71
Go Dutch this winter! in a Holland suedette coat. *Cape Argus Advt.* 25.4.74

Hollander *n. pl.* -s. *hist.* A Dutchman born in Holland: see quot.

A quarter of a century of autonomy had brought the Transvaal to the verge of bankruptcy and anarchy. Civil dissension had been created by the presence of the Dutchmen of European birth or 'Hollanders', whom President Burgers had introduced, and whose advanced political and religious views conflicted with the rigid Puritanism of the old 'Dopper' Boers. These turned to Paul Kruger, the Vice President as their leader. *Cambridge Hist. VIII* 1936 ed. Walker

Hollard Street *n. prop.* Address of Johannesburg Stock Exchange, used to refer to the Exchange itself, *cf. Wall Street.*

It was another golden day for Hollard Street Yesterday. The bullion price continued to zoom and gold shares were again the main feature. *Daily Dispatch* 12.5.73

homeboy *n. pl.* -s. *Afr. E.* Mode of address or reference usu. by urban Africans to others from their own tribe or area; also *home girl.* [*cf. Canad. home boy* sig. an orphan boy from a 'home' in Great Britain, also Colonial use of *home,* England. [*Ngu. umkaya* 'home person', one from home, member of the family]

. . . Ntoane who also came from Lesotho. Ntoane looked up as the Induna brought Monare into his room, and seeing the blanket round Monare's shoulders said: 'Welcome home-boy. Lanham *Blanket Boy's Moon* 1953

I stayed with homeboys in Sophiatown who also got me a job with a garage. *Drum* 22.10.72

homeland *n. pl.* -s. One of those areas set aside under the policy of *separate development* (q.v.) for a particular African people or *nation* (q.v.), being developed with financial support from Government, each with its own *Chief Minister* (q.v.) or Chief Councillor, and a varying measure of self-rule. freq. attrib. ~ *leaders*, ~ *policy* etc.

The Bantu Homelands Citizenship Act of 1970 became operative today in terms of a proclamation in the Government Gazette. As is already the case in the Transkei, the Act provides for certificates of citizenship to be issued to citizens of African homelands. *Daily Dispatch* 27.3.70

The regime has explained that this means real self-determination for the Black man in the so-called homelands. This is the new romantic name for the depressed areas of our country which until recently were known simply as the Native Reserves. *E.P. Herald* 19.8.74

Under the government's cherished *apartheid* program, blacks will become 'citizens' of the nine autonomous homelands, or 'Bantustans', now being established within South Africa. But such a system, even if it should prove acceptable to tribesmen who live in the homelands, would do little for the millions of blacks who live and work in the cities of 'white' South Africa. *Time* 16.8.76

hondjie ['hɔ̃ɪŋkɪ -cɪ] *n. pl.* -s. *Sect. Army.* (*lit.* puppy.) An old motor tyre, part of the training equipment of a *Parabat* (q.v.): see also *marble*. [*Afk. hond cogn.* hound + dimin. suffix -*jie*]

Then there's the 'hondjie' an old car tyre. The instructor roars: 'Take your little dog for a walk!' and off you set for another couple of miles. It's make-or-break in the Parabats. Those who survive the first fortnight with log, hondjie and marble are only the toughest. *Sunday Times* 12.3.72

honey beer *n.* Also *heuningbier*: a type of mead made in S.A. from the earliest times: see quot.

Much wild honey is gathered because honey means honey beer. Coloured people in the Piketberg Sandveld still make it every year, and they prefer honey cakes containing young bees. That, they say, starts the fermentation properly. They use karee moer, a powdered root, as yeast; and when water is added there is a strong drink such as primitive people made in many lands when the world was young. Green *Land of Afternoon* 1949

honeycake *n. pl.* -s. Honey in the comb: see quot. at *honeybeer*. [*translit. Afk. heuningkoek* honeycomb]

honey guide *n. pl.* -s. Also *U.S.*, any of several small birds of the Indicatoridae fam. which lead men and animals to bee's nests usu. in the expectation of collecting a share. ⫽*acc.* C. J. Skead this share consists of the beeswax, not of the honey or the bee grubs. (Various superstitions obtain about those who deny it this, one being that the next time they will be led to a snake's nest instead: see quot.)

This insistent and uncanny brown or grey bird calls on human beings to rob the hive it cannot break into itself. There is no mistaking the honey guide's meaning. Its call is loud and not to be denied. When the honey is reached... the bird remains close at hand teetering with anxiety and greed. Bushmen and Hottentots always leave a share for the honey guide. They say it is a vindictive bird which will lead the way to a snake or a leopard next time if it is cheated. Green *Land of Afternoon* 1949

(h)oogpister [ˈ(h)ʊəxˌpɪstə(r)] *n. pl.* -s. Any of several insects having as a defence-mechanism the ability to squirt acrid fluid some distance often into the eye of its assailant: see also quot. at *piogter*. [*Afk. hoog* high *or oog* eye + *pis* urinate + *agent.* suffix -*ter*]

Hoogpister: The name given to a large beetle, Mantorica, because it ejects to a considerable height (and whether purposely or accidentally, often enough into the eyes of its would-be captor) an exceedingly acrid fluid. Pettman *Africanderisms* 1913

hoogte [hʊəxtə] *n. usu. suffix.* Height(s): found usu. in place names e.g. Voortrekkerhoogte, Helshoogte, Bruintjieshoogte. [*Afk. hoog fr.* Du. *cogn.* high + -*te n. forming suffix*]

If you walk over my farm to the hoogte... you can see Abjaterskop behind the ridge of the Dwarsberge. Bosman *Mafeking Road* 1947

I could almost name the hour when he would emerge from a certain 'hoogte' (rise) with his goatskin bag on a stick slung over his shoulder. Jackson *Trader on Veld* 1958

hoop [hʊəp] *n. abstr. usu. suffix.* Hope found in S. Afr. place names freq. with possessive *n. prop.* e.g. Niekerkshoop, Vorstershoop, Maartenshoop, Keetmanshoop: also Ons (our) Hoop, Hoopstad. [*Afk. hoop cogn.* hope]

hoor hoor [ˈhʊə(r) ˈhʊə(r)] *interj.* 'Hear, hear': exclamation of approval or agreement. [*Afk. vb. hoor cogn.* hear]

... matters of administrative routine, which the Raad approved perfunctorily with gruff rumblings of 'Hoor! Hoor!' Brett Young *City of Gold* 1940

'In spite of what has been said about modern youth, in every respect they are better than we were, in that they will help their neighbours.' 'Hoor Hoor.' *Cape Times* 15.10.73

horries [ˈhɔˌrĭz] *pl. n. colloq.* After-effects of drugs or drinking: see also *babbalas occ.* also used loosely sig. any fear or phobia. *cf. Austral. to have the dingbats,* to be mad or to have *delirium tremens.* [*presum. fr.* 'the horrors', *Afk. horries* delirium tremens]

I watch . . . for a while, and then I get the horries that the rodents might eat me, . . . escape from the soap bubble world of the horries. *Cape Times* 3.7.71

'I can see you don't remember . . . It's the vlam. I can smell you've been drinking the blue-ocean ou pellie' . . . 'You got vlam horries Whitey' . . . 'You wait here, I'll get a borrel dop.' Muller *Whitey* 1977

hotel *n. pl.* -s. Term now not permitted by the Department of Justice for any boarding or rooming establishment not conforming to its minimum hotel grading requirements: see *classify²*, *rotel*.

hotnot [ˈhɔtˌnɔt] *n. pl.* -s. An offensive mode of address or reference to a coloured person: see also quots. at *loop*, *babbie-shop* and *houtkop*, and fourth quot. at (1) *Hottentot*. [*Afk. hotnot* Hottentot]

. . . people who speak of and think in terms of 'hotnots'. These are the racists whether overtly so or covertly so. *Evening Post* 16.7.73

That wretched word 'hotnot' has cropped up again. In a Cape court evidence showed that a White man had used this insulting expression to a Coloured. He was quite rightly reprimanded by the magistrate. But some people never learn. Hotnot, coolie and kaffir should have vanished from our vocabulary years ago. These words can be fighting talk. *Sunday Times* 11.7.76

hotnots- [ˈhɔtˌnɔts- *n. prefix.* Hottentot's: prefixed to numerous plant names ~ *boerboon* see *boerboon*, ~ *brood, Dioscorea elephantipes* see *elephant's foot* (also Kaffir bread tree); ~ *buchu*, ~ *boegoe, Othonna graveolens,* see *buchu*; ~ *koekoemakranka, Gethyllis ciliaris,* see *kukumakranka*; ~ *kooigoed* (bedding), any of several species of *Helichrysum*; ~ *kool, Trachyandra hispida, T. revolutum* or *T. ciliatum* see *hottentot cabbage*; ~ *kool bredie,* a *bredie* (q.v.) made with ~ *kool.*

Hotnotskool Bredie: Use the young un-opened flower shoots of 'hotnotskool' (Anthericos longifolium). They look like asparagus . . . You could use any cauliflower recipe for 'hotnotskool' as it resembles cauliflower in flavour. Gerber *Cape Cookery* 1950

~ *kougoed, Sceletium tortuosum* [*kou* to chew + *goed* substance]: a stimulant plant used as an intoxicant by the Hottentots: see also *kanna* [*cf. Austral. pituri,* a narcotic plant chewed by aborigines, *Duboisia hopwoodii*] ~ *tee, Helichrysum orbiculare,* also called *kaffir tea;* ~ *uintjie, Moraea edulis:* see *uintjie.*

Helichysum orbiculare . . . Usually found along . . . streams or in marshy places. The plants have long been used as tea by the Hotnots [*sic*] whence the vernacular name. The tea has an emollient action and has been used for chest troubles. Smith *S. Afr. Plants* 1966

Hottentot [ˈhɔtnˌtɔt] *n. pl.* -s. **1.** An indigenous people of S. Afr. at the time of the original white settlement.

I was told the Hottentots were uncommonly ugly and disgusting, but I do not think them so bad. Their features are small and their cheekbones immense, but they have a kind expression of countenance. Barnard *Letters & Journals* 1798–1801

For improving the condition of Hottentots and other free persons of colour, at the Cape of Good Hope, and for consolidating and amending the laws affecting those persons – dated 17th July 1828. *Greig's Almanac* 1833

The Hottentots looked upon the coast-belt as theirs. Behind them were marauding Bushmen and a less generous climate. The settlement of white men upon their land was fatal to them. De Kiewiet *Hist. of S.A.* 1941

I prefer to use the term Khoi Khoin because the word Hottentot has become a derogatory stereotype in South Africa, symbolic of the undesirable characteristics attributed to people of Khoi Khoin descent. Carstens *Coloured Reserve* 1966.

2. *no pl.* The Khoisan language spoken by the ~ *s.*

The r does not represent a gutteral in Hottentot as it does in Kaffir. Pettman *S. Afr. Place Names* 1931

I've tried her with every possible sort of sound that a human being can make . . . African dialects, Hottentot clicks, things it took me years to get hold of. Shaw *Pygmalion* 1914

3. Any of several brownish-coloured fish, *Pachymetopon* spp. of the Sparidae fam. *P. blochii,* known as 'hangberger'; *P. aeneum* the blue ~, also known as bluefish or copper bream; and *P. grande,* the *janbruin* (q.v.) known by numerous other common names: see also quots. at *allewêreld* and *doppie³*. [*fr.* (1) *Hottentot*]

Harders are often prepared in this way, but Hottentots, herrings, pilchards, maasbankers, etc. serve equally well. Gerber *Cape Cookery* 1950
In flavour the hottentot beats the roman, and it has a more delicate flesh. The taste is generally described as sweet. *Farmer's Weekly* 18.4.73
4. *prefix n.* numerous plant and other names: also ~'s; see also *hotnots-: ~ apron; ~'s bread; ~ cabbage; ~ fig; ~ god; ~'s tea* all (q.v.).

Hottentot apron *n. obs.* Travellers' term: see also *trassie etym.*
'An excessive development of the labia minora occurring in Hottentot women.' *Webster's Third International Dict.*

Hottentot(s) bread, Hottentot(s) brood *n. Testudinaria elephantipes* of the Dioscoreaceae: see *elephant's foot.*
. . . an extraordinary plant called Hottentots Brood (Hottentot's Bread). Its bulb stands entirely above ground, and grows to an enormous size, frequently three feet in height and diameter . . . The inside is a fleshy substance which may be compared to a turnip, both in consistence and colour . . . The Hottentot informed me, that, in former times, they ate this inner substance . . . cut in pieces and baked in embers. Burchell *Travels II* 1824

Hottentot cabbage *n. Trachyandra revolutum. T. hispida* or *T. ciliatum:* see also quot. at *hotnotskool cf. N.Z. Maori cabbage.*
The flower heads of this plant which thrives abundantly in the deep sands near the seashore, furnish a kind of culinary vegetable, which somewhat resembles asparagus, and is known as *Hottentot's Cabbage* (Hottentot's-Kohl). When stewed and properly prepared, they make no contemptible dish. *Cape of G.H. Almanac* 1856

Hottentot fig *n. pl.* -s. *Carpobrotus* (*mesembryanthemum*) *edulis:* see quot. also *ghokum, sour fig.*
. . . the Hottentot fig grows . . . in large patches. It produces . . . in all seasons . . . a fruit of the size of a small fig, of a very pleasant acid taste, when perfectly ripe. It must, however, be first divested of the outside pulp or coat, which is at all times saltish; and even the fruit, when unripe has a disagreeably saline and austere taste. Its name was given by the first colonists, on account of its form bearing some little resemblance to a fig, and because it is everywhere eaten by the Hottentots. Burchell *Travels I* 1822

Hottentot god *n. pl.* -s. Any of several insects of the Mantidae, the 'preying' or 'praying' mantis.
The Hottentot-gods always attracted him as they reared up and 'prayed' before him;

quaint things, with tiny heads and thin necks and enormous eyes, that sat up with forelegs raised to pray, as a pet dog sits up and begs FitzPatrick *Jock of Bushveld* 1907
Hottentot's tea. See *hotnotstee.*
hou [hɔʊ] *imp. vb.* Hold, keep to: found in place name Hou(w) *Hoek* (q.v.); in expression ~ *moed* equiv. of 'keep your spirits up'; see *moed;* and in signs ~ *links* (left); ~ *regs* (right); also in ~ *jou bek* equiv. 'shut up', see *bek* [*Afk hou Du. houden* to keep, *cogn.* hold]
As after the 1925 floods, so now the women urged their men to *hou moed*, to be courageous in facing the future. Birkby *Thirstland Trek* 1936
What a challenge to . . . the new Prime Minister! Has it ever happened before that a Prime Minister . . . has faced such a baptism? . . . Go to it P. W. – Hou moed. *Het Suid Westen* 6.12.78
hour *n. pl.* -s. A measure of distance in S.A. since C18. 'The reply to a question as to distance generally being "Oh, so many hours" ' (Pettman *Africanderism* 1913.) ¶This usage still obtains, calculated on varying speeds by car usu. 60-80 kph according to the driver: see fourth quot.
The following are the distances on horseback from Stellenbosch to the principal nearest places, viz:
To Cape Town . . . 5 hours+
To Paarl . . . 3 hours
+ By an '*hour*' on horseback, in South Africa is generally understood a distance of about six miles. Greig's *Almanac* 1831
An 'hour' on horseback, in South Africa generally implies a distance of about EIGHT MILES. *Ibid.* 1833
GENERAL DEALER'S BUSINESS . . . 1 hours from nearest railway station. *E.P. Herald* 15.1.1921
Ag plees Daddy won't you take us down to Durban? It's only six hours in the Chevrolet Song, Jeremy Taylor

house camp, house kraal *n. pl.* -s. An enclosure for farm animals close to the homestead: see (1) *kraal.*
household butter. See *butter.*
house mother *n. pl.* -s. The matron or house mistress of a girls' boarding establishment: ¶Also *house-father*, warden of a boys' hostel, also *U.S.* and *Canad.* with several meanings. [*trans Afk. huis moeder/vader* matron, house warden]
House Mother required, April to share supervision of girls in out of school hours and in dormitory.

Apply: Sister Superior . . . School for Girls,
. . . *Farmer's Weekly Advt.* 21.4.72
. . . happy in the hostel where she lived and
was devoted to the house mother and father.
Het Suid Western 2.6.76

hout [həʊt] *n. pl.* -s. *S. Afr. and Rhodesian*
See *houtkop* (1) and quots below.
. . . in Zimbabwe, every black is now a 'hout'
to shoot and it doesn't make things any easier.
Rhodeo 13.10.78
Hout. Afrikaans term of derision for blacks,
meaning woodenhead. Houtie slayer. Rifle.
Time 30.4.79

houtkop [ˈhəʊtˌkɔp] *n. pl.* -pe, -s. Block-
head [*Afk. hout* wood + *kop* head *cogn.*
Ger. Kopf] **1.** An offensive mode of
address or reference to an African.
The 'poison of racism' flowed deep in the
veins of South Africans. Even in the Afrikaans
language racist words such as kaffer, hotnot,
koelie and houtkop, were still being used.
Evening Post 17.2.73
2. Numskull: a stupid person.

howl *vb. intrns. colloq.* Weep: not necessarily noisily. [*translit. Afk. huil* weep]
I howled my eyes out at *Love Story.* Schoolgirl, 17, 1975
. . . he comes storming in the kitchen . . . My
mom reaches out for him, beginning to howl.
Darling 26.11.75

H.S.R.C. [ˈeɪtʃˌesˌaˈsiː] *n. prop.* Human
Sciences Research Council: a statutory
body sponsoring and promoting research projects in the humanities and the
arts: see also *C.S.I.R.* [*acronym*]

huilebalken [ˈhœɪləˌbalkən] *pl. n. rare*
prob. obs. Professional mourners, freq.
slaves, employed to walk weeping in
funeral processions. See also quot. at
tropsluiters [*Cape Du. huilebalk* a blub-
berer, whining person + *pl. suffix -en*]
Her funeral [Lady Charles Somerset: died
11.9.1815] . . . was one of the last attended at
the Cape by *huilebalken* or professional mourners . . . The huilebalken were distinct from the
tropsluiters who were merely employed to
lengthen the procession. Mills *First Ladies* 1952

huisapteek [ˈhœɪsapˌtɪək] *n.* A comprehensive household medicine chest used esp.
in country districts and still commercially available: P*Dutch Medicines Handbook* 1970: see also *Dutch Medicines* and
Old Dutch Medicines. [*Afk. huis* home
+ *apteek fr. Du. apotheek* dispensary]
Every household had a 'huisapotheek', or a
box of Halle medicines, which was so much
used that it was only in very extreme cases
that a doctor was called. Botha *Our S.A.* 1938
. . . a stout little wooden chest kept handy . . .
This was known as 'Die Huis Apteek', the

medicine chest that took the place of the family doctor, . . . usually imported from Germany. McMagh *Dinner of Herbs* 1968

huisbesoek [ˈhœɪsbəˌsʊk] *n. pl.* -e. House
visiting. [*Afk. huis cogn.* house, home +
besoek visit, call] **1.** Parish or district
visiting by clergy or other teachers.
Four times a year, he undertook journeys
through his district for the purpose of holding
these assemblies in various places, for . . . those
whom distance prevented from coming to the
church. These pastoral visits were called *huis-
bezoekings,* or domiciliary visitations. Burchell
Travels II 1824
. . . even if this huisbesoek was not part of
my after-school duties, I would have gone and
visited the parents in any case. Bosman *Unto
Dust* 1963
2. Political canvassing or other house-to-
house visiting.
Nationalist predictions were based on huisbesoek figures and records kept by party
workers at the polls. The . . . predictions proved that . . . many voters who had been canvassed
and had promised to vote for the party had
either abstained or voted for the United Party.
Argus 26.2.72

huis toe [ˈhœɪsˌtʊ] *interj. colloq.* Home-
(wards): equiv. of 'We must be off' to a
person, or 'Go home' to an animal.
[*Afk. huis toe* homewards]
It was very dark and Sanna said 'huis toe.'
We ran home and sat in the dark. Vaughan
Diary circa 1902
This war is not ours. It is time to trek home-
ward. *Huis toe! Huis toe!* Brett Young *City
of Gold* 1940

hunger belt *n. pl.* -s. See *girdle of famine,
lambili/e strap.*
[Umbopa . . . marching along beside me,
wrapped in his blanket, and with a leather belt
strapped so tightly round his stomach, to make
his hunger small as he said, that his waist
looked like a girl's. Haggard *Solomon's Mines*
1888
Hunger belt: A thong of hide . . . worn as a
belt by the Namaqua Hottentots which in
times of scarcity is gradually tightened to
deaden the gnawings of hunger. Pettman
Africanderisms 1913

hunt *vb. trns. and intrns. usu.* To shoot:
similar to *U.S.* usage, not equiv. *Brit.* ~.
~ *ing season* the shooting season: ~ *ing
horse* not a hunter as *Brit.* but sig. a
horse accustomed to being shot over; ~
ing dog usu. a cross-bred pointer-type
dog; ~ *ed out* denuded of game. PThere
is a *hunt* in *Brit.* sense, often a 'drag' or
jackal hunt, not fox hunting, in some
large centres esp. Cape Town. [*S.A.E. in
terms of ordinance given*]

'Hunt' means by any means whatsoever to kill or capture or attempt to kill or capture, or to shoot at, poison, pursue, drive, search for, lie in wait for or wilfully disturb. *Section 21H of the Nature Conservation Ordinance. Ordinance 26/1965*

With the start of the hunting season only about a month away, arms and ammunition dealers . . . are having difficulty meeting demands for large calibre sporting rifles, ammunition and some hand guns. *E.P. Herald* 30.4.76

. . . it was a lovely country of rolling grasslands and vleis, over which buck ran year after year. That was before they were hunted out. Vaughan *Last of Sunlit Years* 1969

. . . excellent ranching veld. Sheltered kloofs, good hunting. *Daily Dispatch Advt.* July 1972

hut *n. pl.* -s. An African dwelling usu. circular, consisting of one room of which several make up a (4) *kraal* (q.v.) either mud-walled with a conical, grass-thatched roof, or of a domed, beehive construction: see also (2) *pondok. cf. Austral. humpy,* also *goondie, gunyah.*

Their huts, all fronting inward to the kraal, are constructed of mats stretched over a frame of sticks in the shape of a bee-hive, and afford but an indifferent shelter in cold weather. Thompson *Travels I* 1827

Seven tribesmen were killed, one injured and 159 huts burned down in a faction fight in the Libode district . . . The trouble started when an impi . . . attacked the Deep Level area and killed two people. Deep Level retaliated when about 80 tribesmen killed five . . . injured one and burned down 43 huts. *Daily Dispatch* 11.5.72

~ **tax** a tax imposed by Government.
. . . although the policy of all governments (according to an official statement) is not to encourage full-time farming in the Reserves, but to impel the Native by the hut-tax to seek a livelihood on the mines or in the Boer farmer's fields, the lure of the white man's cities alone is enough to attract him. Reed *South of Suez* 1950

I

I.D.B. ['aɪ‚di'bi:] *n. prop.* and *n. modifier.* **1.** *I*llicit *D*iamond *B*uying: trade in uncut diamonds by unlicensed persons, forbidden under the Diamond Trade Act which came into force on 1st September 1882, and punishable since then as a criminal offence: hence ~ *Act* familiar term for this Act: see quot. at *Breakwater.* [*acronym*]

Kimberley and the diamond fields coined this new phrase: I.D.B. – illicit diamond buying. The crime was considered to be one of the

most serious that could be committed in South Africa, and the punishment imposed . . . up to seven years hard labour on the breakwater at Cape Town. Klein *Stage Coach Dust* 1937

Police in South-West Africa are investigating one of the biggest and most startling I.D.B. cases in the history of the Oranjemund diamond area. *E.P. Herald* 2.3.70

2. *n. pl.* -s. An illicit diamond buyer: one engaging in the trade of *I.D.B.*: see also *kooper* and *kopje-walloper* [*fr.* (1) ~]

Yet the I.D.B., as the illicit diamond-buyer is called flourishes still. Farini *Through the Kalahari* 1886

. . . the success of the terrible three as illicit diamond dealers was such that other I.D.B.'s took up their quarters at Oliphantsfontein . . . to the trio's great consternation. Cohen *Remin. Johannesburg* 1924

-ie [-ĭ] *suffix.* **1.** *dimin.-forming suffix* equiv. *-djie, -tjie* etc. as in *drogie* (q.v.), *bossie* (q.v.) *etc.*

2. *personif. suffix* affixed to *n.* or *adj.* e.g. *bergie, bangie, dronkie,* all (q.v.) *cf. Austral. broomie, fleecie; Brit.* and *U.S. wierdie, alkie,* etc.

3. *suffix* in children's names for marbles: *alie, glassie, ironie yakkie,* all (q.v.) also *sodie, steelie. bottlie.*

igqira See *iqira*

Ikey ['aɪkĭ] *n. prop. pl.* -s. A student or alumnus of the University of Cape Town: esp. a member of one of its sports teams. [*abbr. n. prop. Isaac*]

I certainly don't hold with what the Ikeys have been doing, but I cannot understand why the police had to behave as they do now. *Daily Dispatch* 3.6.72

Ikeys *n. prop.* The University of Cape Town (U.C.T.) [*see Ikey*]

At Ikeys too, there is a lot of indifference. But that is universal. In our parents' day they had something to fight for. *Star* 8.6.73

ikhaba [ĭ'kaba] *n.* (*pl.* ama-) *Urban Afr. E.* Very young urban African boys and men: see also quot. at *uMac.* [*Xh. -khaba* shoot, sprig, young plant, hence *amakhaba* young men]

The age-set from 15 to about 25 of the tsotsi type are called ikhaba. Wilson & Mafeje *Langa* 1963

ikhankatha [ĭ‚kaŋ'kata] *n.* See *abak(h)wetha, circumcision school, ingcibi,* and quot. below. [*Xh.* warden, guardian]

A certain person with experience in circumcision rites is appointed to take charge of the boys during this period. He is called 'ikhankatha' (teacher). *Voice* 15-21.11.78

imbadada [ˌimbaˈdada] *pl.* izi-, -s. *Afr. E.* Rough sandal with sole and uppers made of strips of motor tyre: *occ. erron.* by white speakers *patats (onomat.) cf. Jam E. jump-and-jives.* [*Zu. imbadada* sandal (as worn by monks or made from old motor tyres)]

But after 12 months probably these Plural-stanians are still girding their loins and tying their imbadadas because they have barely started the long journey. *Drum* Feb. 1979

mbongi [ĭmˈbɔŋgĭ] *n. pl.* -s, izi-. Also *mbongi:* a praise singer to an African chief, whose traditional office it is to chant and recite in honour of his person and his forbears: see also *izibongo.* ⫯Also used occ. of a political stooge or apologist, usu. *'mbongo'* (q.v.). [*Ngu. im-bongi,* bard or professional praiser *fr. vb. bonga* to praise, laud or utter the praises of]

The mbongis are out . . . dancing and singing, shouting 'Bayete' to Mnumzana V— they climb up his ancestral tree singing the praise poems of . . . illustrious forbears. *Drum* 8.3.73

Mr Nelson Mabunu, Chief Kaiser Matanzima's 'mbongi' (Official singer of praises), did a fine job at Dobsonville Stadium during the Transkei celebration for independence. *World* 24.10.76

. . . he is acknowledged as the leading Xhosa imbongi, or praise singer. When he swops his clerk's suit for the imbongi's kaross and assegaais, his whole demeanour changes. He becomes tense and aggressive, and from his lips comes a powerful torrent of completely spontaneous and yet highly sophisticated poetry. *E.P. Herald* 14.6.79

mfe [ˈĭmˌfe] *n. Sorghum dochna:* see *soetriet.* [*Ngu. imfe* sweet cane]

From September to March or April, they lived chiefly on milk, and the large supplies of pumpkins, Indian corn, sweet cane (imfe), and other green crops, which they raise during the summer months. Shaw *My Mission* 1860

mishologu [ˌimĭʃɔˈlɔgu] *pl. n.* See *ama-tongo.* [*pl. Xh. (u)mshologu* ancestral spirit]

mmorality *n.* In S.A. usu. sig. unlawful intercourse between Black and White.

Extension of Immorality law

. . . to introduce a Bill to prohibit unlawful intercourse between European males and native females and other acts in relation thereto. The Bill was read a first time and the second reading was set down for Wednesday next. *E.P. Herald* 4.2.1926

They made love, the woman turned out to be Coloured and they were each sentenced to nine month's jail, suspended, for immorality.

Cape Times 13.1.73

Immorality Act *n. prop.* Act 5 of 1927 and its amendments as Act 21 of 1950 which provides for the prosecution of Black and White who cohabit or attempt to marry, or who by a single actual or attempted act contravene it: *Afr.E. abbr. form Immo Act.* [*fr. immorality (law)* (q.v.)]

The Immorality Act, which bans interracial sex, is still in force, but prosecutions are rare. *Time* 15.10.73

A day earlier he was questioned by a senior police officer, in connection with the alleged breaking of the Immo Act. *Post* 10.10.71

impala¹ [ĭmˈpalə] *n. pl.* Ø, -s. One of the larger and most common of the S. Afr. antelopes *Æpyceros melampus,* bright russet-coloured with a white belly, curved black stripe on the haunch, the male with slender ringed horns. [*Zu. iMpala* pallah antelope]

The reserve is advertising for 10 each of impala, hartebeest, nyala, reedbuck, waterbuck, black wildebeest and blue wildebeest. *Grocott's Mail* 11.4.72

black faced ~

In the west there are also some *black-faced* impalas, one of the rarest antelopes in Africa. *Etosha National Park* 1972

Impala² *n. prop. pl.* -s. A jet-engined military aircraft used by the S. Afr. Air Force. [*presum. fr. antelope*]

Citizen Force squadrons throughout South Africa will be equipped with Impala jet aircraft to replace the Harvard trainers within the next two or three years . . . head of the Light Aircraft Command . . . said today. *Daily News* 8.5.70

impi [ˈĭmpĭ] *n. pl.* -s. Formerly sig. regiment or army as in Chaka's ~*s;* now an armed band of men esp. those engaging in *faction fights* (q.v.) see quots. at *hut, tribesman, Swazi* and (1) *induna.* [*Ngu. impi* regiment, army, military force]

A Zulu impi . . . managed to cut off the chief's cattle and kill some of his followers. *Daily Telegraph* 16.5.1879 *cit.* W. S. Mackie

. . . a massive impi of about 250 men, butchered 19 fellow tribesmen from a nearby kraal. A police reconnaissance patrol which sped to the battlefield was surrounded by the 250-strong assegai-wielding impi. *Daily Dispatch* 2.8.72

improved *vb. partic.* Of or pertaining to a farm or land with *improvements* (q.v.) also *well* ~ : or *un* ~ sig. land which has not been developed in any way.

113

FOR SALE – Well-improved sweetveld farm, approximately 786,4ha in the Elandslaagte district. *Farmer's Weekly Advt.* 21.4.72

improvement(s) *n. usu. pl.* Any dams, fencing or buildings including the homestead (if any) constitute the ~ *s* of a *farm* (q.v.): also *Canad.*

The lessee had no *dominium* in the ground and he could sell or bequeath nothing more than the *opstal* – that is the house, kraals and other improvements. Botha *Our S.A.* 1938

IMPROVEMENTS: Fully fenced. 26 Camps, 4 Strong boreholes 3 with windmills and 1 with Lister diesel engine. *Farmer's Weekly Advt.* 21.4.72

impundulu bird [ˌɪmˌpŭnˈdulŭ] *n.* See *lightning bird, mpundulu bird* and *windbird.*

The *impundulu* is identified with the lightning; thunder is the beating of its wings, while the ɬ ash indicates the laying of its eggs. Hammond-Tooke *Bhaca Society* 1962

A disused Xhosa custom was revived when men stabbed a body during a funeral . . . word went round the home of Mr . . . during his funeral that he was winking. Three men wanted to have the body stabbed because they claimed it was not Mr . . .'s but that of a mystery bird (impundulu) . . . The gossip in the township was that Mr . . . had been changed into a zombie by some witches. *Indaba* 29.9.78

in *prep. substandard.* Used as equiv. of 'At' usu. with 'good' or 'bad' as in 'my son's so bad in Maths but he's quite good in languages': see first quot. at *there-*. [*translit. Afk. goed in* good at]

'She was very good in this job', said Mr — *Het Suid Western* 25.4.74

indaba [ˌɪnˈdɑbə, ˈɪnˌdaba] *n. pl.* -s. **1.** A tribal conference: now extended to mean any conference or parley between politicians or other contracting parties. *cf. N.Z. (Maori) korero, U.S.* and *Canad. pow-pow* [*Ngu. indaba* conference]

A domestic now informed us, that the king was holding an en-daba (a council) with his warriors. Thompson *Travels II* 1827

The indaba ended today . . . informed the chiefs that they must keep their women away from the town. . . . The Chiefs said the women were too much for them, but agreed to the other conditions. *E.P. Herald* 11.1.1897

Diplomatic indabas only rarely produce neatly wrapped solutions to problems. They tend rather to be stages in a process . . . *Sunday Times* 22.10.78

Abakwetha controversy – Iniation schools row. . . . said at an Indaba at the Great Place Tamarha it had been decided that the five years should be reduced to one. *Indaba* 9.2.79

2. Concern, problem: [*presum. figur. fr.*

(1) ~]

Mom refrained from adding that there was also an East wind, which . . . blew, . . . sweeping across the hillside to hit the small cottage bang on. Anyway she felt it was their indaba, put the cheque in her purse, and left them. Vaughan *Last of Sunlit Years* 1969

'Their sex needs are their own indaba. They can have their sex somewhere else,' the matron answered. *Drum* 8.12.72

Indian[1] *n. pl.* -s. *also modifier obs. hist.* An 'Anglo-Indian'. ⟨Formerly used of those who were known and listed in certain directories and *Almanacs* (q.v.) as 'Indian Visitors', persons, usu. British, spending extended periods of sick leave or furlough from India at the Cape: also *colloq. Hindoo(s)* (q.v.): see quots. at *Hindoo* and (1) *Kaapenaar.*

. . . Mr Advocate Cloete, two Indian residents, two medical gentlemen and two English merchants of Cape Town. cit. Meurant *Sixty Yrs Ago* 1885

Take her as your free servant, no longer a slave. Take her as your friend . . . God bless you both, and when you supplicate on high, ask heaven's mercy for Wilkinson the Indian. *Cape of G.H. Literary Gaz.* Vol. I June 1830

. . . at about 6 p.m. arrived at Mr G. du Toit's place where I had the pleasure to meet a large company of Indians, Lord and Lady Doyle* and Major Cloete. Truter *Diary* 26.10.1862 cit. Gordon-Brown 1972

*Sir Charles and Lady D'Oyley.

Indian[2] *n. pl.* -s. A member of the Asian racial group of Indian descent or birth, mostly descended either from indentured Indian labourers, the first of whom arrived to work in the Natal sugar plantations in 1860, or from *passenger* ~ *s* (q.v.), the merchant class who emigrated at about the same time: see *Bombay Merchant, Arab, Indian Council, Indian Reformed Church, Indobond* and *Indian terms,* also second quot. at (1) *black.*

Indian Council, the *n. prop.* An official body constituted by Government consisting of elected and appointed Indian members, to negotiate in all matters concerning the *Indian*[2] community in S.A.: see quot. at *coolie.*

Indian parsley See *Indian terms, dhunia.*

Indian Reformed Church, *n. prop.* See quot. [*as in Dutch Reformed Church* (q.v.)]

This week I spoke to the Rev Edward Mannikon, vice-chairman of the Indian Reformed

Church – the Indian branch of the NG Kerk – about his views on a confrontation between the State and Church. *Sunday Times* 1.9.74

Indian terms *prob. largely reg. Natal* ❡Numerous Indian terms appear in shops, Indian recipe books etc, some of which are for convenience listed here: not exclusively S.A.E., sp. forms vary greatly among language groups. *Clothing:* champals (q.v.), choli (q.v.), mundani (q.v.); khafia (q.v.). *Foods and dishes:* bhajia (q.v.), biriani (q.v.), bunny chow (q.v.), butter-bread (q.v.), butter chilli (q.v.), chilliebite(s) (q.v.), halim/leem (q.v.), kalya (q.v.), naan (q.v.) Indian yeast bread, baked or fried see also quot. at *kalya; papad (puppa/poppadums)* thin fried wafers of pea flour commercially available ; *puri,* thin fried pastry or wheat cake; *roti* (q.v.); *salomi(e)* (q.v.); *samoosa* (q.v.). *Ingredients: chana flour,* see *gram; dhai,* sour milk curds; *dhal, dholl,* various lentils and pulse; *ghee,* clarified butter see quot. at *bhajia ; gram,* pea-flour; *masoor,* brown lentils ; *soojee,* semolina. *Spices : arad,* turmeric see *borrie ; dhunia,* coriander seed, also the green leaves called by some *Indian parsley; elachi,* cardamom/mon seeds ; *hing,* asafœtida, see *duiwelsdrek; jeero,* cumin seed ; *masala* (q.v.) curry spices, also in combination *garum/ghurum/garam* ~ (q.v.); *methi* (q.v.) fenugreek ; *soomf, saunf* fennel or anise seed, see *vinkel; tuj,* cinnamon usu. in stick form. *Sweetmeats* (general terms *mithai, halwa)* burfee (q.v.) ; *goolab jambo,* a syrup-dipped doughnut flavoured with rose water *cf. koeksister* (q.v.) ; *mithai samoosa,* a *samoosa* (q.v.) filled with a sweetened mixture usu. of coconut or other nuts.
See also *Haj, halaal, lounge, goodwill, goodself, Zakaat, ooplang, platoon school.*
indhlunkulu [ˌindɫüŋˈkulŭ] *n.* See *great-*(house).
indicator, honey See *honey guide.*
Indobond *n. prop.* A secret S.Afr. Indian political organization: see quot.
   ... one of the country's best kept secrets – the existence of the Indobond, the Indian version of the very powerful and very secret Afrikaner Broederbond. The Indobond's aims are similar ... to use behind-the-scenes wheeling

and dealing to control the men who control the people. *Drum* 22.9.73

indoena(s) [ɪnˈdüna(s)] *n. usu. pl. Sect. Army.* The top brass: see at *induna.* [*Afk. fr. Ngu. in Duna,* captain, headman]
   Top brass of any designation are *indoenas* ... military English in South Africa is not merely a 'taal' for 'backvelders', neither is it a language for 'lang hare en dik brille' or 'Indoenas'. It is a practical means of communication. Picard *Eng. Usage in S.A.* Vol 6 No. 1 May 1975

induku [ɪnˈdugŭ, -kŭ] *n.* See *battle-stick.* [*Ngu. induku* stick for fighting or a walking stick]
   Tied up with the assegais is commonly a short stout stick, with a large knobbed head called by the kaffirs themselves *indookoo* and by us *keerie.* Alexander *Western Africa I* 1837

induna [ɪnˈdunə, ˌɪnˈduna] *n. pl.* -s. Headman. [*Ngu. in Duna* captain, councillor] **1.** Councillor or headman appointed by a chief: see second quot. at *indaba.*
   ... acting Paramount Chief of the Zulus, addressing chiefs and indunas of the war-like Msinga tribesmen ... urged: 'These are not the times to expend your energies in the wasteful and retrograding exercise of impis but in the rewarding pursuits of knowledge.' *Daily News* 16.9.70
   The lead *induna,* or head man, resplendent in hyena tails and impala, monkey and civet skins, carried an Instamatic ... When the newly enthroned Paramount Chief left the party it was in a new Chrysler. *Time* 13.12.71
2. An African man in charge of any band of workers, farm labourers, cleaners, etc.
   We loaded sacks of mielies on to a lorry. The induna ... kicked and hit us all the time. We do not know what we did wrong. *Sunday Tribune* 14.11.71
3. An African man in charge of a gang of mine workers: see also *boss boy* sometimes a mine policeman, see second quot.
   ... he had obtained the job of Induna or head-boy, when he had decided to leave the mines to better his position. Lanham *Blanket Eoy's Moon* 1953
   Makulu Baas now called an Induna, and requested him to take Monare to the barrack in which lived another policeman, ... who also came from Lesotho. *Ibid.*
4. *Afk. form indoenas pl. Sect. Army:* The 'top brass' *[figur.]*
industrial diamond *n. pl.* -s. A diamond not of gem-stone quality in colour or size, used for industrial purposes: see *grit, melée.*

Nature has produced diamonds that while useless for ornamentation are now essential for industry. That is why they are generally called 'industrial diamonds', a term which includes stones used for an endless variety of industrial, mechanical and scientific purposes. McCarthy *Fire in Earth* 1946

influx control *n. phr.* Government control of entry by Africans into urban areas without workseekers' or other permits.

Lack of influx control led to pools of unemployment, and unemployment to crime; overcrowding again led to sickness and epidemics: and germs were not colour-conscious. Gordon *Four People* 1964
. . . some form of influx control is necessary to prevent a flow of workless and homeless Africans to the cities. *Sunday Times* 12.8.73

ingelegde vis, ingelegde fish ['ɪnxə‚lexdə ¹fɪs] *n.* See *pickled fish.* [*Afk. ingelegde* preserved *fr.* Du. *inleggen* + *vis* cogn. fish]

She . . . made all the old Cape fish dishes . . . the celebrated 'ingelegte vis' fried, curried then preserved with vinegar, onions, mango relish and chillies. Green *Tavern of Seas* 1947

ingcibi [ǐn¹kibǐ+] *pl.* ii- A Xhosa circumcision doctor who performs the operation upon initiates to manhood: see *abakwetha, circumcision school, initiation school, circumcision hut.* ⫽ *acc.* Elderly informant whose brother-in-law is an ~, the traditional charge for circumcision is R2,50 or a bottle of brandy. [*Xh.* circumcision doctor]

. . . suggested the operation could be done by a medical doctor . . . Mr . . . still favoured a traditional ingcibi. . . . Where and how the mkhwetha was circumcised was not the crucial issue at all, said Mr . . . a . . . businessman. *Indaba* 9.2.79

ingubu [(ǐ)ŋ¹gubǔ] *n.* Blanket or garment. [*Ngu. Zu. inGubu* clothing, *Xh.* a blanket] **1.** *prob. obs.* Skin blanket or *kaross* (q.v.).

From the shoulders hangs the ungooboo, kaross, or mantle of softened hide worn with the hair next the body, and fastened with a thong at the neck. Alexander *Western Africa I* 1837

2. An article of clothing, (*Zu.*) or blanket (*Xh.*)*:* see first quot. at *blackjack.*

initiation school *n. pl.* -s. A *circumcision school* (q.v.) in which *abakwetha* (q.v.) learn the arts and are admitted to the status of manhood: see quot. at *circumcision school.*

The fast thinning bush near Port Elizabeth

African townships was the temporary home of more than 200 youths who learnt the art of manhood at the initiation schools. *E.P. Herald* 16.2.74

Inkatha [ǐŋ¹kata] *n. prop.* First founded in 1928 by the Zulu King Solomon* as a cultural and social organisation; recently revived by Chief Gatsha Buthelezi as a body aiming for a single South African State with equality for all. [*Zu. inKhata,* a tribal emblem believed to ensure solidarity and loyalty] *Dinizulu

Inkatha, his latest venture, is termed a Zulu National Liberation Movement . . . Inkatha is seen as a means for Blacks, significantly not just Zulus, unilaterally to determine political policies independently of the White Government. *Sunday Times* 9.5.76

(i)nkona, (i)nkone [(ǐ)ŋ¹kɔna/e] See *Ngoni.*

(i)nkosi [(ǐ)ŋ¹kɔsǐ] *n.* Lord, chief. [*Zu. inkhosi* king, paramount chief] **1.** See *enkosi.*

2. Mode of reference to an African Chief.

Great chiefs (*inkosee incoolo*) are assisted by *amapakati* or counsellors. Alexander *Western Africa I* 1837

3. *poss. reg. Natal* Mode of addressing the master of the house: see *baas* also (*i*)*nkosikazi* and quot. at *nkosaan.*

ink-pen *n. pl.* -s. *colloq. usu. children* Either a fountain pen or a 'dip' pen consisting of nib and penholder, as used now for teaching Italic handwriting, as opposed to more modern writing tools with ball points, felt tips etc.

Mrs B – says my writing would improve if I'd learn to use an ink-pen. *O.I. School child circa* 1972

inkundla [ǐŋ¹kundla] *pl.* izi- ii-. The open area between the huts and the cattle kraal of an African village or *stat* (q.v.) used as a meeting place for discussions, announcements and the giving of directives by the elders. [*Ngu.* meeting place]

The crucial issue is whether the initiant can . . . act according to the instructions he receives from his elders at the inkundla . . . as a fully fledged community man. *Indaba* 9.2.79

insangu [ǐn¹saŋgǔ] *n.* See quot. at *dagga.* [*Zu. insangu, Cannabis sativa*]

. . . and marched off all the smokers, leaving me with a handful of boys charged with more serious matters, such as smoking the weed called *dagga* or *insangu* (called in America marijuana). Paton *Kontakion* 1969

inspan ['ɪn‚spæn] *vb. trns. and intrns.* To harness. [*Afk. fr.* Du. *in* + *spannen* fasten, hitch up, *Ger. spannen* to harness] **1.** To yoke or harness draught animals at the beginning of a journey or after breaking it, also for ploughing: see also *outspan.*

In a few hours we walked out of Pilgrim's Rest and after reaching the waggon and oxen, the latter were duly inspanned, and were soon on the move. Cohen *Remin. of Kimberley* 1911 **2.** *vb. intrns/trns.* to prepare for a journey, or make ready a wagon or other vehicle for travel.

The wind has subsided and they have inspanned as we wish to get to the foot of the Stormberg to-night ... After a short time we inspanned again to go on to the next farm. Dugmore *Diary* 1871

The Marico Boers inspanned their oxwagons and took the road to Zeerust in order to attend the Nagmaal. Bosman *Unto Dust* 1963 **3.** *vb. trns. figur.* To press someone into service to assist or co-operate, or to set to do some particular task. *cf. Brit. to rope in.*

I inspanned the only living being in the vicinity, a Zulu woman named Gertrude, and set about unravelling the vagaries of a paraffin-burning refrigerator. *Personality* 28.5.71
deriv.: ~*ing vbl. n.* The act of (1), (2) or (3) or the time for beginning a journey.

... we proposed to proceed at three o'clock a.m.; but a storm of snow and hail raging at that hour, we delayed *inspanning* till it had blown past, the horses being in the meanwhile put loose into an outhouse. Thompson *Travels I* 1827

She had fallen into a light doze before the inspanning began and throughout it she remained silent and hidden. Smith *Beadle* 1926

intake n. pl. -s Formerly *Sect. Army* now in general use in combinations *1978~*, the *July~*, the *January~*, *sig.* that batch of National Servicemen beginning their period of service at the time indicated: see quot. at *N.S.M.*

international *adj.* Of or pertaining to a rating or status granted to certain hotels and restaurants which are permitted to accommodate or serve black patrons and guests.

... separate entrances are still a ... condition in the granting of all liquor licences. The only exceptions to this are the 60 or so hotels with 'international' status where no separate entrances are demanded at their off-sales outlets ... The whole question of separate entrances at bottle stores is now reported to be under review by the liquor licensing authorities. *Het Suid Western* 20.9.78

International status is giving headaches to an increasing number of South African hoteliers and their black guests. *E. P. Herald* 17.4.79
intombi See *ntombi.*
inyala [(ĩ)n'jalə] *n. pl.* Ø. The S. Afr. antelope *Tragelaphus angasi,* similar to the bushbuck in preferring a thickly wooded habitat.

The small farm is virgin bush ... I would like to get nyala established in the area ... Bushbuck and duiker already run on the farm. *Grocott's Mail* 17.8.71
inyanga [ɪn'jaŋga, ĩn-] *n. pl.* -s. An African medicine man, *herbalist* (q.v.) or *muti man* (q.v.) practising either in a tribal or an urban area: see also *uhlaka* and quots. at *herbalist* and *gumboot dance. cf. Jam. Eng. obeah man/woman, samfi man; Canad. conjuror, medicine man; Austral. koradji coradge; N.Z. tohunga,* native doctor, Maori priest. [*Ngu.* (*i)nyanga* doctor, one skilled in a profession]

The INYANGA is a specialist, a man skilled above his fellows, and the number of different types of INYANGA is very great in the different tribes. Some of them are great tribal officials ... In other cases, the INYANGA is a practitioner at large, putting his skill at the disposal of those prepared to pay for it. Hoernlé *cit. H'book on Race Relations* 1949

An inyanga who sells goods in a shop is required to take out a general dealer's licence Bell *S. Afr. Legal Dict.* 1951

An inyanga this week told how he had been robbed of his bag of herbs and mutiman's diploma ... a fully qualified inyanga – his diploma ... comes from the African Dingaka Association in Pretoria, had spent the day digging herbs for his patients in a field in Kliptown. *Post* 18.1.70
inyongo [ɪn'jɔ̃ŋgɔ̃] *n. Afr.E.* Bile, biliousness: also used of the gall bladder worn by diviners, (see *sangoma, isanusi, twasa*) as a mark of their calling. [*Ngu.* in-yongo bile]

The goat was cut up and eaten and the intestinal fat ... hung round ... 's neck, while the gall bladder (*inyongo*) was attached to a small square of hide and fastened to the head as the distinct insignia of a novice. Hammond-Tooke *Bhaca Society* 1962

Sparkling ... takes away inyongo in seconds. Nothing's faster. And that's why you'll feel better all over. You can tell how quickly ... takes away bile by the fresh, clean taste in your mouth. *Post Advt.* 7.4.68
ipiti *n. pl.* -Ø The blue *duiker* (q.v.).
iqira [ĩ'kĩxa+] *n. pl.* ama-. An African witchdoctor, neither an *isanusi* nor an

inyanga (herbalist,) but one who has some powers of clairvoyance: see also *twasa*. [*Xh. igqira* wizard, medicine man]

The middle order of witchdoctor is the Iqira who possesses the same qualities. The lowest order, the herbalist does not have these psychic qualities. *Panorama* Oct. 1977

ironie [ˈaɪənĭ] *n. pl.* -s. A ball-bearing used as a marble by children esp. esteemed as a *ghoen* (q.v.) on account of its weight: also formerly *steelie*. [*fr.* iron + -*ie suffix* (q.v.)]

Hoffie and the grubby children watched with awe and admiration as the General's 'ironie' scattered the marbles with deadly accuracy. *Argus* 2.4.73

ironwood *n.* Also *ysterhout:* any of several trees having particularly hard wood, also the timber itself: usu. species of *Olea* esp. *O. capensis:* in the Transvaal applied to *Colophspermum mopane:* see *mopane* also *umzimbete* (q.v.)

The iron wood or yezerhout is very common and grows very high. The wood is hard, heavy and of a dark brown colour. Percival *Account of Cape of G.H.* 1804

. . . wheelspokes made of either assegai or kershout yellowwood wheel naves or hubs, yellowwood bed planks for the platform, wheel falloes of hard pear or saffraan and . . . disselboom, foretong, long wagon and other heavy carrying timbers, of black ironwood. *E.P. Herald* 28.5.73

irrigation farm *n. pl.* -s. A farm so described or advertised has a proportion of *lands* (q.v.) under irrigation, the rest being *dryland*(*s*) (q.v.) see quot. at *hanepoot*. [*trans. Afk. besproeiingsplaas, sproei cogn.* spray]

An outstanding irrigation farm along the permanent Sonderend River below the Theewaterskloof Dam. 458 Hectares in extent with approx. 260 hectares irrigable land of which approx. 86 hectares are already being irrigated by means of underground pipes and sprinklers. *Farmer's Weekly Advt.* 27.2.74

isanusi, isanuse [ĭsaˈnusĭ/e] *n. pl.* iz- A diviner, usually a woman, trained in the *smelling out* (q.v.) of witches: see also *sangoma, witchdoctor,* and quot. at *herbalist.* [*Ngu. is-, sing. prefix* + *anuse/i* diviner, smeller out of witches *rel. ukunuka* to smell]

. . . he would be touched by the wand of an Isanusi, as we name a finder of witches. Haggard *Nada* 1895

. . . an Isanusi, or high priest among witchdoctors, is the author of several books. *Sunday*

Times 6.6.76

isibongo [ĭsĭˈbɔŋgɔ] *n. sing. only.* Tribal African clan name: see also quot. at *Unkulunkulu. cf. izibongo.* [*Ngu.* clan name]

Clan membership is determined . . . by the possession by all its members of a clan name (*isibongo*) . . . also used as a polite mode of address or in the place of the usual formula in expressing thanks. Hammond-Tooke *Bhaca Society* 1962

is it? *vb. phr. qn. substandard colloq.* Proforma qn. equiv. of Really? Is that so? expressing astonishment, incredulity or polite interest: ⸿Tense, gender and number are not allowed for. See also *isn't it?* [*translit. Afk. is dit?* is that so?]

Her next words take me right in the guts . . . 'I'm getting hitched, Bloss,' she reckons . . . 'Is it?' I croak. *Darling* 11.6.75

isn't it? *vb. phr. qn. substandard colloq.* General tag-qn. inviting assent, equiv. *Ger. nicht wahr?* ⸿Freq. used in contexts where the tense and number of the *vb.* and the neuter *prn.* are inappropriate: see also *not so? nè?* and *is it?* [*prob. fr. Afk. nie waar nie?* isn't that so? *cogn. Ger. nicht wahr?*]

Shakespeare wrote Macbeth isn't it? British informant quoting S.A. Student 1973

it *prn. substandard sing. prn.* used instead of *pl. them* or *they.* [*prob. fr. confusion of Afk. dit* used as neuter *pl. sig.* they, these, *with sing. dit,* it, this, that]

I gave him two of mine and he gave me a small glassie for it. Vaughan *Diary* circa 1902

'I believe that if there are laws which are not in keeping with our faith, the Church must witness against it. *E.P. Herald* 17.10.74

ithongo [ĭˈtɔŋgɔ] *n. pl.* ama-. An individual's personal guardian spirit: see *amathongo.* [*Ngu. ithongo* ancestral spirit]

The one who sleeps must not be woken suddenly. For then his ithongo will not have the time to return. Louw *20 Days* 1963

izibongo [ĭzĭˈbɔŋgɔ] *pl. n.* Praise song or poetry chanted by an *imbongi* (q.v.) in honour of a chief or king: see also *isibongo.* [*fr. Ngu. -bonga* to praise, laud, extol]

I waited at the entrance of the royal kraal while my Zulu friend and interpreter . . . bowing low with folden hands, yelled out an izibongo, a song in praise of the Zulu kings. *National Geographic Mag.* Dec. 1971

izindlubu [ĭzĭnˈdlubŭ] *pl. n.* Zu. name for *Jugo beans* (q.v.) or *kaffir groundnuts* (*kaffergrondboontjies*) (q.v.) *Voandzeia subterranea* of the *Leguminosae*, an underground nut much cultivated by the Zulus and cooked with *samp* (q.v.) or whole *mealie* kernels (q.v.): see *kaboe mealies.* [*Zu. pl. n. (izin)dlubu* Voandzeia nuts]

izinyanya [ĭzĭ'nja₁nja] *pl. n.* Also *imishologu*, see *amatongo* [*Xh. pl. of isinyanya* ancestral spirit]

J

ja [jɑ] *affirmative interj. colloq.* Yes: *freq.* used in speech: see also *ja-nee* and combination ~ *baas vb. phr.* or *n.* see third quot. [*Afk. ja cogn.* yes]

The 450 workers present shouted a loud 'Ja' when . . . asked if they were prepared to strike if all negotiations failed. *Rand Daily Mail* 18.2.71

It seems that this is a unilingual country, or that you wish it to be,' retorted . . ., and there were shouts of 'Ja.' *Sunday Times* 1.4.73

'I believe I was fired because I am no jabaas' he told DRUM. *Drum* 8.3.76

jaag [jɑx] *vb. usu. intrns. colloq.* Tear about, rush round, as in Don't ~ about like that in this weather: *occ. trns. sig.* hustle, rush or harrass; see also *woel.* [*Afk. jaag* rush, chase, *fr.* Du. *jagen* to hunt, chase]

jaagsiekte ['jɑx₁sĭktə] *n.* Either of two diseases having similar symptoms viz. increased rate of breathing as in a hunted animal. [*Afk. jaag* to chase, pursue *fr.* Du. *jagen* to hunt + *siekte fr.* Du. *ziekte* disease] **1.** Crotalariosis in horses caused by either of two species of *Crotalaria C. dura* and *C. globifera* both known as ~ *bossie.*

'Jaagsiekte' in horses is caused by the plants Crotalaria dura and C. globifera . . . the first noticeable symptom is the increase in the rate of respiration (breathing), which may vary from 100 to 120 per minute. A dry cough is present. *H'book for Farmers* 1937
2. A destructive lung disease of sheep, infectious but of unknown origin: sometimes called droning sickness.

Jaagsiekte has been recognised for a considerable time as specific and very destructive lung disease of sheep in the Union . . . There is no known curative treatment and no case has been known to recover. *Ibid.*

jaap [jɑp] *n. pl.* -s. Also *japie* (q.v.) a crude or inexperienced person freq. a

country bumpkin as in *plaasjapie* (q.v.); see also *takhaar* and quot. at *japie.* [*fr. Jaap*, name *deriv. fr. Jakob*]

jabroer ['ja₁bru:r, -bruə] *n. pl.* -s. A 'yes-man' without opinions of his own: a sycophant or 'stooge' to a politician or big businessman: see also *ja-baas* at *ja.* [*Afk. ja* yes + *broer fr.* Du. *broeder cogn.* brother]

Paul Kruger was sitting . . . on the stoep surrounded by a group of men whom Adrian had called Ja-broers . . . He rose heavily and knocked out his pipe . . . nodded curtly . . . and moved away, followed by his escort of black-coated 'Ja-broers.' Brett Young *City of Gold* 1940

jacht [jax(t)] See *jag.*

jack, half See *half-jack, ha-ja.*

jackal-proof fencing *n.* Also known as 'vermin-proof': fencing with wire netting adequate to exclude predators such as jackals: see also second quot. at *fountain.*

There are 14 land camps; eight grazing camps – jackalproof fencing in very good condition. *Daily Dispatch Advt.* 11.3.72

Veld and Fencing: All boundary lines jakhalproof fenced (except approximately 2 000 yards – stock proof); 16 windmills in the veld and at homestead. *Farmer's Weekly Advt.* 27.2.74

Jacky Hangman *n. pl.* -men. Also *Jacky Hanger, Johnny Hangman*: see *Jan Fiskaal*

. . . and the big brass scale showing how many grasshoppers, mice and lizards are eaten by two shrikes (Jackie Hangman) which in turn feed one hawk, far outweighing the one farmyard chicken which may happen to be caught by the same hawk. *Evening Post* 28.4.73

jacopever ['ja₁kɔp'ïəvə(r)] *n. pl.* Ø. Also *jakopewer*: used of any of several red-coloured bulging-eyed marine fishes including **1.** Either of two species of the Scorpænidae fam. *Helicolenus dactylopterus dactylopterus* or *Sebastichthys capensis*, (now known as 'false ~ ') **2.** See *fransmadam.* See quot. at *maasbanker cf.* Austral. Sergeant Baker, a highly-coloured fish, also *miss lucy* (q.v.) [*fr. n. prop. Jacob Evert(sen)(son) C17* Du. *sea captain*]

Thanks to the old writer Francesci, we know that a pock-marked skipper, Jakob Evertson, had a rubicund face with protruding eyes . . . His crew saw the likeness between man and fish, and Kolbe declared that everyone was 'ravished with mirth in the allusion.' Green *Grow Lovely* 1951

In the very early days of the Colony, the

Dutch East India Company had as captain of one of its ships a certain Jacob Evert. When the local fishermen caught a fish with a red face, bulging eyes and thick lips they named it jacopever. *Farmer's Weekly* 18.4.73

jag [-jax] *n.* Also *jacht.* Hunt: found in S. Afr. place names usu. suffixed to animal name e.g. Ezeljacht, Eseljagpoort, Hazenjacht, Buffeljagsrivier. [*Afk. fr. Du. jacht* hunt]
   ... stumbled ... and in the fall broke its neck. But I must not trouble you with all the particulars of our jagt. A. G. Bain 1834 *cit* Steedman *Adventures II* 1835

jags [jaxs] *adj. slang* (vulgar) Lecherous, 'randy'. ℙ(Afk) Used of an animal on heat, see quot. below (erron. form). *cf. Brit. and U.S. slang 'hot pants'.* [*Afk. jags* lecherous, in season]
   3 Nov. 1850 to put 3 more rams among Tseu's ewes ... Tseu says the rams are not springing well. The ewes are jaging. Lewins *MS Diary transcript*

jakkals [ˈjaˌkals] *n.* Jackal: prefix to various plant names usu. sig. 'something spurious or inferior' (1966 *Smith S. Afr. Plants*) e.g. ~ *bessie,* the fruits of *Sideroxylon inerme* see *melkhout;* ~ *bos/blom* any of several species of *Dimorphotheca;* ~ *kos/food* species of *Hydnora* etc. [*Afk. fr. Du. jakkals cogn.* jackal]

jakkalstrou [ˈjakalsˌtrəʊ] *n. colloq.* A monkey's wedding (q.v.) [*presum. abbr. Afk. jakkals trou met 'n wolf se vrou*]

jakkie [ˈjækĭ] *colloq.* See *yakkie.*

jalappoeier [ˌjaˈla(p)ˌpŭĭə(r)] *n.* Powdered jalap, dried root of a Mexican plant *Exogonium purga* a purgative: one of the *Dutch Medicines* (q.v.) [*fr. Span jalapa :* Mexican place name + *Afk. poeier cogn.* powder]
   Epsom salts, senna and jalop, ipecacuanha and Dover powders were the great remedies of those days. Green *When Journey's Over* 1972

jammerlappie *n. pl.* -s. See at *lappie.*
   Breakfast was served respectably ... Oilcloth table cover and clean, damp jammerlappie made from an old flour-bag ... She handed him the jammerlap to wipe his fingers. Louw *20 Days* 1963

janblom [ˌjanˈblɔm] *n.* The rain frog, *Breviceps parva.*
   The shriek of the loorie and the metallic croaking of the Jan blom frog. *Cape Times* 17.9.1912 *cit.* Pettman

janbruin [ˌjanˈbrœĭn] *n. pl.* ∅. Used of either of two marine fishes of the Spa-

ridae fam., having dull, brownish colouration: *Gymnocrotaphus curvidens,* also known as *blue eye* ~; and *Pachymetopon grande,* see (3) *Hottentot.* (also called *John Brown.*)
   ... the confusion that exists over the popular names of our common fish ... our favourite fish, the jan bruin, is elsewhere a Hottentot, a das, a fatfish, butterfish, bluefish, or bronze bream. *Daily Dispatch* 20.6.72

Jan Compagnie *n. prop. hist.* See *Company, John, Jan.*

ja-nee [ˌjaˈniː, -nĭə] *interj.* An emphatic affirmative equiv. of 'that's a fact', 'that's right' etc.: see also *Yes-no.* [*Afk. as above*]
   Ja-nee, Miems, you need cunning to be a farmer. *New S. Afr. Writing* 4 (no date)

Jan Fiskaal [ˌjanˈfɪsˈkɑːl] *n. pl.* ∅. The black and white fiscal shrike, or butcher bird, *Lanius collaris,* which stores its prey of small lizards etc. in a 'pantry' of a thorn bush or barbed wire fence: also *Jacky/Johnny Hangman, Jacky Hanger, ka(r)nallie* or *laksman,* all (q.v.). [*fr. Afk. fiskaal* fiscal shrike, *or folk etym. as in quot.*]
   For nearly two centuries Cape Town's chief of police was an official with the title of Fiscal – a name which survives in the shape of the Janfiskaal, Jacky Hangman or butcher-bird. Green *Grow Lovely* 1951

Jan Groentjie [ˌjanˈxrŭĭŋkĭ] *n.* **1.** *Nectarinia famosa,* the malachite *sunbird* (q.v.). **2.** Creme de Menthe: peppermint liqueur: see quot. at *Van der Hum* [*Afk. fr. Du. Johan* John, *groen* green + *dimin. suffix -tjie*]

Janpierewiet [ˌjanˈpĭrəˌvĭt] *n. pl.* ∅. **1.** *Malaconotus zeylonus,* the *bokmakierie* (q.v.) shrike. **2.** Title of a well-known Afrikaans folk-song. [*presum. onomat.*]

japie [ˈjapĭ] *n. pl.* -s. *colloq.* See *plaas~, jaap* and quot. at *gawie.*
   ... the insult was either intended for, or intercepted by, another young man on his left, who halted and snarled something about 'bloody Japies'. Walker *Wanton City* 1949

jas [jas] *n. pl.* -te. Coat, overcoat. [*Du. jas* greatcoat]
   ... he was a tall man in a great jas (watch coat), and ate mutton with a crooked knife. Burchell *Travels I* 1822

jawl *vb. anglicized pron. sp. jol* See *jol*[2]
   Let's jawl round to the Hohenort. They're blowing up a storm with Beethoven's Moon-

light Sonata. *Cartoon Caption Cape Times* 26.4.79

jeero *n.* See *Indian terms,* and quot. below.

Marinate in . . . Indonesian sauce made of jeero, dhanya (bought from an Indian food shop), bamboo sesate, ketjap . . . turmeric, a little oil and water. *Sunday Times* 26.3.78

jentoe [ˌdʒenˈtŭ, ˈjɪntŭ] *n. pl.* -s See *gentoo* [*fr. Afk.* form jintoe]

'Hey Maria, you old jentoe! . . .' Maria, not put out by the slur cast on her character, waved back gaily and responded in a likewise manner. *11.41 to Simonstown* Matthews *On the edge of the world* ed. Gray 1974

jerepigo [ˌdʒeriˈpigəʊ] *n.* Various sp.: a sweet fortified wine either white or red. [*fr. Port. cheripiga* an adulterant of port wine, of grape juice, sugar, brandy and colouring]

. . . Some Cherupiga wine for ourselves . . . It is about one shilling and fourpence a bottle here, sweet red wine, unlike any other I have ever drank, and I think very good. It is very tempting to bring a few things so unknown in England. Duff Gordon *Letters* 1861–2

The sweet types [sherries] have their flavour and colour imparted to them . . . by the addition of a quota of jerepigo – a well-matured wine in which the sugar of the original juice has been conserved by fortification with grape spirit. Bagnall *Wines of S.A.* 1972

jigger [ˈdʒɪgə] *n.* The sand flea *Tunga penetrans* of which the female burrows under the skin causing discomfort and irritation: also *U.S.* and *Jam. Eng.* [*orig. Span, fr. chigger, fr. chigoe*]

Sometimes as now they peeled them off their hose

And hacked the jiggers from their gnarly toes, Campbell *Veld Ecologue (Adamastor* 1930)

Jim Fish *n. prop.* An offensive mode of reference to a black man now *prob. rare cf. U.S. Jim Crow, Austral. Billy.*

Safely ensconced in Bloomsbury, he [Roy Campbell] could be as offensive as he pleased to any South African who displeased him. And we all displeased him, from Jan Smuts to Jim Fish. E. Davis 1951 *cit.* Sachs *Bosman* 1971

jintoe See *jentoe.*

jislaaik [ˈjɪsˌlaɪk] *interj. colloq.* An exclamation expressing various moods according to tone, usu. surprise but also dismay or a sense of grievance: see also quot. at *nog 'n piep.* [*etym. unknown: see first quot. poss. fr. first syllable of Jesus, cf. Gee, Jeez* and *Afk. interj.* [jɪəs], *yessus* (q.v.)]

Jislaaik . . . its origin is dubious. Is it a contraction of 'Just like mamma made,' 'Just like the doctor ordered' or some such ungram-

matical comparison? If so, how did it come about that its present meaning approximates to the Cockney 'Crikey' or the American 'Gee'? That it began as a substitute for a popular swear-word is more than likely. *Cape Times* 8.1.72

It's the first time I ever catch the Ed. saying 'Jislaaik' . . . she goes in for more refined stuff such as 'Good God!' *Darling* 27.10.76

job reservation *n.* The restriction of certain kinds of employment to particular racial groups: esp. that of the skilled trades to members of the *white* (q.v.) or *Coloured* (q.v.) groups.

. . . said that already job reservation was a 'dead duck'. He thought that in theory the term job reservation would be retained in order to save face, but in practice it would disappear. *Sunday Times* 7.10.73

In perhaps the most significant modification of *apartheid* since it became national policy in 1948, Prime Minister John Vorster last week virtually abandoned the Job Reservation Act, under which the best jobs in the country have long been reserved for whites. *Time* 15.10.73

Job's tears *pl. n. Also Austral.* The hard dried spikelets of *Coix lachryma-jobi* which are tear-shaped rather than round and still used as beads. ⫽Also *Afk. jobskraaltjies* (little beads) or *jobskrale* (beads). [*fr. Lat. name*]

These good people were very civil to us – gave me . . . Job's tears, a pretty sort of grey seed which the Hottentots string into necklaces – and everything else they could think of. Barnard *Letters & Journals* 1797–1801

Joeys *n. prop. slang.* Also *Johies :* Johannesburg: see also *eGoli* and quot. at *tannie (ticket).*

There are stacks of South Africans at this mining and finance firm I'm with . . . It's like still being in Joeys from 9 to 5. *Personality* 12.7.74

joggie See *petrol joggie.*

John *n. prop.* Mode of address to a black man whose name is unknown (objectionable): see also *boy, Jim Fish.* [*presum. fr. most freq. name*]

On the simplest level (that of housewife and servant) madams used to discuss their maids as though dealing with a sub-species. 'John' seemed a good enough name to call a Black guy if you wanted to draw his attention. *Star* 3.11.73

John(ny), dear See *dear John(ny).*

Johnny *Afr. E.* A soldier: see *amajoni.*

That Johnnies' place where they've got our kehla* sounds like a jail! *O.I. African woman* 7.4.79 *(q.v.)

Johnny hangman Also *Jacky hangman,* see *Jan Fiskaal.*

joiner [ˈdʒɔɪnə] *n. pl.* -s. Abusive term among Afrikaners equiv. of 'traitor': formerly one who went over to the British forces during the *Anglo-Boer War* (q.v.): see also *handsupper* and *bittereinder.* [*fr. Eng.* join + *agent. suffix -er*]

There's . . . who, like the poor, is always with us, calling the members of Verligte Aksie and ASASA 'Joiners'. With a grandfather who fought at Majuba, I had been led to believe that 'joiners' were those who went over to the enemy. Van Biljon *cit. Star* 8.9.73

jointed cactus *n. Opuntia aurantiaca* (*O. pusilla*): known also as 'katjie' in the E. Cape. a sprawling, much-branched succulent. ⟦Identified in 1903 as a noxious weed, ~ today ranks as one of the most serious agricultural problems as an 'invader plant', greater than that of the related species of *Opuntia*, prickly pear, see quots. also quot. at *r(h)enosterbos.*

Jointed cactus – *Opuntia pusilla*, a dangerous weed: it is a near relative of the prickly pear, and threatens to become a pest. Pettman *Africanderisms* 1913

Methods of eradicating jointed cactus came under fire . . . the Jointed Cactus Committee reported on the situation with the persistent and hard-to-eradicate pest, and said present methods of getting rid of the cactus seemed hopeless. *Grocott's Mail* 31.8.71

joking, you('re) *substandard.* See *omissions.*

jol¹ [dʒɔl] *vb.* and *n. slang.* To play, frolic, have fun etc.: See also quot. at *kaal.* [*prob. fr. Du. jolen* to make merry]

. . . plenty of jolling and singing . . . join the rest of the ous as the German band jols off to pastures new. *Darling* 9.10.74

deriv. ~ *ler n. pl.* -s. See second quot. at *ticket tannie.*

jol² [dʒɔl] *vb. colloq.* To go: see also *jawl.* [*unknown prob. fr. jol¹,* by extension of meaning]

Let's slaat it out; let's jol. Let's 'beat it' (make off) South African c. (C. P. Wittstock Letter of May 23, 1946). Ex Afrikaans. Partridge *Dict. of Slang and Unconventional English* Vol. II 1961

So it's Sherman's larney dark blue Merc . . . jolling down to Durbs . . . Complete with . . . padkos . . . biltong, ham sarmies . . . Ouma's soetkoekies . . . Hobbs *Blossom* 1978

jong [jɔŋ] *n. pl.* -s. **1.** *colloq.* An informal mode of address usu. regardless of sex: *cf. man cf. Scottish yonker,* young fellow. ⟦Found also in place name Jongensklip.

[*Cape Du. jongen* boy, lad *cogn.* young(ker)]

We returned to the station and held sweet converse with diverse kêrels, to whom we were 'man' and 'jong' after the nature of the tribe. *E.P. Herald* 25.11.1911

'Pas op, jong,' the prison guard had said to her as she left the gaol one day. 'We watch all the dagga people.' Meiring *Candle in Wind* 1959 **2.** A coloured servant, now *rare prob. obs.* in S.A.E.: see also *boy.* [*fr.* (1) ~]

jongmanskas, jongmanscupboard [ˈjɔŋkɪˌmansˌkas] *n.* Also *jongmanskas:* a fairly small clothes cupboard with two drawers side by side above and cupboard space below, freq. in yellowwood and stinkwood with turned (2) *Cape feet* (q.v.) *cf. Brit. bachelor's chest,* a chest of drawers without cupboard space. [*Afk. jonkman* bachelor + *kas* cupboard]

The restored yellowwood and teak 'jonkmans' cupboard sells at R200. *Fair Lady* 13.6.73

A good quality old Cape yellowwood cupboard, old Cape wakis . . . Jongmanskas in yellowwood and stinkwood . . . Tulbagh Chair. *Cape Times Advt.* 7.6.73

jova [ˈdʒɔvə] *n.* (*pl.* -s). *colloq. prob. reg. Natal* Injection, inoculation or 'shot'. [*Zu.* -*jova* to vaccinate, inoculate: *n. jovo*]

For your trip to the Far East did you have to have jovas for yellow fever as well as cholera? *O.I.* 28.2.79

juffrou [ˈjœˌfrəʊ, ˌjəˈfrəʊ] *n.* Mistress: a form of address, second or *third person* (q.v.) (see second quot.) occ. reference, formerly to married, now only to unmarried women esp. schoolteachers: formerly also titular use with surname, see first quot. [*Afk. fr. Du. juffer, juffrouw* madam, mistress, miss, lady]

In the course of conversation our hostess, the Juffrouw Mare, gave an account of the recent death of one of her relations. Thompson *Travels I* 1827

If Juffrow will but tell me where I can find her I will . . . take her my letter. Smith *Beadle* 1926

'That mark on your cheek, juffrou,' I said, 'Will you tell me where you got it from?' Bosman *Mafeking Road* 1947

Jugo bean *n. pl.* -s. *Voandzeia subterranea :* see *izindlubu,* also *kaffergrondboontjie.* [*unknown*]

In addition . . . are . . . many promising plants which . . . may assume considerable economic importance. Chief amongst these are . . . Dolichos beans, the Kudzu vine and the Jugo bean. The 3 last mentioned crops will be of importance only in the lowveld areas. *H'book for Farmers* 1937

jukskei [ˈjœkˌskeɪ, ˈjək-] *n.* A S. Afr. game similar to *U.S. horse-shoes,* played originally with ~ *s,* yoke-pins or *yoke-skeys* (q.v.) now with bottle-shaped, skittle-type pegs: also the peg itself. [*Afk. juk cogn.* yoke + *skei* pin, skey]
   Jukskei, the popular game played with yoke-pins, is said by some to have its origin in the days of the Voortrekkers. De Kock *Fun They Had* 1955
   Horse-shoe pitching in America originated, like jukskei in South Africa, as a pastime among the pioneers. *Cape Times* 19.4.72

July (the) *n.* Also *Durban July:* the July Handicap held annually at Durban's Greyville Racecourse on the first Saturday in July. ⁋A major social event in S.A. similar to Ascot: in combinations ~ *fashions,* ~ *fever,* ~ *day,* ~ *winners* etc. [*abbr. July Handicap*]
   Held in the middle of Durban's superb winter, the July draws people from all corners of South Africa and Rhodesia . . . Towards the turn of the century the Durban Turf Club came into being and on July 17, 1897 the first Durban July Handicap was run for a stake of 500 sovereigns. *Sunday Times* 25.6.72
   . . . the Rothman's July Handicap, traditionally South Africa's 'Ascot' and prestige race. *Panorama* Aug. 1973

jumat [ˌdʒuˈmat] *n. pl.* -s. Malay charm or amulet: see *goëlery* and quot. at *slamaaier.*

just now *adv. t. phr. colloq.* In a little while: phr. with reference to the immediate future as in 'I'm coming ~ ', occ. also to the immediate past as in 'He was here ~ ', but not the immediate present as in standard 'We have none in stock just now.' *cf. Brit. use of 'presently' sig. 'now, at the moment', also 'in a little while'.* [*trans. Afk. netnou* presently, in a moment, *or with past time reference* 'just']
   Just now you get a bloody good klap. Fugard *Boesman & Lena* 1969

K

k [keɪ] *pl.* -'s. *colloq. abbr.* Kilometre: see also *kg* and *metrication.*
   But it's three years and thousands of k's on in sophistication since she left . . . to become Miss World. *Darling* 15.3.78

kaaiman [ˈkaɪmən, -man] *n.* Used of vari-

ous reptiles usu. alligator, *leguaan* (q.v.) (*Varanus* spp.) or even lizard. ⁋acc. One superstition, a merman. [*Du. kaaiman fr. Port. or Span. caiman* alligator] Also found in place names Kaaiman's River, Kaaimansgat, where the meaning of 'crocodile' is possible. ~ *sblom, Nymphaea capensis* the blue water lily.
   . . . a tremendous ravine called the Kaayman's-gat (Crocodile's hole). This name it has probably received from being frequented by the leguaan, a species of amphibious lizard, growing to the length sometimes of six feet, but quite innoxious. Thompson *Travels I* 1827

kaal [kɑːl] *adj.* Bare, naked: usu. in combinations ~ *blad/blaar,* ~ *gat,* ~ *voet,* all (q.v.): Kaallaagte, Kaalrug, Kaalspruit. [*Afk. Du. kaal* bare]
   The bitter stars I've tasted them, My backside is mos kaal. Clouts *One Life* 1966
   The very thought of a whole klomp of nudists jolling around kaal – I mean, sis! *Darling* 12.2.75

kaalblad, kaalblaar [ˈkɑːlˌblat, -blɑː(r)] *n.* Any of several spineless or almost spineless species of *Opuntia* a near-relation to the prickly pear, see *turksvy,* introduced from America to serve as succulent stock feed in time of drought. [*Afk. fr. Du. kaal* bare + *blaar, blad* leaf]
   The orchard wall was backed by a thick hedge of kaalblad, whose big leaves were like flat, full-green hands. Cloete *Hill of Doves* 1942

kaalgat [ˈkɑːlxat] *adj. slang.* (vulgar). Equiv. of *U.S. 'bare-assed':* often in combination to *swim* ~ : See also quot. at *kaal* [*Afk. fr. Du. kaal* bare + *gat* (q.v.) backside]
   Sitting there in the dust with the pieces . . . Kaalgat! That's what it felt like! Fugard *Boesman & Lena* 1969

kaalsiekte [ˈkɑːlˌsiːktə] *n. Alopecia* or baldness in new born lambs and kids of ewes grazing on *bitterbos(sie)* veld (q.v.): see quot. at *bitter(karoo)bos.* [*Afk. fr. Du. kaal* naked + *siekte fr. Du. ziekte* disease]
   . . . the cause of so-called 'kaalsiekte' . . . in lambs. . . . Kids and lambs develop the disease from 3 days to about 3 weeks after birth . . . They often pull out mouthfuls of hair and swallow it. In severe cases the animal may lose its coat within 24 hours. *H'book for Farmers* 1937

kaalvoet [ˈkɑːlˌfuːt, -fʊt] *adj. colloq.* Barefoot: in combination ~ *rangers,* ~ *brigade* sig. *poor whites* (q.v.), freq.

children. [*Afk. fr. Du. kaal* bare + *voet cogn.* foot]

Would you have thought it was an offence to drive a car kaalvoet? A young friend assured me that it was and . . . I was inclined to believe him, although I have found that barefoot driving allows more sensitivity on the control pedals. *Daily News* 28.5.70

Ka(a)penaar [ˈkɑ:pənˌɑ:r] *n. pl.* -s. One from the Cape. [*Kaap cogn.* Cape + -*enaar personif. suffix*] **1.** *hist.* A Cape Town citizen: see quot. at *Hindoo.*

, . . . at each of which [boarding houses] a stranger is sure generally to find a large party of Indians of every denomination – Qui hy, Mull and Duck – all of whom are distinguished by the 'Kapenaars' or townspeople by the generic name of Hindoos. Polson *Subaltern's Sick Leave* 1837

2. One born or long resident in Cape Town or the Cape Province.

The rage in the Transvaal was declared at . . . Nationalist MPC for Worcester, who was reported as saying that 'we Kapenaars are better brought up than say any Transvaler.' *Argus* 16.6.73

3. Edible marine fish *Argyrozona argyrozona:* see *doppie*, also called *silverfish, carpenter.*

Kaapenaar: Found only in S.A. from Table Bay to Natal, usually in deeper water to 100 fathoms . . . This species is of considerable commercial significance, and the flesh is esteemed. Smith *Sea Fishes of S.A.* 1961 *ed.*

kaapse- [ˈkɑ:psə] *n. prefix.* Cape: sig. either peculiar to the Cape, or the Cape species of a plant which occurs elsewhere: *prefix* to numerous plant names. [*Afk. kaap cogn.* Cape + -*se possessive suffix sig.* of, belonging to]

kaapse draai [ˈkɑ:psəˌdraɪ] *n. phr.* See *draai:* also *prob. obs.* A feat of driving skill: see quot. [*Afk. kaapse* of the Cape + *draai fr. Du. draaien* to turn, twist]

. . . the 'bruidswa' [bridal carriage] . . . headed a long procession; and on arrival at the homestead the driver of the bruidswa reached the height of his skill by making the fancy curve known as a Kaapse draai. It took a fine driver to carry out that 'figure of eight' flawlessly at a full gallop. Green *Land of Afternoon* 1949

kaas(kop) [ˈkɑsˌkɔp] *n.* Also *kaas:* a Hollander: not equiv. S.A.E. *Dutchman* (q.v.): *cf.* Brit. and *U.S. squarehead*, a German. [*Afk. kaas cogn.* cheese + *kop* head *cogn.* Ger. *Kopf*]

My late husband . . . was a Dutch marine engineer . . . trust my luck to marry the only

Kaaskop in the world who wasn't a tightwad. Muller *Whitey* 1977

kabeljou [ˈkabəlˌjəʊ, ˌkabəlˈjəʊ] *n. pl.* ∅. Also *abbr. kob* (q.v.): in general use for the common edible marine fish *Argyrosmus hololepidotus* of the Sciænidae, so named by the early Dutch colonists on account of its likeness to the N. hemisphere cod. See also quot. at *allewêreld.* [*Afk. fr. Du. kabeljauw cogn.* Fr. *cabillaud* cod] ⟨Smith, 1975 *Common and Scientific Names of S. African Fishes* uses *kob* for *Argyrosomus hololepidotus* and ~ only for the cod, Gadidae fam.

Some of the fish, like the Kabeljou (Cape cod), geelbek (the so-called Cape Salmon) and pilchards are of the South African species. Green *Sky Like Flame* 1954

kaboe mealies [kaˈbŭ] *pl. n.* Whole maize kernels, boiled: see *mealie.* [*etym. dub., poss. rel. Zu. iziNkobe* boiled mealie kernels]

. . . an iron pot that a fire had been burning underneath . . . All afternoon it had smelt to me like sheep's inside and *kaboe* mealies. Bosman *Bekkersdal Marathon* 1971

kachie, katchie [ˈkatʃi] *n. pl.* -s. *Sect. Drug users.* A measure of *dagga* (q.v.): See first quot. at *arm.* [*unknown*]

Koppie, tell ou Blare . . . three kachies dagga. Not majat, please. cit. *Staffrider* May/June 1978

Kaffir/Kaffer Note: In S.A.E. this term appears to have six basic meanings or uses. These are (1) Formerly (a sense now *obs.*) a member of one of the three Xhosa-speaking peoples, (*hist*). (2) Their language (*hist.*), (3) An abusive means of address (now actionable) or reference to a black person, (4) A London Stock Exchange term (now *prob. obs.*) (5) An element in S.Afr. place names, (6) A common prefix in names of flora and fauna, etc. The material is therefore presented under six headings with the *relevant* compounds following each alphabetically and numbered accordingly, e.g. (1) *Kaffir War*, (3) *Kafferboetie*, (4) *Kaffir Circus*, (6) *Kaffir honeysuckle* etc. [*Arab. kafir* infidel, unbeliever]

⟨ Non S.A.E.: Muslim term of abuse for non-Muslims. esp. Hindus in use acc. Naipaul in Trinidad: see quots.

'I'm going to tell your father. For a Muslim you ain't got no shame. Going out with a kaffir woman.'

'*You* calling she kaffir. You make yourself

out to be all this religion . . . You aint got no shame. Dog eat your shame.' V. S. Naipaul *Suffrage of Elvira* 1958
. . . 'We could do without the Muslim vote' . . . he lifted his left arm and pinched the skin . . . 'This is pure blood. Every Hindu blood is pure blood . . . Is pure Aryan blood.' Baksh snorted. 'All-you is just a pack of Kaffir . . .' *Ibid.*

Kaffir ['kæfə] *n. prop.* **1.** *hist. (obs.).* A member of the Xhosa, Pondo or Thembu nations, regular C19 use for those speaking the ~ *language*: see (2) *Kaffir,* and occupying ~ *land* (q.v.). In combinations ~ *fair,* ~ *land,* ~ *war* all (q.v.). ¶Also occ. used of any indigenous black S. African, now *rare*: see quot. at (2) *shebeen.*

The word Kaffir, meaning infidel in the Arabic, has been improperly applied to designate the native tribes of South Africa, from Natal to the colonial border. The so-called Kaffirs are divided into three great nations: the Amakosas . . . the Amatembies, or Tambookies, . . . and the Amapondos, or people of the elephant's tooth. Alexander *Western Africa I* 1837

The Caffers and Tambookies were very much pleased at my calling them by their true titles, and it is strange how they should ever have been miscalled. The name of the former is Kosa, plural Amakosa, . . . The latter is Tymba, plural Amatymba. Philipps *Albany & Caffer-land* 1827

1. Kaffir fair A trade *fair* established for barter between (1) ~*s* and Colonists.

Caffer Fair . . . Early the next morning the gun was fired for the commencement of the fair . . . but as there were not many traders expected it was postponed till noon, at which time about a thousand Caffers had arrived . . . several of the Caffer Fair dealers had arrived from Grahamstown. Philipps *Albany & Caffer-land* 1827

1. Kaffirland *n. prop.* Formerly also known as Kaffraria. That part of the Cape Colony comprising Pondoland, Tembuland and the territory eastwards from the Great Fish River: see quot. at *Kaffir War.*

To the eastward of the Great Fish river or Rio de Infante extends a fertile tract of country, broken into hill and dale and diversified with extensive woods of the finest forest trees. This is inhabited by a nation called Kaffers or Caffres, the country being called Caffraria or Kafferland. Ewart *Journal* 1811–1814

1. Kaffir War *n. prop. pl.* -s. Any of the nine (see ¶) wars between the Cape Colonists and the frontier tribes during the period 1779–1877. ¶acc. *Cambridge History of the British Empire VIII* 1936: sources vary as to the number.

I had come to know very little more than I had read in my history books with their rambling account of an endless succession of 'Kafir Wars' and an ever-shifting boundary-line on the dim Kaffrarian border. *Cape Times* 9.11.72

2. Kaffir *n. prop. obs.* The Xhosa language: see also quot. at *Sechuana.* ¶Early language texts were known as (2) ~ *grammars* and (2) ~ *dictionaries* in no derogatory sense: see at *enkosi.*

27 June 1842. I have been very busy compiling a kaffer grammer [sic] . . . I finished it last week and intend to revise and improve it, if spared, next year when I hope to have a little more experience. Appleyard *Journal* 27.6.1842

I requested Mr. Whitworth to preach in English, while I translated what he said into Dutch, from which language one of our people, . . . rendered it into Kaffir, that . . . all might know what the preacher was saying . . . We employed much of the day in collecting words for a Kaffir vocabulary, and on the second night slept very uncomfortably. Shaw *My Mission* 1860

You hear an Englishman speak of dobo grass, dongas, tollies, tsholo, etc. which are pure Kaffir. E. *London Dispatch* 4.9.1912 cit. Pettman

In combination *kitchen* ~ (q.v.) and *mine* ~, both derogatory terms for a 'bastard' language: see *fanagolo* and quot. at *go garshly.*

A wonderful language is kitchen Kafir, a weird medley of dialects, interspersed with English words. *Argus* 2.2.1924

3. Kaffir, kaffer ['kæfə, 'kafər] *n. pl.* -s. A mode of address or reference to an African regarded by most Black and White S. Africans as offensive: now a punishable offence in some parts of Southern Africa. *cf. U.S. nigger.* In combination ~*boetie,* ~ *pak, raw* ~, *white* ~ and ~ *work,* all (q.v.); see quot. at *Dutchman.*

. . . term 'Kaffir' is not a term used by the natives to designate either themselves or any other tribe . . . The border Kaffirs know that the white nations apply this name to them, ¶many of them regard it as a term of contempt. Shaw 28.12.1847 cit. Sadler 1967

100 years ago From the *Cape Times* January 1, 1879:
The year ends not altogether gloomily, although wool is down and in Natal Kaffirs are up . . . We shall probably hear in a few days of war in the adjoining colony. *Cape Times* 1.1.79

¶Zu. *Khafula* 'Term of contempt for a person (black or white) of uncivilized manners (a swear word if used direct to a person).' Doke & Vilakazi *Zulu Dict.* 1948.

A black man recently sued and collected damages from a white who had called him a *kaffir* (the South African equivalent of nigger). *Time* 15.10.73

African wins 'Kaffir' appeal. The Supreme Court ruled yesterday that the word 'Kaffir' was an insult and awarded an African damages of R150 ... Mr Justice ... said ... was fully justified in feeling his dignity had been impaired. *E.P. Herald* 4.6.76

3. kafferboetie [ˈkafə(r)ˌbūtĭ] *n. pl.* -s. An abusive mode of address or reference to a White person thought to be a negrophile, or to one who works for or attaches importance to the welfare of Black people. *cf. U.S. nigger lover.* [*Afk. kaffer* + *boet* brother + *dimin. suffix -ie*)

At that time the Prime Minister accused the hon. member for ... of going through the country and accusing the National party of being 'Kafferboeties'. *Hansard* 16.4.59

The BROEDERBOND-CONTROLLED Action Committee ... an official Nationalist front organisation, has resorted to blatant boerehaat and 'kafferboetie' smears in a desperate attempt to prevent the non-political ... from winning the municipal elections on Wednesday. *Sunday Times* 11.3.73

3. kaffir budgies *n. phr. usu. pl. Army and Rhodesian.* Flies, or any other irritating flying insects, also *blood budgies,* mosquitos. ⁋ Various informants.

3. kafferpak [ˈkafə(r)ˌpak] *n. pl.* -ke. *slang.* A thorough beating, usu. the defeat of a sports team. [*Afk. kaffer* + *pak* beating]

The team, highly delighted over their weekend triumph, seemed totally dejected after the 'kafferpak' they picked up at the hands of S.W.D. police side. *Het Suid Western* 19.4.73

3. kaffir, raw *n. pl.* -s. Derogatory term for an African who has not been exposed to any westernizing influences: see *raw.*

... a raw kaffir, who can't even sign his name, but has got to put a cross at the foot of the things he has said – this raw kaffir is allowed to stand there wasting the time of the court for ten hours. Bosman *Mafeking Road* 1947

3. kaffir, white *n. pl.* -s. An offensive mode of address or reference to a White man: see (3) *Kaffir.* ⁋Not equiv. of *U.S. white trash;* see *poor white.*

3. kaffir work *n.* Unskilled manual labour despised by White persons: see quots. at *poor white,* and *lekker lewe.*

Then there grew up the most unwholesome

tradition that most housekeeping duties were 'kaffirs' work', and that it was the sign and symbol of the ruling race to be quite idle. Bruce *Golden Vessel* 1919

Widespread amongst even the poorest whites was a distaste for 'Kafir work'. There was no corresponding prejudice against all forms of manual and unskilled labour in the other great colonies. De Kiewiet *Hist. of S.A.* 1941

4. Kaffirs *pl. n hist.* London Stock Exchange term for S. Afr. mining shares: also *Kaffers.*

The mines floated on the London Stock Exchange which are classed under the general head of 'Kaffers'. *Nation* 19.12.1895 *cit.* W. *Mackie

4. Kaffir circus *n. prop. hist.* That market of the London Stock Exchange (so-called from the early 1890s) where transactions in S. Afr. land, mining and other shares were effected, also *Kaffir Market.*

... the fame of the Rand was noised abroad, foreign buying, by speculators in London and Paris and New York, who barely knew where the Rand was or what shares they bought, continued to keep it spinning at an even more furious pace. They called the exchange, not unreasonably, the 'Kaffir Circus'. Brett Young *City of Gold* 1940

5. Kaffer-, Kaffir- *n. prefix.* Found in place names poss. reflecting historical events, e.g. Kafferspruit, Kaffersrivier, Kaffir Drift.

6. Kaffir-, kaffer- *n. prefix freq. capitalized.* A common prefix: sometimes derogatory implying inferior as in ~ *dog,* ~ *fowls,* ~ *sheep,* ~ *trader,* ~ *truck,* all (q.v.): more frequently occurring in plant or animal names sig. wild or indigenous e.g. ~*boom,* (also ~*bean tree*), ~ *bread tree,* ~ *crane,* ~ *finch,* ~ *honeysuckle,* ~*melon,* ~ *plum,* ~ *tea,* ~ *wag-'n-bietjie,* etc. all (q.v.): or sig. used almost exclusively by Africans in terms usu. of earlier coinage than (3) *Kaffir* e.g. ~ *almanac,* ~ *beans,* ~ *beer,* ~ *corn,* ~ *eating house,* ~ *groundnuts,* ~ *manna,* ~ *piano,* ~ *pot,* ~ *print,* ~ *sheeting,* ~ *tobacco:* the form *Kaffer* usu. being prefixed to *Afk.* words e.g. ~*baai,* ~*brood(bread),* ~*grondboontjie(groundnut),* ~*vink(finch),* all (q.v.). *cf. Austral. prefix blackfellows'-* sig. aboriginal as in ~*bread,* ~*button:* also *U.S. prefix nigger-* as in ~*daisy,* ~*fish,* ~*toe,* ~*weed* etc.; and *Canad.*

prefix Indian- as in ~ *dog,* ~ *brandy,* ~ *corn,* ~ *blanket* etc.

South African actress – who extols the virtues of rooibos tea on radio and television commercials, quickly regained her composure when told of what the Americans were calling 'her' product. 'Kaffir tea' . . . probably has the same connotations in America as kaffircorn, kaffirpot and kaffirboom have in South Africa. *Sunday Times* 30.7.78

6. kaffir almanac *n. pl.* -s. Either of two species of *Haemanthus see Almanac.*

Kaffir almanac so called in Natal, because the Zulus sow their mealies when this plant is in flower. Pettman *Africanderisms* 1913

6. kafferbaai ['kafə(r)ₗbaɪ] *n.* See ~ *sheeting.*

6. kaffir bean(s) *n. usu. pl. Vigna sinensis :* Cow peas, cultivated usu. by Africans as a vegetable.

Brown Haricot . . . Mixed Cow Peas or Kaffir Beans. *Farmer's Weekly Advt.* 3.1.68

6. kaffir bean tree *n. pl.* -s. *Erythrina caffra :* see ~ *boom.*

This day Henry found the chain we lost at the top of a Kaffre bean tree where the monkey had got fast. Shone *Diary II* 29.8.18

Kaffirboom . . . black pods containing bright red beans with a black spot, known as 'Kaffir beans' or 'lucky beans'. Beeton *Eng. Usage in S.A.* Vol. I No. 2 1970

6. kaffir beer *n.* Also *KB* (q.v.) now freq. called *Bantu beer,* and *rare colloq.* 'white wash' (commercially available in cartons) see *mqomboti; maiza; tshwala; utywala.* [Orig. a home-brewed fermented beverage made from malted grain usu. *kaffircorn* (q.v.) also maize: now brewed on a large scale by municipal breweries for *beerhalls* (q.v.): see also quot. at *suurpap.*

'Kaffir beer' means the drink commonly brewed by natives – from kaffir corn or millet or other grain, and . . . includes fermented liquor made from prickly pears . . . fermented liquor made from honey (commonly called honey beer) and any other fermented liquor which the Governor-General may from time to time, by proclamation in the Gazette, declare to be included in this definition. *Act 21 of 1923 Section 29 Statutes of the Union of S.A.*

It is true that on many occasions in urban areas, for births, ancestor-worship ceremonies, and so on, kaffir beer is bought from the municipal breweries and partaken of with due ceremony and circumspection by many urban Africans. Longmore *Dispossessed* 1959

6. kaffirboom [buəm] *n. pl.* -s. *Erythrina caffra :* also known in the early days as Coral tree or *Corallodendrum. cf. Jam.*

Eng. *coral bean tree, Erythrina corallodendron.* A large deciduous tree with spiny trunk and usu. bright scarlet flowers. The timber is light and cork-like, the leaves thought to be toxic to stock. Its seeds, also scarlet, are called by some *lucky beans* (q.v.), or (6) *Kaffir beans :* see *Kaffir bean tree.* [*kaffir* + *Afk. boom* tree *cogn. Ger. Baum*]

. . . the *Erythrina Cafra* or *Corallodendrum,* known among the Dutch farmers and English colonists as the Kafferboom. This often grows into a large and umbrageous tree, and is sometimes met with standing apart. In the spring season it is covered with innumerable blossoms, of a brilliant scarlet colour, giving it a very gorgeous appearance. Shaw *My Mission* 1860

6. kaffir bread tree, kafferbroodboom ['bruətₗbuəm] *n. pl.* -s. Any of several species of *Encephalartos* with pith rich in starch: see *Hottentot bread, broodboom* [*kaffir* + *brood cogn.* bread + *boom* tree *cogn. Ger. Baum*]

Two plants of the palm tribe were frequently met with, one, the *zania cycadis* or Kaffer's Breadtree, growing on the plains. Barrow *Travels* 1801 *cit.* Jeffreys

6. kaffircorn *n.* Any of several species of *Sorghum* esp. *S. caffrorum* widely cultivated for its grain, which ground is used for porridge, and when sprouted for making (6) *Kaffir beer* (q.v.). Sprouted ~ is known also as *mtombo* (q.v.), or *uitloop* (q.v.): see quot. at *suurpap.*

Mealies, Kafir Corn, Ration Meal, Flour, Mealie Meal, Kafir Corn Meal . . . Transvaal and Boer Tobacco. *Grahamstown Journ.* 20.9.1892

KAFFIR CORN AND MALT Excellent Kaffircorn Malt . . . Best quality Mixed Kaffircorn . . . Sprouted Kaffircorn . . . Fine Kaffircorn Meal. *Farmer's Weekly* 7.11.73

6. kaffir crane *n. pl.* -s. Also *mahem* (q.v.) used of any of several species of *Balearica* (Gruidae fam.) in particular *Balearica pavonina regulorum:* a slate grey crane with a crown or tuft of black plumes, hence also crowned crane: also applied to *Bugeranus carunculatus,* the wattled crane.

6. kaffir dog *n. pl.* -s. Also *kaffir brak,* see *brak :* a mongrel, usu. an ill-kempt, uncared-for dog: see quot. at *brak. cf. Anglo-Ind. pye-dog, pie-dog.*

Long after the trumpet had sounded Young Hopeful would be seen making his way toward the parade, putting his accoutrements on as he came and invariably two or three Kaffir dogs following him. Buck Adams *Narrative* 1884

6. kaffir eating house *n. pl.* -es. *obs.* See quot.: also at *eating house.*

Kaffir eating house . . . is a house in which a person carries on the business of supplying meals to natives. Bell's *S. Afr. Legal Dict.* 1951

6. kaffir finch, kaffir fink, kaffervink *n. pl.* -es, -s. Any of several small S.Afr. birds, *Euplectes* spp. with long black tail feathers: see *widowbird* and *sakabula*: also used of the red *bishop bird* (q.v.) *Euplectes orix.*

. . . the Kaffir finch, whose black and white plumage and red throat were set off by his long streaming tail, the feathers of which are so prolonged that they droop into a perfect arch, and when flying nearly overbalance him. Lucas *Camp Life & Sport* 1878

6. kaffir fowl *n. pl.* -s. A usu. scraggy fowl of indeterminate breed *cf.* (6) *Kaffir sheep.*

When I was a child kaffir fowls were a regular article of commerce on the East London market, called just that. I.O. 1971

6. kaffir groundnut, kaffergrondboontjie [ˈkafə(r)ˈxrɔ̃ntɪbŭĭŋkĭ, cĭ] *n. Voandzeia subterranea*: see *izindlubu*: also called *bambarra* (*ground*) *nut* and *Jugo bean* (q.v.).

6. kaffir honeysuckle *Tecomaria capensis*: see *Cape honeysuckle.*

6. kafferjaghond [ˈkafə(r)ˈjaxɪhɔ̃nt] *n. pl.* -e. A cross-breed hunting dog, licenced under this name in S.A. [*kaffer* indigenous + *jag fr.* Du. *jagen* to hunt + *hond cogn.* hound]

6. kaffir manna *n. Pennisetum americanum* (*typhoides*): large perennial grasses extensively cultivated by Africans for grain which resembles (6) *kaffircorn* (q.v.): see also *babala.* [*Eng.* manna, *Setaria* (*Panicum*) *italica*, Italian or Hungarian millet]

6. kaffir melon *n. pl.* -s. Also *kaffir watermelon, tsamma*(*melon*), *karkoer* and *makataan* all (q.v.) *Citrullus vulgaris*, (*lanatus*) or *C. caffer*, poor varieties, some cultivated as a succulent feed for stock, used also for *waatlemoenkonfyt*, see *konfyt.*

6. kaffir orange, *n. pl.* -s. See *klapper²*.

6. kaffir piano *n. pl.* -s. Also *calabash piano; marimba; mbira* (q.v.): An indigenous musical instrument similar to a xylophone but usu. played with the fingers, often suspended over calabashes,

sometimes containing water, as resonators: hence *calabash* (q.v.) *piano;* see quot. at *mbira.*

. . . the development of African musical instruments, such as xylophones or mbira (so-called kaffir pianos), upon which Africans may play their traditional music and interchange tribal compositions without having recourse to expensive foreign instruments. Hellman & Abrahams *H'book Race Relations* 1949

6. kaffir plum *n. pl.* -s. *Harpephyllum caffrum*, a tall flowering tree with large, sourish fruit resembling a plum, eaten by Africans: also yielding a prized reddish timber.

6. kaffir pot *n. pl.* -s. An almost spherical lidded black iron pot on three legs, used for cooking over an open fire, obtainable from smallest to very large sizes, see quot. at *skerm:* ⟦Also used in white households as coal scuttles, flower containers, or for camp cookery,

Outside, on the stoep, were the single-furrow American ploughs, a pile of yokes, trek-gear, and three-legged Kaffir pots of various sizes, chained together by their handles. Cloete *Hill of Doves* 1942

. . . lard it well with fat bacon and cook it slowly in the good old three-legged kaffir pot. Green *Glorious Morning* 1968

6. kaffir print *n.* Also *German print,* (q.v.) (*Duitse sis*), *African print, 'bloudruk'* : inexpensive cotton material usu. of blue or brown, closely printed with geometrical or floral designs, now also printed in shaped skirt panels.

6. kaffir sheep *n. pl.* ∅. A crossbred sheep inferior for slaughter purposes.

The breeds of sheep mostly encountered in these areas are the Blackhead Persian and the 'kaffir' sheep the former being superior for mutton purposes. The 'Kaffir' breed is scraggy and of poor conformation. Both breeds are very hardy and resistant to heartwater. *H'book for Farmers* 1937

6. kaffir sheeting *n.* Thick soft cotton material, coarsely woven, used for some African dress and much favoured for inexpensive home-decorating (not for sheets). [*Afk. kafferbaai* kaffer baize ; see *linnebaai*]

. . . a coarse white cotton cloth called 'Kaffir sheeting' from which all Qaba dress is made. Broster *Red Blanket Valley* 1967

SUPERIOR HEAVY QUALITY KAFFIR SHEETING . . . fantastic value for every home . . . first quality Kaffir Sheeting with a novelty surface textured effect that goes with every decor. Available in 10 decorator shades. *E.P. Herald Advt.* 27.3.74

6. kaffir tea,[1] **kaffertee** *n.* Certain species of *Compositae* esp. of *Helichrysum* used in an infusion as a medicinal tea for colds and respiratory disorders: see also *hotnotstee.*

6. 'Kaffir Tea'[2] *Non S.A.E.* American Trade name for *rooibos tea* (q.v.) *cf.* See also quot. at (6) *kaffir.*

The American version of Rooibos is marketed . . . They describe it like this: 'Kaffir tea' is the leaves of the rooibusch [sic] shrub which grows on the highlands above Cape Town, South Africa. It has been used . . . there for many generations. 'Kaffir tea' resembles orange pekoe tea in . . . appearance and . . . aroma. Yet it is entirely free from caffeine or theine. *Sunday Times* 30.7.78

6. kaffir tobacco *n.* Early term for (3) *dagga* (q.v.), *now prob. obs.*

When I was a lad dagga was known as 'kaffir' tobacco and had a vast sale at 1s. a 1b. *Sunday Times* 13.2.71

6. kaffir trader *n. pl.* -s. A merchant often in a remote district largely occupied by Africans dealing usu. in foodstuffs, inexpensive goods and (6) *kaffir truck* (q.v.).

6. kaffir truck *n. prob. obs.* Buttons, beads, brass wire, trinkets and cheap soft goods, formerly the somewhat tawdry stock-in-trade of the *kaffir trader* (q.v.), collectively known as ~.

Ayliff and Co. . . . a large and varied assortment of Merchandize . . . Fineries, Clothing, Hardware, Saddlery . . . Kaffir Truck, Breadstuffs, Groceries. *E.P. Directory* 1848

Before long the more sophisticated Xhosas returned to barter their possessions for red clay. They demanded beads, buttons, trinkets and other kaffir truck in return for their goods. Metrowich *Frontier Flames* 1968

6. kaffer wag-'n-bietjie ['vaxņˌbǐkǐ, -cǐ] *n. Acacia caffra :* see (3) *katdoring.*

kahle ['gałe] *interj.* Exclamation enjoining caution, 'Take care', 'Watch out' etc. *cf. oppas :* see also *hamba* ~, *go garshly,* and *go well.* [*fr. Zu.* -*hle, kahle* well, softly]

kaiing(s) [kaıŋs] *n. usu. pl.* Greaves or brow(n)sels from which fat has been rendered down. [*Afk. prob. fr. pl. kaaien Du. kaan* residue of melted tallow]

. . . the food now to be considered is not commonly met and certainly does not appear on any menu . . . Kaiings are little bits of crisp fat which remain when sheep's fat is rendered

down into dripping . . . They must be eaten piping hot and, as they are very rich, should be approached by anyone over the age of sixteen with great restraint. *Farmer's Weekly* 25.4.73

kak [kak] *interj. n. modifier slang.* (vulgar) Excrement: as *interj.* an expletive, also in *vb. phr. gaan* ~, a savage dismissal, equiv. of 'go to hell' etc. ; as *n. modifier* equiv. *U.S.* 'crappy'. ⫿Also used esp. by children as equiv. of 'rubbish' e.g. 'The sermon was the biggest lot of ~ I've ever heard.' *cf. Austral. bullsh, Jam. Eng. caca* filth, excrement, anything dirty: *also Brit. low colloq. cack* [*Afk. fr. Du. kak* excrement *fr. Lat. cacare* to defaecate]

. . . you may remember that a former mayor of Naboomspruit once described Naboomspruit as a kak ou dorpie. Have you any comment on that? *Sunday Times* 12.9.71

And the final big moment came when the visiting actress says 'Goodbye – and in your beautiful Afrikaans – gaan k-.' *Cape Times* 27.1.73

kakebeenwa, kakebeen wagon ['kɑkə-ˌbɪən'va] *n. pl.* -ens. A type of wagon used by early pioneers: see quot. [*Afk. kakebeen* jawbone + *wa fr. Du. wagen cogn.* wagon]

. . . two ox-waggons, their wide heavy bedplanks, worn with much use. They are what is known as 'kakebeen' (jawbone) waggons, because of the high, upright sides, sloping like the jawbone of an ox or a horse. *Drostdy Swellendam* (Catalogue)

kakelaar ['kɑkəˌlɑ:r] *n. pl.* -s. The red billed hoopoe, *Phoeniculus purpureus,* which gets its name from its loud and strident voice ; also called *monkey bird.* [*fr. Du. kakel(en)* to chatter, *cogn.* cackle + *agent. suffix* -aar]

Its voice is harsh and resounding, and has acquired for it the name of *Kackela* among the Dutch, which signifies the 'Chatterer.' Layard & Sharpe *Birds of S.A.* 1875–84 cit. Pettman

kakiebos ['kɑkiˌbɔs] *n.* See (3) *khaki. Tagetes* a source of essential oil: see quot.

Many a South African farmer who fights a losing battle trying to clear his fields of persistent 'kakiebos' would shed tears at the sight of the devotedly cultivated fields of 'tagetes' (kakiebos) . . . Kakiebos would also be regarded by most people as an unlikely raw material for producing fragrant perfumes. *Panorama* May 1974

kak off ['kak ˌɔf] *vb. slang,* (vulgar) *usu. Sect. Army.* To exert oneself to or beyond the limits of endurance: see *afkak parade* and *kak.*

One makes friends very easily here as we are

all in the same boat – H.M.S. Kak-off. *Letter Serviceman* 14.1.79

kakparade See *afkakparade* also *kak off.*

kalander[1] [ˌkaˈlandə(r)] *n. pl.* -s. *Calandra granaria :* the grain weevil which breeds in stored grain [*Afk. kalander* weevil *prob. fr. Lat. name*]
Wheat, maize and other cereals which are stored, are susceptible to attack, particularly by two small insects – the grain weevil or kalander, and the grain moth – and these insects do great damage annually. *H'book for Farmers* 1937

kalander[2] *n. modifier. Podocarpus falcatus*, the Outeniqua yellowwood: usu. in combination ~ *yellowwood.* [*corruption of Outeniekwalander* a dweller in, or native of Outeniqualand, the Knysna-George district]
The highest prices a cubic metre of timber realised at the sales were: stinkwood R1 360, yellowwood R260, kalander yellowwood R230, blackwood R250 and witels R150. *E.P. Herald* 21.9.74

kalkoentjie [ˌkalˈkŭĭnkĭ, -cĭ] *n. pl.* -s.
1. Any of several species of Iridaceae esp. *Gladiolus alatus*, and various *Tritoniae ;* named for their scarlet colouring resembling the wattles of a turkey. [*Afk. kalkoen* turkey + *dimin. suffix -tjie*]
Flowers bloomed everywhere in the warm sunshine – gladioli, ixias, iris, heaths every shade of colour, from the crimson kalkoentjie to the pure white chincherinchees. Fairbridge *Which Hath Been* 1913
2. The S. Afr. bird *Macronyx capensis* [*as at* (1) ~]
To scamper across these Flats is like riding on the top of a Scotch or Yorkshire moor, and only for the scream of some excited kalkoontjie circling over its nest in the heather, the scene is as quiet and subdued as the heart of man can desire. A Lady *Life at Cape* 1870

kalimba [kaˈlĭmba] *n. pl.* -s An African musical instrument adapted to the Western scale by Hugh Tracey from the traditional *mbira* (q.v.) Informant Mr Andrew Tracey, Rhodes University. [*Bantu ka-* dimin. prefix, *limba/rimba* a note]
Grahamstown factory exports kalimbas
The kalimba is a handy, pocket sized instrument, capable when plucked of producing 'sweet and gentle harmonies' from steel reeds attached to a sound box or board. Dr Tracey described it as companionable and uniquely African. . . . There are three types of kalimba: celeste, treble and alto. *E.P. Herald* 18.6.79

kalya *n.* An Indian dish of spiced chicken or mutton cooked with *dhai*, see *Indian terms*, or yoghurt.
Brinjal fritters Chicken Kalya with Naan . . . serve buttered for tea, or with curries, Kalya or with . . . salads. *Fair Lady* 21.7.76

kam(m)assiehout, kam(m)assiewood [kaˈmasĭ(haʊt)] *n.* The timber of *Gonioma kamassi*, very hard and similar to boxwood: called Knysna boxwood or False Cape Box. ¶When fresh there is an alkaloid toxic principle present in the wood which is dangerous to workers. Found in place name Kamassieberg. [*etym. dub. prob. fr. Hott. name*]
Kamassie . . . is often known as Knysna boxwood. It is of particular note as it is one of the two timbers regularly exported from South Africa, although in small quantities. The other timber exported is the Cape box for which kamassie wood is often mistaken. Palmer & Pitman *Trees of S.A.* 1961

kameeldoring [kaˈmɪəlˌdʊərɪŋ] *n. pl.* -s. Also *camelthorn tree* (q.v.) *Acacia giraffae :* see quot. at *camelthorn.* [*Afk. kameel* giraffe *fr. cameleopardalis* + *doring fr. Du. doorn cogn.* thorn]
. . . the 'Kameeldoorn' (camelthorn) . . . is prolific in the desert and carries many nests of society birds, not to mention the tampaan tick, the size of a healthy bug, which descends upon you if you make your resting place under the tree. Jackson *Trader on Veld* 1958

-kamma[1] [ˈkama] *n. suffix.* Stream, water: found in place names Goukamma, Sapkamma, Keiskamma, Kraggakamma etc. See first quot. at (2) *fontein.* [*fr. Hott. kamma/o* stream, water]

kamma[2] [ˈkama] *modifier, slang.* Pretended or put on, e.g. spurious illness, sadness etc. is described as ~ *sick*, ~ *sad* etc. [*Afk. prob. fr. Nama khamo* like, similar]

kamnassie(hout) [ˌkamˈnasĭ(haʊt)] *n. Maytenus cymosa*, a greyish spiny shrub with white flowers (not *kam(m)assiehout* (q.v.)): found also in place names Kamnassieberge(n), Kamnassirivier, Kamnassidam. [*etym. dub. prob. fr. Hott*]

kanallie See *ka(r)nallie.*

kankerbos [ˈkaŋkə(r)ˌbɔs] *n. pl.* -se. Cancer bush: used either of *Euphorbia ingens*, see *naboom*, or of any of several species of *Sutherlandia*, *S. frutescens*, *S. tomentosa* and *S. microphylla :* see *cancer bush.* [*Afk. kanker cogn.* cancer + *bos cogn.* bush]

kanna[1] [ˈkana] *n. pl.* -s, Ø. *obs.* Traveller's

term: *Hott.* name for *eland* (q.v.)
hence ~ *bos* see *khannabos,* see *kanna².*
[*fr. Hott. kanna* eland (elk)]
  Here he saw several herds of *kannas* (or
elands) and quakkas grazing at a distance and
appearing not much to heed the presence of
our party. Burchell *Travels II* 1824
Also ~ *bos,* species of *Salsola:* see
ganna; said to be so called from a belief
that eland fed upon it.

kanna² [ˈkana] *n.* Also ~ *wortel :* Either
of two species of *Mesembryanthemum:
Sceletium tortuosum* or *S. anatomi-
cum :* see *hotnotskougoed.* ⫿ A stimulant
earlier erron. compared with ginseng
root and unrelated to the *Salsola*
(*gannabos*).[*fr. Hott. plant name channa*]
  ... species of Mezembryanthemum, which
is called Channa by the natives, and is exceedingly esteemed among them. Paterson *Narrative of Four Journeys* 1789 cit. Pettman

kanniedood [ˈkanĭˌdʊət] *n.* Popular name
of several species of *Aloe, Gasteria* and
Haworthia which are exceptionally
drought resistant and can live without
food and water, often suspended in a
tree or on a verandah hence the name
'air plant': *also figur.* a diehard. [*Afk.
kannie* cannot + *dood cogn.* dead]

-kant [kant] *n. suffix.* Side: found in
street names e.g. Buitenkant St., Waterkant St. [*Afk. fr. Du. kant* side]

kaparring, kaparrang [ˌkaˈparɪŋ] *n. pl.* -s.
Traditional style wooden sandals worn
by *Malays* (q.v.): see quot., similar to
Japanese *geta.* [*Afk. prob. fr. Javanese
gambarran* sandal]
  When Lady Duff-Gordon saw them the men
wore the toudang – a wide, pointed straw hat –
over a red and white handkerchief bound
turban-wise about their heads, and on their
feet kaparangs or clogs, as the old-fashioned
Malays still wear them. Fairbridge cit. Du
Plessis *Cape Malays* 1944
  By the end of the nineteenth century Oriental
dress had been discarded; but certain characteristic features have been preserved. *Kaparrings*
(probably from the Javanese *gamparan*:
wooden sandals with a knot to push between
the big and second toes) are still in use. Du
Plessis and Lückhoff *Malay Quarter* 1953

kapater [kaˈpɑtə(r)] *n. pl.* -s. A castrated
goat: see also *hamel.* [*Afk. fr. Du.
kapater fr. capade* eunuch *fr. Port.
capado* castrated]
  One poor farmer on the coast decided to go
in for breeding goats, so he went to Grahamstown where he bought a number for this pur-

pose. Proudly he returned to his location with
them, only to be informed by a more knowledgeable friend that they were all kapaters!
Metrowich *Frontier Flames* 1968

kapkar [ˈkapˌka(r)] *n. pl.* -re. *usu. Cape
cart* (q.v.) a corruption of ~.
  ... the tour will stop at the old postcar
station, a resting station for the old 'Kapkar'
that transported post from Knysna to George
during the late 1800's . . . twice a week. *Het
Suid Western* 20.9.78

kappie [ˈkapĭ] *n. pl.* -s. **1.** A large sun-
bonnet usu. of white lawn or linen freq.
tucked and embroidered in intricate
designs, worn by *Voortrekker* (q.v.)
women and still by *volkspelers* (q.v.)
in costume: see also quot. at (1) *Voortrekker* and *moederkappie :* in combination ~ *sis,* see *sis.* [*Afk. fr. Du.
kapje* little hood]
  In the Dutch community, linen caps and
scarves, popular since Van Riebeeck's day,
have developed into the kappie, later to be the
most typical feature of the Voortrekker woman's costume . . . The fichu and the kappie
were typical features of a Voortrekker woman's
dress. The kappies, particularly, were beautifully embroidered in quaint original designs.
Gordon-Brown *S. Afr. Heritage* 1965
2. *colloq. figur.* The circumflex used in
Afk. as in *kêrel* and *môre* indicating a
lengthening and lowering of the vowel:
see pronunciation table. [*fr.* (1) ~]

Karakul [ˈkærəˌkul] *n. prop. pl.* Ø. The
breed name of so-called 'Persian Lamb',
a long-haired fat-tailed sheep somewhat
like the *Ronderib Afrikander* (q.v.): see
Swakara.
  As the word Persian refers to a different
breed of sheep in this country, a certain amount
of confusion is bound to result from its use,
so it is suggested that the true breed name
'Karakul' be retained for general use in South
Africa. *H'book for Farmers* 1937

karamat See *kramat.*

karanteen [ˌkærənˈtin, ˌka-] *n. pl.* Ø. The
small marine fish *Crenidens crenidens :*
in combination *striped* ~ *Sarpa salpa*
also known as *bamboofish* (q.v.) and
stre(e)pie, often used by anglers as bait.
  The Eastern Cape Bamvoosie (derived from
Bamboo fish) is known in Natal as the Karanteen and in other parts as the streepie. *E.P.
Herald* 15.7.71

karbonaadjie [ˌka(r)bəˈnaıkĭ, -cĭ, kar-] *n.
pl.* -s. Various sp. also *carbonaadjie* (q.v.)
and *karmenaadjie**. Mutton cutlets or
'collops' usu. *braai'd* (q.v.) over a fire.

⫷ Also, *rare*, a present of meat* from a neighbour who has slaughtered an animal. [*Afk. fr. Du. karbonade cogn. Fr. carbonade fr. Lat. carbones* grilled meat, *also loan word in Eng. carbonade*, scored, grilled meat]

... I joined half a dozen burgher officers in a tent, and discussed with them . . . the well known South African *karbonatje*. This last consists of pieces of meat roasted on a peeled forked stick, fat alternately with lean, and with a sprinkling of salt and pepper . . . with a field appetite, this is the true way to relish mutton. Vive la karbonatje! Alexander *Western Africa I* 1837

Doctor! Doctor! (*Oom Doors halts the doctor's progress...*) A karmenaadjie. We slaughtered yesterday. Fugard *Guest* 1977

kareehout, kareeboom, kareemoer See *karree*

karem [ˈkarəm] *n.* A Cape Malay game related to billiards, played on a board with a cue and discs, not balls, now gaining popularity. [*unknown* numerous spellings see *kerim*]

Mr. Ishmail Floris (84) . . . invented the game Karem many years ago and he still makes boards for this game as he has done for 25 years. The game has been played by Cape Malays for the last quarter of a century. It came into being one day, when a visiting Indian asked Mr. Floris . . . to copy a board that was used for a game in which people flicked discs with their fingers. *Panorama* Aug. 1974

ka(r)koer [ka(r)ˈkuːr, -ʊə] *n. pl.* -s. *Citrullus lanatus* or *C. amarus:* see *kaffir melon*, also *tsamma*. [*etym. dub. poss. Hott. fr. Bantu cakulo*]

. . . only one water-melon was on sale. Rube had a passion for them, and the dorp's licensed comedian suborned a 'Coolie' to bid him up. At 7s. 6d. the Indian dropped out and Rube seized the prize, cut it greedily at once and found that it was only a 'karkoer', the 'Kafir pumpkin' which grows as a weed amongst the mealie-crops. Prance *Tante Rebella's Saga* 1937

ka(r)nallie [ˌka(r)ˈnalɪ] *n. pl.* -s. One of the names of the fiscal shrike or butcher bird: see *Jan Fiskaal*. [*Afk. fr. Du. kanalje cogn. Fr. canaille fr. Lat. canis* dog]

Karoo See *Kar(r)oo*.

kaross [kəˈrɒs, kaˈrŏs] *n. pl.* -ses, -es. **1.** Traditional clothing of certain tribes made of skins; also combinations *fore* (*voor*) ~ and *hind(agter)* ~: see also quot. at *ngubu*. [*Afk. fr. Hott. (k)caro-s* skin blanket]

. . . but the rest of their body was quite uncovered, except by a bundle of small greasy leathern aprons . . . These aprons, which they distinguish into fore-kaross and hind-kaross, and which are tied just over the hips, are their only permanent clothing: for the large kaross or cloak, is only worn, or thrown off, agreeable to the weather or the fancy of the wearer. Burchell *Travels I* 1822

2. A blanket of sewn skins used on a bed or occ. on the floor. [*fr.* (1) ~]

On leaving, Mr Webster made me a present of a very handsome kaross – a number of skins sewn together. Buck Adams *Narrative* 1884
Shall I take the blanket from the Baas's bed? Shall I bring the Baas's best karos, the one of silver jackal skins, or the rooikat? Cloete *Hill of Doves* 1942

karree [kəˈriː, -ɪə] *n.* Also *kiri: honey beer* (q.v.): also a drink prepared fr. prickly pear syrup: see quot. at *kaffir beer*. In combination ~*moer* (*kirimoer*) see quot. at honey beer: the powdered root of any of several plants esp. *Trichodiadema stellatum* used as the ferment in making ~. ⫷Used acc. Pettman by Kar(r)oo housewives as leaven for baking bread. [*Afk. fr. Hott. karib*]

Detective Ferreira said that the sample produced was 'karriemoer' with water. It was not exactly kaffir beer. *E.P. Herald* 27.8.1925

karreehout, karreeboom [kaˈriːhəʊt/bʊəm, kəˈrɪə-] *n.* Any of several species of *Rhus* more esp. *R. lancea* (bastard willow) and *R. viminalis* and their timber: some of which are also known as *taaibos* (q.v.). ⫷Their fruits, *karree berries*, are used for making *mampoer* (q.v.). [*prob. fr. Karoo acc.* Smith *S. Afr. Plants* 1966]

Because the Sandveld is a semi-desert, the local carpenter's wood resources were stunted trees and shrubs. He drew not only from the farm orange grove, but from the wild olive and mulberry in his environs, the syringa and bastard willow (karree) that grew next to the river. Baraitser & Obholzer *Cape Country Furniture* 1971

There are some people that will try to tell you that kareebessies make the best *mampoer*, but I . . . agree with Oom Daan the best is the sort that is made from peaches that have been fermenting for eight days until they taste sharp, but not too sharp. *Star* 17.1.79

Kar(r)oo [kəˈruː, -ʊə] *n. prop.* The arid plateau or semi-desert regions: the *Great* ~ and the *Little* ~; *cf.* use in second quot. [*Afk. fr. Hott. karo* dry]

Betwixt the Zwarteberg and Nieuveld or last great chain, extend the immense arid deserts called by the Hottentot name of Karroo, where no human creature has yet attempted to fix a habitation. Ewart *Journal* 1811–14
. . . throwing up clouds of dust from the arid ground, which is here quite a karroo, and miserably parched and poor. Thompson *Travels I* 1827
. . . the Karoo, the semi-arid but healthy plateau that covers nearly one-third of South Africa's interior, and which supports the world's second-largest sheep population. *Panorama* Sept. 1971

Also in combinations ~ *area*, ~ *climate*, ~ *coal* (q.v.), ~ *soil*, ~ *veld*, see (2) *veld*, ~ *bush* (q.v.) etc. and *prefix* in very numerous plant species indicating their habitat hence *deriv.* *Kar(r)oid* of or pertaining to the (1) *Kar(r)oo*.

Kar(r)oo bush, Kar(r)oobos *n.* Generic name of numerous varieties of scrublike bush, some toxic or semi-toxic, or noxious weeds: others useful as grazing. *bitter* ~ *Chrysocoma tenuifolia*, see *bitterbos;* ~ *thorn Acacia karroo* regarded as an undesirable species, or 'invader plant'; see *plant migration:* ~ *buchu/boegoe* an aromatic shrub *Diosma oppositifolia* similar to the genuine medicinal buchu *Agathosma betulina*, unpalatable as grazing: *sweet* ~ or *goeie(good)* ~ *Phymaspermum parvifolium* a valuable fodder-pasture for sheep, also *Pentzia incana* (*ankerkaroo*).

On we tramped silently as shades through the night and in the heavy sand. The Karoo bushes caught our feet . . . and the sand worked into our veldtschoens. Haggard *Solomon's Mines* 1886
Most of the valuable Karoo bushes give off a sweet aromatic perfume when bruised . . . and the flavour appears to be agreeable to the palates of herbivorous animals. The mild degree of bitterness . . . is rather appreciated than objected to by them. The plants which animals neglect are deficient in these characteristics. Wallace *Farming Industries* 1896
Unpalatable thorny types such as bitter bush, thorny figs, buchu Karoo bushes and Karoo thorn types are to be found on the western Karoo route [of plant migration]. *E.P. Herald* 28.2.73
Abundant water . . . 13 excellent camps with water, outstanding sweet karoo grazing carrying 1 200 sheep. *Farmer's Weekly Advt.* 21.4.72

Kar(r)oo caterpillar *n. pl.* -s. A voracious pest which causes rapid deterioration of the (2) *veld* (q.v.) by destroying the grazing: see *rusper*.

. . . non-participants whose veld had deteriorated. There was a bad drought and a karroo caterpillar plague between the first and second stages. *E.P. Herald* 21.12.74

Kar(r)oo coal, *n.* Dried *mis* (q.v) used as fuel: see also *Free State* (*coal*).
. . . it's no joke starting a fire with only dry dung for fuel. Karoo coal we call it down at the Cape. Van Alphen *Jan Venter* 1929

Kar(r)oo encroachment, *n. phr.* A serious veld and soil conservation problem freq. the result of over-grazing: see quot., also *Stock Reduction Scheme.*

Karoo encroachment is a serious matter, meaning as it does in plainer words, a spreading outwards of the desert and semi-desert that so limits the farming potential of the inland plateau. *E.P. Herald* 14.6.73

Kar(r)oo veld *n.* (2) *Veld* supporting vegetation typical of *Kar(r)oid* areas: see (2) *Kar(r)oo*, also *Kar(r)oo bush.*

Although there is no doubt about it that the perennial bushes form the staple, and at times indeed the only, means of subsistence in the Karroo veld, grasses and many annual 'weeds' are by no means to be despised as food for stock. *H'book for Farmers* 1937

Kar(r)oo wilg, kar(r)oo willow [vɪlx] See *pepper tree.*

kas [kas] *n. pl.* -te. **1.** Cupboard: *usu.* suffixed to *n.* e.g. *hoek* (q.v.) ~, *jonkmans* ~ (q.v.), *klere* ~ clothes, *kos* ~ (q.v.), *linne* ~ linen. [*Afk. kas fr. Du. kast* chest, wardrobe, cupboard]
The Dekenah kas is characteristic of the Riversdale District. These cupboards are typically made of yellowwood and teak and are of the wardrobe type with two drawers at the bottom. Baraitser & Obholzer *Cape Country Furniture* 1971
Country koskas with one gauze door. *Ibid.*

2. ~, *the n. prop. slang* Gaol: as in 'he landed up in the ~ ': see also *tronk.* [*fr.* (1) ~]

kas [kas] *n. pl.* -te. *Sect. Army.* **1.** A regulation army locker. [*Afk. kas* cupboard]
'The steel locker is a *kas*, a tin trunk is a *trommel* and the mess tray is a *varkpan.* J. H. Picard in *Eng. Usage in S.A.* Vol. 6 No. 1 May 1975

2. Prison: see *kas* 2.
An arrest is referred to as 'put him in the kas'. *Ibid.*

kat [kat] *n.* Cat: suffixed to names of various animals, e.g. *rooi* ~ (q.v.) *Felis caracal; muskeljaat* ~ (q.v.) *Genetta* spp. *Genetta tigrina; kommetjiegat* ~, *Atilax paludinosus,* (also octopus, *see* ~ .) Prefixed to the names of certain plants, e.g. ~ *doring* (q.v.), ~ *bos, Lycium hirsutum,* also species of Asparagus, see *katdoring;* ~ *pisbossie, Acalypha angustata;* ~ *naels, Hyobanche sanguinea* etc.: also *dimin.* ~ *jie,* (see *jointed cactus*), ~ *piering* (gardenia) (q.v.). Also found in place names Katrivier, Katberg, Katteriver. [*Afk. fr. Du. kat cogn.* cat]

katdoring ['kat₁duərɪŋ] *n .pl.* -s. **1.** *Scutia myrtina:* see *bobbejaantou* and *droog-my-keel.*
2. Several species of *Asparagus,* the thorns of which do damage to sheep and goat fleeces.
3. Sometimes applied to *Acacia caffra* the *kaffer wag-'n-bietjie* (q.v.). [*Afk. fr. Du. kat cogn.* cat + *doorn cogn.* thorn]

katel ['katl̩] *n. pl.* -s. **1.** A lightweight bedstead, portable and often thonged with *riempies* (q.v.), see also *trekbed* and quot. at (2) *trek:* part of the essential *trek gear* (q.v.) on wagon journeys: *cf. Anglo-Ind. cot.* [*Afk. fr. Malay katil fr. Hindi/Tamil kattil* a bedstead]
... and were weary enought at night to sleep soundly on their reim-bottomed kaatles without either feather-beds or curtains. Dugmore *Remin. Albany Settler* 1871
... the African Trek Wagon was really a caravan in which people lived as they travelled. This meant the addition of several features unknown to the European wheelwright: the katel or bed... is the most interesting of these, a word... that is said to come from Hindustani. This was a wooden frame on which rawhide thongs were interwoven. It was carried under the wagon-tilt by day and brought out at night. Morton *In Search of S.A.* 1948
2. The bier used by Cape Malays in the ceremonies for their dead. [*Malay fr. Hindi/Tamil as above*]
Three sheets of linen or *kaffang* are now wrapped round the body, which may be perfumed with aromatic oils and rose petals. The body is then transferred to the stretcher or *katil.* Du Plessis & Lückhoff *Malay Quarter* 1953

katjiepiering ['kaɪkĭ₁pïrɪŋ] *n. pl.* -s. Any of several species of *Gardenia* including *G. jasminoides, G. thunbergia* and *G. florida.* [*Afk. katjie* kitten + *piering fr. Malay piring* saucer (*prob. by folk etymology*) *Malay katchapiring,* the Cape jasmine]
There are no formal speeches, no costly gifts, but friends bring the first daphne of the year, the richest purple violets, the sweetest

katjepierings, until the house is heavy with the perfume of flowers. Fairbridge *Which Hath Been* 1913

katkop ['kat₁kɔ̆p] *n. colloq.* Used in various districts sig. bread loaves of different kinds – sometimes doughy 'batch' type bread, cottage loaf or even occ. a standard loaf baked in a tin; said to originate from prison slang: see quot. *acc.* H. C. Davies, *ex-Serviceman, Letter* 2.2.79 ~ has been adopted into army usage; also *sig.* bread. *cf. N.Z. barracouta* long 2 lb. loaf, *Jam. Eng. womanbreast* small soft loaf etc. [*Afk. kat cogn.* cat + *kop* head *cogn. Ger. Kopf*]
Katkop ... originates from *Central* [Gaol] where bread is cooked in distinctive small brown loaves, looking like a 'cat's head.' Itinerant 'bandiete' have universalised the term now. *S.Afr. Gaol Argot Eng. Usage in S.A.* Vol. 5 No. 1 May 1974

katonkel [kə'tɔ̆ŋkəl] *n. pl.* ∅. Also sp. *katunker: Scomberomorus commerson* the 'king mackerel' or Spanish mackerel: also applied loosely in the Cape to *Sarda sarda* the Atlantic or blue bonito: both species being of the Scombridae fam. [*etym. dub. Afk. prob. fr. Malay fish name ketung or katjang*]
... recalled the days, not so long ago, when katonkel (barracuda) abounded just off Port Elizabeth. *Evening Post* 27.5.72
Somewhat similar to the albacore ... is the katonkel, although he is, I believe, more nearly related to the tunny, the largest of all our fish. *Farmer's Weekly* 18.4.73
... found that while the tunny in those waters took feather lures, the so-called barracuda family, to which the katonkel belongs, would not ... but occasionally took bright spinners. *E.P. Herald* 1.8.74

kattebak ['katə₁bak] *n. pl.* -s. *colloq.* The dickey seat of an old-fashioned motor car. *cf. U.S. rumble seat.* [*Afk. fr. Du. kattebak* boot or seat attached to a vehicle]
My mother sat in front with my father, the two babies on her lap and the other three of us were strapped in with a big leather strap where the kattebak should be. Guy Butler 25.11.73

katunker [kə'tœŋkə(r)]. See *katonkel.*

kav [kav] *n. Sect. Army abbr.* Kavalleris; see quot. at *troepie.* [*Afk. kavalleris* trooper, cavalryman]

k(a)ya ['kaɪa] *n.* See *kia.*
Inanda's Kraal was a cluster of kyas and rondawels, shaped in a half-moon. Buchan *Prester John* 1910

KB ['keɪˌbi] *n.* See *Kaffir beer.* [*acronym*]
In that city, boys have solved the day-to-day financial problems of the parents by buying gallons of KB from the council's beer hall depots, and selling it to guzzlers. *Drum* Apr. 1965

keelvol ['kɪəlˌfɔl] *adj.* *slang* Fed up, disgusted: also *gatvol* (vulgar). [*Afk. keel* throat, *vol cogn.* full]
I am 'keelvol' of the constant 'gekyf'* in your columns about the dubbing of English language TV programmes into Afrikaans. *TV Times* 26.3.78 *bickering, wrangling

keer [kɪə(r)] *vb. trns.* and *intrns.* **1.** *Sect. Farm.* To direct or check animals being herded [*Afk. keer* herd, direct, *etc.*]
The scrum parted and heaved and broke, striped jerseys scattering like a break of quail and one was in the lead crouching low . . . There was a spatter of shouts: 'Here Farrie!' To me!' Keer hom!' Bennet *Mister Fisherman* 1964
2. *usu.* ~ *om:* turn round: found in place names, Keerom, Keerom Koppie, Keerom Street: see also *draai.* [*Afk. keer* to turn (+*om* round)]

kehla [keɬa] *n. pl.* -s. An elderly African man, often a councillor, wearing a *headring* (q.v.): see quot. at *headring.* [*Ngu. khehla* a man with a headring, elderly man]
I observed, however, that he was a 'Keshla' or ringed man. Haggard *Solomon's Mines* 1886
The old Zulu 'khehlas' and councillors felt that it was improper for a king of the Zulu nation to marry a divorcee. *Drum* Nov. 1964
Afr.E. Endearing mode of address to a young boy equiv. of 'old chap'.
Come kehla let me brush your hair. O.I. (Zulu) 12.3.74

'Kei *n. prop.* See *Transkei.*

Kei apple [kaɪ] *n. pl.* -s. Also *Dingaan's apricot: Dovyalis caffra* esp. abundant in the Transkei area, a densely spiny shrub used for hedges, or its fragrant but acid fruit which is used for jam and jelly. [*fr. Kei name of river*]
Charles and I got out together. It was the Ky apples. They make all people to vomit . . . one must not throw up in a Church. Vaughan *Diary circa* 1902
The kei-apple is known to many gardeners as a spiny shrub which makes a stout impenetrable hedge . . . The fruit is roundish . . . velvety, and bright yellow when ripe, with a very acid flavour. Palmer & Pitman *Trees of S.A.* 1961

kenner ['kenə(r)] *n. pl.* -s. A knowledgeable person, connoisseur. [*Afk. ken fr.*

Du. kennen to know *cogn. Fr. connaître* + *agent. suffix -er*]
. . . we found the champion red wine beautifully delicate but neither full bodied nor deep coloured . . . one of the *kenners* told me this could possibly be because the young vines . . . hadn't yet got their roots down to the sub-soil. *Cape Times* 30.8.73

kennetjie ['kenǐkǐ, -cǐ] *n.* The 'tipcat', a small wooden peg tapered at each end and with which the game ~ is played: also the game itself, see quot. at (2) *boeresport.* [*etym. dub. poss. fr. Frisian kunje fr. Lat. cuneus* wedge]
The kennetjie itself was a piece of wood about five inches in length, and an inch and a half in diameter, tapered off at both ends. It was laid on the ground, the player then struck one end smartly with his stick, and if it rose high enough he was able to beat it away as it fell, and run . . . But . . . the law intervened, and kennetjie was pronounced a dangerous nuisance in the streets. De Kock *Fun They Had* 1955

kêrel ['ke:rəl] *n. pl.* -s. **1.** A mode of address *usu. pl.* equiv. of 'chaps', unless *ou* ~ 'old chap': see quot. at (1) *jong.* [*Afk. kêrel cogn. O.E. ceorl,* also *carl*]
'Go right through, kêrels,' he said, 'the dancing is in the voorhuis. The peach brandy is in the kitchen.' Bosman *Mafeking Road* 1947
2. A chap, fellow, sometimes in the sense of boy-friend; often in the combination *slim* ~ (q.v.), a tricky, cunning fellow. [*as above*]
. . . the general's excellent defensive arrangements . . . soon put a stop to their proceedings; but not before the 'slim carles' had played . . . an ugly trick. Alexander *Western Africa II* 1837

kerim See *karem* and quot. below. ⸙ This name has several sp. forms: also *kerm* and *karm.*
ᴋᴇʀɪᴍ board complete with cues and chips, as new R8. *Advt. Cape Times* 11.12.78

Kerk [kerk, kɜk] *n. prop.* The *Dutch Reformed Church* (q.v.): also a church building, usu. *D.R.C.* (q.v.). *cf. Scottish kirk,* and in place names Kerkstraat, Kerkplein; for ~ *huis* see *nagmaalhuis.* [*Afk. kerk* church]
. . . the Dutch farmers . . . in consequence of my having come with the 'English Boors'* . . . are very respectful to me, and always honour me with the appellation which they give their own minister, viz. 'Predicant' . . . and many . . . have expressed to me their thankfulness that they shall now have an opportunity of attending Kerk. Shaw Sept. 1820 *cit.* Sadler 1967
* The Settlers of 1820
The Dutch farmers were tearing home from

Kerk in their carts – well-dressed, prosperous-looking folks, with capital horses. Duff Gordon *Letters* 1861–62

In combination *Moeder~*, mother-church.

Further along Drostdy Street is the neo-gothic Dutch Reformed 'Moederkerk', which dates in its present form from 1863, although some walls were part of an earlier church built in 1722. *E.P. Herald* 14.6.79

Kerkbode, Die [dǐ'kerk₁buədə] *n. prop.* The official organ of the *D.R.C.* (q.v.) in which Church news is published, and the names of newly qualified *dominees* (q.v.) are brought to the notice of *gemeentes* (q.v.) that may wish to 'call' them. [*Afk. kerk* church + *bode* messenger]

Afrikaans churches agreed with at least two amendments to the draft Bill on Publications and Entertainment, according to an editorial in the latest issue of Die Kerkbode, official organ of the NG Kerk. *Evening Post* 4.3.74

Kerkraad ['kerk₁rɑt, 'kɜk-] *n.p l.* -s, -en. The parish council of a *Dutch Reformed Church* (q.v.) also a member of the council; see quot. at (2) *heemraad*. [*Afk. kerk* church + *raad* council *also* councillor]

At yesterday's meeting of the Dutch Reformed Church Synod . . . a delegate asked what actions should be taken in cases of members of the Kerkraad who participated in dances and races. *E.P. Herald* 4.5.1906

Although both their husbands were members of the kerkraad Mrs . . . discouraged a friendship with Mrs . . . *Sunday Times* 10.9.72

-kerrie, knob- ['nɒb₁kerǐ]. See *kierie*.

kersbos(sie) ['kers₁bɔ̃s(ǐ)] *n. pl.* -s. Also *Bushman's candle* or *Boesmankers*: *Sarcocaulon burmannii* also *S. patersonii* and *S. rigidum*, all species of dwarf wax-secreting shrubs which burn like a candle even when fresh, not to be confused with *kers(ie)hout*. ¶*S. burmannii* and *S. patersonii* were used for the treatment of diarrhoea by Hottentots, and the ground roots by colonists as substitute for a mustard plaster. *S. rigidum* in decoction was used as an abortifacient. Found also in place name Dwarskersbos. [*Afk. kers fr. Du. kaars* candle + *bos(ch) cogn.* bush]

kersie ['kersǐ] *n. pl.* -s. Also *kersogie*: used of various species of *Zosterops* including *Zosterops capensis* and *Z. annulosa*, see *white eye*, also *witogie*. [*Afk. kers* candle + *dimin. suffix -ie*]

kersogie ['kers₁uəxǐ] *n. pl.* -s. See *kersie*. [*Afk. kers* candle + *oog* eye + *dimin. suffix -ie*]

kettie ['ketǐ] *n. pl.* -s. *Sect. Army.* A rifle: see quot. at *bokkie* and second quot. at *tankjokkie*. [*presum. fr. cattie* (q.v.)]

kettle *n. pl.* -s. Used freq. of a tea or coffee pot: usu. in combination *coffee* ~. [*translit. Afk. ketel*]

We had three days rashons to carrey: our coffe Kittle: our Kanteen for the water or git not one drop for twenty four hours. Goldswain *Chronicle II* 1838–58

The coffee maker had his own fire burning, his flat-bottomed enamel coffee kettles, their flannel coffee-bags bulging, all ready and bubbling. McMagh *Dinner of Herbs* 1968

keurboom ['kǐœ(r)₁buəm] *n. pl.* -s. *Virgilia oroboides (capensis)*: an evergreen shrub or tree with scented pale-pink to rosy lavender flowers, growing esp. freely on river banks: see quot. at *assegai wood*: found in place names Keurboomsrivier, Keurboomstrand. [*Afk. keur* choice + *boom cogn. Ger. Baum* tree]

The vernacular name appears to have been inspired by the floral beauty of the plants, thus the 'pick of all' (Afr:keur) The name was first recorded by Kolben at the Cape in 1705. Smith *S. Afr. Plants* 1966

kg [₁keɪ'dʒi] *n. pl.* -s. Spoken *abbr.* of 'kilogram' see also *k* and metrication. ¶ *abbr.* 'kilo' is rare.

kgotla ['xɔ̃tla] *n. pl.* ma-. The meeting place of a tribe where gatherings and discussions are held, also the tribal court: see also *lekgotla, makgotla. cf. N.Z. (Maori) marae.* [*Sotho (le)kgotla* courtyard, meeting place]

. . . the kgotla serves another significant purpose. It gives all Bantu, even the very humblest among them, the right to bring their grievances to the notice of their superiors. The Kgotla is almost sacred ground. It is the place where a man can 'Let down his hair' . . . without fear of reprisal . . . Over the centuries Bantu Law and justice has been controlled by what is known as the kgotla-system. Becker *S.A.B.C. Bulletin* 23.2.70

khafia ['kafɪə] *n. pl.* -s. The usu. white skull-cap worn by conservative Muslim men. [*presum. Arab.*]

Bearded . . . won't take off his kafia (hat) for anyone. He wears it all the time, except in bed . . . 'It is a kafia, neither a cap nor a hat, and that is that.' *Indaba* 29.10.76

khaki ['kɑkǐ] *n. pl.* -(e)s. **1.** *Afk. sp. kakie. Boer* name for a British soldier during the *Anglo-Boer War* (q.v.) on account

of the khaki-coloured uniform. see also second quot. at *drift*. [*Hindi khaki* dust-coloured *fr*. *Persian khak* dust]
Willem said here is a kaki. The Tommy rode fast up to us. Vaughan *Diary circa* 1902
2. *colloq*. *rare*. An English speaking non-Nationalist South African: see also quot. at *smeerlap* and second quot. at *mbongo*. [*fr*. *earlier use*]
Among other things the article says: 'We are fighting the English. The fight has been declared against the khakies – the battlefield is wide open. But first let us dig up the old-old hate against the English.' *Sunday Times* 11.6.72
In combination ~ *gevaar:* see also *swartgevaar, Roomsegevaar*
. . . predicted the three 'emotional' fronts the National Party would employ were 'Khakigevaar, swartgevaar, and rooigevaar.' 'Khakigevaar,' he said, was merely a new name for 'Boerehaat'. *Argus* 16.9.72
3. *adj.* or *n.* Also *kakie*, prefixed to several usu. alien species, mostly noxious weeds: ~ *bos/bush*, ~ *weed/onkruid*, ~ *klits*, usu. sig. that they were introduced from elsewhere, in forage or equipment for British troops see (1) *khaki*. ⫞Assumed by Pettman to be *fr*. colour of foliage: *Tagetes minuta* a proclaimed weed, see quot. at *kakiebos; Inula graveolens*, a pest in grain lands; *Alternanthera pungens* and *Bidens bipinnatus* both of spiny habit. [*prob. fr*. (1) *khaki sig*. alien, *poss. fr*. *colour*]

khakibos, -bush, -weed See (3) *khaki* and *kakiebos*.

Khalifa [ˌkaˈlifa, ˌxa-] *n.* and *n. modifier.* Also *Chalifa:* a Cape Malay sword-dancing ritual accompanied by chanting and drumming on the *ghomma* (q.v.) so called from the 'Khalif' or leader, chosen for his spiritual and stoical qualities: see quot. at *goëlery*. [*fr. Arab Khalif* successor to the Prophet, leader]
The Malay sword dance known as the *Chalifah* should really take place on the 11th day of Rabi-I-achier in honour of Abdul Kadir Beker . . . but its original religious implications have been modified with the result that the *Chalifah* now amounts to a skilful exhibition of sword play . . . Chalifa or Khalifa . . . is the name of the central person conducting the ceremony. Du Plessis *Cape Malays* 1944
Khalifa performances have no religious significance. Most Moslem religious leaders discourage the art as creating a misleading Malay image, and would like to see it die out. *Cape Times* 6.6.70

khanna [ˈkana] *n.* See *kanna* also *ganna*.

khaya See *kia*.

Khoi-Khoin [ˈkɔɪˌkɔɪ(n)] *n. prop.* Also *Khoekoen:* the self-styled name of the Hottentots: [*Hott. khoii* a man + *khoin* the men]
deriv. Eng. adj. form. *Khoisan* of or pertaining to the Hottentot and Bushman races or their languages: See fourth quot. at (1) *Hottentot* and second at (2) *Bastard*.
The Hottentots, or, as they were pleased to designate themselves, the Khoi-Khoin (men of men), driven before the flooding Bantu tide. Pettman *Africanderisms* 1913

khotla [ˈxɔtla] *n.* See *kgotla*.

kia [ˈkaɪə] *n. pl.* -s. Properly an African hut or house, see quot. at *k(a)ya:* usu. loosely used of detached servants' quarters on white properties esp. in Natal. ⫞ occ. *slang* for any house: see first quot. [*Ngu. khaya* home, house]
I'm half wild to be back in our kia and to grope through the sweet scented trees on the river bank. Stormberg *Mrs P. de Bruyn* 1920
He toddled off to his kia and from the tin trunk, which held his treasures, withdrew the parcel the boss had given him the day before. *Argus* 24.2.73
In combination *piccanin(ny)* ~ : see *P.K.*

kiaat(hout) [kĭˈatˌ(h)əut] *n.* Name assigned to several species for their resemblance to teak but esp. to *Pterocarpus angolensis* of the N.E. Transvaal and Natal, and to the hard reddish timber of this tree: also called *bloodwood* (q.v.) *Transvaal teak, Rhodesian teak*. [*Afk. fr. Du. kajaten* teak *prob. rel. Malay kaju* wood]
Kiaat is a medium sized tree, usually not exceeding 13m in height . . . the wood has a light brown colour and because it exudes, when cut, dark red juice, it is also known as bloodwood. Baraitser & Obholzer *Cape Country Furniture* 1971

kick-the-tin *vb. phr.* as *n.* Children's game: see *skop-die-blik* ⫞Not equiv. *Austral*. ~ to stand drinks. *cf. U.S.* (children) *kick the can*.
When it is full moon we play I spy kick the tin at the market bell. Vaughan *Diary circa* 1902

kiepersol [ˈkĭpə(r)ˌsɔl, -sɒl] *n. pl.* -s. Any of several species of *Cussonia* esp. *C. thyrsiflora:* also called *sambreelboom* (umbrella tree) from the shape of the terminal crown of leaves: the roots are chewed for their sap: also in place name Kiepersol. [*Afk. and Du. kiepersol fr*.

Indian kittisol fr. Port. quita-sol excluding the sun]
   . . . there were no less than six kippersol trees . . . she . . . took a sharp stone, and cut at the root of a kippersol, and got out a large piece, as long as her arm, and sat to chew it . . . It was very delicious to her. Kippersol is like raw quince, when it is very green. Schreiner in *Tales fr. S.A.* 1967 ed.

kierie [ˈkìrì] *n. pl.* -s. Also *knobkierie, -kerrie :* see also *induku :* a fighting club or stick usu. with a knobbed head. *cf N. Z. mere* Maori war club, *Jam. Eng. coco-makka* bludgeon, *Austral. nulla-nulla* aboriginal club, *Canad. casse-tête.* [*Afk. fr. Hott. kirri/keeri* stick]
   Their arms consist of a long spear called a Hassagai which they throw in the manner of a javelin, with great certainty to the distance of fifty or sixty paces, and a small club called a keerie which they use when closely engaged. Ewart *Journal* 1811–14
   . . . while they were arguing Mr. . . . attacked him with a kierie. He grabbed the kierie from Mr. . . . and struck him. *Post* 14.3.71
In combination *snoek ~ ,* fisherman's club used for dispatching *snoek* (q.v.) which are savage when landed.

kiewietjie [ˈkìvĭ₁kĭ, -cĭ] *n. pl.* -s. The crowned plover or peewit *Vanellus coronatus,* which has a loud and persistent cry esp. if defending its young. [*presum. onomat.*]
   . . . a peewit, or as it is called here, a *Keewit;* in moonlight nights they are constantly crying on the wing, and they are the harbingers of sunrise. Philipps *Albany & Caffer-land* 1827
   Among the plovers that love the coast are the Kittlitz sand plover and the crowned plover or Kiewietjie. Green *S. Afr. Beachcomber* 1958
~bene [-bɪənə] *colloq.* One who is excessively thin: or a description of very thin legs.
   She's dieted herself down to such a skeleton we call her kiewietjiebene. O.I. Schoolgirl 1974

kikuyu [kɪˈkujŭ] *n.* and *n. modifier. Pennisetum clandestinum,* a coarse and hardy grass species used in gardens as a lawn grass, and in certain areas as pasture. [*fr. Kikuyu, E. Africa*]
   . . . growing a garden in Lyndhurst is an almost hopeless business, and ours consisted of a young wilderness of coarse kikuyu grass and a few beds of petunias. Jacobson *Long Way fr. London* 1953
   60 Acres under irrigation & further 100 under dry land cropping. Large areas established kikuyu & Eragrostis. *Farmer's Weekly Advt.* 3.1.68

Kimberley Club *n. prop.* A sherry-type aperitif produced in S.A. before the importation of European *flor* culture. [*prob. fr. name of Club founded in Kimberley in 1881*]
   To the K.W.V. must go the credit for first making flor sherry here. Before that the Cape developed a wine of its own, different from the Spanish, with a sherry character which was and still is widely used as an aperitif. Much of this type is sold as Kimberley Club sherry . . . Some brands of Kimberley Club . . . although they are dry have as much colour as some of the sweet sherries. Beck *Cape Wines* 1955

Kimberley Train *n. prop. hist.* **1.** Title of a World War I folk song of the coloured people: also of a play on the colour question by Lewis Sowden.
2. A train from Kimberley, the departure of which signified an attempt, successful or unsuccessful, to *'try for white'* (q.v.) for *Coloured* (q.v.) people.
   ☞ It was said to be easier to pass for white in the rougher conditions of Kimberley than it was at the Cape.
   The Kimberley Train is coming tonight,
   It comes in black and it goes out white.
   Song : A. E. Voss 1962

kimberlite [ˈkɪmbə₁laɪt] *n.* The *blue ground* (q.v.) of the diamond fields which occurs in cylindrical pipes: usu. weathered yellow at the surface: see *yellow ground,* also *diamond.* [*fr. place name Kimberl(ey)* + *ite* + *adj. forming suffix -ic*]
   Alluvial diamonds were probably released from kimberlite by erosion. They are found along water courses and along the Atlantic Ocean. It is possible that kimberlite pipes exist beneath the sea. *Panorama* Dec. 1972
deriv. ~ ic; of or pertaining to *~ .*
   They found two Kimberlite pipes, the sort of soil structure named after the diamond city of Kimberley . . . an inconclusive scattering of Kimberlitic minerals. *Daily Dispatch* 14.7.72

kind [kɪnt, kɪndərs, -z] *n.* usu. *pl.* -ers. Child. [*Afk. kind(ers) fr. Du. kinderen* children]
   Distance to them is no consideration; the boor puts his vrouw and kinders into the wagon, lights his pipe and sets off to travel five hundred miles with as much ease as we should ten in England. Philipps *Albany & Caffer-land* 1827

kinderbewys [ˈkɪndə(r)bə₁veɪs] *n. Sect.* Roman Dutch Law: a legal deed made by the surviving spouse of a couple married in *community of property* (q.v.): see quot. [*Afk. fr. Du. kinder* child + *bewijs* warranty, deed]
   Kinderbewys . . . a deed of hypothecation or

mortgage passed by a surviving spouse married in community of property, for the purpose of securing to the minor children of the marriage the portions due to them from the estate of a deceased spouse. *Bell's Legal Dict.* 1951

kinderspeletjies [ˈkɪndə(r)ˌspɪəlïkïsˌ-cïs] *pl.* -s. Mere play, children's games [*Afk. kinder(s)* children, *speel* game, *-etjie dimin. suffix* + *pl.* -s]
This is the most serious crisis in South Africa's history. It makes America's Watergate look like a smudged pastel sketch. Britain's Profumo scandal was kinderspeletjies by comparison. *Het Suid Western* 6.12.78

kingklip [ˈkɪŋˌklɪp] *n.* Used of either of two edible marine fishes of the Ophidiidae fam., the eel-like *Xiphiurus capensis*, an esteemed table fish, and the related 'false' ~ *Hoplobrotula gnathopus.* ⟦Also applied loosely in the Knysna area to two fishes of the Serranidae fam., the 'koester' or soup-bully *Acanthistius sebastoides*, and the rock cod *Epinephelus andersoni.* [*Afk. fr. Du. koning cogn.* king + *klip* rock + *visch cogn.* fish]
South Africa: also dish Hottentot; Kingklip; Kabeljou; Musselcrackers; Red Roman; Steenbras; Stockfish. Nidetch *Weight Watchers' Cookbook* 1972
Grapefruit adds a piquant flavour to kingklip steaks, baked crisp with buttered crumbs. *E.P. Herald* 27.5.75

king protea, *n. pl.* -s, Ø. Also *giant protea : Protea cynaroides*, a common S. Afr. emblem appearing on the obverse of the twenty cent piece, formerly on the *tickey* (q.v.).
He has between 2 000 and 3 000 of the Giant (or King) protea alone. . . . one of the King (Cynaroides) proteas . . . This variety is much sought after overseas and a favourite among growers. *Cape Times* 13.9.74

kippersol *n.* See *kiepersol.*

kirimoer [ˈkïrïˌmuːr] *n.* See *karree.*

kis(t) [kɪs(t), kɪs(t)] *n. pl.* -s, -te. Also *kis :* chest, usu. a lidded box not a chest of drawers: also *Scottish kist.*
A heavy kist, or dower chest, was almost always to be found in one of the reception rooms. It was of stinkwood, when locally made, and decorated with fine brass-work. Gordon-Brown *S. Afr. Heritage* 1965
In combination: *bruids* ~ (q.v.) ; *~ klere* (q.v.) ; *wagon/wa* ~ (q.v.): also *agter* ~ , 'hind chest' in a wagon and *voor* ~ the 'fore chest' serving also as box-seat for the driver. [*Afk. kis fr. Du. kist cogn.* chest]
. . . several features unknown to the Euro-

pean wheelwright: the katel, or bed, and the voorkis and agterkis, the front and rear chests or kists, in which clothes and possessions could be carried. Morton *In Search of S.A.* 1948
The Voorkis (the chest in front of the wagon) is bigger and has sloped front and sides where the clothes and possessions could be carried. The agterkis (the chest at the back of the wagon) is smaller and square. Baraitser & Obholzer *Cape Country Furniture* 1971

kisklere [ˈkɪsˌklïərə] *pl. n.* Sunday-best clothes of country or farming people kept in a *kist* (q.v.) and only brought out for *Nagmaal* (q.v.), funerals and other solemn occasions. See third quot. at *Nagmaal.* [*Afk. kis fr. Du. kist* + *klere* clothes]
. . . at least one suit of *kisklere*, a carefully tended Sunday best. Mockford *Here are S. Africans* 1944
For special occasions like Church services and weddings, men kept at least one suit of 'kisklere.' Telford *Yesterday's Dress* 1972

kitchen *n. prefix.* Low: ⟦Pettman observes that ~ was used of low Latin 300 years ago. In combinations: ~ *Dutch* see (3) *Dutch; cf. Canad. monkey French*, E. *Afr. kisetla* 'Settler language'. ~ *Kaffir* see quots. at (2) *Kaffir, Fanagalo* and *garshly :* also used of a mixture of English and some African language by those speaking to servants, (usu. derogatory, but sometimes self-deprecatory) cf. *Chilapalapa* (q.v.).
Sir Henry and Umbopa sat conversing in a mixture of broken English and kitchen Zulu in a low voice, but earnestly enough. Haggard *Solomon's Mines* 1886.
The Dutch is easy enough. It's a sort of kitchen dialect you can learn in a fortnight. Buchan *Prester John* 1910

klaar¹ [klɑː(r)] *adj.* Finished: see *finish(ed)* and *klaar.* [*Afk. klaar* finished, ready, *cogn.* clear]

klaar-² [klɑː(r)] *adj. prefix.* Clear: found in place names e.g. Klaarwater, Klaarstroom: see also *helder-.* [*Du.* or *Afk. klaar cogn.* clear]

klaar out [klɑ(r) aʊt] *vb. phr. Sect. Army.* To finish the period of military service and leave the army: also occ. to leave one camp for another. [*Afk. klaar* finish, *poss. rel. Eng. colloq.* 'clear out' *sig.* leave]
When we klaar out we're going to run down with our trommels – you keep all your kit except what was in the trommel. *O.I. Serviceman* 15.4.79

klab(ber)ja(a)s [ˈklab(ər)ˌjas, -ˌjɑs] *n. prop.*

and modifier. See *klawerjas*: various sp. forms.

kla on ['klɑ ˌon] *vb. phr. Sect. Army.* To lay a charge against someone: see quot. at *spook*[3]. [*prob. translit. Afk. aanklagte doen/maak* make a charge, arraign *Afk. vb. kla* complain]

> [I told her . . . if I did she wold go to the Magrest [Magistrate] and clar of my beating her . . . Goldswain *Chronicle II* 1838-58]

> If you get injured while you're on pass or overdo the sunburn or even if you shave off your moustache, once you've been allowed to grow one, they can kla you on for damage to state property. *O.I. Serviceman* 9.3.79

klap [klap] *n. and vb. slang.* Usu. among children sig. smack e.g. 'I gave him a good ~', or 'I'd had enough so I ~ed him': roughly equiv. of clout or 'clobber': see quot. at *just now* and *moerava*. [*Afk. klap* cuff, smack *etc.*]

> The two . . . men closed in on me . . . The one on the left had frying-pans for hands. He said 'If you move I'll clap you dead with these.' Matshikiza *Chocolates for Wife* 1961

klapbroek ['klapˌbrük] *n.* Trousers closing at the waist with a flap buttoning horizontally: see also quot. at *Dopper*. [*Afk. fr. Du. klap* flap + *broek* trousers *cogn. breeks*]

> . . . the flap-fly, which earned the garment the name of 'klapbroek' . . . Trousers were of the 'klapbroek' type, and had in place of the centre fly opening, a flap (hence the name) fastened at the sides with buttons. Telford *Yesterday's Dress* 1972

klapper-[1] ['klapə(r)-, 'klæp-] *n. and n. prefix.* General term for numerous species usu. prefixed to *-boom* or *-bos* sig. plants with bladder-like fruits or pods which break with a sharp snap, and in which the seeds rattle: also fruits of these species esp. *Nymania capensis*. [*Afk. fr. Du. klapper* rattle]

> Passing through the Karroo in spring, travellers often pause in astonishment at splashes of pure and vivid colour . . . This is the famous klapper or Chinese lanterns, *Nymania capensis*, which grows among scrub on dry rocky soil, . . . The flowers . . . develop into inflated fruits with a papery covering like that of a large gooseberry.* Palmer & Pitman *Trees of S.A.* 1961
> *see *Cape gooseberry*

klapper[2] ['klapə(r), 'klæp-] *n. pl.* -s. The fruit of *Strychnos pungens* and *S. spinosa* called also *wooden orange, monkey orange* (q.v.) and *kaffir orange*. [*Afk. fr. Du. klapper* coconut *fr. Malay kelapa*]

> We had a capital lunch from some wild fruit

about three times the size of an orange, called a clapper. It has a hard shell outside, which one must batter against a tree to crack or break. Baldwin *African Hunting* 1890 *cit.* Pettman

> Low-slung and depending upon the tight waistband and buckles klapbroeke were sometimes called nierknypers (kidney pinchers). Strutt *Clothing in S.A.* 1975

klawerjas ['klɑvə(r)ˌjas] *n. or n. prop.* A card game 'pam' in which the Jack or Knave of Clubs is the highest card and all knaves are trumps: corrupted to *klabberjas, klabjas* and *klaberjaas*. [*Afk. fr. Du. klaver* club *cogn.* clover + *Du. jas* knave of trumps, ('pam' *fr. Gk. abbr. pamphilos* beloved (*philos*) of all)]

> Players of Klabberjas will tell you (as will bridge players of their game) that it is the *only* card game. . . . In the bistros of Paris they call it 'Belotte', the Dutch call it 'Klaviarsz' and in South Africa it's often called just 'Klabjas'. E.P. *Herald* 10.2.72

as *modifier:*

> The Eastern Province Klaberjaas championships . . . Klaberjaas is . . . similar to bridge or German whist, and is believed to have originated in Holland. It is usually played by four people, playing in two pairs. *Ibid.* 18.7.74

kleilat(jie) ['kleɪˌlat, -ˌlaɪkǐ, -cǐ] *n. and vb. trans. lit.* 'Clay twig': a game played by children, the missile being a ball of mud or clay propelled by flicking it off a springy twig: as *vb* to bombard some place or person with a ~. [*Afk. fr. Du. klei cogn.* clay + *lat* twig (+*dimin. suffix -jie*)]

> When no other sport presented itself, young South Africans often amused themselves by fighting with sticks and, more especially, with the so-called kleilat or clay-stick. De Kock *Fun They Had* 1955

klein [kleɪn(ə)] *adj. prefix.* Small: used sig. 'lesser' in the names of about sixty S. Afr. plant species, opposed to *groot* (q.v.): also found in place names e.g. Klein Brakrivier, Kleinbegin, Kleinpoort, Kleinmond, and with *attrib. inflect.* Kleinemonde. In combination ~ *huisie* (q.v.). [*Afk. klein* small]

kleinhuisie ['kleɪnˌhœïsǐ] *n. pl.* -s. *colloq.* Privy: usu. an outdoor earth or bucket closet, occ. loosely, any lavatory, *cf.* 'the smallest room': see also *P.K. cf. Austral. dunny(can), dyke, shouse; Canad. honey bucket etc.* [*Afk. klein* small + *huis* house + *dimin. suffix -ie*]

> The N9 may be the smiling Gateway to Cape Town; the main line is surely the way out

of the back door, past the dustbin and the kleinhuisie. *Cape Times* 28.2.72

First to go were the cluster of uithuisies, kleinhuisies, groothuisies and the tacked-on corrugated iron stoep. *Ibid* 16.8.73

kleintjie [ˈkleɪŋkǐ, -cǐ] *n. pl.* -s. Little one, child. [*Afk. kleintjie fr. Du. kleintje* little one]

For three years, ever since I was a kleintjie, I serve the Vrouw Smedinga. Fairbridge *Which Hath Been* 1913

kleios [ˈkleɪˌɔ̃s] *n. pl.* -se. Child's plaything: an ox modelled in clay used by S. Afr. children from the early days: see (1) *dolos* [*Afk. klei cogn.* clay + *os* ox]

klinker [ˈklɪŋkə, ˈklɪŋkə(r)] *n. pl.* -s. **1.** Extra-hard brick used from the earliest days for exposed work, steps etc., also called *geel(yellow)* ~s and *klompies* (q.v.), used today for fireplaces etc.: also *faggot* (q.v.). [*Afk. fr. Du. klinker* brick (*klinken* to sound, ring + *agent suffix -er*)]

In the days of the Company small hard yellow bricks, 'geel klinkers' – locally called 'klompjes' – were imported from Holland in large quantities and used for all exposed work, and for reinforcing arches over openings. There was still a market for them at the Cape as late as 1829, and they were even proposed for the steeple of St. George's Church as an alternative to English bricks. Lewcock *C19 Architecture* 1963

Brick klinkers and wood make an attractive combination. Like face bricks their biggest advantage is that they do not require plastering or painting. *Panorama* Sept. 1971

2. *prob. figur.* Hard ship's biscuit, hard tack. *cf. Canad. shanty cake/biscuit*, hard tack. [*prob. fr.* (1) ~]

klip [klɪp] *n* Stone. [*Afk. klip* stone *fr. Du. klip* rock] **1.** *colloq.* A diamond: see (2) *blinkklip*, also *stone*.

2. Stone-, rock-: common *prefix* to names of flora and fauna usu. sig. preference for a rocky habitat e.g. ~ *fish* (q.v.); ~*springer* (q.v.); ~*vygie*, see *vygie*: or having stony texture or appearance ~*kous* (q.v.).

3. Rock: stone found in place names as prefix or suffix. e.g. Klipfontein, Klipheuwel, Platteklip, Henley on Klip.

klipfish, klipvis(sie) [ˈklɪpˌfɪs(i)] *n. pl.* -es, -se, -s. Any of numerous species of Clinidae usu. frequenting tidal rock pools. [*Du. klip* rock + *vis(ch) cogn.* fish + *dimin. suffix -ie.*]

. . . the klipfish seldom attains a length of even 12 inches . . . Hilda Gerber refers to it as growing up to two pounds. Her informant must have been a genuine fisherman, for eight ounces would be nearer the mark. The klipfish is fried in butter. *Farmer's Weekly* 18.4.73

klappertert [ˈklapə(r)ˌte(r)t] *n.* A tart filled with a sweetened coconut mixture [*Afk. fr. Malay kelapa* coconut, *tert cogn.* tart]

The baskets of food contained two traditional New Year items to relish – curried pickle [sic] fish and *klappertert*. *Cape Times* 13.1.73

klip-klip [ˈklɪpˌklɪp] *n.* Children's game 'five stones', 'Jacks' or 'fives'. ⫿ Not equiv. 'fives' played in a fives court, *Afk. kamertennis* or *kaatspel*. [*Afk. klip* stone]

One moment I would be playing 'klip-klip' with the young children, the next asking permission to go dancing with guys who were . . . at university. Van Biljon *cit. Sunday Times* 24.12.78

klipkous [ˈklɪpˌkəʊs] *n.* One of the largest S. Afr. shellfish: any of several species of *Haliotis* esp. *H. midae* also known as *perlemoen* (q.v.): also *H. spadicea, siffie* (q.v.), known as elsewhere as 'Venus ear' from the beauty of the pearl-lined shell. See quot. at *ollycrock U.S. abalone.* [*Afk. fr. Du. klip* rock + *kous* stocking]

A rarer delicacy is the perel-le-moen or klipkous, largest of all the Cape shellfish. This monster grows far out on the rocks, and can be reached only at very low tides with the aid of a spear. It has a beautiful shell, but the meat must be scrubbed and put through a mincer. Green *Few are Free* 1946

The perlemoen or klipkous (stone stocking) a species of shellfish . . . The shells are lovely with a mother-of-pearl lining. The fish is most delicious if properly cooked. *Duckitt's Recipes* ed. Kuttel 1966

klipspringer [ˈklɪpˌsprɪŋə(r), -sprɪŋ-] *n. pl.* -s. The mountain antelope *Oreotragus oreotragus* similar to the chamois of Europe. [*Du. klip* rock + *spring* jump, bound + *agent suffix -er*]

. . . the klipspringer, which is closely allied to the chamois of Europe, and coney-like, has its house on the mountain top, being furnished with singularly coarse hair, imparting the appearance of a hedgehog. Harris *Wild Sports* 1839

I shall always remember these hills as the home of the klipspringer. No other antelope could find a foothold where the klipspringer lands with ease. Green *Glorious Morning* 1968

klipsweet [ˈklɪpˌswɪət] *n.* See *dassiepis.* [*Afk. klip* stone, rock + *sweet cogn.* sweat]

klipvissie ['klɪpₙfɪsĭ] *n. pl.* -s. See *klipfish/-vis.*

... gazed into rockpools full of starfish, slowly waving sea plants and green-and-gold *klipvissies* drifting lazily from crevice to crevice – with here and there a small red crab. Krige *Dream & Desert* 1953

klomp(ie)[1] ['klɔmp(ĭ)] *n. pl.* -s. *colloq.* Group or cluster of trees, animals or persons. *cf. Canad. bunch* (trees, horses, oxen etc.): see second quot. at *kaal.* [*Afk. fr. Cape Du. klomp* a lot, cluster, *cogn.* clump (+ *dimin. suffix -ie*)]

A dry, arid, barren waste ... of queer shaped jagged edged or flat topped kopjes; of dried up watercourses and dancing dust devils, with every now and then a klompie of trees, very green against the . . . brown veld. *The 1820* Sept. 1973

klompie[2] ['klɔmpĭ] *n. pl.* -s. Also *klompje :* a small hard brick in use from the days of the *Dutch East India Company* (q.v.) to the present, for outdoor work and fireplaces: also called *faggots :* see quot. at (1) *klinker* [*Afk. fr. Du. klompje; klomp* lump, nugget + *dimin. suffix -je*]

... the small red Amsterdam bricks were the neatest and the glazed Dutch tiles the most artistic. Large shipments of bricks arrived until late in the eighteenth century. You find them mainly in stoeps and face works, with the yellow klompjes that have weathered so well. Green *Grow Lovely* 1951
It is not known that 'klompjes' were even used to face an entire building at the Cape as they were in Holland. Lewcock *C19 Architecture* 1963

klonkie ['klɔŋkĭ] *n. pl.* -s. Also *klong :* a young coloured boy: see *kwedin* and *umfaan.* [*Afk. klong, contraction of klein-jong* servant boy + *dimin. suffix -kie*]

... the klonkies there, the small black boys, having learned it from the soldiers who camped in Venterspan during the war of 1939, saluted him. Paton *Too Late Phalarope* 1953

kloof [klu:f, klʊəf] *n. pl.* -s. A deep ravine or valley, usu. wooded ; gorge between mountains. *cf. Scottish cleugh. Canad. draw, gulch,* gully or ravine. [*Afk. fr. Du. kloof* ravine, gorge *cogn.* cleft]

I have since been informed ... that the woody kloofs, or ravines in this range, contain many of the forest trees and other plants which, according to common opinion, are only to be found in more distant parts of the colony. Burchell *Travels I* 1822
Also found in S. Afr. place names as prefix Kloofeinde, Kloofnek; alone,

Kloof (Natal) or as suffix Fonteinskloof, Waterkloof, Baviaanskloof, Langkloof.

klopjag ['klɔpₙjax] *n.* A police raid or round-up. [*Afk. klopjag* police drive round up, *fr. Du. klopjacht* battue]

Non-Nationalists shrink away in revulsion and anger . . . when there is another Security police klopjag with a whole lot of decent, ordinary citizens being raided at dawn. *Rand Daily Mail* 28.8.71

kneehalter ['ni:ₙhɔltə] *vb. trns. occ. n.* To restrict the movement of a horse turned out to graze by tying the bridle to the foreleg, allowing the animal to graze freely but preventing its running away: see quot. at *riem. cf. Austral. legrope n.* and *vb.* [*translit. Afk. kniehalter*]

No time was now to be lost. Our horses were hastily *knee-haltered* (i.e. tied neck and knee to prevent their running off) and turned to graze till the night closed in. Thompson *Travels I* 1827

knickerje ['(k)nɪkə(r)jə] *n. obs.* Marble or game of marbles. [*fr. Du. knikkeren* to shoot at, or play, marbles]

Mr. Rex made me a present of a very fine golden cuckoo. After dinner played knickerje . . . the game of knickerje learned from his Dutch neighbours must have been new and amusing to him. Duthie *Diary* 22.11.1830 *cit.* Metelerkamp *Geo. Rex* 1955
Also *knikkertjie.* Very hard, grey coated seed also called *seeboontjie* washed up on the S. Afr. coasts used for ornaments or for infants' teething, so called because of the likeness to a marble. [*cf. Jam. Eng. nicker* a marble, ar.d two species of *Caesalpinia,* grey nicker and yellow nicker, the seeds of which are used as marbles.

knobkerrie, knobkierie *n. pl.* -s. Also *knob/knopstick* (rare): see *kierie cf. Canad. casse-tête* war club.

Lt. . . . had his arm injured by a battle axe attack when police restored order and Sgt. . . . was hit on the head with a knobkerrie. *Daily Dispatch* 6.10.71
Many South Africans know the wood of Umzimbeet (which word is a corruption of the Xhosa and Zulu umSimbithi, meaning ironwood). Until its use was protectively stopped, it was the favourite wood for the tribal craft of making knobkieries. *E.P. Herald* 12.1.72

knobwood *n.* Also *knophout, knobthorn* and *perdepramboom,* see *pram :* any of several species having spine-tipped knobs or horn-like protuberances on the trunk esp. *Fagara capensis :* also its hard, yellow timber formerly used for making household utensils. see quot. at *boekenhout.*

Amongst the most unusual of our trees are the . . . knobwoods, easily identified by the knobs which stud the stems. These are sometimes up to 3 inches long . . . and often tipped with thorns. A tall knobwood, with a brown trunk studded with knobs like small stout horns, is an extraordinary and unforgettable sight. Palmer & Pitman *Trees of S.A.* 1961

knorhaan [ˈknɔ̃(r)ˌhɑn] *n. pl.* -s.
1. See *korhaan.* [*Afk. fr. Du. knorren* to scold + *haan* cock]
2. Also *knoorhaan.* Gurnard: any of several species of *Trigla : T. capensis. T. kumu,* etc. also used of *Pomadasys commersonni, tiger, spotted grunter :* see *grunter* ⫽All these fish are named after the *k(n)orhaan* (q.v.) or 'scolding cock' because of the noise they make when taken from the water. [*fr. bird name*]

kob [kɒb] *n. pl.* Ø. Also *cob : abbr. kabeljou* (q.v.): *Argyrosomus hololepidotus.* [*abbr. kabeljou* cod, *cogn. Fr. cabillaud*]

One of the highlights in an otherwise quiet period of fishing along the Natal coast was the recent catch . . . at Tugela Mouth . . . landed a giant kob which tipped the scales at 36,9 kilograms. This is the biggest kob caught by any rock and surf angler during recognised competitive events this year. *Sunday Times* 3.9.72

koedoe [ˈkŭdŭ] See *kudu.*

koedoekos [ˈkŭdŭˌkɔ̃s] *n. Haworthia viscosa* a succulent-leaved plant of rocky habitat thought to have been a favourite food of the *kudu:* also *kudukos.* [*Afk. koedoe, kudu* (q.v.) + *kos fr. Du. kost* victuals]

kocjawel (*Afk.*) see *guava.*

koekoek [ˈkŭˌkŭk] *n. pl.* -s. A S. Afr. breed of fowl developed in Potchefstroom at the Cattle and Dairy Research Institute, registered with the S. Afr. Poultry Union under the name *Potchefstroom ~.* [*Afk. poss. fr. Du. koekoek,* cuckoo, *more prob. Ngu. nkuku* fowl]

Purebred Koekoek. – Not the old-fashioned Barred Plymouth Rock with its small eggs. A pure breed bred by Potchefstoom Agricultural College as a tablebird with stamina and high egg production. *Farmer's Weekly Advt.* 30.5.73

South Africa has produced its very own kind of fowl – a breed of its own, called . . . the Potchefstroom koekoek . . . it produces chickens

which can readily be identified as boys or girls immediately they hatch out. *E.P. Herald* 6.7.76

koekoemakranka See *kukumakranka.*

koeksister [ˈkŭkˌsɪstə(r)] *n. pl.* -s. A traditional Cape confection poss. of Malay origin: a deep fried twisted or plaited doughnut immediately dipped in syrup. *cf. U.S. cruller, Canad. twister :* see also *goolab jambo* at *Indian terms.* [*Afk. fr. Du. koek cogn.* cake + *sissen* to sizzle, hiss + *agent suffix -ter*]

They bought green-fig and watermelon preserve, and koeksisters, the plaited and syrup-dipped doughnuts that are a Cape delicacy to this day. McMagh *Dinner of Herbs* 1968

koelie[1] [ˈkŭlĭ] *n. pl.* -s. *obs. hist.* An Anglo-Indian from Calcutta: see quots. at (1) *Kaapenaar, Hindoo* and *Indian.* [*poss. fr. Qui-hy, Qui-hi fr. Koi hai?* Is anyone there? *mode of calling a servant : see* quot. at (1) *Kaapenaar :* see also *coolie*]

. . . I must explain to you that the Hindoos is a soubriquet applied by the local residents to all visitors from India – whether they be Koelies from Calcutta, Mulls from Madras, or Ducks from Bombay. *Cape of G.H. Literary Gaz.* 15.9.1830 *cit.* Pettman

koelie[2] An offensive mode of address or reference to an Indian: see (2) *coolie* and quot. at *houtkop.* [*Afk. koelie* coolie *fr. Tamil, Hindi, kuli* labourer, *also Chinese*]

koester [ˈkŭstə(r)] See *kingklip.*

koffie [ˈkɔ̃fĭ] *n.* Coffee: found in place name Koffiefontein, and as *prefix* in combinations ~ *huis,* coffee house, old style restaurant; ~ *ketel,* see *kettle;* ~ *geld* (*hist.*) see quot. [*Afk. koffie cogn.* coffee *geld* money]

The President still sat on the verandah of his house early in the morning, ready to meet visitors and offer them the cup of coffee, for which the Government allowed him 'Koffie-geld' of £500 a year. Rosenthal *De Wet* 1946

kofia See *kufija.*

koggelmander [ˈkɔ̃xəlˌmandə(r)] *n. pl.* -s. Any of several species of *Agama* or rock lizard: best known is the *bloukop ~ Agama atra,* of slightly dragon-like aspect with a large triangular blue head. [*Afk. (blou cogn.* blue + *kop* head +) *koggelmander* rock lizard]

kokerboom [ˈkʊəkə(r)ˌbʊəm] *n. pl.* -s. *lit.* 'Quiver tree': *Aloe dichotoma,* a much-branched tree-aloe of up to 10m high the timber of which is porous and corklike. ⫽Formerly branches were hollowed out by Bushmen and Hottentots for quivers hence the vernacular name. Found also in place names in Namaqualand Kokerboom, Eenkokerboom. [*Afk. fr. Du. koker* quiver + *boom cogn. Ger. Baum* tree]

Those who lived in houses built of Kokerboom trunks stood back and watched as the matches were struck and the flames leapt and crackled through the wooden structures. *E.P. Herald* 28.1.74

kokkewiet [ˈkɔkəˌvĭt] *n. pl.* -s. The bush shrike *Malaconotus zeylonus:* see *bokmakierie.*

Kolbroek [ˈkɔlˌbrŭk] *n. pl.* -s. A popular breed of pig, the first of which were rescued from the wreck of the ship *Colebrooke* lost off the S. Afr. coast in 1778. [*Afk. translit. Colebrooke*]

On most farms . . . it will be found profitable to keep a few pigs for home consumption. The most suitable breed is probably the Kolbroek . . . The Kolbroek . . . does not require so much protein and, being of a sluggish nature is less mischievous when allowed loose around the house. *H'book for Farmers* 1937

kole (*Afk.*) See *coals.*

kolk [kɔlk] *n. prefix* and *suffix.* A deep pool, or water hole usu. the deepest point in a *vlei* (q.v.): found in S. Afr. place names Kolkfontein, Granaatboskolk (pomegranate), Tontelboschkolk (tinder bush), Kootjieskolk. [*Afk. kolk* abyss, pool, pothole]

All over the Calvinia district you find names ending in *kolk* – Granaatboskolk, Abiquas Kolk, Korannakolk, Wildehondekolk, Leeukolk, Klipkolk. A *kolk* is a pan that holds water after rain. Green *When Journey's Over* 1972

kolstert [ˈkɔlˌstert, -ˌstɛət] *n. pl.* Ø. See *blacktail:* also called *dassie* [*Afk. kol* spot + *stert* (*fr. Du. staart*) tail]

. . . has registered the largest blacktail (kolstert) taken by a club angler for at least two seasons. The fine specimen he landed weighed 1,644kg. *Grocott's Mail* 25.6.71

kolwyntjie [ˌkɔlˈveɪŋkĭ, -cĭ] *n. pl.* -s. Small cake of the type formerly baked on St. Columbine's day. ~ *pans* usu. of copper with shaped depressions similar to *Fr. madeleine* tins, are prized by collectors of antiques. [*fr. Du. kolombijntje*]

Tart-pans and kolwyntjie-pans: these are possibly the most typical of all old Cape kitchen utensils, and the most eagerly sought after by collectors. Cook *Cape Kitchen* 1973

kombers [ˌkɔmˈbeːrs, -bɛəs] *n. pl.* -e (-en). *obs.* Blanket usu. suffixed as in *vel* ~

(q.v.) *cf. karos(s),* also *wol(wool)* ~ [*Afk. fr. Du. kombaars* (ship's) blanke]

The qualities of the linebayi and the wo kombersen were elaborately discussed. Du more *Remin. Albany Settler* 1871

komfoor [ˌkɔmˈfʊə(r)] *n. pl.* -s. See *kon foor.*

Brass Kettles and Comfores; Superior sing and double barrelled Guns made to order fc the Cape Market. *Cape of G.H. Almanac Adv* 1841

kommetjie [ˈkɔmĭˌkĭ, -ˌcĭ] *n. pl.* -s. A sha] low circular depression in the ground *cf. Austral. ghilgai.*

His stretcher-bearers thankfullly deposite him in a kommetje (a round saucer-like d pression in the ground from which the pla takes its name) while they rested their wear limbs. Metrowich *Frontier Flames* 1968 Shallow basin-like depression found i place names Kommetjie, Kommetjie Flats. ¶The form 'Committee' in ~ Drift and ~s Flats is thought by som to derive from this. [*Afk. fr. Du. kom metje* a small cup or basin]

. . . We came upon a plain full of strange hol like basins: hence this plain is called Commat Flats. One of the guides . . . explained the caus of these scientifically: 'The earth here' he saic 'had once been much soaked with rain, an had got the shivers.' Alexander *Western Afric II* 1837

konfoor[1] [ˌkɔnˈfʊə(r)] *n. pl.* -s. A warme for a tea or coffee pot or kettle, usu made of brass and containing a *tessĭ* (q.v.) or firepan for coals. [*Afk. fr. Dt komfoor* chafing dish, brazier]

Brass chafing-dishes (konfoors), coffee po and smaller articles in brass or silver were mad by the Cape craftsmen of the 17th and 18t Centuries; . . . Three old Brass Konfoors ir tended to contain burning coals to keep a kett] hot. Gordon-Brown *S. Afr. Heritage* 1965

konfoor[2] Also *voetstofie* or *stofie* (q.v.): footwarmer-cum-footstool consisting o a perforated wooden box also containin a *tessie:* see quot. at (1) *Cape Dutc] [prob. fr. Fr. chauffoir* a footstove]

To preserve warmth the Dutch women us an apparatus to set their feet upon called *komfoor.* It is a square box, with a few hol cut through the top, and closed only half-wa up the front. Into the inside a few hot charcoa embers are introduced from time to time in a iron basin. Backhouse *Narrative* 1844 *cit.* Pet man

konfyt [ˌkɔnˈfeɪt] *n. pl.* -s. Preserves o conserve, either of whole fruit or larg pieces in syrup. [*Afk. fr. Du. konfi, preserves, conserve]

A delightful Afrikander custom . . . the offer to the visitor, during the forenoon, of a cup of tea, with a liberal supply . . . of some beautifully preserved home-made *confyt*. Wallace *Farming Industries* 1896
. . . spread quickly on the whole biscuits and then press on them those which have the centre cut out, filling up the cavity in the centre with crystallised cherries, or other candied konfyt. *Duckitt's Recipes ed.* Kuttel 1966

In combination ~*jar*.

Konfyt Jars. These imported crystal jars both open and covered were among the most gracious ornaments on the table. Housewives took a particular pride in their own home-made preserves. Fehr *Treasures at Castle* 1963 *teewater* ~ : see *teewater*: also quot. above (Wallace 1896). *waatlemoen/ watermelon* ~. A traditional preserve of watermelon rind: see *kaffir melon*.

They had supper. Babotie and rice, sweet potatoes and pumpkin, followed by a milk tart and watermelon konfyt. Cloete *Rags of Glory* 1963

onsistorie [ˌkɔnsɪsˈtʊəri] *n. pl.* -s. The vestry of a *Dutch Reformed Church* (q.v.) ; ~*stoel/chair*, a high-backed chair of the type traditionally used in the ~. ⫽Also used of the governing body or 'vestry' of the church: see *kerkraad*. [*Afk. fr. Du. konsistorie* consistory, vestry]
. . . Elder Landsman tiptoed out of the church and went round to the Konsistorie where the Nagmaal wine was kept. Bosman *Bekkersdal Marathon* 1971

ontant [ˌkɔnˈtant] *n.* Cash. [*Afk. fr. Du. kontant* ready, available (money)]
The slave was knocked down to the Stranger, and the auctioneer demanded – kontant – cash. *Cape Literary Gaz.* Vol. I June 1830
It was not always sheer ignorance that led people to keep large amounts of money on their farms. 'Kontant', hard cash, was the rule of the veld, . . . and men buying farms and stock in distant places simply had to have the money close at hand. Green *Land of Afternoon* 1949
Cash (kontant) for your clean used car. *E.P. Herald Advt.* 5.5.73

oolkop [ˈkʊəlˌkɔp] *n. pl.* -pe. See *SSB*. [*Afk. kool* cabbage + *kop cogn.* Ger. *Kopf* head]

oper [ˈkupə, ˈkʊəpə(r)] *n. pl.* -s. A buyer of diamonds during the early days of the Diamond Fields before the introduction of legislation, see *I.D.B.*, in 1882. [*anglicization kooper fr. Du. koper* buyer.]
. . . it came under earnest debate whether to burn the proprietors' tents . . . or to sack the stalls of the koopers. Little groups of the dangerous class gathered round each diamond buyer's tent, loudly threatening. Boyle *Cape for Diamonds* 1873

koornvreter [ˈkʊərn̩ˌfriətə(r)] *n. pl.* -s. *Plocepasser* spp.: see *koring/koornvreter/ voël*.

kop [kɔp] *n.* **1.** A prominent hill or peak, domed, pointed or flat-topped: *freq.* in *dimin. form* ~*pie* (q.v.). *cf.* Canad. *coteau*. [*Afk. fr. Du. kop* (head) peak, hill]
Me that 'ave watched 'arf a world
'Eave up all shiny with dew,
Kopje on kop to the sun . . .
Kipling *Five Nations* 1903

2. *n.* suffix *pl.* -en. Hill, peak, found in place names usu. as *suffix* e.g. *Boskop* (q.v.), Kranskop, Daskop, Spioenkop (spy), Driekoppen, Witkoppen. [*as* (1) ~]
Behind the town of Swellendam are four conspicuous peaks which form a natural sundial, and for more than two hundred years these have been known to local farmers as 10 uur, 11 uur, 12 uur and 1 uur Kop. Gordon-Brown *S. Afr. Heritage* 1965

3. *n. pl.* -s. *slang.* A butt or blow with the head. [*fr. Afk. kop* head *cogn.* Ger. *Kopf*]
. . . just swore at him. So the CID man gave him three quick kops (butted him in the face). *Sunday Tribune* 16.7.72

kop-en-pootjies [ˌkɔpɛnˈpuĭkĭs, -cĭs] *n.* Head and trotters – a stewed dish made usu. of sheep: see also quot. at *pens-en-pootjies* [*Afk. kop* head + *en* and + *poot* foot of animal + *dimin. suffix -jie*]
Concertina and guitar music goes on until dawn with few intervals, but with many rounds of *vaaljapie* and brandy. You remember the candle light, the *vastrap* music, the supper of *kop-en-pootjies* (head and trotters) or *bloedwors*, the black pudding made from the blood of a pig. Green *Giant in Hiding* 1970

kopje [ˈkɔpĭ, ˈkɔpĭ] *n. pl.* -s. Dutch form of *koppie* (q.v.). ⫽ Rhodesian *sp.* form, *obs.* in S.A.
Then mock not the African kopje,
Especially when it is twins,
One sharp and one table-topped kopje,
For that's where the trouble begins.
Kipling *Five Nations* 1903
Salisbury is like a city under siege . . . There is a . . . curfew on and around the park-like kopje, a popular view-point just out of the city centre. *Sunday Times* 8.4.78

In combination ~ *walloper* (q.v.), ~ *walloping* (q.v.).

kopje walloper [ˈkɔpĭ-, ˈkɔpĭ-] *n. pl.* -s.

hist. See also *kooper.* A diamond buyer usu. Jewish, buying directly from diggers on their claims in the early days before the *I.D.B.* (q.v.) legislation of 1882. [*cf. Canad. gravel puncher*, miner with primitive equipment. [*fr. kopje walloping,* below]

... the occupation of dealing in diamonds. the slang camp term indeed for this was 'kopje walloper' ... from the circumstance that in the earliest days the diamonds were obtained from a number of kopjes or small hills in the neighbourhood ... and the dealers travelled on foot from one to the other purchasing the finds as they were turned out at the sorting tables. Raymond *Barnato Memoir* 1897

kopje walloping *vbl. n.* or *partic.* Buying diamonds in this way: see *kopje walloper*, also *kooper.* [*presum. fr. kopje* (q.v.) + wallop to thrash]

... I was kopje-walloping in Kimberley, feeling more depressed and dispirited than usual when in the course of my rounds I came across a Dutchman sorting at his table. Cohen *Remin. of Kimberley* 1911

koppie ['kɔpĭ] *n. pl.* -s. A hillock, flat topped or pointed, a common feature of the S. Afr. (1) *veld* (q.v.): see also *kopje,* and quot. at *klompie.* [*Afk. kop* hill, peak + *dimin. suffix* -(*p*)*ie*]

When I faked death on the Serengeti one large male lion climbed a koppie that was very close by and studied me for a long time. *Sunday Times* 12.8.73

Also found in place names e.g. Koppiealleen, Koppies, Weskoppies.

kop toe ['kɔp ˌtŭ] *modifier. Sect. Army.* Swelled-headed: usu. in phr. *a ~ ou* a conceited fellow, or in 'portmanteau' term *koptoeraal* a corporal whose recent promotion has 'gone to his head' (kop toe). [*Afk. kop cogn. Ger. Kopf* head, *toe* closed, or to (direction towards)]

korhaan [ˌkɔrˈ(h)ɑn] *n. pl.* -s. Any of several species of *Otis,* (Otididae fam.) the S. Afr. bustard: its name derives from its noisy call esp. if disturbed: see also *knorhaan.* [*Afk. and Du. korhaan* bustard]

I saw only a few of the larger ... species ... the *pouw,* which is a sort of large bustard, and very delicate eating; the *korhaan,* a smaller sort of bustard, also prized by epicures; cranes, Namaqua partridges, and white-necked crows. Thompson *Travels I* 1827

Also found in place name Korhaan.

koringkriek ['kuərɪŋˌkrĭk] *n. pl.* -s. An insect pest *Eugaster longipes,* which feeds on various crops. [*Afk. koring fr. Du. koorn cogn.* corn + *kriek* cricket]

Koringkrieke ... Grey or black, long-legged, clumsy, spiny long antennaed grasshoppers. Common in drier parts of Orange Free State and Bechuanaland. *H'book for Farmers* 1937

koringvreter, koornvreter, koringvoël, koornvoël ['kuərɪŋˌfrɪətə(r)/ˌfʊəl] *n. pl.* -s. Used of numerous species of sparrows, sparrow weavers (*Plocepasserinae*) and other seed eaters, particularly the common *Passer melanurus,* see *mossie,* which feed in grain lands. [*Afk. koring fr. Du. koorn cogn.* corn + *voël fr. Du. vogel cogn.* fowl, *or vreter* eater *fr. Du. vreten* to eat greedily]

8 August 1811 ... a troublesome bird to farmers, and well deserving the name they have given it, of Koornvreeter (Corn-eater). It has very much the manners of the common sparrow of Europe. Burchell *Travels I* 1822

korrelkop ['kɔrəlˌkɔp] *n. (modifier).* One who has *pepercorn/peperkorrel* (q.v.) hair. [*Afk. korrel* grain, granule *abbr. peperkorrel* peppercorn + *kop* head *cogn. Ger. Kopf*]

Hans is a korrelkop bushman. Vaughan *Diary circa* 1902

kos [kɔs] *n.* Food: *usu. as suffix* e.g. *pad~* (q.v.), *veld~,* see (1) *veld* compounds, or *prefix ~kas,* a store cupboard; see also quot. at *lekker* [*Afk. fr. Du. kost* food, victuals]

Yellowwood Table; Yellowwood Washstand; Yellowwood Koskas. *Evening Post Advt.* 27.10.73

Yellowwood Wakiste ... Yellowwood Koskaste. *Grocott's Mail* 25.5.73

Also suffix in plant names *koedoe~* (q.v.), *Haworthia viscosa; olifants~,* *Portulacaria afra,* see *spekboom; slang~* (snake), *Amanita phalloides,* the death cap/cup, a highly poisonous fungus, etc.

koster ['kɔstə(r)] *n. pl.* -s. The verger of a *Dutch Reformed Church* (q.v.). [*Afk. fr. Du. koster* verger]

A little further on lived a Koster, that is, a sexton, a set of people that are more respected by the Calvinists than with us. *trans.* Sparrman *Voyages I* 1786

kotch [kɔtʃ] *n. and vb. intrns.* Puke, vomit: see *cotch cf. Austral. perk* [*fr. Afk. kots* vomit (*vulgar*)]

kouvoël ['kəuˌfʊəl] *n. pl.* -s. *Aquila rapax :* see *lammervanger* [*prob. fr. Hott.* to catch, hold + *voel fr. Du. vogel* bird *cogn.* fowl]

kraai [kraɪ] *n.* Crow: usu. in compounds: ~*bos* (q.v.), *withals* ~ (q.v.): or in place names Kraaibos, Kraayenkraal, Kraaifontein, Kraaipan. [*Afk. fr. Du. kraai cogn.* crow]

kraaibos [ˈkraɪˌbɔs] *n. Diosporos austroafricana* and other species of *Diosporos*, a small bushy aromatic shrub used medicinally in various forms by indigenous tribes now regarded as an 'invader' plant of the Kar(r)oo region. Also a place name, Kraaibos. [*Afk. kraai cogn.* crow + *bos cogn.* bush]

The black berries of a bush called *kraijebosch*, or crow-bush, were greedily devoured by the crows at the Cape. Thunberg *Travels I trans.* 1795

kraak [krak] *vb. slang prob. usu. motor cyclists.* To speed, 'send it'. [*prob. fr. Afk. (gaan) dat dit so kraak*, go like blazes, hell etc.]

... he can only move that iron ... That first arvey we kraak out along the Ben Schoeman highway, and it does a ton easy, no kidding ... s 'true 's bob there 's something about going fast, it gets you. Hobbs *Blossom* 1978

And kraaking out to Harties on the Sunday breakfast run, then sending it ... *Darling* 26.10.77.

kraal ¶ Note: This term has several basic meanings or uses (see quot. below), viz. (1) An enclosure for farm animals; (2) A game sanctuary (rare); (3) *vb.* To drive animals into a kraal; (4) A village or settlement of an indigenous tribe; (5) A cluster of huts occupied by one family or 'clan'; (6) The seat of a chief (usu. *royal kraal*); (7) An element in S. Afr. place names; (8) A bead: (in combination only). Combination forms for (1) *kraal* are listed with it. [*Afk. fr. old Du. koraal fr. Port. curral/corral*, fold, pen]

kraal [kra:l, krɔ:l] *n. pl.* -s.

The Dutch word *Kraal*, as used in the Colony, has three different significations: – a string of beads, a cattle fold, and a native horde or encampment. Thompson *Travels I* 1827

1. *n. pl.* -s. An enclosure, pen or fold for farm animals: in combinations *cattle* ~; *sheep* ~; *goat* ~; *skut* ~ (see *skut*); *lambing* ~; *house* ~, a ~ close or attached to the homestead; *out* ~, a ~ set at a distance; ~ *manure*, sheep or goats' manure collected from the ~, much favoured as garden fertiliser. *cf. Jam.*

Eng. crawl (*fr. Colonial Du. kraal*) 'an enclosure, pen or building for keeping hogs', also *pig crawl, turtle crawl etc.*

... out-houses occupied one side; and beyond them were *kraals* (enclosures) of thorn bushes for the cattle. Alexander *Western Africa I* 1837

Goodly and great was the yard and a run was around it. The swineherd Built it himself for the swine in the years when his master was absent, ... Stones he had dragged from afar for the kraal and with thorns he had fenced it. Cottrill *trans. Odyssey* 1911

40 Morgen newly established Eragrostis. Dip with equipped kraals. 2 Well-built Native quarters ... *Farmer's Weekly Advt.* 20.3.74

Also figur. In a political sense: a separated group, racial or political.

As for the accusation that the English Press are trying to chase the Afrikaners, like cattle, into separate 'verligte' and 'verkrampte' kraals, I fear that my friend seriously overestimates the power of the Press. *Sunday Times* 24.8.69

2. *rare:* a game sanctuary on private property. [*fr.* (1) ~]

The buck made off for the 'kraal' the sanctuary which most farmers had and where the animals soon realised they would be safe. Even wounded animals were not followed into the kraal. *Daily Dispatch* 22.7.72

3. *vb. trns.* To drive animals into a (1) ~. *cf. Brit. to fold sheep; Canad. corral*, to enclose animals in a circle of wagons, see *laager*. [*fr.* (1) ~]

I found the lady of the mansion kraaling her flocks and herds, her lord being absent. Thompson *Travels I* 1827

... do not kraal the animals but leave them in the open veld ... If they must be kraaled a wire kraal ... is the best. *H'book for Farmers* 1937

Because of past attacks by killer dogs, the Bouwers kraal their sheep every night. *E.P. Herald* 9.8.74

also figur. usu. ~ *off*, to separate racially, linguistically or politically into groups. see quot. at (1) *kraal.*

How can children from these different societies ever hope to understand each other when they are systematically kraaled off into different schools and raised in isolation from each other? *Het Suid Western* 19.4.73

4. *n. pl.* -s. A village or settlement occupied by an indigenous tribe, formerly Hottentot or African, now usu. African: see also *stat/d* and quot. at *tribesman. cf. Canad. rancherie*, an Indian village.

... the beautiful Tyumie valley besides, is now dotted with the houses of European farmers, winklers, or missionaries, surrounded by numerous kraals or villages of Fingoes, located there after the war. McKay *Last Kaffir War* 1871

The entire population of one kraal attacked by an impi of 250 tribesmen on Sunday have fled and it may be several days before the first of them filter home and life at the kraal returns to normal. *Daily Dispatch* 3.8.72

5. A cluster of huts occupied by one family or 'clan': see quot. at *k(a)ya*.

They are divided into a great number of independant clans, or *kraals*, as they are termed. Thompson *Travels II* 1827

in combination ~ *head*, see also *headman*.

Although the system of family government was patriarchal, the will of the kraal-head was tempered and, to a great extent, influenced by the individual and collective views of his wife or wives, and the male adult members of his household. Longmore *Dispossessed* 1959

6. *usu.* in combination *royal* ~, the official place of residence of a tribal king or paramount chief called also *Great Place* (q.v.) see quot. at *izibongo*.

Evidence was that Chief Zulu was building a royal kraal for himself at Nongoma. *Daily Dispatch* 8.8.72

7. (4), (5) and (6) ~, *usu. hist.*, found in place names e.g. Chaka's Kraal, Hammanskraal, Kommandokraal, Jan Fourie's Kraal.

8. *kraal, pl.* krale. Bead(s): in combination *Jobskrale* (see *Job's tears*) *kraalogie* (*ogie* little eye), *Zosterops capensis*, see *witogie:* [*Afk. kraal* bead] also found in place name Kraletjies Baai, where beads from wrecks are still washed up. [*dimin. form*]

krag [krax] *n. colloq. slang* Energy: see also *woema*. [*Afk. fr. Du. kracht* energy, power, strength]

That's the altar, chancel steps and font – who's got krag to arrange some at the West Door? *O.I. Grahamstown Cathedral* 8.12.76

kragdadig [ˌkraxˈdɑdɪx] *adj.* Forceful, usu. figur. in a political sense. *cf. Canad. hist. 'direction action (man)'*. [*Afk. kragdadig* forceful, (*krag* force, might + *daad* deed + *adj. forming suffix -ig*) + *-heid* (q.v.) *equiv.* -ness]

Instead of considering public concern seriously Mr. Vorster adopts a kragdadig attitude to the subject and introduces a red herring. *Sunday Times* 31.10.71

~ *heid, n.* Forcefulness, also *figur.* of political 'steamroller' tactics.

. . . a smokescreen to cover a great retreat from apartheid . . . It was, I reckoned, a deliberate display of kragdadigheid both to compensate for this great retreat *Evening Post* 6.10.73

The dour, stocky political patriarch of South Africa, . . . has the ironfistedness his fellow Afrikaners call *kragdadigheid. Time* 28.6.76

kramat [ˌkraˈmat] *n. pl.* -s. Also *karamat:* The shrine, usu. the tomb, of a holy man, often a place of pilgrimage for *Malays* (q.v.), see also *tuan* [*Afk. fr. Malay keramat*, sacred, hallowed (place)]

Near by are two white-walled karamats, which are to-day places of pilgrimage to all true believers. *Cape Times* 30.11.1929 *cit.* Du Plessis 1944

This karamat is one of a series which stretches round the Cape Peninsular to form a rough circle. Every Muslim believes that all followers of the prophet who live within this circle are safe from fire, famine, plague, earthquake and tidal waves. Du Plessis *Cape Malays* 1944

There is a legend that, when the body was removed . . . one of its little fingers could not be found, and the present tomb was built as a *kramat* or holy place, for this relic of the saint. *Cape Herald* 14.9.76

krans [krɑns] *n. pl.* -s. Also *krantz, kranz:* an overhanging sheer cliff-face or crag, often overlooking a river or river bed. *cf. Canad. rampart* precipitous river bank: see also quot. at *mopane.* [*Afk. krans* sheer rock wall poss. *fr. High Ger. Kranz*]

But 'e wasn't takin' chances in them 'igh an' 'ostile kranzes –
He was markin' time to earn a K.C.B.
Kipling *Five Nations* 1903

Also found in place names as *prefix* and *suffix* e.g. Kranskop, Kransberg, Klipkrans, Bloukrans, *Swartkrans* (q.v.).

krans athlete [krɑns] *n. pl.* -s. *slang. figur.* A baboon: an offensive mode of reference to an *Afrikaner* (q.v.); also *abbr. kransie*, thought by some to be the origin of *crunchie* (q.v.): see also *rock (spider), hairy(back).*

krans bee *n. pl.* -s. Also *black* ~: the wild, indigenous S. Afr. honey bee: *cf. Anglo-Ind. dingar*, large wild honey-bee.

kransie [ˈkrɑnsĭ, ˈkrans-] *slang.* See *krans athlete.*

kreef [kri:f, krɪəf] *n. pl.* Ø. *Jasus lalandii* the Cape rock lobster or Cape spiny lobster (*Fr. langouste*): see also *crayfish.* [*Afk. kreef* rock lobster, crawfish *fr. Du. kreeft* lobster]

'Crayfish' or 'kreef' is also plentiful all through the summer. We also call it Cape Lobster. Hilda, *Diary of a Cape Housekeeper* 1912 *cit.* Pettman

... **Cape Independence**
What sold me was a demand that the export of smoked snoek to the Transvaal and the export of kreef to any part of the world be prohibited. The motto would be 'A kreef in every pot'. *Cape Times* 26.4.79
In combination ~ *boat* a fishing vessel carrying baskets, like *Brit. lobster pots*, for catching ~. *Potted* ~, home made slightly spiced paté of ~.

kremetart(boom) [ˌkreməˈtartˌ(buəm)] *n. pl.* -bome. *Adansonia digitata*, the baobab tree, also called *cream of tartar tree* and *lemonade tree* (q.v.) from the fruits which contain citric acid, and from the flavour of the drink sometimes made from their pulp. [*Afk. fr. Du. kremetart* cream of tartar]
Bush and koppie, withaak and kremetart and kameeldoorn were dreaming languidly under a cloudless sky. Bosman *Unto Dust* 1963

kreupelboom [ˈkrɪœpəlˌbuəm] *n. Leucospermum conocarpodendron* (Proteaceae): a small tree with showy flowers, also *goudsboom*, formerly used for firewood hence *brandhoutboom* (firewood tree). ⫽The bark was formerly used both medicinally and for tanning hides. [*Afk. fr. Du. kreupel* cripple *fr. the bent lower branches* + *boom cogn. Ger. Baum* tree]
12 Dec 1810. A long broad walk, or avenue ... leads from the street called the Heeregragt towards an uncultivated plain extending to the foot of Table Mountain, and, in some parts, abounding in low scrubby trees of Kreupelboom, much used for firewood. Burchell *Travels I* 1822

krimpsiekte [ˈkrɪmpˌsiktə] *n.* Also *nenta: Cotyledonosis*, a disease of domestic animals esp. goats thought to be from eating various species of *Cotyledon* esp. *C. cacaliodes, C. eckloniana harr.* and *C. ventricosa burm.* all known as ~ *bossie*. ⫽There are two forms *dun* ~ (thin) and *opblaas* ~ ('blow up', distended) i.e. the chronic form from habitual eating of small quantities of the plants, or the acute from sudden consumption of a large quantity: see *nenta* (*bossie*). [*Afk. krimp fr. Du. krimpen* shrink, diminish + *siekte fr. Du. ziekte* disease]
All our domestic animals are susceptible to krimpsiekte, but goats suffer chiefly as the krimpsiekte areas are mostly suitable for goat farming. *H'book for Farmers* 1937

kroes(ie) [ˈkrŭsĭ] *adj.* Frizzy, crisp, over-curly. [*Afk. kroes* frizzy (+ *-ie fr. Eng. adjective-forming suffix* -y)]
... the British Government wanted to give the vote to any Cape Coloured person walking about with a kroes head and big cracks in his feet. Bosman *Unto Dust* 1963

kroesblaar [ˈkrŭsˌblɑː(r)] *n.* Leaf curl: a virus disease of tobacco similar to leaf curl in peach trees. [*Afk. kroes* curly + *blaar* leaf]
... Kroesblaar ... very damaging in some years. Leaves appear crinkled ... and develop frilled leaf-like outgrowths from the undersurface of veins. *H'book for Farmers* 1937

kroeskop [ˈkrŭsˌkɔp] *n.* Someone with frizzy or over-curly hair: see *korrelkop; peperkorrels* and quot. at *kroes*. [*Afk. kroes* frizzy + *kop* head *cogn. Ger. Kopf*]

krom [krɔm] *adj.* Crooked: found in place names e.g. *Kromdraai* (q.v.), Kromkloof, Kromrivier. [*Afk. krom* crooked]

Kromdraai(ape)man [ˈkrɔmˌdraɪ] *n. prop. pl.* -men. A primitive man, *Paranthropus robustus* or *Australopithecus robustus*, with an extraordinarily massive jaw, known from skull and other fragments discovered at Kromdraai, Transvaal, S.A. *cf. Swartkrans ape man:* see also *Boskop man*, and (2) *Stellenbosch* [*fr. place name Kromdraai*]

kromnek (disease) [ˈkrɔmˌnek] *n.* A virus disease of tomatoes, tobacco and other plants spread by feeding of thrip or white fly. *U.S. tomato streak* [*Afk. krom* crooked + *nek cogn.* neck *poss. sig.* stalk]
Kromnek: a very destructive disease which attacks plants at any age. Growth suddenly stops. *H'book for Farmers* 1937

Kruger Rand See (4) *rand.*

kruidjie-roer-my-nie [ˈkrœĭkĭˈru:rˌmeɪˌnĭ, ˈkrœĭcĭ-] *n.* Any of several species of *Melianthus* esp. *M. major* the leaves of which are highly toxic to animals. However, honey is produced from the copious nectar of the flowers hence the name *heuningblom* (honey flower). ⫽Used in *Afk.* of a touchy or 'stand-offish' person. [*Afk. fr. Du. kruidjie* little herb, plant + *roer* touch + *my* me + *nie* not]

kruis [krœĭs] *n.* Cross: found in place names Het Kruis, Kruisfontein, Kruis Vallei. [*Afk. fr. Du. kruis cogn.* cross]

kruithoring [ˈkrœĭtˌhuərɪŋ] *n. pl.* -s.

Powder-horn. [*Afk. fr. Du. kruit* gun-powder + *horing fr. Du. hoorn cogn.* horn] **1.** A horn usu. of a cow, with a lid or cap, formerly used for the carrying and storage of gunpowder.

Each man had three rifles, ready primed with small bags of slugs at hand and the horns (kruithorings) filled with dry powder. *Daily Dispatch* 16.12.71

2. The emblem of the *Nationalist Party* (q.v.).

krummelpap [ˈkrœməlˌpap] *n.* A crumbly dry-cooked *mealie-meal* (q.v.) 'porridge' served both as a breakfast food and as an accompaniment to meat and gravy or (2) *braaivleis* (q.v.): see also *putu*. [*Afk. krummel* crumb + *pap* porridge]

Mieliepap for every Taste. Have you tried it with meat & gravy? A real old time South African favourite – perfect at braais. Make the firm stywe pap or the dry and crumbly Krummelpap. *Fair Lady* 30.10.68

kudu¹ [ˈkʊdŭ, ˈkŭdŭ] *n. pl.* ∅. Also *koedoe: Tragelaphus strepsiceros,* a large species of antelope with magnificent curling horns: shot for venison, *biltong* (q.v.) and the hide: see ~ *leather, skin.* ⫽A traffic hazard in the E. Cape: see quot. [*Afk. koedoe fr. Hott. prob. fr. Nama kudu-b* or *kudu-s*]

At night, the car lights attract the kudu, which then tends to jump on to the vehicle. With these animals weighing up to 800 lbs, severe damage can be inflicted and lives endangered. *Daily Dispatch* 2.9.71

Farmers in five areas of the Eastern Cape have been allocated hunting permits to shoot 1 409 kudu during the June-July hunting season. *E.P. Herald* 12.6.73

~ *feet.* A style in furniture making. *cf. Brit. hoof feet* both in silver and furniture.

. . . lounge suites made in ball and claw, Queen Anne or Kudu feet. *Farmer's Weekly* 21.4.72

~*skos. Haworthia viscosa :* see *koedoe-kos.* ~ *leather/skin.* The hide of the ~ sometimes used for making *vel(d)skoen* (q.v.) whips etc.

GAME LEATHER SHOES, (veldskoene), Strong Kuduleather Uppers; feathercrepe soles, comfortable and durable, natural yellow or brown suede. *Farmer's Weekly Advt.* 21.4.72

Stockwhips from plaited Koodoo skins R3,30; heavier R5,30. *Farmer's Weekly Advt.* 20.3.74

kudu² *n. pl.* -s A mine-proof armoured military vehicle.

Here in front of a Kudu mine-protected

vehicle at Victoria Falls are two members of the AANS. *Argus* 29.6.79

kudu milk *n. colloq. poss. reg.* E. Cape. Brandy, also *tiger/tier milk/melk* (q.v.) *cf. Canad. moose milk, wolf juice.*

Kudu milk is the Albany for brandy. Guy Butler 1971

kufija This Malay word has numerous spellings: *qufi(y)a* (q.v.) *kofia khofia khafia kafia kofija* etc., see at *khafia.* ⫽ This small, white crocheted skull-cap is called by its Persian name *topee, topi* by Muslim Indians in Natal, and sometimes *fez* by Cape Malays.

kugel [ˈkʊgəl] *n. pl.* -s. Sect. Jewish. A young girl of the wealthier class, whose interests are men, money and fashion, speaking a recognizable drawling dialect developed within the group. [*Yiddish kugel* a heavy suet pudding sometimes with raisins *cf. Ger. Kugelhopf*]

. . . a *kugel* (*kugel* is a Yiddish word meaning pudding, but in this country it now also means daughters of wealthy parents whose only interest in life is their appearance and how to spend more money. Kugels have developed a jargon and accent of their own.) *Personality* 29.1.71

Kuhne meal [ˈkunĭ] *n.* Coarsely crushed wholewheat flour used for *Kuhne/ Kuny* (q.v.) *meal bread,* usually a health bread: see also *boermeel.* ⫽Thought to be the name of a German doctor who developed the 'health loaf' at the turn of the century, 'to clear up the illnesses prevalent at the time'.

5 cups unsifted brown flour (Kuhne meal) . . . 1 cup crushed wheat . . . ½ cup sunflower seeds. *Darling* 28.4.76

kuier [ˈkœiə(r)] *vb. intrns. colloq.* To visit, usu. to make a stay as opposed to a call. [*Afk. kuier* to visit *fr. Du. kuieren* to walk, stroll]

They soon began to feel themselves at home as they were allowed to visit, or as they call it, kuyer, at the kraal. Burchell *Travels I* 1822

When the graze . . . runs out, we use up some of Sherman's extra petrol on kuiering with other parties up the road where the beer never stops flowing and the braais never go out. Hobbs *Blossom* 1978

kuif [(keɪf) kœïf] *n. pl.* -s. *colloq. usu.* Among young boys: also *kyf* a brushed back quiff of hair as opposed to a forward brushed fringe. [*Afk. kuif cogn.* quiff]

'For heaven's sake Ma don't make my hair into a kuif.' O.I. Boy 12 1973

. . . all curly blonde hair with a kuif to it and lekker mooi sidies. Hobbs *Blossom* 1978

kuil [ˈkœil] *n. pl.* -s. Pool, hole, pit: found in place names e.g. Kuilsriver, Kuilfontein. [*Afk. kuil* water-hole]

. . . we reached a spot known to my guides, called a *kuil* or pit, where we found a small natural reservoir of tolerable water. Thompson *Travels I* 1827

kuku [ˈkŭikŭ] *n. pl.* -s. *colloq.* A fowl: see *koekoek : cf. Austral. chook*(*y*). [*Ngu. nkuku* a fowl *prob. onomat.*]

kukumakranka [ˌkŭkŭmaˈkraŋka] *n. pl.* -s. Any of the species of *Gethyllis :* usu. sig. their fragrant club-shaped fruit which are dried for scenting rooms or cupboards, eaten, or infused as ~ *brandy :* see second quot., also quot. at *buchu* (*brandy*). [*Hott. name. Afk. form koekoemakranka*]

. . . a celebrated little plant which still preserves its original Hottentot name . . . kukumakranki. It has . . . a bulbous root close to which is produced a long, yellow, soft fruit, of the length and size of a lady's finger, its top just appearing above the ground. The taste of it is somewhat pleasant, but its smell is delightful, having a perfumed odour of ripe fruit. Burchell *Travels I* 1822

. . . a jar . . . half filled with brandy, each koekemakranka placed in the liquid and the jar tightly screwed. After a period of fermenting, the liquid could be used as a remedy for stomach complaints. *Argus* 29.5.71

kulu [ˈkulŭ] *n. poss. reg. E. Cape* Familiar term for grandmother among children. See also *makulu, omkulu* [*fr. Xh. makulu* granny ('big mummy')]

Beloved wife of . . . mother of . . . Loving kulu of . . . and . . . There never was a wife or mother lovelier. *E.P. Herald* 4.3.78

kuni, kuny (meal) [ˈkunĭ] *n.* See *Kuhne meal.*

It's supposed to be made from something called kuny meal but I just used flour. *Sunday Times* 22.12.74

Also ~ bread.

. . . if you're just feeling peckish, how about honey and sunflower seeds between slices of kuny bread? *Darling* 18.8.76

Kupugani [ˌkupŭˈganĭ] *n. prop.* A non-profit making organization established in 1962 for supplying nourishing food at below-retail prices esp. to Africans: *attrib.* ~ *soup kitchens,* ~ *biscuits* etc. [*unknown :* said to mean '*help yourself*']

Their slogan is 'Nutrition is our Business'. They are Kupugani, a non-profit making organisation which aims at providing nourishing food at low cost to the needy. *Fair Lady* 13.11.74

kurper [ˈkɜpə] *n. pl.* -s. Any of several species of freshwater fish of the Anabantidae esp. *Ctenopoma* spp. and *Sandelia* spp. including varieties of climbing perch and the *Cape* ~ *S. capensis.* ⫽ ~ is loosely applied to numerous species of the Cichlidae fam., esp. of the *Tilapia* and *Sarotherodon* genera: in particular *S. andersoni, S. mossambicus* and the banded *Tilapia, T. sparrmanii :* see quots. at *moggel* and *oog*. [*Afk. fr. Du. karper* carp]

Loskop Dam, . . . the scene of record-breaking kurper catches . . . is at last delivering some nice fish, mostly from the kurper shoals congregating in the Olifants River inflow . . . had limit bags of blue kurper over a weekend outing. *Sunday Times* 27.10.74

kurvey [ˌkɜˈveɪ] *vb. intrns. obs.* To convey goods by oxwagon, usu. over long distances, from the coast up country. ~ *or n. pl.* -s. A carrier: see also *transport rider/ing cf. Canad. packer.* [*fr. Afk. karwei* to cart goods *prob. by folk etym., fr. Du. karweien* to do a job]

. . . according to local legend a karweier of the early days, an English transport rider nicknamed John Bull, drove his horse too hard up that hill, so that it died. Green *These Wonders* 1959

~ *ing. vbl. n.* the trade of a *kurveyor.*

I tried a trip at kurveying, I took a load to Fort Willshire. Stubbs *Remin.* 1876

kussingslaan [ˈkœsɪŋˌslɑn] *n.* Traditional game: see also *boeresport.* [*Afk. kussing* pillow + *slaan* to strike, hit]

Amid cheers of 'very good, Boetie!' these youngsters had a go at 'kussingslaan' (a pillow fight on a beam) – one of the traditional boeresport items. *Panorama* Oct. 1975

kustingbrief [ˈkœstɪŋˌbrif] *n. pl.* -ven. Roman Dutch Law A special type of mortgage *bond* (q.v.) passed simultaneously with registration of title securing the purchase price of the property. [*Afk. fr. Du. kusting* mortgage + *brief* letter, note]

A kustingbrief is a particular variety of special mortgage bond. It is a bond intended to secure the purchase of land and passed simultaneously with the transfer thereof. Wille *Law of Mortgage & Pledge* 1961

kwaai [kwaɪ] *adj. colloq.* Bad tempered, harsh: vicious or savage if used of an animal: in combination ~ *vriende,* see also *bad friends,* and in place name Kwaaihoek. *cf. Austral. fr. dial. Eng. maggoty,* ill-tempered, also *niggly, le-*

mony etc. [*Afk. fr. Du. kwaad* bad]
The *Daily News* commenting on President Kruger's recent speech, says it is all to the good, when there is any question of tampering with the paramount position of Great Britain, that the Queen should be found 'een kwaai vrouw'. *E.P. Herald* 15.3.1897
The second race I take a chance on a big grey job with a *kwaai* look about him, and guess what? It pays R45. *Darling* 24.12.75
. . . don't say anything. Just sit still. Pretend we're still *kwaai vriende*. Fugard *Boesman & Lena* 1969

kwaal [kwɑl(ə)] *n. pl.* **kwale.** *slang.* Ailment, complaint, usu. *pl.* [*Afk. kwaal* complaint]
Don't ask her how she's feeling, she's always full of kwale and you'll never get away. O.I. Schoolgirl, 1973

kwagga [ˈkwaxa] *n. pl.* Ø, -s. Also *obs.* *kwakka*, see quagga.

kwashiorkor [ˈkwaʃĭɔˌkɔ] *n.* A protein-deficiency disease prevalent among undernourished African children characterized by wasted limbs, distended belly, oedema and a reddish depigmentation of the hair. ▯ First identified by a team from the University of Cape Town Medical School doing research in W. and Central Africa in the 1940's. The name ~ is thought to be from one of the languages of Ghana and is said to mean 'red boy'. *Informant* Professor J. Brock [*c.* 1947.]
He had been struck by the paradox of wards full of cases of malnutrition and kwashiorkor in cities 'which must be the wealthiest in Africa'. *Rand Daily Mail* 22.10.70

KwaZulu [ˌkwaˈzulŭ] *n. prop.* The *homeland* (q.v.) of the *Zulu* (q.v.) consisting of territory in Zululand and Natal. [*Zu. kwa-* adv. locative formative equiv. 'at' or 'in' + *Zulu,* sky, heaven (*praise name*) see *izibongo*]
The people of KwaZulu want to live in one land and not the Bantustan pieces. *Drum* 8.9.72
Discussions continued between officials and representatives of the Zulu people, however, and in 1970 the latter agreed to accept the Government's scheme. Chief Gatsha Buthelezi was unanimously elected Chief Executive Officer of the territorial authority that was created shortly afterwards for the area in future to be known as KwaZulu. Horrell *Afr. Homelands in S.A.* 1973
On March 30, 1972, a legislative assembly was established for KwaZulu. The assembly was constituted on the same lines as the former Zulu territorial authority (in accordance with

the Bantu Authorities Act of 1951). *Panorama* July 1973

kwedin(i) [ˌkweˈdin(ĭ)] *n. pl.* -s. A young boy. usu. Xhosa (see *Zu. umfaan*) strictly, only before circumcision: see *abakwetha.* [*contraction Xh. inKwenkwendini,* vocative case of inKwenkwe uncircumcised boy]
One White trader complained that kwedins (boys) had destroyed more than 1 000 white telephone cups in the Idutywa area by shooting at them with catapults. *Daily News* 18.12.70
YOUTH GETS SIX CUTS FOR KILLING . . . A 16-year-old kwedini was this week sentenced to six cuts for culpable homicide. *World* 9.11.76

kweek [kwɪək] *n.* Also *quick* (q.v.): a name generally applied to any of several lawn or pasture grasses referring to their habit of growth i.e. propagating by underground or surface runners, which root at the nodes: orig. *Cynodon dactylon, fyn* ~ called 'fine quick', also *Stenotaphrum secundatum; growwe* ~ known as *coarse quick.* Also in numerous compounds e.g. *khaki* ~ see (3) *khaki; rooi* ~ (red), *strand* ~ (seashore): see also *uintjie* ~ and third quot. below. [*Afk. fr. Du. kweek* dog's grass *poss. rel. kweken* to cultivate, grow]
There is a large number of pasture grasses which develop dangerous amounts of prussic acid when they are wilted. The most dangerous of these are the 'quick grasses' or 'kweekgras' different species of Cynodon. *H'book for Farmers* 1937
'The mealie-planter doesn't seem to work well on the lands, behind a plough, going over kweekgras sods and turned-up anthill, as it does on the smooth stoep of Policansky's hardware store.' Bosman *Jurie Steyn's P. O.* 1971
How can I get rid of kweek from my skaapplaas lawn? *S. Afr. Garden & Home* Mar. 1975

kwela [ˈkwelə, ˈkwe:la] *n.* or *n. prop.* African *pennywhistle* (q.v.) music: see first quot. [*Ngu. khwela* (*vb.*) climb on, mount]
The Xhosa verb – kwela is in this context used to mean to begin, to get moving. The word, which originally meant to climb, assumed this different meaning during the time of the penny whistle music. Before the famous pennywhistle group, the Black Mambazo . . . began playing, their leader . . . to notify his band that . . . they should begin . . . would say 'Kwela' meaning begin. From this, then, the name 'Kwela music' was given to pennywhistle music. *Drum* Dec. 1965
Here is Lemmy Special Mabaso who put kwela and the penny-whistle on the map and

took kwela music around the world. *Drum* Mar. 1971

kwela-kwela [ˈkwelə-ˈkwelə] *n. pl.* -s. *Afr. E.* A police pick-up van. [*Ngu. khwela* climb on, in, mount]

On the left is the kwela-kwela van used on April 2 when three prisoners died. The picture was taken when the judge trying the driver and his assistant . . . inspected the van. *Post* 27.4.69

A young cop in a kwela-kwela who happened to be passing by, gave chase. *Drum* 8.11.72

kweperlat [ˈkwɪəpə(r)ˌlat] *n. pl.* -te. A young quince twig, much in favour as an instrument of punishment for the allegedly keen smart it inflicts: see also *kleilat(jie).* [*Afr. kweper* quince, *lat* twig]

To the south ran a long quince hedge. Real intelligent planting, those quinces. A 'kweperlat' was the sanctified instrument of correction for Karoo boys, and every school of any pretensions grew its own supply of 'latte'. Butler *Karoo Morning* 1977

kwêvoël [ˈkwe:ˌfuəl] *n. pl.* -s. The *go-away bird* (q.v.) or grey *loerie* (q.v.) *Corythaixoides concolor,* occ. known as *groot* (*great*) *muisvoël* from the similarity of appearance, see quot. at *spooring.* [*onomat. fr. bird's alarm call*]

K.W.V. [ˈkeɪˌdʌbljuˈviː, ˈkaˌviˈfi] *n. prop.* Co-operative wine producers union often used as *prefix* e.g. K.W.V. 10 year Old (Brandy) etc.: see quot. at *Kimberley Club* and at (1) *hanepoot.* [*acronym Kooperatiewe Wijnbouwers Vereniging*]

The ultimate result was the now powerful KWV, the Ko-operatieve Wijnbouwers-Vereniging van Zuid-Afrika . . . The main functions of the KWV are to promote the interests of South African wines. *U.C.T.* Vol. 6 1976

kyf [keɪf] *n. pl.* -s. *erron. form kuif* (q.v.).

L

laager [ˈlɑːgə(r)] **1.** *n. pl.* -s. *hist.* An encampment of wagons lashed together for the protection of the people and animals within, and as a barricade from which to fire on attackers; the regular defence of the *Voortrekkers* (q.v.) and other pioneers; occ. loosely a refuge from attack by hostile tribes; see second quot. *cf. Canad. corral n.* a circle of wagons. [*Afk. poss. fr. Ger. Lager cogn. Eng.* lair]

. . . the Voortrekkers began to see the folly of the lonely *laager* in a country ravaged by savage hordes. A laager was formed by the wagons being drawn up, end to end, to form a circle, square or triangle, according to the nature of the terrain. The spaces between the wheels were barricaded with rawhide *riems* and thornbushes. Mockford *Here are S. Africans* 1944

This farmhouse had formed a valuable laager and refuge for other neighbouring members of the Bowker family during these Kaffir Wars, as it was best suited for protection purposes. *E.P. Herald* 9.2.73

2. *vb. trns. freq. pass. or as partic.* To form into a (1) ~ [*fr.* (1) ~]

. . . advanced into their country with 5 weak columns, did not laager the camp at Isandula [Isandhlwana] and this occasioned the fearful disaster there. Alexander *MS. Sketch Book* 1878

The transport wagons were to be laagered and left behind. Cloete *Rags of Glory* 1963

3. Found in Afk. form *laer,* in place names Laerfontein, Laersdrif, Laerwag.

4. *n. figur.* A protective environment usu. in a political sense sig. an ideologically impenetrable enclosure, see quots. at (1) and (3) *kraal.* cf. *Brit. figur. fold n. esp. in ecclesiastical sense.* [*fr.* (1) ~]

. . . ex-servicemen returning to South Africa from the war had a more cosmopolitan outlook. People wanted to read what was happening outside the *laager* and turned to the overseas magazines which poured on to the market. *Personality* 11.6.71

. . . 's main assignment, after all, is to bring as many Afrikaners as possible back into the laager. *Sunday Times* 24.9.72

Back to the laager

South Africa has a word for it – laager housing. Property developers prefer it to simplex housing . . . simplex or laager housing had got off to a bad start with the . . . belief it was a poorer type of housing. Basically laager living was a single-level ground floor development . . . Elderly people were not faced with stairs. In an emergency they were living in a community where help was readily at hand. *S.A. Digest* 19.10.78

laagte [ˈlɑːxtə] *n. pl.* -s. Also *leegte;* a low lying area, valley, dip or depression. *cf. Brit. and U.S. bottom(s),* low lying grass land, *also Canad. and U.S. river bottom.* [*Afk. fr. Du. laagte,* valley, low ground, *alternate form, leegte*]

A herd of wildebeeste and zebra were grazing in a leegte. Cloete *Turning Wheels* 1937

Two leopards . . . trotted down the easy slope of a three-mile-wide laagte. Stander *The Horse* 1968

Also found in place names e.g. Hollaagte, Elandslaagte, Langlaagte, Vol-struisleegte (ostrich).

la(a)itie [ˈlaɪtĭ] *n. pl.* -s *Urban Afr.* A boy: also *lightie* (q.v.) [*fr.* light]

Mine dump, one room dirty streets, dirty laaities, no lights, no hot water, clevahs, Jesus man. *Staffrider* May-June 1978

'You'll look really nice . . . my laiti . . .' 'Solly, you've cleaned us out . . . Solly my laitie, make 'n laas daar!' 'Ok gents, let's go and have a drink.' Dike *First S. African* 1979

laan [lɑːn] *n. suffix.* Avenue: found in street names e.g. Eikelaan, Stellenberg-laan. [*Afk. fr. Du. laan cogn.* lane]

laani See *larn(e)y.*

laatlammetjie [ˈlɑtˌlamĭkĭ, -cĭ] *n. pl.* -s. *colloq.* A child born long after the others in a family; *cf.* Eng. 'afterthought'. [*Afk. laat cogn.* late + *lammetjie* lambkin]

She told me she was born when her mother was 40. 'I was a laatlammetjie – twelve and sixteen years behind my two brothers and I cried and nagged for a little brother or sister to play with.' *Fair Lady* 17.9.75

As modifier:

Births
. . . To Nancy and Willie a 'laatlam' daughter and sister to Jeanine *E.P. Herald* 27.4.79

Labarang [ˌlaˈbaraŋ] *n. prop. poss. reg. Cape* Malay term for Eid-ul-Fitr, the festival celebrating the end of Ramadhan.

Special Offer for Labarang!! Special Offers for Eid (Labarang) Discount of 10% on cash purchases over R30. *Cape Herald Advts* 14.9.76

THE NIGHT OF TERROR DURING LABARANG
The eve of Labarang became a night of terror when a man tried to run down 76-year-old Mr . . . threatened to knife him. *Sunday Times* 3.10.76

lady *n.* A mode of address usu. among *Afk.* speakers, more courteous than 'madam': see *mevrou* and quot. at *bry.* ¶Sometimes misinterpreted as an impertinence: see first and third quots. [*trans. Afk. dame* lady]

I do wish the man on the telephone exchange wouldn't call me 'lady', I can't bear it. O.I. 1970

That's quite all right lady. I jus' come to visit. Don't worry about me. Maclennan *The Wake* 1971

I shall miss . . . Table Mountain but it will be good to see bowler hats sloshing their way to work once more, and I do prefer 'luv' to 'lady'. Letter, English Visitor 1.1.75

ladyfish See (2) *moonfish.*

laeveld [ˈlɑəˌfelt] See *lowveld.*

laer [ˈlɑər] See (3) *laager*: for combination ~*plek* see quot. at *gatjaponner.*

lager [ˈlɑːgər] See (1), (2) *laager.*

laksman [ˈlaksˌman] *n. pl.* -ne, Ø. Executioner, hangman: name for the Fiscal Shrike, *Lanius collaris*: see *Jan Fiskaal* ¶Also in combinations applied to numerous species of the Laniidae e.g. *boslaksman* (bush shrike). [*Afk. fr. Malay laksamana, title of high ranking state official, hence equation with Fiscal* (q.v.) *one of whose concerns was corporal punishment*]

It is commonly known as the 'Jack-hanger' and 'Butcher bird' in the Cape, the 'Jacky-hangman' in Natal, and the Lachsman in the Transvaal. Haagner & Ivy *S. Afr. bird Life* 1908 cit. Pettman

lala kahle [ˈlala ˈgaɬe] *vb. phr. Afr. E.* Sleep well, rest well: see *kahle* and *hamba kahle.* ¶ A child being told by an African to lie down is told to *lala*: 'sleep' and 'lie down' are thus sometimes confused in children's minds: see *sleep*[1] [*Ngu. ukulala* to sleep, *kahle* well, softly]

You are still in the mind of your daughter . . . Rest in holy peace. Lala kahle 'ndlovu' elihle. *Daily Dispatch* 27.3.75

lambili, lambile strap [ˌlamˈbilĭ, -e] *n. pl.* -s. See *girdle of famine, hunger belt.* [*Ngu. Xh. ukulambile* to be hungry, *Zu. lambile perfect of lamba* become hungry]

The leathern strap worn round the waist is called by the savages a lambele strap, or hunger-girdle. Harris *Wild Sports* 1839

lammergeier [ˈlamə(r)ˌxeɪə(r)] *n. pl.* -s. Used of S. Afr. species of ~ : *Gypaetus barbatus* (Aquilidae fam.), the largest European bird of prey: see *lammervanger.*

lammervanger [ˈlaməˈ(r)ˌfaŋə(r)] *n.* Also *lammergeier* (q.v.) used of the tawny or golden eagle of S.A. *Aquila rapax* (*kouvoël*) and of the rare *Gypaetus meridionalis* (Aquilidae) which destroy lambs and poultry. [*Afk. fr. Du. lammer* lambs + *vang* catch + *agent. suffix* -*er*]

. . . the big eagle, the lammervanger, who is sometimes seen sitting on a rock brooding, or more often gliding about the sky on unmoving wings. Morton *In Search of S.A.* 1948

MALTAHOHE, Thursday – A 13 year-old White youth captured a long taloned golden eagle (lammervanger) with his bare hands here. *Daily News* 4.6.70

lamp oil *n.* Paraffin, kerosene: *cf. fish oil* (*visolie*) [*fr. Du.* and *Afk. lampolie* (lighting) paraffin]
... and misfortunes (the effect of Pawell's bad packing) ... the bottle of lamp-oil having been broken by the jolting of the wagon, and a bag of raisins finely soaked with it. Barnard *Journal* May 1798

lamsiekte [ˈlamˌsɪktə] *n.* A paralytic disease of cattle caused by eating toxic organic material, esp. bones, in phosphate deficient areas where 'osteophagia', bone craving, is frequent: ⫽Dosing with bonemeal is sometimes adopted as a preventive measure: also *gal~* (gall, bile) bovine parabolutism, as above, complicated by anthrax or splenic fever: see also *styfsiekte.* [*Afk. fr. Du. lam* paralytic *cogn.* lame + *siekte fr. Du. ziekte* disease]
... lost 6 cows from lamziekte ... they say that they are even more convinced than ever that the disease is caused through the cattle eating putrid bones. *E.P. Herald* 25.1.1921

land¹ *n. pl.* -s. A cultivated field, usu. fenced, see *camp vb.* and quot. at *kweek. cf. Brit. field*, similarly combined *wheat ~, mealie ~* etc. [*Afk. fr. Cape Du. land* cultivated or arable field]
... Lands – Excellent for all kinds of grain; at present there are lands for 16 bags of mealies, but more lands can be made. *The Friend Advt.* 25.8.1930
About fifty years ago the grave was dug up by some farmers while they were making a land. Metrowich *Frontier Flames* 1968
In combination *dry ~*, see also *irrigation farm.*
... well improved mixed farm ... 120 Hectares dry lands and 5 hectares under irrigation. *Farmer's Weekly Advt.* 21.4.72
Irrigation. 12 ha. Magnificent dry lands. Balance thick bushveld teeming with game. *Ibid.*
~ camp a fenced cultivated area as opposed to fenced or enclosed grazing: see also *camp.*
The farm is divided into four grazing camps and 10 land camps – well fenced and watered. *Daily Dispatch Advt.* 14.8.71

land² See also *Ons Land.* ⫽ *~* in S.A. is commonly used in the somewhat biblical style as in 'the land of Cana' *sig.* 'country'. [*prob. fr. Afk. Ons Land* (q.v.)]

Land Bank *n. prop.* An autonomous institution responsible through Parliament to the Minister of Finance: its main business being to advance money to farmers, and agricultural co-ops. ⫽ *~ loans* are of four types, mortgage loans, charge loans, hypothec loans and loans in the form of cash credit accounts. [*abbr. Land and Agricultural Bank of South Africa*]
... such a scheme would have to be partly or largely financed by either the Land Bank or some other financial organisation. Where else may one obtain the finance required and at what rates of interest? ... Money is very scarce and expensive. The Land Bank's rate of interest is six per cent, plus an insurance premium of between 1½ and two per cent. *Farmer's Weekly* 21.4.72

landdrost [ˈlan(t)ˌdrɔst, ˈlænˌdrɒst] *n. pl.* -s. *hist.* A magistrate having jurisdiction over a particular magisterial district or *drostdy* (q.v.) superior in rank to the *field/veld cornet* (q.v.) *cf. Canad.* (*hist.*) *district warden*: see also first quot. at *Volksraad.* [*Du. land* country + *drost* sheriff, bailiff]
District of Cape Town His Majesty's Fiscal Officiates as Landdrost for this District. *Afr. Court Calendar* 1807
13 Oct 1830 Whereas the office of landdrost was abolished on the 31st day of December, 1827. *Statutes of the Cape of G.H.* 1862
Before civil commissioners were appointed for the Cape districts, the landdrost was the chief officer in each; and his residence was styled 'the *drostdy*'. Alexander *Western Africa* II 1837

lang [ˈlaŋ(ə)] *adj. prefix.* Long: found in place names Langkloof, Langverdriet (sorrow), Langberg, Langklip; also with *attrib. inflect.* Langeberg, Langebaan or as *adv.* in Langgewens (long desired). [*Afk. fr. Du. lang cogn.* long (+ *attrib. suffix* -e)]

langbek [ˈlaŋˌbek] *adj./adv. colloq.* Used of one who is sulky, pouting or depressed: see also *dikbek. cf. Brit. down in the mouth.* [*Afk. lang cogn.* long + *bek* (q.v.) mouth]

lang hare and dik brille [ˈlaŋ ˈharə n ˈdɪk ˈbrɪlə] *n. phr.* Sect. *Army lit.* 'long hair and thick spectacles': the S.Afr. equiv. of *Brit.* 'boffins' or 'backroom boys', designers, etc.: see also quot. at *indoenas.* [*Afk.*]
Specialist staff are often referred to as 'lang hare en dik brille' ... Picard in *Eng. Usage in S.A.* Vol. 6 No. 1 May 1975

lang tande [ˈlaŋ ˈtandə] *pl. n. phr. colloq. lit.* 'long teeth': usu. in phr. 'to eat with ~ ' *sig.* with distaste or unwillingness. ⫽ *metaph.* use as in 'I agreed to go on a trip to Europe with long teeth'. (O.I. 1977) is *prob.* idiosyncratic. [*Afk. met lang tande eet*, to eat without relish, toy with one's food]

The greenhorn ... through an ill placed shot may destroy most of the edible parts and be left with a bloodied mess of hind- or forequarter and bone fragments which even a dog would eat with *lang tande*. *Farmer's Weekly* 9.5.73

Language Movement See *Taalbeweging*.

lanie See *larn(e)y.*

lank [læŋk] *adj. colloq. usu. children's slang* Either a term of general approbation *cf. mooi, lekker* etc. e.g. He drives a ~ car; It was a ~ film etc. : or in fixed expression *a ~ age, sig.* 'a long time' e.g. He kept us waiting a ~ age before he rocked up (*O.I.s children*). [*presum. fr. Afk.* lank *cogn.* long]

lap [lap]. See *lappie.*

lapa¹ [ˈlapə, -a] *adv./modifier colloq.* 'Here', or *demon.* 'this' as in 'lapa side'. In combination ~ *language*, an appellation for *fanagalo* (q.v.): see also *Chilapalapa*. [*Ngu.* lapa here, this side]

lapa² *n.* The courtyard of a cluster of Ndebele houses: see quot. [*Ndebele*, courtyard]

The houses ... are ... built on the simplest rectangular plan. But each complex is integrated in an individual fashion. In some, the units are grouped about a centralised forecourt – or 'lapa' – with entrance through a gateway in the decorated 'lapa' wall. *Panorama* May 1974

lap(pie) [ˈlap(i), ˈlæpi] *n. pl.* -s. *colloq.* Any cloth or rag for cleaning, dusting, patching etc. See also quot. at *verneuk.* [*Afk.* lap rag, cloth + *dimin. suffix* -(*p*)ie]

Someone on foot gave the man who held the letter a white lappie tied to a stick, a white rag on a stick – a flag of truce. Cloete *Hill of Doves* 1942

... his one pair of jeans is beginning to look like last year's polishing lappie. *Darling* 21.1.76

In combination: ~*pop* a rag doll; *smeer* ~ (q.v.)

jammer ~ (*pie*) [ˈjamə(r)-] a damp cloth used for spills, accidents or for wiping the hands at a (1) *braai* (q.v.) [*Afk. jammer* sorry]

Table napkins were seldom provided in those days, but a special wet cloth known as a

jammerlappie was passed around after a meal so that guests could wipe their hands. Green *When Journey's Over* 1972

~*pie-legs* Used of one with poor or sloppy gait.

... his socks were usually round his ankles. He was nicknamed 'sigaret beentjies' or 'Lappy legs'. Butler *Karoo Morning* 1976

~*piesmous* A soft goods pedlar, see *smous.*

tjanga ~*pie colloq.* see *beshu, mutsha.* (cf. *tanga*, 'a string' bikini) [*prob. fr. Ngu. iThanga*, fleshy part of the thigh]

larn(e)y [ˈlani] *n. pl.* -s and *adj.* **1.** *Afr. E.* usu. *la(a)ni(e)*, also *Ind.E. lahnee.* See quot. at *bra.* A White person. [*see first quot.*]

As I sat down ... I heard a murmur 'Laanis' the *tsotsi* word for White men. Sampson *Drum* 1956

The lanies we get on with. They are just like us ... now they are bulldozing the place down. Jiggs *Doornfontein* (Poem 1975)

2. *colloq.* Posh, classy, dressed-up [*prob. fr.* (1) ~ *or poss. Fr. l'ornée* the decorated one]

... blue eye shadow, orange blusher, pink lipstick, her purple swade wedgies ... and her best diamante drop ear-rings. Talk about larny! Auntie Vilma would put a rainbow to shame any day. *Darling* 31.8.77

las [las] *n. slang Sect. Army.* Money as in phr. 'make me a ~ ' *sig.* 'lend me some money' (*acc. O.I.s* common in the forces) Also *to make a ~*, to contribute or to pool resources. [*fr. Afk. colloq. vb.* las to augment, increase.]

They were thirsty. Old Chris looked around meaningfully and said: 'Well, gentlemen, between us we should be able to make a *las* for a *dop*,' meaning we should have enough money to buy a drink. *Family Radio & TV* 23-30.1.77

laventel [laˈfentəl] *n. usu. Rooi ~:* lavender drops, one of the *Dutch Medicines.* [*Afk.* laventel *cogn.* lavender]

'A doctor,' Kobus suggested helplessly, 'Mustn't we get a doctor now?' 'The *laventel* ... ' Alida begged weakly. Burgess in *Edge of the World* ed. Gray 1974

lawaai [ˌlaˈvai] *n. slang.* Noise, disturbance, row: see also *bohaai* and quot. at *skop.* [*Afk. fr. Du. prob. ex Low Ger.* lawaai noise, row]

Then this morning in all the lawaai and mix-up – gone! I wanted to look, but Boesman was in a hurry. Fugard *Boesman & Lena* 1969

lay-by *n. and n. modifier.* The reservation of an article by payment of a deposit followed by regular instalments, or the article itself: also Austral. *lay-by* n. U.S. *to lay away: also occ. as vb trns.* to reserve an article in this way.

LAY-BY SHOPPING HAS PITFALLS FOR UNWARY Shopping on the lay-by system has pitfalls for the unwary buyer. If a person finds he cannot continue paying instalments on the article reserved for him in the shop, he seldom gets his money back. *E.P. Herald* 11.6.73

leader *n. pl.* -s **1.** *hist.* A small boy, often a Hottentot, employed to lead a team of draught oxen for those travelling by wagon: see quot. at *togt²* also *voorloper, touleier.* [*trans. Du. and Afk. leier cogn. leader, prob. fr. touleier* (q.v.)]
. . . little Hottentot boys, who usually run before and guide them . . . The attachment of the animals to their little leaders is very great, and sometimes you will see them look about for them and keep bellowing and uneasy until they come to their heads. Percival *Account of Cape of G.H.* 1804
23 Feb 1812 Old Hans had engaged a Half-Hottentot, named Daniel Kaffer, and his son, to be the driver and leader of my waggon from the river bed to Klaarwater. Burchell *Travels I* 1822
2. *hist.* Used of the head of an 1820 Settler *party* (q.v.).
The plan of the large joint-stock parties was ill devised, and proved a fertile source of disunion. The heads or leaders were in many instances merely nominal, and neither in property nor intelligence superior to their followers. Thompson *Travels II* 1827

lead water *vb. phr.* To irrigate, usu. by means of *sloots* (q.v.) or *furrows* (q.v.) from a public supply in towns which have *water erven* (see *erf*) from farm dams or other irrigation schemes: freq. in form *lead out water:* see quot. [*trans. Afk. water lei*]
On examining the banks I observed with regret the impracticability of leading out the water for irrigating the adjoining lands by dams and ditches – the usual and only method of cultivating the soil in the interior of Southern Africa. Thompson *Travels II* 1827
It's mainly a matter of water. Piet has spent most of his money on that; he's made furrows, built dams. The whole farm is a network of furrows that follow the contours; he can lead water anywhere. Brett Young *City of Gold* 1940

leadwood see *hardekool* and quot. at *boekenhout*

leaguer [ˈliɡə] *n. pl.* -s. An old Dutch liquid measure of 5,82 hectolitres, about 150 gallons, sometimes a vessel containing this, or a greater amount: ⸙*presum. not obs.* see second quot. at (1) *hanepoot.* [*fr. Du. ligger, legger*]

. . . the wine is kept in very large vessels somewhat shaped like the hogshead . . . Each of these butts or reservoirs, which they call leagers though an inapplicable term, as a leager is a measure of one hundred and fifty gallons, will contain from six to seven hundred gallons. Percival *Account of Cape of G.H.* 1804
. . . the roads are bad, lying chiefly through deep sand, and require eighteen oxen to convey two leaguers of 152 gallons each, occupying two or three days to perform the journey. Thompson *Travels II* 1827

learn *vb.* **1.** *vb. intrns. substandard.* Used mainly by children as equiv. of 'swot' or 'study' as in 'I've been learning for three hours' or 'I must go and learn'. *poss. fr. Afk. leer (vir)* to study]
2. *vb trns. substandard.* To teach. *cf. Brit. slang 'I'll l(e)arn you':* see quot. at *fanagolo.* [*trans. Afk. leer* to teach, *learn cogn. O.E. laeran* to teach]

Lebowa [ˈleɪbɔ̃wa] *n. prop.* The *homeland* (q.v.) of the North Sotho peoples, the *N. Ndebele* (q.v.) and *Pedi* in the N. Transvaal.
A new constitution for Lebowa was adopted at a meeting of the Legislative Assembly in July 1972. In terms of Proclamations 224, 225 and 226 of 29 September this homeland became a self-governing territory within the Republic as from 2 October with Seshego as the seat of Government. Horrell *Afr. Homelands of S.A.* 1973

lechwe [ˈletʃwe] *n. pl.* ∅. *Kobus leche*, a small African antelope preferring a marshy habitat. [*Bantu rel. Sotho letsa,* lechwe]
The reserve hopes soon to get a number of lechwe a type of waterbuck. *Grocott's Mail* 8.6.71

leegte [ˈliəxtə]. See *laagte.*

leeningsplaats [ˈliənɪŋsˌpla(t)s] *n. pl.* -en. See *loan place, loan farm* and quot. at *opstal.*

leervis, leerfish [ˈliə(r)ˌfɪs] *n. pl.* ∅. Also abbr. *leerie/y: Lichia amia* called *garrick* in Natal, an edible S. Afr. game fish, thought to be named for its leathery skin: *cf. Austral. leather jacket.* [*Afk. fr. Du. leer, leder* leather + *vis fr. Du. visch cogn.* fish]
. . . more leervis should be caught in our waters. This is a wide ranging fish round the Eastern Cape where it is found wherever salt water occurs. Leervis are caught at sea, in the surf and in estuaries right up to the highest point of tidal influence in some rivers. *E.P. Herald* 2.5.74

leeu [lju:] *n.* Lion. [*Afk. fr. Du. leeuw,*

lion] **1.** Found in place names e.g. Leeukop, Leeublad, Leeukraal, Leeufontein. **2.** As *prefix* in plant names esp. ~ *bekkie* (little mouth) *Antirrhinum majus,* also *Nemesia capensis;* ~ *gras, Aristida marlothii,* see *twa(a)grass;* ~ *doring, Harpagophytum procumbens.*

left hand (wife) See *right hand (wife).*

leguaan [ˈleguˌɑn] *n. pl.* -s. Also *Afk. likkewaan :* various spellings: the iguana, a species of the monitor lizard *Varanus* usu. *V. niloticus* the *water* ~ which attains a maximum length of between 2 and 3 metres in S.A.: also *V. albigularis* the *rock* ~. *Austral.* go(h)anna, iguana, *Varanus :* see quot. at *kaaiman.* [*Du. leguaan fr. Fr. l'iguane* the iguana]

　. . . the Leguan is not a crocodile at all. It is, indeed, an animal of the Lacerta class, and amphibious, but perfectly harmless. Lichtenstein *Travels I* 1812

　. . . . pointed out a leguan – a giant, five foot lizard – swimming slowly up against the current with the water forming a tight, green ruff around his sinuous throat. Collins *Impassioned Wind* 1958

lekgotla [ˌleˈxɔ̃tla] *n. pl.* (a)ma-. A tribal court, now urban: see *makgotla,* also *kgotla.* [*Sotho (le)kgotla* courtyard, court]

　An indication of local concern with crime has been the re-emergence of the traditional lekgotla, or tribal court, . . . The lekgotla tries offenders and dispenses the rough justice of public floggings with a sjambok on Sunday mornings in Naledi township. *Sunday Times* 27.10.74

lekker¹ [ˈlekə(r)] *adj. colloq.* A term of general approbation esp. among children sig. pleasant, excellent, delicious etc. Sometimes rendered as *adv.* + *-ly,* also inflected ~ *est :* see also *mooi* and quots. at *nog-'n-piep* and *Terry.* Found in place names Lekkerdraai, Lekkerlag and Lekkerrus. *cf. Austral. beaut.* [*Afk. fr. Du. lekker* pleasant, nice tasting, dainty *etc.*]

　Lekker kost as much as you please,
　Excellent beds without any fleas.
　Inn Sign – Farmer Peck's – near Muizenberg, *circa* 1840 *cit.* Gordon-Brown 1965

　. . . makes lekker padkos . . . packs to feed any number of travellers . . . tastiest take along food for people on the move. *E.P. Herald Advt.* 28.5.76

　The lekkerest ladies in London . . . in . . . leading model agencies . . . the upper crust vowels

are rapidly being replaced by . . . international drawls that include a great deal of South African. Lekker hey! *Sunday Times* 24.4.77

freq. in combination ~ *ou,* see *ou;* also ~ *jeuk,* ~ *lewe,* ~ *ruik.*

　To rank as a lekker ou at Bishops, Michaelhouse or St. Andrews is surely not to have lived in vain no matter how savagely the inwardly delighted recipient of this accolade lashes out at corruptions of English pure and undefiled . . . 'He was a lekker ou' Jislaik! *Cape Times* 1.8.72

~ *jeuk* [jœk] [*jeuk* itch] *colloq.* Scabies: a skin disease caused by parasites.

　. . . would appear to have something called lekker-yuk, and this can be cured with antiscabies lotion. *Post* 21.6.70

~ *lewe* [~ lɪəvə] [*lewe cogn.* life] *cf. Italian la dolce vita.*

　The Boer's idea of what he called the *lekker lewe,* the sweet life, was the free and easy life of the frontier . . It would be pleasant to report that they had attained the *lekker lewe.* But the *lekker lewe,* of course, is one more mirage; it is not a state that men can ever achieve, or were meant to achieve. White *Land God Made* 1969

　We have developed an attitude of 'we can't do that, it's Kaffir's work.' Yes, we prefer 'die lekker lewe'. *Evening Post* 21.10.72

prefix ~ *ruik* [~ rœik] [*Afk. ruik* to smell] pleasant-smelling: prefixed to certain plant and flower names: ~ *bossie* either of two species of *Lippia;* ~ *gras Elyonurus argenteus;* ~ *heide/heath,* sweetly scented species of *Erica;* ~ *pypie Cyrtanthus suaveolens,* see *brandlelie.*

lekker² *adj. slang.* Tipsy, merry *cf. Brit.* slang '*Nicely, thank you.*' [*Afk. lekker* nice (tipsy)]

　There was no shortage of food and drink . . . an endless supply of snacks with plenty of cheese. The cheese helped keep everybody just 'lekker' and not too stoned to make them rowdy. *Post* 12.10.78

-lelie- [ˈlɪəlĭ] *n. pl.* -s. Lily, suffixed to certain names of flowers e.g. *vark* ~ (q.v.), *berg* ~ (q.v.), *blouwater* ~ see *kaaimansblom,* also *paddapreekstoel, brand* ~ (q.v.). Also found in place names e.g. Leliesfontein, Blouleliesbos. [*Afk. fr. Du. cogn.* lily]

lemonade tree *n. pl.* -s. *Adansonia digitata :* the baobab tree: see *kremetartboom.*

　In these pods is the white pulp that gives the baobab yet another name, the 'lemonade tree', for when this is mixed with water it makes a refreshing drink with the cream of tartar flavour. Green *Glorious Morning* 1968

emoenhout, lemonwood *n. Xymalos monospora* also known as *borriehout:* see also *sandveld chair* and *borrie* [*Afk. lemoen* orange, *hout* wood]

. . . some confusion exists because a small tree found in the Eastern province called the wild lemon . . . produces a wood called lemoenhout; but numerous chairs found in the sandveld made of lemoenhout were probably made from the light yellow, fairly soft wood of the orange tree and other citrus trees found . . . in those parts. Baraitser & Obholzer *Cape Country Furniture* 1971

end *vb. trns. substandard* Borrow: see also quots. at *borrow.* [*trans. or translit. Afk. leen* borrow, *also* lend]

I could lend Fanie's bakkie off him so's we can take tents . . . and braai stuff. It'll only cost the petrol. *Darling* 9.6.76

'I'll borrow you a lend of my clean blouse to get home with, Trix,' . . . *Ibid.* 29.9.76

ength *n. substandard.* Height, as in 'Give your length and waist measurement.' *Advt. Sunday Times* (no date). [*translit. Afk. lengte* height (of a person)]

Leopard *n. pl.* -s. A member of the African S. Afr. national rugby team, black equiv. of rugby *Springbok* (q.v.).

The burly Leopards prop . . . was fired when he reported back from Italy at his job as a labourer at a factory . . . here yesterday. *E.P. Herald* 29.5.74

eopard crawl *n. and vb. Sect. Army.* A particular method of stalking, moving through low cover keeping flat propelling oneself forward by the elbows and knees while balancing the rifle held above the ground in both hands: as *vb.* to stalk in this manner *O.I. Serviceman* (demonstrated 14.4.79)

epel, (lê) ['lɪəpəl ˌle:] *n. or vb. phr. colloq. slang lit.* 'spoon' *usu. vb. phr. to lie, lê lepel, sig.* to lie in bed facing in the same direction fitted into each other, like spoons beside each other in a drawer. [*Afk. lê cogn.* lie *lepel* a spoon]

Patsy and Jen slept together in one bed with Patsy's legs warmly tucked into Jen's. Lepel, they called it. Roberts *Outside Life's Feast* 1975

êsiekte ['le:ˌsɪktə] *n.* See also *krimpsiekte* and *nenta: Cotyledonosis,* a wasting disease of stock esp. goats. [*Afk. lê cogn.* lie + *siekte* disease]

Lêsiekte according to the book on medicinal plants not only affects small stock . . . There are

recorded deaths of fowls dating back to 1909, and horses and bovines. *E.P. Herald* 6.3.78

In combination ~ *veld*

. . . it is thought that goats are most susceptible because conditions in lêsiekte veld are suitable practically only for goat-farming. *Ibid.*

Lesotho [lə'sutŭ, le-] *n. prop.* The country formerly known as Basutoland: see also *Botswana, Kwazulu,* see quot. at *makulu.*

let *vb. trns. substandard.* To cause to do, to make, not equiv. 'permit' or 'allow': as in 'Order the things and let the shop deliver them'. [*translit. Afk. laat* to cause to do, make ; *cogn. O.E. forlaetan*]

I don't want to give up work – I won't until my doctor lets me. O.I. (pregnant) 1969

leting ['leˌtɪŋ] *n.* A Sotho drink, not as strong as *Kaffir beer* (q.v.) [*Sotho leting*]

Before the day had ended a bag of green mealies and a large pot of leting, which is a milder drink than kaffir beer, found its way to our doorstep. Pohl *Dawn and After* 1964

liedjie ['lĭkĭ, -cĭ] *n. pl.* -s. An Afrikaans folk song: see also *boeremusiek,* second quot. at *babbie shop* and quot. at *meisie* [*Afk. fr. Du.* lied song *cogn. Ger.* Lied(er) + *dimin. suffix* -jie]

The songs, in the Afrikaans language, were plainly old folk-songs, known in South Africa as 'liedjies'; but not one of the listeners could recognise either tunes or words. Green *Where Men Dream* 1945

Afrikaans liedjies and some of the Bantu singing that made an impression during the Royal Tour would be a feature of a £25,000 'Meet South Africa' exhibition due to open in London on March 18. *E.P. Herald* 6.3.48

lieg [lĭx] *vb. colloq. (vulgar)* To lie. [*Afk. cogn.* lie *vb. also n.*]

'No . . . This time you lieg.'

'Don't say to me I lieg! . . . I know what I'm doing.' . . .

'Lieg your soul into hell for enough to live.' Fugard *Boesman & Lena* 1969

lightie, lighty ['laɪtĭ] *n. pl.* -s -ies *Slang.* A young boy, also *la(a)itie* (q.v.): also Naval usage, [*fr.* light]

Witnesses further alleged that prisoners . . . had rank systems . . . and the lowest ranks called 'lighties' were used by senior convicts for acts of sodomy. *E.P. Herald* 8.8.79

lightning bird *n.* See *impundulu* and *mpundulu bird.*

He went on to speak of the Lightning Bird and other things which caused trouble. Vaughan *Last of Sunlit Years* 1969

likkewaan ['lɪkəˌvɑn] *n. pl.* -s. See *leguaan* and quot. at *kaaiman.* [*Afk. fr. Du. leguaan*]

lily *n. pl.* -ies. Also *piss* ~: a temporary plastic urinal in use on the Border or in desert conditions, bowl-shaped on a stem, hence name. (Various Informants). *cf. Austral. Army* rosebowl, desert rose. See quot. at *chicken parade*.

line fish *n. s.* or *pl.* Any of different varieties of fish caught by rod and line as opposed to trawling.

Bourride: 3 kg mixed line fish (reserve heads and carcasses) *Fair Lady* 8.6.77

links [lɪŋks] Left: see *hou* ~. [*Afk. links* left (direction)]

linnebaai [ˈlɪnəˌbaɪ] *n. obs.* A fabric of wool and linen used for clothing in the early days. *cf. Brit. and U.S. archaic linsey-woolsey:* see also quot. at *kombers.* [*Afk. linne fr. Du. linnen* linen + *baai* flannel, baize]

. . . plain and printed Muslins, Voerchits, Linnebaay, Sail Canvass. *Grahamstown Journ.* 1.8.1833

liretlo [ˌlĭretlɔ̃] *n.* Ritual killing as practised among the Basotho: see quot. also *ritual murder* and first quot. at *muti.* [*Sotho liretlo* ritual killing]

. . . the Moruti, the priest . . . denounced all killing as evil, and Liretlo, the ritual killing, as the greatest evil of all. . . . the clever lawyer . . . had surely made plain to the court that Liretlo was not murder . . . where men took life for some petty personal motive . . . That which had been done had been done for the good of all Lesotho – and was the proof . . . not clear for all to see? The crops were good and the land fertile. Fulton *Dark Side of Mercy* 1968

litre, liter *n. pl.* -s. A measure of cubic capacity in S.A. see *metrication.* [*prob. fr. Fr. litron obs. measure of capacity*]

210 litre freezer R10,32 monthly 210 litre refrigerator R9,42 monthly. *Advt. circular* received 28.5.74

Deep-Freezers Big '15' 428-litre *Pretoria News* 3.10.74

little *adj.* Used redundantly in S.A.E. as direct transference of *Afk.* dimin. form of noun following *klein* (small) e.g. 'n klein blik*kie* a small tin. [*fr. Afk.* dimin. *suff.* -tjie, -jie, -ie etc.]

You think I haven't got secrets in my heart too? That's mine. *Sies!* Small little word, hey. *Sies.* But it fits. Fugard *Boesman & Lena* 1969

What's the use of buying a small little tin like that? O.I. *Supermarket* 1972

loan place, loan farm *n. pl.* -s. *hist.* A tract of land, originally granted at an annual rent of twenty-four *rixdollars* (q.v.), the

tenure being a lease in perpetuity; som͏ ~*s* became quitrent estates: see als͏ quot. at *improvement.* [*trans. Du. leen ing(s)* loan + *plaats* farm, place]

Land Tax or Rent of Estates, which (wit͏ the exception of a few freeholds) are all hel͏ of Government under the denomination o͏ Loan Lands or Places, and at a rent for eac͏ place of twenty four rixdollars per annum, an͏ the regular payment of this rent insures ͏ perpetuity of the lease. Ewart *Journal* 1811–181͏

Every holder of a loan place, on his makin͏ application by memorial to the governmen͏ for the purpose, shall have a grant on his place on *perpetual quitrent. Cape Statutes* 6 Aug 181͏

In 1717 the Company decided to halt th͏ issue of freehold land. Instead the farmer coul͏ obtain a 'loan farm' in return for an annua͏ rent. De Kiewiet *Hist. of S.A.* 1941

lobola [ləˈbɔʊlə, ˌlɔˈbɔla] *n.* The brid͏ price, usu. in cattle, paid by an Africar͏ man to the parents or guardians of hi͏ prospective wife: see quot.: used attrib͏ in ~ *system,* ~ *cattle.* [*Ngu. ukulobol͏* to give dowry]

My girlfriend and I want to get married. My problem is that I do not have cattle for lobola *Drum* August 1971

According to tradition it is the divine righ͏ of kings for the Prince to take whoever h͏ pleases to the royal place, Kethomthandayo ͏ which means choose the one you fancy . . Zulu monarchs are exempt from paying lobola It is the nation that has to pay it. *Drum* 18.4.7͏

Do we also need to obey these laws? Do we have to subsidise the lobola of chiefs or chief tainesses who marry at the drop of a kaross? *Voice* 8-14.11.78

location *n. pl.* -s. **1.** *hist.* The land grantec͏ to a *party* (q.v.) of Settlers usu. namec͏ after the party, e.g. James's Party; als͏ used of the land of an individual settle͏ see first quot. *cf. Canad. location,* parce͏ of Government land applied for anc͏ granted for settlement; *Austral. block section* (*also N. Z.*) lot into which land i͏ divided for settlers by Government [*fr. Lat. locatus fr. locare* to place]

Monday 25 (December 1820): Expected M͏ Bird, it threatened rain, he did not come – th͏ D(algairns) went to Mr Bailey's location to ͏ dance. Sophia Pigot *Diary* 1819–1822

. . . could not fail to give a spur to the exer͏ tion of the Settlers. As the nights had consider ably shortened, and the Caffers had ceased t͏ harrass the Locations the people were allowe͏ by degrees to separate and settle in smalle͏ parties. *Greig's Almanac* 1831

2. A segregated area on the outskirts o͏ a town or city set aside for black, usu͏ African, housing and accommodation͏ which whites require a permit to enter: see also *township*. [*fr. Lat. locatus fr. locare* to place. ⟦*poss. by transference from* (1) ~ *see Shaw* 1847]

The Kaffir Chiefs have entire confidence in him; and although a great soldier, he is a sincere lover of peace. I go to Kaffirland again on the 7th of January, at the final arrangement of native locations, Missionary-stations etc. in British Kaffraria. Shaw 28.12.1847 *cit.* Sadler 1967

'Is the Moruti staying in Johannesburg?' 'Yes, I stay at Orlando Township – that big location we are now approaching. Lanham *Blanket Boy's Moon* 1953

A Bantu Council for a specific residential area, such as a location, may be established at the instigation of the local authority, but can also be requested by a Bantu Advisory Board. *Personality* 23.10.70

ocks *pl. n.* Also *lox*: Woolbroker's term, the lowest grade consisting of shearing shed sweepings, dung-soiled parts of fleeces. *cf. Austral. dag(s), usu. pl.* (lock of wool clotted with dung); also *locks* small pieces of wool that fall off during shearing.

Locks: the lowest or most inferior line of wool made when wool classing and includes sweatlocks, stained wool, second cuts and shankings. *Vet. Products H'book* (I.C.I.) (no date)

ocust bird *n. pl.* -s. Also known as *sprinkaanvoël* (q.v.) any of several species which destroy and feed on locusts including the European white stork, the *great* ~ (Ciconiidae fam.).

We are happy to be enabled to announce that the Locust Bird has at last visited this district in such numbers, that there is every prospect of a deliverance from the Locusts now in their larval state. *Grahamstown Journ.* 24.2.1832

erie ['luərĭ, 'lu:rĭ] *n. pl.* -s. Afk. and usu. S. Afr. form of lourie, lory: any of several of the Musophagidae (*Loriinae*): parrot-like touracoes of brilliant plumage: esp. the *Knysna* ~ of the Knysna forests, the *purple-crested* ~ of Natal and the *grey* ~, *kwevoël* or *go-away bird* (q.v.). ⟦In combination also *vlei* ~, the *rain bird* (q.v.) *Centropus superciliosus burchellii* of the Cuculidae, and the *bush* ~ *Narina trogon* of the Trogonidae: also *Austral. lory*. [*Afk. fr. Du. ex Malay luri, form of nuri*, parrot]

In the aviary, I saw the Touracoo, called Loeri by the colonists. Burchell *Travels I* 1822
. . . the lovely 'lourie' wing, which he tells me is obtained from a rare bird in the Knysna

forests . . . When flying in the sun, they glitter all over like burnished metal with the lustrous green and deep claret hues of their feathers. A Lady *Life at Cape* 1870

Also found in place names Loerie, Loeriesfontein.

At night we came to Lory's River, so called from a species of parrot, which is found there. Masson *Botanical Travels* 1776

Long Cecil *n. prop.* The long-range gun built by De Beers in Kimberley during the siege of the city in the Anglo Boer War. [*fr.* Cecil Rhodes, who ordered it.]

. . . the 28-pounder gun, the famous 'Long Cecil' improvised by De Beers engineers. Every year a military ceremony commemorates the raising of the siege. *Panorama* Aug. 1977

longdavel ['lɒŋˌdavəl] *n. pl.* -s. A *rondavel* (q.v.) lengthened by a rectangular section between two semicircular ones, resulting in a building of roughly oval shape. [*'portmanteau word' long + rondavel*]

'RED-TOP' PREFABRICATED STEEL RONDAVELS complete with door and window . . . 'RED-TOP' STEEL LONGDAVELS comp. 1 door, 2 windows *Farmer's Weekly Advt.* 20.3.74

long drop [lɒŋ drɒp] *n. pl.* -s. *colloq.* A pit privy: *cf. Austral. outhouse*, etc. Also *Sect. Army 'longdrop* – toilet on the border [self explanatory]': *Informant* H. C. Davies *Letter* 2.2.69. ⟦ This term cannot be traced in any available dictionary American or English, slang or conventional, except in the C18 sense *the drop*, sig. a gallows.

In October 1972 they dug up the old pit lavatory or 'long drop' near the cottage . . . But it turned up no bones or any further clues. *Het Suid Western* 7.2.79

longsiekte ['lɔŋˌsĭktə] *n.* See *lungsickness*. [*Afk. long cogn.* lung + *siekte fr. Du. ziekte* disease]

looi [lɔɪˌlŭĭ] *vb. Sect. Army.* To thrash, beat, get the better of. [*Afk.* lash, lick, thrash]

Another word is looi (thrash). You 'looi' the enemy, you 'looi' a job. *Panorama* Jan. 1978

loop [luəp] *vb. usu. imp.* Go! Run, *imp. sig.* 'go away', or else, 'march': used by wagon drivers to their teams: see also *hamba. cf. Canad. mush (fr. marchez).* [*Afk. fr. Du. lopen* walk *cogn.* lope]

Without delay, the drivers clap their long whips and . . . loudly call out to the oxen, Loop! and instantly the whole of the caravan are again in motion. Burchell *Travels I* 1822

I was furiously assailed by dogs and the shrill

voice of an old woman intimating that the master was not home, and desiring me also to 'loop', or take myself off. Alexander *Expedition I* 1838

Then this morning: *Loop, Hotnot!* Just had time to grab our things . . . I didn't even have on a *broek* or a petticoat when we started walking. Fugard *Boesman & Lena* 1969

looper ['lʊəpə(r), lupə] *n. pl.* -s. Large sized buckshot or slugs. [*Afk.* slug]

A Corporal Walker, who sat among the number, remarked 'I will put some loopers' (ounce ball cut into four parts) 'into my musket, and if any Kafir comes within range, he will not run far afterwards. MacKay *Last Kaffir War* 1871

. . . he, forgetting that he had taken the big 'looper' cartridges from his gun and reloaded with No. 6, fired. FitzPatrick *Jock of Bushveld* 1907

los [lɔs] *adj.* and *vb. slang.* Loose: in combinations ~ *my* equiv. of Let go!; (*vulgar*) *a piece of* ~ a promiscuous girl or woman; and ~ *hotnot* (q.v.). [*Afk. los* loose *adj.* and *vb.*]

'I reckon it should come up higher,' he says, sliding his fingers upward . . . 'I'm warning you, Charlie, los my . . .' I screech. *Darling* 27.10.76

los hotnot ['lɔsˌhɔtnɔt] *n. pl.* -s. *colloq.* One who is 'footloose and fancy free', without responsibilities, unoccupied. [*Afk. los hotnot* one free of responsibility, holiday maker, also free-lance, grass widower *etc.*]

Erika Theron happy to be a 'los hotnot' . . . their gentle and matronly Chairman [of the multiracial Erika Theron Commission] wistfully mentioned how much she was looking forward to the day that she'd be a 'los hotnot' again. *Sunday Times* 9.5.76

losieshuis [lŭ'sïsˌhœïs] *n. pl.* -s. *colloq.* A lodging or rooming house: used by *Eng.* speakers of hotels; 'More like a third grade ~ than a three star hotel.': also of persons continually required to accommodate all and sundry house guests; 'Her friends/the firm/the community/use her as a full time ~.' [*fr. Afk. loseer* to board, lodge, *huis cogn.* house]

Mother and daughter were talking in a room in Ben's Losieshuis, the boarding house that Mrs Clarke ran. Bosman *Willemsdorp* 1977

loskop ['lɔsˌkɔp] *adj.* and *n. slang.* Crazy, scatty, forgetful: see also *malkop.* [*Afk. lit. los cogn.* loose + *kop cogn. Ger. Kopf* head]

Please accept my apologies with a solemn undertaking not to be so loskop next time.

Letter 19.4.77

lossie ['lɔsï] *n. pl.* -s. *slang.* A floozie, see at *los.* [*fr. Afk. los cogn.* loose + *dimin. suffix*]

'. . . ? That's the club where tired-out business men go to pick up lossies' *O.I. Girl*, 20. 30.11.78

lounge *n. pl.* -s. *Ind. E. poss. usu. reg. Natal.* A restaurant (Indian) is so designated e.g. Bhagat's Vegetarian Lounge, Peter's Lounge, Victory Lounge (Durban). ⫽Now also Albany Lounge Grahamstown. [*unknown*]

Now that the building is finished they're making a director's lounge on the top floor so we don't have to go out for meals. O.I. (Indian) Durban Nov. 1972

Louis *n. prop. Sect. Army.* Pet name for the train which takes National Servicemen home after completing their period in the forces.

Louis – the mindae train. *Various O.I.s Servicemen* Jan.-Mar. 1979

love gap *n. pl.* -s. See *passion gap.*

lowveld [felt] *n. Afk. laeveld:* the subtropical area of the N. and E. Transvaal where malaria is endemic: the ~ *climate* of intense heat and damp is generally considered unhealthy. [*trans. Afk. laeveld*]

The swamp-miasma and the tsetse-fly killed off his sheep, cattle and horses; while the lowveld mosquito gave malaria to the members of his clan. Mockford *Here are S. Africans* 1944

The lowveld is sub-tropical . . . Altitude about 500 to 2,000 feet. Frostless. Temperatures up to 120°F in the shade during summer. Rainfall 15 to 20 inches in north increasing to 20–25 in southern portion. Area occupied by dry savannah-like type of bush consisting mainly of Acacias. King *Tree Planting* 1951

lucky bean *n. pl.* -s. The black-eyed scarlet seeds of *Erythrina caffra* (Kaffirboom), sometimes called *Kaffir beans* (q.v.) and those of *Abrus precatorius* which are used as charms, are both so designated. ⫽The former are freq. sewn like beads into mats, necklaces or other ornaments.

luilekker- [ˌlœï'lekə(r)] *adj. prefix.* Easeful, luxurious: prefixed to other items ~*lewe* (see *lekkerlewe*) ~*land* the mythical country of *ho autonomos bios.* ⫽ ~*land* is portrayed in Breughel's painting of the same name. [*Afk. fr. Du. lui* lazy + *lekker* pleasant, sweet (*lewe cogn.* life)]

He had to work in order to live, but his work was so uncommonly like play that, not without reason, the district was named 'Luilekkerland'. Marais 1914 *cit. Road to Waterberg* 1972 'Don't overdramatize. Snap that defence mechanism of yours out of top gear. Drinking may be part of your *luilekkerlewe* (sweet life) but you're not drunk.' Jenkins *Bridge of Magpies* 1974

lungsickness *n.* Pleuro-pneumonia, a highly infectious disease of cattle, and horses. [*trans. longsiekte* (q.v.)]

There is a terrible murrain, called the lungsickness, among horses and oxen here, every four or five years, but it never touches those that are stabled, however exposed to wet or wind on the roads. Duff Gordon *Letters* 1861–62

Lung sickness, 'longziekte', or *pleuropneumonia*, a highly infectious disease in cattle, is one of the most severe stock scourges in the Colony. Wallace *Farming Industries* 1896

lus [lœs] *adj.* and *n. slang* Having a longing for, a strong desire. [*Afk. lus vir* desirous of, wanting]

Being with you and talking Japanese again gives me a terrific lus for it – to go back there, for the language. *S. Afr. Lawyer* (*ex London*) 25.4.78

lynx See *rooikat*[1] and *rooikat*[2]

M

Ma- [ma] *prefix. Afr. E.* 'Mother of' or 'Mrs', prefixed to a woman's name. ¶In *Sotho*, used both as 'Mrs' and as a clan prefix. In *Zulu*, often prefixed to the maiden name of a married woman. Thus *Ma-M sig.* 'daughter of the M-family' though her married name may be Mrs N. In English-language African fiction, usu. simply 'Mrs' or 'mother of' see third quot.

Our eldest boy being named Robert, Mrs Livingstone was, after his birth, always addressed as Ma-Robert, instead of Mary, her Christian name. Livingstone *Missionary Travels* 1857

George had been engaged to play the piano for the two days by Auntie Ma-Ndlovu. The drinks, skokiaan and other concoctions were sold in a room adjoining. Dikobe *Marabi Dance* 1970

. . . it was because of Ma-Ndlovu and Mkhulu that Thembi's mother came to know the whole story . . . Ma-Thembi was wise every time to have a good look at these gentlemen. *Drum* 22.9.73

maagbom ['maːxˌbɔm] *n. pl.* -me(n). *prob. obs.* Flour dumpling. See *stormjaer* (2), *cf. Jam. Eng. stick-in-the-middle* (dumpling), *Austral. buck-jumper.* [*Afk. fr. Du. maag* stomach + *bom cogn.* bomb]

maak gou [ˌmakˈxəʊ] *imp. vb. phr. colloq.* Make haste: equiv. 'hurry up', 'buck up' etc. [*Afk. maak* do *cogn.* make + *gou* quick(ly)]

'Kaartjies, tickets. Come on you black skelms. Maak gou.' Fulton *Dark Side of Mercy* 1968

maar [ma(r)] *adv. colloq.* Just; only; merely. Often redundant in S.A.E.: see quot. at *ask*. In combination *toe* ~ (q.v.). [*Afk. maar* but, yet, only, just]

'Ag, ja wat, lady. It's all right. I'm just *maar* thinking what it is that you must send to the Transvaal for.' Kavanagh *Merry Peasants* 1963

If they want more land, why don't they maar apply to the Land Bank. *Daily Dispatch* (*Cartoon Caption*) 6.5.72

maas [mas] *n.* Also *amasi, amaas:* thick, naturally soured milk, a favourite food-beverage of Africans: now a commercially available dairy product. *cf. Jam. Eng. bani* (*kleva*). Also used in cookery (see *dhai*) and for making ~*kaas* (*kaas cogn.* cheese) cottage cheese, also obtainable commercially under this name. [*Ngu. amasi* sour milk]

. . . Umslopogaas brought Nada the Lily maas to eat and mealie porridge. She ate the curdled milk, but the porridge she would not eat. Haggard *Nada* 1895

. . . it seems a great pity that the urban African should be deprived of the cheap, healthy nourishing 'maas' he has been able to make in the past from naturally soured raw milk. *E.P. Herald* 11.5.72

If you like yoghourt but find it rather expensive, try making maas as the Zulus do. *Darling* 28.4.76

maasbanker [ˌmasˈbaŋkə(r)] *n. pl.* Ø, -s. The horse-mackerel *Trachurus capensis*, a marine fish economically important in S.A. and used instead of herring for making kippers etc. also for *bokkems* (q.v.) [*fr. Du. marsbanker* horse mackerel]

Some harders and maasbankers were hung to dry as 'bokkoms' in the wind and sun on lines stretched between posts near the fishing villages. Grindley *Riches of Sea* 1969

maaskaas ['masˌkas] See *maas.*

maat [mat] *n. pl.* -s. *colloq.* Friend; mate; chum. Often *ou* ~ (old): see also *chommie, pellie, (ou)boet: cf. Austral. mate.* [*Afk. fr. Du. maat* friend, comrade, *cogn.* mate]

Many of the Bechuanas selected maats or comrades, after their manner, from among their allies, presenting, in a formal manner, an ox to the individual pitched upon. Thompson *Travels I* 1827

'. . . ou maat,' 86-year-old Mr. . . . said as he nearly dislocated my shoulder with his hand-shakes. *Drum* 8.12.72

mabela [ˌmaˈbeːlə, -la] *n.* Millet, *Sorghum caffrorum*, usu. called *Kaffircorn*, (q.v.) extensively cultivated orig. by Africans for brewing of beer from the malted grain: also available ground as ~ *meal*, malted or plain, for making porridge. [*Ngu.* amabele (*pl.*) grain sorghum, millet] ⁋ *acc.* Zu. legend the grain was discovered growing wild and fed to a woman by a jealous rival who wished to poison her. Instead she grew plump and better looking than ever, and its nutritive properties were recognized.

At length she asked her if the Amabele was nice. She replied 'Nice indeed!' And from that time the women cultivated Amabele and it became an article of food. Callaway *Religious System of the Amazulu* 1884

madala [maˈdɑla, -lə] *n. Afr. E.* Old one: a mode of address among workmen to the oldest among them (not necessarily the chief): *cf. Brit. gaffer.* See also *abadala.* [*Ngu.* -dala old, aged]

Now I myself have become old, you do not give me, an old madala, time to stop working, because without work, and you, I myself would die. *Forum* Vol. 6 No. 2 1970

madoda [maˈdɔda] *pl. n. colloq. Afr. E.* Mode of familiar address, *abbr. amadoda* (q.v.) men.

One man a father of three teenage daughters said: 'Madoda, it is doubtful that a Miss Sexy contest could be the idea of a woman. The whole thing smirks [sic] of mischievous men . . .' *Indaba* 6.4.79

madolo [maˈdɔlɔ] *n. Afr.E. prob. slang.* Wine: *cf. Austral. plonk.* [*unknown poss. fr.* Zu. amadolo knees *cf. sgomfaan*]

. . . he has a nipinyana of 'madolo' which is the name non-voters prefer to call wine and which wine he buys . . . at the bottle store. . . . begins to partake of the 'madolo' and I do the same with the ndambola. *Drum* 8.3.74

madressa [məˈdresə] *n. pl.* -s. Not exclusively S.A.E. A Muslim school in which Arabic and the Quran are taught by the *moulvie* (q.v.). [*Arab. madrasah*]

This will be flanked by a double storey Madressa, consisting of nine classrooms, a library, caretaker's residence and offices. Mur-al-Islam Masjeed Appeal, Lenasia, Nov. 1970

Since the beginning of this year, our local Madressa has been held in a classroom. What-ever the children are taught in Arabic, is also translated for them into Afrikaans. *Drum* June 1971

madumbi [maˈdumbĭ] *n. pl.* -s. *Arum esculentum:* a native plant of the East yielding starchy tubers cultivated by Africans in Natal and Mozambique since before the coming of Europeans to S.A. [*Zu. (a)ma pl. prefix* + (*i*)*Dumbi* tuber]

Madumbies (*Colocasia esculanta* (L) *Schott*) are grown in warm humid areas with a good rainfall . . . It has underground organs similar to a canna or artichoke. The mother madumbi is fibrous and is usually used as seed while those radiating from it are eaten. *Farmer's Weekly* 7.11.73

maffick [ˈmæfɪk] *vb. intrns.* To celebrate with noise and rejoicing. [*back formation from vbl. n.* mafficking *fr. Mafeking*]

We trust Cape Town . . . will 'maffick' today, if we may coin a word, as we at home did on Friday and Saturday. *Pall Mall Gaz.* 21.5.1900 cit. O.E.D.

The news that Mafeking had fallen reached London at 9.30 p.m. on Friday, May 18 (1900). The siege of seven months was over, and all England was in an uproar. London had gone mad and the verb 'to maffick' was born. Cloete *Rags of Glory* 1963

mafufunyana, mafufunyane [ˌmaˈfʊfʊnˌjana, -e] *n.* Condition of extreme hysteria, usu. in women, freq. self-induced by belief in witchcraft. ⁋Often cited as an extenuating circumstance in criminal cases. *cf. Canad. piblokto*, 'Eskimo madness': hysteria esp. among women in the dark winter. [*Bantu:* Zu. ufufunyana/e disease causing delirium and mania]

. . . a girl . . . who was suffering from Mafufunyana, a type of hysterical condition, which residents in the Tsomo district believed to be induced by witches. The court was satisfied that the murder was committed because of the belief of the people who did it that . . . was a witch who was causing the girl to suffer from 'mafufunyana'. Such a belief has been repeatedly found to constitute an extenuating circumstance. *Daily Dispatch* 3.3.70

Mr . . . said his sister . . . had suffered from mafufunyana since last November . . . they approached a sangoma Miss . . . to help and drive away the evil spirits. *Indaba* 6.4.79

mafuta [ˌmaˈfuta] *n.* (anglicized *pl.* -s). Fat(s) [*Bantu pl. n.* (*a*)*mafutha*, fat] **1.** Mode of address to a fat person, cf. 'Fatso', or 'Fats': see also *vetsak.*

She glanced back over her shoulder into the kitchen. 'Mafuta!' 'Nkosikazi'. A round black face, split by a dazzling smile, peered out

of the pantry. Collins *Impassioned Wind* 1958
2. *pl.* Fats of various real or mythical
creatures e.g. lion, *likkewaan* (q.v.) *tikoloshe* (q.v.), crocodile etc. part of the
stock-in-trade of the *witchdoctor* (q.v.)
or *herbalist* (q.v.). *cf. Jam. Eng. balm
oils; oil of consolation, oil of Virgin
Mary* etc.
  MEDICINAL AFRICAN Herbal College
for 20 years has taught the people how to use
African Mutis and Mafutas and also wonderful
imported Herbs by correspondence. Low fees.
Easy terms. *Post Advt.* 23.5.71

magageba [ˌmagaˈgeba] *n. Urban Afr.
slang.* Also 'magegebes': money. prob.
flaai taal, fly taal (q.v.) [*unknown, see
mali*]
  Kid Fall is in the right Xmas mood – or
rather spirit . . . and boasts that we will never
finish his boodle. He explains this by saying
that he is in the magegebes. Casey Motsitsi
cit. Drum 8.1.74
  . . . a stranger to the tsotsi's dangerous world
could still save his throat if he has some knowledge of basic words and phrases: . . . Magageba
– money. Venter *Soweto* 1977

Magaliesberg [maˈxalisbɜg, -berx] *n. prop.
usu. as modifier.* A Virginian tobacco
grown in the Magaliesberg area. [*fr.
name of mountain range*]
  The pipe tobacco manufactured from Magaliesberg leaf might be described as a mild smoke.
Taylor *Tobacco Culture* 1924 *cit.* Swart *Africanderisms Supp.* 1934
  – all the time that I was filling my pipe from
a quarter-pound bag of Magaliesberg tobacco;
the sort with the picture of the high-bounding
blesbuck on it. Bosman *At his Eest* 1965

mageu See *mahewu.*

magtig, magtie [ˈmaxtɪx, -ti] *interj. colloq.*
An exclamation *usu.* of surprise *equiv.*
'Good Lord', 'Lawks' etc. poss. *abbr.*
to avoid any suggestion of blasphemy:
see also *Here.* [*abbr. Afk. allemagtig
cogn.* almighty]
  . . . an old Hottentot sat down . . . and swore
with a round oath (almagtig) that he would not
go back. Alexander *Western Africa II* 1837.
  And I'm afraid of a mad person, ou Baas.
Magtig, but I am afraid! Meiring *Candle in
Wind* 1959.

mahem [ˈmeɪˌhem, ˈmɑ-] *n.* The crested
Balearic crane *Balearica pavonina regulorum:* one of those birds known as
bromvoël (q.v.), named from the sound it
makes. [*Xh.* (a)*ma pl. prefix* + (i)*Hemu*
crested crane, *onomat.*]
  . . . they hold in high estimation a beautiful
crane . . . they call it maahoom from the noise

it continually makes. Philipps *Albany & Caffer-
land* 1827.
  . . . and if a person kill by accident a *mayhem,* (or Balearic crane) . . . he is obliged to
sacrifice a calf or a young ox in atonement
Thompson *Travels II* 1827
In combination ~ skuif [*Afk. kuif* top-
not] a grass *Rendlia altera.*

mahewu [maˈxeŭ] *n.* A drink made of
thinned, slightly fermented mealie-meal
porridge: various *sp.* (Now commercially available in cartons, see third quot.;
also as a 'mix'.) ⫼ The *h* is frequently
rendered *r,* in this word, see quot. at
mqombothi, and the sp. *maheu* is also
used. [*Bantu: various sp. Zu.* (ama)*hewu*
fermented porridge drink]
  . . . they drank mamaghew which is made
from mealie meal and water . . . Mamaghew is
made and left overnight before it is drunk.
Daily Dispatch 18.8.71
  . . . the children had interrupted their lessons
at about 10 am on Wednesday for the traditional drink of 'maheu' the vernacular name
for their nutritional drink. *Evening Post* 2.11.73
  The number one food for health and strength
MAGEU NUMBER 1 – mieliemeal, wheat and
sugar. *World Advt.* 24.10.76
INSTANT POWDERED MAHEWU
  It's the fantastic new thirst-quenching Mahewu that is ready to mix and drink immediately
. . . It fills and satisfies! *Indaba* 26.1.79

mahog(a) [maˈhɔg(a), -hɒg-] *n. Afr. E.
slang.* Brandy or poss. other brown-
coloured spirit. *cf. Brit. dial. mahogany,*
strong brandy and water; (Cornish for
gin and treacle). See also quots. at *ha-ja*
and *straight.* [*prob. abbr. mahogany*]
  . . . each time he has a few slugs of mahog
mixed with be-ah under his belt he begins falling all over the place. *Drum* 8.1.74

mailship *n. pl.* -s. The weekly, formerly
passenger ship of the company contracted to the S. Afr. Government to
carry mails to and from Britain. ⫼The
~ constituted S.A.'s main link with
Europe before air travel became general:
see second quot. at *overseas.*
  The departure of the mailship on Wednesday
afternoons never fails to arouse an answering
thrill in the hearts of spectators crowding the
Ocean Terminal to watch this majestic sight.
Panorama July 1971
  And as the mailships are dying out so are the
people who sailed in them to summer in South
Africa . . . ninety percent . . . are dead. And the
newer generation jet in on far briefer visits.
Sunday Times 16.5.76

mainman [ˈmeɪnˌman] *n. pl.* -ne, -men.
slang. usu. a local hero in a school or

university etc. on grounds of sporting prowess or other prestige: also *mynman* (q.v.) *cf. manne, die*

Goodness yes, I remember him – he was one of the mainmanne when I was a student. He won a donkey race because his legs were long enough to do the walking! Really one of the 'big bulls'. *O.I.* 9.4.79

maiza [ˈmaɪza] *n. Afr. E.* African beer (see *Kaffir beer, tshwala, mqomboti, KB, Bantu Beer*), made from maize, see *mealie*, for consumption usu. in *beerhalls* (q.v.): see also quot. at *ai-ai*. [*presum. fr.* maize]

I went to a beerhall . . . As bold as a bachelor on pay day, I walked right in to one of those boozing temples and supped my maiza with the best of them. *Drum* Oct. 1970

Maiza is the greatest source of revenue for the city council. *Drum* 8.10.73

maizena [məˈziːnə, ˌmeɪ-] *n.(prop.)*. Cornflour (trade name). [*fr.* maize]

. . . thicken the soup with a little maizena dissolved in water, and a lump of fresh butter. Gerber *Cape Cookery* 1950

Mix the maizena, salt and lemon juice. Then add the orange juice and stir over boiling water until mixture thickens. *Evening Post* 27.10.73

majat [maˈdʒat] *n. Sect. Drug users* The lowest grade of *dagga* (q.v.): see quot. at *kachie, katchie* [*unknown, poss. Malay*]

The first grade is called 'Rooipoortjie', second grade is 'Jong Dagga' or 'Pieper' and the third grade is called 'Majat' in the Cape. *Drum* Aug. 1969

mak¹ [mak] *adj.* Prefixed to numerous plant names *sig.* a variety which is not wild, also occ. 'harmless' e.g. ~ *dagga*, ~ *boerboon*, ~ *ghaap* etc.: see *dagga, ghaap, boerboon*. [*Afk. fr. Du.* mak tame]

mak- in compound vernacular names literally has the opposite meaning of *wild(e)*, used in the sense of 'domesticated' or 'cultivated' or 'of cultivation' or 'growing near dwellings'. Smith *S. Afr. Plants* 1966

mak² *colloq.* Docile, easy-going, tractable *lit.* 'tame'. [*Afk. mak,* tame]

The Police Commandant might be so 'mak' and friendly as almost to entitle him to rank as . . . one of us; but . . . Prance *Tante Rebella's Saga* 1937

In combination ~ *Afrikaners,* ~ *Engelse*
For years the Progs have been accused of attracting only 'mak Afrikaners' to their ranks. *Sunday Times* 13.11.77

One of these days the so-called 'mak-Engelse' (tame English) will tell the likes of . . . where they get off. *Ibid.* 19.3.78

makataan [ˌmakəˈtɑːn] *n. pl.* -s. *Citrullus*

166

lanatus : see *kaffir melon, tsamma melon* and *karkoer*. [*Tswana makataan*]

Makataans, stock or kaffir melons and Tsamma are types of non-saccharine melons used for stock feed. This group is more drought resistant and does better in poor sandy soils than pumpkins. *H'book for Farmers* 1937

make *vb. trns. substandard.* Do, sig. prepare, cook ; of a raw material as opposed to a dish, as in 'I'm ~ing chicken, ~ing rice, ~ing sweet potatoes' see quots. *cf. Canad. make land, timber, fish* etc. sig. 'work'. [*poss. translit. Afk.* maak do]

. . . our South African dishes – things like sosaties, the real bredies, sweet potatoes as we make them, pumpkin too . . . *Sunday Times* 3.3.74

Spoil your husband. He'll love to taste a real old-fashioned *fresh* farm chicken like his mother used to make on Sundays. *E.P. Herald Advt.* 18.11.76

make a plan *vb. phr. prob. substandard.* In common use 'think of, arrange, fix up something', or 'work it out' even in quite trivial matters. [*translit. Afk.* 'n plan maak devise a scheme] ⁋This usage was noted by John Buchan in *Prester John* 1910. 'I grew angry . . . "Ek saī 'n plan maak", I told myself in the old Dutchman's words. I had come through worse dangers, and a way I should find.'

We will make a plan to show that uitlander a thing or two, nè? *S. Afr. Writing* Vol. 4 (no date)

'We gotta make a plan somehow,' Mr Baumgartner mutters, looking all hassled. Sales have dropped off something terrible since Christmas.' *Darling* 17.3.1976

make sections *vb. phr. Sect. Army.* '~ mate': a request for a share of another man's cigarette, *cf.* 'Give us a pluk, mate', asking for a single draw. *Informant Serviceman* 15.2.79

makgotla [ˌmaˈxɔtla] *pl. n. Afr. E.* 'Bush courts' based on the tribal courts of former times, now operating in urban areas allegedly in the interests of law and order to stamp out violence, gang warfare and other *tsotsism* (q.v.): see also quot. at *lekgotla*. [*fr. sing. (le)kgotla* (q.v.)]

Councillor . . . told the police officers that the residents had shown their full support of the makgotla. He said it was the parents who brought their children for flogging to the makgotla. *World* 13.5.74

'We formed the first lekgotla here in Naledi during December last year. There are now 20 makgotla in Soweto . . . More are being formed weekly.' How do the makgotla work? 'A person comes to the lekgotla to report that his son is giving him trouble. We send word to the boy to be here on Sunday when the lekgotla sits. At the court each side has his say, and the men of the lekgotla discuss the case and if the boy is found guilty, he is sjamboked. All in African tradition.' *Drum* 22.8.74

makoti [ˌmaˈkɔ̃tĭ] *n.* Mode of address or reference by the husband's people to the daughter-in-law in a Zulu household whatever her age, as the one who has married into it: see also *hlonipa*. [*Zu. makoti* bride]
. . . they sang the whole night, and had by custom to be fed by the people of *makoti* (the bride). Longmore *Dispossessed* 1959
. . . I am in love with a guy . . . Every Saturday he takes me to his parent's home. When his father and mother see me they say, hello Makoti. *Drum* 8.10.75

makulu[1] [məˈkulŭ, ˌma-] *adj. colloq.* Big, great: usu. in combination ~ *baas*, also ~ *trouble* etc.: see also quot. at *great*. [*Bantu -kulu* great]
Now was the heart of Monare lightened, for the Makulu Baas turned out to be the sort of white man who spoke his language, Sesotho, and who had travelled in his homeland, Lesotho. Lanham *Blanket Boy's Moon* 1953

makulu[2] See *atshitshi*, *zoll*(1)[4], *dagga* and *skuif(ie)* sk*yfie* (also *makulu*[1]).
My makulu (yeah, another name for dagga) is smooth like honey, no headaches with this stuff – not worried of doing time in the poky so long as there's makulu, I'm not worried of niks . . . and anyway there's more zoll inside the cell than outside.' *Darling* 8.11.78

Malay *n. pl.* -s. *usu. Cape* ~ : Muslims (Mohammedans) of Asian descent, living largely in the Cape (see quots.), now officially classified as *Coloured*, see *classify*[1].
The Malays form no inconsiderable part of the population of Cape Town. These have been brought at different periods from the Dutch settlements in Java and neighbouring islands. Ewart *Journal* 1811–14
In Cape Town the terms 'Malay' and 'Mohammedan' are often used as synonyms; but strictly speaking 'Malay' stands for that section of the local *Muslim* community in which the descendants of Malay slaves and political exiles are to be found. Du Plessis *Cape Malays* 1944

malgas [ˈmalˌxas] *n.* The common Cape gannet, *Sula bassana capensis* (Sulidae fam.) found on all S. Afr. coasts: also in place names Malgas and Malgasrivier. [*Afk. fr. Du. malagas, mallegas,* prob.

abbr. fr. Port. mangas de velludo sleeves of velvet *fr. black wing tips*]
. . . another mystery joins the sardines – the Cape Gannet commonly known as Malagas, 'Moras Capenses'. These gannets come all the way from West Africa. How then do they know the exact day of the arrival of the sardines? *Daily Dispatch* 25.9.71

mali [ˈmɑlĭ, ˈma-] *n. colloq. now poss. rare. Afr.E.* Money. *cf. Austral. moolah, splosh; U.S. dough.* [*Ngu. iMali* money]
. . . promised that a Kaffir runner leaving with despatches should bear it; but intelligent 'Ikona Mali' (No Money), and two others of his fraternity . . . were but broken reeds on which to depend. Du Val *With a Show II* 1882
Mali – a word in constant use among the natives, and frequently heard among the colonists also, for money. Pettman *Africanderisms* 1913

malkop [ˈmalˌkɔp] *n. pl.* -s. *slang.* A fool, one who is 'mad in the head', also a term of abuse *cf. houtkop domkop.* [*Afk. mal* mad + *kop cogn. Ger. Kopf* head]
'Do you know where this path leads?' 'Didn't the boy say?' 'He's a *malkop* – just that it went to the west' McClure *Rogue Eagle* 1976
'Come on, the officer's finished with you *malkop*' . . . 'Can't you see that?' *Ibid.*
In combination ~*siekte* (disease), the staggers in horses and other stock.

malombo [maˈlɔ̃mbɔ̃] *n. Afr.E.* A jazz idiom combining tribal rhythms with urban jazz: see quot. Combinations ~ *drum*, ~ *sound* etc., see also *mbaqanga*. [*fr. Venda see first quot.*]
Seldom have jazzmen been so reticent . . . Try as DRUM would they kept mum and said: 'Look, dad, the term "Malombo" comes from the Venda and means spirit. You've heard people . . . speaking of "soul" in jazz. Well, ours is based on tribal rhythms blended of course with urban township sounds.' *Drum* Nov. 1964
In our Black society, when the ancestral spirits speak through somebody, we beat the malombo drums as accompaniment for the spirit voices. I just happen to be a musical medium . . . We introduced the malombo sound successfully and now hope to make a record when we reach Europe. *Drum* 22.3.73

malpitte [ˈmalˌpɪtə] *pl. n.* The poisonous seeds of the *stinkblaar* (q.v.) *Datura stramonium*, which contain belladonna and which, when swallowed or smoked, are hallucinogenic: see quots. [*Afk. fr. Du. mal* mad + *pit* pip + *pl. suffix* -(*t*)*e*]
. . . stinkbaar . . . a weed which must be treat-

ed with respect however. Generations of South African school boys have known these seeds as *malpitte* because of the queer behaviour and delirium they produce. Green *Land of Afternoon* 1949
. . . a nightmare four-day hallucinatory drug trip induced by a common garden weed, known as 'malpitte' (mad pips), which grows wild . . . is the latest craze among South African teenagers and drug-takers . . . the pips . . . affected the brain and could do mental as well as physical harm. An overdose was sufficient to kill and the drug was as hallucinatory as the drug L.S.D. *E.P. Herald* 27.4.73

Angl. form *malpita:* see quot.
. . . chewing morning glory seeds and malpita which was originally imported . . . to prevent soil erosion. Then the power of the little black seeds was discovered, and it started eroding minds . . . Malpita can derange you for ever . . . You see people that aren't really there. You do things without knowing what you're doing. *Darling* 22.11.78

mama [ˈmama] *n. Afr. E.* Mother: a mode of address for older women used by younger Africans: see also *ma-, tata* and quot. at *A.N.C.*[1] *cf. Jam. Eng. Ma,* respectful address, title. [*Bantu form of address uMama*]

mamba [ˈmămbə, ˈmamba] *n. pl.* -s. Venomous snake of the genus *Dendroaspis* with a deadly bite: usu. in combination *green* ~, *D. angusticeps; black* ~ *D. polylepis.* [*Bantu iMamba* snake]
On ploughing through an ant-heap he cut a large mamba in two. Unfortunately the mamba, before dying, bit Botha on the leg. Botha expired within two hours. *E.P. Herald* 11.3.1926
. . . he tells me, the black mamba is by no means the world's most poisonous snake. 'He's in the top 10, I suppose, in about eighth or ninth position.' *Ibid.* 10.4.73

mamba, green *n. pl.* -s. **1.** See *mamba* **2.** See *green mamba.*

mambakkie *erron.* See *mombakkies.*

mamlambo [ˌmamˈlambɔ] *n.* A river snake mythical or actual, about which there are varied beliefs. ⍰That it is a beneficient fertility spirit for wealth and prosperity; a love amulet (*see etym.*), a malevolent force even causing death: see quot., also *tokoloshe* and *impundulu bird.* [*Ngu. umamlambo* river snake. *Zu.* 'water-snake *kept in the hut by women of a polygamous household to ensure the husband's favour (this is a current Native belief)*' Doke & Vilakazi *Zulu Dict.* 1948]
. . . these fairies of Africa are sinister . . . Sexually depraved and of insatiable appetite . . .

they exercise great powers of magic, cast spells and can change themselves into animals. The most common are Thikoloshe, Mamlambo and Mpundulu, . . . Mamlambo the mother of the river is a mythical snake that lives under the water . . . She escapes from the river as a beautiful woman and in this guise a man falls in love with her. Immediately he is in her thrall she takes up her abode in his kraal and causes his father, brother or uncle to die. Broster *Red Blanket Valley* 1967

mammajoor [ˈmămaˈjʊə(r)] *n. pl. Sect. Army.* Sergeant Major. ⍰ This is not a new notion: *cf. Brit.* World War II Song 'Kiss me goodnight Sergeant Major.' ['portmanteau' word, *mamma +* *Afk. majoor* major]
Should they fail to find everything in order their *mammajoor* (sergeant major) gives them *storings* or 'troubles', but if everything is in order they will feel *bakgat.* Picard en *Eng. Usage in S.A.* Vol. 6 No. 1 May 1975

mampara [ˌmamˈpara] See *mompara.*

mampoer [ˌmamˈpuːr, -ˈpʊə(r)] *n.* Home distilled spirit often called *peach brandy* made of the juice of peaches and numerous other soft fruits including wild *marulas,* medlars (*mispél*), prickly pears (*Opuntia*) and *karree* berries (all q.v.) [*Afk. acc. some fr. Mampuru, name of a Pedi (Sotho) chief*]
. . . the mampoer you make from kareeberries. . . . But karee-mampoer is white and soft to look at, and the smoke that comes from it when you pull the cork out of the bottle is pale and rises in slow curves. Bosman *Mafeking Road* 1947
The canteens carried a surprisingly large range of goods – from Cape 'dop' and 'mampoer' to imported liqueurs. *Sunday Times* 23.4.72

man[1] [mæn] *interj.* **1.** *colloq. interj.* Used for emphasis. *cf. U.S. (Oh) Boy!*
It's working perfectly, Man! I mus' say I like these old stoves. Maclennan *The Wake* 1971
2. *colloq.* A mode of address regardless of the sex of the person addressed and often redundant as in 'Come on man, Ma, we're all waiting for you', or replacing a proper name: equiv. use in *Jam. Eng.* see quots. at *skyfie*[1], *toe maar* and first quot. at *jong.*
'. . . you needn't screech like that, Myrtle, I'm not deaf, man.' *Darling* 3.9.75

man[2] [mæn] *n. pl.* men. Husband. [*Afk. fr. Du. man* husband]
In this house I saw the first trait of female industry, the vrows being employed in making clothes for their 'men'. Barnard *Letters & Journals* May 1798
The country lady strolls in to buy a hat . . .

tries on everything in the shop . . . tells them to put it down to her 'man's' account . . . Pay next year when the wool comes in. *Stormberg Mrs P. de Bruyn* 1920

manel [ˌmaˈnel] *n. pl.* -s. A black frock coat worn by elders (see *ouderling*) *diakens* (q.v.) and other officials of the *Dutch Reformed Church* (q.v.) usu. with a *white tie* (q.v.): see also *gatjaponner.* [*Afk. fr. Du. manille fr. Span. manilla* frock coat]
. . . no matter what sort of clothes I had on either. Even if I was wearing my black *Nagmaal manel.* I wouldn't fancy my chances much . . . on Judgement day on any part of your farm. Bosman *At His Best* 1965
. . . when Elder Landsman came back into the church he had a long black bottle half hidden under his manel. Bosman *Bekkersdal Marathon* 1971

maningi [maˈnïŋgï] *intensifier. colloq. usu.* Equiv. of 'very' etc. occ. as equiv. of 'lots of'. [*Zu. pl. prefix* (*a*)*ma* + *-ningi* plenty, many]
The gardener arrived at my study window and said it was 'maningi hot' and could he have shorts like the cook? *Sunday Times* 21.2.71

manitoka [ˌmænïˈtɒkə] *n. pl.* -s. *Myoporum insulare,* a native plant of Australia naturalized in S.A. esp. as a plant used for hedges near the sea. [*prob. Austral. aboriginal name*]
. . . the caravans were standing . . . with hedges of fleshy, narrow leaved manitokkas bushing out between the caravan sites. Philip *Caravan Caravel* 1973

manna [ˈmænə] *n. usu. suffix.* Various millets usu. in compounds, *boer* ~ *Setaria italica,* also called Italian millet; *red/rooi* ~, *white/wit* ~, *yellow/geel* ~ all *Kaffer* ~ (q.v.) *Pennisetum americanum* also *babala grass.* [*Eng.* manna species of grass *Setaria : Shorter O.E.D.*]
SEEDS – We are buyers for Manna and Millet. . . . MACHINE CLEANED AND AIR WASHED Imported Red Millet. Red Manna . . . Yellow Manna . . . Golden Millet . . . White Millet . . . Babala . . . *Farmer's Weekly Advt.* 20.3.74

manne, die [dǐ ˈmanə] *pl. n. phr.* Urban African slang lit. 'The men'; 'the bulls' 'mainmen' (q.v.) or inner circle of township gang leadership: see also *bra* [*Afk. die* the, *man*(*ne*) *pl.* men]
There was this . . . ou . . . who used to bully me, take my money, beat me up . . . He was one of 'Die manne' and I had no alternative . . . I also realized that if you're one of 'die bra's' no one else is going to mess you around. *Drum* 22.9.74

manslag (*Afk.*) See *culpable homicide.*

manyano [ˌmanˈjanɔ̃, -jɑ-] *n.* (*pl.* -s). An African women's church association of several sects, meeting on Thursday afternoons throughout S.A. each having a distinctive uniform: see also quots. at *doek* and *stockfel.* [*n. fr. Xh. vb. ukumanya,* to join, unite] ⟦Men's church unions are also known in some parts as ~, the uniform being usu. a waistcoat of the same colour as the jacket of the women's ~ of the same denomination.
On this particular occasion, the male equivalent of the women's Manyano is also present, although the number of male members is insignificant in comparison. Brandel-Syrier *Black Woman* 1960
Mia Brandel-Syrier . . . the author of a penetrating study of the *manyano* church women's associations (Black Woman in Search of God). *Sunday Times* 26.9.71
An energetic churchwoman, . . . has been chosen as the president-elect of the Grahamstown District Methodist Women's Manyano at the annual conference in Port Elizabeth. *E.P. Herald* 30.6.73

marabaraba See *morabaraba.*

marabi [maˈrabï] *modifier. Urban Afr. E.* Of or pertaining to a township music mode which preceded jive and rock-and-roll: usu. in combinations ~ *music,* ~ *band,* ~ *tunes,* ~ *dance.* [*prob. fr. place name Marabastad*]
She likes the Rolling Stones, Elvis . . . Satchmo . . . township jazz, and believes that nostalgic teenagers like herself are rekindling a national interest in marabi music . . . Here they danced to marabi tunes . . . earthy music. Venter *Soweto* 1977

marble *n. pl.* -s. *Sect. Army.* A concrete slab, part of the training equipment of a *Parabat* (q.v.): see also *hondjie* and *Aapkas.*
Like everybody else he had his 'marble' – a slab of concrete weighing 31,7 kg (70 lb), which is a Parabat's constant companion. *Sunday Times* 12.3.72

marewu See *mahewu.*

marimba [maˈrïmba, -rïm-] *n. pl.* -s. Any of numerous African xylophones. [*Bantu ma-* pl. prefix *rimba/limba* a note]

market master *n. pl.* -s. A municipal official in charge of the market in each town: the ~ *'s* report on current prices is freq. published in the press. see quot. at (2) *mevrou.*
UITENHAGE MARKET REGULATIONS The proceedings of the Market to be under the

controul of a Market Master . . . The following Tariff of Fees shall be exacted for the purpose of paying the Market Master a salary . . . One Farthing sterling on every Rix-dollar under the sum of Twenty-five Rix-dollars . . . *Greig's Almanac* 1831

MARKET REPORT. The Market Master, . . reports as follows on the sales held at Grahamstown for the week ending April 5. *Grocott's Mail* 11.4.74

maroela [ma'rŭla] *Afk.* form of *marula* (q.v.).

marsbanker See *maasbanker.*

marsh rose *n. pl.* -s *Orothamnus zeyheri* a very rare flowering species of the Proteaceae. [*fr. habitat and shape of flower*]

. . . the marsh rose a plant so rare that since its discovery it was twice thought to be extinct. According to Dr Pappe the Cape botanist who named the flower in 1848 Orothamnus zeyheri was discovered by that intrepid collector Carl Zeyher, 'in marshy places on the summit of the Hottentot's Holland Mountains in the month of July.' *Argus* 9.4.77

martevaan ['matə¡vɑːn] *n. pl.* -vanen, -s. Large earthenware storage jar used aboard Dutch East Indiamen: *Anglo-Ind. Martaban jar.* [*fr. place name Martaban*]

Martavanen were large earthenware jars of Chinese origin used on board Dutch East Indiamen for carrying oil and wine. They were mostly shipped from Pegu in the Gulf of Martaban, Burma, of which the name is a corruption. Gordon-Brown *S. Afr. Heritage* 1965

. . . a magnificent pair of Martevaans in brown glaze decorated with panels of flowers R625 the pair. *Cape Times* 26.1.73

marula [ma'rŭla, -ə] *n. pl.* -s. The deciduous tree *Sclerocarya birrea* (*S. caffra*) common in the hotter parts of S.A.: also a place name Maroelaboom. [*fr. Sotho name*]

The marula, with its straight bole and dense, graceful foliage, is a common sight in the lowveld of Natal, Swaziland and the Transvaal, and in Rhodesia. Palmer & Pitman *Trees of S.A.* 1961

Its fruit also known as ~*s* or ~ *berries*, used for jam and jelly and for making ~ *beer* and *mampoer* (q.v.), and which in the fermented state intoxicate birds and animals: see also quot. at *monkey orange.*

It was the season of the ripening of the marula, that yellow plum-shaped fruit which the natives call *umgana* or 'friend'. . . . the women and children of the village stooped below, gathering the fallen fruit into baskets and gourds for the brewing of marula-beer. Brett Young *City of Gold* 1940

I hardly saw any animals in the Kruger Park – except an elephant that had got drunk on marula berries. *E.P. Herald* 10.4.73

Mary *n. pl.* -ies. **1.** An Indian woman (now regarded as offensive by many): see quot. also (1) *Sammy.* [*prob. fr. n. prop. Mariamma*]

Because 'Mariamma' was a common name among Indian women and 'Munsamy' among the men, we were referred to as 'Marys and Sammies'. This was considered to be insulting. *E.P. Herald* 22.9.73

2. An Indian girl or woman fruit and vegetable hawker (*prob. reg.* Natal): see (2) *Sammy.* [*prob. fr.* (1) ~]

Their women folk, each called Mary, just as the Indian males were known as Sammy, hawked fruit and vegetables in flat baskets the design of which they had brought from their native land . . . half a century before. McMagh *Dinner of Herbs* 1968

3. Loosely: a Non-White woman. *cf. Austral. mary* (also *gin* or *lubra*), an aboriginal woman. [*unknown: poss. Austral. poss. fr.* (1) *and* (2) *Mary*]

Look at the colour-scheme that Mary's wearing – you and I couldn't get away with it. O.I. White Woman 1971

masala *n. pl.* -s. *Ind.E.* Curry spices: see *Indian terms.* [*prob. fr. Hindi mussalla* ingredients, curry stuffs]

Saffron, Spice, Masalas, Brassware, Ornaments. *Grocott's Mail Advt.* 18.6.76

Cumin seeds . . . used in curry powder (masala) mixture . . . Coriander seeds, powdered and used for a base for every mixture of masala. *Fair Lady* 21.7.76

Mashona [ma'ʃɔna] *pl. n. anglicized pl.* -s. A group of Bantu peoples of Rhodesia, speaking a number of closely related languages with a single literary form, see *Shona:* also used, prob. only by whites, of a single member of these peoples. In combination ~*land* [*pl. prefix* (*a*)*ma* + *Shona*]

The Mashona, however does not appear to be incapable of becoming an efficient miner . . . 'it was stated . . . that when the Mashonas are willing to work for longer periods, they become quite as efficient as the boys from outside the territory.' *Natives of S.A.* 1901

Mashona piano See *mbira.*

mason *n. pl.* -s. In S.A. a bricklayer, *Brit. stonemason*, poss. equiv. *U.S. mason* (who builds with stone or brick). [*see second quot.*]

Town Mason. H. Schutte *Afr. Court Calen-*

dar 1807

Mason – This word, influenced probably by the Dutch use of the word metselaar for both mason and bricklayer, is used all through South Africa, where, in England, the word 'bricklayer' would be employed. Pettman *Africanderisms* 1913

. . . work therapy provides the prisoner with a wide variety of types of work . . . masons, carpenters, electricians and many others. *Farmer's Weekly* 20.3.74

masonga [maˈsɔŋga] *pl. n.* A variety of caterpillar, the grubs of which are dehydrated and eaten by Pedi tribesmen. ▶ Their vernacular name relates them in habit to the various species known as *army worm* (q.v.)

The masonga . . . tastiest and most nutritious of all . . . are found by the million on the mopane, marulla and . . . syringa trees. The vernacular name for them is masotju (soldier) because they look like soldiers on the march. MacDonald *Transvaal Story* 1961

masoor *n. Ind.E.* Brown lentils used in making *biriani* (q.v.) see also *Indian terms. Anglo-Ind. mussoor.* [*presum. Hindi*]

GHEEWALA'S MASALA, CINNAMON, CLOVES, BORRIE, MASOOR . . . CHANA, DHAL, DHUNIA. *Advt. Argus* 29.6.71

mass *n.* Weight: under the S.Afr. system of *metrication* (q.v.) ~ is calculated in grammes and kilogrammes, and is not equiv. of *volume* (q.v.) see also *litre* and quot. at *middelskot.* [*S.Afr. Bureau of Standards term*]

Due to its light mass and flexibility the . . . pipe is ideal for irrigation systems. *Farmer's Weekly* 4.7.73

I am 20 years of age, 1,5m tall, mass of 75kg, have black hair, brown eyes. *Ibid.* 'Hitching Post' 11.7.73

over ~ overweight.

Are you overmass? . . . Test yourself by taking hold of a piece of fleshy skin over the lower ribs . . . Make a little roll of it . . . If the roll is more than 3cm thick you should lose mass. *E.P. Herald* 1.5.74

Master, the *n.* Regular *abbr.* for the Master of the Supreme Court.

All persons claiming to be creditors under this . . . Estate, are required to take notice, . . . that the Master has appointed the third meeting to be held before the Resident Magistrate. *Grahamstown Journ.* 2.3.1832

-master *n. suffix. sig.* 'Man in charge of . . . ' in combinations such as *market* ~ (q.v.) and *pound* ~ (q.v.). [*poss. trans. Afk. fr. Du. -meester* (q.v.)]

mat *n. pl.* -s. substandard *colloq.* Loosely.

sig. carpet and in *Afr. E.* linoleum also. [*poss. fr. Eng.* mat *or Afk.* mat floor rug]

Carpets: of course you have one in the lounge but remember to call it a mat. *Personality* 5.6.69

sleeping ~ a grass mat used by tribal Africans woven usu. of *Cyperus textilis:* see *matjiesgoed.*

Matabele [ˌmætəˈbilĭ, ˌmataˈbeːle] *pl. n.* anglicized *pl.* -s. The *Ndebele* (q.v.) of Matabeleland, a people of *Zulu* stock living in Rhodesia, speaking a dialect of Zulu known usu. as *Ndebele* also *Sindebele* or *Sitebele, Sitebela.* See also *Mashona.* [*fr. Sotho letebele*] ▶So called by the Sotho-Tswana people, the name meaning 'disappearing ones' from the size of their shields behind which they sank down (*teba*) when fighting.

. . . the British South Africa Company, whose very existence had recently been imperilled by a fierce struggle with Lo Bengula's trained regiments of Matabele (1893), followed by the rebellion and murders in Mashonaland. Headlam *Milner Papers* 1897–99, 1931

Matie [ˈmatĭ] *n. pl.* -s. A student or alumnus of Stellenbosch University, freq. used in reference to a member of the University's rugby team: see also *Ikey, Tukkie, Uppie.* ~ *land, colloq.* Stellenbosch. [*acc. some. abbr. Afk.* tamatie tomato *fr. red colour of Stellenbosch University's rugby jersey, acc. others fr.* matie *cogn.* matey, friendly]

. . . 23-year-old Stellenbosch senior education diploma student, has been elected cheer leader of the Maties for this year's intervarsity at Newlands on May 15. *Argus* 4.5.71

matjiesgoed [ˈmaɪkĭsˌxŭt] *n.* Also *mat rush* (*obs.*): *Cyperus textilis,* a reed used esp. by Africans for thatching, mats and baskets: *cf. Jam. Eng. basket withe.* [*Afk. fr. Du.* mat + *dimin. suffix* -je + goed material]

Our guides . . . brought me . . . to a part of the river where the mat-rush grew in great abundance; expecting this to persuade us to stop here several days, till the rush-gatherers had finished their work. Burchell *Travels I* 1822

Cyperus textilis . . . A rush 2 or 3 feet high, which grows in marshy localities and in the beds of rivulets. From it baskets and mats are manufactured by the natives who call it matrush (matjiesgoed). *Cape of G.H. Almanac* 1856

The roofing material was still generally thatch, made from 'matjes goed' reed. Lewcock *C19 Architecture* 1963

matjiestou(w) [ˈmaɪkĭsˌtəʊ, -cĭs-] *n.* Cord

made from *matjiesgoed* (q.v.) formerly used for tying or binding thatch. [*Afk. fr. Du. mat* + *dimin. suffix -jie* + *tou*(*w*) rope]

matjiestou(*e*). *Cyperus textilis* . . The name applied to the rope (Afr: tou) made from the culms [of matjiesgoed]. Smith *S. Afr. Plants* 1966

Matric *n. prop. abbr. Matriculation* (q.v.).

Matriculation *n. prop.* Usu. abbr. *Matric,* the examination written at the end of *Standard* (q.v.) ten, ~ *exemption* qualifying a student for University entrance. ¶*Conditional exemption* is granted to certain students wishing to begin their University studies, conditional upon their completing the requirements for ~ *exemption* before having credit for any university courses passed. A *school-leaving certificate* may be awarded to a student not satisfying the full ~ requirements, on the same examination or on a combination of subjects not constituting a full 'University Matric'.

In the various provincial examinations, Cape, Natal, Transvaal and Orange Free State Senior Certificate, a certain standard must be achieved for ~ exemption.
JOINT MATRICULATION BOARD (constituted under the provisions of Section 15 of Act No. 61 of 1955 to control and conduct the Matriculation Examination of the Universities of South Africa. *Matriculation Examination H'book* 1973

mat rush *n. obs.* Travellers' term: see quots. at *matjiesgoed.*

mawo [ˌmaˈwɔ̃] *interj. Afr.E.* An exclamation of dismay, surprise or sympathy used by Xhosa-speaking Africans: see also *hau.* [*Xh. interj.*]

The wonder became great, and the Kaffirs exclaimed 'Mar-whow'. Bisset *Sport and War* 1875

The kaffirs, finding escape hopeless and shouting 'Mah Wo', formed themselves into a compact body and steadily waited our approach. Buck Adams *Narrative* 1884

Mayibuye i Afrika [ˌmaɪiˈbuje(i) ˈafrɪka] *vb. phr. interj.* 'Come back Africa', also abbr. *Mayibuye.* Cry or motto of the now banned African National Congress (A.N.C.): also the *A.N.C.* song. See *A.N.C.*[1] *cf. E. Afr. uhuru* (q.v.), now also in S.A.E. [*Ngu. mayibuye* come back]

. . she raised her glass then she sipped slowly, . . . She said, 'Heil, Gesondheid, Mayibuye.' Matshikiza *Chocolates for Wife* 1961

Africa emerging from the colonial haze, poised between two oceans . . . her face impressive but evocative as the rapacious sun, Mayibuy'i Afrika! Poem: *Drum* 8.5.73

mbaqanga [mbaˈkaŋ(g)a +] *n.* An African music mode which originated in Soweto, Johannesburg: see also *malombo* and quot. at *patha-patha.* ¶Acc. purists not jazz: see second quot. [*unknown poss. rel. Zu. umbaqanga* thick porridge *poss. sig.* 'mixture']

. . . there has been the Zulu-idiom, which forms the basis of what is loosely termed 'Mbaqanga', or pop jazz. But the Malombo sound is unique. It is hard to pinpoint. *Drum* Nov. 1964

And don't sell people a pack of lies telling them that it's going to be a jazz festival, when you know that it'll be a hodgepodge of jazz, soul and mbaqanga. If you call it a jazz festival, let it be a jazz festival. *Drum* 8.7.1974

mbira [mˈbira] *n. pl.* -s. Any of several musical instruments found in S.W. and Central Africa called variously *kalimba* (q.v.) *likembe,* '*Mashona piano*', '*calabash piano*', '*Kaffir piano*' (q.v.), played with the thumbs sometimes over or inside a *calabash* (q.v.). *Informant* Mr Andrew Tracey, Rhodes University. [*fr. Bantu vb. -imba* to sing, here a metathesis of *rimba*/*limba* a note, see *marimba*]

Posselt might have remained longer at Zimbabwe but he listened to a native playing the *mbira* or kaffir piano with such exquisite melancholy that he became homesick and decided to return. Green *Glorious Morning* 1968

. . . one of the Mbiras which he has designed. This instrument is often wrongly called a thumb piano. *Panorama* May 1973

mbombela [(m)ˌbɔ̃mˈbeːla] *n. Afr.E.* African term for a train carriage of the open *sit-en-kyk* (q.v.) pattern: see also *bombella train.* [*unknown poss. onomat. poss. rel. Xh. mbomba*(*loza*) (g)rumble: *also poss. rel. Zu. bambela,* catch, hold on to, *also poss. Zu. mbombozela* '*applied form*' *of vb. sig.* resound or rumble]

But hundreds of them . . . hastily bundled their blankets, clothing and other belongings and caught the 'mbombela', or workers train, home to the Transkei. Gordon *Four People* 1964

mbongo [mˈbɔŋgɔʊ] *n. pl.* -s. *colloq.* An offensive mode of reference to a political 'stooge' or apologist. [*fr. Ngu. imbongi* (q.v.) praise singer]

Government supporters have danced round their ministers like dutiful, adoring and disciplined mbongos. [*cit. O.E.D.*] *Argus* 16.9.48

Never call an MP a mbongo! ... a fair sample of expressions which have been disallowed in the House of Assembly including 'he has a yellow streak', 'agitator', 'gangster', 'stooge', 'mbongo', 'khaki pest', 'fat-head'. Van Biljon cit. *Sunday Times* 3.10.76

(-)meal *n. and n. suff.* Also *meel:* ground grain: suffix to the name of any grain, coarsely or finely ground e.g. *mealie ~ kaffircorn ~, mabela ~* all (q.v.). *cf. oatmeal:* also in combinations *boer ~* (q.v.) brown wheat flour containing bran; *Kuhne/kuny/i ~* coarse crushed wheat; see quots. at *samp, Kuhne ~* and *kuny/i ~*.

He has a bedstead and a bed, with clean bedding, a table and some chairs; a chest in which he keeps his clothes; and in some sacks a supply of meal, sugar, coffee etc. Shaw 1860 cit. Sadler 1967

... none of this white-flour business either. 'Meel', it said, which I translated as 'meal' and measured 9kg into a great tub. *Fair Lady* 7.3.73

mealie ['milĭ] *n. pl.* -s. Maize, Indian corn, *Zea mays: U.S. and Canad. corn. Afk. form mielie* (q.v.): extensively cultivated and used in numerous forms in S.A. [*fr. Afk. mealie prob. fr. Port. milho fr. Lat. milium* grain *esp.* millet]

Though it is not indigenous to South Africa, the mealie could almost qualify as our national vegetable. *Evening Post* 27.10.73

green ~s, fresh *~s* boiled and served like sweetcorn on the cob as a separate dish or as a vegetable.

White maize, at the semi-sweet stage, known as the good old 'green mealie' has been sold as a substitute for sweet corn in this country for many years. *Farmer's Weekly* 13.6.73

Made into *green ~ bread,* also called *~ bread* similar to *U.S. corn pudding* or *corn pone; Jam. Eng. dokuno* (various sp.).

... green mealie bread made for him by a trader's wife in Pondoland. She put the mealies through a mincer three times, added baking powder and salt, and boiled the mixture in a dish cloth. It looked like a suet pudding, but served hot with butter it was delicious. Green *Land of Afternoon* 1949

in combinations *~ cruncher* (q.v.); *~ meal* (q.v.); *~ pap* (see *mieliepap*); *~ rice,* dried maize kernels ground to resemble fine rice grains, used as a substitute for the more costly rice; see also *samp, stamp ~s* and *kaboe ~s*.

Serve hot with mealie rice. *Drum* 8.11.72
MAIZE PRODUCTS Samp 90kg R5,46
Mealie Rice R4,98

Farmer's Weekly Advt. 21.4.72

~ pip, mielie pit, colloq. Maize grains or seed usu. for planting.

The yellow fish greedily ate at the porridge, but they tricked themselves with a single mealie pip. *Farmer's Weekly* 12.5.71

~ stalks, mieliestronke. The stalks of already harvested *~s* used as green feed or as stover for stock, also milled for ensilage: *~ (stalk)borer Afk. mielie-rusper, Calamistis fusca* a pest which attacks maize crops.

When his cattle had the heart-water or his sheep had the blue-tongue, or there were ... stalk-borers in his mealies, Webber would look it all up in his books. Bosman *Mafeking Road* 1947

mealie cruncher *n. pl.* -s. An offensive mode of reference to an Afrikaner of the lower class: see also *crunchie, hairy* (*back*), *krans athlete, kransie, rock spider.*

mealie meal Finely ground, granular maize meal, white or yellow, the staple food of much of the S. Afr. population. ⫫Not equiv. cornflour, see *maizena,* equiv. *Italian polenta, U.S. hominy grits, Jam. Eng. corn-corn:*

straight run ~, yellow unrefined *~,* containing maize germ.

'I'd much rather the children didn't have so much mealie meal and rice,' ... Mrs. Dladla said, 'But I have to keep their tummies full.' *Fair Lady* 24.11.71

Why is it so difficult to buy yellow unrefined mealie meal? ... several appetising recipes for straight-run mealie meal, their name for unrefined mealie meal. *Ibid.* 19.4.72

mealy bug *n.* Non S.A.E. A scale pest of the Pseudococcidae occ. known in S.A. as *Australian bug* (q.v.), also a type of scale.

Mealy bugs ... These insects belong to the same family as the scale insects and Australian Bug. Eliovson *Gardening Book for S.A.* 1960

Meat Pie See *Durban Beach.* Also, *Border usage* a turtle.

mebos ['mǐbɒs, -bɔ̆s, 'meɪˌbɒs] *n.* A confection of salted and sugared dried apricots readily available commercially, *cf. Jam. Eng. dolce, dosi* (of guava or mango). [*Afk. prob. fr. Japanese umeboshi* pickled plum (*ume* plum)]

I have bought some Cape 'confyt'; apricots, salted and then sugared, called 'mebos' – delicious! Duff Gordon *Letters* 1861–62

medora [mə'dura] *n. pl.* -s. A Malay headdress worn on special occasions only, such as the return from *Haj* (q.v.) or a wedding, and consisting of a scarf brought from Mecca, heavily embroidered with gold.

When Sadia changed into her . . . wedding gown, the bridesmaids who had worn dark pink headdresses called mishfals, changed to white medoras, adding matching pink capes. *Sunday Times* 18.2.73

(-)meel [mɪəl] *n. suffix* See (-)*meal* and quot. at *semels*.

meelbol [ˈmɪəlˌbɔl] *n.* Traditional infant cereal made of flour: see quot. ⟦Commercially available. [*Afk. fr. Du. meel* flour + *bol cogn.* ball]

. . . after infants had been weaned they were given 'meel bol'. This was flour that had been tied in a pudding cloth and boiled for several hours. When the cloth was opened the flour was found to have become a solid cake which was grated and boiled in milk. McMagh *Dinner of Herbs* 1968

When Baby needs *more* than milk, build him up with wonderful . . . Meelbol. The balanced food that always agrees. *Argus Advt.* 13.5.71

meercat¹, meerkat [ˈmɪəˌkæt, mɪə(r)kat] *n. pl.* -s. Any of several small S.A. mammals of the Viverridae similar to the mongoose, *Suricata tetradactyla* (also applied to other species *Cynictis pencillata, Suricata suricatta, Mangusta levaillantii*). ⟦~s are sometimes tamed as pets but are subject to, and the commonest carriers of rabies. The main types are the slender tailed or *stokstert* ~ and the *waaierstert* ~, see *waaier-*, or bushy-tailed ~: also *graatjie* ~ corruption of *gharretjie*. [*Afk. fr. Du. meerkat* type of monkey: *meer* sea + *kat cogn.* cat *hence an imported creature*]

. . . a sick meercat . . . the prettiest, but most ferocious little brute I ever saw. Of these animals there are several kinds, only to be classed together by their habit of living in holes upon the veldt, and standing on hind legs to survey the prospect, in a very droll manner. Boyle *Cape for Diamonds* 1873

Of the confirmed cases, nine were meercats and one a wild cat . . . warn their children . . . not to catch wild animals especially meercats. He said sick animals were usually docile and easily caught and such animals could have rabies. *E.P. Herald* 15.9.71

meercat² A newly-developed all-purpose vehicle for underground use in mines.

The Meercat is powered by a twin-cylinder diesel engine, is fully automatic and all hydraulic, and can be fitted with a light or heavy duty bucket, a roof bolter drill, a scissors lift, a mobile crane, and a personnel carrier. *S.A.*

Digest 19.1.79

meester [ˈmɪəstə(r)] *n. pl.* -s. Teacher. [*Afk. fr. Du. meester* teacher, (school)-master]. **1.** Mode of address to a teacher or other scholarly person.

. . . various titles are given to people holding special positions in the community. All teachers and former teachers are addressed as *Meester*. Carstens *Coloured Reserve* 1966

The station porter called me 'meester' – I suspected him of irony. O.I. University Professor 1972

2. *hist.* A resident tutor in a Dutch family, or an itinerant school-master, often an ex-soldier moving from farm to farm: see quot.

7th July 1811 The daughters . . . were under the tuition of an itinerant tutor, or Meester, as he was called, who had been for several months an inmate of the family. Burchell *Travels I* 1822

There were many men of his kind, old soldiers, English and German, attached to the households of the frontier Boers in those days. For the most part they acted as 'meesters' being allotted the task of instructing the children of the family in reading and writing and ciphering. Brett Young. *Seek a Country* 1937

meid [meɪt] *n. pl.* -e. A non-White girl or woman usu. Coloured, Cnless in combination *kaffer* ~. [*Afk. cogn.* maid]

It's mos funny, Me! *Ou meid* being *donnered!* . . . You've made it worse for yourself. Dead *Kaffer* and a *Hotnot meid* with bruises . . . and Boesman sitting by with no skin on his knuckles. Fugard *Boesman & Lena* 1969

meisie [ˈmeɪsĭ] *n. pl.* -s. A young girl: freq. in combination *mooi* ~ a pretty girl, found also in place name Mooimeisiesfontein. [*Afk. fr. Du. meisje cogn. Ger. Mädchen* maiden, girl]

And the girls who provided tasteful items in the programme in the intervals between burgher 'liedjies' and soldier-songs – they were a refreshing novelty in Afrikander life, not shy like the backveld 'meisje'. Prance *Tante Rebella's Saga* 1937

Any girl hereabouts who resents being called a mooi meisie must be singular of her kind even if she is of Norman blood and the line of Vere de Vere. *Cape Times* 8.1.72.

-mekaar [məˈkɑ:(r)] *prn. suffix.* Each other, one another: found in place names as object of imp. vb., Helpmekaar and Soekmekaar. [*Afk. fr. Du. malkander,* one another, each other; (*Du. zoeken* seek, search for)]

melée [ˈmeˌleɪ] *n. Sect. Mining.* Diamond trade term, non S.A.E.: small diamonds: ⟦Authorities vary as to the exact size which constitute ~: see quot. 'A small diamond cut from a fragment of a larger stone and usu. less than one-eighth carat in weight.' *Webster's Third International Dict.* [*unknown poss. fr. Fr. se mêler* to mix, mingle, *or mêlee* a scramble]

There are also special terms for diamonds classified by weight, 142 carats being reckoned to the ounce. Small diamonds of less than a carat are known as 'melée', while broken stones are described as 'chips' if of melée size, and 'cleavages' if larger. White *Land God Made* 1969

'Stones' are usually over one carat (a carat being 200 milligrams). Anything smaller falls in the 'melee' category. *Panorama* Dec. 1972

melk [melk] *n. lit.* Milk: prefixed to numerous plant names sig. the presence of white latex or 'milk' which can, esp. in the Euphorbiaceae, be poisonous, and corrosive on the skin, e.g. ~ *boom*, ~ *hout*, ~ *bos* all (q.v.). [*Afk. fr. Du. melk cogn.* milk]

melkbos ['melk₁bɔs] *n.* Any of numerous species of *Euphorbia* (*also Austral. milk bush*) and species of *Aesclepiadaceae*, all containing white latex, some of which are poisonous e.g. *A.glaucophylla* (*blou* ~), whereas *E.mundii, skaap* (*sheep*) ~ is much favoured by sheep. Found in place names Melkbospoort and Melkbosstrand. [*Afk. fr. Du. melk cogn.* milk + *bos cogn.* bush]

melkboom ['melk₁buəm] *n.* Any of several trees with white latex in the bark and esp. the arborescent species of *Euphorbia*: see *melkbos*, also *naboom*. [*Afk. fr. Du. melk cogn.* milk + *boom cogn. Ger. Baum* tree]

melkhout(boom) ['melk₁(h)əut₁buəm] *n.* Used of several trees with milky latex, esp. in the bark and esp. of *Sideroxylon inerme* and/or its closely grained timber: see *milkwood*. Found in place names Melkhoutkraal, Melkhoutfontein. [*Afk. fr. Du. melk cogn.* milk + *hout* wood (+ *boom cogn. Ger. Baum* tree)]

melksnysels ['melk₁sneɪsəls, -z] *pl. n.* A traditional Cape dish containing 'dough-cuttings' similar to noodles, see also *snysels*. [*Afk. fr. Du. melk cogn.* milk + *snysel(s)* slice(s) *fr. Du. snijdsel*]

And melk snysels, cooked with milk, sugar and cinnamon. Cloete *Watch for Dawn* 1939

melktert ['melk₁tert] *n. pl.* -s, -e. Tradition-al Cape custard tart usu. dusted with cinnamon, nutmeg or other spice: see quots. at *poffertjie* and at *sea-cat*. [*Afk. fr. Du. melk cogn.* milk + *tert cogn.* tart, *Fr. tourte*]

The Cape Malays excelled in the art of baking, especially the melktert with crushed cardomons and cinnamon slightly strewn on the egg and milk filling. *Argus* 5.6.71

It is wrong to talk of milk tart – an insipid, anaemic-sounding name for a really delicious old Cape delicacy. WHAT IS WRONG WITH MELKTERT? *Ibid.* 10.7.71

meltsieke ['melt₁sïktə] See *miltsiekte*.

meneer [məˈnɪə(r)] *n.* Sir, gentleman. [*Afk. fr. Du. myn/mijn* my + *heer* lord (hence gentleman) sir]

1. Also *mijn-, mynheer:* a mode of address equiv. 'Sir': in titular use with surname or Christian name and surname as equiv. Mr (*abbr. Mnr*); and used as a mode of address in the *third person* (q.v.): see first quot.: see also *mevrou, third person address,* and second quot. at *tannie*

Her mind, her heart, her soul – all these were now his and he would do what he would with them. 'Mijnheer knows,' she said again. 'Let him do as it seems right to him.' Smith *Beadle* 1926

2. A mode of reference (also *mynheer*) now poss. archaic *equiv.* 'the gentleman'; also occ. 'the *dominee*' (q.v.).

. . . good looking cheerful Frouw, two neat daughters and the mynheer a very capital specimen of the best class of Dutch Boer. *Traveller's Journal* 1821–23

. . . when Meneer die Predikant turned up in person on his rounds – . . . there was a period of intensive worship. Jackson *Trader on Veld* 1958

Meraai [məˈraɪ, me-] *n. prop.* The female character in *Gammatjie Jokes*. see quot. at *Gammat*. [*presum. corruption of Maria*]

'Sowaar, Meraai – I'm suffering from an energy crisis – I just can't get up!' Gammatjie Cartoon Caption *Drum* 8.1.74

Met, the *n. prop. abbr. Metropolitan Handicap,* the major summer race meeting of the year held in Cape Town in January: combinations ~ *fashions* etc. See also *July*, (*the*).

Its best not to smile at the races. Not unless you are leading in the Met winner, and then only briefly. *Capetonian* May 1979

met(h)i *n.* Fenugreek: see *Indian terms.*

. . . (Fenugreek is obtainable in Indian shops under the name of meti.) Gerber *Cape Cookery* 1950

metrication *n.* Conversion to the metric system of weights and measures in S.A. in combination ~ *Board:* see *litre, mass, volume, cumec, k* and *kg. [adj. metric + n. forming suffix -ation: term suggested to the S. Afr. Bureau of Standards, see S.A.B.S. by Oxford Dict. editors: also adopted in Australia]*

METRICATION Important Notice. The Executive Council of the Newspaper Press Union of South Africa in collaboration with the Metrication Board has agreed that a ban be placed on the use of Imperial Measures and Figures in all advertisements. *Evening Post* 2.3.74

mevrou [mə'frəu] *n.* Mistress. *[Afk. fr. Du. mevrouw* madam, mistress]
1. A mode of address to married women, esp. to *Afk.* school teachers (see also *juffrou*), equiv, of 'Madam': also in titular use with surname ~ *X* as *equiv.* Mrs *(abbr. Mev.)* and used as a mode of address in the *third person* (q.v.): see quot. See *lady* and second quot. at *tannie.*

And does Mevrouw like it too and – and your daughter? Fairbridge *Which Hath Been* 1913
2. A mode of reference to a married woman, also to the mistress of a household.

. . . preparations for these guests . . . kept Mevrouw busy in her kitchen with Andrina in constant attendance upon her there. Smith *Beadle* 1926

Mevrouw appealed to her husband when she found him at her side, to back up her statement that the Market Master had assessed her berry wax. Fitzroy *Dark Bright Land* 1955

mfezi [m'fe:zi] *n. Naja nivea:* the Cape Cobra, see *rinkhals.* [*Zu. imFezi* cobra, *rinkhals*]

'The imfezi spits in the eye of the Amandebele and blinds them . . . They nourish it in their kraals thinking it to be an ancestor and they cannot see truly . . . Drive out the imfezi and all will be well.' Heaton Nicholls *Bayete* 1923

middlemannetjie ['mɪdl̩ˌmanĭkĭ, -cĭ] *n. pl.* -s. Also *middelmannetjie:* a hump, usu. continuous, between the wheel ruts in an unsurfaced rough road or track. [*Eng.* middle *or Afk. middel + mannetjie (man + dimin. suffix)*]

Remember when mud and middlemannetjies made motoring a continuing adventure? Well, not all romance has left the road with tar. *E.P. Herald* 4.3.74

middelskot ['mɪdl̩ˌskɔt] *n.* Intermediate payment made to farmers for their crops between the *voorskot* (q.v.) and the *agterskot* (q.v.). *[Afk. middel* middle *+ skot cogn.* shot: intermediate payment]

According to a statement issued by the Board in Pretoria, the 'middelskot' will be paid on the net mass of buckwheat delivered by producers to agents of the Board during the period January 1 to October 31, 1972. *Cape Times* 9.11.72

mielie- ['mĭlĭ-] *n. pl.* -s *and n. prefix.* See *mealie:* ~ *gif,* see *rooiblom;* ~ *pit* see *mealie pip/pit;* ~ *stronke,* see *mealie stalks. [Afk. fr. Port. milho* millet *fr. Lat. milium* grain *esp.* millet]

mieliepap ['mĭlĭˌpap] *n. Mealie (maize) meal* porridge of various types: staple food of the African people, but used by most S. Africans: see also *putu, krummelpap, stywe pap,* and *pap* also quot. at *vleis. cf. Canad. sagamite, supon: U.S.* cornmeal mush, *Jam. E. turn-turn.* *[Afk. mielie* (q.v.) + *pap]*

'All I'm longing for now is a big dish of well cooked mealiepap and a chunk of meat. I have been missing them,' he greeted us. *Drum* 22.9.73

The Zulus [in uMabatha] . . . Their diet will be typically English and instead of mieliepap they'll have to settle for oatmeal porridge. *E.P. Herald* 15.6.73

The typical South African food that still means as much to us today as it meant to our fathers and grandfathers . . . Mieliepap for breakfast. Mieliepap with meat and gravy. Mieliepap with dry beans or other vegetables . . . Mealiepap. It's the power food of our people. *Bona* Mar. 1974 (*Maize Board Advt.*)

mies(ies) ['mĭˌsĭs] *n.* Mode of address and occ. reference with definite *art. the* ~ to the mistress of a household: also to or of other White woman *usu.* by non-Whites: equiv. of 'madam'; also in form *missus* (q.v.) *cf. Hong K. tai-tai. Anglo-Ind.* memsahib. [*presum. fr. Eng. dial.* missus]

. . . they have an affectionate way of saying 'my missis' when they know one, which is very nice to hear. Duff Gordon *Letters* 1861–62

. . . asked for a telephone directory, and phoned for an ambulance, explaining: 'The miesies is sick.' *E.P. Herald* 30.8.74

miggie See *muggie.*

mijnheer See *meneer.*

mik [mɪk] *n. pl.* -s. *colloq.* The forked stick of a catapult: see *cattie/y. [Afk. mik(stok)* forked stick]

. . . has chosen such a killer great mik for my new cattie that I won't be able to pull it. O.I.

Child 1974

milkwood *n. pl.* -s. See *melkhout(boom)*
white ~ /*wit melkhout*, timber of *Sideroxylon inerme; red* ~ /*rooi melkhout*, timber of several species of *Mimusops esp.
M. obovata.* [*trans. Afk. melkhout*]

milt [mɪlt] *n.* Also *melt:* spleen, usu.
butcher's term. [*Afk. fr. Du. milt*
spleen]

 . . . the dust that the Bechuana with his head
under his arm raised . . . seemed to become
part of and reach beyond the Milky Way that
shone through his milt and was also a road.
Bosman *Jurie Steyn's P.O.* 1971

miltsiekte [ˈmɪltˌsĭktə] *n.* Also *meltsiekte.
gif(t)siekte:* anthrax in stock, highly
infectious and communicable to man.
❡Cases caused by eating ~ contaminated meat are *usu.* fatal [*Afk. fr. Du.
milt* spleen + *siekte fr. ziekte* disease]

 Anthrax, locally termed 'meltziekte' or
'gift-ziekte', is a disease which, in its well-
known erratic and spasmodic way, appears
from time to time in all districts of the Colony,
and among all classes of stock. Wallace's
Farming Industries 1896
 . . . the year after the drought, the miltsiek
broke out. The miltsiek seemed to be in the
grass of the veld . . . the water of the dams, and
even in the air the cattle breathed. Bosman
Mafeking Road 1947

min dae [ˈmɪnˌdaə] *n. and n. modifier.
Sect. Army. colloq.* 'Few days': used
sig. obligatory National Service is
almost over: as modifier ~ *smile,*
~ *salute* sig. careless or insubordinate:
see *ouman* and quots. at *varkpan* and
blougat. [*Afk. min* few, little + *dae*
days (*dag* + *pl.*)]

 To the oumanne of E. Squadron . . . min
dae and vasbyt from Dave. *Radio S.A.* Forces
Favourites 1.4.72
 A favourite form of greeting is 'Min dae'
when they have forty days or less to serve, and
'forty days' also means something to them.
Letter Dr Amy Jacot-Guillarmod 2.1.70

In combination ~ *sign n. pl.* -s A 'safe
sign' (index and little fingers pointed,
third and fourth folded back, waving
slightly from side to side) exchanged by
those *oumanne* (q.v.) who have *mindae,
mindays* left to serve: ❡The index and
little fingers are brought closer together
as the days run out, when the two (if
possible) meet.

 You know they're on forty days when they
start going around flashing min-day signs at
each other. *Serviceman* 4.2.79

mine-dump *n. pl.* -s. A large 'hill' of solidified crushed quartz from which the gold
has been extracted, a dominant feature
of the *Rand* (q.v.) (Reef) landscape.

 . . . the mine dumps, dominating the scene and
continually changing colour, reminded me of
brown pyramids, grey flat-topped hills in the
Karoo, the ochre foundations of gigantic old
temples never completed . . . – how striking at
times are the prosaic old mine dumps of the
Witwatersrand. Krige *Dream & Desert* 1953
 It has been estimated that the mine dumps of
the Witwatersrand are among the largest man-
made landmarks, comparable perhaps to the
pyramids . . . or to Manhattan Island. *Panorama*
May 1972

minesweeper *n. pl.* -s. *Rhodesian.* A truck
used to clear airstrips and runways of
land mines for aircraft. *cf. S. Afr. spook²*
and *sprinkaan³* [*presum. fr. orig.* naval
use]

 A truck drove slowly up the runway – a
'minesweeper' Robin called it – and we landed
in its tracks. *Fair Lady* 16.3.77

mis [mɪs] *n.* Dung, manure [*Afk. fr. Du.
mest* manure (+ *koek cogn.* cake)]
1. Dried cattle dung which burns to
white hot coals, used for fuel, even
braaiing (q.v.) where wood is scarce, see
quot. at *hardekool.* See also *Free State
coal* and *Kar(r)oo coal:* ~*koek* see quot.
cf. Canad. buffalo-chip, prairie chip, dung
used as fuel.

 Here . . . we burn a strange kind of coal. It
is dung . . . It is called mis. Vaughan *Diary
circa* 1902
 . . . the manure was well trampled into the
ground, forming a layer three to four inches
deep, which was periodically cut out with spades
in square cakes of about 18 inches. Packed on
top of the kraal walls, where it soon dried
ready for burning, these cakes were sold in
adjacent villages at fourpence each, and were
called miskoeke. Jackson *Trader in Veld* 1958
2. Fresh cattle dung used in various
mixtures with ash, blood, mud or water
for *smearing* (q.v.) or dressing floors in
country districts, also for making floors,
see ~*vloer:* also form *mist,* see quots.
and *etym.*

 The floors . . . she smeared regularly with a
mixture of cowdung and ashes called mist.
The little house smelt always of mist, of strong
black coffee. Smith *Little Karoo* 1925
 Following traditional Cape practice, such
composition floors were sometimes smeared at
regular intervals with a mixture of cow dung
and water known to the settlers as 'mist'.
Lewcock *C.19 Architecture* 1963
3. Dry blocks of stamped and cut up ~
used for building purposes as well as

for fuel: see quot. at (1) *mis*.

. . . in the drier pastoral districts, where manure is not much in demand for fertilising purposes . . . it is the practice to cut the material, which resembles a light fibrous peat and is known as 'mist', into slabs 3 to 4 inches thick, and about 16 inches long and 12 broad. In this form it is used for the building of kraal fences or as fuel. Wallace *Farming Industries* 1896

mishfal [ˈmɪʃˌfal] *n. pl.* -s. The draped flowing headscarf usu. oblong or square in shape and embroidered at the edges, still worn by many Malay women: see quot. at *medora*.

miskruier [ˈmɪsˌkrœïə(r)] *n. pl.* -s. The dung-roller beetle or 'tumble-bug', a beetle of the Scarabæidae. *cf. Jam. Eng. tumble-turd.* [*Afk. fr. Du. mest* dung + *kruier* a porter]

Mistkruier – . . . The not inappropriate appellation of the various dung-rolling beetles. Pettman *Africanderisms* 1913

mispel [ˈmɪspəl] *n.* Medlar: found in several plant names e.g. that of *Vangueria infausta* the fruits of which resemble the cultivated medlar and are used for making *mampoer* (q.v.).

misrybol [ˈmɪsˌreiˌbɔl] *n. Amaryllis belladonna* and *Haemanthus coccineus* which flower March/April, when the vineyards are being manured: see *ride*. [*Afk. fr. Du. mest* dung + *rijden* to cart + *bol* bulb]

A variety of *Haemanthus*, which appears to have received this inelegant appellation because it happens to be in flower just about the time that the mest (manure) is being carted, or in South African English 'ridden' on to the vineyards. The name is also applied to a fragrant Amaryllis. Pettman *Africanderisms* 1913

miss lucy *n. pl.* Ø. The red *stumpnose* (q.v.) *Chrysoblephus gibbiceps*: see also quot. at *allewêreld*. [*unknown poss. fr. n. prop. (cf. jacopever) see second quot.*]

Jan-bruin were plentiful and of a good size. Several romans and daggerhead were caught, and a couple of Miss Lucy (red stump-nose). *Grocott's Mail* 30.6.72

If you do not know this fish as miss lucy you might call it a red stumpnose or magistraat or bont dageraad or miggel or rooistompneus or stump . . . I rather like the name miss lucy. Maybe in the days of the early fishermen there was a colourful character ashore known . . . as Miss Lucy and this fish was named after her. *E.P. Herald* 22.9.76

missus *n.* Mistress, madam: in S.A. usu. *mies(ies)* (q.v.). [*presum. fr. mistress, Brit. dial.* missus]

You can already dress like a white missus,

and Jo'burg girls dress like that. Dikobe *Marabi Dance* 1970

misvloer, misfloor [ˈmɪsˌfluːr] *n.* Also *dung-floor* (q.v.), see quot. at *pepper tree* a floor made of stamped, dried *mis* (q.v.) usu. cattle dung: found in old *barns* and *waenhuise* (q.v.). [*Afk. fr. Du. mest* manure + *vloer* cogn. floor]

The living room had a misvloer, but the floor of the wagon formed the roof of the room. Green *Land of Afternoon* 1949

His three-roomed house was of clay, floored with *mis*, and had no ceilings. Fitzroy *Dark Bright Land* 1955

mithai See *Indian terms*.

mist rain *n. pl.* -s. Drizzle, light rain. *cf.* 'Scotch mist', *Jam. Eng.* dew dew, ju ju [*translit. Afk.* misreën drizzle]

. . . They said that it was now the commencement of the mist rains at Walvisch Bay . . .

The people said there was plenty of mist (or small rain) in the cool months, which would bring forward the vegetables. Alexander *Expedition II* 1838

As *vb. substandard*; see quot.

The weather looks as if it is going to clear. . . . I don't like it when it is just mist raining it must either rain or it must clear. Letter schoolgirl 17, 30.9.74

mkaya [(u)mˈkaɪa] *pl.* aba-. See *abakhaya* and *homeboy*.

mk(h)onto [mˈkɔntɔ] *n.* Spear, (*assegai* (q.v.)): see also *wash spears*. [*Ngu.* umkhonto spear]

. . . they charge with a single umconto, or spear, and each man must return with it from the field, or bring that of his enemy, otherwise he is sure to be put to death. Thompson *Travels II* 1827

(u)mkhonto ka Shaka [(u)mˈkɔntɔ ˌga ˈʃaga, -ka] *n. prop.* The Zulu opposition party, 'The spear of Shaka' (Chaka), also *Shaka's Spear* (q.v.), also occ. *The Spear* (q.v.). [*Zu.* umkhonto spear *ka* (*possessive*) of + *Shaka*]

King Sobhuza had heard about the formation of the Zulu opposition Umkhonto wa [*sic*] Shaka (Shaka's Spear) and expressed 'deep concern' about it. Last year King Sobhuza outlawed political parties in his kingdom [Swaziland]. *E.P. Herald* 6.2.74

(u)mlungu [(ŭ)mˈlŭŋgŭ] *n.* and *n. modifier*, *pl.* -s, -abelungu. *Afr.E.* A mode of address or reference to a white man by a black one: now usu. ironic as in term ~ *stan. cf. U.S. whitey; honkey: Anglo-Ind. Gora*, Englishman; also some uses of *Jam. Eng. backra: Canad. kabloona, mooneas*, also *saganash* white man esp.

an Englishman: *Hong K. gwai-lo* Chinese mode of reference to a European: [*Ngu.* (*u*)*mlungu* a white man]
Then his head fell back, his eyes, however, never leaving my face. 'Hau . . . umlungu . . .' he groaned. Krige (*Death of the Zulu*) *Dream & Desert* 1953
. . . when Mlungu journalists were barred from the Transkei Hotel at Umtata where our Bantustan leaders had their first summit meeting. *Drum* 22.12.73
The Chief was not being flippant when he called on bantustanians to convert their mlungus . . . mlungustanians need a refresher course. *Drum* 8.11.73
In combination ~ *stan cf. Bantustan:* see third quot. and quot. at *-stan.*

mngqusho *n.* See *samp.*

mnumzane, mnumzana [m|'nŭm₁zǎna, -e, -ə] *n.* A mode of address, usu. written equiv. Mr. ❡At one time adopted for official reference or correspondence to avoid the use of 'Mr' to or for a black man, with the salutation 'Greeting(s)' to obviate 'Dear Sir'; abbr. *Mna, Mnu.* see quot. at *imbongi.* [*Ngu.* (*u*)*mnumzane/a* gentleman, *also* kraal-head]

modder ['mɔdə(r)] *n. prefix.* Mud: found in place names Modderfontein, Moddergat, Modderrivier: see quot. at (2) *voetsek.* [*Afk. fr. Du. modder* mud, muddy]
The farm where we now stopped is named Modder-Fonteyn (Muddy Fountain) an appellation so common in the Colony that I have visited, I believe above a dozen places of that name. It is strange to observe the barrenness of fancy of the boors in giving names to places . . . This may, perhaps be ascribed to the sameness and monotony of South African scenery: it however occasions much inconvenience and confusion to the traveller. Thompson *Travels I* 1827

Modimo [mɔ'dimɔ] *n. prop.* Sotho mode of address or reference to God: see also *Tixo, Qamata, Unkulunkulu* ❡ Forms vary: *Molimo, Mlimo* etc. in historical writers, esp. missionaries.
He stood . . . on a platform of sorts and he seemed to be praying in the Sotho language . . . of which she had small knowledge. She bowed her head reverently, but beyond the occasional Modimo – God – she heard little. Louw *20 Days* 1963

modjadji, (the) [mɔ'dʒadʒĭ] *n. prop.* The Chieftainess and goddess of rain of the N. Sotho *Pedi:* see also quots. at *Rain Queen and tickey;* combination ~ *palm, Encephalartos transvenosus:* see *kaffir*

bread tree, Hottentot bread: also in place name Modjadjisberg. [*Sotho n. prop.*]
It has been my privilege to stand in the swirl of cloud and mist among the age-old stems of the Madjadja Palms on the hill below the Kraal of the Rain Queen and feel something of the magic of rain in Africa. Giddy *Cycads of S.A.* 1974 cit. *E.P. Herald* 13.11.74

moed [(xə)|mŭt] *n. abstr.* Courage, spirit. [*Afk. moed* courage *cogn. O.E. mōd* heart, spirit]. **1.** Also *gemoed: courage,* heart: found in various compounds in place names Moedverloor, Welgemoedboven, Moedig, Moedwil.
2. Courage: in combination *hou* ~, found also on the coat of arms of the Orange Free State: see quot. at *hou.*

moederkappie ['mŭdə(r)₁kapĭ] *n. pl.* -s 'Granny-bonnet', any of several species Orchidaceae the flowers of which are not unlike a bonnet in shape, including some of the *Disas, Disperis capensis, Bonatea speciosa* and *Pterogodium catholicum:* also *disa, oupa-en-sy-pyp, begging hand.* [*Afk. fr. Du. moeder cogn.* mother + *kappie Du. kapje* bonnet (hood) *kap* + *dimin. suffix -je, -ie*]
moederkappie – . . . The whole flower bears a striking resemblance to the old Dutch 'kappie' (bonnet) and the lateral sepals enhance the effect by their suggestion of ribbons for tying the 'kappie'. Smith *S. Afr. Plants* 1966
'. . . Seventeen varieties . . . There were pale yellow hoods with little faces inside them –' 'Moederkappies', said Mrs le Seur, nodding. Fairbridge *The Torch Bearer* 1915

moeg [mŭx] *adj. colloq.* Tired, weary. [*Afk. moeg* tired *fr. Du. moe* weary]
I'll keep on walking. I'll walk and walk . . . until you're so bloody moeg that when I stop you can't open your mouth! Fugard *Boesman & Lena* 1969

moegoe ['mŭxŭ] *n. pl.* -s. *Afr. Urban slang:* A 'square', bumpkin: see also (*i*) *moogie* [*unknown*]
Another man said: 'If you wear [Trade Name] shoes people will know you are not a moegoe! *The Reader* Issue 1 30.6.78
You see that's a small time clever . . . maybe he thinks I'm going to sukkel to sell this jacket. But he forgets this location is full of moegoes. Dike *First S. African* 1979

moenie worry nie, moenie panic nie ['mŭnĭ] *vb. phr. colloq.* Don't worry/panic – sometimes with *alles sal regkom* (q.v.). [*Afk. moenie* don't *fr. Du. moet niet,* must not (+*nie,* no(t) in *Afk. double neg.*

structure)]

'Man, we're never going to be the same again!' And the Afrikaner nodded sagely and replied 'Moenie worry nie. Alles sal reg kom.' *Drum* 1.1.72

'I promised I wouldn't go out anywhere in public whilst he's gone.' 'Moenie panic nie, Lorn. You couldn't hardly call a Saturday morning show . . . public . . . more like feeding time at the zoo. *Darling* 21.7.76

moepel ['mŭpəl] *n. pl.* -s. The red milkwood *Mimusops spp.* see *milkwood:* also *melkhoutboom.*

It was queer that willows should be in that place, when all around there was just African bush, soet-dorings and moepels and withaaks and kameeldorings. Bosman *Willemsdorp* 1977

moer[1] [mu:r] *n. and interj. obscene:* as an abusive mode of address or expletive of rage or disgust. [*Afk. and Du. fr. moeder cogn.* mother *sig.* womb, dregs, lees *etc.* (+ *jou cogn. you: see first quot.*)]

. . . staring up at a bird . . . *Jou Moer!* . . . shakes her fist at it *Jou Moer!!* '. . . the ultimate obscenity. Contraction of *Jou ma se moer* your mother's womb.' Fugard *Boesman & Lena* 1969, Glossary to 1973 *ed.*

In combinations ~ *of a, modifier;* also *the ~ in,* see *hell in.* slang ~*ava* usu. among children equiv. '*hell of a*', see *helluva sig.* tremendous, 'almighty', used of a blow, quarrel etc.

He tripped me up on purpose so I gave him a moerava klap. O.I. Child 12, 1973

moer[2] *n.* Yeast, ferment: as *prefix or suffix* to plant names: in combinations ~*plantjie,* any of several species of *Anacampseros* the dried roots and stems of which afford yeast for baking or brewing; ~ *wortel*(*tjie*) (root + *dimin. suffix*); *Glia gummifera* and other species from which a ferment is made esp. for *honey beer* (q.v.); *karee* ~ (q.v.); also *kiri* ~. [*Afk. moer fr. Du. moeder, here sig.* yeast, leaven]

Moor-wortel is an umbelliferous plant, from the root of which and honey the Hottentot's make, by fermentation, an intoxicating liquor. trans. Thunberg *Travels II* 1795

moer(a) ['mʊr(ə), 'mŭ-] *vb. trns. slang.* Murder, prob. by sound association, beat up, thrash, *cf. donder vb.* [*fr. Afk. moor* murder]

We'll go on out and moera all the outjies next door. Song '*Ag Plees Deddy*' Jeremy Taylor

moes [mŭs] *n.* A puree of some cooked fruit or vegetable e.g. *applemoes* apple sauce. *cf. U.S. mush, corn-mush* etc.

[*Afk. fr. Du. moes* pulp, puree *prob. fr. Fr. mousse*]

. . . mutton sosaties on skewers, tasting of dried apricots and onions and orange-leaves and curry; with pumpkin 'moes', rich with butter, sugar and cinnamon, and yellow rice, and new peas. M. Weiner 1969

PAMPOENMOES

Pumpkin . . . sugar . . . salt, stick cinnamon . . . margarine . . . Use pumpkin which is not quite ripe and has a tender peel. *Fair Lady* 8.6.77

moffie ['mɔfĭ] *n. pl.* -s. Homosexual, sometimes a male transvestite: see *trassie* also *bof cf. Austral. punce/ponce, poofter; Canad. burdash, Jam. Eng. mampala, U.S. fairy* etc. [*poss. fr. Afk. or Du. slang mofrodiet, Brit. seaman's slang mophy 'a delicate well-groomed youth'* (*Partidge Dict. Slang & Unconventional Eng.*) *fr. substandard slang 'mophrodite'* hermaphrodite]

The life of Edward – described as a beautiful moffie with a sweet soprano voice – was a strange affair from its beginning to its weird end last week. *Post* 23.5.71

In combination ~ *dom.*

Susie, Rodney and Pauline relax on the Rhodes Memorial, next to the University of Cape Town. They are just three of the moffies who have been engaged in meticulous research to trace their ancestors right back to the days of Van Riebeeck. Jackie Heyns gives DRUM readers an exclusive insight into the early days of moffiedom. *Drum* July 1977

mog [mɒg] *n. pl.* -s. *Sect. Army. abbr.* 'Unimog' a four-wheel drive rough-terrain vehicle. ℙ When stripped of superstructure for action known as a *vasbyt* (q.v.) ~ on the *Border* (q.v.)

moggel ['mɔxəl] *n. pl.* ∅. Fresh water sand barbel *Labeo umbratus* a type of carp. [*etym. dub. poss. fr. Du. mokkel* plump child or woman]

These simple plants could be nourished in quantity by the excreta of the high-value eels and thus provide food for the weed-eating fish such as the moggel (sand fish) or tilapia (kurper) and these then could provide cheap protein food for the masses. *Farmer's Weekly* 13.6.73

molen [-mʊələn, -məʊlən] *n. prefix and suffix.* Mill, found in place names Oude Molen, Molenrivier. [*Du. molen* mill (*Afk. meul*)]

mombakkie(s) [ˌmɔmˈbakĭ(s)] *n. pl.* -s. Carnival mask. [*Afk. sing. mombakkies* mask *fr. Du. mom cogn.* mummer, mask, guise + *Du. bakkes* face]

This month is to be Guy Fox [*sic*]. We are making in our street our own G. Fox with all our friends helping. The big men make a great G. Fox with 1000 crackers in its stummick. They put a mambakkie on its face. We all buy the cheap mambakkies in the shop to wear for walking the Guy. Vaughan *Diary circa* 1902 ~*sblom(metjie)* (little flower) *Disperis capensis, Pterygodium catholicum* (Orchidaceae): see *moederkappie.*

mombakkiesblom(metjie) . . . the vernacular name is derived from the suggestion of a mask . . . conveyed by the peculiar shape of the flowers. Smith *S. Afr. Plants* 1966

mompara [ˌmɔmˈpara, ˌmɒmˈpɑrə] *n. pl.* -s. Also *mampara* : a fool, greenhorn or incompetent: also a mode of address (abusive) *cf. E. Afr. (m)pumbavu.* ¶Formerly used of a *raw* (q.v.) African labourer: see first quot. [*etym. dub. poss. fr. ma-* Bantu *pl. prefix + form of Afk. baar* raw, inexperienced *fr. Malay baharu* new(ly)]

The shepherd must be, not the recently engaged 'mompara' whose only language is unintelligible – not the 'piccanin' who spends the day catching meerkats or sleeping, but the most intelligent native available. *Farmer's Annual* 1914

'It makes me sick,' said Mostert angrily, 'When I think of those smug mamparas sitting there in Parliament six thousand miles away from the nearest bullet, making a decision like that.' Krige *Dream & Desert* 1953

-mond- [mɔnt] *n. prefix and suffix. pl.* -e. (River) mouth: found in place names Mondplaas, Gouritzmond, Kleinmond, Kleinemonde. [*Afk.* mond mouth]

mondfluitjie [ˈmɔntˈflœɪkĭ,-cĭ] *n. pl.* -s. *colloq.* A mouth organ, harmonica: also *bekfluitjie.* [*Afk.* mond mouth + *fluitjie* whistle, pipe, mouth organ.]

In a touching ceremony during which some of the participants had to be helped to their feet four or five times . . . the raft was officially handed over . . . to the strains of the Hallelujah Chorus played on a mondfluitjie. *Sunday Times* 4.3.79

monk [mʌŋk] *n. pl.* ∅. Also *angler-(fish).* *Lophius* spp. ¶Sold frozen commercially as a substitute for *crayfish* (q.v.) as *fillets of* ~ for seafood cocktails. [*unknown poss. fr. shape or colouring*]

Monk, Anglerfish, Broad head with large mouth and recurved teeth. Brown or slaty grey, lighter on belly . . . A repulsive and rapacious fish. Smith *Sea Fishes of S.A.* 1949 *ed.*

monkey bomb, monkey snuff *n. pl.* -s. *colloq.* usu. children. *Lycoperdon hyemale*: see *devil's snuff box.*

monkey bird *n. pl.* -s. The red billed hoo-

poe ; see *kakelaar.*

monkey(face)stone *n. pl.* -s. *poss. reg.* Cape. Natural weirdly-shaped stone usu. from Table Mountain, sold or collected for rockeries and other garden work. [*unknown, prob. fr. shapes*]

Garden and Orchard MONKEYFACE rockery stone R35 per load. *Argus Advt.* 4.7.75

Rockeries, waterfalls, fish ponds built with monkeyface stone. *Ibid.* 9.7.75

monkey orange *n. pl.* -s. The fruit of several species of *Strychnos* esp. *S. innocua* or *S. pungens*: see *klapper²* also called *kaffir orange.*

There is food in the veld – not only maize, but marula, custard apple, monkey oranges and manketti . . . says . . . the C.S.I.R. journal Scientiae, after an investigation of the nutritional value of wild plants. *Argus* 1.1.77

monkey rope *n. pl.* -s. Any of several species of *Cynanchum, Secamone, Rhoicissus* etc. all of which are of a lianous habit of growth: see *bobbejaantou* and quot. at *wait-a-bit.*

monkey rope . . . see BOBBEJAANTOU, a name less appropriate than 'monkey rope' as baboons . . . very seldom scamper up the lianes, whereas monkeys frequently do. Smith *S.A. Plants* 1966

It was thick bush with big old yellow-wood trees festooned with creepers as thick as a man's wrist. We called them monkey ropes. *Fair Lady* 14.6.72

monkeys' wedding *n.* Simultaneous rain and sunshine poss. so called from the incongruous combination: see *etym.*: also *jakkalstrou cf. Jam. Eng. devil rain* (while the sun is shining) and *Du. Kermis in de Hel.* [*poss. trans. Zu. for this phenomenon, umshado we Zinkawu* a wedding for monkeys]

As I speak to you we have a burst of late sunshine making this into a monkey's wedding and a half . . . let's hope this sunshine does sweep the rain away. Radio *S.A.* 20.11.71

monkey thorn *n. pl.* -s. Any of several species of *Acacia* affording cover for monkeys: see *apiesdoring, anaboom/tree.*

moochi, moocha [ˈmutʃi,-a, -ə] *n. pl.* -s. See *muchi, mucha.*

Yesterday she had been here, clad in a moocha like a man and bearing a shield. Haggard *Nada* 1895

(i)moogie [ĭˈmuxĭ] *n. pl.* ∅ -s. *Urban African slang.* A fool, clot, raw bumpkin: see also *moegoe* (and *sp. mugu*) [*unknown presum. tsotsi-* or *'fly-taal'*]

He was sent home . . . by his senior homeboys whom he referred to as *imoogie* – country

bumpkins. Wilson & Mafeje *Langa* 1963
... a stranger to the tsotsi's dangerous world
could still save his throat if he has some knowledge of basic words ... A mugu – an idiot.
Venter *Soweto* 1977

mooi [mɔɪ] *adj. colloq.* Pretty, good, nice
etc. general term of approbation, often
among children even in *adv.* form as in
'My new bike goes ~*ly*'; see quot. at
meisie. [*Afk. mooi* pretty]
He said it was just a 'mooi Klipje' ... He
certainly had no idea of its being any more
valuable than any other Mooi Klip (pretty
stone) found among the water-washed pebbles
in the river. *E.P. Herald* 3.3.1921
For the young people this was a chance to
do some courting, to seek a *mooi meisie*, a pretty
girl. Cloete *Rags of Glory* 1963
Also found in place names e.g. Mooi
Uitsig, Mooimeisiesfontein, Mooi Rivier,
Mooifontein.

mooi loop [ˈmɔɪ ˈluəp] *imp. vb. phr.* 'Walk
pleasantly', '*Go well*' (q.v.) farewell
greeting to someone leaving: see *hamba
kahle*. cf. *Jam. Eng. walk good* also a
form of farewell. [*Afk. mooi (adv. m.)*
pleasantly, well + *loop* walk, go]

mooi-moois [ˈmɔɪˌmɔɪz] *pl. n. Sect. Army
lit.* 'pretty-pretties': military '*step-out*'
(q.v.) uniform: see quot. at *Ride-safe
sign*, cf. *Austral. glamour gowns*, khaki
dress uniform. [*Afk. mooi* pretty, nice]
The step-out uniform is known as mooi-
moois while the barrackroom is referred to as
varkhok.* Picard *Eng. Usage in S.A.* Vol. 6 No.
1 May 1975 *pigsty

mooinooi(en)tjie [ˌmɔɪˈnɔɪɪŋkĭ, -cĭ] *n. pl.* -s.
Pretty girl. [*Afk. mooi* pretty + *nooi(en)-
tjie* young girl – *nooi* (q.v.) + *dimin.
suffix*] **1.** *Sarpa salpa*: the *bamboo fish*
(q.v.) (*bamvoosie*) or *stink fish* also
stre(e)pie, see *karanteen*. cf. *Jam. Eng.*
names *Miss Pretty, None-so-pretty,
Nancy Pretty* for several colourful fish.
He defended the bamboo fish, or mooi
nooientjie (pretty girl) also known as the stink
fish, explaining that its diet of seaweed gave
out a peculiar smell when the fish was cleaned.
Nevertheless, it is sound food. Green *S. Afr.
Beachcomber* 1958
2. *Cleome rubella*, an annual weed with
mauve flowers *usu.* avoided by stock
and once 'suspected of being the cause
of gallamsiekte'. Smith *S. Afr. Plants*
1966 (see *lamsiekte*).

moonfish *n. pl.* ∅. **1.** Also *moonie/y* either
of two species of the Monodactylidae,
the Cape moony *M. falciformis*, or the

Natal moony, *M. argenteus* both also
known as Cape lady, sea-kite, kite-fish.
[*fr. colour and shape*]
... landed three moonie ... The lucky fish
which took the sealed award was a moonie
of the exact weight of 2lb 7oz. *Albany Mercury*
29.1.70
2. Either *Trachinotus rusellii* also *ladyfish*, or *T. blochii*, both known in S.A.
and elsewhere, as the *pompano*.
... caught two cob, a jan bruin, a black steenbras, a zebra and a moonfish ... three white
steenbras, two cob, two galjoen and a moonfish ... *Daily Dispatch* 29.5.74
3. *Lampris regius* the opah or kingfish.

mootjie [ˈmŭĭkĭ, cĭ] *n. pl.* -s. Fillet of fish:
see quot. at *pekelaar*. [*Afk. fr. Du. moot*
slice, fillet (of fish) + *dimin. suffix* -*jie*]

mopane(tree) [ˌmɔˈpane, -ĭ] *n.* The shrub
or tree *Colophspermum* or *Copaifera
mopane*, sometimes called *ironwood*
(q.v.), an important fodder tree in low
rainfall areas, giving its name to ~ *veld*,
where it is the chief vegetation, and to
the ~ *worm*, a black spotted caterpillar
which feeds upon it and is eaten by
Africans: *various sp. mopaane, mopani/e*
etc. [*mopane* Setswana (*Tswana*) *plant
name*]
The vegetation consists largely of Mopane
trees, the soil is sandy, and the rocky outcrops
and krantzes are ideal as sleeping quarters for
the baboons. *Panorama* May 1971
But then, when we were in the bundu on a
safari, everything was served with *stywepap* –
even mopani worms. *E.P. Herald* 6.5.74

moppie(s) [ˈmɔpĭ(s)] *n. usu. pl.* Humorous, often teasing, Malay street song.
[*cf. Jam. Eng. banter-sing*, a work song
often satirically alluding to local people,
as do African work songs in S.A. [*Afk.
fr. Du. mopje*, a tune, ditty]
... street-songs or moppies ... Moppies are
little songs (often of doubtful content) sung in
order to challenge, deride or irritate the listener, or merely as foolery. When singing a mop-
pie, the singer often includes a person's name,
and if the person referred to cannot respond in
similar vein he is laughed at by all present.
Du Plessis & Lückhoff *Malay Quarter* 1953

moraharaba [mɔˈrabaˌraba] *n.* Also *marabaraba*: a game allied to draughts
played by Africans, freq. on a sketched
or otherwise improvised checkerboard.
[*Sotho* name of game]
Mkize looked up from his game of morabaraba and answered him. Langham/Mopeli Paulus *Blanket Boy's Moon* 1953

môre- [ˈmɔːrə-] *n. prefix.* Also *morgen:* morning, found in place names e.g. Môreson, Môregloed, Morgenster. [*Afk. môre fr. Du. morgen cogn.* morning, morrow]

môre is nog 'n dag [ˈmɔːrə ɪs ˈnɔxə ˌdax] *idiom. colloq.* Tomorrow is another day: equiv. of *Span. mañana.* [*Afk. môre* tomorrow + *is* + *nog* another, still + *'n fr. Du. een* a (*indef. art.*) + *dag cogn.* day]

Only in South Africa, I thought, where 'môre is nog 'n dag', could this happy atmosphere attend a roaring railway engine and coaches running through a small village. Vaughan *Last of Sunlit Years* 1969
Will we Whites sit back, pour out another drink and say: 'Môre is nog a dag?' *Sunday Times* 18.2.73

morena [ˌmɔˈreːna] *n.* Sotho title of respect accorded to chiefs, *cf.* (*i*)*nkosi*, also used as equiv. 'sir' or 'master'. [*Sotho morena* lord, sir]
... the Chief is highly respected by his subjects, who call him Morena. Lanham *Blanket Boy's Moon* 1955
Morena, give my greetings to the other Morena. Tonight I shall come to the Morena's truck and we shall speak again of Mokhotlong. Krige *Dream & Desert* 1953

morgen¹ [ˈmɔːgən, ˈmɔ(r)xən] *n. pl.* ∅. A Dutch land measure used in S.A. until the adoption of the metric system, see *Metrication* ╟A ∼, roughly regarded as the amount of land which could be ploughed in a morning, is just under a hectare, and just over two acres. 1 morgen = 0.856 ha. [*fr. morgen²* (q.v.)]
Dutch farmers ... sometimes make a deficiency of water a pretence for asking for an additional grant of land, a full-sized farm is reckoned from two to three thousand *morgen,* or double that number of acres. Alexander *Western Africa I* 1837
On a piece of ground 4½ morgen (4 hectares) in extent, situated on the eastern side of the town, 71 jukskei pitches were laid out. *Panorama* June 1970

morgen² Morning: see *môre.* [*Du. morgen* morning]

morning gown *n. pl.* -s. *Afr. E.* and *Ind. E.* A dressing gown: see also *gown.* [*unknown, poss. as opposed to nightgown/-clothes*]
'I have even bought a morning gown and slippers. Things I never thought I would ever need,' he said as his wife looked at him proudly. *Drum* 22.9.73

morogo [mɔˈrɔxɔ] *n. Afr. E.* Any of se-

veral species of edible plant sometimes called 'wild spinach' cooked and eaten as 'greens' by Africans. [*Sotho morog* greens]
We found Mr Soko eating mealiepap and morogo out of a pot outside ... a tin shack. *Drum* Oct. 1977

Morolong See *Barolong.*

morsdood [ˈmɔ(r)sˌduət] *adj. colloq. slang.* Dead as mutton/a doornail etc.: see also *mossie.* [*Afk. fr. Du. mors/murs* absolutely, completely, *intensifying prefix,* also *poss. rel. Lat. mors,* death + *dood cogn.* dead]
Ja! he's dead ... He's dead Boesman ... *Morsdood?* Ja. Fugard *Boesman & Lena* 1969

moruti [ˌmɔˈrutĭ] *n. pl.* baruti. Priest, minister: see also *umfundisi* and *fundis:* see quot. at *liretlo.* [*Sotho (mo)ruti* priest, *ba- pl. prefix*]
Yes, Moruti, I was baptised during childhood at the French Mission, by a Moruti who bore your name. Lanham *Blanket Boy's Moon* 1953
... tell me, you who consort with Baruti, how come you to this pass? I should rather have thought to see you following behind the white men who beat the big drum and wear the red caps. *Ibid.*

mos¹ [mɔs] *adv. particle. colloq.* Indeed, in fact, actually, but: used redundantly in S.A.E.: see also quots. at *meid, bry.* and *kaal. cf. maar.* [*Afk. fr. Afr. and Du. im(mers)* indeed, in fact: *thought by some to be abbr.* Yiddish *mozel,* luck]
A person can't live on ice cream and with these teeth in my mouth I can mos only eat soft stuff like ice cream. Philip *Caravan Caravel* 1973
In combination *slang* usu. among children as in 'I did it *for* ∼, or *for* ∼*sie, sig.* for no special reason, 'for the hell of it', 'just because', etc. *cf. sommer.*
It was darem high time some-one did for South African English what Mr Fowler did for the language in Britain. It obviously was not a sommer-for-mossie job and one can only be grateful ... for their splendid contribution. Van Biljon *cit. Sunday Times* 2.11.75
... this hang-gliding. Stone crazy ...You wouldn't catch me jumping off a hill jis for moz like that. *Darling* 4.8.76

mos-² [mɔs] *n. prefix.* Must, of grapes, usu. in combinations ∼*bolletjie,* ∼*beskuit,* ∼*konfyt,* ∼*konfyt jar* all (q.v.). [*Afk. fr. Du. most cogn.* must]

mosbeskuit [ˈmɔsbəˌskœit] *n.* See *mosbolletjie.* [*Afk. mos* must + *beskuit* rusk *cogn. Eng. and Fr. biscuit*]

mosbolletjie [ˌmɔsˈbɔlĭkĭ, -cĭ] *n. pl.* -s. Bun

made of dough leavened with a ferment of must, *mos*[2], instead of yeast, baked together and broken apart: also oven-dried as rusks known as *mosbeskuit* or *boerebeskuit* (q.v.), similar to *U.S. zwieback* (q.v.). [*see first quot.*]

> Mosbolletjies – so-called from 'Mos', the juice of the grape in its first stages of fermentation, and 'Bolletjie', a bun. During the wine-making season the freshly fermented grape-juice is commonly used instead of yeast by the country people. *Duckitt's Recipes ed.* Kuttel 1966
>
> Hilda Duckitt started the day on the farm by making rusks from mosbolletjies and serving them with the early coffee. She used fermented grape juice in these buns instead of yeast. Green *Tavern of Seas* 1947

mosdoppie [ˈmɔsˌdɔpǐ] *n. pl.* -pe. *Sect. Army.* The inner plastic heat-protector lining of a military helmet or *staaldak* (q.v.) also known as a *doibie* (q.v.) See quot. at *balsak*. [*unknown: poss. fr. Afr. doppie* a grape skin: see *mos*, or *fr. dop*, head, shell (empty cartridge, eggshell, husk *etc.*) or motor cyclist's *dop*[5] (q.v.), a crash helmet (or *poss. dophoed* bowler hat)]

moskonfyt [ˈmɔskɔnˌfeɪt] *n.* Thick syrup prepared from grapes: used in cookery, see quot. at *patat*, also second quot. below. [*Afk. fr. Du. most* must + *konfijt* conserve]

> Cream butter and sugar and add coffee essence, egg and moskonfyt. (Use golden syrup if you cannot get moskonfyt.) Gerber *Cape Cookery* 1950
>
> . . . success was achieved with the export of moskonfyt or grape syrup . . . because there was a good demand for it in Britain for the preparation of so-called 'British Wines'. Opperman *Spirit of Vine* 1968

In combination: ~ *jar/bottle* early glass, now collectors' pieces: see quot. at *konfyt*.

> Cape made glass water jugs and most konfyt jars. Fehr *Treasures at Castle* 1963

Mosotho [mɔˈsutǔ] *n. pl.* Basotho. A citizen of *Lesotho* (q.v.) see quots. at *makulu* and *Sesotho*. [*Basotho (Basuto) the pl. form is freq. erron. used for one person. cf. Bantu (pl. form)* (q.v.)]

mossie[1] [ˈmɔsǐ] *n. pl.* -s. The common S. Afr. sparrow *Passer melanurus* similar to the English variety, also called *koringvreter* (q.v.) in the grain season. [*Afk. fr. Du. musch* a sparrow]

> . . . the Bible says, not even a little mossie

can fall to the ground without our heavenly father knows it. Maclennan *The Wake* 1971

> . . . wakes at 6.30 a.m. or so loudly claiming . . . that he's a mossie and carrying on ear-splitting, vulgar conversations in mossie-language with the mossie hoi-polloi outside. *Cape Times* 15.1.73

as dead as a ~ stone dead: see also *morsdood* [*trans. Afk. so dood soos 'n mossie*]

to live on mieliepap and ~*s*, to be on short commons.

> We're thrilled to be building our own place in the sun, even though we have to live on mieliepap and mossies to do it. Letter Sept. 1972

mossie[2] See *mos*[1].

M.O.T.H. [mɒθ] *acronym pl.* MOTHS *M*emorable *O*rder of *T*in *H*ats, a S.Afr. ex-servicemen's organization founded in Durban in 1927 for social activities and the raising of funds for charities. ℙ ~ s are organized into 'Shellholes' with regional leaders known as 'Old Bills': ~ wives are known as ~*was*, and their leaders as 'Lady Billies'.

> The M.O.T.H. organization is not an importation from overseas. Like the . . . protea it is a plant indigenous to South Africa. 31.3.45 *cit.* Franklin *This Union* 1949
>
> 400 Moths Mothwas and Moth children set out on a big march . . . to raise funds for Moth charities. *Rand Daily Mail* 16.3.77
>
> A number of new Moths who recently completed Border service and joined the order will also be competing. *Daily Dispatch* 11.6.79

mother, small, big *n. pl.* -s. *Afr.E.* In African family structure the *big mothers* are the mother's elder sisters and the *small* her younger sisters: see also *mama*, *tata* and quot. at *father*.

motivation P.T. *Sect. Army.* Euphemism for Punishment Drill [Straf P.T.], see also *afkak parade*.

> '. . . gave us bloody motivation P.T. until 10.30 at night the day before we were due to go off on pass – I nearly didn't make it for inspection. *O.I. Serviceman* 10.3.79

moulvie [ˈmulvǐ] *n. pl.* -s. *Ind. E. not freq* in *S.A.E.* An Islamic scholar who gives religious instruction in a *madressa* (q.v.) and is usu. in charge of a mosque; also *mullah* (*rare*). [*Hindi maulvi fr. Arab mawlawi*]

> Only last week a learned moulvie brought to me the lines which I translate for you. They were written by Abul Fazl, the friend and adviser of the great Akbar. Fairbridge *Which Hath Been* 1913
>
> My name is Abdul Wahid, I am Moulvi

at a mosque near Cato Manor. Lanham *Elanket Boy's Moon* 1953

mousebird *n. pl.* -s. Any of several species of *Colius* (Coliidae fam.): see also *muisvoël*. [*prob. trans. Du. muisvogel*]
Two very singular birds, the Caffer finch and mousebird (the latter so called from being the colour of a mouse) have two very beautiful marks on their wings; the rest of the plumage is dark grey. At the commencement of spring their tails begin to grow, and get to such a length as to appear to be an absolute incumbrance. Philipps *Albany & Caffer-land* 1827

(i)mpundulu (bird) [(ĭ)m₁pŭn¹dulŭ] *n.* The *lightning bird* (q.v.) see also quot. at *impundulu* (*erron. form mpundu.*) An evil spirit, about which there are various beliefs. ⸿It is invoked by witches, and freq. cited before judges as the instigator or cause of crime, see quot. at *mamlambo. cf. Canad. hohoq,* thunderbird. [*Ngu. i-mpundulu* – 'bird supposed to be used by women in witchcraft'. *Doke & Vilakazi Zulu Dict.* 1948]
The evil spirit most feared and the most potent is Impundulu the lightning bird . . . Like a vampire it feeds on blood. The people believe that lightning is caused by the Mpundulu streaking across the sky to strike the earth and lay eggs of thunderbolts. Broster *Red Elanket Valley* 1967
In addition to the tokoloshe there was the mpundu bird, which 'kicks a person in the chest' and causes TB. *E.P. Herald* 20.7.73

mqomboti [mkɔm¹bɔtĭ +] *n.* African beer usu. brewed with sprouted *kaffircorn,* (q.v.) see *mtombo:* also *kaffir beer, (u)tshwala, maiza.* [*Xh. mqombothi* beer]
Mdantsane residents may brew mqomboti (African beer) in unlimited quantities provided it was for home consumption and not for sale. *Daily Dispatch* 4.8.72
'Afraid we can no longer afford bottled liquor – will you try a can of homemade marewu, or would you prefer mqombothi?' *Daily Dispatch* 27.6.78
Sorghum not Jabulani, MP told
. . . asked what the department was doing about the sale of jabulani which was now sold under the name 'Mqombothi' and 'Xhosa Beer' in cartons although it had been banned by the Assembly . . . In 1976 the Minister of Justice [Transkei] banned the commercial brewing of Xhosa beer, Jabulani . . . Traditional home brewing of beer was not affected. *Indaba* 6.4.79

mrezan [m¹xezan₁m¹re-] *n. pl.* -s. *Prob. tsotsi language or fly-, flaai-taal* (q.v.). Girlfriend, woman: see quot. at *worry* [*unknown*]

msobosobo See *sobo-sobo.*

(u)mthakati [(u)m₁ta¹gatĭ] *n. and adj. Various sp.:* see *tagati.*

mtombo [m¹tɔmbɔ] *n. Afr. E.* Malt, sprouted grain usu. *kaffircorn* (q.v.), see *uitloop* for brewing *kaffir beer* (q.v.) ; also commercially available: see quot.: ~ *mmela* (beer brewing mix). [*Ngu. (u)mthombo* sprouted grain]
. . . Mtombo-Mmela Home Brew. Stronger, smoother, richer guaranteed pure and healthy . . . mtombo-mmela the strongest home brew. *Drum Advt.* Jan. 1971

much better *adv. m. or degree colloq. poss.* non-S.A.E. *equiv.* 'Very much', 'like anything' etc. See quots. at *snik* and *poegaai.* [*unknown*]

muchi, mucha [¹mutʃĭ, -a] *n. pl.* -s. A loincloth of strings of hide or animal's tails worn by African tribesmen: see also *beshu, moocha.* [*Zu. umutsha* loincloth]
Cetywayo never appeared in European clothes, but wore simply his mutya, and a necklace of lion's or tiger's claws. Buchanan *Pioneer Days* 1934
. . . the menfolk stride along bare-thighed, in swinging umutshas of monkey-tails and hidestrips instead of trousers. Walker *Kaffirs are Lively* 1948

muggie [¹mœxĭ, ¹mǝ-] *n. pl.* -s. *colloq.* Midge, gnat: used of irritating flying insects generally: see also *brommer, gogga, nunu:* erron. *miggie* (pron. sp.) [*Afk. fr. Du. mug* gnat + *dimin. suffix -(g)ie*]
Joan was playing frightfully bad tennis but she suddenly swallowed a muggie and flew into such a rage her game improved 100%. O.I. 1974

mugu See *(i)moogie,* also *moegoe.*

muhle [¹muɬe] *adv. poss. reg. Natal.* Nicely, well: see *mush(ie),* also *kahle* and *hamba kahle.* [*Ngu. (umu)hle* good, well]

muid [mju:d] *n. pl.* -s. A Dutch measure of capacity consisting of four *schepels* (q.v.) and weighing just under 200 lbs (91 kg). In combination ~ *sack. cf. Canad. minot* (1.07 bushels) also called *Canada bushel.* [*Afk. fr. Du. mud cogn. O.E. mydd* bushel, *and Lat. modius* peck]
CORN MEASURE 4 Schepels, equal to 1 Muid – 10 Muids, equal to 1 Load. The Muid of Wheat weighs on an average, about 180lbs. Dutch, being somewhat over 196lbs. English. *Greig's Almanac* 1833
Muid sacks are commonly used in this coun-

try for harvesting [cotton]; on both sides of the opening a riem or cord is fastened and the sack is then hung over the shoulders of the picker. *H'book for Farmers* 1937

muishond ['mœĭs₁(h)ɔ̃nt] *n.* pl. -s, ∅, -e. Skunk: *Ictonyx striatus* of the Mustelidae fam. which emits a fetid smell if disturbed. [*Afk. fr. Du. muishond* weasel, polecat]

Our first victim was a muis-hond, a destructive little animal of the weasel species and very numerous. Philipps *Albany & Caffer-land* 1827

also (*figur.*) Something which smells bad *lit.* as cheese or tobacco, or *figur:* see quot.

. . . it remained the *muishond* of the world with whom nobody wanted anything to do. *News/Check* (12–25th) June 1970

muisvoël ['mœĭs₁fʊəl] *n.* pl. -s. Also *mousebird* (q.v.): any of several of the Coliidae, fruit eating birds, with soft hairy greyish plumage and long tails: mouse-like both in colouring and in their habit of creeping among the branches of trees: *groot/great* ~, see *kwêvoël*. [*Afk. fr. Du. muis cogn.* mouse + *vogel cogn.* fowl]

. . . whole colonies of little birds the Dutch call *muizvogels.* The name signifies mouse bird, they have an appearance almost of grey fur with a tufted crest and long tails. Papa not partial to them since they pierce the fruit. Fitzroy *Dark Bright Land* 1955

multi-national *adj.* In recent S.A.E. official equiv. multi-racial sig. 'mixed' (black and white): see *nation* as *n.* ~*ism.*

Turning to sport . . . said the Government 'has been dragooned by world pressure into creating a facade of multiracial sport under the guise of multinationalism'. *E.P. Herald* 9.11.73

mundani [₁mʊn'danĭ] *n.* Ind. E. rare in S.A.E. The broad ornamented border on that end of a sari which hangs over the shoulder. [*prob. Urdu or Hindi*]

SPECIAL OFFER! India Silk Hindipur SARRIES . . . Narrow and Wide Borders AND MUNDANI. *Graphic Advt.* 18.7.69

munt(u) ['mʊnt(ŭ)] *n.* pl. -s. An offensive mode of reference to an African: ⫿Recent source of racial 'jokes' see second quot. cf. *Austral. boong,* aboriginal. [*Bantu uM(u)ntu* a person pl. *aBantu*]

Words like 'munt' and kaffir' are never heard. Racial discrimination and intolerance have all but disappeared from the capital. *Daily News* 13.5.70

If a . . . midwife is a confinemunt . . . and a . . bank manager is an embezzlemunt . . . then a . . . sex-kitten must be an enticemunt. *cit. Het*

Suid Western 14.12.72

mush(ie) ['mʊʃ(ĭ)] *interj. and adj. colloq.* Great, super: a general term of approbation. cf. *Austral. beaut.* [*fr. Zu. umuhle* good, well *etc.*]

Really . . . is getting on my nerves. If I said 'I've got cancer all over and all my kids have been killed in a smash' she'd still say 'Oh mush!'" O.I. Mar. 1974

muskeljaatkat [₁mœskəl'jɑt₁kat] *n.* pl. -s. Musk-cat, any of several species of genet esp. *Genetta tigrina.* [*Afk. muskeljaat* musk + *kat cogn.* cat]

ONE MUSK CAT (muskeljaatkat), full grown, R10 or nearest offer. *Farmer's Weekly Advt.* 3.1.68

musselcracker, musselcrusher *n.* pl. ∅, (-s) A marine fish of the Sparidae fam., *Sparodon durbanensis,* with exceptionally powerful jaws and teeth, also known as silver steenbras, steenbras, brusher: ⫿~ is also applied to the *poenskop* (q.v.) *Cymatoceps nasutus.* [*fr. powerful jaws*].

Zululanders can expect best results operating out of Mtunzini, where good catches of deepsea reef-fish – including some big musselcracker have been reported. *Daily News* 16.4.71

must *vb. equiv.* of 'Shall' and 'should' in qns ~ unless stressed does not sig. obligation in S.A.E. or *Afk.* also characteristic of (prob. translated) S.A.E. 'officialese', e.g. 'This form must be filled in' etc.; see second quot. and quots. at *beneek, laventel* and have to. ⫿Also acc. R. W. Burchfield now in use in *Brit.* [*prob. trans. moet* shall, should *cogn.* must]

The idea of compulsion in the word *must* is frequently absent in the Cape idiom. Herrman *Note on Cape Idiom* 1959

Anybody who can help identify the woman must phone Det/Sgt . . . at 61111. *E.P. Herald* 22.10.75

What must I wear? Denim boy, dandy or diehard conservative doubts assail them constantly . . . asked three top male fundis for the answers. *Darling* 9.6.76

muti ['mutĭ] *n.* pl. -s. African medicines, spells and herbs, parts of animals or even human bodies, used in therapeutic or pseudo-therapeutic treatment, or in witchcraft or magic. cf. *Canad. medicine* and compounds, *Jam. Eng. obeah.* [*Zu. uMuthi* tree, shrub, hence herb]

It was alleged that I killed Shibongile to use parts of her body for muti. *Post* 28.6.70

'One man . . . told me: "I was drugged, ▶

couldn't see a thing in there. The sangoma kept giving us large doses of strong muti," ' said Lieut. Moloto. *Sunday Times* 9.1.77

In combination ~*man*/*woman*, an African *inyanga* (q.v.), also *herbalist*, (q.v.) *cf. U.S. and Canad.* (*Indian*) *medicine man, Jam. Eng. obeah man*/*woman*/ *samfai man*/*woman* etc., *Austral. koradji*/ *coradgee :* see also quot. at *inyanga*.

Muti men and women receive high payments . . . Muti men find a ready market in gullible sports administrators and sportsmen who think they cannot achieve anything without the use of herbs. *Daily Dispatch* 9.3.72

figur. colloq. sig. 'the doctor' as in 'If he's not better by the morning we'll have to call the ~ *man.*' ~ *shop*/*house*, the retail shop of a ~ *man* or *herbalist* (q.v.) see quot. at *inyanga*, supplying African ~*s* and *mafutas* (q.v.). ⫫Several are run as large scale mail-order businesses even supplying spells against *mamlambo* (q.v.) and (*i*)*mpundulu* (q.v.) and love charms by post.

You may buy all these things only a kilometre or so from the computered heart of Johannesburg. The muti shop is in colourful Diagonal Street. *Sunday Times* 6.2.72

muurkas [ˈmyːrˌkas] *n. pl.* -te. A wall cupboard, usu. built-in: see also *hoekkas*. [*Afk. muur* wall *cogn. Lat. murus* + *kas fr. Du.* cupboard *cogn.* case (+ *dimin. suffix -ie*)]

Downstairs in the reception room, with its stinkwood muurkas and pieces of Delft. Green *Land of Afternoon* 1949

. . . where eighteenth-century furniture was often built-in (the 'muurkassie') or at least kept small and neat (the Georgian book-case) the new furniture was heavy and dominant. Lewcock *C.19 Architecture* 1963

mynheer *n. pl.* -en. See *meneer* and second quot. at *riempie*.

mynman See *mainman* and quot. below.

'That ou's just thinking he's a mynman.' replied the drunk, and added '. . . he mustn't start with me again.' *Forum* Vol 6 No 2 1970

mynpacht [ˈmeinˌpaxt] *n. rare.* The agreement by which one tenth of the claims on an area proclaimed a goldfield were reserved for the use and profit of the owner of the land: see also quot. at *bewaarplaats.* [*Du. mijn cogn.* mine + *pacht cogn.* pact, agreement]

. . . took care that the Mijnpacht, the tenth share reserved for the owner, and the sixty Vergunnings, or Preference Claims . . . should be grouped compactly round this rich area which he had assayed. Brett Young *City of*

Gold 1940

No contract of sale or cession in respect of land or any interest in land (other than a lease, mynpacht or mining claim or stand) shall be of any force. Statute Act No. 68 24.6.57

N

naafi(e), naffy [ˈnafǐ, ˈnæfǐ] *adj.* and *n. Sect. Army. Non S.A.E. acc.* Partridge *Dict. of Slang* '~ used as a pejorative adjective connotes shirking. Services ca 1940.' ⫫ In S. Afr. forces ~ is used of those National Servicemen who slack when they have *mindae* (q.v.), also *forty days* (q.v.) left to serve: also used of or pertaining to a *naf* (q.v.). [*acc. O.I.s, Servicemen*, acronym *no* ambition and *f*-all interest, *acc.* Partridge *fr.* 'Navy's naffy rating, a shirker. Here the initials N.A.A.F.I. are interpreted as standing for *no* aim, ambition or *f*-ing initiative']

naafie – term used for oumanne with min dae – cocky and feel nothing. *Informant* H. C. Davies Letter 2.2.79

naan *n.* See *Indian terms* and quots. at *kalya* and *halim.*

. . . Bakery. Manufacturers of Bread, Cake, Pies and Naans. *Muslim Digest Advt.* Dec. 1967

naar [nɑːr] *adj. colloq.* Queasy, sick. ⫫Also sickening, rare in S.A.E.: *cf. Brit. and U.S.* 'sick' as a term of disapprobation. [*Afk. naar* nauseated (or nauseating) *fr. Du. naar* disagreeable]

I find that that cough mixture often makes my patients a bit naar, try another. O.I., Doctor, 1971

One day your turn. One day mine. Two more holes somewhere. The earth will get naar when they push us in. And then it's finished. Fugard *Boesman & Lena* 1969

na(a)rtjie [ˈnɑtʃǐ, ˈnarkǐ, -cǐ] *n. pl.* -s. *Citrus nobilis :* the soft, loose-skinned tangerine or mandarin orange, used as the flavouring in *Van der Hum* (q.v.) and preserved as ~ *konfyt.* [*Afk. fr. Tamil nartei prob. rel. Arab naranj and cogn. Span. naranja* (*Du. mandarijntje*)]

Van der Hum. You detect the naartjie flavour at once . . . made in the right way, from flavouring extracted from the peel of the Cape naartjie (and not from tangerine essence) it can be very wholesome and satisfying. Green *Tavern of Seas* 1947

Nartjie konfyt Mandarin orange preserve. My Grandmother's Dutch Recipe. *Duckitt's Recipes ed.* Kuttel 1966

~ *girls* (*poss. obs.*): Jocular S. Afr. Airways ground hostesses.

South African Airways 'naartjie girls' – ground hostesses in their bright orange uniforms – have been given a new look . . . Replacing the 'naartjie' hat is a chic bowler. *Daily Dispatch* 16.10.71

naboom [ˈnɑˌbʊəm] *n*. Also *melkboom* any of several species of tree *Euphorbia*, some of which contain rubbery latex, including *E. ingens*, a highly toxic species with corrosive latex said to be used by Zulus in minimal doses, often with fatal results, for cancer (hence *kankerbos* (q.v.)) also for dipsomania and as a purgative. [*Afk. fr. Hott. gnap* powerful + *boom cogn. Ger. Baum* tree]

All the tree Euphorbias described are commonly known as 'naboom'. According to the authors of 'The Succulent Euphorbiaceae' 'na' is a corruption of a Hottentot word 'gnap' meaning strong or energetic, and emphasizes the vigorous habit of growth of the trees. Palmer & Pitman *Trees of S.A.* 1961
Also in place names Naboomspruit, Naboomkoppies.

nadors [ˈnɑˌdɔ(r)s] *n*. Excessive thirst after drinking: see also *babbala(a)s*. [*Afk. na* after + *dors* thirst]

This is the time of the year, alas, when people are prone to wake up feeling tired and listless with a headache, maybe even a little sick, and of course, with a great *nadors. Cape Times* 14.12.72

naf [næf] *n.* pl. -s. *slang.* In S.A. used for an ineffectual person, a fool, 'drip' or 'wet', see also *naafi(e), naffy* ¶ *acc.* Partridge *Dict. of Slang* 'gnaff or n'aff a petty thief or informer low Glasgow mid C 19-20', '*cf.* Parisian s. *gniaffe* a term of abuse for a man: *prob.* of the same origin as gonnof.*'

*Hebrew gannav, thief

Talk about insults . . . to tell you the truth . . . I never felt such a total and utter naf in my entire life. *Darling* 29.3.78

nagana [naˈgɑnə, -a] *n*. A disease of domestic animals esp. cattle and horses, almost always fatal to *unsalted* (q.v.) animals, caused by a trypanosome transmitted by the bite of the *tsetse fly* (q.v.) see quot. at *salted*. [*Zu. u-nakane* cattle sickness caused by tsetse fly, *also a variety of grass, believed by some to cause the disease*]

It was DDT that removed much of the insect threat to health . . . In Natal, particularly, it overcame the dreaded nagana by wiping out nearly all tsetse fly. *Daily News* 16.10.70

. . . involved in countless anti-tsetse fly . . . culling operations . . . 'I refer to the anti-Nagana shooting campaigns of 1929-30 and 1942-50 which virtually exterminated all the large game species' *Tribune* 3.6.79

nagapie [ˈnaxˌɑpĩ] *n. pl.* -s. *lit.* 'Night-ape': also known as bush baby: a lemur of nocturnal habits [*Galago spp.* usu. *G. maholi*] [*Afk. nag fr. Du. nacht cogn.* night + *aap* monkey + *dimin. suffix -ie*]

A young scientist is studying the world of the bushbaby, or nagapie . . . it sounds like a rather cross baby crying. *Radio S.A.* 9.4.73

I nearly cried and I made my eyes all big to absorb the tears. That's when they took the picture . . . I look very nervous . . . My dad said I looked just like a nagapie. *Darling* 20.12.78

nagmaal [ˈnaxˌmɑl] *n.* (*pl.* -s). The sacrament of Holy Communion celebrated quarterly in the *Dutch Reformed Church* (q.v.) an occasion for large numbers of country people to visit the town, some occupying their ~ *huise* (q.v.) others wagons in the *outspan*² (q.v.) see second quot. also quot. at (2) *inspan*. In combinations ~ *tent*, see quots. ; ~ *clothes*, see *kisklere*, also quot. at *haak-en-steek;* and ~ *wyn* (q.v.). [*Afk. nag fr. Du. nacht cogn.* night + *maal cogn.* meal]

. . . the great season for traffic is the period for the quarterly administration of the Sacrament (or *nachtmaal*) on which occasion large numbers of farmers . . . assemble, and the place assumes the appearance of a fair rather than an assemblage for the celebration of a solemn religious ordinance . . . an attendance at church at such seasons is considered an imperative duty . . . they will often travel hundreds of miles with their families to perform it. *Cape of G.H. Almanac* 1843

But both before and after the actual *nagmaal*, the night-meal, the supper, the Lord's Supper, there was much visiting in the outspan, relatives and friends rejoicing in this rare occasion of reunion. Mockford *Here are S. Africans* 1944

. . . about 300 people who had come from miles around in cars, in bakkies . . . wept into the sleeves of seldom-worn black suits bought decades before to last a lifetime of weddings, Nagmaals and funerals. *Sunday Times* 24.11.74

Green Waterproof Canvas . . . Nagmaal tents 6ft x 7ft x 6'3" R28 *Farmer's Weekly Advt.* 3.1.68

A common sight from mid-19th century to early 1930s was reconstructed at Rustenburg – Nagmaal tents pitched beside the church for a week-end of worship and socialising. *Pano-*

rama Oct. 1975

nagmaalhuis [ˈnaxˌmɑlˌ(h)œïs] *n. pl.* -e.
A small town-house owned by a farmer
from the outlying district, also *kerkhuis:* and *town house²* (q.v.) see quot.
[*Afk. nagmaal* (q.v.) + *huis cogn.*
house]

A group of old 'nagmaal huise'... the oldest
buildings in Alexandria were built by the early
Boers ... to live in when they came together
for 'nagmaal'... This house was originally the
'nagmaal huis' of the Muller family ... Her
last purchase is the 'kerk' or 'nagmaal huis'
of the Scheepers family. *Evening Post* 2.11.74

nagmaalwyn [ˈnaxˌmɑlˌveɪn] *n.* Sweet fortified wine, similar to that used in the
administration of the Sacrament, sold
commercially under several brand
names: see quot. at *konsistorie.* [*Afk.
nagmaal* (q.v.) + *wyn cogn.* wine]

His ... wine was in great demand as a
'*Nagmaal*' or Communion Wine ... Oom
Abram Mouton ... of the historic farm Brak-
fontein ... made a '*Nagmaal*' wine almost but
not quite equal to the Worcester wine. Leipoldt
300 Yrs of Cape Wine 1952

naloop [ˈnɑlʊəp] *n.* The 'faints' or last,
weak runnings of a brandy still, opposed
to the 'heads' or *voorloop²* (q.v.). ⫿ It
is said by many to be of medicinal value:
see second quot. as *modifier.* [*Afk. fr.
Du. nalopen* to follow, run after]

Alcohol meters ... have no place in the
tradition. Every farmer has his own infallible
method of telling when his brew's reached the
right strength and the *naloop* begins. *Family
Radio & TV* 19-25.9.77

Tannie ... has developed her herbal *naloop*
practice to the point where she can administer
a blend for any complaint. 'It works for everything,' she says. *Ibid.*

Nama(qua) [ŋəˈmækwə, naˈmakwa] *n.
prop. pl.* -s. **1.** A member of the Hottentot people giving their name to ~ *land*,
speaking *Nama*, a dialect of *Hott.* [*name
of language* + *suffix -qua,* man]

The Namaquas are a race of Hottentots ...
They are a pastoral people, resembling the
Korannas, and the aboriginal tribes of the
Colony, in their general characteristics; living
chiefly on milk, addicted to a roaming life; and
of a disposition mild, indolent, and unenter-
prising. Thompson *Travels II* 1827

2. *prefix.* Found also as ~ *land* prefixed to names of flora and fauna,
~ *gousblom/marigold, Venidium fastuo-
sum;* ~ *daisy, Dimorphotheca sinuata*
(*aurantiaca*); ~ *patrysbossie, Walafrida
saxatilis* in which the ~ *partridge,*

Pterocles namaqua shelters (see quot. at
korhaan); also ~ *dove* (q.v.) and ~
sheep (q.v.).

Namaqua dove *n. pl.* -s. *Œna capensis,* of
the Columbidae: a common species in
arid areas.

16 Sept 1811 ... a small bird called the Namaquas duif (Namaqua Dove) a name which
is also given to the Columba capensis. Burchell
Travels I 1822

Namaqua sheep *n. pl.* ∅. A variety of the
indigenous *fat-tailed* or *Afrikander sheep*
(q.v.) somewhat flat ribbed cf. *ronderib*
(q.v.) hardy and resistant to arid conditions.

The Namaqua Afrikaner Sheep ... This is a
non-wooled sheep ... long legged and a good
rustler. Owing to its hardiness and capacity
for travelling long distances it can stand up
better than any other to the droughts. *H'book
for Farmers* 1937

Namibia [naˈmɪbɪə] *n. prop.* Name adopted by the United Nations Organization
for South West Africa in 1966 in
anticipation of its becoming independent. See also quot. at *Herero.* [*fr. Namib
a desert on the S.W. coast of Africa*]

United States influence in South Africa must
be used to persuade the South African Government into opening direct negotiations with
Swapo as soon as possible, Mr Sean MacBride,
United Nations Commissioner for Namibia,
said in Lusaka yesterday. *E.P. Herald* 30.8.76

derivatively ~ *n adj. or n,* an inhabitant
of ~ .

nanny *n.* A mode of address to African
women now often objectionable and
becoming *rare, sisi* (q.v.) having in
some areas taken its place. ⫿ *Also occ.*
a mode of reference to a general servant,
not a nursemaid (*cf. Brit. also S.A.E.
nanny*) poss. to avoid term *girl* (q.v.) or
even 'servant'.

The woman in the shop kept calling me 'Nanny' – surely she could see I wasn't somebody's
servant? O.I. African Medical Student, Durban
1953

When I see the native girls walking ... I
usually stop. 'Do you want a lift nanny?' I ask
them, but apparently 'nanny' is a term that
native girls hate – the same as their being called
'my girl' or 'Mary'. Gough Berger *Where's
Madam* 1966

naras [ˈnaras] *n. pl.* ∅. *Acanthosicyos
horrida* which grows in the Kalahari
and South West Africa bearing the
spiny ~ fruit which inside resembles a ·
melon and serves as both food and drink

in these arid areas. [*Hott.*]

... the ripe 'naras fruit, which served them for food and water ... The bay people catch and eat fish after the 'naras is out of season, and the carcases of whales, killed by the crews of whaling ships, afford them savoury repasts in the months of May, June, July, and August. Alexander *Expedition II* 1838

nartj(i)e See *na(a)rtjie.*

nas(s)ella [nəˈselə] *n. Stipa trichotoma* Also ~ *grass* and ~ *tussock* or *polgras:* a grass introduced from S. America in the early 1900s: now, being non-nutritive, an encroaching plant pest found from the Cape to the Drakensberg. [*fr. former generic name Nassella.*]

A national campaign should be launched by the State for the eradication of nasella grass ... He said nasella completely overpowered natural plant growth, it was a threat to grazing, flourished in high rainfall areas and the seeds remained active for 15 years. *Daily Dispatch* 28.7.78

nas(ter)gal(bossie) [ˈnas(tə)(r)ˌxalˌbɔsĭ] *n. pl.* -s. *Solanum nigrum* or black nightshade also known as *nagskade, nagtegaalbossie* (with numerous corruptions and sp. forms): also as *sobosobo* (q.v.). ⫽The berries were used for making ~ *ink* in the early days. [*corruption of Du. nacht cogn.* night + *schaduw cogn.* shadow, shade]

nasgal(bossie) ... a native of Europe and introduced into South Africa about 1652 ... a common garden or rural weed. The globose purple-black berries are eaten by children, poultry and birds and ... made into jam. So far apparently no cases of the berries being toxic to humans have been recorded ... the green berries of somewhat wilted plants are fatal stock poisons. Smith *S. Afr. Plants* 1966

Nat [næt] *n. pl.* -s. A member of the National(ist) Party: *attrib.* of or pertaining to practices or policies of the party: in combination ~ *and Sap* [nat ən sap] (see *Sap*): the ~*s* the Nationalist Party, also the Government: see second (*figur.*) quot. at *off-load.* [*abbr.* national(ist)]

He wanted consensus on matters of common interest where Nat and Sap would stand together. *E.P. Herald* 13.6.73

He compares the United Party to the Nats. He cannot give any good reasons to back up his nonsensical statement because the United Party is far superior to the Nats or any other political party. *cit. Evening Post* 19.5.73

Natal whisky *n.* see *Cane spirit.*

nation *n. pl.* -s. **1.** *hist.* An African or other indigenous tribe.

... some extraordinary rumours had reached him a few days ago, respecting an immense horde, or nation, who were said to be approaching from the north-east and who were laying waste the country. Thompson *Travels I* 1827

2. Used as equiv. of race or ethnic group: see quot. at *XDC.*

During the year there has been increasing emphasis by National Party spokesmen on the 'multi-national' (as opposed to the 'multiracial') nature of the population of South Africa. There were considered to be a white, a Coloured, an Asian, and eight distinctive Bantu nations. *S.A.I.R.R. Survey* 1969

3. In combination e.g. *white* ~ prob sig. group, race.

... also made it plain in his speech on Monday that when the Broederbond Government speaks of the 'White nation of South Africa' it means only the 2 000 000 Afrikaners or 'Boerevolk' as he calls them. *Evening Post* 9.9.72

National Road *n. pl.* -s. A major road built and maintained by the Government-appointed National Transport Commission, numbered N1, N2 etc. ⫽There are no toll-roads in S.A. *cf. district road.*

The national road from the east to Cape Town runs through Riviersonderend, Caledon, Botrivier and Grabouw, and is linked by tarred roads to Bredasdorp in the east, Villiersdorp in the north-west and the coastal resorts. *Argus* 16.9.72

... snapped up a piece of ground ... only a quarter of a mile from the national road ... Ian ... steered his car along the bumpy dirt road that led off the national road to his farm and factory. Louw *20 Days* 1963

~ *camp.* A semi-permanent community of ~ workers and their families usu. in prefabs., caravans, *rondavels* (q.v.) or occ. tents: see *padkamper.*

National Scout *n. pl.* -s. *hist.* A Boer deserter who joined the British forces: see *joiner* and first quot.

The *joiners* ... defected openly and accepted remuneration ... from the enemy ... as Town Guards ... and National Scouts ... Du Toit ... decided to join forces with the British: he felt his duty lay in this way to help bring the unhappy and destructive war to an end. He ... joined the National Scouts, a military movement consisting of Republican deserters. These and similar movements developed from so-called Burgher Peace Committees ... formed as early as 1900. Brits *Introd.* (1974) Du Toit *Diary of a National Scout* 1900-1902

'Robbers, murderers, adulterers and fornicators the Lord and I can forgive – but National Scouts? Never!' 'So' said Oom Koos ... 'my old friend Jakob was sentenced to life imprisonment . . .' Welch *The Brothers* (*unpub. story*) 1978

National Service *n.* As elsewhere obligatory military training for young men. ~ *men*, army trainees undergoing their ~ training, not career soldiers, see also *P.F.*, *blougat* (*blouie*), *ou man*, *vasbyt*, *min dae*, and *varkpan*.

native *n. pl.* -s. and *modifier.* A member of any of the *Bantu* (q.v.) tribes of S.A. but never sig. a *Coloured* (q.v.) South African. Still in freq. use esp. by elderly White speakers, though superseded in Governmental use by *Bantu* (see fourth quot.) and in the Press by *African* or *Black.* ⫽Now a pejorative term to many speakers both black and white. See also *African*, *Kaffir.*

Baptized five Adult Natives on their profession of faith in our Lord Jesus Christ. Shaw *Diary* 1826
To prove its simplicity a Native will do the shearing. *Friend* 25.8.30
Native means a person who in fact is or is generally accepted as a member of any aboriginal race or tribe of Africa. *Population Registration Act. no. 30 of 1950*
Subject to the provisions of this Act, there is hereby substituted for the words 'native', 'Native', 'natives' and Natives' wherever they occur in any law, the words 'Bantu' 'Bantu', 'Bantu' and 'Bantu' respectively. *Subsection 100 of Bantu Laws Amendment Act No. 42 of 1964*
Their concept of a Non-White person is that he is a 'Kaffir' a 'Native' a 'boy' a 'girl' or a maid. Neame *History of Apartheid* 1962

Also *attrib.* and in combinations: ~ *area*, (see *reserve*); ~ *beer*, *Kaffir beer* (q.v.); ~ *boy*, see *boy*; ~ *cattle*, (see *nkona*); ~ *girl*, (see *girl*); ~ *law*, traditional African law; ~ *location*, see *location*; ~ *reserve*, see *reserve*. *blanket* ~, an African adhering to tribal dress; *school* ~, an educated African.

natuurlik [na'ty(r)lɪk] *adv.* Naturally, of course. [*Afk. natuurlik*, naturally]

At last! South Africa's own keyboard. From . . . Natuurlik. . . . a truly South African typewriter . . . all the accents found in Afrikaans plus the typical ''n'. *Sunday Times* 9.4.78

ndambola [n̩ˌdam'bɔla] *n. Afr. E. prob. urban slang:* African beer, liquor: see quot. at *madolo* [*unknown*]

I fritter away my dough at the . . . beerhall sipping Ndambola laced with wine. *Drum* 8.8.74

Ndebele [nde'beːle] *n. pl.* ama-, -s
1. A people of the N. Transvaal noted

for their *beadwork* (q.v.) and style of building; see *lapa²*; also the *Matabele* (q.v.): a member of either of these peoples.

. . . Chieftainess Esther Kekana, head of the 10,000 strong Mandebele tribe in the Northern Transvaal. *Drum* Oct. 1970
The Ndebele beadwork . . . considered by connoisseurs to be the best in the world . . . has attracted interest from at least four overseas countries. *Rand Daily Mail* 5.4.71

2. The language spoken by the ~, known also as *Sindebele*, a dialect of *Zulu.*

My mother language is Shona and I can speak English and Ndebele. *Drum* 18.1.73

ndiblishi [n̩'dɪbˌlɪʃi] *n. Afr.E.* prob. *obs.* A penny. ⫽Also, Xh. a farthing (*obs.*) [*Zu. fr. Du. dubbeltje* (q.v.)]

Dubbeltje – the Dutch name for a penny; the word is, however, sometimes used for money generally. This word is corrupted by the Natal native into 'Deeblish'. Pettman *Africanderisms* 1913

nè [neː] *qn. particle.* 'Isn't that so?' a tag qn. or qn. particle inviting the assent and participation of the listener *cf. Ger. nicht wahr?* ⫽Similar in sound, placing and meaning to the *Japanese* tag qn. *nee?* See also *not so?* and quot. at *make a plan.* [*poss. contraction of nie waar nie? cogn. nicht wahr?* isn't that true/so?]

Take for instance the story of the ark. You know about Noah's ark, nè? Maclennan *The Wake* 1971

neder ['nɪədə(r)] *adj. prefix.* Low(er): found in S. Afr. place or other names e.g. Neder Paarl, Nederburg. *cf. Brit. Nether Wallop.* [*Du. low(er) cogn.* nether]

Nederduits(c) Gereformeerde Kerk ['nɪə-dərˌdœits(ə) xəˌrefɔr'mɪərdə 'kerk] *n. prop.* Also known as *Ned. Ger. Kerk*, *N.G.K.*, *the Kerk*, *Dutch Reformed Church* (q.v.) and *D.R.C.*

. . . theologian and former editor of Die Kerkbode, mouthpiece of the Nederduitse Gereformeerde Kerk, the most powerful of the three Afrikaans churches. *E.P. Herald* 7.10.74

nee, ja- ['jaˌni, -nɪə] See *ja-nee.*

neef [nɪəf] *n. pl.* -s. Nephew, cousin: mode of address or occ. reference used by uncles and aunts to nephews, or older people to boys not related to them: also *nefie* (q.v.) *cf. swaer* (q.v.) and *Brit. 'coz'.* [*Afk. fr. Du.* cousin *cogn.* nephew *fr. Gk. nepos*]

... photos of relatives were shown them, ... This is your Oom Davie, whom the English murdered at such a place; this is your Tante Sara and your poor little neef Koos, whom the English murdered at some other place. O'Connor *Afrikander Rebellion* 1915

nefie ['nɪəfĭ] *n. pl.* -s. Cousin, nephew: an affectionate mode of address or reference usu. from an older man to a younger: also as in *neef* (q.v.) used to or of a boy by his older relatives: see also *niggie.* [*Afk. neef* cousin *cogn.* nephew + *dimin. suffix -ie*]
It is seldom now that I meet a man twice my age, but there it was and he called me nefie. Green *Grow Lovely* 1951

negative, uses of Various uses of *neg.* are characteristic of S.A.E. esp. *no* (q.v.) and regular use of the *neg.* interrogative in making requests e.g. Won't you sit down so long? Won't you lend me a pencil? etc. where *neg.* form is used as a signal of tentative request, or to avoid the apparent imperative in 'Will you ....'

negotie winkel [nə'xʊəsĭ, vɪŋkəl] *n. pl.* -s *hist.* Trading store, retail shop. [*Du. negotie (Afk. negosie)* trade, commerce + *winkel* shop]
Old Bertram had a Store and had over his door Negotie Winkel – the Settlers thought it was his name, and always called him Old Nigerty Winkle. Stubbs *Remin. I* 1876
Shop windows played a small part in commerce until after 1850, and a visitor to Cape Town in 1823 wrote that all shops were in private dwelling houses, distinguishable only by the notice 'Negotie Winkel' over the doors. Gordon-Brown. *S. Afr. Heritage* 1965

nek [nek] *n. pl.* -s. A raised narrow ridge or strip of land usu. between and connecting two mountains. [*Afk. fr. Du. prob. fr. Eng.* neck 'isthmus, cape, promontory or mountain pass.' *Webster's Third International Dict.*]
... the sea once covered the large areas of the Cape Peninsula, including both the narrow nek of land at Kommetje, and the low coast where Grotto beach lies. Green *Few are Free* 1946
Also found in S. Afr. place names e.g. Brooke's Nek, Nicholson's Nek, Kloof Nek, Qacha's Nek.

nek-strop ['nek₁strɔp] *n. pl.* -s. See *strop* and quot. at *yokeskey.* [*cogn.* neck strap]
The yokes are straight and pierced ... to receive ... two straight pegs, one on each side of the ox's neck, and having notches on their outer sides to receive the *nek-strop* (neck strap). Burchell *Travels I* 1822

nenta(bossie) ['nenta₁bɔ̆sĭ] *n. pl.* -s. Hottentot name for various species of *Cotyledon* esp. *C. ventricosa, C. cacalioides, C. wallichii* and *C. decussata,* causing *krimpsiekte (Cotyledonosis)* in stock. [*Hott. (t)nenta* plant name]
The Hottentots called by the name of *Nenta,* a plant ... which was said to be poisonous to sheep. Thunberg *Travels II* trans. 1795
The disease itself is also so called. *Krimpsiekte or Nenta.* H'book for Farmers in S.A. 1937

nerine, nerina [nə'rin(ə)] *n. pl.* -s. *Nerine sarniensis* or Guernsey Lily or *N. bowdeni* a genus of the Amaryllidaceae also called *berglelie* as is *Vallota speciosa,* the George or Knysna lily. [*fr. Lat. name*]
Nerine sarniensis ... the tradition attached to the species being that a vessel bound for Holland from the East Indies via the Cape had a quantity of bulbs on board and was wrecked on the Guernsey (formerly Sarnia) coast in the early part of the 17th century. Smith *S. Afr. Plants* 1966

nerves, on my, on his, on her *predic. adj. phr. substandard.* Tense, edgy. [*trans. substandard Afk. op my/sy/haar senuwees* (nerves)]
I'm always 'on my nerves' ... when I'm expecting visitors. O.I. (Woman) 1969

neuk [nĭœk] *vb. trns. slang.* To thrash, beat up: also harrass, pester; *poss. rel. Scottish neuk vb,* to corner, outwit, deceive, humble. [*Afk. neuk* to thrash, plague *poss. cogn. knock*]
They wound up by asserting that they would straightaway 'nierk' him. 'Nierking' not being a very pleasant proceeding for the 'nierked', Barnard asked 'What for.' 'Never mind. We will nierk you,' was the only reply vouchsafed. *E.P. Herald* 24.5.1910

never *adv. or pseudo vb. substandard.* A substitute for *didn't* esp. among children as in such exchanges as 'You did!' 'I never!' but also in other S.A.E. speakers where ~ *sig.* standard 'not ever' would be inappropriate: see quots. [*prob. fr. Afk. nooit* never, used as emphatic negative]
I appeared on the BBC ... just sitting there staring into the camera. I never blinked once! *Darling* 27.10.76
She was expecting her mother at a flat she was decorating and became concerned when she never arrived. *E.P. Herald* 19.11.76

N.G.K. ['en₁xɪə'ka, 'en₁dʒi 'keɪ] *n. Nederduits(e) Gereformeerde Kerk* (q.v.), the

Dutch Reformed Church. [*acronym*]
It was attended by eight WCC member churches, including the NG Kerk and NH Kerk, who, among other things, decided that there were no Scriptural grounds to prohibit mixed marriages. *Sunday Times* 1.9.74

ngoma [ŋ'gɔma] *n.* An African drum: see quot. [*Ur-Bantu ngoma* drum *cf. ghomma* (q.v.) *Ngu. Zu.* (*iz*)*ingoma* 'Dance song performed at festivals esp. that of the first fruits, royal song, national anthem'. *Doke & Vilakazi Zulu Dict. 1948: also Swahili for* 'song']
. . . the baritone of the mirumba and the deep bass of the ngoma, thud out an intoxicating symphony in praise of sex and its mystic role. Bulpin *Lost Trails* 1951
In combination ~ *dancing*, African tribal dancing esp. by mine workers, often arranged as entertainment for visitors to a mine: see also *gumboot dance. cf. Canad.* (*Indian*) *drum dance.*
Prof. Kirby states that the word ngoma is used by the Arabs trading on the East African coasts to describe their dances. Du Plessis & Lückhoff *Malay Quarter* 1953
Parading proudly down Durban's Golden Mile in the procession were the traditional 'Zulu Warriors', the Ngoma dancers. *Panorama* 1974

ngoni [ŋ'gɔnĭ] *n.* **1.** A form of *Nguni* (q.v.) **2.** *poss. erron.* form of *iNkone/a* (q.v.) indigenous native cattle (see also *Zulu*) of various colours including red and black. [*fr. Ngu. inkone* 'a beast black, brown or red with white patch on the back.' *Doke & Vilakazi Zulu Dict.* 1948]
The Ngoni cattle of Red Nkona pattern have been improved over the past 20 years for beef, milk and hornless; bred and reared on controlled heartwater and redwater veld. *Daily Dispatch Advt.* 15.5.73

ngubu [ŋ'gubŭ] *n. rare* African clothing formerly the (1) *karos*(*s*) (q.v.) see quot. at *blackjack* used of a blanket esp. among Xhosa speakers; among Zulu speakers sig. clothing in general. [*Ngu. ingubu Xh.* blanket, *Zu.* clothing]
From the shoulders hangs the *ungoobo*, kaross, or mantle of softened hides, worn with the hair near the body, and fastened with a thong at the neck. Alexander *Western Africa I* 1837

Nguni [ŋ'gunĭ] *n.* A group of S.E. Bantu peoples comprising *Zulu, Xhosa* and *Swazi* all (q.v.), also the group of languages of these peoples. [*prob. fr. Zu. nguni sig.* one of ancient stock]
. . . the Nguni – in which tribes like the Zulus,

Swazi and the Xhosa are historically and traditionally welded together – form the largest ethnic group in the townships. Venter *Soweto* 1977
The youth came closer. Three scars formed a triangle on his left cheek – the markings of an Nguni tribe. It was incongruous, an ancient sign on the face of a modern youth in denims. *Ibid.*

N.H.K. ['en₁ha'ka, 'en₁eɪtʃ'keɪ] *n. prop.* *N*ederduits *H*ervormde *K*erk: see quot. at *N.G.K.* [*acronym*]

nie-blankes ['nĭ₁blaŋkə(s)] *n. usu. pl.* Non-White(s): freq. seen on notices, doorways etc. usu. in combination ~ *alleen* (only): see *blanke*(*s*). [*neg. prefix equiv.* non- + -*blanke* white person *cogn. Fr. blanc*]
The sad thing is it IS true . . . as witness a letter I received recently, postmarked London and starting: 'Dear Juby, I miss the sunshine and the signboards reading "Nie-Blankes Alleen" . . .' *Drum* Sept. 1968

nierknypers See *klapbroek.*

niet [nĭt] *neg. particle.* Not: found in S. Afr. place names in combination with *imp. vb.* or *vb. partic.* e.g. *Twist* (q.v.) -*niet* (quarrel), Terg-niet (tease), Niet-verdient (q.v.) (deserved). [*Du. niet* not]

nigger-ball *n. pl.* -s. Large hard spherical black aniseed flavoured 'suck sweets' which change colour in successive layers. *cf. Austral. black-ball* usu. a humbug. [*fr. shape and colour*]
'Life is like a nigger ball, hard but nice,' . . . Life, in its blacker moments, appears so hard that the sweet doesn't come through for a long time. Yet, like a nigger ball, the sweet will eventually come through as will the layers of colour. *S. W. Herald* 14.5.71

niggie ['nĭxĭ] *n. pl.* -s. A mode of address or reference to a female cousin or niece, also to a young girl unrelated to the speaker. [*Afk. fr. Du. nicht* niece, cousin + *dimin. suffix* -*ie*]
Ouma du Preez on the riempie settee . . . said to me: 'Is dit nie mooi nie? – excuse – I mean toch isn't it beautiful mij niggie?' Stormberg *Mrs P. de Bruyn* 1920

niks [nĭks] *neg. prn.* Nothing: *cf. U.S. nix*, no, 'nothing doing': in combination ~ *doen* (do), a policy of inaction, also idleness. [*Afk. fr. Du. niets* nothing]
. . . 'What's the matter with your eyes?' Frank asked. 'Niks' the man replied sniffing. Jacobson *Dance the Sun* 1956
. . . white civilization . . . was in danger, they said, because Smuts stood for *niksdoen*, for

'letting the situation develop', which meant letting white civilisation drift on to the rocks. Hancock *Smuts II* 1968

nipinyana, nipinyane [ˈnĭpĭnˌjana, -e] *n. Afr.E. prob. urban slang.* A 200 ml 'nip' bottle of liquor: see also quot. at *madolo.* [*Eng.* nip *quarter bottle* + *Bantu dimin.* suffix *-inyane sig. usu.* 'nice or dear little . . .']
Now this cherrie who first asks Kid Fall to buy her a nipinyana comes up with a good suggestion that we should . . . dump him in the bedroom . . . to snooze it off. *Drum* 8.1.74

njebe See *ntshebe.*

(i)nkona, (i)nkone [(ĭ)ŋˈkɔna] *n.* Also *ngoni, nguni* (q.v.) Indigenous breed of cattle. [*fr. Ngu -nkone, -nkona* white patched or spotted (on the ridge of the back)]
(i)nkone . . . a beast black, brown, or red with a white patch on the ridge of the back. Doke & Vilakazi *Zulu Dict.* 1948

(i)nkosana [(ĭ)nkɔˈsana] *n.* See *(i)nkosi.*

(i)nkosazana, -zaan [(ĭ)ŋˈkɔsaˌzɑn(a)] *n. Afr.E.* A mode of respectful address by Africans to an unmarried woman of the upper class, black or white. [*Ngu. inkosazana* mistress (unmarried)]
A native umfaan in a white suit of kitchen clothes came trotting down the path. 'Tea time, nkosazana' he told Jessica, showing twenty eight perfect teeth. Collins *Impassioned Wind* 1958

Soon after his nineteenth birthday, Solomon . . . decided to get married. 'Nkosazana,' he told the girl of his dreams, 'I must have you for my wife.' 'I am not your princess,' teased . . . Maria. 'If you are serious, discuss it with my parents.' Venter *Soweto* 1977

(i)nkosi [(ĭ)ŋˈkɔs(ĭ)] *n.* **1.** A mode usu. of address to a male superior sig. Master, Chief, Lord: also *dimin. (i)nkosana* (q.v.) [*Ngu. (i)kosi* chief, lord, sir]
–Inkosana? That's little inkosi, isn't it?
–It is little inkosi. Little master, it means. Paton *Cry, Beloved Country* 1948
2. Thank you, sig. obligation or respect to the giver, see *enkosi.*

(i)nkosikazi [(ĭ)ŋˈkɔsĭˌkaz(ĭ)] *n.* Lady, Madam: a mode of, usu. address, also reference to a woman of the upper class, and esp. in Natal equiv. 'madam' used to White women: see quot. at (2) *mafuta.* [*Ngu. inkosikazi* married woman of higher class, often a wife of a chief]
Fastened to his arm by a thong . . . was the great axe . . . and each man as he came up saluted the axe, calling it '*Inkosikaas*', or

Chieftainess. Haggard *Nada* 1895

Nkosi Sikelel'i Afrika [ŋˈkɔsĭˌsĭkəˈleːl(ĭ) ˈafrĭka] *n. prop.* God Bless Africa, a hymn rapidly being accepted as the national anthem of the *Black* (q.v.) people: see quot. at *Stem, die.* [*Xh. inkosi,* lord, chief + *sikelela* bless + *iAfrika* Africa]
Nobody could fail to be stirred by the stately *Nkosi Sikelel' iAfrika,* which has been chosen by the Transkei as its national anthem. *Daily Dispatch* 11.11.75
The national anthem of Transkei is 'Nkosi Sikelel' iAfrika, a well-known Xhosa song in Southern Africa. The first verse was written in 1897 by a member of the Tembu tribe, Enoch Sontonga. . . . He intended publishing 'Nkosi sikelel' iAfrika, but died about 1901 before he was able to do so. It is not known who wrote the music. *Panorama* July 1976

no *particle.* See also *negative, uses of:* ~ is used in various non-standard ways in S.A.E. [(1) and (2) *prob. fr. Afk. nee* as *sentence initiator esp. in formulating replies to qns.* (3) *fr. ja-nee* (q.v.)]
1. ~ sig. 'yes' or a willingness to comply.
Why do South Africans all say 'No' when they mean 'Yes'? O.I. (British Immigrant) 1968
No, that'll be fine. We can do that for you easily. O.I. (shop assistant) 15.10.74
2. ~ used redundantly e.g. Q: How are you? A: No, I'm fine now. ~ See quot. at *boykie, boytjie.*
'But why do you ask?' Thereupon Gysbert van Tonder said no, it was nothing. It was just something Jurie Steyn had been mentioning . . . Bosman *Jurie Steyn's P.O.* 1971
3. In combination *Yes-no* (q.v.). An emphatic affirmative e.g. Yes-no, she's completely better now.
Yes-no, rhenosterbos isn't a problem here thank goodness. It really can damage your veld. O.I. (Farmer) 1972

noem-noem-(bessie) [ˈnŭmˌnŭmˌbesĭ] *n. pl.* -s. Also *num-num :* see quot. at *amatungulu* the fruit of either of two species of *Carissa, C. macrocarpa,* see *amatungulu,* or *C. bispinosa,* the 'small amatungulu' [*prob. onomat. sound of enjoyment* + *bessie* berry]

nogal [ˈnɔxˌal] *particle. colloq.* Also *nog:* in S.A.E. equiv. of 'what's more' or 'into the bargain' etc. [*Afk. nogal* rather, quite, fairly, *poss. confusion with nog,* still, also, as well]
This wall chart, attractive . . . and tweetalig nogal, is to be distributed to all . . . institutions which could benefit and help save victim's

lives. *Cape Times* 21.6.73
A rise and a whole string of compliments,
nog. Flattery gets people everywhere. *Darling*
17.3.76

nog 'n piep ['nɔx ə ˌpĭp] *interj. colloq.
equiv.* 'And another one' usu. heard
after the third cheer to encourage a
fourth. [encore! more! *Afk. lit. nog* still,
another + '*n* a(*fr. Du. art. een*) + *piep*
cheep, squeak]
   . . . it only breaks up well after midnight
with the traditional three cheers for oupa.
'And nog 'n piep,' Uncle Fanie reckons.
'Jislaaik, but the ole man always did like a
lekker party, hey? Rest he's soul.' *Darling*
15.10.75

non-black *Afr. E.* see *black.*

non-European See *European,* also *White.*

nonna ['nɔna] *n.* Mode of address or
reference usu. by a servant to or of the
mistress, *poss. archaic:* see also *nooi:*
also dimin. *nonnie.* [*prob. fr.* Malay
*nonya respectful form of address to a
woman of the upper class: poss. fr. Port.
dona,* lady, *fr.* Lat. *domina,* mistress]
   . . . an old blind fellow . . . devoted to his
nonna, and to keep him happy and occupied
she gave him little tasks that required her
presence. . . . 'My nonna, where is my nonna?
We must fly . . . the river he comes down, my
nonna.' McMagh *Dinner of Herbs* 1968
   Nonnie was a diminutive of *Nonna* the
polite word for mistress used in Francois's
world. The daughters . . . inevitably become
Nonnie (little mistress). Van der Post *Story
Like Wind* 1972

non-voter *n. pl.* -s. *Afr. E. colloq.* An African sig. that he is without a vote.
   Some non-voter whose house and stokkie . . .
this turns out to be, comes out of the bedroom.
Drum 8.3.74
   You see, the Johannesburg city fathers
recently met and remembered that we non-
voters were so inconvenienced. *Ibid.* 22.3.73

Non-White *n. pl.* -s. also *modifier.* A member of any *Black* (q.v.) racial group: a
term resented by many Blacks and now
freq. avoided in Press usage. See also
African.
   . . . Three bowling greens, two golf courses
and various other facilities for the convenience
of the Non-Whites. *Panorama* Nov. 70
   The S.A. Institute of Race Relations wants
the 'offensive' term 'non-white' replaced by a
more acceptable term. *Sunday Times* 28.11.70
   Non-White is a terrible word to use for a
description of us. It makes us appendages of
Whites. If there were no Whites, there wouldn't
be non-Whites. *Drum* 6.2.72
   The Cape Peninsula Publicity Association
has a full list of accommodation available

for Non-Whites (capital N, mind you) in the
Cape Peninsula. *Sunday Times* 21.11.76
As *modifier, attrib.* See also at *Black.*
   Perhaps the most alarming feature of these
stirrings is the growth of the Black Power
idea at non-white universities. *News/Check*
10.11.70

no objection permit *n. pl.* -s. A permit to
bring stock from one area to another,
to be produced at a stock sale.
   IMPORTANT: Buyers from other districts
must bring 'No Objection' permits, where
necessary. *Farmer's Weekly Advt.* 20.3.74
   A 'no objection permit' is issued by a local
State veterinarian or stock inspector to con-
firm that there is no objection for cattle to
be moved to the farm of destination. It also
states whether quarantine facilities are available
at the destination, should this be required.
Ibid. 20.3.74 (Correspondence)

Noodhulpliga [ˌnʊət'hœlpˌlɪxa] *n. prop.*
An Afrikaner First Aid organization:
see also quot. at (2) *Voortrekker* [*Afk.
noodhulp* emergency aid + *liga cogn.*
league]
   Noodhulpliga workers gave him mouth to
mouth resuscitation and oxygen . . . The
Noodhulpliga chief commandant said seven
spectators had heart attacks during the game.
E.P. Herald July 1974

nooi [nɔi] *n.* See also *nonna:* Mistress: a
mode of address or reference to the
mistress of a household, and used as a
term of address in the third person with
article: see second quot. see also quot.
at *sieur.* ‖Also used of a young girl:
esp. in combination *klein ~,* young mistress, the daughter of a household. [*Afk.
prob. fr. Mal. nyonyah/njonjah rel. Port.
dona fr.* Lat. *domina,* mistress]
   . . . Abdol, who was diligently cutting up
berry-wax preparatory to blending it with raw
spirit, and thereby producing such a floor-
polish as would satisfy even this new and particular nooi. Fairbridge *Which Hath Been* 1913
   'I am pardoned by the *nooi?*' she said half-
timidly. *Ibid.*

nooit [nŭit, nɔit] *adv.t.* Never: found in
S. Afr. names esp. of farms prefixed to
vb. partic. e.g. Nooitgedacht, Nooitverwacht [*Afk. nooit* never]

noord [nʊə(r)t] *n.* North: found in S. Afr.
place names usu. as prefix as in Noord-
einde, Noordkaap, Noordoewer. [*Afk.
noord* north]

nophepha bag See last quot. at *shebeen.*
[*Xh. no- fem. prefix,* (see *auntie* (2)) +
pepha fr. paper]

normalize *vb.* To make non-racial; used

exclusively of sport and *prob.* now obs. *cf. international* and *multi-national. [presum. fr. Afk.* official term *normaliseer.*]

Now Rhodes normalizes rugby ... the club's officers would not enter into any discussions ... about the colour issue ... the decision to normalize rugby at Rhodes is merely in accordance with the spirit of the present policy of the South African Rugby Board ... Commenting on the decision to go non-racial, Rhodes rugby coach ... said 'I am very excited about the latest developments.' *Oppidan* Mar. 1977

not so? *qn. interj.* Tag question following positive statements, equiv. isn't it? etc. inviting the agreement, assent or participation of the hearer. *cf. Ger. nicht wahr? Japanese soo desu nee?* see also *isn't it* and *nè. [presum. fr. Afk. nie waar nie?* isn't that so? or its abbr. *nè?* (q.v.)]

Lena	It's a hard life for us brown people, hey.
Boesman	He's not brown people, he's black people.
Lena	They got feelings too. Not so, Outa.

Fugard *Boesman & Lena* 1969

nou [nəʊ] *adj. prefix.* Narrow: found in place names e.g. Noupoort, Noukloof. [*Afk. nou* narrow]

now *modal adv. substandard.* Misplaced or used redundantly in S.A.E. ; see first quot.: also used in past tense sentences (see second quot.) freq. for 'then' as in 'He was now eighteen years old': see quot. at *doppie²*. [*fr. Afk. uses of nou cogn.*now]

'*There's* a piece of ground for you now. For somebody ... that isn't afraid of a bit of hard work Catholic or Protestant, *there's* now a ...' Bosman *Jurie Steyn's P.O.* 1971
'What about the Volksraad member's words?' Jurie asked. 'There was something for you now ...' *Ibid.*

now-now *adv.t. colloq.* In S.A.E. usu. *sig.* immediate future, shortly, at once, right away. *cf. just now.* [*translit. Afk. nou-nou* in a moment (*with present tense*), *also* a moment ago (*with past tense*)]

I'm going to town now now. O.I. 1970
Also used referring to immediate past.
It was now, now, that he went. Paton *Cry, Beloved Country* 1948

NSM [ˈenˌesˈem] *acronym. abbr. n. phr. National Serviceman/men* (q.v.)

As a national serviceman I wonder if you can explain a gross injustice perpetrated on myself and thousands of other NSM of the 1977 in-

take ... As a NSM I do not feel free to disclose my name. *Letter cit.* Steenkamp *Cape Times* 25.6.79

-ntaba- [(ĭ)nˈtaba] *n.* Mountain: in various forms in S.Afr. place names Entabeni, Intabakandoda, Thaba 'Nchu, Thaba Bosiu, Thabazimbi. [*Bantu thaba* mountain]

(i)ntombazaan [(ĭ)nˈtɔmbaˌzɑn] *n. pl.* ama-, -s. A mode of address or reference to a little girl *cf. ntombi.* [*Ngu. intombazana* little girl]

(i)ntombi [(ĭ)nˈtɔmbĭ] *n. pl.* -s. Mode of address or reference to a virgin of marriageable age. [*Ngu. intombi* marriageable girl]

... we off-saddled to rest, the Kaffirs coming out to stare as usual, the young intombes (Kaffre maids), like their white sisters, curious to see the strangers, came out. Anderson *25 Yrs. in Waggon I* 1887

ntshebe [nˈtʃeːbe] *n. interj. Zu.* greeting to a bearded man. [*Zu. -ntshebe* beard]

Impassive faces of Dingana's spearmen
Under municipal oilskins; travellers
On the same road as I.
... I take
A widening eye, a flash of teeth, the shout
Njebe! as communication
Wm. Branford, UNISA Eng. Studies *Poetry* 1974
People always make remarks about his beard ... The Africans stop in the road and yell 'ntshebe'! *Darling* 30.3.77

num-num-(bessie) See *noem-noem-(bessie)*.

nunu [ˈnunŭ] *n. pl.* -s. *colloq.* Any insect. *cf. gogga Jam. Eng. tichi. E. Afr. dudu.* [*Zu. inunu* uncanny animal, monster, bogey]

... did not know what had stung her, all she knew was that a nunu had walked across her arm. O.I. Durban 1970

Nusas [ˈnjuˌsæs] *n. prop.* The National Union of *S. Afr.* Students founded in 1925. [acronym]

PROTEST CALL BY NUSAS JOHANNESBURG: The National Union of South African Students (Nusas) has called for nation-wide student protests *Evening Post* 3.10.70
He exhibited a faith and an interest in students ... His banning along with seven other Nusas leaders was a blow to all students. *Cape Times* 1.10.77

Nuwejaar, Tweede [ˈtwɪədəˌnyvəˈjɑː(r)] See *Tweede Nuwejaar*.

nyala See *inyala*.

nyanga *n. pl.* -s. See *inyanga*.

O

O.B. [ˌɔʊˈbi] *n. prop.* See *Ossewa Brand-wag* and quot. [*acronym*]

... there was another and more important rival to Malan's party among the Afrikaner people – the Ossewabrandwag, or O.B. This ... was formed in 1938 as a cultural movement inspired by the emotions of the Great Trek centenary ... it became an avowedly National-Socialist organization, working for a totalitarian republic and the abolition of the party system. Keppel Jones *Hist. of S.A.* 1948

oblietjie [ɔˈblĭkĭ, ŭ-, -cĭ] *n. pl.* -s. A thin wafer cooked over a fire in an ~ *pan* or 'wafer iron' and rolled, like a brandy snap: brought to S.A. by the French Huguenots. [*Afk. fr. Fr. oublie* sacramental wafer *fr. Lat. oblatus partic. sig.* offered]

She also served the rolled wafer tea cakes called oblietjies, made with cinnamon and white wine – a Huguenot contribution to Cape Cookery. Green *Tavern of Seas* 1947

Oublie-pans are called by the ironmongers 'wafer-pans', and can be obtained in all the English shops. *Duckitt's Recipes* 1966 *ed.* Kuttel

An old fashioned deal chair was sold for R38 and an 'oblietjie' pan for R50. *Argus* 27.11.72

oefening [ˈŭfənɪŋ] *n. hist.* A prayer meeting, retained in form ~*shuis.* [*Afk. fr. Du. oefening* prayer meeting (*huis cogn.* house)]

In the evening was held an *oeffning* (or meeting as distinguished from the regular church service.) Burchell *Travels II* 1824

In combination ~ *huis*

Apart from the Drostdy the town boasts of several more historical monuments. They include the old ... Meeting House (Oefenings-huis), which was built in 1838 for the holding of 'religious services and prayer meetings and the education of the heathens' ... and the old Residency. *E.P. Herald* 23.1.79

~ *houder* a lay preacher

Such instructors, catechists or lay preachers (oefeninghouders) ... *Cape Statutes* 1843 (1862)

deriv. (*fr. vb. oefen* to exercise) ~*aar* a lay preacher: *cf. Scottish: exercises,* prayers.

Is care taken that no lay preachers ('oefenaars') perform divine services ... without the consent of the minister? *Ibid.*

oes [ŭs] *adj. slang.* Out of sorts, seedy, under the weather: see *verlep,* (2) *pap :* also blanket term of somewhat pitying

disapprobation of a poor specimen or a poor show. *cf. Austral. crook/cronk,* ailing, out of sorts ; *E.Afr. shenzi (adj.)* low grade. [*Afk. oes* out of condition, feeble, shabby *etc.*]

I'm feeling really oes today. Guy Butler 1971 How oes that ... girl always looks! O.I. (British born) 1972

oester [ˈŭstə(r)] *n.* Oyster: found in place-names Oesterbaai, Oestersteeg etc. [*Afk. fr. Du. cogn.* oyster]

off *n. pl.* -s. *Afr.E.* Used by African speakers sig. time or day ~, also in *pl.* [*pressum. abbr. time-off*]

... the watch boy ... hadn't come back from his 'off' which is the native abbreviation for 'day off'. Gough Berger *Where's Madam* 1966 What shall we do about our offs this week? O.I. African servant 1970 I get my off once a week, on Sundays. O.I. African Nurse, Cape Town 21.6.71

off-colour *adj. Sect. Mining.* Used of or pertaining to a diamond neither pure white nor a definite colour: see *fancy.*

When in a general way one talks of 'yellow' stones, one means 'coloured' of that tint, not 'fancy'; in the Fields we incorrectly call them *off-colour.* The true *off-colour* has no distinct tinge at all. Boyle *Cape for Diamonds* 1873 *cit.* Pettman

A good many of these bigger ones, however, we could see by holding them up to the light, were a little yellow, 'off coloured' as they call it at Kimberley. Haggard *Solomon's Mines* 1886

off-load *vb. trns.* To unload, now widely used outside S.A. *esp.* of cargo. [*translit. Afk. aflaai* unload]

Dock workers at Durban harbour spent all day yesterday off-loading the long-awaited consignment of New Zealand butter. *Daily Dispatch* 1.6.71

Also *figur.* to sell or dispose of usu. unwanted or poor quality goods: see quot. at (2) *queen.*

... the indifferent quality which they off-load on to local markets would not even fetch a return as pig swill on the competitive and discriminating markets of Europe. *Farmer's Weekly* 11.7.73

The Nats can say goodbye to my vote. For two years I have been trying to offload my clay deposits on to the Government. *Sunday Times* 12.3.72

offsaddle *vb. trns. and intrns.* To unsaddle a horse, also to break a journey: see also *opsaal, afsaal,* and quots. at (4) *vaal-* and *ntombi.* [*translit. Afk. afsaal* unsaddle]

The travelling paces of South Africa are a

canter on level road, and a walk for the declivities; with frequent 'off saddling', to let the horses roll and refresh themselves. Alexander *Western Africa I* 1837

At the foot of the garden Jan Botha offsaddled his horse, knee-haltering it and allowing it to stumble away in search of grazing. Brett Young *Seek a Country* 1937

. . . here, there, and everywhere, riding, fighting, retiring, advancing, saddling up and offsaddling. Cloete *Rags of Glory* 1963

off-sales *n.* The *bottle store* (q.v.) attached to or owned by an hotel, in which liquor is sold by the bottle for consumption off the premises, also ~ *department:* ~ *licence:* equiv. *Brit. off-licence* and *U.S. off sale (adj.)*. See quot. at *international.*

Hoteliers here rely largely on off-sales and liquor sales in the hotels to make a profit. . . . but liquor sales both in the bars and the off-sales had dropped compared with last year. *E.P. Herald* 18.2.72

oke [əʊk] *n. pl.* -s. Also *okie:* see *ou, outjie* and quot. at *smaak. [fr. outjie* (q.v.)]

. . . and a sexy foreign accent . . . I never been out with a reel continental okie before. *Darling* 9.10.74

Old Dutch Medicines *pl. n.* See *Dutch Medicines:* and quot.

It is clear that Dr. Juritz was making up the mixtures that later became 'Old Dutch Medicines'. Turlington for chest complaints, Jamaica Gemmer Essence, Dr. Stahl's Versterkende Druppels – all these were on the list. Dr. Juritz also announced that he had the original recipe for Napoleon's Borstpillen, . . . Cajoputi oil for pain in the limbs, buchu azyn, . . . and Hoffman's Druppels for headaches. Pynstellende druppels, zinkings druppels (own recipes) koorts* druppels and oogwater all speak for themselves. Green *Land of Afternoon* 1949 [*Du fever]

old man's beard *n.* Either of two lichens of the Usneaceae, *U. barbata* or *U. florida* are so called in S.A. from their greyish dangling growth. Also *U.S.* used of *Spanish moss (Tillandsia usneoides)* and species of *Clematis.*

Old man's beard . . . very common lichens in the South coastal forests where they form long festoons on the branches of *Podocarpus elongatus* (geelhout) and other trees. Cosmopolitan in distribution. Smith *S.Afr. Plants* 1966

Old Year's Night *n. phr. and modifier.* New Year's Eve is freq. so called in S.A., as in ~ *party [trans. Afk. Oujaarsnag/aand* old year's night/evening]

olifants¹ [ˈŭlĭˌfant(s)] *n. prefix.* Elephant

[*Afk. olifant cogn.* elephant.] **1)** Found in numerous plant names *sig.* either giant or very large species, e.g. ~ *riet, Restio giganteus;* ~ *melkbos* (q.v.), *Euphorbia hamata;* or having the appearance of the elephant as ~ *oor* (ear), *Eriospermum capense;* ~ *stert* (tail), *Vellozia retinervis;* ~ *voet, Dioscorea elephantipes;* see *Hottentots bread;* or 'eaten by the elephant' as in ~ *kos* (food) *Portulacaria afra,* see *spekboom;* ~ *doring, Acacia giraffae,* see *kameeldoring.*

2. Found in place names presum. referring to actual events or habitat e.g. Olifantsfontein, Olifantshoek, Olifantsbrug.

olifant² *Sect. Army.* The S.Afr. tank in use in the Armoured Corps, see *SSB*, of the *SADF* (q.v.)

ollycrock [ˈɒllĭˌkrɒk] *n. pl.* -s. See *ali/arikreukel. [corruption of alikreukel]*

If the poenskoep are feeding they will take . . . redbait, whole seacat . . . chokka, ollycrock and siffie (venus ear). *E.P. Herald* 1976

-om- [ɔm] *adv./prep.* Round, about (direction): found in place names e.g. Omdraaisvlei, Keeromstraat *(keer* turn). [*Afk. fr. Du. om* round, around]

omgekrap [ˈɔmxəˌkrap] *adj./partic. slang.* Upset, disaffected, 'scratchy': equiv. *Brit. rubbed up the wrong way. [Afk. omgekrap* irritable]

We've had to give up going to the . . . Church because when my husband gets up he's so omgekrap he's not fit to live with for the rest of the day. O.I. 1965

omissions Certain items are frequently omitted in S. Afr. speech and occ. in writing: not all these omissions are exclusively S. Afr. usage but are here recorded as commonly occurring: see also *redundancies.*

1. -ed (q.v.) 'alveolar suffix' of *past partic. modifiers*; see quot. at *-ed: three bedroom home* etc. and at *klappertert.*

2. -'re [ə] reduced form of *are.* freq. omitted in speech before *prns.* we, you, they, e.g.

We working on it now. You looking tired. O.I. They coming next week. O.I. You joking *(poss. fr. Afk. jy jok,* you're fibbing, romancing etc.). O.I. including Radio S.A.: see also *quot.* at *drol.*

. . . while you tapping yore feet in the gutter waiting . . . they all pally blues . . . it doesn't look like you trying to . . . if you talking about long hair. *Darling* 9.10.74

3. *articles* or other noun-determiners e.g. his, her, their, our, are omitted before nouns *esp. in adv. prep. phr.* in speech and writing.
> We came back from holiday last week.
> They're on honeymoon.
> He walks to office.
> Let's go to flick. O.I.
> Learn to play guitar, piano . . . in 2 weeks. *Advt.*

4. *nouns* or *pronouns* omitted after certain *vbs.* normally requiring complements (*vbs.* of 'incomplete predication') e.g. *have, got;* or requiring objects (*vb. trns.*) e.g. *find, learn* (q.v.) *fetch.* e.g.
> A. 'Would you like another cup?'
> B. 'No thanks, I still have.'
> 'These come from East London: we can't get from P.E. any more.'
> A. 'I was looking for some shoes in town.'
> B. 'And did you find?'
> 'They coming to fetch just now.' OI

5. *nouns* omitted leaving modifiers alone e.g.
> Come to my twenty-first.
> She is nursing at the Mental.
> He wants his Christmas/New Year.
> The lowest priced automatic in South Africa. *E. Province Herald Advt.* (car) 5.12.74
> How automatic is your automatic? *Sunday Times Advt.* (washing machine) 17.10.76
> 1968 Vauxhall Viva with roadworthy. *Advt. Grocott's Mail* 15.8.75
> '. . . making the tea and forging the roadworthies.' *Darling* 28.4.76

6. *to, at, in,* etc. omitted after such verbs as 'explain' 'reply' or 'lecture'.
> Viv explains me the assignment then I write it. O.I. B.A. Student
> Granny didn't reply me. O.I. Child 10 yrs.
> It is no use moralising and lecturing them. Newspaper article
> What are you lecturing this term? O.I. University teacher

7. Various *particles* omitted (usu. by children):
> Be careful your feet. O.I. Child
> I'd laugh he fell off. O.I. Child
> I wouldn't mind that stove was mine. O.I. Housewife

cf. U.S. out the window etc.

8. *Possessive suffix* 's occ. dropped in speech.
> We beat the . . . Women Hockey Club 4–nil. O.I. Schoolgirl 1974

Also dropped as in *U.S. barber shop, florist shop* etc.
> An application from a businessman to open a chemist shop in . . . one of the city's biggest medical centres was refused. *E.P. Herald* 14.10.76

9. [*Afk. speakers usu.*] -s, -es third person singular present tense marker :
> I'm no musician but the wife play. O.I.
> . . . always come when a person isn't expecting him. O.I.
> A shoal of geelbek approach, one of them spot the strepie . . . *Farmer's Weekly* 4.12.74

omkulu See *great* (*wife*) (quot. ex Dugmore diary), also *kulu* and *makulu*.

ompad ['ɔmˌpat] *n.* Detour: seen on road signs ; also a roundabout way, long way round. [*Afk. ompad* roundabout way]

-onder- [-ˌɔn(d)ə(r)-] *adj.* Under, below: found in place names as *prep.* prefixed or suffixed e.g. Waterval Onder, Onder-Papegaaiberg, Onder-Smoordrif: as modifier, prefixed to *n.* e.g. Onderplaas: see also *onderste, bo(ven)* also *under*. [*Afk. fr. Du. onder* below, lower]

onderdeur ['ɔndə(r)ˌdiœ(r)] *n.* The lower half of a *stable door* (q.v.) or *bo-en-onderdeur* [*Afk. fr. Du. onder,* lower + *deur* cogn. door]
> . . . when any stranger arrived . . . Tante . . . would lean over the onderdeur and inform the visitor in most vigorous language that if he happened to be a 'digger' his presence anywhere on the farm was most unwelcome. MacDonald *Romance of Rand* 1933

onderste ['ɔn(d)ə(r)stə] *adj.* Lowest: found in place names Onderstepoort, Onderstedorings. [*Afk. onder* lower + *superlative suffix -ste* lowest]

onderstebo *slang colloq.* Upside-down, topsy turvy: see also *deurmekaar*. [*Afk. onderste* lowest, *bo* above, hence upside down]
> 'T't, t't, t't,!' clicked old Mrs van der Merwe sympathically. 'My but this woman has turned Vredendorp upside down – all ondersdaboo, eh?' Fairbridge *The Torch Bearer* 1915

only *Intensifier, adv. of degree. colloq.* Various non-standard uses are found in S.A.E. 1. *with adj. or other modifier.* 'Really' or 'completely' (also Irish) sometimes placed before *n. phr.* with or without redundant *but :* see quot. at *tom.* [*unknown*]
> Since he didn't know what our talk was about the lorry driver's assistant looked only mystified. Bosman. *Bekkersdal Marathon* 1971
> 'Yessas she is but only a mooi pop.' quoted by Guy Butler *cit. Daily Dispatch* 29.10.74

2. *with vb.* 'Really', with or without redundant *but.* [*unknown poss. trans. Afk. net* only]
> Yias but we were only moving! O.I. Yachtsman George Lakes 8.12.74
> It was an exclusive party. We *didn't* invite . . .

Liz Taylor, van der Merwe or Idi. But Des Olivier brought along the . . . and that party only swung. *Darling* Advt. 13.4.77

3. 'Just' [*prob. trans. Afk. maar* (q.v.) just, only *as in Hy is maar klein*] Shame, he's only small. O.I. 1972

4. *intensifier equiv.* of '*really*' *in phr.* ~ *with pleasure, it's* ~ *a pleasure, sig.* complete willingness. [*poss. trans. Afk. slegs* merely, nothing but: *poss. maar* but, only]

It's only a pleasure to fit him, . . . only a pleasure. O.I. 14.2.75

on the moment *prep. phr. substandard.* At the moment, at present ⫫Not *equiv.* 'on the instant' *sig.* 'at once'. [*prob. trans. Afk.* '*op die oomblik*' (*Afk. op usu. sig.* on) at the moment, at present]

We've no places left for that flight on the moment, but there may be a cancellation. O.I. Jan 1974

ons land [ˌɔns ˈlant] *n. prop.* 'Our country': a patriotic phrase sometimes used as a toast at formal dinners: see also *land²* [*Afk. ons* our, *land* country]

Boetie saw only the romance of it, of all these strange men united in a common purpose – to defend their land. Ons Land as they called it. Cloete *Rags of Glory* 1963

oo- *pl. prefix:* see at *u-* for *ooclever, ooMac, ooscuse-me.*

oog [ʊəx] *n.* The *eye* (q.v.) or source of a stream, river or *fountain* (q.v.). In place names Die Oog, Molopo Oog, Malmane Oog. [*Afk. oog* eye]

At the Molopo Oog, where there is a modern holiday resort, and where kurpers, geelvis, carp and black bass are bred on a large scale. *Panorama* Oct. 1973

oogpister [ˈʊəxˌpɪstə(r)] *n. pl.* -s. *Anthia*, an insect also known as *hoogpister* (q.v.) and erron. '*piogter*'. [*Afk. oog* eye + *pis* urinate + *agent. suffix* -(*t*)*er*]

There are also the piogters, black with white stripes, when you touch them they shoot nasty burning stuff in your eyes. Vaughan *Diary circa* 1902

oom [ʊəm] *n. pl.* -s. Uncle: mode of address or reference to an uncle or older man unrelated to the speaker, with or without a name following, used by children esp. in the country as a mark of respect: occ. of a national figure e.g. *Oom Paul* (Paul Kruger), *Oom Jim* (President Fouché): see also *uncle, auntie, tannie, tante* and quots. at *oubaas, neef, alles van die beste.* [*Afk.*

fr. Du. oom uncle.] ⫫Also '*man* (*used by children referring to male white adults*)' Bosman, Van der Merwe and Hiemstra, *Tweetalige Woordeboek* 1972 ed.

Old Lucas, or as he was more familiarly called, Oom Hans (Uncle Hans), . . . Burchell *Travels I* 1822

. . . we are all a primitive people here – not very lofty. We deal not in titles. Every one is Tanta and Oom – aunt and uncle. Schreiner *Afr. Farm.* 1883

. . . Oom Jakob in middle-age was firmly established as 'Oom' to every burgher young enough to be his nephew or his son. Prance *Tante Rebella's Saga* 1937

oondskop [ˈʊəntˌskɔp] *n. pl.* -pe. A long-handled scoop by means of which bread was inserted into and removed from the depths of a *bakoond* (q.v.) [*Afk. oond* oven + *skop* shovel *cogn.* scoop]

The loaves were then inserted one at a time on the floured flat end of a wooden oondskop . . . a sort of shovel with a handle 1.5 m to 2m long. Cook *Cape Kitchen* 1973

ooplang [ˈuplaŋ] *n. Ind. E.* Businessmen's term for profits not disclosed on the balance sheet. [*Gujerati ooplang* assets kept secret.]

With the rise in taxes most of us need our ooplang to live on. *Indian Informant* 3.7.78

oord [ʊə(r)t] *n.* Used as equiv. of 'place' in street names e.g. Cavan-Oord: also seen on road signs in combinations: *vakansie* ~, a holiday resort; *strand* ~, beach resort. [*Afk. oord* place, region]

oorlams [ˈʊərˌlam(s)] *modifier and n. hist* Shrewd, knowing, experienced esp. (as noun) 'old hand', a black or coloured person long exposed to white or Western civilization: see quots. [*Afk. fr. Du oorlam fr. Mal. orang* person + *lama* long (of time) old (long standing)]

The word has come down from the days of the Dutch East India Company; the men who had seen considerable service were called oorlammen (Mal. *orang lami*, old person), while the recruits were called Baren (Mal. *orang barn*, new hand). Pettman *Africanderisms* 1913

They were the first truly 'detribalized' Natives; and they and their descendants, the so called 'Oorlams,' became so completely assimilated in a hereditary master-servant relationship, into the structure of Afrikaner society that their very mother-tongue became Afrikaans. Hoernlé *S. Afr. Native Pop.* 1939

oorskiet [ˈʊərˌskit] *n. pl.* -s. *colloq.* Left-overs, bits and pieces, remainders: lately derogatory mode of reference to an 'outsider', among schoolboys (*slang*). [*Afk* *fr. Du. oorskiet* remnant, leftovers]

These roses are oorskiets really . . . my dad's started pruning. O.I. 1971

op [ɔp] *predic. modifier. colloq.* Finished, all gone: usu. of some material substance rice, tea, paper, drink etc. as in 'I'm sorry, the beer's op'. *cf. Brit. time's up.* [*trans. Afk. op* finished, exhausted, *in phr. 'dit is op'*, *equiv.* there's nothing left, it's all gone, *etc.*]

opblaas krimpsiekte *n.* See *krimpsiekte (acute)*

op die kop [ˌɔp dï ˈkɔp] *adv. t. prep. phr. colloq.* Precisely, usu. of time *cf. on the dot, Jam. E. bap(s).* [*Afk. fr. Du. op den kop* precisely]

At the next stroke of the gong it will be exactly – op die kop – a quarter past seven. *Radio S.A.* 21.5.73

See you at seven, then, op die kop. O.I. 1973

opdraand [ˈɔpˌdrɑnt] *n. colloq.* Rising ground, usu. on a road or path e.g. 'We'll stop at the top of the next opdraand and change drivers': also *figur.* as in *Eng.* to have 'uphill', tough going etc. [*Afk. opdraand* uphill path, acclivity]

There's a sort of road going off. You go along it towards the river . . . there's a river just on the other side of the opdraand. Jacobson *Dance in the Sun* 1956

Operational Area Official designation of any sector in which S. Afr. forces are serving: in everyday parlance referred to as *'The Border'* (q.v.) and at times by men in the army as *'The Red Area'* (q.v.)

The South African soldier acquits himself exceptionally well in the Operational Area. He need not stand back for any other soldier in the world. *Paratus* Jan. 1979

opgaaf [ˈɔpˌxɑf] *n. hist.* Certain income and property taxes formerly levied by Government. [*Du. opgaaf* statement, account (*Afk. poll tax*)]

In Cape Town, and in some of the districts, the opgaaf taxes have been assessed upon the principle of a property tax, calculated upon a valuation of all stock and property of whatever description. *Reports of the Commissioners II* 1827

. . . INCOME. – Two pounds per cent, on all whose income exceed £30, except from Farming Stock, chargeable to Opgaff. *Greig's Almanac* 1833

The custom at that time was that the Chief Clerk of the Civil Commissioner travelled annually in an ox-wagon to collect the 'opgaaf' (taxes), Albany and Somerset were then one

fiscal district and this collecting journey took up several months. Meurant *Sixty Yrs Ago* 1885

oppas [ˈɔˌpas] *interj. colloq.* Warning exclamation equiv. of Watch out! Careful! see also *pas op* and quot. at (2) *bloubaadjie*. [*Afk. oppas* look out, mind, be careful *fr. Du. oppassen* mind, beware, be on guard]

Oppas! You'll go too far one day. Death Penalty. Fugard *Boesman & Lena* 1969

opper [ˈɔpə(r), ɒp-] *n. pl.* -s. A stack of grain bundles: see quot. [*Afk. fr. Du. opper* (hay)cock, hay rick]

After the crop is cut the bundles . . . are then packed into 'shocks' or 'oppers'. Shocks are heaps of about 16 bundles standing upright on the straw. Oppers are heaps of about 50 bundles packed flat, like a small stack. *H'book for Farmers* 1937

ops [ɔpsˈɒps] *vb. colloq. slang* usu. children To swop, exchange, lend or give. [*unknown*]

. . . the primarily schoolboy 'ops' meaning give as in 'Hey ops me some cake'. Gordon Jackson cit. *To the Point* 14.4.78

. . . he whispers to me and Lorna and Charmaine under the gum tree where we . . . opsing each others sarmies. I crave Lorna's marmite ones and she goes for my fish paste. Hobbs *Blossom* 1978

opsaal [ˈɔpˌsɑl] *vb.* See *upsaddle.* [*Afk. fr. Du. op cogn.* up + *zadelen* to saddle]

That night Alexander passed the word round to 'inspan' and 'opsaal,' and with all their kit and gear the little commando took the road. Klein *Stagecoach Dust* 1937

opsit [ˈɔpˌsɪt] *vb.* To court a girl by 'sitting up' with her after the family have gone to bed: see quot., also *sit up, cf. Austral. to track with,* equiv. of *Brit. walking out.* [*Afk. fr. Du. opzitten* to sit up]

Among the farming community a direct method of courting was observed. It was called *Opsit.* And it allowed a girl to indicate very clearly what her feelings were towards a suitor. If she favoured him a long candle would be produced. De Kock *Fun they Had* 1955

In combination ~*kers* (candle).

If acceptable to the parents, he was given, as nightfall approached, a piece of 'opsitkers' (sitting up candle) . . . On the table flickered the 'opsitkers' and as long as this bit lasted he was allowed to sit with her, talking sweet nothings. Jackson *Trader on Veld* 1958

~ *bank* (bench) equiv. of *U.S. love-seat,* bench or sofa for courting couples. *cf. Jam. Eng. couple bench* (in church).

opskepper [ˈɔpˌskepə(r)] *n. pl.* -s. *Sect. Army.* Someone who is required to dish up for others in the mess: a despised job. [*Afk. opskep* to ladle, dish up + *agent. suffix -er*]

The farmer gave the order to 'schenk een zoopje (pour out a dram), and then to *skep op* (set the victuals on the table). Alexander *Expedition I* 1838

op skoot, to bring/put a gun [ˌɔpˈskʊət] *vb. phr.* To set the sights of a gun. [*fr. Afk. vb. phr. op skoot bring*]

. . . the boys were putting their guns op schot, or sighting them, to go after lions the next day. Bisset *Sport and War* 1875 *cit.* Pettman

opskop [ˈɔpˌskɔp] *n. colloq.* An informal dance, 'hop', also *sheepskin* (q.v.) *velskoen* (q.v.) and *opskud* (q.v.): *cf. Brit. a 'knees up'.* [*Afk. opskop* lit. kick up (a noise *etc.*)]

opskud [ˈɔpˌskœt] *interj.* **1.** *colloq.* Buck up! Get cracking! of *Anglo-Ind. jildi,* be quick. [*Afk. opskud* to buck up, hurry, *lit.* 'shake up']

'Opskud kêrels! I heard. But it was not Serfina who gave that command. Bosman *Unto Dust* 1963

2. Also *opskop* (q.v.) a *sheepskin* (q.v.) dance with *boeremusiek* (q.v.) *Jam Eng. buck up* a social gathering; *Brit. a 'knees up'.* [*prob. fr.* (1) ~]

In combination ~ *en uitkap :* a dance.

. . . he could do a vastrap, that man, non-stop, on all strings, at once . . . Polka, tickey-draai, opskud en uitkap ek sê . . . that was jollification for you . . . Fugard. *Blood Knot* 1968

Vastrap, Tickey-Draai, Opskud en uitkap: traditional South African folk dances. Fugard *Glossary* to *Boesman & Lena* (1969) 1973 *ed.*

opslag¹ [ˈɔpˌslax] ˋ*n. hist.* Auctioneering term: preliminary upward bidding to 'run up' the price before selling by downward bidding: see *afslag.* [*Afk. fr. Du. opslag* price increase]

The method of selling a house or farm was unusual. First it was put up by opslag, that is by advancing bids in the usual way. But the highest bidder to whom it was knocked down did not wish to be the purchaser; his function was to help the auctioneer to increase the final selling price and for this he received a bonus or strykgeld. . . . The final figure was generally well in advance of the opslag bid, but should it fail to reach that amount the man who had received the strykgeld was compelled to become the buyer at that figure. Gordon-Brown *S. Afr. Heritage* 1965

opslag² *n. and n. modifier.* Used of any self-sown crop e.g. ~ *potatoes,* ~ *barley* etc. Also profuse secondary plant growth (chiefly annuals) after rains esp. on *braakland* (q.v.) stubble and *ouland*

(q.v.): see also quot. at *tramp out.* [*Afk. fr. Du. opslag* self sowing, regrowth]

The Karroo veld consists of the various types of Karroo bush and a secondary plant growth (opslag) . . . It is especially these 'opslag' plants . . . which provide the stock with an early green succulent. *H'book for Farmers* 1937

They are especially glad that the first young shoots have started growing . . . stock are voracious feeders on this *opslag.* Cape Times 6.5.70

opstal [ˈɔpˌstal] *n.* Buildings or other *improvements* (q.v.) on a farm (excluding the land.) See *loan-place* [*Afk. fr. Du. opstal* building, superstructure]

The lessee had no *dominium* in the ground and he could sell or bequeath nothing more than the *opstal* – that is the house, kraals and other improvements. Most of the farms in the interior were held under this form, and the farm was generally known as a *leeningsplaats,* a loan farm. The stock farmers owned such places. Botha *Our S.A.* 1938

opstand [ˈɔpˌstant] *n. hist.* Rebellion. [*Afk. fr. Du. opstand* insurrection]

But politically the Union of South Africa was a house divided against itself. The suppression of the 'opstand' of 1914 was followed by an increase of political opposition among the Afrikaners to the Botha Government. *Cambridge Hist. VIII* 1936 *ed.* Walker

opstoker [ˈɔpˌstʊəkə(r)] *n. pl.* -s. Inciter agitator: ~*ery* stirring up strife or ill-feeling: see also *voorbok (figur.)* [*Afk. opstoker* agitator]

That well-known Boerehaat opstoker, . . . is at it again. This week he warned . . . the English Press to stop mocking and cursing the Afrikaner and the things that were holy to him. *Sunday Times* 22.10.72

. . . the Afrikaner-dominated National Party, which controls 123 of the 171 seats in the House of Assembly, remains committed to *kragdadigheid* (forcefulness) against all *opstokers* (troublemakers). *Time* 2.5.77

Orange Express *n. prop.* Also *colloq. Natal* 'Marmalade Express': express train, partially orange in colour, running between Durban and Cape Town via the Orange Free State: see also *Blue Train, White Train* and *Apple Express.*

South Africa's special trains are the Blue Train which travels between Pretoria and Cape Town; the Drakensburg, also a luxury train which operates between Johannesburg Durban and Cape Town once a week; the Orange Express. *Panorama* Dec. 1977

orange wine, *n.* Designation, not officially permitted, of alcoholic beverages, of which there are several varieties, including a 'sherry', made by fermenting oranges.

Spend Friday night at the Golden Valley Hotel at Muden. While you are there, taste some of the locally made orange wine or aperitif. *The 1820* Christmas 1973

oranje- [ŭˈranjə, -jĭ] *adj.* or *n. modifier.* Orange, found in place names usu. as modifier of *n.* Oranjefontein, Oranjemund, Oranjekrag; alone, Oranje (Bastion of Castle at Cape Town) and Eng. Orange (River). [*Du. oranje (fr. House of Orange*)]

Order of Ethiopia, *n. prop.* See *Ethiopia, Order of.*

Order of Good Hope, *n. prop.* See *Good Hope, Order of.*

ore [ʋərə] *pl. n. slang.* The police: see also *gattes, die; boere* (5) [*Afk. oor,* ear + *pl.* -e]

And now they are bulldozing the place down here's hardly anyone left; except the whores and the oere [sic]. Jiggs *Doornfontein cit. New Classic* Vol 1. 1975

oribi [ˈɒrɪbĭ] *n.* Any of several species of small antelope with small straight horns of the genus *Ourebia,* esp. *O. ourebi,* usu. inhabiting open plains. [*Afk. prob. fr. Nama orab*]

The dogs gave chace to and killed an *orabi,* a very pretty kind of buck, smaller than any we had seen. Philipps *Albany & Caffer-land* 1827

There are fewer than 600 oribi left in the Cape, of which none will be found in either national parks or provincial nature reserves. *E.P. Herald* 7.9.74

ossewa [ˈɔsəˌva] *n. pl.* -s -ens. An ox-wagon: see also *trek wagon* ‖ Now usu. *figur.* a symbol of the *Voortrekkers* (q.v.) and hence of the Afrikaners or the Nationalist party: see also *kruithoring.* [*Afk. os cogn.* ox *pl.* -se + *wa fr. Du. wagen cogn.* wagon]

South Africa can no longer be . . . excited by the belief that the verligtes will one day hijack the party ossewa and drag it into the future. *Sunday Times* 31.7.77

Ossewa Brandwag [ˈɔsəˌvaˈbrantˌvax] *n. prop.* Also *O.B.* (q.v.) (see quot. at *O.B.*) [*Afk. ossewa* ox wagon + *brandwag* sentinel]

On the eve of the war, a party existed in South Africa which was dedicated to the Nazi cause and ideology. It came into existence in 1938, and called itself the Ossewa Brandwag – the 'Sentinels of the Ox-Wagon'. White *Land God Made* 1969

ostrich *n. pl.* -es. ‖ Classed as farming *stock* (q.v.) in S.A.

A bird [ostrich] of fifteen months is worth about R25 and provides quality feathers, duster feathers, skins used mainly in the manufacture of shoes and handbags, biltong, cattle-feed and bone meal. *Panorama* Dec. 1973

When is the driver of a vehicle compelled to stop?. . . At the request of a person leading, riding or driving any cattle, horse, ass, mule, sheep, goat, pig or ostrich. *Road Traffic Ordinance* 1966; *Examination of applicants for drivers [sic] licences. Questions and answers.* Amended Jan 1977.

~ **skin** the de-feathered hide of the ostrich used for shoes, handbags and numerous S. Afr. curios esp. in the S.W. Cape, as are the shells of ostrich eggs. ‖ The meat is used as pets' food or making the highly esteemed ~ *biltong* (see quot. at *stellasie*), and the eggs used for cakes, omelettes and other cookery, see quots. at *poffertjie* and *zeekoe-spek.*

A considerable traffic might also be carried on in ox hides, either dried or salted, to which may be added the skins of the leopard, panther and tiger cat, as well as the valuable ones of the ostrich. Ewart *Journal* 1811–1814

~ **palace** Any of several elaborate houses built in Oudtshoorn, Cape Province at the time of the boom in ostrich feathers, also Feather palace.

. . . Ostrich Farm (Illustration) Ostrich Feather Palace. Traditional Ostrich farm and Private Nature Reserve. One of the finest in existence. See the Ostrich Feather Palace. *W. and S. Cape Telephone Directory* 1976/7, (Oudtshoorn) 1977/8.

otherside [ˈʌðəˌsaɪd] *prep. substandard.* usu. among children. 'On the other side of' as in 'The Post Office is otherside the road.' [*trans. or lit. rendering of Afk. anderkant or trans. oorkant* (over) on the other side of]

. . . and where is the Cat River? Antjie: In the Settler Country, otherside Grahamstown. Butler *Cape Charade* 1968

'It's what other people are saying, also. About your sitting alone with Jack Brummer in his car under the thorn-trees otherside the bridge.' Bosman *Willemsdorp* 1977

otherwise *adj. and adv. colloq.* Contrary, difficult, troublesome. [*etym. dub. poss. fr. Afk. eiewys* contrary, *or trans. anders* otherwise]

I didn't want to be otherwise about it but I did feel that I had to complain. O.I. 1968

Otherwise = 'perverse' as in 'he's acting otherwise' also used adjectivally, 'he's an otherwise ou'. *S.Afr. Gaol Argot* in Eng. *Usage in S.A.* Vol. 5 No. 1 May 1974

ou¹ [əʊ] *n. pl.* -s. *colloq.* Chap, guy, fellow: a common mode of address or reference to a man or boy as in *come on you ~s*, equiv. of 'come on chaps'; also in dimin. form *outjie* (q.v.) and occ. *oke* (q.v.): see quots. at (2) *bioscope, lekker, otherwise* and *pellie:* various combinations e.g. *bike ~*, motor cyclist. See quots. at *breker*. ⁋ In combinations, usu. puns. manifesting as the '*~joke*', which enjoyed a vogue in the 1950s, e.g. a commercial traveller is a Roamy-~ a drug addict is a pill-~ etc. [*Afk. ou* fellow *poss. fr. Du. ouwe* an elderly man, *cf. Brit. old chap, Fr. mon vieux as modes of address*]

Sissy: Well then do something!
Shorty: But What? What must a ou do? I slog.
Fugard *People Living There* 1969
'Have any of you ous got a radio?' he demanded. *Forum* Vol. 6. No. 2 1970
poss. Sect. Army pl. -ens.
I am about to go to church with the other English 'ouens' (plural for ou in the forces) . . . This is a huge camp . . . about 1550 ouens. *Letter Serviceman* 14.1.79

ou² [əʊ] *adj.* Old. [*Afk. fr. Du. oud cogn.* old] **1.** Used jocularly as in *~ maat* (q.v.) *colloq.* also as part of a *dimin.* structure '*a kak ~ dorpie* (a rotten little (old) dump), see quots. at *kak* and *eis, on my. cf. U.S. little old me.*
Ou Heath's stupid, hey – why doesn't he use the Suppression of Communism Act. *Daily Dispatch* Cartoon Caption 17.2.72
2. Found in S. Afr. place names e.g. Oukraal, Ouwerf.
3. Old (usu. in appearance): prefix to plant names e.g. *~ doring* (thorn), *Acacia robusta; ~ hout* (q.v.), *~ bos* (q.v.), *~ koe Cotyledon reticulata.*
4. Old, elder: regular prefix in modes of address or reference to older people e.g. *~ boet, ~ baas, ~ sus(sis), ~ man, ~ ma* and *~ pa;* also *outa* all (q.v.).

oubaas [ˈəʊˌbɑs] *n.* and *n. prop.* Old master [*Afk. fr. Du. oud* old + *baas cogn.* boss, master] **1.** Mode of address or reference to the master, if he is an elderly man, or occ. to the father of the master or mistress of a household.
My father's always '*Oom Jan*', though mother calls him the '*ou baas*,' and she's *Tante Lisbet*. Brett Young *City of Gold* 1940
2. God: referred to as *the ~ up there, ~ in the sky* etc., often in relation to rain. *cf. Austral. Hughie*, God of weather,

in 'Send her down Hughie!'
'The oubaas has got it up there for us, I know.' O.I. (elderly farmer) 1973
Jackson Mamba . . . who saved the life of . . . when she was almost electrocuted in her home, told me . . . that it was not his doing but that of the 'Oubaas in the sky' that the girl was not dead. *Het Suid Western* 25.4.74
3. Affectionate mode of reference to General Smuts (Field Marshal J. C. Smuts).
With General Smuts' departure the United Party will . . . forfeit the votes of many people who had supported them because of a personal loyalty for the 'Oubaas'. *Press Digest* No. 3 1948
Those who enter the gates of 1977 in trepidation should discover resolution in the words of General Smuts so long ago . . . in a favourite philosophic saying of the Oubaas: 'The dogs bark but the caravan moves on: *Argus* 1.1.77

ouboet [ˈəʊˈbŭt] *n.* Mode of address or reference to an elder brother, also to an old friend; see also *swaer* and *ou maat*. [*Afk. fr. Du. oud* old + *boet fr. broeder cogn.* brother]
. . . and Kleinboet would call him Ouboet, eldest brother. Ouboet . . . It was such a good name, it had such a full round sound. Krige *Dream & Desert* 1953

oubos [ˈəʊˌbɔ̃s] *n. Leucosidea sericea:* also *ouhout* and (3) *geelhout*, an 'invader plant' see *plant migration:* found in place name Oubosstrand. [*Afk. ou cogn.* old + *bos cogn.* bush]

ouderling [ˈəʊdə(r)ˌlɪŋ] *n. pl.* -s. An elder in the *Dutch Reformed Church* (q.v.) distinguished by his dress; see *manel* and *white tie*, also *gatjaponner. cf. Canad. (hist.) marguillier* churchwarden. [*Afk. fr. Du. ouderling* elder]
We had no predikant there; but an ouderling, with two bandoliers slung across his body, and a Martini in his hand, said a few words. Bosman *Mafeking Road* 1947.
He was unhappy about the whole thing. As chief ouderling, he was also in trouble with his church. *Cape Times* 8.11.72

oudstryder [ˈəʊtˌstreɪdə(r)] *n. pl.* -s. Veteran, ex-soldier. [*Afk. oud* former + *stryder* combatant fighter]
The unveiling . . . was attended by a gathering of more than 1 000, amongst whom were many oudstryders (veterans) of the Anglo-Boer War – some of whom were present at the battle being commemorated. *Panorama* Nov. 1970
. . . the uncompromising Boer outstryder . . . told the Queen that he could never forgive the British for fighting against the Boers. The Queen was all sympathy; as a Scot, she said,

she understood his feelings perfectly. *E.P. Herald* 13.10.75

ouhout [ˈəʊˌ(h)əʊt] *n.* Any of several species of shrub in which either the name is accounted for by the fact that the flowers are borne on the old wood, as in species of *Halleria*, or by the old or brittle appearance of the wood esp. *Cordia caffra* and *Leucosidea sericea*, (*oubos*, (3) *geelhout*). [*Afk. ou cogn.* old + *hout* wood]

The vernacular name OUHOUT said by some to be in allusion to the usually crooked branches which may suggest an aged . . . appearance but is more probably derived from the fact that the trunk burns like rotting wood (Afr.: ouhout). Smith *S.Afr. Plants* 1966

Oujaarsdag [ˈəʊˈjɑrsˌdax] *n.* '*Old Year's Day*' (q.v.), used for 'New Year's Eve' in S.A. [*Afk. ou* old, *jaar cogn.* year + *dag cogn.* day]

Ou-Jaarsdag. The usual thing. No thunder and no danger. Plaatjes *Diary* Date missing (1973)

ouklip [ˈəʊˌklɪp] *n.* 'Pudding stone,' gravel stone, honeycomb-rock usu. decomposed dolomite. [*Afk. ou cogn.* old + *klip* stone: gravel stone or pudding stone]

A few months ago it was discovered by a farmer living in the vicinity that large beds of Ou Klip (honeycomb gravel rock) on the farm were literally saturated with mercury in the native state. *Grahamstown Journ.* 20.9.1892

A strata of ouklip, through which we had not picked, was an added problem, causing deeprooted trees to wane and fall to the winter gales. *Argus* 8.5.71

ouland [ˈəʊˌlant] *n.* Unsown or fallow land: see also quot. at *opslag*. [*Afk. ouland* unsown land, fallow land]

The grazing obtainable from 'braakland', 'ouland' and stubble land may be regarded as a type of natural pasture . . . it may be braaked in July-September, or it may be left lying for a year or longer as ouland. *H'book for Farmers* 1937

oulandsgras [ˈəʊˌlantsˌxras] *n.* Also *ouland(e)gras*: any of several pasture grasses flourishing on *ouland* (q.v.) esp. *Eragrostis curvula* (*soetgras*). [*Afk. ouland(s)* fallow land + *gras cogn.* grass]

Sweet grass and Eragrostis (oulandsgras) are both fetching up to 50c and veld grass up to 40c. *Farmer's Weekly* 12.5.71

oulap [ˈəʊˌlap] *n. colloq.* A penny: see also *ndiblishi*. [*Afk. ou cogn.* old + *lap* (q.v.) rag, clout *fr. Du. seaman's term*, coin]

. . . the 'tickey' is the old Cape Town slang. Some . . . coloured phrasemaker early last

century fixed the penny forever as an 'ou lap' – valueless old rag. Green *Tavern of Seas* 1947

ouma [ˈəʊˌma] *n. pl.* -s. Grandmother: a mode of address or reference to a parent's mother, also, with surname, to an old lady: in combination ~ *grootjie*, great grandmother; ~ *kappie* 'Granny Bonnet'; several species of terrestrial orchids, see *moederkappie.* See also *oupa*, quot. at *niggie* and second quot. at *tannie*. [*Afk. fr. Du. oma* (*children's language*) grandmother, *also ou cogn.* old + *ma* mother]

I'd like to order 1 million gift vouchers to give all the South African oumas, oupas, moms, dads, swingers and groovers, who've been good boys and girls this year. *E.P. Herald Advt.* 5.12.74

. . . Ouma K . . ., dearly beloved granny . . . and great-granny . . . Rest in Peace. *Ibid.* 8.5.74

ouman [ˈəʊˌman] *n. pl.* -ne. *lit.* 'Old man', an old hand: a *National Serviceman* who has completed six months or more of his compulsory military training: see quot. at *blougat*. [*Afk. ou* old + *man*]

To the oumanne of E Squadron . . . min dae and vasbyt from Dave. *Radio S.A.* 1.4.72

When a national serviceman becomes an 'ou man', he's served on white plates which are washed for him afterwards. *Sunday Times* 1.4.79

oumeidsnuif(doos) [ˈəʊˌmeɪtˌsnœifdʊəs] *n. Lycoperdon hyemale:* lit. old coloured woman's (see *meid*) snuff (box): see *devil's snuff box.*

oumeidsnuifdoos . . . A rounded puffball which with its dark brown contents . . . recalls a snuffbox . . . used by old native women. Smith *S.Afr. Plants* 1966

oupa[1] [ˈəʊˌpa] *n.* Grandfather: a mode of address or reference to a parent's father, also occ. with or without a proper name, to an elderly man: see also *ouma* and second quot. at *tannie*. [*Du. opa* grandpa]

'Come in then, Ou-pa' she said . . . and see the little grandson that you have with his round bald head. Smith *Beadle* 1926

. . . that only other death he had ever known, the death of his grandfather three years ago. Oupa Kotze had died in his sleep. Krige *Dream & Desert* 1953

~ *grootjie* Great grandfather. [*Afk. fr. Du. grootje* grandparent]

His four nearest sires on the male side are all preferent bulls. In other words, this young bull's proven records date from beyond his oupagrootjie. *Farmer's Weekly* 27.2.74

oupa[2] [ˈəʊˌpa] *n. pl.* -s. *Sect. Army.* A National Serviceman of the two-year service period, who has served longer than the *ouman* of the one-year service period. ⫿ ~ is a new term which may or may not become fully assimilated: see ~¹.

ousie [ˈəʊˌsɪ] *n.* A mode of address to an African girl or woman by Whites or other Africans, by whom it is regarded as showing admiration; see second quot. *cf. sisi.* [*Sotho ausi fr. Afk. ousus, ousie* elder sister]

 One night a person knocked on her door while she was in bed . . . a voice said 'Ousie, ousie open the door I want to give you something.' *Post* 7.12.69
 . . . is a very well proportioned and prettiful ousie . . . in the same street. *Drum* 22.9.75

ousus [ˈəʊˌsœs, -ˌsəs, -ˌsɪs] *n.* Mode of address to an elder sister or sometimes an older woman friend. [*Afk. ousus* eldest/elder sister]

 'Drink without food will make you so drunk . . do you think your sister knows nothing about you?' 'Never mind', Louw said, 'Go on, ousis. Bring the bottle here. Jacobson *Dance in Sun* 1956
 My boet lets forth this hollow larf . . . 'You have gotta be joking ousus.' *Darling* 16.2.77

outa [ˈəʊˌta] *n. pl.* -s. A mode of address or reference to an elderly usu. coloured or African man, often by children to an elderly servant as a mark of respect for age: also as a title with usu. a Christian name: see also *aia* and quot. at *not so. cf.* U.S. use of '*uncle*' to an elderly negro; also Hong K. courtesy prefix Ah- with surname e.g. *Ah-Chim, Ah-Poon* to menservants or to amahs. [*Afk. ou cogn.* old + *Ngu. ta(ta)* (q.v.) father *prob. fr. Du. dial. tate* father]

 When she started imitating Outa Adoons or Aia Rosie, they would laugh (including Outa Adoons and Aia Rosie) till the tears came. Krige *Dream & Desert* 1953

outers [ˈaʊtəz] *n. slang.* The 'domicile' of vagrants: doorways, bus shelters, bushes etc. See quot. at *outie*, and *cf. bergie* [*prob. fr. Austral. on the outer,* penniless]

 I was quickly introduced to the language of their world beyond the fringe . . . They aren't hoboes, they are outies, and their domain is the outers. I never learned whether this was an abbreviation for outside, out of luck, out of respectability or a combination of all these . . . in the words of old Chris: 'Anything, but anything, is better than the outers.' . . . two newcomers to the outers . . . had been sleeping in parking garages, bus-shelters and behind bush-

es. *Family Radio & TV* 23-30.1.77
 . . . she didn't smell all violets – smelled, in fact, as if she had been living on the outers for a couple of weeks. Muller *Whitey* 1977

outie [aʊtɪ] *n. pl.* -s. *slang.* A down-and-out: an inhabitant of the *outers* (q.v.): see also quots. at *vlam, outers,* and *goffel.*

 A hardened outie becomes resigned to sleeping on cardboard in shop doorways but he doesn't enjoy it . . . The behaviour of my own outie companions dramatically illustrated . . . within the realm of the outies, everyone is everybody else's friend and there are no social bounds. Where there are no social strata there is no apartheid either. *Family Radio & TV* 23-30.1.77
 It didn't take long to find out what an outie's biggest problem is . . . that one overwhelming thought in an outie's mind: where is the next dop coming from. *Ibid.*

outjie [ˈəʊˈkɪ, -cɪ] *n. pl.* -s. *dimin.* form of *ou* (q.v.) sig. 'little chap' referring to a child, or slightly contemptuously to an adult as in 'those outjies up in Johannesburg don't know which side their bread's buttered.' see quots. at *moera* and *oke.* [*Afk. ou* fellow + *.dimin. suffix -tjie*]

 . . . when I reached the first pondokkies and the thin dogs, the wind turned and brought the stink from the lake and tears, and a clear memory of two little outjies in khaki broeks. Fugard *Blood Knot* 1968

outkraal [ˈaʊtˌkral] *n. pl.* -s. A *kraal* (q.v.) at some distance from the homestead or *werf* (q.v.) [*Eng.* out + *Afk. kraal* (q.v.)]

 . . . if there be cattle to see to or outkraals to visit. Schreiner *Thoughts on S.A.* 1891

outspan¹ [ˈaʊtˌspæn] *vb. trns. and intrns.* To unyoke oxen from a wagon or plough or unharness other draught animals, also to break a journey, now used even *figur.* of a break in a journey by car. [*translit. Afk. uitspan uit cogn.* out + *span fr. Du. spannen* to hitch or put to]

 Having outspanned – that is, unyoked the oxen – the driver . . . gave me an iron kettle and told me to follow the oxen because they would be sure to find the water. Buck Adams *Narrative* 1884
 . . . a place called Mohaddah, a little place of only two or three houses. This was the half way place, and we 'outspanned' here all day. *Argus* 13.4.1878 cit. Du Plessis 1944

outspan² *n. pl.* -s. An area formerly specifically set aside for travellers by ox wagon to rest and refresh themselves and their animals, now also used of areas of commonage used for grazing: see quots. at *eaten out, Divisional Council* and *nagmaal.* [*fr. Afk. uitspan vb.*]

'. . . complaints from farmers in the areas of the over-utilisation and over-grazing of these outspans . . .a natural consequence of hiring out veld,' . . . The Council has been concerned about the condition of the outspans for some time. Members have called them remnants of ox-wagon days no longer necessary today. *Grocott's Mail* 3.10.72
[*presum fr. Afk. uitspanplek* (~ place) *fr. Du. uytspanplaats*]

outspan³ A rural servitude: the right of 'outspan'.

Outspan, uitspan: the right to allow one's draught animals to rest, graze and water on the land of another. Wille *Principles S.Afr. Law* 1945 *ed.*

ou-vrou-onder-die-kombers ['ɔʊ ˌfrɔʊ ˌɔn-(d)ə(r) dĭ ˌkɔmˈbɛə(r)s] *n. prob. obs. lit.* 'Old-woman-under-the-blanket.' see quot. [*Afk. ou,* old; *vrouw,* woman (under the) *kombers,* a ship's blanket]

This is the humorous designation given by the Dutch to a dish consisting of Carbonatjes . . . baked in dough or batter – not unlike the English dish known as 'Toad-in-the-hole'. Pettman *Africanderisms* 1913

Ovambo ['ɔʊˈvæmbɔʊ, ŭˈvambŭ] *n. pl.* -s. Also *Owambo* (q.v.): a negroid Bantu people of South West Africa, or a member of this people.

Police in South West Africa's model Owambo homeland are on stand-by this morning after a day of violence in which 3 000 Owambos clashed with armed constables, incurring several dozen casualties. *E.P. Herald* 16.8.73

over *prep. substandard.* About, at: see also quot. at *stokkies-draai.* [*trans. Afk. oor* about, over]

The Chief complains over the Goes for having taken their baken (or landmark) away. *Colonial Office Documents* 1827 *cit.* Theale Parents anxious over boy, 16. *E.P. Herald* 7.6.74

overmass *adj.* See quots. at *mass, metrication, volume.*

overseas *n. n. modifier and adv. p.* In S.A.E. equiv. of *Brit.* 'abroad', also occ. *n. sig.* the outside world (see second quot.), freq. as modifier of *n.* ~ *interference* an ~ *trip,* ~ *visitors,* ~ *markets.* ⫿Use similar to that in B.*O.A.C.* [*trans. Afk. oorsee,* abroad: *also in OWO ons was oorsee, cf. W. Afr.* 'a been to'

(*Europe*) *n. phr.*]

. . . Fuerte avocadoes arrive in good condition at overseas markets if the fruits are pre-cooled to 4,5 deg. C. *Farmer's Weekly* 12.5.71

During the week-end we relaxed into the quietude of our peaceful existence, untroubled by thoughts of 'overseas' until at six o'clock on Monday a Union-Castle liner brought to Cape Town the illustrated papers, the merchandise and the passengers that formed our link with the Mother countries. Henshilwood *Cape Childhood* 1972

Owambo [ŭˈvambŭ] *n. prop.* Official designation of the *homeland* (q.v.) of the Ovambo people: see quot. at *Ovambo.*

owerheid ['ʊəvə(r)ˌheɪt] *n.* The authorities, 'powers that be' – usu. in a political sense, or referring to superiors at work or in business. *cf. the* (*top*) *brass.* [*Afk. owerheid* authority *fr. Du. over cogn.* over + -*heid* (q.v.), *n.-forming suffix*]

I can see the owerheid won't give me any more staff. *O.I.* 1971

Owl Run *n. prop.* See quot. at *take the gap, to.*

P

paapies ['papĭs, -z] See *papies.*

paarde- [pɑː(r)də] *n. prefix.* Horse: found in place names Paardeberg, Paardevlei, Paardefontein, Paardeneiland; also in plant names e.g. ~*pis* (urine) *Vepris undulata,* white iron wood; and in ~ *ziekte* (q.v.) more commonly in *Afk.* form *perd*(*e*) (q.v.). [*Du. paard* horse]

paardeziekte ['pɑː(r)dəˌzĭktə, -ˌsĭk-] *n.* Horse sickness *Œdema mycosis.* [*Du. paard* horse + *ziekte* disease]

. . . one of the few situations in this part of the country where horses are not liable to the *Paardeziekte* (Horse distemper) which rages during the summer season, and annually carries off great numbers. Burchell *Travels I* 1822

Horse-sickness or 'paard-ziekte', *Oedema mycosis,* is a deadly epizootic disease which has been known in Cape Colony since 1719.
. . . its full force about the beginning of February, and disappears for the season on the advent of frosty weather. Wallace *Farming Industries* 1896

paarl [pɑː(r)l, ˈpeːrəl] *n.* Pearl: found in place names Paarl, Paarlshoop, Paarlberg, Paarl Rock and freq. in the names of wines. [*Du. paarl cogn.* pearl]

We had a pleasant ride from the Fransche Hoek to a village called the Paarl or Pearl, from an immense bare rock perched on the summit of an insulated mountain at the foot of which the village is built. Ewart *Journal* 1811–1814

packhouse [ˈpækˌhaʊs] *n. pl.* -es. A packing shed usu. on a farm freq. for handling fruit or other soft crops, also fleeces etc. [*prob. translit. Afk. pakhuis* packing shed *fr. Du. pakhuis* warehouse]

The Cape Provincial Administration has given permission for the Bathurst Farmers' Union to turn its disused Park Road citrus packhouse, a 'non-conforming usage' in a residential area, into a bulk store. The old packhouse will be used to store general merchandise. *Grocott's Mail* 2.2.73

pack in *vb. trns.* **1.** *substandard.* To pack clothing, books etc. [*translit. Afk. inpak* to pack (into some container)]

There is certain . . . wear you can't possibly do without. So pack it in. *Sunday Times Advt.* 1.9.74

2. *slang.* freq. among children to ~ (*laughing*), also *pack out* (*laughing*) or *pack up* (*laughing*) *sig.* to burst into sudden laughter as in 'We all packed in/out/up (laughing).' [*unknown*]

Wouldn't . . . fit . . . a two-ton ouma. That's when the whole lot of us pack up – larf! . . . we was really rolling around. *Darling* 10.12.75

pack out *vb. trns. substandard.* To unpack goods, clothing etc. from containers: also in combination ~ (*ing*) *out parade Sect. Army.* Check involving each individual's unpacking and spreading out his equipment, clothing etc. for inspection. [*translit. Afk. uitpak* unpack]

I turned around to pack out the night clothes from the suitcases. *Sunday Times* 9.3.69

I have had great difficulty keeping the shop fully stocked. As fast as the staff can pack out goods they are bought up. *Albany Mercury* 23.12.69

pad [pat] *n.* Road: usu. in combination Nasionale Pad, see *National Road*, wa ~ (q.v.) wagon track, trail etc., ~ *kos* (q.v.). In combinations also in law of servitudes – rights of way: see second quot. [*Afk. fr. Du. pad cogn.* path]

He asked me why I had stopped his Waggon on the Govt. Padt. Stubbs *Remin.* 1874–6

Right of way include . . . foot path *voetpad:* the right of walking across the land of another. Bridle path *rijpad:* the right of riding on horseback across the land of another. . . . Trek path, *trek pad:* the right of driving cattle, including large flocks of sheep, across the land of another, and also of allowing the stock to graze as they travel. Wille *Principles S. Afr. Law* 1945 *ed.*

padda [ˈpada] *n. pl.* -s. Frog, prefix to certain plant names e.g. ~ *snuifdoos*

Lycoperdon hyemale, see *devil's snuff-box,* ~ *preekstoel* (pulpit) *Nymphaea capensis,* see *kaaimansblom,* ~ *kombers* (blanket) green slime on ponds (*Spirogyra* etc.). Also in combination *brul* ~ (roar) *Pyxicephalus adspersus,* bullfrog. [*Afk. fr. Du. pad,* toad *cogn. Ger. Padde*]

padkamper [ˈpatˌkampə(r)] *n. pl.* -s. Derogatory term for a worker, or a member of his family, in a *national road camp* (q.v.) *cf. spoorie.* [*Afk. pad* road *cogn.* path + *kamper cogn.* camper]

They look like a lot of ruddy padkampers. O.I. 1974

padkos [ˈpatˌkɔs] *n.* Provisions: *lit.* 'road-food': a picnic, sandwiches or other snacks for consumption on a journey: also *figur.* 'Take ~ for your car!' (*Oil Advt.*): see also quot. at *lekker.* [*Afk. pad* road *cogn.* path + *kos* food, victuals]

I was not allowed to depart without a good supply of pat-koss, and other comforts provided by the kindness of the parishioners. Gray *Journal II* 1851

. . . tins of petrol – apart from 'padkos' – were essential to a feeling of confidence as a motorist set out on a long journey. *Evening Post* 26.2.72

pagter [ˈpaxtə(r)] *n. pl.* -s. *obs.* A farmer having the monopoly of selling wine and spirits by retail over a certain area or district. [*Du. pachter* farmer *Du. pagt, pacht cogn.* pact (*Lat. pactus* agreement)]

There were also a town butcher and baker, and a pagter (pakter), or retailer of wine and brandy; who are appointed by licence from the landdrost. Burchell *Travels II* 1824

painted lady *n. pl.* -ies. Any of several species of *Gladiolus* esp. *G. blandus* 'the vernacular name is in allusion to the markings on the lower segments of the flowers' Smith *Names S. Afr. Plants* 1966. *cf. Brit.* ~ a type of Dianthus or carnation-pink. ⁋Also a migratory butterfly *Vanessa cardui* found in the S.W. Cape: not exclusively S. Afr., also *Brit.* and *U.S.*

In the Spring arum lilies and orchids and delicate gladioli, which South Africans called Painted Ladies, grew wild in the black soil, and tiny sunbirds glittered over golden, and rose, and white proteas. Spilhaus *Doorstep Baby* 1969

paintstone *n.* Travellers' term *prob. obs.* Stone containing powder-pigment much commented on by early travellers: see quot.

> I send you a specimen of the paint-stone, which you may break if you like – you will find within a fine impalpable powder which when mixed with oil serves all the country people here as a paint for their houses, waggons &c. They are found of every possible colour except green, and what is very extraordinary they know the colour without breaking the stone, tho' on the outside they are all exactly of the same tint. The blue is the most rare, and is the Native Prussian blue. Barrow *Letter* 4.6.1797

paling ['pɑlɪŋ] *n. pl.* -s, Ø. Any of several species of marine eels of the Muraenidae: *bont* ~ used of several species of *Echidna* and *Gymnothorax* (Muraenidae fam.) known as 'moray (eel)' or 'kwatuma'. [*Afk. fr. Du. paling* eel]

> . . . landed a moray eel, the 'bont paling'. They are not common in our area and one this size (one metre) is exceptional. These creatures can inflict a nasty bite. *Grocott's Mail* 4.3.72

palmiet [ˌpalˈmɪt] *n. Prionium palmita,* a plant common on river banks which frequently spreads into and chokes the flow of such watercourses. ⁋ The young shoots were formerly used as a vegetable and the stiff, sharp-edged leaves used for weaving into straw hats: see third quot. Also found in place names esp. of rivers, Palmiet River(s), Palmietfontein. [*Afk. fr. Du. palmiet fr. Sp. or Port. palmetto*]

> Most of the rivers which we passed . . . are choked up with the plant called Palmiet by the colonists, and from which this one derives its name. Some notion of the appearance of these plants, may be gained by imagining a vast number of ananas, or pine-apple plants, without fruit, so thickly crowded together as to cover the sides and even the middle of the stream. Burchell *Travels I* 1822
>
> It is strange to observe the barrenness of fancy of the Boers in giving names to places. In every quarter of the colony we find Brak River, Swart River, Palmiet River, Baviaan's Kloof and so forth. Thompson *Travels I* 1827
>
> . . . the beaver gave way to the home-made palmiet or coffee straw, and the tiger-skin cap. Dugmore *Remin. Albany Settler* 1871

pampelmoes¹ [ˌpampəlˈmŭs] *n. pl.* Ø. Also *pomelo, pompelmoes. Citrus grandus* or shaddock. A citrus fruit related to the grapefruit with drier pulp and loose skin: used loosely in S.A. of grapefruit of any variety; see quot. at *withond. cf. Fr. pamplemousse,* shaddock or grapefruit.

[*Afk. pampelmoes fr. Du. pompelmoes prob. fr. Malay pumpulmas* ⁋ 'Later called Shaddock by the English after the captain who first imported the seeds]

> And there were always dishes piled with fruit on the table – peaches, nartjes or tangerines, pears, grapes, pompelmoes or grapefruit, spanspek or muskmelon, figs, pomegranates. Mockford *Here are S. Africans* 1944
>
> . . . and the pampelmoes, which is not unlike the citrus fruit called shaddock. Green *Grow Lovely* 1951

pampelmoes² *n. pl.* Ø. An edible marine fish *Stromateus fiatola,* bluish or purplish in colour also called butterfish, *Cape Lady* (q.v.), bluefish and blue butterfish. [*unknown: poss. fr. pompano* also a *butterfish*]

> Pampelmoes . . . An Atlantic species . . . quite common in 30–80 fathoms from the Cape to Natal, also sometimes taken from the shore. A beautiful fish, flesh delicate and tasty but does not keep well. Smith *The Sea Fishes of S.A.* 1961 ed.

pampoen [ˌpamˈpŭn] *n. pl.* -s. *Cucurbita pepo,* pumpkin: in combination ~ *moes* (q.v.). [*Afk. fr. Du. pompoen*]

> . . . they rarely have anything given to them but bread; at some of the farmhouses they are even worse off, getting the fourth part of a raw pampoon, a sort of pumpkin or bad melon which they carry into the fields with them when they have cattle to tend, – it must last them for the day. Barnard *Letter* May 1798

figur. ~ *(kop) slang.* A fool: *cf. U.S. punkinhead.* ⁋In *dimin. form (Afk.)* ~ *tjies,* mumps.

> Fancy that pampoen Whyte-Whyte putting up for Parliament again. Walker *Wanton City* 1949
>
> *Pampoenmoes:* Pumpkin . . . sugar . . . salt, stick cinnamon . . . margarine . . . Use pumpkin which is not quite ripe and has a tender peel. *Fair Lady* 8.6.77

Also found in S. Afr. place names Pampoenskraal, Pampoenpoort.

pan [pæn] *n. pl.* -s. A nearly circular depression in which water accumulates *cf. vlei,* also *Austral. clay pan* 'natural hollow of clay soil retaining water after rain' and *g(h)ilgai.* See quot. at *kolk.* [*Afk. fr. Du. pan* hollow among dunes]

> It was the year of the big drought, when there was no grass, and the water in the pan had dried up. Bosman *Mafeking Road* 1947
>
> . . . the entire surface of the pan is never completely covered and the water is only a few centimetres deep. *Panorama* Sept. 1976

Most freq. in combination *salt* ~ not only S.A.E.

These salt pans . . . are merely pools of water, collected in different parts of the arid plains during the rainy or winter months, and which imbibing a portion of the saline matter with which the ground is strongly impregnated, deposit . . . when the water evaporates, a considerable quantity of fine white salt. Ewart *Journal* 1811–1814

Also found in place names Du Toits Pan (see quot.), Etosha Pan, Soutpansberg, Panbult, Pansdrif.

At some distance on our left, the pan, which gives its name to all this famous spot, was seen in the shape of a tiny puddle. Boyle *Cape for Diamonds* 1873

panga¹ [ˈpæŋgə] *n. pl.* -s. A large broadbladed knife used for heavy cutting of bush, cane etc. [*etym. dub. Ngu ukuphanga* rush, rob, gobble up (food), *various meanings. poss. rel. Zu. phanga* shoulder blade: (*'native name in East Africa'. Webster's Third International Dict.*)]

Knives, daggers and pangas were found in many of the . . . derelict cars removed from open spaces, streets and lanes in the African townships. *Evening Post* 30.9.72
. . . a gang of four men armed with pangas ransacked their home last week. . . . 'They carried the longest pangas I've ever seen and wore red head bands.' *Cape Herald* 12.1.74

panga² *n. pl.* Ø, -s. *Pterogymnus laniarius*, an edible marine fish of the Sparidae, also called *dikbek(kie)* (q.v.) and '*reds*' a common and abundant fish with 'furry lips and flaring canines' (Smith *Sea Fishes of S.A.* 1961 ed.) [*Afk. prob. fr. Malay ikan* fish *pangerang* prince

Similarly it is remarkable that those skilful fishermen the Malays only bestowed one fish name of clear Malay origin. That is the panga, which resembles the 'ikan pangirang' (prince) or silver fish of the East Indies. Green *Grow Lovely* 1951

pannekoek [ˈpanəˌkŭk] *n. pl.* Ø. A rolled pancake usu. sprinkled with sugar and cinnamon, or served with lemon. ¶Similar to *Fr. crêpe*, but larger and somewhat thicker; thinner than *U.S. pancakes* which are usu. served flat. [*Afk. fr. Du. pannekoek cogn.* pancake]

THE OLD AND THE NEW That was the story of the Blood River Monument – old and new in transport. Pannekoek and coke. Voortrekker dresses, bonnets and minis mingled at gleaming car park and laager alike. *Sunday Times* 19.12.71

pansy shell *n. pl.* -s. The ornamental shell of the mollusc *Echinodiscus bisperforatus* (*colloq.* 'pansy') which is peculiar to the S. Afr. coast between Keurbooms River and Mossel Bay, particularly at Plettenberg Bay where it is used as an emblem and sold in large numbers in curio shops: For this reason the 'pansy' is becoming regarded as an endangered species. ¶ A similar species is found in California where they are known as *sand dollars*.

The pansy shells are named for their decorative under-side which, in varying shades of purple and white strangely resembles a pansy. . . . The shells have been exported to Australia and America to conchologists who know that this is the only part of the world where they are found . . . When the pansies die, it takes about a week for them to be bleached white by . . . the salt and sun. If they are not left in the sun they will retain their purple colour. *Het Suid Western* 8.9. 76

pantihose *pl. n.* S.A.E. term for *Brit. tights, U.S.* (and elsewhere) *pantistockings.* [*unknown, poss. rel. archaic* 'trunk-hose']

There was only one bed in the house so he and . . . shared it. He was naked and she wore pantihose. He did not touch the girl because she asked him not to. *Cape Times* 8.1.75

pap¹ [pap] *n.* Porridge, usu. of *mealie meal:* in combinations *krummel~, mealie/mielie~, stywe~,* all (q.v.) see also *putu* and quot. at *wors.* [*cogn. Eng.* pap, soft, pulpy food (*Zu. mpuphu* fine meal)]

With typical childlike imitation all native labourers began to find boiled mealies disagreed with them and plumped for the more refined 'impoopu' or 'pap'. Tait *Durban Story* 1961
. . . For the first three weeks I was lost in Australia. I terribly missed my favourite dish of pap and steak. *Post* 28.2.71

pap² [pap] *adj. colloq.* Flabby or feeble of persons or substances, exhausted; see also *verlep, oes.* [*Afk. fr. Cape Du. pap* weak, soft]

Snoek is one of the fish which must never be hung up to dry in the moonlight, or it will become pap. . . . I think it is the dew that spoils the snoek rather than the moon. However, pap snoek . . . seems to bring out rashes and cause headaches in some people. Green *S. Afr. Beachcomber* 1958

There's another side to the problem of African widows trying to find work. 'They're too pap, washed-out, burnt-out. They have lost all incentive. Life's kicked them so hard they no longer have the initiative.' *Evening Post* 10.6.72

In combination ~ *nat*(wet), soaked,

sodden.

Remember that night the water came up so high? When we woke up pap nat with all our things floating down to the bridge. You got such a skrik you ran the wrong way. Fugard *Boesman & Lena* 1969

Also ~ *snoek* (*figur.*) Softie, coward: ~ *broek* (*trousers*) (q.v.) below.

papad See *Indian terms*.

papbroek ['pa(p)₁brŭk] *n. pl.* -s, -e. *colloq.* A coward. [*Afk. fr. Du. pap*² (q.v.) flabby + *broek* pants]

Boys will be boys, and none liked to be called 'pap-broek' or nincompoop, so there was soon a lot of money and 'Good-For' in the pool . . . Prance *Tante Rebella's Saga* 1937

He's such a helluva papbroek no one wants him in the house team. O.I. Child 1973

deriv. ~*ig adj.* Cowardly or pertaining to a *papbroek* (q.v.). [*Afk. papbroek* coward *lit.* 'flabby pants', -*ig adj.* forming suffix equiv. Eng. -y]

He added that Mr . . . was scared to release the evidence. 'I say you are too papbroekig to do it' . . . the . . . Minister shouted. *Daily Dispatch* 10.2.79

papaw ['pɔ₁pɔ] *n. pl.* -s. *freq.* Pawpaw, the subtropical fruit of the tree *Carica papaya* usu. known elsewhere as papaya. ⫿Used in *U.S.* for both tree and fruit of *Asmina triloba* the N. American ~. *Jam. E. papaw* and numerous compounds. [*fr. Span and Port. papayo, papaya*]

The breakfast papaw had been grown in a nearby garden. This fruit is like a cross between a melon and a marrow, and is eaten all over the Union with sugar and a slice of lemon. Morton *In Search of S.A.* 1948

However, the cooler conditions at this time have led to the development of many small 'cocktail' papaws, or papayas as the Americans call this fruit. *Farmer's Weekly* 18.4.73

papegaai [₁papǝ'xaɪ] *n. lit.* Parrot: the traditional form of target used for marksmanship competitions at the Cape from the early days: see also at *wapenschouw*. [*Afk. fr. Du. papegaai cogn.* popinjay]

He emphasized the importance of good marksmanship during the gay fornight of his birthday. The principal target of the practice was a clay bird called a parrot or papegaai. Mills *First Ladies* 1952

In combination ~ *shoot* see quot.

Members of the South African Historical Arms Association take part in the traditional 'papegaai' shoot at the festivities marking the birthday of the Governor Simon van der Stel in Stellenbosch today. *Argus* 14.10.72

papies ['pɑpĭs, -z] *pl. n.* Bots [*Afk. fr. Du. paapje* cocoon + *pl. suffix*] **1.** The maggots of the gadfly *Aestrus* (*or Gastrophilus*) *equi* harboured for 10–12 months in the digestive tract of horses from their ingestion of newly hatched eggs irritating the skin.

2. Found in place name Papiesvlei poss. referring to (3) *papies*.

3. *Typha australis* and *T. capensis*, plants also known as ~.

paapkuil: Sometimes contracted into PAPIES because it was believed that horses developed 'papies' (bots) when grazing in vleis or marshy places where the plants grow. Smith *S. Afr. Plants* 1966

Parabat ['pærǝ₁bæt] *n. pl.* -s. A member of the *Para*chute *Bat*talion also *the 'Bats*: see second quot. at *vasbyt*, also *marble* and *hondjie*. ['portmanteau' word]

A 29 year old Parabat company commander jumped to his death . . . yesterday when he landed in a farm dam and drowned. *Daily Dispatch* 8.9.78

paraffin¹ *n. poss. reg.* E. Cape: Insignificant rain, insufficient to benefit crops and which serves simply to burn young seedlings when the sun comes out. [*Afk. paraffien presum. fr. volatile nature of paraffin oil* (*kerosene*)]

Had any rain?

Nothing but paraffin for the last six months. O.I. E.Cape farmers 1973

paraffin² *n. Afr. E. Urban slang* Gin: see ~¹. ⫿ Also acc. servicemen, anything stronger than beer. [*presum.* because colourless]

. . . cases of mahog and paraffin which is the name some non-voters prefer to call gin. *Drum* 8.7.74

Paratus [pǝ'rɑtǝs] *n. prop.* 'Prepared': the S.A. Defence Force magazine, also a television serial on Army life. [*Latin paratus* prepared]

'Paratus' named after the official magazine of the South African Defence Force, centres on the fictional staff of that magazine, and each selfcontained segment will concern a story in which these journalists become involved. *Sunday Times* (TV) 7.8. 77

parmantig [₁par'mantɪx] *adj., adv. m. colloq.* Cheeky, impudent. [*Afk. fr. Du. parmantig* proud, arrogant, conceited]

. . . we have no need for active persecution, beyond making an example, now and then, of an 'impudent' (*permantig*) non-European who does not remain in his place. Hoernlé *S. Afr. Native Pol.* 1939

Magtig! You spoke 'parmantig' like that to Mr. Mountjoy? Butler *Take Root* 1970

As to being 'parmantig' or conceited, I believe we must bring the Griqua and Hottentot in as guilty, at any rate from a caste standpoint. Mackenzie *North of the Orange River* 1871 (1971)

particles, omission of See *omissions.*

party *n. pl.* -ies. Term used to designate a group of British Settlers of 1820 of indefinite number, officially placed under a head or *leader* (q.v.) and assigned to a particular *location* (q.v.) [*official term*]

. . . has not left the Party, he is removed from being head of the party and has not been allowed to draw the rations for the party for some time. Hancock *Notebook* June 1820

. . . portion of the Location of the Salem Party (sub-division No. 6) and Portion 23 of Salem Commonage, all situate in the Division of Albany. *Grocott's Mail Advt.* 24.4.71

Still found occ. in the E. Cape in place names of former *locations* (q.v.) of settler ~ *ies*, e.g. Deal Party, James's Party, Nottingham Party.

At the distance of four miles, is a village called *Ebzn Ezer* and sometimes James's Party: The greater part of this location lies upon an elevated ridge of limestone formation. *Cape of G.H. Almanac* 1843

pasbrief ['pasˌbrïf] *n. obs. hist.* A document carried by any slave or other servant at the Cape indicating his master's identity and willingness to his being abroad: also a pass for permitting a Hottentot to move from one area to another. [*Du. pas* pass + *brief* letter]

1st Jan 1811 Every slave or even Hottentot, who is found at a distance from home, without a pasbrief, or passport, signed by his master or some responsible person, is liable to be taken into custody as a runaway or vagabond; and this precaution alone is a powerful check to keep a slave from absconding. Burchell *Travels I* 1822

pasel(l)a See *bonsella.*

pas op ['pasˌɔp] *vb. phr. interj. colloq.* Warning interjection sig. 'be careful', 'watch out', also *oppas* (q.v.). See quot. at (1) *poenskop.* [*Afk. fr. Du. oppassen* to be on guard]

'Burghers, burghers!' she screamed. 'Pas op! Look out! The Tswana are coming up the hill!' Gibbon *S. Afr. Stories* 1969 ed.

pass *n. pl.* -es. **1.** Also *passbook* now known as *reference book* (q.v.) and in *Afr. E.* 'book-pass', *domboek* (q.v.) and *dompas* (q.v.), an identity document carried by all Africans with details of

employment and other personal data: see also *pasbrief*, and quots. at *reference book* and *aid centre.*

The Chief asked the Government to reconsider the whole question of passes. He pointed out that any policeman, no matter how junior, could 'humiliate me by demanding this thing from me at any time.' *Cape Times* 18.1.73

Freq. in combination ~ *law* (*system*) (*offences*) (*arrests*) etc.

. . . nowadays an offender is less likely to be automatically jailed for a pass-law transgression or other minor infraction – partly because of the work of a string of government 'aid centres' that have been established to help blacks cope with the law. *Time* 15.10.73

The last resolution urged that the Republican Government '. . . repeal pass laws and influx control regulations, to enable the Black people to sell their labour wherever required.' *E.P. Herald* 2.11.72

Also ~ *burning* the burning of passes as a gesture of defiance.

2. *hist. obs.* A written document usu. from the head or leader of a settler *party* (q.v.) carried by any settler employed by him when not on his *location* (q.v.).

To be obliged to procure a 'pass', in order to go merely from the location to Graham's Town, without incurring the risk of getting a night's lodging in the 'tronk' on arriving there, and even when applying for one, to be flouted by petty official insolence, . . . chafed the minds of Englishmen in a manner that soon gave a pledge of the downfall of the system. Dugmore *Remin. Albany Settler* 1871

3. *Sect. Army abbr.* of 'leave pass': combinations *weekend* ~, *seven day* ~, etc. and *prep. phr. on* ~.

They . . . started washing, drying and ironing their clothes, bathed did P.T. on the beach, wrote more letters and went out on pass. 'Twede in Bevel' *Piet Kolonel* 1944

. . . leave and leave passes are granted as follows . . . After completion of nine weeks two weekend passes a month are allowed . . . A pass is a privilege and *not a right* . . . A pass is only given when a member is entitled to it. *Circular to Parents* 13.1.79

The first pass I get is on the 2nd of March and I am not sure how long it is – probably only three days. *Letter Serviceman* 14.1.79

passenger Indian *n. pl.* -s. An *Indian* (q.v.) of the merchant class, often a Muslim, who came to S.A. not as an indentured labourer but independently: see *Arab*, and *Bombay merchant.*

The Gujeratis who alighted as 'free' or 'passenger' Indians, were on a higher economic rung. *Indian Delights ed.* Mayat 1961

Free, or passenger Indians as they came to

be called, followed in the wake of the indentured to Natal, but White colonists became alarmed by the competition offered by these merchants . . . By 1913 Indian immigration was generally prohibited by law. Meer *Portrait of Indian S. Africans* 1969

passion gap *n. pl.* -s. *slang.* Also called 'love gap': the space left by the extraction of up to four upper front teeth: see quot.

. . . our record 'haul' at one session now stands at 53 extractions . . . relief of pain is our sole objective. We do not cater for 'passion gap' dentistry . . . achieved by the extraction of the four top middle incisor teeth which is reported to increase sex appeal when kissing . . . *False Bay Hospital Assoc. W.O.* No. *1133 1972-5*

pastei [ˌpasˈteɪ] *n. pl.* -s. A pie, usu. of meat or chicken: in combination *vleis* ~ (*meat*), *hoender* ~ (*chicken*), skilpad ~ (water tortoise). [*Afk. fr. Du. pastei* pastry, pie *cogn.* pasty]

In the kitchen the elaborate preparation of curries, frikkadels, and pasteis had occupied the attention of a troop of slave girls . . . The *schildpad pastei* in a large, round blue and white dish was so clearly the pride of mevrouw's heart. Fairbridge *Which Hath Been* 1913

Do you like roast beef and Yorkshire pudding? Or turkey and cranberry sauce? Or hoenderpastei and geelrys? *E.P. Herald* 15.11.72

pastorie [ˌpastŭˈrĭ] *n. pl.* -s. The parsonage or manse of a *Dutch Reformed Church* minister, the *dominee* (q.v.) or *predikant* (q.v.). [*Afk. fr. Du. pastorie* parsonage *fr. Lat. pastor* shepherd, *hence* priest]

On the stoep of his pastorie sat the minister, Henricus Bek, watching Dorp Straat come to life again after its midday sleep. Fairbridge *Which Hath Been* 1913

Eventually, on the Methodists acquiring land elsewhere, the house was granted to the Dutch Reformed Church as a 'pastorie', which function it served until well into the present century. Lewcock *C19 Architecture* 1963

patat [pəˈtat] *n. pl.* -s. Sweet potato: the tuberous root of *Ipomoea batatas* of the Convolvulaceae: similar to the yam, see quot. at *pocket. N.Z. kumara,* sweet potato (*Maori*). [*Afk. fr. Span. batata(s)*]

. . . recipe for Stowepatats . . . Mrs M's Stewed Sweet Potatoes. If available the fat yellow *borriepatat.* Peel and slice them . . . stew gently . . . Add a tablespoon of *moskonfyt. Darling* 16.3.77

In combination ~*salf* one of the *Dutch Medicines* (q.v.) used for the treatment of *veld sores* (q.v.)

patha-patha [ˈpataˈpata] *n.* An African dance music mode originating prob. in the 'sixties. [*Ngu. patha,* touch]

'Cats' in Windhoek . . . are raving mad with mbaqanga, the Soweto originated song-dance mood which came soon after patha-patha (touch-touch) a dance tempo popularised by . . . Miriam Makeba. *Post* 15.6.69

Soweto families prefer to visit a beer garden for 'Bantu beer,' or a shebeen (speakeasy) for stronger drink and the sensuous local music called *patha patha.* Shebeens are unlawful, but police tolerate them as pressure valves. *Time* 28.6.76

pauw See *pou.*

pawpaw [ˈpɔˌpɔ]. See *papaw.*

payday *n. Afr. E.* The day on which disability or old-age pensions are paid to Africans.

For the old and disabled, it was another 'pay day', which meant getting up early to stand in the queue, but for the . . . Women's Club . . . it was another day to serve hot soup . . . to the pensioners. *World* 28.9.77

peach brandy *n.* Home-distilled spirit made from fermented peaches: see also *mampoer* and quot. at *kêrel.*

What fetched us all to Abjaterskop in the end was our knowledge that Willem Prinsloo made the best peach brandy in the district. Bosman *Mafeking Road* 1947

Beside the canal an old man fed logs into the oven of a home-made peach-brandy distilling plant. A young boy scooped fermenting peaches into a bucket from a large cow-hide container suspended from four stout poles. Becker *Sandy Tracks* 1956

peach leather *n.* See *perskesmeer.*

But didn't he say anything else about the peach leather? . . . He bit off a piece, chewed it, and said: 'Peach leather? . . . Well, its well named, for sure.' Butler *Take Root* 1970

peach-pip floor *n. phr. pl.* -s. Also *perskepitvloer.* A floor of peach pips embedded in clay still occ. found in the Cape. [*trans. Afk. perskepitvloer*]

Fruit pips provide floors in the Cape, though the mixture of ant-hill clay and cow-dung is far more common. Nevertheless, the peach-pip floor is still to be found on some farms. It takes thousands of pips to cover even a rondavel floor, and the surface is liable to be slippery. Such floors are burnished with ox-blood, and are relics of more leisurely days. Green *Land of Afternoon* 1949

peanuts *n.* or *pseudo prn. colloq.* 'Little or nothing' usu. applied to remuneration, (see quots.) or importance, as in 'It means peanuts to me.' Also *U.S.* [*unknown*]

. . . artists, . . . have to be treated decently . . . Pay them peanuts and you will get them staggering in hours after the show was to have started. *Drum* 8.7.74

One builder . . . estimates that to safeguard the entire town's decayed Settler houses would cost no more than R30 000 'and this is peanuts in today's money market' he added. *Star* 30.9.74

pear *n.* Also *peer:* any of several native timber trees so named on account of some resemblance to the pear tree or its timber: usu. in combinations *hard* ~, also *rooibessie* (q.v.), *red* ~, *white* ~, etc.: see quot. at *saffraan*.

Other Cape woods used in old furniture, but seldom met with or recognized were olive wood, witte els, rooi els, camphor wood and pearwood. Gordon-Brown *S. Afr. Heritage* 1965

Pedi ['pedï] *n. pl.* Ba-. A N. Sotho people: see *Lebowa*.

peer ['pɪə(r)] *n.* See *pear*. Combinations *hard(e)* ~, *rooi* ~, *wit* ~, *wilde* ~. [*Afk. fr. Du. cogn.* pear]

peetsho ['pitʃɔ̃] See *pitso*, also *indaba*.

pekel- ['pɪəkəl] *n.* Pickle: prefix as in ~ *balie* pickle jar, vat ; ~ *aar* (q.v.). [*Afk. fr. Du. cogn.* pickle]

kitchen shelves . . . held crockery and items such as a candle mould, pudding moulds, 'pekelbalie' (tub for pickling meat) . . . *Panorama* Jan. 1975

pekelaar ['pɪəkə,lɑ:r] *n.* Salted fish, fish pickled in brine: usu. in combination *snoek* ~. [*Afk. fr. Du. pekelharing* pickled herring]

Snoek Pekelaar. Have the snoek cut into mootjies. *Miss Hewitt's Cookery Eook* 1890
'Snoek *pekelaar* is the name we give to fillets of snoek slightly salted and sundried.' Hilda's *Diary of a Cape Housekeeper* 1902 *cit.* Pettman

pelile [ˌpe'lile] *partic. usu. colloq.* Finished, dead beat: as in 'I've worked on the thing until I'm just about pelile', see also quot. [*Ngu.* (*ndi*|*ngi*)*phelile* (I am) finished: *qualificative form fr. ukuphela* to finish]

. . . as nothing compared with their ancient African wisdom. '*Ngipelile* . . . (I am finished . . .)' Krige *Dream & Desert* (*Death of the Zulu*) 1953

pellie ['pelï] *n. pl.* -s. *colloq.* Pal: mode of address or reference to a friend, esp. *ou* ~, or ~ *blue:* see quot.: see also *chommie*/*tjommie*. [*presum. fr. Eng.* pal]

. . . along with four other ous . . . They all pally blues. . . . travelling around the world together in an old clapped out Kombi. *Darling* 9.10.74

Vernon's he's ou pellie blue from back home. *Ibid.* 12.2.75

penkop ['pen,kɔ̃p] *n. pl.* -s, -pe. A youth, esp. used of one of below military age

during the Boer War. [*Afk. penkop* youth, youngster, cub]

A cavalcade approached – old men riding side by side with young boys: just as it was in the Boer War – greybeards and *penkop* youngsters taking up their rifles against the English. Bosman *Willemsdorp* 1977

Peninsula, The *n. prop. reg. Cape Town abbr.* The Cape Peninsula: see quot at *tuan*.

pennyline *n. pl.* -s. *Afr. E.* Cheap prostitute *cf. tickeyline.*

'I do not ask your husband to make love to me. Talk to your husband and leave me alone' . . . 'Ja you Shoeshoe women ! You are terrible "pennylines" !' Longmore *The Dispossessed* 1959

penny-whistle *n. pl.* -s. The leading instrument in *kwela* music: see quot. at *kwela.*

A penny whistle wailed a wistful, halting melody. Penny-whistle people, Richard was thinking, fluting and strumming in the dark – no matter what happens. Thompson *Richard's Way* 1965

pens-en-pootjies ['pe:nsn̩'puĭkĭs, -cĭs] *n.* A stewed dish of tripe and trotters made usu. of mutton: see also quot. at *kop-en-pootjies*. [*Afk. fr. Du. pens* belly + *en* and + *poot* animal's hoof or foot + *dimin. suffix* + *pl.* -jies]

Kop en pootjies (sheep's head and trotters) and pens en pootjies (tripe and trotters) are two more traditional dishes of the Cape which may be classified as stews – but what stews! *Farmer's Weekly* 25.4.73

people See *family* ℙ ~ is also freq. used in S.A. *sig.* 'guests' *e.g.* I'm going to be busy I've got people . . . for the weekend . . . for dinner . . . this evening, etc.: see also quot. at *goffel.*

. . . an old woman died of hunger in this house when her peopel [sic] went to Nagmaal and forgot her. Iris Vaughan *Diary* circa 1900.

peperboom ['pɪəpə(r),buəm] See *pepper tree.*

peperkorrels ['pɪəpə(r),kɔ̃rəls, -z] *pl. n.* 'Peppercorns': usu. *peppercorn hair*, or *korrelkop* both (q.v.). [*Afk. fr. Du. peper cogn.* pepper + *korrels* granules]

. . . pure White and pitch Black, straight hair and peperkorrels. Guy Butler *cit. Forum* Oct. 1961

peppercorn hair *n.* Also *peppercorns*, *peperkorrels:* sparse tufts of tightly curled, woolly hair, characteristic of the Hottentots and Bushmen: see also *korrelkop.*

These yellow skinned people, small of stature, their round heads dotted with 'pepper-corns' of wool, ugly in feature, active and agile as monkeys, skipping and strutting about to the tune, leaping, clapping their velt-schooned feet in the air. De Kock *Fun They Had* 1955
As a very small boy I was shown over the mine compounds, and learnt to identify a Bushman by his wizened face and patches of peppercorn hair. Green *These Wonders* 1959

pepper tree *n. pl.* -s. Also *peperboom* and *kar(r)oowilg*, *kar(r)oo willow*, *Schinus molle*, an exceptionally hardy evergreen tree cultivated for shade in the Kar(r)oo. ℙIts drooping clusters of red fruits have a pungent taste hence 'pepper', and its pollen is said to be a cause of hay fever: a 'characteristically' S. Afr. tree, see quots., but a native of S. America. Also known in S.A. as *peppercorn tree*.
Those which most attract the attention are, the red pepper tree, the castor oil shrub, the silver tree. Percival *Account of Cape of G.H.* 1804
At last here's a local film that looks and smells like the Africa we know. You almost get an aromatic whiff of pepper trees, red dust, burnt wood, manure, kraal, coffee pots and cow-dung floors. *Sunday Times* 4.7.71

perlé ['pɔleɪ] *adj.* and *n.* Either of or pertaining to, or a name for slightly sparkling or pétillant wines. ℙ Since ~ was taken over as a popular trade name, the term *crackling* (q.v.) has replaced it in the terminology of certain manufacturers. Neither refers to sparkling wines made by the 'champagne' process, which description is not permitted of S. Afr. wines. [*Fr. perlé* beaded, set with pearls, of jewels, fabrics etc. (not a description of wine)]
Light and strong sherries, Lachrymae Christae [sic] (Frontignac, white and red muscadel or Perle Constantia) and genuine Constantia Wines . . . all warranted genuine produce of the grape . . . *Advt. Cape of Good Hope Almanac for* 1845
He's a smart bird now . . . but every now and then he gets a hankering for a bit of boerewors . . . or a drop of perlies . . . all these . . . worms and goggas every day don't agree with him. Brink & Hewitt *The Birds* 1973
Perlant French term for slightly sparkling wine.
Perlé Wine The same as perlant, slightly effervescent as distinct from sparkling wines . . .
Perlwein German term for white semi-sparkling wine made in a pressure tank which retains carbon dioxide generated by fermentation. Bolsmann *S.Afr. Wine Dictionary* 1977
. . . [Trade Name] Late Harvest and Perlé . . .

Advt. EP Herald 30.3.79

perlemoen ['pe:rlə‚mŭn, 'pɛə(r)-, 'pɜ-] *n. pl.* ∅. Also *klipkous* (q.v.): used of several edible species of *Haliotis* esp. *H. midae*. U.S. *abalone :* see quot. at *klipkous*. [*etym. dub. prob. fr. Afk. fr. Du. perlemoer* mother of pearl]
Miss Duckitt knew all about the perlemoen or klipkous, largest and most beautiful of Cape shellfish. The flesh is so tough at first that some people put it through a mincer; but she preferred to boil it with wood ash, then beat it with a mallet. Green *Tavern of Seas* 1947
The perlemoen were being canned and sent to the East to restore the failing virility of wealthy orientals. *Ibid. S. Afr. Beachcomber* 1958

perde- ['pɛə(r)də-, 'pe:(r)-] *n.* Horse: prefixed to numerous plant names usu. sig. the association of a particular plant with horses, e.g. ~ *gras* several varieties of *Aristida ciliata* hay and pasture grasses ; ~ *pisbos/boom, Clausena anisata;* ~*vy*, see *Hottentot Fig;* ~ *pram, Fagara capensis* (see *knobwood*) so called from the likeness of the protuberances to a mare's teat, etc. [*Afk. fr. Du. paard* horse]

perskepitvloer ['pɛə(r)skə‚pɪt‚flu:r, 'pe:-] See *peach pip floor*.

perskesmeer ['pɛə(r)skə‚smɪə(r), 'pe:-] *n.* A thin layer of pounded peaches, dried in the sun and rolled: also *smeerperske*, *peach leather*, occ. also *tammeletjie rol(l)*. [*Afk. fr. Du. perske* peach + *smeer* spread]
. . . and hard and flavourless peach which was pulped after cooking, dried, sprinkled with sugar and rolled to form the sweetmeat known as *perskesmeer*. Green *Glorious Morning* 1968

Permanent Force *n. prop.* See *P.F.*
. . . last week a unit of the Permanent Force killed and wounded a number of terrorists. One member of the S.A. Permanent Force . . . lost his life. *E.P. Herald* 7.74

permantig See *parmantig*.

person, a *pseudo-prn. substandard.* Equiv. of *Eng.* one, *indef. prn.* for which 'you' is freq. substituted both *Brit.* and S.A.E. See quot. at *mos*[1]. [*trans. Afk. 'n mens fr. Du. een mensch* one, someone]
'Come Boetie, don't be offended'. 'But you make a fool of a person.' Stormberg *Mrs. P. de Bruyn* 1920
There is also severe scarring to his leg which is now an ugly sight and offends a person's aesthetic sensibilities. (Quoted from a judgement.) *Daily Dispatch* 20.8.69

'. . . A person can get very lonely living by herself.' She often refers to herself as 'a person' preferring perhaps the anonymity of the word . . . '. . . he said . . . that he was going to marry someone else. I mean what can a person do?' *Darling* 10.1.79 Also used as equiv. of *someone, somebody* the 3rd person indefinite *prn. see* quots. at *makgotla* and *mpundulu bird.*

Peruvian *n. pl.* -s. *Sect. Jewish.* An E.European Jewish immigrant to S.A.: as used by S. Afr. Jews sig. E. European Jews retaining their foreign accents, customs and characteristic eating habits. ⫶Also used loosely by anti-Semitic persons for anyone Jewish. [*prob. fr. acronym P.R.U.* Polish and Russian Union]

. . . illicit liquor traffic was in full blossom, and mostly in the hands of 'Peruvians' (lowest type of Polish Jews). Cohen *Remin. Johannesburg* 1924

Among Dutch-speaking Afrikaners Meninsky was usually known as 'the Peruvian' – by heaven knows what confusion of ideas, for he had come to South Africa from Poland by way of Whitechapel. Brett Young *City of Gold* 1940

In Kimberley during the early days, there was an organisation calling itself the Polish and Russian Union, which had its own Club. This became shortened into 'PERU' . . . the expression of 'Peruvian' originated. Sonnenberg 1957 *cit* E. Rosenthal

petrol joggie ['jɔxĭ] *n. pl.* -s colloq. A petrol pump attendant. [*Afk. joggie* boy, lad, caddy]

. . . he was allowed by the court to get a holiday job as a petrol joggie and to go to church on Sunday. *Het Suid Western* 9.3.77

peuloog ['pĭœl̩ʊəx] See (2) *Cape lady* and *fransmadam.*

phutu See *putu.*

P.F.[1] ['pi̩ef] *n. prop. Permanent Force*, or one of its members: the S. Afr. standing army formerly *U.D.F.* (q.v.) [*acronym*]

. . . They found themselves subjected to what they felt to be the indignity of being treated as recruits by P.F. instructors of limited qualifications, whose sole claim to fame rested on their parrot-like knowledge of Pamplet 2, Lesson 3. 'Twede in Bevel' *Piet Kolonel* 1944

P.F.[2] *Rhodesian.* Patriotic Front. [*acronym*]

The magazine *Illustrated Life Rhodesia* says the forces of the Patriotic Front (PF) are 'succeeding in their objective, the erosion of order' . . . There is a fundamental difference between . . . all-party talks and simply demanding that the PF join the traditional government . . . the most probable effect of which would be to split the PF. *Financial Mail* 18.8.78

216

phatha-phatha See *patha-patha* and quot. below.

'Perhaps your daughter wants to phatha-phata,' the Black Jack said, using the coarse expression for sexual intercourse. . . . a stranger could still save his throat if he has some knowledge of basic words and phrases: . . . Phataphata – sex. Venter *Soweto* 1977

pheasant *n. pl.* -s, ∅ Applied to several of the Phasianidae family particularly certain francolins. [*translit. Afk. fisant*]

. . . a bird styled 'pheasant', though about as like a pheasant of England as a Dutch Boer is to a Bond Street exquisite. Polson *Subaltern's Sick Leave* 1837

There is no true pheasant . . . found in Africa . . . Several species of Francolins belonging to the same family are . . . known as pheasants. Wallace *Farming Industries* 1896

phumula See *pumula.*

phuza See *puza* and *pusa.*

piccanin ['pɪkə̩nɪn] *n. pl.* -s. Usu. a small African child: not exclusively S.A.E. (also *Austral.* aboriginal child) but used esp. in *Afr.E.*, see quot. and quot. at *mompara :* also in compound ~*ny kia*, (see *P.K.*); and in derived form *pik* or *pikkie* [*fr. Port. pequeno* small + *-ino dimin. suffix*]

Picannin I am so short that my friends call me picannin and I don't like this. Please tell me what I must do to grow taller. I am 19 and only 4 ft. tall. *Drum* 22.3.74

pickled fish *n.* Traditional Cape dish see also *ingelegde vis :* fish prepared with onions in a vinegar sauce flavoured with curry powder, *borrie* (q.v.) and other spices: obtainable commercially in tins ; *cf. pekelaar* in which brine or dry salt is used for pickle. [*Afk. fr. Du. ingelegde* preserved + *visch cogn.* fish]

Pickled Fish. For this old Cape favourite choose any firm fish such as snoek, Cape Salmon, Kabeljou, albacore etc. Gerber *Cape Cookery* 1950

pick out *vb. phr. substandard.* Reprimand, scold, not equiv. of 'select' *vb. trns.; cf.* to pick on someone: also *U.S. bawl out, Austral.* give curry, go crook at. [*poss. fr. Afk. uitskel* abuse, call names, *skel cogn.* scold]

'. . . okay Mummy, I started it. It's my fault.' Her mother picks her out and tells them to play nicely. *Fair Lady* 13.4.77

pick up (weight) *vb. phr. substandard.* To gain or put on weight. [*etym. dub. substandard Afk. gewig optel, lit.* 'pick up

weight', *poss. fr. S.A.E.*]
.... he had a normal figure until . . . he suddenly picked up weight for no apparent reason. *Sunday Times* 21.8.77

pick-up van *n. pl.* -s. Also *kwela kwela* (q.v.) a police van, 'Black Maria'.

There comes the big van . . .
They call it the *pick-up van*
Where's your pass?
Walker *Kaffirs are Lively* 1948
Sergeant Van Rooyen of the Marabastad Police Station lifted a hand and ordered him to accompany him to an awaiting pick-up van. Dikobe *Marabi Dance* 1970

piesangblom [ˈpǐsaŋˌblɔ̌m] *n. pl.* -me. *Strelitzia reginae*, also *S. alba*, lit. 'banana flower' so called because of the resemblance of the plant and its flower to the banana: see *crane flower*. [*Afk. piesang fr. Malay pisang* banana + *blom cogn.* bloom *n.*]

piet-my-vrou [ˈpǐtˌmeɪˈfrəʊ] *n. pl.* -s. The migratory red chested cuckoo *Cuculus solitarius*. ⫽Its noisy and distinctive call is heralded, as is that of the cuckoo in Europe, as the first sign of summer: it is found in S.A. fr. October to March. [*onomat.*]
. . . the most reliable weather forecaster known to man – the piet-my-vrou or red-chested cuckoo. . . . the nearest relative of the European cuckoo to reach this corner of the African continent. His repetitious call, which gives him his name, is one of the most typical sounds of Africa. *Argus* 16.9.72

pig *n. pl.* -s. *Afr. E.* A Saracen armoured vehicle. [*prob. fr.* shape of bonnet]
At Nyanga the policemen were spinning tops with the Native children and giving them rides in the Saracens – nicknamed 'the pigs' by them. Louw *20 Days* 1968

piggyback heart operation *n. pl.* -s. An operation developed at Groote Schuur Hospital, Cape Town, in which a 'donor' heart is implanted as a support or auxiliary to the patient's existing heart, which is not removed. [*prob. fr. abba(hart)* see *abba*]
I think it is shameful that heart transplant and piggyback heart patients cannot find employment. *Sunday Times* 29.10.78
Piggy-back heart man dies
. . . two years after receiving a second heart in a 'piggy-back' operation . . . he died last Thursday when both hearts stopped beating. *Daily Dispatch* 19.6.79

pig-lily *n. pl.* -s. Also *varklelie, varkblom* (q.v.) *varkoor :* the arum lily *Zantedes-*

*chia aethiopica (Richardia africana)*which grows in large numbers in damp situations such as *vlei ground* (q.v.). ⫽The rhizomes are said to be much rooted up and enjoyed by pigs. [*trans. Afk. varklelie*]
They always laugh at my enthusiastic love of flowers, and especially smile at my passion for the 'arum' which grows in all the ditches under the title of pig-lily and reaches an enormous size. A Lady. *Life at Cape* 1870

pignose grunter *n. pl.* ∅. *Lithognathus lithognathus*, also *varkbek*, the *white steenbras* (q.v.) see quot. *cf. Jam. E. sow mouth* (fish).
. . . the white steenbras. In the Eastern Cape between the Sundays River and the Keiskama, this fish is frequently called the pignose grunter. The 'pignose' part is descriptive of the elongated head and the thick rubbery lips resembling a pig's snout, but why the 'grunter'? The steenbras does not grunt as does the spotted grunter. *Grocott's Mail* 16.8.72

pik(kie) [ˈpɪkǐ] *n. pl.* -s. Mode of reference to a child: often used among children to or of each other as a mark of contempt – as in 'I don't want to play with her she's only a pik.' [*fr. Afk. pikkie* little chap *poss. fr. picannin fr. Port. pequeno* small *or fr. Du. piek* chicken, later small child]
Magtig! But he's grown, eh, Kleinhansie? Last time I saw you, you were just a little pikkie, so high! Meiring *Candle in Wind* 1959

pik [pɪk] *n. and vb. trns. colloq.* Among children: the pointed metal end or spike of a child's top. [*Afk. pik* peck, bite, *as n.* pick, sharpened tool]
My top's pik's come out, now it won't spin. O.I. Child
As *vb.*, to knock another top over with a throw of one's own.

pill *n. pl.* -s. A *dagga* cigarette: see first quot. at *arm*, and *zoll*[1](2). [*unknown poss.* because 'rolled']
. . . he and the constable each then started rolling a 'pill' . . . started to smoke his but the constable said he would . . . wait to smoke his until after supper. *Het Suid Western* 17.10.74

pincushion *n. pl.* -s. Any of several colourful varieties of the Proteaceae which are so called from their having long stamens which look like pins stuck in a little cushion.
The whole protea family thrives here, including the delicate 'blushing bride' and the gay pin-cushion proteas. *Evening Post* 5.6.71

pinkie [ˈpɪŋkǐ] *n. pl.* -s. The little finger, not exclusively S.A.E. (pinky) [*Afk. fr. Du.*

pink – little finger + *dimin. suffix -ie*]
Heavens! Look at this one! X-Ray left
pinkie. O.I. *X-Ray Department, Settler's
Hospital, Grahamstown*

pinotage [ˈpïnəʊˌtɑʒ, -nə-] *n. and n. modifier.* **1.** A red grape developed in S.A.,
the cultivar being a cross between the
Pinot Noir and Hermitage varieties,
produced by Dr. A. I. Perold of the
K.W.V. (q.v.) see quot. at *stein* [*'portmanteau word' pinot* + *hermitage*]
A young pinotage vineyard is flourishing on
a portion of land once regarded as near useless
. . . 16 000 vines – Pinotage on Richter 99 –
were planted in plastic mulch. *Farmer's Weekly*
14.4.76

2. A dry red table wine made from ~
grapes: on labels usu. preceded by the
estate name: see first quot.
. . . with his well known Simonsig Pinotage.
(Caption). *Ibid.* 14.4.76
Pinotage, grown and made under ideal
conditions, could be fat, full, fruity and just
as elegant as any other great wine. *Cape Times*
7.7.76
My pretentious wine-loving friends . . . are
bored stiff . . . with constantly pointing out
that the pinotage is full-bodied and robust, but
scarcely so generous or philanthropic as the
Cabernet Crus. 'Passing Show' *Sunday Times*
17.10.76

piogter Erron. form of (*h*)*oogpister* (q.v.)
see quot. at *oogpister.*

pirate *vb.* See *pirate taxi* and quot. below.
'What's the car got to do with business?
You'll just waste your time if you apply for a
taxi permit.' 'I'm going to pirate', Joe smiled.
. . . 'I'm not going to bugger around like the
other pirate taxis, begging people with my
hooter. I want regular fares. The same people
every day.' Venter *Soweto* 1977

pirate taxi *n. pl.* -s. An unlicensed usu.
African- or Indian-owned taxi plying
for hire illegally: common in large cities.
cf. Hong K. pak-pai [*prob. fr. idea of
'piracy' of trade or customers from
legally operated vehicles*]
. . . I bought a Chev for R700 and went into
the pirate taxi business in Leopold Street,
Durban. At that time the Crimson League
was operating and time and time again I had
to fight for my passengers. But business was
good, and soon I bought a de Soto and hired a
driver. *Drum* 22.7.74
In combination ~ *driver.*
A war is raging between licensed taxi drivers
and pirate taxis . . . A pirate taxi driver said
that one of the reasons for the friction was the
issue of four licenses to men operating as pirates
on the Mdantsane King William's town routes.
Indaba 6.4.79

pit [pɪt] *n. pl.* -s. *vb. trns.* Pip or stone of
fruit or vegetable: occ. a *vb. sig.* stone,
equiv. of 'remove stones'. In combination *denne*(q.v.) ~ *s* 'pignoli' or pine
nuts: not exclusively S.A.E., see quot.
cf. U.S. pitted cherries (stoned.) [*Afk. pit*
pip stone]
Why Pit, not Pip? . . . I would like to ask . . .
why he consistently uses the word 'pit' throughout his short story . . . when the word he
obviously means is 'pip'. . . . ['Pit', according
to the English dictionary, is a 'fruit stone',
it is also the Afrikaans word for 'pip'. This
was a South African short story and the use of
the word 'pit' lent an added South African
flavour to it.] (editor) *Argus* 13.12.69

pitso [ˈpïtsɔ̃] *n.* A Sotho tribal conference
gathering in the *kgotla* (q.v.) *cf. indaba*
(q.v.) see also *lekgotla* and *makgotla.*
[*Sotho pitso* conference]
A large Pitso was held on Monday by the
paramount chief Letsie at Matsieng. *E.P.
Herald* 12.4.1906
. . . the Chief had summoned the men of the
clan to the Khotla and opened the pitso by
telling those assembled that the season had
been poor, the crops bad, the cattle infertile.
Fulton *Dark Side of Mercy* 1968

P.K. [ˈpiˈkeɪ] *n. pl.* -s. *Picannin*(*y*) *kia :*
usu. an outdoor earth closet but loosely
used of a water closet also: see *kleinhuisie cf. Brit.* 'smallest room'. *Austral.
dunny, outhouse etc.* [*acronym: kia fr.
Ngu. khaya* house, home]
Please don't tear my precious **PK** down.
There's not another like it in a village or a
town. Des Lindberg (Song) *Folk on Trek* 1960's
. . . the old brick and corrugated-iron roofed
houses and such romantic survivals as the
'piccanin kias' rolled back the years. (I am
not suggesting that many 'piccanin kias' are
still in use but they are there like historic
monuments in suburban gardens.) Green
Glorious Morning 1968

-plaas- [plɑs] *n. prefix and suffix.* Farm:
usu. in combinations ~ *japie* (q.v.),
~ *seun* a country boy, not in a derogatory sense ; ~ *boer* Afrikaans speaking
farmer; *leeningsplaa*(*t*)*s*, see *loan farm.
cf. Austral. selection.* [*Afk. fr. Du. boeren-
plaats* farm]
I don't want to change the running of this
country because I'm no blerry politician, but
just a plain plaas boer. *Drum* 8.4.74
Place, farm: found in place names Veeplaas, Mondplaas, Boplaas.

plaasjapie [ˈplɑsˌjɑpï] *n. pl.* -s. See *jaap,
japie,* also *backvelder :* usu. derogatory
unlike *plaasseun* (see *plaas*): *cf. U.S. hick,
hayseed, etc.*

plaats [plɑ(t)s] *pl.* plaatzen. *hist. largely obs.* Place, farm. See *place.* Found in combination *leening ~ , bewaar ~* (q.v.), or in place names e.g. Boomplaats, Waaiplaats. [*Du. plaats* property, estate, *cogn.* place]

He possessed eleven plaatzen, or farm properties. Pringle 1822–3 *cit. Cape of G.H. Almanac* 1843

placaat [plə'kɑt] *n. pl.* -en, (-s). *hist.* An edict or proclamation set up for the information of the people. [*Du. placaat* edict, ordinance]

This Placaat, fixing the Tariff of Stamp Duties on Public Instruments. *Cape Statutes* 1862

The Council of the Dutch East India Company issued a 'Placaat' forbidding the importation of Malay slaves into the Cape. Green *Few are Free* 1946

place¹ *n. pl.* -s. *hist.* Farm: usu. in combination *loan ~farm* (q.v.), *request ~* (q.v.), *full ~* see quot. [*anglicization Du. plaats Afk.* plaas farm *cogn.* place]

They are the proprietors of the farms, or places as they call them, which they occupy in right of ownership. Bird *State of Cape of G.H.* 1823

The farms here, and indeed throughout all the frontier districts except Albany, are of the average extent of 6 000 acres; this large extent only being considered a *full place.* Thompson *Travels I* 1827

For releasing the place Roodebloem from the Entail of *Fidei commis* dated 10th July 1826. *Greig's Almanac* 1833

place² *n. colloq.* Used as equiv. of space, room etc. [*trans Afk.* plek room, space, (*also* place)]

They said we should bring . . . too: do you think there'll be place for him? O.I. 1974

place of, in *prep. phr.* Instead of: see quot. at *beshu, klapbroek.* [*translit. Afk. in plaas van* instead of]

For variation, use fresh or tinned mangoes in place of apricots. '. . . *Cookbook*' (received 1974) *Advt.*

plak [plak] *vb. trns. slang.* To stick, slap on. [*Afk.fr. Du.* plakken to paste, glue]

First I scribbles out a ticket
Tears it quickly out and lick it
Then I plaks it on the car with all my might
Pip Freedman (Song) 1971

plakkie ['plakï] *n. pl.* -s. Also *plakker:* any of several species of *Cotyledon* (q.v.) esp. *C. orbiculata* called kouterie(bos)'*, and *C. leucophylla* the leaves of which are used as plasters or poultices for plantar warts, corns and abcesses. [see quot.]

plakkie . . . the leaves of several species were applied (Afr.: plak) to abcesses . . . apparently first applied to *Cotyledon orbiculata*. Smith *Common Names S. Afr. Plants* 1966

*kouterie (bos) . . . for the softening and subsequent removal of hard corns. *Ibid.*

plakkies *pl. n.* See *slip slops, beach thongs:* also army term for any sandals. *U.S. flip flops.*

Beach thongs, sandals made of rubber, are very popular among all beach-goers and have a great many names here sloppies, slip-slops, plakkies etc. *O.I.* 1971

. . . plakkies – always for sandals. *O.I. Serviceman* 9.3.79

plan, to make a See *make a plan.*

plant migration *n.* The spreading tendencies of undesirable 'invader' plants, a problem to soil conservationists, farmers and agricultural officials: see quot., also *Kar(r)oo encroachment.*

Plant migration includes the natural tendency of woody types of vegetation to spread in definite directions and along distinct routes . . . said . . . there were five such routes in the Karoo region. *E.P. Herald* 28.2.73

plat¹ [plat] *adj.* Vulgar, unrefined usu. of or pertaining to accent or other linguistic manifestation: *cf. Plat Deutsch.* [*Afk. fr. Du. plat* unrefined, vulgar *cogn. flat*]

She's quite a nice girl but she talks the most frightfully plat Eastern Cape English. O.I. 1970 Plat SAP *Sunday Times* 11.7.76

plat(te)² [plat(ə)] *adj.* Flat, low lying: found in place names both in the *attrib. inflect.* (te) and uninflected forms e.g. Platrand, Platberg, Platteklip Gorge etc. [*Afk.fr. Du. plat cogn.* flat]

Flat: found in certain plant names sig. usu. a low habit of growth or other flatness of feature e.g. *~kroon, Albizia adianthifolia* see flatcrown; *~ turksvy, Opuntia aurantiaca* see *turksvy; ~ dubbeltjie, Tribulus terrestris,* see *dubbeltjie; ~doring, Asparagus plumosus,* see *katdoring.*

platanna [ˌplaˈtana, -ə] *n. pl.* -s. *Xenopus laevis,* the clawed toad, the spawn of which formerly much used in tests for early pregnancy: also occ. *plattie,* see also *platanner.* [*corruption of Afk.* or *Du. plat* flat + *hander* handed one]

On the banks if you go quietly you can see the platans and lakavans and sometimes great awful crabs in holes in the bank . . . The willow branches hang far in the river. Vaughan *Diary* circa 1902

platanner [ˌplaˈtanə(r)] *n. pl.* -s. Ø.
Squatina africana: the African angel-shark, also known as monk fish, angel shark. ▮In Afk. the term *platannavis* is used of the frogfishes or 'toadfishes' of the Batrachoididae fam. [*as above: prob. fr. large flat pectoral fins*]

platoon school *n. pl.* -s. A school usu. Indian, operating on account of large numbers in two sessions, morning and afternoon, for two separate bodies of pupils latterly with half the pupils indoors and half out of doors. ▮Not equiv. *U.S. platoon school.*
   This lack of school accommodation led to the introduction of 'platoon' classes as an emergency measure to accommodate more pupils . . . While educationists acknowledge the fact that the platoon school system is both undesirable and educationally unsound, one must be mindful of its achievements in Indian education. *Fiat Lux* Oct. 1974

platsak [ˈplatˌsak] *adj. colloq.* 'Broke', out of funds, *cf. archaic Brit.* purse-pinched, *Austral.* flyblown, *Canad.* on skid row. [*Afk. fr. Du.* plat flat + *sak* pocket]
   . . . where do you think we are going to find food and money? we are platsak.' Meiring *Candle in Wind* 1959

platteland [ˈplatəˌlant] *n. and n. modifier.* The country districts, rural areas: as modifier ~ *farmers, the* ~ *vote,* ~ *towns. cf. U.S. Canad. the sticks, Austral.* backblocks, outback: see also *backveld.* [*Afk.* platteland country districts]
   The play 'Brakwater', has as its theme the decay of village life on the platteland. *Evening Post* 9.9.72
   A retired doctor in Port Elizabeth said the shortage had become progressively worse as the Platteland was depopulated over the past 10 years. . . . 'As the Platteland is depopulated the number of doctors is decreasing,' he said. *E.P. Herald* 15.5.73

plattelander(s) [ˈplatəˌlandə(r)s, -z] *n. usu. pl.* Country people, those from the *platteland* usu. regarded as unsophisticated and often narrow-minded. *cf.* plaasjapie, backvelder: *Austral.* backblocker. [*Afk.* platteland country + *personif. suffix.* -er roughly equiv. -ite]

play *vb. intrns. substandard.* Used of a film (movie) being shown: as in 'I hear there's a good film playing this week'. [*fr. Afk. die stuk/film speel* the piece/-film is being given, shown; *speel* play]

play-play [ˈpleɪˌpleɪ] *modifier. colloq.* Freq. among children: equiv. of make-believe, not real, as in *a* ~ *battle, a* ~ *house* etc. [*poss. erron. Afk. speel-speel* (lit. play-play) *sig. very easily, with no trouble, as if child's play*]

play sport *vb. phr. substandard* (but in general use). To take part in games or other athletic sports: freq. in qn. form 'What sport do you play?' see quot. also *Austral. to play sport.* [*unknown: poss. fr. Austral. prob. confusion between 'play games' and 'do' or 'take part in sport'*]
   . . . the horrible phrase 'playing sport' of which I am sorry to say some of our best newspapers are guilty. One can play cricket, rugby, tiddly-winks, or the piano; but what of other sports like yachting, motor-racing . . . mountaineering, even huntin' and fishin'? How can one possibly PLAY athletics, for example? *Sunday Times* 9.12.73

play white *vb. phr.* **1.** To (try to) cross the colour line: successfully achieved at times by light skinned *Coloureds* (q.v.) see *Kimberley train, try for white: cf. U.S. crossing the line.*
   It is a well known ploy: if you want to play white, you make your mother into your maid. Mackenzie *Dragon to Kill* 1963
2. *n. pl.* -s. more usu. as modifier. A Coloured person who has crossed the colour line.
   The time has come for the Coloured man in the street, to decide once and for all whether he is going to align himself with Black consciousness or the play-White system . . . What about Black consciousness and the 'play-white' coloured? Adam Small *cit. Sunday Times* 10.10.76
deriv. ~ *ism.*
   In the front rows the coloured kids converse with their white counterparts with no sham of play whitism. It is common . . . to hear 'Hey whitey, move up one seat.' *Drum* June 1976

plein [pleɪn] *n. pl.* -s. Equiv. of square, plaza etc. open space surrounded by buildings see quot. at *stadhuis:* found in urban names Groentemarkplein, Plein Street. [*Afk. fr. Du.* plein square]
   Byl, David Retail Shop 1 Boereplein . . . Lowrie, Gesina, 7 Kerkplein. *Afr. Court Calendar* for 1815
   All that was left of Kruger Day were orange peels and banana skins and pieces of paper littering the church plein. Bosman *Willemsdorp* 1977

plural *n. pl.* -s. A Black person: in facetious use only: e.g. ~ *stanians* see quot.

at *imbadada, rural* ~ *mural* (a Bushman painting) etc. ⫽Now also a column in *Drum* 'Plural Stan' (Formerly Bantu Stan and Black Stan) Hence *deriv.* '*Singular*' (q.v.) a White person. [*fr.* Plural Relations and Development see *P.R.D.* and quot. at *B.A.D.*]

. . . this Pluralstan scribe has no time for the . . . I-told-you-so brigade. But while talking of the BophutaTswana first uhuru anniversary I find myself perilously close to qualifying . . . because I was right there in that Pluralstan haven a year ago. *Drum* Feb. 1979

Singular sign at PE bottle store.
Off-sales apartheid has taken a new twist . . . with the display of a 'reserved for plurals only' sign at a . . . hotel bottle store. The owner . . . said the use of the term 'plurals' was an attempt to find a palatable name for customers who were not white. *E.P. Herald* 19.4.79

Don't say plurals says Nat
. . . the bottle store . . . that put up a sign reading 'Plurals-Plurale' . . . He said this was a clear attempt to impugn the dignity of blacks. *E.P. Herald* 13.6.79

plus-minus *quantifier.* Also ±: approximately, about, 'more or less', an indefinite measure of area, number, time, amount, age etc. [*Afk. fr. Lat. plus-minus* approximately]

125 Dorper and cross-bred Lambs (plus/minus six months). *Evening Post* 27.2.71
I use the deep patty pan. Bake in oven 425°F for plus/minus 10 minutes. *Ibid.* 6.10.73
PLUS MINUS Thirty Thousand bags of mixed milled Feed at 50c per bag. *Farmer's Weekly* 30.5.73

pocket *n. pl.* -s. Sack or bag of varying size in which various commodities are sold, used as a measure for trade purposes. [*poss. fr. Afk. sak* pocket in clothes, *sakkie* bag, pocket (q.v.)]

SWEET POTATOES, moderate demand . . . Borrie 150 to 300; others 100 to 150 per sgr pkt. [sugar pocket] GREEN BEANS, well presented, demand for quality, prices showed a decline. *Farmer's Weekly* 21.4.72
Most of the residential abodes are covered over with mealie sacks. A few . . . have their walls and roof constructed of sugar-pockets bearing the trade mark of Messrs . . . Close examination shows that the hessian in a sugar pocket is of finer texture and better woven than that in a mealie-bag. Bosman *Cask of Jeripigo* 1972

poegaai [ˈpŭˌxaɪ] *adj. slang.* Dog tired, exhausted: also drunk e.g. 'He kept us all training until we were poegaai with exhaustion,' or 'By the end of the evening he looked absolutely poegaai.' *Austral. stonkerd.* [*Afk. fr. Du. poechai*

fuss, bother]
I had a ball that night – go-going on the table for this ring of chuckling ole dads in paperhats, most of them *poegaai*, all of them egging me on much better. Clapping, whistling, ogling and all. *Darling* 29.1.75

poe(n)skop [ˈpŭnsˌkɔp] *n. pl.* -s. **1.** A polled or hornless animal, formerly an elephant cow without tusks. [*Afk. poena* a polled animal *prob. fr. Du. poen(a)* dumpy, thickset (person or object) + *kop* head]

Of all the things mind cow elephants without tusks; they are not common, but if you do come across a poeskop like this, 'pas op' (take care). Drayson *Sporting Scenes* 1858 *cit.* Pettman
Usually the dehorning process works and they're poenskop for life; occasionally they grow little stumps. O.I. Farmer 1973

2. *Cymatoceps nasutus* the *mussel-cracker/crusher, black steenbras* or *biskop* (q.v.): acc. Smith *Sea Fishes of S.A.* 1961 *ed.* 'one of the premier angling fishes.' See quots. at *ollycrock* and *biskop.* [*fr. Afk. poena* polled animal *presum. fr. appearance*]

The South African records do contain a Musselcracker, but it is no. 719 which is our Poenskop. The Eastern Province records list it as a Poenskop. *E.P. Herald* 15.7.71

poep- [pŭp-] *n. usu. prefix vulgar* Fart, usu. in comb. ~ *hol* an abusive form of address or reference, or ~ *scared fr. Afk.* ~ *bang* (frightened): also as *n.* equiv. of 'fool' *cf.* poop. [*Afk. poep* to break wind, *also n.*]

That tickey deposit heart of his is tight like his *poephol* and his fist. Fugard *Boesman & Lena* 1969
Talk about nipping straws. OK so they give up at last, but . . . never been so poep-scared in my whole entire life before or since. *Darling* 17.3.76

poes [pŭs] *n. slang* (obscene). The female genitals: see *puss.*

pofadder [ˈpɔfˌadə(r)] *n. pl.* -s. *Puffadder* (q.v.), also a place name Pofadder, N. Cape Province. [*Afk. cogn.* puffadder]

poffertjie [ˈpɔfə(r)kĭ, -cĭ] *n. pl.* -s. Light fritter (beignet, aigrette): traditional Cape Dutch sweet dish, also occ. used of choux pastry. [*Afk. poffer* puff (*n*) + *dimin. suffix -tjie*]

Jeanette van Duyn demonstrated the old Cape recipes at Wembley. When one of the ostriches there laid an egg, she was able to bake a cake and prepare melktert and poffertjies for Queen Mary and her ladies-in-waiting. Green

Land of Afternoon 1949
No matter that 'Les Beignets aux Ananas' is only a pynappel poffertjie in disguise, the guests were having none of it, as they vastrapped across the concrete to the strains of 'Sarie Marais'. *E.P. Herald* 16.5.74

political commissioner *n. pl.* -s. *hist.* (*obs.*) Senior Official of the (Dutch) Reformed Church.
Reformed Church P. J. Truter, Sen. Esq. Political Commissioner *Afr. Court Calendar* 1819
SYNOD OF THE REFORMED CHURCH POLITICAL COMMISSIONERS The Hon Sir John Truter, LL.D. D. F. Berrange, esq. LL.D. *Greig's Almanac and Directory* 1831

pomelo [ˌpɔ́ˈmɪəlŭ, ˈpɒmələʊ] *n. pl.* -s. Used in *Afk.* as equiv. of grapefruit: more properly 'shaddock'; see *pampelmoes*[1].

(m)Pondo [ˈpɒnˌdəʊ, ˈpɔndɔ́] *n. pl.* -s or ama-: A Xhosa-speaking Nguni people of Pondoland in the E. Cape Province, also a single member of the ~ people or the AmaMpondo: see quot. at (1) *Kaffir* [*Ngu. ama(M)pondo*]
The Pondo are one of the tribes living between the Indian Ocean and the Drakensberg Mountains . . . The tribe is composed of a nucleus of 46 related clans which trace descent from a common ancestor, Mpondo. Many members of 21 other clans, unrelated to the descendants of Mpondo have subsequently accepted the authority of the paramount chief and have so become members of the Pondo. *Family of Man* Vol. 6 Part 80 1976

pondok(kie) [ˈpɔ́nˌdɔ́k, ˌpɔ́nˈdɔ́kĭ] *n. pl.* -s. Hut, shack [*acc. some Afk. dimin. fr. Malay pondok* hut, leaf roofed shelter, *more prob. Hott. pondok, pontok,* hut] 1. A crude hut or shelter made of scraps of wood, corrugated iron etc. *cf. Austral. humpy,* shack (from aboriginal hut), *Jam. Eng. wappen-bappen,* slum house made of bits and pieces, *Canad. shack/ie*): *Anglo-Ind. pandal,* temporary shed, booth or arbour: see also quots. at (2) *-rand-, bakgat* and *outjie.*
. . . she endeared herself to many thousands of poor Coloured people, mainly among the shacks and pondokkies of the Cape Flats. *Cape Times* 8.6.70
2. In S.W. Africa a grass-built hut of indigenous tribesmen. Also place name Pondok Mountains so called fr. beehive shape of peaks.
Most of these traditions were dropped during the 19th century when Christian Herero adopted European dress . . . The traditional dome-shaped hut or *pontok*, made from curved

branches plastered with a mixture of cow-dung and clay, gave way to rectangular houses . . . Poverty-stricken tribesmen have tended to build shanties from discarded sheets of metal. *Family of Man* Vol. 3 42. 1975

pont [pɒnt] *n.* -s. A flat bottomed ferry-boat, like a moving bridge, *cf.* pontoon, worked on chains or ropes to convey passengers, animals, wagons or even motor cars across African rivers: also in place names Norvalspont, Pontdrif. *cf. Canad. pont* a solid bridge of ice packed across a river. [*Du. pont fr. Lat. ponto* punt *cogn. pons* bridge]
Ponts, or floating bridges, are used with great success on the Berg and Breede Rivers. Thompson *Travels II* 1827
Fishermen, traders and hunters await the arrival of the pont to take them and their belongings across the river. Telford *Yesterday's Dress* 1972

Pontac [ˈpɒnˌtæk] *n.* A sweet red dessert wine made from Pontac grapes, a variety introduced prob. fr. S. Europe in the earliest days. [*fr. name of grape*]
. . . he has always on sale, a choice of old Cape Wines of superior quality consisting of Madeira, Sherry Hock, Aromatic Hock, Pontac, equal and in many respects preferable to the Port wine generally imported into this colony: sweet wines of very excellent quality and fine flavor, to wit: Pontac, Frontignac, Muscadel etc. not to be equalled. *Cape of G.H. Almanac Advt.* 1843

poort [pʊə(r)t] *n. pl.* -s. Narrow pass or defile through mountains usu. along a stream bed: *cf. nek* [*Afk. fr. Du. poort* pass(age) *cogn. Lat. porta* gate]
It is romantically diversified by gentle undulations, by precipitous woody ravines or kloofs, by stupendous poorts or passes through the mountains and by clumps of elegant evergreens. *Greig's Almanac* 1831
Also found in place names, freq. as suffix e.g. Onderstepoort, Noupoort, Howison's Poort, Seweweekspoort.
. . . our road lay through a narrow defile, which opened upon more extensive scenery. This defile Frederick thought fit to name 'Thompson's Poort,' (i.e. Gate or Pass,) in honour of the narrator. Thompson *Travels I* 1827

poor white *n. pl.* -s. A member of the most indigent white group often living as a *bywoner* (q.v.) on another man's land. *cf. U.S. white trash, red neck, Jam. Eng. white jeg :* also often derogatory as in 'live like ~*s*', 'dress like a ~'.
Here, as the child of poor-whites and as the

mother of poor-whites she had drifted for seventy years from farm to farm in the shiftless, thriftless labour of her class. Smith *Little Karoo* 1925

. . . the poor white is psychologically handicapped by his tradition of membership of the master-class, expressed in contempt for 'kafirwork' and unwillingness to undertake it, especially in public labour-gangs. Hoernlé *S. Afr. Native Pol.* 1939

[The woodcutters] . . . made very little money (the term 'poor Whites' was coined about then), and yet from one generation to another they clung jealously to their trade in the dark heart of the forest. Hjalmar Thesen *cit. E.P. Herald* 28.5.73

poppie ['pɔ̆pĭ] *n. pl.* -s. *colloq.* Familiar mode of address or reference to a young girl, equiv. of 'girlie' etc. [*Afk. pop* doll + *dimin. suffix -ie*]

Ag, shame, now that I come to think of it it's that mauve tin . . . poppie. Over there. *Darling* 25.6.75

Porra ['pɔ̆ra] *n. pl.* -s. *Sect. Army.* A Portuguese: also *Porro,* prob. reg. Johannesburg, a Portuguese market gardener. [*corruption of Portuguese*]

The Porra that we get our beer from has been shot. *Letter SWA, Informant* Aug. 1975

We used 'Porro' on the Border for a captured Portuguese truck – they were in real demand – no speed governors on them. *O.I. Ex-Serviceman* 15.2.79

Port Jackson (willow/wattle) *n. pl.* -s, Ø. *Acacia cyanophylla* and *A. longifolia:* originally a native of Australia introduced about 1857, now a plant pest in many parts: see also quot. at *rooikrans.*

Port Jackson Acacia . . . introduced from Australia as a useful sand binder on the Cape Flats and from there it spread. It proved an aggressive antagonist of the native flora when it spread further afield. Smith *S. Afr. Plants* 1966

-pos(-) [pɔ̆s] *n. usu. suffix.* Also post: a former military post found in place names e.g. Sannaspos, Rykaartspos, Krugerspos, Venterspos: occ. prefixed in *Eng.* Post Retief, Post Chalmers, and in *Afk.* Posdrif.

Potchefstroom koekoek See *koekoek*

potwan(a) ['pɔ̆t₁wan(ə)] *n. pl.* -s. *reg.* E. Cape: *sugar pocket* (q.v.) 100 lb. (45kg) bag, half a *muid sack* or *bag* (q.v.) [*Xh. unompotwana,* half bag]

pou(w) [pəʊ] *n. pl.* -s, Ø. The S. Afr. bustard: any of several of the Otididae fam. see *korhaan* (q.v.); also in combination *gom* ~ the Kori bustard *Otis* kori, and in place name Poupan. [*Afk. pou fr. Du. pauw* bustard, *also* peacock]

I saw only a few of the larger and more hardy species . . . the *pouw* which is a sort of large bustard, and very delicate eating; the korhaan, a smaller sort of bustard, also prized by epicures. Thompson *Travels I* 1827

I saw that the birds were a flock of pauw or bustards, and that they would pass within fifty yards of my head. Haggard *Solomon's Mines* 1886

Roast gompou was a rare delicacy . . . this largest of all game birds – a flying turkey indeed, with enough luscious meat on it to allow a second helping for everyone. Green *Last Frontier* 1952

poundmaster *n. pl.* -s. The official in charge of a pound, (also poundmistress) in S.A.: also *Brit.* and *U.S.* [*trans. Du. schut* pound + *meester cogn.* master]

Impounded in the Municipal Pound, and if not previously released will be sold . . . W. R. Dixon Poundmaster. Notice 45 – 25th May 1971 *Grocott's Mail* 28.5.71

praise singer/praise singing *n.* See *imbongi, izibongo.*

The installation of the new Chancellor . . . was a dignified affair but it is difficult to understand what relevance a Xhosa praise singer had to the function. Letter *cit. E.P. Herald* 27.4.77

-pram- [pram] **1.** Breast (shaped): found in names of mountains or hills e.g. Pramberge, Pramkoppen, Verlepte (withered) Pramberge (Namaqualand). *cf. Canad. mamelle,* a breast-shaped hill. [*Afk. fr. Du. pram* a woman's breast]

The mountains of the Karreebergen . . . among them one in the form of a depressed cone surmounted with an additional summit, was distinguished by the name of Pramberg. Burchell *Travels I* 1822 *cit.* Pettman

2. Breast: found in various plant names usu. sig. breast-shaped protuberances e.g. ~ *boom/bos/doring* also called *perde* (q.v.) ~ *Fagara capensis:* see *knobwood/ knophout, ouma-se-* ~ (grandmother's) *Nycteranthus rabiei,* a dwarf succulent. [*as above*]

Different good Mesems like ouma-se-pram (Aridaria Rabiei) grow in the shade of Lycium Suada. *Farming in S.A.* Nov. 1933 *cit.* Swart *Africanderisms: Supp.* 1934

praying mantis *n. pl.* -es. Also *preying mantis* fr. its predatory habits: see *Hottentot God.*

Praying mantises have acquired a bad name as husband-eaters and voracious insects of prey . . . true of course. A female does sometimes eat her husband and they do have tremendous appetites. But gardeners and farmers can be thankful . . . for the amount of pests they eat is prodigious. *Panorama* Feb. 1975

P.R.D. [ˈpiˌɑˈd] *acronym.* Plural Relations and Development: ⫷ In 1979 redesignated Department of Co-operation and Development. See quot. at *BAD*, also *plural.*

predikant [ˈprɪədəˌkant] *n. pl.* -s. A minister of the *Dutch Reformed Church* (q.v.), also *dominee;* see also *pastorie*, and quots. at *kerk* and *fiemies.* [*Afk. fr. Du. predikant* preacher, minister]

They cannot believe that a predikant would walk . . . It is vain to tell them that our Lord and Master and His holy apostles walked. It may have been so. But they know that predikants don't walk. Gray *Journal II* 1851

. . . it was there that the earnest young Presbyterian minister from Aberdeenshire took up his first and only appointment as *predikant* of the Dutch Reformed Church, where he laboured until his death some 40 years later. *Cape Times* 25.6.72

preekstoel [ˈprɪəkˌstŭl] *n. pl.* -s. The pulpit in a *Dutch Reformed Church* (q.v.) centrally placed and normally larger than in other churches. [*Afk. fr. Du. predikstoel* pulpit]

. . . and high and remarkable rock, which, on account of its resemblance to a pulpit, is called by the herdsmen the *Predikstoel.* trans. Lichtenstein *Travels I* 1812

2 April 1817 Finished the Preek stoel to day and occupied it in the evening. Barker *Diary* 1815–28

. . . the platform stood on its edge . . . so it would look more like a *preekstoel*, the place where Dominee Welthagen was to stand. Bosman *Jurie Steyn's P.O.* 1971

In combination *padda* (q.v.) ~ (*frog*), *Nymphaea capensis :* see *kaaimansblom.*

prepositions Non-standard prepositions and phrases containing prepositions peculiar to S.A.E. including those which are components or formatives in place names appear as separate entries in the text: they are listed here as cross references only: see *agter-, bo-, -boven, by, in, for, onder, otherside, over, under, with;* also *veld, off the; on his/her/my nerves; on the moment; (ride) on water, in the/out of the (road); beaten out; eaten out, shot out (hunted out); tramped out; through the face;* see also *omissions* and *redundancies* for reference lists of these.

Pretoria [prəˈtɔrɪə, prɪ-] *n. prop.* The central government or its departments used similarly to *Brit. Whitehall, U.S. Washington.* [*Pretoria, the Administrative Capital of the Republic of S.A.*]

The resolutions list some of the things we Bantustanians would like to have. There is nothing unreasonable about the leader's pleas. . . . Please remove this and that, they hurt. It is Pretoria which will have the final say. *Drum* 22.12.73

The question I want to explore is whether South Africa is prepared to separate its own future from Rhodesia and Namibia (the Pretoria-ruled territory, also known as South West Africa, that wants independence). *Time* 28.6.76

In terms usually reserved for denunciations of Pretoria, Zambian Ambassador, Dunstan Kamana accused France of 'naked aggression' *E.P. Herald* 28.10.76

preying mantis *n. pl.* -s. Usu. *praying mantis* (q.v.) fr. gesture of lifted forelegs: see *Hottentot God.*

Pride of- *n. phr.* As elsewhere a prefix to several plant and flower names, ~ *India* or ~ *China*, the Cape lilac, see *sering* (*boom*) ; ~ *India, Lagerstroemia indica;* ~ *de Kaap, Bauhinia galpinii* a showy, red flowered bush or climber ; ~ *Table Mountain, Disa uniflora,* see quot. at *disa;* ~ *Franschoek, Serruria florida,* see *blushing bride.*

prison farm *n. pl.* -s. A farm maintained by convict labour, employed during working hours and confined under prison conditions at night. ⫷Formerly a system employed by private farmers: now one of 20 non-profit making farms owned by the Department of Prisons operated by usu. long term prisoners: also *U.S.*

Where Cops and Robbers farm together. Recognised as one of the leading establishments of its kind in the world, this prison farm . . . rehabilitates its inmates using 'agricultural therapy' . . . an environment has been created in the form of prison farms to provide a 'work climate' that will bring out the best in the prisoner with . . . training which can be utilised successfully after release. *Farmer's Weekly* 20.3.74

private school *n. pl.* -s. Fee paying, English-speaking, non-Government schools, in S.A. often owned and run by churches: equiv. *Brit. public school* and *U.S. private school, Canad. separate church school.*

. . . a small minority of English children is sent to 'private' schools in South Africa. Hoernlé *S. Afr. Native Pol.* 1939

Several local teachers have lost faith in 'coloured education' and . . . are sending their children to exclusive, private schools in Cape Town . . . uniforms are costlier than the public schools, books have to be paid for and school fees . . . *Sunday Times (Extra)* 8.7.79

proclaimed weed *n. pl.* -s. Any of numerous species classified as plant pests or noxious weeds the presence of which on any farmer's land is a punishable offence: see *boetebos(sie), satansbos, hakea* and *jointed cactus* and quot. below.

The control of noxious weeds is vested in the Union Department of Agriculture and Forestry. A list of proclaimed noxious weeds is published and farmers are required by law to eradicate such weeds from their farms. The Division of Plant Industry acts in an advisory capacity to the Department, and recommends whether or not any particular plant should be added to the list of proclaimed weeds. *H'book for Farmers 1937*

Dangerous invader now gazetted as proclaimed weed . . . Sesbania punicea, has at last been proclaimed a noxious weed. The proclamation in terms of the Weeds Act appeared in the Government Gazette of March 9. *Grocott's Mail* 18.4.79

pro deo *adv. phr.* or *modifier. Legal. lit.* 'For God': used of the means of representation of (or of the advocate who represents) someone accused of a capital offence, at a nominal fee, duly instructed by the court. In combinations ~ *work,* ~ *advocate,* ~ *representation* or in *adv.* sense *appeared~,* see quot., *represented* ~ etc. ▮ *acc.* Hon. Mr Justice J. P. G. Eksteen the abolition of capital punishment and free legal aid under the welfare state has caused the term to fall out of use in Britain, while it is regularly retained in S. Afr. law. [The term ~ is also used loosely, usu. by lay persons, of free defence in non-capital cases, and even of free surgical or medical treatment beyond the means of the patient]

Two to hang for murder
Mr . . . appeared pro Deo for . . . Mr . . . appeared for the State. *E.P. Herald* 28.4.78

Prog. [prɒg] *n. pl.* -s. A member of the Progressive Party formed in 1959. [*abbr. Progressive*]

A man stands as an Independent and then everybody sets out to prove that he's a crypto-Nat or a Prog in disguise or somebody pretending to be UP while he is really a card-carrying member of the Broederbond. *Sunday Times* 27.2.71

Progressive Federal Party *n. prop.* Also *P.F.P. (acronym)* political party formed in 1977: *abbr. Progfed* a member of ~.

JOHANNESBURG. A new verligte opposition party – the Progressive Federal Party – was launched here yesterday with a strong commitment to shared political power by way of a national convention. *E.P. Herald* 6.9.77

Progressive Reform Party *n. prop.* Also *P.R.P.:* an opposition party formed in 1975 by a merger between the *Reform Party* formerly the *'Young Turks'* (q.v.) of the *United Party* (q.v.) and the Progressive Party, see *Prog.:* also form *Progref,* a member of the ~.

The Progressive Reform Party . . . has shown that even in opposition, a party can be dynamic . . . In the nature of things, it will not be possible for the PRP to ignore the UP in the Parliamentary debates. Sharp clashes between the two groups can be expected. *Progress* Jan-Feb 1976

pronk [prɔŋk] *vb. intrns.* **1.** To curvet or prance usu. of springbok displaying themselves, see first quot., also *figur.* of someone in new clothes etc.: in combination ~ *bok, Antidorcas marsupialis;* see (1) *springbok.* [*Afk. fr. Du. pronken* to strut, parade, show off]

There was a herd of springbok in the distance. They were pronking, jumping high in the air, their white manes glinting in the sun. Cloete *Hill of Doves* 1942

When Pascal's [poodle] been clipped and has a new hairdo he pronks about for days. O.I. 1969

2. Found in plant names sig. usu. 'showy' as in ~ *ertjie* (pea) *Lathyrus odoratus,* the garden sweet pea and ~ *gras Pennisetum setaceum.* [*fr.* (1) ~]

pronoun, omission of See omissions.

prop(ped) *vb. usu. parile, colloq. figur.* 'Stuffed'. [*prob. fr. Afk. intensive form propvol* absolutely full *fr. Afk. prop* to stuff]

Young Vermaak, propped to the ears with raw book learning, with which he is only too eager to overawe his collocutors. Abrahams Introduction to *Jurie Steyn's P.O.* 1971

protea [ˈprəʊˌtɪə] *n. pl.* -s. **1.** Any of several of the genera and species of the Proteaceae esp. of the showy flowered varieties incl. *P. cynaroides* the *giant* or *king* ~ (q.v.); *P. mellifera,* see *suikerbos, sugarbush; P. barbigera,* the bearded ~ ; *P. arborea (grandiflora)* also *waboom* (q.v.). ▮The *silver tree* or *witteboom* (q.v.) *Leucadendron argenteum* is also of the Proteaceae as is *hakea* (q.v.). [*fr. Gk. god Proteus*]

The name protea was first given by Linnaeus in honour of the Greek god Proteus . . . The Protea family is of special interest to botanists, some of whom see in it possible proof of a former close connection between South Africa and Australia, for this large family is abundantly represented in these two countries . . . It does not occur at all in Europe. Palmer & Pitman *Trees of S.A.* 1961

Professor Shigeo Suga . . . Japanese expert on flower arranging, spoke about the beauty of the South African protea on his arrival in Johannesburg yesterday . . . *Rand Daily Mail* 2.11.70

2. *n. pl.* -s. A member of the *Coloured* (q.v.) S. Afr. Rugby team[*fr.* (1) ~]

[A] surprise announcement that the Coloured Proteas could play the rugby Springboks came in reply. *E.P. Herald* 15.10.74

Protected Village *n. pl.* -s. *Rhodesian.* African villages fenced and guarded for the protection of civilians from terrorist attack. ⫽ Occupancy of these villages is voluntary.

The Protected Villages (they looked hideously like concentration camps when they were first shown on television) are Security fenced, and villagers returning home are searched – even children. *Fair Lady* 16.3.77

Proto *n. and n. modifier* usu. capitalized. Trade name of breathing apparatus used for rescue work in mines: usu. in combination ~ *team*, a team specially trained for underground emergencies.

Look . . . ring Mr Nolte *now.* Tell him to get the Proto teams to No. 3 shaft at once. Get the fire plan folder. *Unpublished playscript* 1974

Ten of the dead . . . were brought to the surface from 60 metres underground . . . Two more bodies were found in the haulage by proto teams last night. The proto teams continued their search early this morning for further possible victims. *Cape Times* 4.1.74

Province of South Africa *n. prop.* One of the Provinces of the Anglican Church, see *Church of the* ~.

P.R.P. See *Progressive Reform Party.*

P.R.U. [ˌpəˈru] See *Peruvian.*

pruimpie [ˈprœïmpï] *n. pl.* -s. A quid or 'chew' of tobacco. [*Afk. pruim* plum + *dimin. suffix -pie*]

. . . there were rolls of Transvaal tobacco, for every farm hand expected and was given a weekly 'pruimpje' of the strong black stuff. McMagh *Dinner of Herbs* 1968

puffadder *n. pl.* -s. The common S. Afr. viper *Bitis arietans*, thick and slow of movement but inflicting a poisonous

bite. [*see first quot. Afk. pofadder, pof* puff, blow up + *adder*]

The puff-adder is often met with: it is so called from its swelling itself out to a great size when enraged . . . it is nearly as thick at the tail as at the head. Percival *Account of Cape of G.H.* 1804

The slow-witted puffadder is short and sluggish . . . Their markings are often beautiful. If you step on one by mistake . . . the puffadder immediately becomes aggressive. Puffadders bring forth their young alive; and even a tiny puffadder can inflict a terrifying bite. Fortunately their aim is often erratic. Green *Tavern of Seas* 1947

pula [ˈpuˌla] *n. interj.* Rain [*Bantu pula,* rain] **1.** A greeting or invocation.

Mattebe then made the same movements with his assegai as at the commencement after which he waved the point towards the heavens, when all called out 'Poola!' i.e. rain or a blessing . . . Thompson *Travels I* 1827

Pula. In the middle of Southern Africa there is a country whose coat-of-arms bears, instead of some Latin tag boasting power and glory, the single word: rain. . . . In Botswana there is always the possibility of rain. The hope of rain. Rain *is* hope: *pula* means fulfilment as well as rain. *London Magazine* Feb/Mar 1973

2. The currency of Botswana introduced 23.8.76 [*fr.* (1) ~]

Botswana's new currency, the pula, will be introduced today . . . Botswana's new coins, thebe, will replace South African cents from September 6. *E.P. Herald* 23.8.76

pumula [pŭˈmula] *n. and vb. colloq. prob. reg. Natal* Rest [*Zu. ukuphumula* to rest, -*phumulo n.* rest]

I'm feeling pretty tired, think I'll turn in. Yes, go to bed and get a good phumula. *O.I.* July 1964

. . . looking for a horse . . . so that I can go home to pumula . . . very tired. McMagh *Dinner of Herbs* 1968

pundus [ˈpŭnˌdŭz] *pl. n. colloq. prob. reg. E. Cape.* Buttocks, backside: see quot. at *takes me up to.* [*Xh. impundu,* buttock]

Let her get up; otherwise her pundus will get sore from sitting in bed too long. O.I. Doctor, Grahamstown 1975

punt [pœnt] *n. suffix.* Point, promontory: found in S. Afr. place names usu. in trans. e.g. Seepunt, Groenpunt, Kaappunt. [*Afk. fr. Du. punt cogn.* point]

puri See *Indian terms.*

pusa *vb.* See *puza n.*

Say, Auntie, what about a popla?' That means you would like a beer. Or, 'A dop of moonshine, sister!' . . . for a tot of whisky. Or even, 'I just want to pusa.' Meaning you want to fly high on something and you are not particular. Venter *Soweto* 1977

push beat, to *vb. phr. Sect. Army.* To do guard duty. ⫿ Various manifestations: see quot. at *chicken parade.*

. . . so you've finished . . . cleaning your gatt, squaring off your bed . . . checking out your mohair, pushing beat, sweating at PT, moaning under your breath about the S.M.; bribing a tiffy to fix your cab. *Scope* 10.1.75

puss Objectionable in S.A.: see *poes.*

put [pœt] *n. pl.* -s. A well, water hole. [*Afk. fr. Du. fr. Lat. puteus* well, *cogn.* pit, *O.E. pytt* well]

Where Kabalonta lives there are no rivers; the people and cattle drink from puts*. Smith *Diary* 29 July 1835   *wells
Found in S. Afr. place names as *prefix* and *suffix* e.g. Putsonderwater, Swartputs, Noenieput, Broedersput.

Putco [ˈpʌtˌkəʊ] *acronym.* Public *U*tility *T*ransport *Co*rporation: the Transvaal bus company which carries over 200,000 commuters from the African townships to the reef towns daily in green buses commonly called '*green mambas*' (q.v.).

The children refused to budge . . . they were playing a new game called 'Soweto Soweto'. 'How . . .?' 'Easy . . . Sissy is a PUTCO bus and I'm setting her on fire' Venter *Soweto* 1977
The bus will always be an important form of transport. Putco carries 96,000 commuters a day into Johannesburg from Soweto, Alexandria and Tembisa . . . its suburban service inside Soweto carries 27,000 blacks a day . . . another 70,000 daily commuters into Springs and Boksburg. *Panorama* Mar. 1978

putu [ˈputŭ] *n.* Traditional African preparation of *mealie meal* (q.v.) cooked until it forms dry crumbs: equiv. of *krummelpap* (q.v.) eaten by Africans with meat and gravy etc. or with *calabash milk* (q.v.) or *maas* (q.v.) ⫿It is also a popular breakfast food among whites, served instead of porridge; see second quot. [*Ngu. uphutu* crumb porridge, anything crumbly *e.g.* earth]

Soon one of Mazibe's wives . . . entered on her hands and knees. Permitted to kneel but not to stand, she adroitly balanced a baby strapped to her back . . . while serving us roast chicken and uphuthu, a kind of hominy that has been a staple Zulu dish for centuries. *National Geographic Mag.* 6.12.71
. . . her first move is to the kitchen where Lukas her cook is preparing breakfast . . . Whatever else might be in the offing a large pot of putu – crumb mealie meal porridge – will be ready for eating. *Fair Lady* Jan. 1972

puza [ˈpuza] *n.* Drink, liquor. [*fr. Zu. ukuPhuza* to drink]

'You are thirsty, headman' . . . 'I have

brought you white man's puza.' Umbalose . . . closed his eyes as if to shut out the vision of happiness the bottles had conjured up. 'I know white men,' he said . . . 'and I know that you do not bring me this puza expecting nothing in return.' Greene *Adventure Omnibus* 1928

pylhout [ˈpeɪlˌ(h)əʊt] *n. Grewia occidentalis,* see *assegaibos.* [*Afk. fr. Du. pijl* arrow + *hout* timber]

-pypie [ˈpeɪpĭ] *n. lit.* 'little pipe' suffix to vernacular names of numerous species usu. of the Iridaceae with long tubular flowers e.g. the *Afrikaner* (q.v.) and other *Gladiolus, Watsonia* and *Hesperantha:* see *aandblom.* [*Afk. fr. Du. pijp cogn.* pipe + *dimin. suffix -ie*]

The term 'pypie' appears to have been first applied to one or other species of Gladiolus or Watsonia from the suggestion of a miniature long-stemmed pipe . . . conveyed by the flowers. Smith *S. Afr. Plants* 1966

Q

Qaba [ˈkaba+] *n. prop. pl.* ama-. An ancestor-worshipping tribal African: see quots. at (1) *red* and *kaffir sheeting.* [*Ngu. qaba,* heathen, pagan]

Christians are called by pagans *amakholwa* (believers) . . . while pagans are referred to as *amaqaba* (those smeared with red ochre). Hammond-Tooke *Bhaca Society* 1962

quacha [ˈkwaxa] *n. Equus quagga;* see *quagga:* found in place name Quacha's Nek.

quagga [ˈkwaxa] *n. pl.* -s, ∅. The now extinct wild ass of S.A. *Equus quagga,* related to the zebra but differing in its markings: *hunted out* (q.v.) through indiscriminate destruction from the early days; see second quot. [*Du. quagga fr. Hott. qua-ha onomat. fr. braying of the species, also poss. rel. Ngu. iQwara,* zebra, something striped]

19 April 1811 This beautiful animal has been hitherto confounded with the naturalists with the Zebra. When these were first described . . . the Quakka was considered to be the female Zebra, while both that and the true Zebra bore in common, among the colonists, the name Quakka. Burchell *Travels I* 1822
. . hunting the wild game, to save the consumption of their flocks, and feeding their Hottentot or Bushmen servants, with the flesh of the Quagha, or wild ass. Thompson *Travels II* 1827

In combinations *berg* ~ , *bont* ~ .
The old pioneers used 'quagga' rather loosely

for the mountain zebra (bergkwagga), Bur-chell's zebra (bontquagga) and the quagga itself, so that it is difficult to say how far the range of the extinct species extended north. *Daily Dispatch* 22.7.72

Found as *prefix* to several plant names e.g. ~ *kos, Pleiospilos bolusii* a succulent said to have been a favourite food of the ~ ; ~ *couch grass Danthonia purpurea;* ~ *kweek* see *kweek, Cynodon hirsutus,* etc.

quaestor ['kwaɪstə(r)] *n. pl.* -s. *Quaestor Synodii:* treasurer of the *Dutch Reformed Church* Synod (q.v.). [*Lat. quaestor* revenue collector]
Synod of the Reformed Church Moderators Rev. T. Herold *President* Rev. J. Spyker *Secretary* Rev. A. Faure *Actuarius* and *Quaestor* S. *Afr. Almanac* 1834

(u)Qamata [(u)ka'mata+] *n. prop.* Traditional pre-Christian Xhosa name for God. see at *Tixo.* [*Xh.*]
. . . he is also a strong believer in the traditions and customs of his people, 'They do not necessarily conflict with Christianity' . . . Then he points out that the Xhosa have always spoken of 'uQamata' the God above, and Uvuko or the re-awakening. His people have never been heathens he maintains. *Bona* Oct. 1978
. . . *the bones* [of Maqoma] *were on Robben Island**! She told him she had spoken to Qamata Who said He would help her find the exact spot. *Ibid.* June 1979 *(q.v.)

quarter evil *n.* Non S.A.E. ; also black-leg, black quarter ; see *sponssiekte.*
These sheep have been entered by the following well-known breeders . . . (all sheep inoculated for Quarter Evil, Pulpy Kidney and Blue Tongue). *Daily Dispatch Advt.* 11.3.72

quawal(i) [ka'wal(i)] *n. pl.* -s. *Ind. E. various sp.* Muslim religious singing usu. in commemoration of some historically significant event, inducing a high degree of spiritual excitement. [*prob. Arabic qawwal* singer]
. . . in commemoration of the martyrdom of Hassan the grandson of the Prophet . . . a number of persons fell into a trance while listening to various quawali singers recanting the death of the Islamic laminary. *Leader* 5.3.76

queen¹ *n. pl.* -s. **1.** An infertile or barren animal usu. a cow. ⫽Also acc. Pettman hunters' term for a barren elephant cow. [*fr. Du. kween* a barren cow]
Old cows and especially queens (barren females) will be found more difficult to deal with than the bulls. Nicoll & Eglinton *Sportsman in S.A.* 1892 cit. Pettman
Afrikander Heifers and Tollies . . . old cows . . . trek oxen . . . 1 Queen Heifer, Shorthorn

type. *Daily Dispatch Advt.* 16.10.71

queen² *n. pl.* -s. *abbr. Shebeen* ~ : see also quot. at *skokiaan.*
A nip of Ai-Ai sells in Soweto shebeens for 25 cents normally and some queens have been offloading it at 15 cents. *Post* 16.2.69
Shebeens . . . a long way from the traditional African beerbrewing days . . . and the people sell liquor instead. 'We must have an income somehow', one queen said. *E.P. Herald* 14.10.76

queen³ *n. pl.* -s. *colloq.* among children: a glass marble without trace of colour: also *glassie.* ⫽Also formerly *bottlie* or *sodie, fr.* soda water bottle top: once used for making imitation diamonds ; see quot. at *schlenter.* [*unknown*]

queen's tears *n. Afr. E.* Also *obs.* formerly 'tears of the King of England': European type liquor, usu. spirits. cf. *gologo,* (q.v.) among Africans.
'Brandy: they call it tears of the King of England'. Kuper *Witch in my Heart MS.* 1953

qufiya See *kufiya: acc.* Malay informant, preferred *sp.* is *q* as in *Quran (Koran).*

quick *n.* Any of several varieties of *Cynodon* used as lawn or pasture grasses: combinations *fine* ~ and *coarse* ~ : see *kweek(gras).* [*prob. translit. Afk. kweek* (q.v.)]
. . . all the quick Grass and much of the other destroyed by the Locust. Collett *Diary II* 30.4.1841
Quick grass. Various stoloniferous grasses which spread rapidly when once established and often become troublesome in gardens. Smith *S. Afr. Plants* 1966

quitrent ['kwɪt‚rent] *n.* That system of land tenure introduced in 1732 by which occupancy was given on a lease of fifteen years after which it had to be renewed. ⫽*cf.* The yearly renewal of a *loan place* (q.v.). The system of perpetual ~ by Proclamation 6.8.1813 gave leave to holders of *loan places* to convert these titles.
In the early days of the Cape Settlement, although freehold tenure was not unknown, the commonest form of holding agricultural land was by precarious tenure, the holding being known as a 'loan place'. Gradually however there grew up a system of 'perpetual quitrent', the rights and obligations under which were substantially those of the Roman emphyteusis; and . . . the precarious nature of the loan tenure was modified by customs in the direction of permanence and heritability. *Cambridge Hist. VIII* 1936 ed. Walker

Qwaqwa ['kwa‚kwa] *n. prop.* Also *Basotho* ~ : the *homeland* (q.v.) of the S. Sotho

(*Shoeshoe*) people in the Witzieshoek area of the Orange Free State.

Qwaqwa would have to resort to violence . . . if the Transkei continued refusing to allow the more than 40 000 South Sotho living in the territory to be incorporated in the Qwaqwa homeland, the Chief Minister of Qwaqwa . . . said here at the weekend. *E.P. Herald* 25.8.76

R

R1, R4 *n. pl.* -s. S.Afr. service rifles as issued by the *S.A.D.F.* (q.v.): see quots. at *G5* and description below.

Some R4s had already been issued and further issues depended on circumstances . . . the introduction of the R4 did not mean the abandonment of the well-tried R1 South Africa's service rifle for 17 years . . . asked if the R4 was capable of providing 'burst fire' as well as automatic fire . . . said the R4 could be given such a feature . . . the R4 can be fitted with a bayonet and . . . is slightly cheaper to manufacture than is an R1. *Cape Times* 26.4.79

raad [rɑːt] *n.* **1.** A council, often in combination *Kerk* ~ (q.v.) also a councillor as in *heem*~ (*pl. -en*) (q.v.) see also ~*saal*. [*Afk. fr. Du. raad* council, councillor, counsel]
2. *hist.* The legislative Assembly of one of the Boer Republics before the establishment of British rule in S.A.: see also *Volksraad*.

The gentlemen . . . obtained their grant under solemn seal and bond of the Transvaal Parliament or raad. Boyle. *Cape for Diamonds* 1873
Even when the Raad was not sitting he felt it his duty to stay in Pretoria and stand by the President. Brett Young *City of Gold* 1940

Raadhuis [ˈrɑtˌhœis] *n.* See *Raadsaal*. [*Afk. fr. Du. raad* council + *huis cogn.* house]

So they started work on their capital, opened up water by blasting, made gardens and finally erected a small 'Raadhuis' or Parliament House. The first Volksraad was elected in 1872. Green *Last Frontier* 1952

Raadsaal [ˈrɑtˌsal] *n.* The council or parliament house in which the legislative assembly, see (2) *raad* met in the time of the Republics of the Orange Free State and Transvaal: occ. *Raadhuis* (q.v.) [*Afk.fr. Du. raad* council + *saal* hall *cogn. O.E. sele*]

The harmony of Bloemfontein is in a great measure due to buildings erected in the proper style for South Africa. The perfect little Raadsaal gave me great pleasure. Morton *In Search of S.A.* 1948

raadsheer [ˈrɑtsˌhɪə(r)] *n. pl.* -here. Also *alderman.* ⫽Not equiv. of Brit. alderman who does not fight for re-election, but one who will go to the polls like other city councillors: see quot. [*Afk. raad* (q.v.) council + *heer* gentleman]

A new word in our Civic vocabulary these days is aldermen or raadshere, a term of respect now being used for councillors with more than 20 years' service, or former mayors. *Cape Times* 13.9.74

raak [rɑk] *n.* and *vb.* A hit, to hit with a missile or bullet. [*Afk. fr. Du. raken* to hit]

. . . bang went Hendrik's, who shouted 'Dats raak!! – I've hit him!' Mackenzie *10 Years* 1871
If you miss your cap go on shooting until you raak. Mitford *Aletta* 1900

rainbird *n. pl.* -s. **1.** *Vlei loerie*, see *loerie*, *Centropus superciliosus burchellii* and other species of the *Centropodinae* thought to be weather prophets whose frequent call heralds rain: in the *U.S.* any of several species of the Cuculidae. **2.** The turkey buzzard or *bromvoël* (q.v.) *Bucorvus leadbeateri*, the drowning of which is thought by some African tribes to bring rain.

rainmaker *n. pl.* -s. A *witchdoctor* (q.v.) who in addition to healing and other witchcraft, practices rainmaking or bringing rain by medicines or incantations, and, acc. some, by watching the weather: see quot.: also *U.S.* term for Amerindian medicine man with similar functions.

Rainmakers and witches have great influence in some tribes: the former are often consulted, not only regarding the weather, but also on other matters. When rain will not come they send off the young men to catch a baboon, (a difficult matter) for the purpose of gaining time. Alexander *Western Africa I* 1837

rain queen, the *n. pl.* -s. The Mojadji (Mujaji) or hereditary queen of the Lobedu, a Sotho people, who by divine right of queenship has affinity with the elements and special medicines inherited from her predecessors. ⫽Informant Eileen Krige. See also quot. at (1) *tickey* and *Modjadji*.

Last of the great 'rain queens' in the Transvaal was Mujaji, that withered and famous old woman who was known to Rider Haggard. . . . For centuries the rain queen was expected, in her old age, to pass on her secrets to a daughter or younger woman, and then to commit ritual suicide by taking poison. Mujaji was pre-

vailed upon by missionaries to break this savage tradition, and she died of old age. Green *These Wonders* 1959
... the new assembly [of Lebowa] ... of these 59 would be chiefs, designated by the chiefs in the district ... concerned, one would be a representative of the Chieftainess of the Lobedu (i.e. the 'Rain Queen'), and 40 would be elected. Horrell *Afr. Homelands of S.A.* 1973

ramkie(tjie) [ˌramˈkĭkĭ, -cĭ] *n. pl.* -s. Formerly a primitive stringed instrument of the Bushmen and Hottentots as described below: now usu. in *dimin.* form ~*tjie*, a home-made instrument esp. among coloured children, usu. with four strings and sound box made of a 5 litre oil can, now somewhat ousted by the cheap guitar. [*Afr. fr. Nama ramgi-b poss. via Port. + dimin. suffix -tjie*]
In the evening we were entertained by a Bushwoman ... playing on the *Raamakie* – an instrument about forty inches long by five broad, and having the half of a calabash affixed to the one end, with strings somewhat resembling those of a violin. With this instrument she produced a dull monotonous thrumming, in which my ear was unable to trace anything like regular melody. Thompson *Travels I* 1827

rampi(sny) [ˈrampĭˌsneɪ] *vb. phr.* as *n. prop.* The Malay Feast in celebration of the Prophet's birthday: see quots. [*fr. Malay rampai* mixture + *Afk. sny* cut]
... the Feast of the Orange leaves is held. On this occasion the women go to the mosques on Saturday afternoon from two o'clock till sunset. They sit on the carpeted floor on which the men assemble for prayer and on which the women and children, on other occasions, look down from the galleries. Here the afternoon is spent cutting up orange leaves, dipping them in costly, sweet-smelling oils, and tying them up in sachets (*Rampi's*, from the Malay rampai: a mixture) Du Plessis *Cape Malays* 1944
... cutting up orange leaves on small boards using special knives. The pieces were put on trays and sprinkled with rare oils. The practice is known as *rampi-sny*. *Cape Times* 18.5.70

ramsammy [ˈræmˌsæmĭ] *n.* A common first name for Hindu men: see *sammy*. combination: ~ *grass:* either of two grass species of Natal *Stenotaphrum secundatum*, also known as *buffalo grass*, *buffelsgras* (q.v.), or *Imperata cylindrica*. [*fr. Hindi Rama* god + *swami* lord]
ramsammy grass ... the first word of the vernacular name is a corruption of Rama swami ('Lord Rama') and is used as a sort of generic name for Indians in Natal, contracted in the Cape to 'Sammy.' Smith *S. Afr. Plants* 1966

rand [rănt, rant] *n. pl.* -e. **1.** Also *rant:* a

ridge of mountains or hills, a typical feature of the S. Afr. landscape: see also *bult*, *koppie* and quot. at *ribbok*. [*Afk. fr. Du. rand* ridge (+ *dimin. suffix -jie*)]
I shall never forget the scene ... in the early morning, when there were still shadows on the rante, and a thin wind blew through the grass. Bosman *Mafeking Road* 1947
The veld consists of iron stone rante, vleis and sweet karoo flats. *Farmer's Weekly* 3.1.68
freq. in *dimin.* form ~*jie* [raɪŋkĭ, -cĭ] a low ridge, in place names Randjieslaagte, Randjie Alleen.
Here and there a few kopjes relieved the monotony of the view, and every few miles, randjes, or low stony hills, stretch across the plains. Brinkman *Breath of Karroo* 1915
2. -*rand*- *n. prefix and suffix.* Ridge: found in place names e.g. Witwatersrand, Bosbokrand, Randburg, Randfontein. [*as at* (1) ~]
He had staked out six thousand acres of land on the grassy slopes of the Witwatersrand, the Ridge of White Waters, and built for his delicate wife a pondakkie hut in which, anxiously, they awaited the birth of their firstborn. Brett Young *Seek a Country* 1937
3. *Rand, the, n. prop.* The gold mining area of the Transvaal, also known as the *Reef* (q.v.) of which Johannesburg is the chief city: see quot. at *Kaffir circus*. [*abbr. Witwatersrand see quot. at* (2) ~]
The social scale on the Rand was a money weighing machine and nothing more. Cohen *Remin. Johannesburg* 1924
... there is hardly a family in many of the reserves without a member who is working or has worked on the Rand. Goold-Adams *S.A. To-day* 1936
In combination ~*lord* one of the great tycoons of the Rand of the 1890s and after: see quot. *cf. Hong K. tai-pan, Anglo-Ind. nabob.*
'The Nineties' says a historian, 'were the high and balmy days of the great Randlords. Johannesburg seemed nearer to London than any English town.' Roberts *Churchills in Africa* 1970
The Golden Age of the great Randlords ... 1890–1910 ... They came to conquer and conquer they did. They were a glamorous and exciting fraternity, those giants of the past ... Round all of them was an element of magnificence ... Where are the Randlords today? Dead. All dead ... And I doubt we'll see their like again. *Personality* 8.1.70
4. *rand* [rănt] *n. pl.* ∅ or [rænd] *n. pl.* ∅ or -s. The unit of S. Afr. currency consisting of 100 cents, usu. paper money, *one* ~, *two* ~, *five* ~, *ten* ~ notes,

though one ~ coins do exist: ⫿The *Kruger* ~ is of gold (one ounce) first struck in 1967 and marketed by the S. Afr. Mint at R31 each, since when the price has risen to around R200 apiece. [*fr.* (3) *Rand*]
> The informer bought three – packets of dagga with the one Rand note and received 25 cents change. *S.W. Herald* 14.5.71
> Already millions of rand have been spent to prepare Damaraland for separate development. *E.P. Herald* 28.9.72
> South Africa's Kruger rand is the most sought-after gold coin in the world today. Its overseas sales have earned around R270-million in foreign exchange – and there's more to come. *Cape Herald* 14.9.74

rant [rant] See (1) *rand* and quot. at *ribbok*.

Rapportryers [raˈpɔrtˌreɪə(r)z, -s] *pl. n.* An Afrikaner political organization: see also quots. at (2) *Voortrekker* and (2) *Afrikaner*. [*Afk. rapportryer* dispatch rider]
> The Broederbond's concern about membership of the Rapportryers arises from the fact that the Rapportryers are one of the Bond's front organisations and can be said to be controlled and dominated by the Broederbond. *Sunday Times* 8.10.72

ratel¹ [ˈrɑtəl] *n. pl.* -s. The Cape badger *Mellivora capensis* of the Mustelidae: omnivorous but called 'honey badger' from its liking for honey and for robbing wild hives. [*Afk. ratel fr.* Du. *ratelmuis fr. heuningraat* honeycomb]
> A ratel, the destroyer of bees' nests, crossed our path; the Tambookie dogs soon overtook it, and kept it at bay until the assegais put an end to it. Its stench when killed was intolerable far exceeding that of the pole cat. Philipps *Albany & Caffer-land* 1827
> ... it is a long time since a ratel, or honey badger, was seen in the Peninsula ... A ratel is a vicious opponent, often more than a match for a pack of dogs. Green *Grow Lovely* 1951

ratel² [rɑtəl] *n. pl.* -s. Sect. Army. A six-wheeled high-speed armoured vehicle, designed and built in S.A., primarily in use as a support weapon, ⫿ The ~ carries a crew of three plus eight infantrymen. [*Afk. ratel* honey badger: *prob. fr. Afk. saying 'so taai soos 'n ratel'*, as tough as a ratel]
> Here the Ratel is seen climbing a very steep embankment. This vehicle is so versatile that there is virtually no terrain through which it cannot travel. Even water up to the depth of 1.2 metres ... Shown here the Ratel is seen forcing its way through dense bush and shrub-

bery. Marks *Our S. Afr. Army Today* 1977

rather *adv.* Used as an equiv. of 'instead' freq. placed at the end of a sentence, e.g. Let's jol on the grass rather (child). [*unknown, poss. fr. Afk. liewer, liewers* rather, instead]
> 'But that's what I've been saying also,' Oupa Bekker persisted. 'I say why doesn't Jurie rather go in his mule-cart?' Bosman *Jurie Steyn's P.O.* 1971

rather very *intensifier, substandard.* Somewhat, a bit: usu. with adj. [*trans. Afk. bietjie baie*, excessively somewhat, (*lit.* 'a little, a lot' *fr. Malay banyak* a lot, much)]
> The work is rather very difficult. African student 1952
> Since I saw you I've been rather very ill. Karroo informant Letter 1960

ratiep [raˈtĭp] *n.* A Malay Muslim religious sword-dancing ritual in which the performer demonstrates his faith by piercing himself with swords, skewers and other sharp instruments without drawing blood. *various sp.*: see also *Khalifa*
> The ritual called ratiep, is performed under the guidance of a spiritual leader known as a Khalifa. Enthusiasts say their only protection from certain death is faith in ... Allah, and the ceremony cannot be performed without a Khalifa praying throughout. *Argus* 13.1.77

rat pack *n. phr. pl.* -s. Sect. Army abbr. *ration pack:* issued to men in the bush or on the Border. ⫿ The entire contents mixed together have been called 'Angolsh goulash' in mimicry of the Portuguese language.

R.A.U. [raʊ] *n. prop.* The Randse Afrikaanse Universiteit, Johannesburg. [*acronym*]

raw *adj.* As in *Brit.* '~ recruit', inexperienced: in S.A.E. usu. applied to an African who has not been exposed to civilization, town life etc. freq. derogatory.
> My other maid doesn't know Monday from Friday – she's raw, but reliable. *Cape Times* 13.4.73

reclassification See under *reclassify*.

reclassify *vb. usu. pass.* To assign to another group an individual already *classified* (q.v.) in terms of the Population Registration Act as belonging to a specific racial group. ⫿Under certain circumstances, *e.g.* wishing to marry outside the assigned racial group, or always having lived as a member of another, such persons can apply for *reclassification*. [*fr. classify* (q.v.)]

A Durban Coloured woman . . . who was acquitted in January of contravening the Immorality Act with her White boyfriend, is still battling to be re-classified as a White. *Daily News* 9.6.70

rector *n. pl.* -s. Also *rectress :* the principal of an *Afk.* university or other college. [*fr. Du. rector* headmaster, principal, *fr. Lat.*]

VACANCY FOR RECTRESS Department of National Education *Sunday Times Advt.* 5.9.71

Hostess at a recent garden party in Alice was . . . wife of the Rector of Fort Hare [University]. With her is one of the guests of honour. *E.P. Herald* 24.5.74

red *adj. or n.* **1.** Of or pertaining to the Xhosa *AmaQaba* (q.v.) people who use red clay or ochre on their bodies and/or blankets: see second quot.

'Do all the people in the Transkei believe these things?' 'No, not all the people but the Red people.' 'Red people?' 'The people who smear clay on their bodies.' Gordon *Four People* 1964

Blacks who wear tribal beadwork and dress are called in Xhosa 'amaQaba', a name which signifies that they worship their ancestral spirits . . . those who adhere to this belief are recognised by the red ochre or clay which they apply to body, blankets and clothing. This is the colour beloved by the ancestral spirits and so their followers are called 'Red People'. The colour of the ochre varies . . . from palest orange to deepest red. Broster *cit. Panorama* Dec. 1974

2. Loosely used, often by Africans, of any reddish brown or russet colouring esp. of cow, dog or other animal: see quot. at *ngoni :* also a variety of *Africander* (cattle).

His entire herd of 175 Outstanding Red and Yellow Africanders . . . 81 Young Red Cows – All with calves 5 Old Red Cows – All with calves 8 Dry Red Cows . . . *Farmer's Weekly Advt.* 20.3.74

3. *Sect.* (*World War II*) *:* of or pertaining to the oath taken by S. Afr. servicemen to signify willingness to go anywhere in Africa, or to the tabs or flashes worn on their uniforms as evidence of this undertaking: see second quot.

. . . a soldier with the red tabs of the South African Army . . . on his unbuttoned tunic. Walker *Shapeless Flame* 1951

He took the red oath, which meant that he would go anywhere in Africa, and they gave

him red flashes to put on his shoulders. But the red oath, to those who would not take it, meant only one thing, that the wearer of it was a Smuts man, a traitor to the language and struggle of the Afrikaner people, and a lick-spittle of the British Empire and the English King, fighting in an English war that no true Afrikaner would take part in. Paton *Too Late Phalarope* 1953

Red Area *n. Sect. Army.* Danger zone of the *Operational Area* (q.v.). ⫽ Also *red road* one in which land mines have been laid [*fr. red sig.* danger]

You get Bush leave at the rate of one day for two weeks in the Red Area. *O.I. Ex-Serviceman* 13.2.79

redbait *n.* Also *rooi-aas; Pyura stolonifera* a type of sea squirt, of the Ascidiae enclosed in a thick cartilaginous covering and popular among fishermen for bait: see quot. at *ollycrock*. [*prob. trans. Afk. rooi-aas* red bait]

. . . how about some redbait? This humble sea-squirt is not good company when old, and many an angler's wife has had reason to complain bitterly of the lingering aroma. Fresh red bait, however, can be used as an ingredient in fish soup. Green *S. Afr. Beachcomber* 1958

redfish *n. pl.* ∅. General term for fish of red skin colouring esp. the *dageraad* (q.v.), *miss lucy* (q.v.) (red stumpnose) (q.v.): also *dikbek(kie)* known as 'reds': see *panga*[2].

I'm so sorry no kob today – only redfish. Mind you the reds are tasty even if they are a bit bony. Fishmonger *O.I.* 1972

redgrass *n.* Usu. *rooigras* any of several good pasture grasses with a reddish tinge, including species of *Cymbopogon, Hyparrhenia* and *Themeda*. [*trans. Afk. rooigras*]

100 good camps and excellent red grass grazing with carrying capacity 50 head of cattle and 500 sheep. *Farmer's Weekly Advt.* 21.4.72

red hot poker *n. pl.* -s. Any of several species of *Kniphofia* with cylindrical flame coloured flower heads like those of aloes with which they are sometimes confused, also called *soldier, torch lily :* also *Brit.* see quot.

red hot poker . . . the vernacular name was apparently first coined for *K. uvaria* in English gardens, and subsequently (probably) after 1820 applied to this species in the field from the suggestion of a glowing poker conveyed. Smith *S. Afr. Plants* 1966

red ticket *n. Sect. Mining.* See quot. at *skip.*

redundancies Items used redundantly in S.A.E. are of several kinds: most appear as head words of entries in this text. See also omissions and prepositions.
(a) interpolated *Afk.* particles, see *darem, maar, mos, sommer;*
(b) items usu. translated, redundant in various structures, see *again; already; article, redundant* (also *third person address*); *busy; but* (see *only* (1), (2)); *little; only; now; so; still; what; with; yet.*
(c) *negatives* (q.v.) ; see also *no.*
(d) tautologous phrases e.g. *horse riding, sugar diabetes, now-now* (q.v.), *yellow jaundice;* also *finish(ed) and klaar* (q.v.).

redwater *n. and n. modifier.* **1.** A febrile disease of cattle similar to *gallsickness* (q.v.) caused by a parasite transmitted by the blue tick which destroys the red blood corpuscles: ‖These wastes are converted to excess bile, some of which causes jaundice as in *gallsickness* and some excreted by the kidneys causing red colouration of the urine: see first quot. at *salted.*
The Division of Veterinary Services reported that redwater was the most troublesome disease in cattle herds in most parts of the country during the past year. *E.P. Herald* 9.12.74
As *modifier :*
Cattle that have lived for several generations in redwater areas become less susceptible to the disease than are freshly introduced cattle. The calves of such cattle get a mild form of the disease, although some may die of it, the rest are then immune. *H'book for Farmers* 1937
In combination ~ *veld* areas in which ~ *fever* is endemic, see quot. above.
The animals run on virulent redwater and gallsick veld and are extremely healthy. *Farmer's Weekly* 20.3.74
2. *colloq.* Bilharzia, which is also characterised by haematuria.

reds *n. Dikbek(kie),* or *panga²* (q.v.) see also *redfish.*

reebok [ˈrɪəˌbɔk, riː-] *n.* **1.** See *rhebuck/bok, ribbok.*
2. Rhebuck: in place name Reebokrand, also Rhebokfontein.
3. *prefix* to plant names ~*blom,* see *rhebokblom; ribboksuring, Rumex angiocarpus.*

Reef, the *n. prop.* The heavily urbanised and industrialized gold mining area centering upon Johannesburg as the largest of several large cities usu. known as the ~ *towns :* the Witwatersrand: see also (3) *Rand* and second quot. at *banket.* [*presum. Eng.* reef: *a deposit of ore*]
˹The newcomer to the Reef will wonder too what all the talk of pollution is about. *Panorama* May 1972

reference book *n. pl.* -s. An identity document carried by all Africans containing details of domicile, employment and other personal data: introduced by law in 1952 repealing the *pass* (q.v.) laws and providing for ~*s* instead: see also *domboek, dompas* and quots. at *aid centre* and *citizenship certificate. cf. Book of Life* (q.v.).
The movement of Bantu work-seekers and labourers is controlled by their having to carry a personal document, the reference book that may be compared to a passport. This system of issuing reference books has certain advantages over the former 'pass' system. Tomlinson *Commission* 1955
Chief Gatsha Buthelezi of KwaZulu yesterday denounced reference books as symbols of oppression and the greatest cause of resentment between Whites and Africans. Waving his own passbook in the air, the Chief reminded the special session of the KwaZulu Legislative Assembly in Nongoma that he had been arrested several times for not carrying it. *Cape Times* 18.1.73

Reformed Church *n. obs. hist.* The *Dutch Reformed Church* (q.v.) while the Cape was under British rule: see also quot. at *Political Commissioner.*
Reformed Church
Wilhelm Bussinne Esq. Political Commissioner
Elders . . .
Deacons . . . *Afr. Court Calendar* 1809

reggie [ˈrexi] *n. pl.* -s. *slang. abbr. regmaker* (q.v.).

regmaker [ˈrexˌmɑkə(r)] *n. pl.* -s. **1.** *colloq.* A drink or other stimulant, or medication, taken as a cure for a hangover: *abbr. reggie,* (vulgar), see *babala(a)s. cf. Brit.* 'hair of the dog' etc. [*Afk. fr. Du. recht* right + *maker maak* + *agent. suffix -er*]
. . . the search for the ideal 'regmaker' goes on. We decided to ask those potentially best qualified to know – the dispensers of the potential hangover – the barmen . . . a barman from Kensington . . . says there is nothing better than a pint of cold beer as a . . . 'regmaker'. *Cape Herald* 22.9.73
2. As *n. prop.* the trade name for anti-

hangover tablets: also the official journal of Alcoholics Anonymous.

A charming true story culled from the pages of Regmaker, the official journal of Alcoholics Anonymous. *Star* 6.10.72

Stop leading a dog's life, take Regmakers. Every Regmaker . . . contains enough (non-habit forming) caffeine to put you back in human shape again. Regmakers the extremely fast, highly effective means to rid yourself of a hangover. *Personality Advt.* 18.10.74

Rehoboth(er) See *Basta(a)rd, Baster.*

release *vb. trns.* See quot. *cf. zone.*

The farms in the district were 'released' (declared Black territory) some time ago but there had been no movement as yet. *E.P. Herald* 2.8.74

remskoen ['rem₁skŭn] *n. pl.* -s, -e. A lockshoe of heavy timber used to brake the rear wheels of a wagon before the invention of the screw-operated brake: see first quot. ❘Wagons lacking a ~ paid higher toll than those so fitted. see second quot. [*Afk. fr. Du. remmen* to brake + *schoen* shoe]

30 April 1811 The remschoen (lockshoe or skid) is a log of wood, generally about eight inches square, and nearly two feet long, having a groove in it to receive the felly of the wheel; and is furnished in front with a stout loop of twisted raw hide. Burchell *Travels I* 1822

Upon every wheel of every four wheeled vehicle not provided with a wooden shoe (remschoen) or an iron shoe not less than eight inches broad . . . 3d. *Cape Town Directory* 1866

figur. One who impedes progress, an obstructive person *cf. Brit. stick-in-the-mud*: ~ (*party*) *politics*, ultra-conservative or obscurantist policies.

Those arguing against the Act were 'sukkelars', and formed them a 'remschoen'. He [General Botha] asked them to co-operate in making the Act a success. *E.P. Herald* 27.10.1911

renoster- [re'nɔ̆stə(r)-] *n.* Rhinoceros: [*Du. renoster* rhinoceros] **1.** Found in place names Renosterkop, Renosterspruit.

2. *prefix* to plant names esp. ~ *bos* (q.v.) also ~ *gras, Eragrostis curvula;* ~ *kweek, Cynodon dactylon* said to have been favoured by the rhinoceros.

renosterbos(sie) [re'nɔ̆stə(r)₁bɔ̆s(i)] *n. pl.* Ø, -s. *Elytropappus rhinocerotis,* a troublesome weed: a greyish blue-green shrub which characterizes large tracts of veld where it has encroached on the grazing or fallow wheat lands (see quot. at *Swartland*) often as a direct result of

veld burning (q.v.) which encourages the growth of 'invader' plants, see *plant migration;* also as a result of 'having been carried through the Colony by the brandy distilling Boers of old time, who used it as dunnage in packing the casks on their wagons'. (Wallace *Farming Industries* 1896): ❘Farmers in the C18 believed the invasion of their land by ~ was in retribution for their sins. ~ is known in the E. Cape as *boeboes, fr. Xh. ibhubhusi:* [*Afk. fr. Du. renoster* rhinoceros + *bos cogn.* bush]

Rhenoster bosch (Rhinoceros bush) and said to have formerly been the food of the huge rhinoceros, till those animals fled before the colonists, as these gradually advanced over the country where the shrub grows. Burchell *Travels I* 1822

. . . Tumbleweed, prickly pear, rhenosterbos, and jointed cactus invaded the territory of the edible grass and nutritious plants. De Kiewiet *Hist. of S.A.* 1941

Rent-a-bakkie *n. prop.* Any of several branches of commercial organizations specialising in hiring out *bakkies* (q.v.) or other light delivery vehicles.

repatriate *vb.* To send or return urban blacks to the *homelands* (q.v.) of their particular nations. ❘ This can imply that a town-born black may be ~*d* into a country in which he has never been: see quot. at *bachelor quarters.*

Philip . . . is smiling again after 2½ years of living in fear of being 'repatriated' to a strange country although he has spent nearly 40 years in South Africa. *Star* 31.5.73

Republic, the *n. prop. abbr.* the *Republic of South Africa* used in speech and writing: see *R.S.A.* also *Union.*

In the Republic we might talk with pride of our veld but it is doubtful if such pride is justified on today's picture. *Farmer's Weekly* 20.3.74

request farm, request place *n. pl.* -s. *hist.* A farm not exceeding 3 000 *morgen* (q.v.): see also *place, full* granted on application. *cf. Austral. hist. free selector/ selection* of Crown Land, *Canad. location* [*fr. Du. rekwest* application]

A boor, upon discovering water on a sufficient quantity of unoccupied land, forwards, through the secretary of his district, what he terms a 'request' for a place, – that is a memorial, asking for a grant of 6 000 acres. Thompson *Travels II* 1827

The Rents hitherto received by the Colonial Government have not yet exceeded a third of the sum likely to accrue to the Revenue when the survey of the numerous 'Request Farms' now in progress shall have been completed and the occupiers put in possession of their title deeds. *Greig's Almanac* 1831

reserve(s) *n. usu. pl. obsolescent:* usu. *Native* ~: lands occupied by Africans usu. under tribal conditions; now replaced by *homelands* (q.v.) see second quot., also quot. at *donga. cf. U.S. (Indian) Reservation, Canad. (Indian) Reserve.*

Under existing laws in South Africa the ownership of land by Africans is limited to the areas known as Native reserves. Outside these areas . . . Africans may not own land, and within them ownership in most cases is by tribal tenure and not by private title. Hellmann and Abrahams *H'book on Race Relations* 1949

One step towards this goal should be the gradual extension of home rule to the Natives in the areas they now occupy, namely the Reserves, or, as some say, 'the Bantu Fatherland'. Brummer *Problems & Tensions* 1955

Some persons in the government hope to meet this problem by offering inducements and exerting pressures upon new industries to locate near the Reserves.* *Ibid.*

*see *Border Area, Border Industry*

resettle(ment) *n. and n. modifier, also vb. trns.* Removal of Africans, esp. those *endorsed out* (q.v.) of urban areas from one area to another, usu. in an African *homeland* (q.v.) in terms of legislation providing for this; as modifier ~ *camp,* ~ *areas,* ~ *removals.* [*fr.* Native Resettlement Acts]

The new native resettlement bill of the Nationalist government provides for a natives' resettlement board which would compel and override city councils which refuse or fail to carry out the provisions of resettlement legislation. De Kiewiet *Fears and Pressures* 1954

Appealing to the Government to halt all removals of Africans to homeland 'resettlement' areas, Mrs. Suzman said the film [*The Dumping Grounds*] had exaggerated in saying there were four million Africans to be removed, but otherwise appeared to be factual. *Daily Dispatch* 10.5.71

resin bush *n. pl.* -s. Any of numerous species of *Euryops;* see *harpuisbos:* an 'invader plant'; see *plant migration.*

rest *vb. and n. Sect. Farming.* Usu. in combination *to* ~ *a camp;* to refrain from using a *camp* (q.v.) for grazing in order to allow the (2) *veld* (q.v.) to recover; to practice rotational grazing, see *veld management* and *stock reduction scheme:* as *n.* the non-use of a grazing camp for a certain period: see quot.

Similarly, when deciding . . . on the time a

camp should come out of rest it was not wise to rely on the calendar or the overall look of the veld. *E.P. Herald* 3.12.74

rest camp *n. pl.* -s. Accommodation for visitors and holiday makers in Game Reserves. ⫽A ~ consists usu. of a cluster of *rondavels* (q.v.) in a fenced enclosure.

The Game Reserve provides a rest-camp with a swimming bath and rondavels where visitors can spend the night. *Panorama* Jan. 1974

LIONESS OCCUPIES REST CAMP LOO

An elderly lioness has taken up residence in the only ablution block of a rest camp in the . . . Reserve and is jealously guarding her new home from intrusion by visitors to the camp. *E.P. Herald* 13.1.75

reverend *n. pl.* -s. *Sect. Jewish.* A Jewish scholar, who lacks rabbinical qualifications, employed to minister to a congregation: see second quot.

There's such a small Jewish community here that there isn't even a real rabbi here we just have a reverend. *O.I.* 1968

. . . the use of the word 'Reverend' as a noun, i.e. 'a person employed as a "Reverend" by a congregation.' In practice this applies to partly qualified Jewish ministers, i.e. those without a Rabbinical diploma. *Letter* Eric Rosenthal 23.5.72

R.F. ['ɑ(r)₁ef] *acronym.* Rhodesian Front: see *P.F.*[2]

'They do not trust it' . . . 'Everyone in the transitional Government is talking the language of the Rhodesian Front . . . not . . . the language of their people' . . . said the army was an RF force. *Drum* Dec. 1978

rhebuck, rhebok ['ri₁bʌk, -₁bɔk] *n. pl.* ∅. Either of two species of small S. Afr. antelope, also *ribbok,* see quot. the *red/rooi* ~ with curved horns *Redunca fulvorufula,* or the *vaal/grey* ~ *Pelea capreolus* with straight horns [*4fk fr Du. ree* roe, hind + *bok cogn.* buck]

Other buck in the reserve are reed-buck, vaalrheebok, rooirheebok, bush-buck. Grocott *Mail* 11.4.72

Graaff-Reinet: Open season for buck (ordinary game) – May 31 to July 31. Blesbok, red rhebuck and steenbok declared ordinary game. . . . Humansdorp wards 5 and 7: Grey ribbok protected. Daily bag limits: Steenbok one. *E.P. Herald* 19.3.73

rhebokblom ['ri₁bɔk₁blɔm] *n. pl.* -me. Either of two species of (2) *Afrikaner* (q.v.) *Gladiolus grandis,* and *G. tristis.*

rhenosterbos(ch) See *renosterbos/sie.*

Rhodes (grass) *n. Chloris gayana:* perennial, hardy grass species which makes excellent hay and pasture, introduced by

Cecil Rhodes from Rhodesia at Groote Schuur, his residence in Cape Town. [*fr. n. prop.*]

Soil recently reclaimed from bush, now planted to eragrostis, Rhodes and kikuyu pastures, contoured and divided into 19 camps. *E.P. Herald Advt.* 10.5.74

Rhodesian teak *n. Guibortia coleosperma* (*rooisering*): see also *kiaat*.

RhodZim *n. prop.* Variant on *Zimbabwe-Rhodesia* (q.v.) prob. ephemeral.

. . . he said he hoped RhodZim wouldn't become a bankrupt banana republic. I thought it already was one. *Sunday Post* 3.6.79

rhogun ['rəʊₗgʌn] *n. pl.* -s. See quot. at *Uz(z)i.* [*presum. 'portmanteau word' Rhodesian + gun*]

ribbetjie ['rıbıkı, -cı] *n. pl.* -s. Loin of lamb or mutton, usu. in combination *braai* ~ grilled over an open fire, or *sout* ~ (q.v.). [*Afk. rib* rib, loin + *dimin. suffix -(be)tjie*]

Her range of culinary art was limited to serving us with 'braairibbetjies' (grilled mutton chops) without vegetables for breakfast, lunch and dinner. Jackson *Trader on Veld* 1958

This consists of 4–6lb of mutton or lamb . . . Keep the ribbetjie in one piece, but chop through the bones so as to cut it into convenient serving pieces when grilled. *Evening Post* 17.10.70

ribbok ['rıbₗ(b)ɔk] *n. pl.* ∅. See *rhebuck/bok.*

Give me just a plain piece of ribbok – just roasted on the ashes. . . . Seeing that Jurie Steyn's was the only farm in those parts where you could get an occasional ribbok in the rante. Bosman *Bekkersdal Marathon* 1971

ride *vb. colloq. substandard.* Also ~ *in,* ~ *on* (*water*) (q.v.): equiv. in S.A.E. of convey, cart. *cf. transport riding/er,* and see quot. at *misrybol* [*translit. Afk.* (*in*)*ry* convey, bring in]

One of our neighbours has put a fence across his road which we have been using for riding mealies to the railway siding for the last fifteen years. Van Alphen *Jan Venter* 1929

Farmers have to ride in feed. *Grocott's Mail* 15.12.72

ride on water *vb. phr. Sect. Farming.* See also *ride.* To transport water, usu. for stock in time of drought. [*translit. Afk. water aanry* to convey, transport water]

It's terrible these days. I spend all my time riding on water and the red cats are getting at my sheep. O.I. Farmer 31.1.70

'Ride Safe' *n. prop.* Also *Project Ride Safe:* see quots.

Project Ride Safe, the give-a-troopie-a lift scheme is now well under way . . . Thanks to the Ride Safe scheme our troopies' hitchhiking is rather better regulated than it is in Israel. Steenkamp *cit. Cape Times* 20.1.79

'Ride safe' scheme not going well.

The Ride Safe scheme launched last year by the Defence Force to get National Serviceman safe lifts home . . . is not working satisfactorily. The Defence Force admits this – and so do spokesmen for pickup points in Port Elizabeth . . . A project now successfully operating in the Transvaal . . . called in Afrikaans 'Bel en ry na Ma' (Phone and ride to mother) was advertised on radio. *E.P. Herald* 1.3.79

~ **sign** *n. pl.* -s. A sign displayed by garages which arrange lifts for National Servicemen on leave.

If you see a garage displaying a red-and-yellow Ride Safe sign call in, in case there is a soldier waiting for a lift. Steenkamp *cit. Cape Times* 20.1.79

The Ride Safe sign is red and yellow – yellow background with a National Serviceman in mooi-moois in red. *O.I. Serviceman* 4.2.79

ridgeback *n. pl.* -s. A large smooth-coated dog, originally bred in Rhodesia, brindled to pale brown in colour with a narrow strip or 'ridge' of hair about 30 cm long growing upwards and crossways along the back: also *Rhodesian* ~.

The finest type of Bushman hunting dog, a light brown ridgeback mongrel with dark stripes and a trace of the greyhound in his appearance, is now verging on extinction. Green *Where Men Dream* 1945

riem [rĭm] *n. pl.* -s, -e. **1.** A thong of softened raw hide used for numerous purposes instead of rope, (see quot. at *muid*), tethering domestic animals, leading oxen etc.: use *riempie, brey/* In combination *os*(ox) ~: see quot. at (*stoel*)*riempie* ǀǀAlso in phr. *brey, brei* ~s see quot. also *breipaal.* [*Afk. fr. Du. riem* strap *cogn. Ger. Riemen* stripes, bands]

. . . turned the cart over. We found it considerably damaged, but Ludwig, who is a most invaluable and indefatigable man, bound it together with 'riems'. Gray *Journal II* 1851

We then 'kneehalted' them, which is done by fastening the head to within about 18 inches of the knee by means of a Rhiem – a strip of hide – just allowing them sufficient length to enable them to reach the grass to eat. Buck Adams *Narrative* 1884

The Breying of Riems. The general practice is to stretch the skin directly the hair has been removed, and cut it into riems. . . . such a riem is 70 to 80 yards in length. It is looped over a breying-pole. *H'book for Farmers* 1937

Whips/Whipsticks/Riems. Boermaak Riems.

10ft. R5,00; Strops R3,00. *Farmer's Weekly Advt.* 7.11.73

figur. Link or fastening.

Then the headman [of the Riemvasmakers] spoke. He said: 'the riem that was made fast 60 years ago was now being torn loose.' . . . 'I don't know why we have to give up this land,' the headman said. 'I don't know why the Government wants us to tear the riem loose.' *E.P. Herald* 28.1.74

[*see also resettlement*]

By transference of meaning used of actual rope. [*Afk. vang* catch, *kalf cogn.* calf]

VANGRIEME ideal for Show purposes, in 10 metre lengths. Also available in cotton, KALFRIEME in Nylon only. Length – 2 metres. Diameter – 12mm. *Farmer's Weekly* 20.3.74

2. Thong, fastening: found in place names Riemvasmaak (see quot. at (1) *riem*) and Riemland (Bethlehem district.) [*fr.* (1) ~]

There were more than enough hides for 'riems' (straps), hence the name 'Riemland'. Many of the early inhabitants made their living by selling riems, and the beams and door-frames of their houses were made out of bundles of reeds held together by blesbuck riems. *Panorama* May 1974

riempie ['rïmpï] *n. pl.* -s, and *n. modifier.*

1. A fine narrow *riem* (q.v.) of softened hide used for thonging the backs and seats of chairs, *rusbank(s)* (q.v.) and *katel(s)* (q.v.), and even for shoe-laces: see quot. at *swartwitpens:* cf. *Canad. shaganappi,* rawhide thong. [*Afk. fr. Du. riempie* leather thong]

Thongs, called 'rimpies', made from the fine soft skins of bucks, are extensively employed by people living on the veld for pointing whip lashes, mending harness, and for the common purposes for which twine is generally employed in more densely populated places. Wallace *Farming Industries* 1896

'Be seated, mynheeren,' said the hospitable minister, pushing forward the stout teak chairs with riempies seats, which stood on the stoep. Fairbridge *Which Hath Been* 1913

In combinations ~ *chair/stoel, stoel~s* (q.v.) see quot. at *niggie.*

In this room the Englishman now had his bed, his bath, his guns, two rimpie chairs, a yellow-wood table littered with papers, pipes and tobacco jars. Smith *Beadle* 1926

In combination: *stoel(chair)* ~.

Best Boermaak Osriems, 10 ft. . . . Stoel-riempies, R1,60 pound, . . . *Farmer's Weekly Advt.* 3.1.68

2. *vb. trns.* usu. *pass.* or as *partic.*

To weave ~s, usu. in an open criss-cross basket work pattern for the seats and backs of chairs etc. ; also bedsteads: see *katel.* [*fr. n. riempie*]

The seat is riempied with fine riempie in the pattern of caning. Baraitser & Obholzer *Cape Country Furniture* 1971

riet [rït] *n.* Reed [*Afk. fr. Du. cogn.* reed]

1. Found in place names usu. with connotations of water, e.g. Rietfontein, Rietvlei, Rietkuil, Rietbron.

. . . a place called Rietfontein, occupied by a Griqua. Thompson *Travels II* 1827

Rietbron, Riethuis, Rietkop, Rietkuil, Rietvlak, Rietpan, Rietwater, And sixteen Rietfonteins. A. E. Voss Song *S. Afr. Place Names* 1962

2. Suffixed to the names of several grasses e.g. *dek* ~ (q.v.), *soet* ~ (q.v.) ; see also *biesie.*

rietbok ['rït,bɔk] *n. pl.* ∅. Also *reedbuck :* any of several small S. Afr. antelopes of the genus *Redunca,* esp. *R. arundinum,* which frequent reed beds and marshy *vlei* (q.v.) areas, and whistle shrilly if alarmed. [*Afk. riet cogn.* reed + *bok cogn.* buck]

A bushbuck barks like a dog and a reedbuck whistles like a bird. And if you watch a reedbuck when he makes his whistle, you'll see it doesn't come from his throat, but from a gland-like opening on his flank when he contracts his thigh muscles. *Sunday Times* 12.8.73

rietbul ['rït,bœl] *n. pl.* ∅. Knysna name for larger specimens of *kabeljou* (q.v.) *Argyrosomus hololepidotus* usu. preferring an estuarine habitat. O.I. Hjalmar Thesen 1974. [*Afk. riet cogn.* reed + *bul cogn.* bull]

right hand (wife) *n.* The wife, also known as *owasekunene* 'the one of the right hand' second in authority to the *Great Wife* (q.v.) (*omkulu*) of a chief. ⫿ The hut of the ~ is to the right of that of the Great Wife and her descendants form the 'right hand house' as opposed to the 'great house' of the family; the 'left hand' wife and her descendants, forming the next in seniority.

. . . inferior positions are then assigned to the descendants of two other of the wives under the denomination of the 'right hand' and 'left hand' of the family, . . . apparently taken from the relative situations occupied by the huts of these wives respectively. Shaw *My Mission* 1860

The hut of the right hand wife is placed to the right side, that of the left hand wife to the left of that of the 'inkosikazi'. Matthews *Incwadi*

Yami 1887

right, not *adj. phr. colloq.* Not in one's right mind, mentally abnormal. cf. *Brit., U.S. not right/straight in the head* and *Jam. Eng. no(t) righted*, mad; *Austral. dipped.* [*Afk. nie heeltemal reg nie* not in full possession of the senses; *reg* right, sane]

The parson's name is Damp . . . He is very strange sometimes. Charles says he is not right, but Hester our cook, says who is not right. All have a little mad in them. Vaughan *Diary circa* 1902

rinderpest ['rɪndə(r)ˌpest] *n.* A virulent highly infectious cattle disease long known in Europe but with particular historical significance in S.A. where it broke out in 1896; described as 'the greatest shock ever sustained by the agricultural community of S.A.' *Farmers Annual* 1914: see quot., also quot. at *blue tongue* [*Ger. Rind(er)* cattle *(pl.)* + *pest fr. Lat. pestis* plague]

The settlers were, for the most part, simply beggared by war and rinderpest, the latter from every point of view the most terrible calamity imaginable. *Milner Papers* 1931 Letter to Mr Chamberlain 1.12.1897

People today can have no idea of the terror that now unfamiliar word carried to the South Africa of the nineties. Rinderpest was a cattle plague, deadly, implacable, moving faster than the railway, through lands where there were no railways, without a cure, without regard to political or other boundaries. Jackson *Trader on Veld* 1958

Also used as a date landmark (*cf. 'flu*), and as an informal expression of time. *before the* ~ sig. a long time ago, *cf.* in the year *voetsak* (q.v.). See quot. at *'flu.*

My Dad bought this farm in the year of the rinderpest. O.I. Farmer 1973

ring [rɪŋ] *n. pl.* -e, -s. A number of Dutch Reformed parishes, see *gemeente*, joined together in a manner similar to that of the English diocese or Scottish presbytery. [*Afk. ring* presbytery]

. . . The Public Morals Commission of the Parow ring. Mr . . . is a member of the congregation of Parow North, which falls under the jurisdiction of the Parow ring. *Rand Daily Mail* 28.1.74

ringhals See *rinkhals*.

ringhalskraai *n. pl.* -e. *Corvus albicollis*, see *withalskraai*.

rinkhals ['rɪŋkˌ(h)als] *n. pl.* Ø. Also *ring-hals, spuugslang, bakkop, Hemachatus*

haemachatus or 'ringnecked cobra' closely related to the cobra (*Naja*), which seldom strikes but spits or sprays its venom, which can cause blindness, when disturbed: hence term *spuugslang* (q.v.). [*Afk. fr. Du. ring + hals* neck]

The Rinkhals so called because of a white narrow band across its throat, is an especially dangerous snake. *E.P. Herald* 1.11.1911

I rushed out to see a hooded rinkhals rearing its head and swaying backwards and forwards *Star* 22.9.71

rissie ['rɪsï] *n. pl.* -s. A chilli or other hot pepper. [*Afk. fr. Malay via Du. ristjes* cayenne pepper]

Bosman found just over one hundred Afrikaans words with clear Malayo-Portuguese origins; . . . Those hundred words included many vivid glimpses of the orient: . . . *rissies* (red pepper) *sambalbroek* (wide trousers) . . . and such every day words as *baie* (many) and *nooi* (girl). Green *When Journey's Over* 1972

ritual murder, ritual killing *n. pl.* -s. Murder or killing practised among certain tribes in which parts of the body are used for ritual purposes, witchcraft, medicine or spells: see *liretlo*, also first quot. at *muti*.

Our chiefs are educated men who understand the vagaries of climate. They are not likely to resort to ritual murder to fill the ancient medicine horns. Fulton *Dark Side of Mercy* 1968

Nine people are now known to have been burned alive by angry mobs in a wave of ritual slaughter in Lebowa . . . All the people killed have been branded as witches or wizards by witchdoctors. *Sunday Times* 9.1.77

~ *er*, ~ *killer.*

'Ritual' killer sought. . . . policemen are hunting for a ritual-type murderer following the brutal killing of a woman on Tuesday. The mutilated body . . . was found on Wednesday . . . certain parts of her body had been cut out. *E.P. Herald* 27.9.74

rixdollar ['rɪksˌdɒlə] *n. pl.* -s *hist.* The monetary unit at the Cape first issued by the Dutch East India Company; see quots., also *schelling, skilling, stuiwer.* [*anglicization Du. rijks* imperial + *daalder cogn. Ger. Thaler*]

. . . the colonial paper rix-dollar of the Cape, first issued by the Dutch East India Company in 1781, was declared to be equal to forty-eight full weighted pennies of Holland, (about 4s sterling), and which, under all its fluctuations, had generally been considered to be its normal value. . . . The value of the rix-dollar gradually sunk in exchange, till in the year 1825 it appears to have reached its lowest point of depression, viz. below 1s.5d. Thompson *Travels II* 1827

Accounts are kept either in Pounds, Shillings, Pence and Farthings, or Rix-dollars, Skillings and Stivers.

1 Stiver equal to ¾ of a Penny
6 Stivers   2¼ Pence, or 1 Skilling
8 Skillings   18 Pence, or 1 Rix-dollar
Greig's Almanac 1833

. . . an early Cape shopkeeper kept his account in rix-dollars, skillings and stuivers, . . . In 1735 the value of a rix-dollar was 2½ guilden or 50 stuivers, but in 1770 it was changed to 48 stuivers equalling four English shillings. In 1806 the official value of the rix-dollar was two shillings or 24 stuivers, but by 1824 it was only worth 1s.6d., and it was decided to introduce British currency. Gordon-Brown *S. Afr. Heritage* 1965

road camp Also *Canad.*: see *National Road* and *padkamper*.

road, in the, out of the ~ *adv. pl. sub-standard*. In the way, out of the way. [*prob. fr. Afk. in die pad* in the way, *pad* (q.v.) road]

The old man told me that he was living with his son and his family but he was afraid of being 'in the road'. I didn't understand him for a moment – he meant he might be in the way. *O.I.* 1972

Robben Island *n. prop.* (*hist.*). An island in Table Bay, variously a penal colony, leper colony and mental asylum: now a place of detention for political prisoners hence ~ *er*. [*fr. Du. rob(ben)* seal(s)]

May 8th 1855 The Kaffir Chief Seyolo conveyed to Robben Island, his strange conduct at Wynberg – having, it was said, evidenced an aberration of the mind. *Cape of G.H. Almanac* 1856
General Infirmary Robben Island Lunatic keeper and Matron, J. Nutt and wife. *Cape Town Directory* 1866
One day in 1921 we flew across Table Bay and made history by landing on the beach at Robben Island. The island, at that time, was a leper and convict settlement. Green *Where Men Dream* 1945
And memories of six hard years on Robben Island, where he was sent after being found guilty. *Drum* 8.12.71
. . . at the installation of Chief Maqoma of the Amajingqi tribe at the Great Place of the first Chief Maqoma (The old chief died soon after he was exiled to Robben Island in 1874 by the Colonial Government) . . . 'There is one thing I wish to make clear' Mr. Sebe said, 'Robben Island was not introduced by the Afrikaner. It was here in the time of Lord Charles Somerset.' *E.P. Herald* 16.12.74

~ *stone n.* Bluish stone also called ~ *slate* used for paving and monumental purposes in the early days at the Cape.
J. Fitzpatrick, Stone Cutter . . . Begs leave

to inform the public that he has constantly on hand polished Robben Island stones of various sizes; and that he undertakes to cut out Inscriptions on Tomb stones at very moderate prices. *Cape of G.H. Almanac Advt.* 1841

robot [ˈrəʊbɒt, -bəʊ] *n. pl.* -s. Traffic light. cf. *Brit.* and *U.S. robot*, automaton. [*etym. dub. poss. because automatic*]

Between Cape Town and Port Elizabeth, if you take the direct route through Mossel Bay there are only four robots – two in George and two in Knysna. *Het Suid Western* 3.10.74

rock lobster *n.* The name under which S. Afr. *langouste*, see *crayfish* and *kreef* is marketed abroad: also known as *spiny* ~.

rock rabbit, *n. pl.* -s. Usu. *dassie* (q.v.) *Procavia capensis.*

I also shot several Dasses (or the Cape Hyrax) known by the name of Rock Rabbit in the Colony. Leyland *Adventures* 1866

rock(spider) *n. pl.* -s. *slang*. An offensive mode of reference to an Afrikaner, freq. in form *rock*: see also *hairy(back)*, *kransie, krans athlete, mealie cruncher, crunch(ie)* cf. *Brit. rock scorpion*, a resident of Gibraltar, usu. a policeman. *Austral. rockspider*, a mountain-climbing enthusiast, also a petty thief in a park. [*unknown, poss. fr. Brit. use*]

He said a professor at the university who abhorred any form of censorship took offence at the use of the word 'rock spider'. He gave the definition of 'rock spider' as a slang word used for an Afrikaans person. *Star* 11.4.73
The committee held that the terms 'rock-spider' and 'hairyback' . . . were derogatory of and maligned Afrikaners as a whole. It also found the words to be grossly racialistic and grossly insulting overall to the Afrikaans community. *Cape Times* 4.5.73

roer [ruːr, rʊə(r)] *n. pl.* -s. Old fashioned, heavy gun. [*Afk. fr. Du. roer* long-barrelled gun]

His gun or roer as the Dutchman calls it, is his never failing accompaniment. Philipps *Albany & Caffer-Land* 1827
She turned . . . hurrying straight to the rack where the guns were hung at the back of the *voorhuis*. They were all there: a great elephant *roer*, two old-fashioned muskets. Brett Young *City of Gold* 1940

In combination *Boer en(met) sy* ~, the prototype of the colonial Dutchman.

The new uniform is completely South African and even if it does not have any direct link-up with the keen-eyed, straight-shooting 'Boer en sy roer' it does have in it this element of

239

basic simplicity to which has been added the colour and gold braid so beloved of the great military powers of Europe. *Farmer's Weekly* 12.5.71

This was the roer that created a South African legend and the first image of that heroic figure Die Boer met sy roer. *Daily Dispatch* 22.7.72

roes(t) [rœst] *n. hist.* See *rust*. [*Du. roest* plant disease causing orange or red discolouration]

The Dutch call the blight which has destroyed the corn the roest or rust it begins about the time of its coming into ear. Hancock *Notebook* 9th Feb. 1826

roker [ˈruəkə(r)] *n. pl.* -s *colloq. lit.* 'smoker': one who smokes *dagga* (q.v.) also known as a '*boom* (q.v.) boy'. See *gerook*, also quots. at *goffel* and *skuif*. [*Afk. fr. Du. roken*, to smoke + *agent. suffix -er*]

In the garden at Bethany the 'dakka rookers', or smokers of the intoxicating and deleterious leaves of hemp, had an opportunity of filling their pouches. Alexander *Expedition II* 1838

He is not yet a confirmed 'roker' having smoked the stuff about ten or twelve times at random over a period of three years. *Drum* 27.8.67

rolbos [ˈrɔlˌbɔs] *n. pl.* -se. Also *roll-bush* applied usu. to *Salsola kali* the stems of which break off from the rootstock as the plant dies off, after which it rolls, sometimes at great speed over the veld in the wind: occ. a fodder crop, see second quot. *cf. U.S. tumbleweed, Austral. roly-poly (grass).* [*Afk. fr. Du. rollen cogn.* roll + *bos cogn.* bush]

. . . piled up against the fences the up-rooted rolbosse (tumbleweed) – that strange round mass of ashen-grey twigs, weightless, powerless, – driven hither and thither by the wind. *Cape Times* 30.7.73

. . . farmers are planting a variety of crops on dry lands as grazing for their small stock. On one farm a dense stand of rolbosse on 50 hectares of dryland easily carried 200 dorper ewes from lambing time until the lambs were weaned. *E.P. Herald* 9.9.74

roll-bush *n. pl.* -es. See *rolbos*.

Far and near the skeletons of roll-bushes careered over the veld. Pohl *Dawn and After* 1964

roman *n. pl.* ∅. Also *red roman*, a marine fish of the Sparidae, *Chrysoblephus laticeps*, called in Natal *daggerhead* poss. by confusion with *C. cristiceps* see *dageraad*: see also quot. at *allewêreld*. [*fr. Afk. rooi fr. Du. roode* red + man, *or poss. as in first quot.*]

. . . there is also one peculiar to Simons Bay called roman fish, from being first found near

a rock of that name in the entrance of the bay; it is of the size and shape of the silver fish, and of a beautiful deep rose colour. Ewart *Journal* 1811–14

In the piping days of plenty, the red roman was not rated among the finest of fish. Its name is a corruption from 'rooi man' or red man. To refer to it as 'red' is, therefore, just a piece of tautology. *Farmer's Weekly* 18.4.73

rondavel [ˌrɒnˈdɑvəl] *n. pl.* -s. Also *rondawel* (*Afk.* form): a circular house, usu. of one room with a conical thatched roof, resembling an African hut in shape, often a guest room, office etc. next to a farm house or holiday cottage, now prefabricated in kits in asbestos or steel and used as tool sheds, workshops or on *National road* (q.v.) camps: see also quots. at *zinc, kaya* and *peach pip floor.* [*etym. dub. poss. fr. Du. rondeel* round bastion *or fr. rond cogn.* round + *Malay de wala* wall, *poss. Port. roda* ring, wheel + *vallo* wall]

Her home was made of two rondavels (the circular one-roomed huts with a conical thatched roof which the white man has adapted from the Native kraal) joined together by a small kitchen and bathroom. Reed *South of Suez* 1950

COTTAGES, Rondavels available, Blue Sea Cottages, Ifafa Beach, Natal. *Farmer's Weekly Advt.* 21.4.72

deriv: longdavel (q.v.) and *rare, squaredavel.*

rond(e) [ˈrɒnd(ə), ˈrɔn-] *adj.* Round, circular: found in S. Afr. place names in *inflected attrib.* form: Rondebosch, Rondevlei.

rond(e)ganger *hist.* A watchman or roundsman whose job was to sound the hour in the Castle at Cape Town in the early days: see first quot. at *rondloper.* [*Afk.* roundsman, watch]

During September, 1716, a 'rondeganger' or roundsman left the guard room to sound the hour . . . the bell did not sound, nor did the man return . . . another was sent to enquire . . . and as he climbed the ladder he received a slap on his cheek from a hand which swung in mid-air . . . The missing man had hanged himself. Laidler *Tavern of Ocean* 1926

ronderib [ˈrɒndəˌrɪb] *n. pl.* -s. *Africander* (q.v.) sheep also called *blinkhaar* (q.v.) one of the original, indigenous fattailed sheep so called from its wellsprung ribs. [*Afk. fr. Du. rond(e)* round + *rib*]

The Ronderib is . . . a hardy breed and thrives under semi-arid conditions . . . Though distinctly heavier than the Persian, the Ronderib also may not be regarded as a desirable mutton breed for export purposes . . . there is an undesirable localization and abundance of fat. *H'book for Farmers* 1937

Africanders FOR SALE . . . Ronderib Blinkhaar Africander rams R30 and studs R75. Also young ewes . . . The mothers of these sheep weigh up to 180lb. live weight. *Farmer's Weekly Advt.* 20.3.74

rondloper ['rɔnt₁luəpə(r)] *n. pl.* -s. A gadabout, tramp, hobo, or one with wanderlust: also *rondganger*, and occ. *vb. rondloop*(ing) *cf. Austral. swagman, U.S. bum, Canad. pack-sack citizen.* [*Afk. rond cogn.* round + *loop fr. Du. loopen* to go + *agent. suffix -er*]

. . . the old bell which tolled the hours, and where story has it, a rondganger hanged himself for love of a secunde's daughter, still hangs there. Vaughan *Last of Sunlit Years* 1969

I got to know this country well when I was a child because my uncles and aunt were great rondlopers and holidaymakers and would pack up and go at the drop of a hat. *Sunday Times* 7.11.71

rood [ru:d] *n. pl.* -s. Dutch linear measure equal to 12.396 feet (3.78 metres) formerly used in S.A., also *square* ~ Dutch land measure 148.752 square feet (14.08 square metres), now obsolete: see *metrication.* [*Du. measure*]

roode- [ruəde] *n. or adj.* Red: see also *rooi:* Du. prefix found in names of red wines, and older forms of plant, bird and animal names e.g. *roodebekkie,* see *rooibekkie; roodebok,* see *rooibok; roode els,* see *rooi els:* also in place names Roodekrantz, Roodepoort, Roodebank. [*Du. rood* red + *inflect. suffix -e*]

roof See *roofie, rowe-* and quots. at *blougat* and *blouie.*

roofie ['ruəfi] *n. pl.* -s. National Serviceman of the newest intake: see quots. at *blouie* and *blougat* [*unknown, poss. rel. Afk. roof* scab]

rooi- [rɔi] *adj. prefix.* Red [*Afk. fr. Du. rood* red] **1.** Found in place names e.g. Rooiberg, Rooikraal, Rooipan, Rooiwal.

2. *prefix* to names of over 120 plant species e.g. ~ *afrikaner* (q.v.) see *aandblom,* also *rhebokblom; ~boegoe, Diosma rubra,* see *buchu; ~gras* (q.v.).

3. *prefix* to the names of various fauna, ~*kat* (q.v.), ~*meerkat* (see *meerkat*), ~*vink* (q.v.), ~*valkie,* see *valkie* etc.:

also ~*bekkie,* ~*bok,* ~*lip* all (q.v.).

rooi aas ['rɔi ₁as]. See *red bait.* [*Afk. rooi* red + *aas* bait, carrion]

He had been left undisturbed, eking out a simple healthy existence on a tiny pension augmented by what he caught in the sea or . . . by collecting bait in advance for a few of the regular anglers who came down to collect their rooi aas, mussels or chokka from him. Dodd *S. Afr. Stories* (no date)

rooibaadjie ['rɔi₁baikĭ,-cĭ] *n. pl.* -s **1.** Redcoat: a British soldier: also *Canad.* [*Afk. rooi fr. Du. rood* red + *baadjie fr. Malay baju* jacket]

There were not more than 50 Kaffirs guarding them and as soon as they caught sight of us they raised the cry 'Rooi Badjies' – Red Coats – and away they went as fast as their legs would carry them. Buck Adams *Narrative* 1884

The time was just after the first Anglo-Boer War of 1881. At that time Kipling's Private Thomas A. was the sturdy Victorian redcoat . . . 'a rooibaadjie' in South Africa. *Daily Dispatch* 18.2.71

2. 'Red-Jacket': a locust at the *voetganger* (q.v.) stage. ⟨Also acc. Pettman *Acridium purpuriferum* a large red and green locust. [*fr. colour*]

. . . you see the very earth become alive with diminutive insects, which develop . . . from day to day, first into a moving mass of black minutiæ, and from that increasing in size and becoming the colour of the brightest red. At this stage they are called the Rooi baatyes or red soldiers. Bisset *Sport and War* 1875

rooibekkie ['rɔi₁bekĭ] *n. pl.* -s. The common waxbill *Estrilda astrild:* the name is also given to one of the widow birds *Vidua principalis.* [*Afk. rooi fr. Du. rood* red + *bek* cogn. beak + *dimin. suffix -kie*]

I will try to bring home some cages of birds – Cape canaries and 'roodebekjes' (red bills), darling little things. Duff Gordon *Letters* 1861–62

rooibessie ['rɔi₁besĭ] *n. pl.* -s. Any of several species of *Olinia* esp. *O. capensis:* see *hard pear, pear:* and *O. cymosa,* large forest trees so called after their pinkish-red fruits. [*Afk. fr. Du. rood* red + *bessie* berry]

rooibessie(boom) . . . the timber of *O. cymosa* is valued on account of its durability and was formerly largely used for fencing posts and waggon wood, for railway sleepers and telephone poles. Smith *S. Afr. Plants* 1966

rooiblaar ['rɔi₁bla:(r)] *n.* See *rooibos.*

rooiblom ['rɔi₁blɔm] *n.* Also occ. *rooibossie: Striga asiatica,* also *S. elegans*

and *S. forbesii*, an annual plant with
bright scarlet flowers, parasitic upon
the roots of mealies and other cultivated
and wild grasses, also known as *witchweed* and *mieliegif* (poison) [*Afk. rooi
fr. Du. rood* red + *blom cogn.* bloom]
    She was like a weed that spread over the
land; like the Rooiblom that was so beautiful,
but which lived by sucking the sap from the
mealies. Cloete *Watch for Dawn* 1939
rooibok [ˈrɔɪˌbɔ̌k] *n. pl.* ∅. See *impala*.
[*Afk. rooi fr. Du. rood* red + *bok cogn.*
buck]
    On the banks of this stream I observed a
species of antelope, that I had not previously
seen. It is called by the Bechuanas *Paala*,
and Mr. Burchell has described it under the
name of the red buck. Thompson *Travels I*
1827
    Antilope Melampus. The Pallah. Rooye-bok
of the Cape Colonists. Harris *Wild Sports* 1839
rooibos [ˈrɔɪˌbɔ̌s] *n.* Also *rooi blaar*
(leaf): a timber tree, any of several
species of *Combretum* esp. *C. apicula-
tum*, also applied to other unrelated
species. [*Afk. rooi fr. Du. rood* red +
bos cogn. bush]
    'rooibos: . . . in the case of species of *Com-
bretum* the vernacular name is generally applied
from the colouration of the fruits when these
ripen, or in some, the autumn colouration of
the leaves. Smith *S. Afr. Plants* 1966
rooibostee, rooibos tea [ˈrɔɪˌbɔ̌sˈtɪə, -ˈtiː] *n.*
The dried leaves of *Aspalanthus con-
taminata* (*corymbosa*) or *A. cedarbergen-
sis* or the tea made from them, thought
to have tonic and other medicinal properties: see first quot. ⫿The red colour
is developed during the sweating process. See quot. at *Kaffir tea²*. [*Afk.
rooi fr. Du. rood* red + *bos(ch) cogn.*
bush + *tee* tea]
    . . . the amazing discovery that Rooibosch
tea alleviated milk allergies in infants . . . so
important that it made front page news
throughout South Africa, and was even noted
overseas. *Personality* 26.1.73
    Ever heard of iced Rooi (red) tea, Rooi tea
cool drink, Rooi tea 'Collins' or Rooi tea
Rhine wine? Or do you know it only as Rooibos tea, the tea with the distinctive taste? This
plant, known as Rooibos tea or Rooi tea. . . .
was cultivated by the indigenous populations
from the earliest times. *Panorama* Apr. 1973
rooi els [ˌrɔɪ ˈels] *n. Cunonia capensis*, a
large evergreen forest tree or its red-
coloured timber: which, being moisture
resistant, was formerly used in the construction of water mills: also *witels* (q.v.)

see quot. at *pear* (wood). [*Afk. rooi fr.
Du. rood* red + *els* alder]
    rooi els . . . the vernacular name is derived
from the colour of the wood and the resemblance to the European alder. Smith *S. Afr.
Plants* 1966
rooigevaar [ˈrɔɪxəˌfɑː(r)] *n. colloq. lit.*
'Red-peril' a reference to Communism.
cf. swartgevaar (q.v.) also *Brit.* and *U.S.
yellow-peril*. [*analg. swartgevaar* (q.v.)]
    . . . predicted the three 'emotional' fronts
the National Party would employ were 'Khakigevaar, swartgevaar and rooigevaar.' *Argus*
16.9.72
rooigras [ˈrɔɪˌxras] *n. and n. modifier*. Also
redgrass (q.v.) any of several species
are so called, esp. species of *Themeda*,
T. triandra and *T. burchellii*, excellent
densely foliaged pasture grasses: as modifier ~ *hay*, ~ *veld*, ~ *pastures;* see
quot. at *steekgras*. [*Afk. rooi fr. Du.
rood* red + *gras cogn.* grass]
    A farmer, for example, having steekgras
dominating his veld cannot follow the same
system of grazing as the farmer who has rooigras dominant. *H'book for Farmers* 1937
    . . . ideal ranching veld of high-carrying
capacity, mainly rooigras, sheltered kloofs
and good hunting. *Daily Dispatch Advt.* 22.7.72
rooihout [ˈrɔɪˌhəʊt] *n.* Any of several
species with reddish timber: *Phyllogeiton
zeyheri*, see quot.; *Trichocladus grandi-
florus*, *Erythrophloeum suaveolens* and
Ochna arborea or Cape plane, see quot.
[*Afk. rooi fr. Du. rood* red + *hout*
timber]
    'rooihout: . . . Heartwood pink to crimson
and of a remarkable bright red when dry, but
darkening with oiling. . . . excellent for work
requiring great strength, being chiefly employed
for yokes, felloes, and knobkerries . . . one
of the hardest of South African woods, even
more than *Umzimbiet*. Smith *S. Afr. Plants* 1966
    No self-respecting woodcutter would have
the handle of his axe made from any timber
other than 'rooihout' (Cape plane) a reddish
close grained wood. *E.P. Herald* 28.5.73
rooikat¹ [ˈrɔɪˌkat] *n. pl.* -s ∅. The lynx,
Felis caracal, a fierce and troublesome
predator among sheep etc., see quot. at
ride on water. [*Afk. rooi fr. Du. rood*
red + *kat cogn.* cat]
    Of vermin, a large variety existed, especially
the jackal, . . . small tiger-spotted cat, 'rooikat'
(lynx), hyena, wild dog, and lastly the aardvark
(antbear). Jackson *Trader on Veld* 1958
rooikat² *n.* also *lynx:* a multi-barrelled
multi-calibred S. Afr. manufactured
hand gun.

Various versions of the Lynx/Rooikat showing different barrel lengths and several finishes for the local market. (Caption) *Farmer's Weekly* 27.6.79

rooikeurtjie [ˈrũĭ‚kĭœ(r)kĭ‚-cĭ] *n. Sesbania punicea*, red sesbania an alien plant of S. American origin, now a *proclaimed weed* (q.v.) [*Afk. rooi* red + *keurtjie* (a wild sweet-pea) *keur* choice, –*tjie* dimin. *suffix*)].

In the Government Gazette of 9 March Sesbania punicea (red sesbania or 'rooikeurtjie') was declared a weed throughout the Republic in terms of the Weeds Act of 1937. *Evening Post* 28.4.79

rooikrans (willow) [ˈrɔɪ‚krăns-] *n. Acacia cyclops:* a variety introduced from Australia as a sand binder: a useful fodder plant, but regarded by many as a plant pest encroaching upon natural bush. [*Afk. rooi fr. Du. rood* red + *krans* wreath, garland; *fr. red aril surrounding the seed*]

. . . Port Jackson willow (Acacia cyanophylla) and Rooikrans (Acacia cyclops) can be a menace by ousting natural vegetation in some areas of high rainfall . . . In the . . . low rainfall area . . . where grazing control is a delicate matter and droughts frequent . . . the Rooikrans flourishes and is a useful fodder and shade tree as well as an essential tool in stabilising shifting sand dunes. *Farmer's Weekly* 4.12.74

rooilip [ˈrɔɪ‚lɪp] *n. pl.* -s. *Crotaphopeltis hotamboeia:* also *herald snake*, a poisonous Afr. snake named for its red upper lip: a back-fanged variety. [*Afk. rooi fr. Du. rood* red + *lip*]

rooinek [ˈrɔɪ‚nek] *n. pl.* -s. An Afrikaner or Boer name for an Englishman, originally on account of the susceptibility of the British troops to severe sunburn in S.A., formerly derogatory: see also *khaki*, now any Englishman, *cf. Austral. pommy* etc. [*Afk. rooi fr. Du. rood* red + *nek cogn.* neck]

Often the Boers . . . chuckled at the incredible behaviour of the Rooineks (red necks) – a term ever after applied to the English because their untanned skins often burned red in the brilliant South African sun. *Panorama* July 1970

'Oh, I am proud and so is my wife.' . . . 'It's quite a thing being appointed commodore of a South African fleet, especially as I'm a rooinek.' *E.P. Herald* 8.9.73

As a mode of address occ. *verdomde* (q.v.) ~ and as modifier as in 'he married a ~ woman' etc.

Only a few of the most bitter mumbled that this was judgment – that God had taken this young Boer maid to his bosom rather than see her lie in wedlock with this rooinek soldier. Cloete *Rags of Glory* 1963

rooi tea See *rooibostee, rooibos tea.*

rooivink [ˈrɔɪ‚fɪŋk] *n. pl.* -s. *Euplectes orix*, the red bishop bird, also called *kaffir finch* (q.v.) of which there are several sub-species: the males have bright puffed-out scarlet plumage in the breeding season. [*Afk. rooi fr. Du. rood* red + *vink cogn.* finch]

Roomse gevaar [ˈruəmsə xəˈfɑ:(r)] *n. prop.* Occ. used of the Roman Catholic Church and/or its influence in S.A.: see also *swartgevaar, rooigevaar. cf. Brit.* and *U.S. yellow peril.* [*analg. swartgevaar* (q.v.) *Roomse* Roman + *gevaar* danger]

I stood there with childhood memories of 'Roomse Gevaar' still echoing albeit softly, through my mind, with my arms linked . . . with Catholic Pentecostals, Anglicans, nuns and laity. Willie Marais *cit.* Crozier Nov. 1975

roosterkoek [ˈruəstə(r)‚kŭk] *n. pl.* -e (-ies). Dough cakes either leavened or unleavened, baked on a gridiron over a fire. [*Afk. fr. Du. rooster* grid(iron) + *koek cogn.* cake]

. . . they then filled their pockets with 'rooster koekies' (baked scones), took their overcoats, rifles, and bandoliers and climbed the embankment to the great steel bridge. Klein *Stagecoach Dust* 1937

Both Vetkoek and Roosterkoek were made either with unleavened dough or with a dough leavened with suurdeeg. Combine all ingredients to stiff dough and allow the dough to rise. Gerber *Cape Cookery* 1950

. . . my family's favourite braai side dish along with *Roosterkoek!*

. . . Roosterkoek. This traditional, untranslatable flat bread, baked on or over hot coals has as many variations as there are South Africans. Basically it's plain bread dough . . . *Darling* 1.2.78

rotel [‚rəuˈtel] *n.* Usu. a non-*classified* (see *classify²*) motel, therefore not entitled to the term, but which nevertheless offers accommodation and meals to travellers by car: also occ. what was formerly a 'private hotel', an unlicensed therefore unclassified boarding or rooming house. [*presum. fr. hotel poss. ro fr. road or room, analg. with* motel]

Fraser's Camp Rotel. Fraser's Camp, Cape Province Bathurst House Rotel Grahamstown.

roti *n. pl.* -s. Flat bread or *chupatti* cooked on a flat surface: see also *naan, Indian terms.* ⫸Known as salo(a)mi/e (*reg.* Cape Town) when stuffed with curry and rolled. [*unknown prob. Urdu or Hindi*]

Roti is unleavened bread made of either white flour or unsifted boermeal with a little shortening, but instead of being baked they are toasted over griddles until they are freckled gold. *Indian Delights ed.* Mayat 1961

roukoop ['rəu̯ˌkʊəp] *n.* Roman Dutch law term in law of sale: forfeiture of moneys deposited as part of the purchase price by a buyer who wishes to withdraw from a sale already agreed upon: see also *uitkoop.* [*Afk. fr. Du. rouw* regret + *koop* purchase]

... where a sum is given as a deposit in part payment of the purchase price can be clearly proved ... to be *arrha,* roukoop or earnest, it is forfeited to the vendor if the sale falls through due to default on the part of the purchaser, and forms a portion of the purchase price if the sale goes through. Belcher *Law of Sale in S.A.* 1961

row [rau̯] *vb. trns. slang.* Rebuke, reprove, among children.

... a boy laughed in church becos he was reading I must not marry with my grandmother ... and Mr Damp rowed him in front of all. Vaughan *Diary circa* 1902

rowe- ['rʊəvə-] *prefix. Sect. Army prefix deriv. fr. Roof*[*ie*] *usu.* derisive as in ~*cut* first, very short army haircut, ~*raal* newly promoted corporal, and ~*rant* a lieutenant who has just received his star. *cf. Austral. Army 'newly wed'* for a newly promoted serviceman.

R.S.A.[1] [ˌɑresˈeɪ] *n. prop.* Republic of South Africa, more freq. the *Republic* (q.v.), formerly called the *Union* (q.v.). [*acronym*]

The need for a single comprehensive and authoritative work of reference on the RSA has long been experienced, not only in South Africa, but particularly abroad. *Panorama* Nov. 1974

R.S.A.[2] *n. prop.* Radio South Africa. [*acronym*]

ruggerbugger ['rʌɡəˌbʌɡə] *n. pl.* -s. *slang.* A recognizable, aggressively masculine type fanatical about sport and usually partial to all-male gatherings.

The boy's father may have been in local parlance a ruggerbugger who did nothing more strenuous in the house than call for another beer. *Fair Lady* 1.10. 75
... says the trend for greater recognition of scholastic achievement ... still continues, 'There is far less of the 'ruggerbugger' hero worship now. Even our choice of prefects today reflects the growing academic influence.' *Retiring Headmaster cit. Grocotts Mail* 23.3.79

rugstring ['rœxˌstrɪŋ] *n.* See quot. The backbone of a carcass, usu. a sheep, with the meat on either side of it, is called the ~. ⫸ In the case of a buck the ~ is usu. the long fillet which stretches from neck to tail on either side of the spine and is the cut most favoured among many farmers' wives for making *biltong* (q.v.) [*Afk. rugstring* spinal column, backbone]

MODE OF CUTTING UP SHEEP ON A FARM
... Whereas the butcher divides the backbone into several divisions, the farmer only cuts off the ribs on each side. The whole part from neck ... to loin chump end remains as one piece and is called the backbone (rugstring) Because the marrow is present this piece does not keep and is used after the entrails. It is generally used for stewing. Higham *Household Cookery for SA* 1916 (1939 ed)

ruigte ·['rœix̯tə] *n.* General term for rushes, reeds and sedges which grow densely near water esp. standing water as in *vleis* or *pans,* or near sluggish streams, including *Cyperus textilis* (see *matjiesgoed*) also *palmiet:* found in place name Ruigtevlei. [*Afk. ruigte,* undergrowth *fr. adj. ruig* thickly grown]

Chopping out of the ruigte and drainage of the vleis checks its growth and spread. At Elsenburg, drainage and chopping out of ruigte and valueless bushes are the means of changing a useless vlei into a valuable pasture. *H'book for Farmers* 1937

Ruiterwag ['rœit̯ə(r)ˌvax] *n. prop.* See quot. also *Broederbond* [*Afk.* mounted guard, *ruiter* horseman + *wag* guard]

... former Springbok rugby captain and until recently chairman of the Ruiterwag – the junior Broederbond – has called for the selection of multi-racial South African cricket and rugby teams. *Sunday Times* 1.9.74

ruk[1] [rœk, rək] *n. and vb. colloq.* Tug or jerk as in 'Give it a ruk and it'll come loose' or 'Ruk it off, ruk it out man.' [*Afk. ruk* jerk *or* tug *n. and vb.*]

ruk[2] [rœk, rʌk] *vb. S. Afr. sp.* form of international rugby term *ruck* sig. to form a loose scrum. [*fr. ruck*]

Too Much Rukking in S.A. Rugby (headline) Informant Eric Rosenthal

'Rum Run' *n. Sect. Army acc. Paratus* (q.v.) the transport run of DC3 aircraft which carry supplies to men in outlying

parts of the *Operational Area* (q.v.).
[*fr. earlier term* ~*ing*, smuggling of
liquor]
[. . . the bases that have no suitable runways
for large transport aircraft are served by Dako-
as (DC3s) . . . on this run to the remote areas as
they can do up to ten take-offs and landings . . .]
Paratus Jan. 1979

rusbank [ˈrœsˌbaŋk] *n. pl.* -e, -s. A wooden,
open-armed settee without upholstery
often made of *stinkwood* (q.v.) usu. with
riempied (q.v.) back and seat or occ.
caned: see also *bankie* cf. *Scottish bunker*,
bench or chest used as a seat. [*Afk. rus
fr. Du. rust cogn.* rest + *bank* bench]
How alike all these Boer houses were. Each
had the same rough, home-made riempie-
seated rusbanks, the same beds, the same tables.
Cloete *Watch for Dawn* 1939
At all times the rusbank has been a 'Multiple
chair' in which the form was copied from the
single chair of the time. Atmore *Cape Furniture*
1965
Yellowwood and Stinkwood Rusbanke with
riempie seats. *Grocott's Mail Advt.* 19.9.72

rusper [ˈrœspə(r)] *n. pl.* -s. *Heliothis
obsoleta*, a striped, greenish to dark
velvety brown caterpillar destructive to
fruit, crops and (2) *veld* (q.v.) cf. *U.S.
bollworm*: see also *kar(r)oo caterpillar.*
[*Afk. rusper* caterpillar *fr. Du. rups*]
The grass cover is probably better than it
has been for 20 years . . . although there are
areas where the Karoo bush has still not recovered from the long dry spells and attack
from ruspers. *E.P. Herald* 16.2.74

Russians *n. usu. pl.* **1.** South *Sotho* (q.v.)
blanketed gangs who terrorized the
townships from the 1940s onwards.
Those Russian cases during the 40s and 50s
were the most difficult I ever had . . . Rival
Russian factions were reluctant to talk because
of reprisals. Elias Xaba *cit. Drum* June 1969
Three men were shot dead . . . when two
rival gangs of blanketed 'Russians' fought it
out yesterday. *Post* 24.10.71

2. Soweto term for S. *Sotho* (q.v.) people.
. . . a legendary 'Famo' or South Sotho
party. The South Sotho known in Soweto as
the 'Russians' because of their distinctive
blanket dress have built a reputation for wild
parties. *Sunday Times* 20.10.74
. . . her parents living the rural life and clinging to the old ways . . . 'He says my mother's
people were real Russians. That is what we call
the South Sotho people, because of the way they
wear their blankets. Venter *Soweto* 1977

rust¹ [rʌst] *n.* Non-S.A.E. Any of several
plant diseases manifesting in reddish

spots or colouration: *hist.* the blight
which destroyed successive crops of
the Settlers of 1820: see also quot. at
roest. [in S.A. fr. Du. roest (q.v.)]
The blight or rust, though also prevailing
here of late years, has never been so universal
or inveterate as in Albany and other tracts
along the sea-coast. Thompson *Travels II* 1827
They were successively afflicted by rust in their
wheat; by floods which swept away their houses
and crops. Alexander *Western Africa I* 1837

rus(t)² [rœs(t)] *n. abstr.* Rest: found
in place names e.g. Nelsrust, Oden-
daalsrus, Volksrust, Rust-en-*Vrede* (q.v.)
Rust-en-Vreugd(joy): also as *adj.* Rus-
tig. [*Du. rust cogn.* rest + -*ig adj.
forming suffix cogn.* -y]

ry versigtig [ˈreɪˌfə(r)ˈsɪxtɪx] *imp. vb. phr.*
Drive carefully: either a warning road
sign or a mode of bidding farewell.
cf. *hamba kahle, go garshly, mooi
loop,* cf. *Jam. Eng. drive good*, a farewell, also *walk good.* [*Afk. ry fr. Du.
rijden* to ride + *versigtig* carefully]

S

sa [sɑ] *interj. imp.* Used when setting
a dog on something or someone equiv.
of 'after him' cf. *U.S. sic'em.* [*Afk. fr.
C16 Du. sa, tsa, tza fr. Fr. ça, Xh.* (*poss.
fr. Afk.*) *ukuTsatsa* run quickly]
It was very well to punish the dogs, but what
was to happen to the owner of the dogs who
stood by urging them on and crying *tsaa?*
FitzPatrick *Transvaal from Within* 1900 cit.
Pettman
Then you must run. It will chase you too. Sa!
Fugard *Boesman & Lena* 1969

S.A.A.A.S *n. prop. S. Afr.* Association
for the Advancement of Science, usu.
known as S₂A₃ (q.v.). [*acronym*]

S₂A₃ [ˈesˌtuˈeɪˌθri] See *S.A.A.A.S.*

saaidam [ˈsaɪˌdam] *n.* As *n.* see second
quot.; also ~ *system,* a method of
irrigation by initial flooding: see quots.
[*Afk. fr Du. zaaijen* to sow + *dam*
dam, reservoir]
The so-called 'saaidam' system is practised.
Flood water is guided over level terraces surrounded by low walls, in which the water is
allowed to stand and soak into the soil. When the
soil has been sufficiently soaked the remaining
water is guided on to a lower terrace. As soon
as the soil is workable it is ploughed over and
sown. The crop then often grows to maturity
without further irrigation. *H'book for Farmers*
1937

The *saaidam* is simply a low embankment thrown across a flat valley or plain to delay the flow of flood water and ensure sufficient moisture in the soil for the germination of a crop. Green *Karoo* 1955

saamie ['sɑmĩ]. See *sarmy, sarmie.*

S.A.B.C. ['esˌeɪˌbi'si] *n. prop.* Also *Sabc*: *S. Afr.* Broadcasting Corporation. [*acronym*]

It is understood in London that the SABC has already invited tenders for its key television consulting service. *Daily Dispatch* 20.5.71

S.A.B.S. ['esˌeɪˌbi'es] *n. prop. S. Afr.* Bureau of *S*tandards: also ~ *mark*, a guarantee of quality or standard maintained: see quot. at *canopy.* [*acronym*]

In collaboration with the South African Bureau of Standards (SABS) it will also attempt to get an international compulsory hallmark, the SABS mark in connection with the hallmark of gold. *Panorama* Dec. 1973

S.A.D.F. ['esˌeɪˌdi'ef] *acronym.* South African Defence Force: see also *U.D.F.,* *P.F.*

The name 'South African Defence Force' describes our aims and intentions. We have no expansionist ambitions, but we will defend our country and all its peoples against any military threat. Lt. General C. L. Viljoen *cit.* Marks *S. Afr. Army Today* 1977

safari [sə'fɑrĩ] *n. pl.* -s. Not S.A.E. but in use in S.A. as elsewhere sig. a hunting trip usu. after big game: also a commercially organized long-distance sightseeing tour. In combination ~ *suit* popular light-weight washable outfit for summer wear consisting usu. of a short sleeved 'bush jacket' with short or long trousers. [*fr. Arab. safariy* of a journey (+ suit)]

Insistence on rigid dress on the campus had been replaced by permission for men to wear safari suits and women to wear slacks. *Evening Post* 8.5.71

Safari Suit a hit on Broadway. The American male has discovered the safari suit and the bush jacket. *E.P. Herald* 27.6.75

safe *adj. slang* esp. among children: poss. non-S.A.E. A regular term of approbation equiv. usu. of smart, good-looking etc. ⫽This may originate among surfers, baseballers or motorcyclists. [*unknown*]

I am not . . . a raving beauty, . . . but I'm not exactly Dracula neither. Give me a safe new ensemble and a new hairstyle and a drop of My Sin behind each ear, and I can slay the ou's with the rest of them. *Darling* 29.9.76

Safe-Ride Scheme See *Ride-Safe.*

saffra(an) ['sæfrən, sə'frɑːn] *n.* **1.** The hard yellow timber of *Cassine crocea* (*crocoxylon*) [*Du. zaffraan fr. yellow colouring cf. borrie* (q.v.)]

The other woods most in request, and found in Albany are – Red and White Milk, Red and White Else, Red and White Pear, Saffran. *Greig's Almanac* 1831

A typical wagon of the Great Trek period would have had . . . wheel falloes of hard pear or saffraan. *E.P. Herald* 28.5.73

2. The tree itself: a tall forest tree of 15m or more.

He points to a large pear tree, a Dutch saffraan, as the oldest inhabitant of the Cape Town gardens. Green *Grow Lovely* 1951

SAI [saɪ] *n. pl.* -s. *acronym.* South African Infantry: ⫽ When used referring to a specific unit, preceded by a numeral 1 ~, 6 ~ etc.: used also of a member of the S. Afr. Infantry: see *bokkies.*

We're not bloody S.A.I.s – we're the black berets*. *Armoured Corps. see *SSB. O.I. Serviceman* 6.4.79

Looking pleased to be back after a stint of duty in the operational area . . . The men of 6 SAI arrived in Port Elizabeth yesterday. *E.P. Herald* 11.6.79

sail *n. pl.* -s. Any tarpaulin or canvas sheet covering a wagon or other load etc., also in combination *buck* ~, a covering for a *buckwagon* (q.v.) *car* ~, *tent* ~ and ~*cloth*, canvas. [*prob. translit. Afk. seil* canvas, tarpaulin]

The wagon in which I was had seventy-two stabs in the *sail*. Bird *Annals of Natal* 1888 *cit.* Pettman

. . . took refuge on the second wagon, drawing a tent-sail over them. Haggard *Nada* 1895

Protect your car. New car sails at factory prices. *Farmer's Weekly Advt.* 3.1.68

sakabula [ˌsakə'bulə, -a] *n. pl.* -s. The long-tailed widow bird or 'flap' *Euplectes progne:* see quot. at *kaffir-finch.* [*Zu. iSakabuli* widow bird]

They wore upon their heads heavy black plumes of Sacaboola feathers like those which adorned our guides. Haggard *Solomon's Mines* 1886

. . . the scarlet-chested sunbirds, the turquoise-and-mauve rollers and the long-tailed black widow birds commonly known as sakabullas. *Evening Post* 28.4.73

Sakekamer, the ['sakəˌkamə(r)] *n. pl.* -s. An Afrikaans Chamber of Commerce. [*Afk. fr. Du. zaken* business, affairs, *saak* + *pl.* + *kamer cogn.* chamber *Lat. camera*]

The Bond has just as much right as the Sons of England or the Chambers of Industries or the Sakekamer to make representations to the Government. *Sunday Times* 24.9.72

sakkoffie ['sɑk(k)ɔfĭ] *n.* Coffee made by infusing ground coffee in a linen bag suspended in boiling water, or by pouring boiling water through it: see also second quot. at *kettle* [*Afk. sak* bag + *koffie cogn.* coffee]

... an excellent cup of coffee from steeping ... the basic principle in the old sakkoffie, when the grains are suspended in a fine linen pouch ... and boiling water poured over them ... Aunt Kate ... and the real thing: sakkoffie, great boeretroos in her Malvern kitchen. *Darling* 11.5.77

sala(ni) kahle [ˌsala(nĭ) ˈgaɬe] *imp. vb. phr.* Stay well: the reply to *hamba kahle* (q.v.) a farewell greeting to the person(s) staying behind: see also *stay well.* [*Ngu. -sala* stay + *pl. suffix -ni* + *kahle* well]

'I'm glad you are going to stay. Your God will look after you. Salani kahle ...' he quickly disappeared into the forest. *Voice* 11-17.9.78

salomi [sə'lɔumĭ] *n. pl.* -s. *prob. reg. Cape.* A *roti* (q.v.) spread with curry stuffing and rolled ⫿ The ~ is said to be of Malay origin, *poss. orig.* a trade name.

... a spread worthy of any rajah's table – a selection of curries, salomis, rotis, samoosas, salads – exotically presented. *Capetonian* May 1979

salted *partic.* Of or pertaining usu. to a horse which is immune to horse sickness or other disease through having had it and recovered from it: see quot. at *skimmel.* Also *U.S.* [*trans. Afk. gesout* immunized]

What is more, this lot were thoroughly 'salted', that is, they had worked all over South Africa, and so had become proof ... against red water, which so frequently destroys whole teams of oxen when they get on to strange 'veldt' or grass country. Haggard *Solomon's Mines* 1886

Hunters who went down to the Bushveld prized most of all a 'salted' horse which had worked up an immunity to the horse sickness that made summer hunting impossible. *Daily Dispatch* 22.7.72

Also in *neg.* form *un* ~.

Nagana is almost always fatal to unsalted horses. Stander *The Horse* 1968

sambal ['sæmbəl, 'sam-] *n. pl.* -s. A highly seasoned relish or chutney, usu. of raw vegetables and/or fruit with spices. [*fr. Malay sambal, sembal* condiment]

The spicy stew of meat and vegetables is enhanced by various *sambals* (condiments): sliced onion sprinkled with fine pounded chili, fresh grated quince mixed with pounded chili, and a highly seasoned *sambal* of mint leaves and chili pounded together and moistened with vinegar. Du Plessis *Cape Malays* 1944

Sambals or side dishes and strong Indian pickles complete the meal. Sambals: some of the following side dishes may be served with the main course: *Argus* 29.5.71

In combination ~ *broek*, wide trousers: see quot. at *rissie.*

sambok ['sam₁bɔk] *n. pl.* -s. See *sjambok.*

I ran to the major, to give him the assistance of a *sambok*, or rhinoceros-hide whip. Alexander *Western Africa I* 1837

sammy *n. pl.* -ies. **1.** An objectionable mode of reference to an Indian man: see quot. at (1) *Mary.*

2. *prob. reg.* Natal: An Indian fruit and vegetable hawker usu. with a van. ⫿Objectionable among certain groups, though prob. seldom intended as offensive. [*etym. dub.* (*poss. fr. swami*) *see quots.* at (1) *Mary* and *ramsammy*]

... different in temper and disposition from the shopkeepers ... and widely removed in looks from that other Indian tribe which the white man calls 'Sammies' and whose members sell fruit and vegetables to the white women. Lanham *Blanket Boy's Moon* 1953

samoosa [sə'musə] *n. pl.* -s. A fried Indian pastry, finely folded as a small triangular parcel, usu. containing spiced minced meat: in combination *mithai* ~ a ~ containing a sweetened preparation usu. of coconut or other nuts. ⫿Ready-prepared ~*s* are commercially available deep frozen, also cooked in certain cafés. Acc. some, ~*s* are Malay in origin. [*prob. Urdu/Gujarati, poss Malay* (*mithai* sweetmeat)]

Beginners and non-Indians are often put off attempting samoosas as they have heard that it is a laborious and complicated procedure. Admittedly it is time consuming ... Samoosas are the tastiest of savouries, made from the most economical dough ever devised ... Therefore the art of samoosa making should be a must for who-so-ever wishes to learn Indian cookery. *Indian Delights* ed. Mayat 1961

samp [sæmp] *n.* Also *stampmealies* (q.v.) maize kernels stamped and broken but not ground as fine as *mealie rice* or *mealie meal* (q.v.) cooked freq. with dried beans as a staple food among the Africans; known as *mngqusho*: see also *kaboe mealies. Canad.* and *U.S. samp.*

U.S. *hominy*. [*Narraganset nas(a)ump* 'coarse hominy or a boiled cereal made from it.' *Webster's Third International Dict.*]

Most African shops specialise in merchandise like paraffin, candles, bread, tea, coffee, sugar, milk, soap, samp, mealie meal, meal, fats and vegetables. *Daily Dispatch* 11.10.71
MAIZE PRODUCTS Samp 90kg . . . R5,46
Mealie Rice . . . R4,98½
Farmer's Weekly Advt. 21.4.72

In the meantime the men [Transkei diplomats] say they don't like American food and both families are missing their mngqusho, or samp. *Argus* 8.11.75

sandloper ['sant₁luəpə(r)] See (2) *strandloper.*

sandveld ['sant₁felt] *n. prop.* **1.** A geographical area characterized by light sandy soil: see quot. at ~ *grainworm*: [⌐*r. sandy soil + veld* terrain, area]

Most of the rye is grown in the sandveld north of the Berg river. In the sandveld the production of a grain crop is not easy – the soil is a light sand, not particularly fertile, which blows readily. *H'book for Farmers* 1937

2. *prefix* in several plant names sig. a preference for this habitat, esp. ~ *lelie* the pink (2) *Afrikaner* or *sandpypie* (q.v.) ; ~ *bitteraar(bossie)* any of several species of *Oligomeris* see *bitterbos;* ~ *kapokbossie,* either of two species of *Eriocephelus :* also in ~ *chair* (q.v.), and ~ *grainworm* (q.v.)

sandveld chair *n. pl.* -s. Very rare S. Afr. chair made in the *sandveld* (q.v.) area of lemon and other woods indigenous to it: see quot. at *karreehout.*

. . . complemented by the 30 'Sandveld' chairs in the room. These chairs are made of wild lemon wood, and no more than approximately 500 are still extant in South Africa. *Panorama* Mar. 1974

sandveld grainworm *n. pl.* -s. *Apophylia duvivieri:* an insect pest of winter grain crops in the *sandveld* (q.v.).

The sandveld grainworm . . . occurs in the sandveld only, the infestation being particularly severe in certain sandy areas such as Saldanha, Vredenburg, Velddrift and Graafwater where it causes extensive damage to grain lands annually. *H'book for Farmers* 1937

sandvygie ['sant₁feɪxĭ] *n. pl.* -s. *Dorotheanthus bellidiformis,* the *Bokbaai vygie* see *vygie.* [*Afk.* sand + *vy fr. Du.* vijg fig + *dimin. suffix* -(g)ie]

sand vygie . . . generally and more commonly known as *Bokbaai vygie* from the little bay where the species grows very commonly. Smith

S. Afr. Plants 1966

sandworm *n. pl.* -s. *Tunga penetrans :* see *jigger U.S. chigoe,* also *jigger.*

sangoma [₁saŋ'gɔma] *n. pl.* -s. An African *witchdoctor* (q.v.), usu. a woman often claiming supernatural powers of divination: see quot. ; also *smeller out* (q.v.) of witches: see *twasa, isanusi* and quots. at *muti* and *smell out.* [*Ngu. iSangoma, pl. itangoma,* a diviner, usu. a woman]

Banquo's ghost becomes the Tokoloshe (a Zulu evil spirit), the witches are three bone-throwing Sangomas (witchdoctors). *Sunday Times* 5.4.72

Several sangomas declared that they had indeed helped solve some murders that have puzzled the Police and other local authorities in the losses of articles and missing persons . . . is a fully fledged sangoma in her class. With her were three student-sangomas. *Bona* Mar. 1974

In combination ~ *land*

. . . the police eventually called in sangomas and witchdoctors to augment them . . . First the local sangomas . . . Then came others older and more experienced . . . Then the police asked for 'Mpapane', . . . the most powerful sangoma in the Transvaal. He came in . . . armed with bones, flywhisks, all the blessings from sangomaland and a bold heart as tall as the mountains of his native village. *Pace* Dec. 1978

sanna ['sana] *n. pl.* -s. Early flintlock musket: *cf. Brown Bess, Big Bertha, Long Tom* etc. [*abbr. ousanna Afk.* ou, old *(su)sanna*]

. . . the old-fashioned Sanna (Susan), the flint-lock musket which we used alike against Kaffir and wild beast. Blore *Imperial Light Horseman* 1900

We had to spend a good bit of money on defence, in those days. Gunpowder and lead, and oil to make the springs of our old sanna's work more smoothly. Bosman *At his Best* 1965

Sanna 77 *n.* A semi-automatic 9 mm hand carbine, based on the design of the Czech Vz25, modified and manufactured in S.A. ℙ The ~ was used earlier in its fully automatic version in Rhodesia, manufactured in Salisbury, and known as the GM-15: see also *rhogun* and quot. at *U(z)zi* [*presum. fr. orig. Sanna* (q.v.)]

The Sanna 77, said to be the first fully South African made, semi-automatic hand machine-gun, has been launched in Durban. *S.A. Digest* 29.9.78

The latest in . . . South African weaponry is the Sanna 77 – a light, handy weapon . . . and the basic design could be adapted to provide rapid fire . . . this weapon could prove to be one of the farmer's best friends. . . . The manufacturers of the Sanna maintain that the . . . fixed striker blowback operated gun has an effective range of 300 m. But considering it fires a 9 mm round I would regard 200 m as an optimum range. *Farmer's Weekly* 21.3.79

Santas [ˈsæntəz] *pl. n. Sect. Army slang abbr.* 'Santa Marias', voluminous army-issue underpants [*poss. fr.* whiteness or *fr.* great modesty: *unknown*].

Santa Marias – army issue underpants, usually hang down around the knees. H. C. Davies *Letter* 2.2.79

We all get our Santas – they have to be folded in a perfect rectangle for inspection – but nobody wears them – they come up to here! *Serviceman* 10.3.79

S.A.P. [ˈesˌeɪˈpiː] *n. prop.* South African Police: see also *S.B.* [*acronym*]

If we weren't such respectable people, respectable farmers and so on, Johnny Coen said, we wouldn't mind if even a dozen mounted policeman . . . came marching into our voorkamers with S.A.P. on their shoulders. Bosman. *Bekkersdal Marathon* 1971

Sap [sap] *n. usu. capitalized pl.* -pe. The South African Party, later *United Party* (q.v.) or one of its members: see quot. at *smelter.* ⫿The name South African Party (SAP) has been revived as that of a new party formed in August 1977 [*acronym S.A. Party*]

He says it is obvious I am a Sap (United Party Supporter) and many Sappe are good guys. *Cape Times* 3.7.71

In combinations: *Nat and* ~ ; see also quot. at *Nat.*

In other words . . . non-consensus is more 'on' among Sap and Nat rank and file than consensus itself is. *Argus* 21.4.73

bloed [blŭt] ~ a staunch, ultra-conservative member or supporter of the United Party usu. of long standing. [*Afk. bloed cogn.* blood]

Another startling claim by . . . is that some UP verkramptes (he calls them 'Bloedsappe') are getting ready to leave the United Party. *Sunday Times* 30.7.72

It is interesting to recall that some verkrampte Nationalists, who pose now as super Afrikaners, were once 'bittereinder bloedsappe . . .' Another super Afrikaner these days is . . . In the 1950s he, too, was a Sap. *Ibid.* 27.10.74

S.A.R. [ˈesˌeɪˈɑː] *n. prop.* South African Railways ; *Afk.* form *S.A.S.* (*spoorweë* railways). [*acronym*]

'Then it couldn't have been in the Union,' Chris Welman shouted out, trying to be *really* funny. 'You couldn't have been travelling on the S.A.R.' Bosman *Jurie Steyn's P.O.* 1971

sarmie, sarmy [ˈsɑːmĭ] *n. pl.* -s. *colloq.* A sandwich, esp. among children. *cf. Austral. sang.* [*unknown poss. fr. Afk.*

saam, together]

. . . North of England . . . renowned for a delicacy known as the chip butty (sort of sarmy with chips in the middle). *Darling* 12.3.75

sasa(a)tie [səˈsɑtĭ] *n. pl.* -s. Also *sassaa(r)-tjie :* see also *sosatie* and quot. at *sea cat.* [*Afk. sosatie* (q.v.) *fr. Du. sasaatje*]

There existed in those days what we termed 'Sasaatje and Rice' houses, places where a favourite Dutch dish called 'Sasaatjes' was served in the evenings. . . . two sassaatjes (diamond-shaped inch-sized pieces of mutton, curried and about half a dozen stuck upon a bamboo skewer, and then roasted upon a gridiron). Meurant *Sixty Yrs Ago* 1885

SASO [ˈsæsəʊ] *n. prop.* The South African Students' Organization: see also quot. at *Azania.* [*acronym*]

After Sharpeville in 1960 the Government drove the ANC and PAC underground and tried to force all blacks to accept ethnic separation. The mid-sixties were disastrous years for black nationalism, but with the founding of SASO in 1967 younger blacks took a new direction. *Oppidan* Oct. 1976

sassaby, sassabi [ˈsæsəˌbĭ, ˈsas-] *n. pl.* -s, ∅. *Damaliscus lunatus :* also *tsessebe :* see *hartebeest.* [*Tswana tshêsêbe*]

satansbos [ˈsɑtansˌbɔs] *n. Solanum eleagnifolium,* named ~ by R. H. Murray of Kendrew: the silver-leafed bitterappel, a proclaimed noxious weed, which being deeply rooted, infests irrigable land and chokes other plant growth: ⫿Originally from the Southern *U.S.* the *white (silver) horse nettle,* and S. America *trompillo.* Informant E. Brink [*Afk. satans* Satan's, devil's + *bos fr. Du. bosch cogn.* bush]

Satansbos . . . has taken over valuable irrigable land in the Kendrew ward . . . Satansbos was introduced to South Africa in pig food which was fed to animals at Kendrew and . . . has spread to all four provinces and South West Africa. *E.P. Herald* 18.4.75

Satan's dung *n. prop.* Asafoetida, see *dui-welsdrek* one of the (*Old*) (q.v.) *Dutch Medicines* (q.v.) [*fr. duiwelsdrek* (q.v.)]

The 68-year-old Coloured mystic went . . . with a prominent . . . business man after an appeal for help from the minister of the 'bewitched' congregation. He said that he found eight bottles of Satan's dung (duiwelsdrek) hidden in the roof of the manse. Three more bottles of the potion were unearthed in the Church building. Unity and harmony in the congregation were restored after the discoveries. *Sunday Times* 4.5.69

Saterdagaand-se-kind [ˈsɑtə(r)ˌdaxˈaːntsə-ˈkĭnt] *slang.* An illegitimate child: [*Afk.*

Saterdag Saturday, *aand* evening, *se possessive suffix, kind* child].
. . . the very expressive language . . . Coloured people have at their command. Within a few seconds I was given half a dozen words or expressions for 'illegitimate child' . . . best of all *Saterdagaand se kind.* Van Heyningen & Berthoud *Krige* 1966

S.A.W.A.S. [ˈsɑwəz] *acronym South African Women's Auxiliary Services:* a voluntary organization of women all over the country, usu. housewives, who provided hospitality and recreation for troops in most of the towns and cities of S.A. during World War II.
The S.A.W.A.S. memorable Christmas hospitality in 1941 to the Battalion in Windhoek was remarkable . . . S.A.W.A.S. spared no pains in catering and caring for the troops, whether in sickness or in health. 'Tweede in Bevel' *Piet Kolonel* 1944

S.B. [ˌesˈbi] *n. prop. Special Branch* of the S.A. Police, see *S.A.P.* for the investigation of political as opposed to criminal or civil matters: see also *B.O.S.S.* [*acronym*]
TOP COP WARNS OF FAKE S.B. MEN South Africa's top policeman, . . . this week warned the Labour Party to beware of people who might be posing as Security Branch cops . . . if there was any doubt about the identity of the S.B. men, anyone had the right to demand his name, number, rank and station. *E.Cape Post* 22.6.69

S.C. *acronym. Senior Consultus*, equivalent of the letters patent of K.C. or Q.C., granted to *advocates* (q.v.) of the the Supreme Court of South Africa. ℙ The right to elect was granted to those who had already taken silk before the referendum of 1960, following the precedent set by the Republic of Eire: there are thus still Q.C.s in S.A. see quot. [*Lat. senior consultus* (therefore correct in *abbr.* in both *Eng.* and *Afk.*) not *Senior Counsel*]
The hearing of the application brought by Mr . . ., S.C. and Mr . . ., S.C., lasted four days. One of the documents . . . handed in by Mr . . ., Q.C. . . . was a copy of the judgment. *E.P. Herald* 17.3.79.

scaf [skaf] See *skaf.*

scale *n. pl.* -s. A drinking vessel, usu. of plastic, holding about 600ml, used as a measure for *Bantu beer, KB, tshwala* all (q.v.) both in beer halls and the *shebeen* (q.v.) trade: see *skaal,* also quot. at *ai-ai.* [*fr. Eng.* scale a drinking bowl or cup,

O.E. scealu, Du. schaal, Ger. Schale cogn. shell]
Old Ma Plank sat over a huge vat in the yard and doled out scales of beer and collected shillings in return . . . Xuma put the scale to his lips. Abrahams *Mine Boy* 1946
. . . a year-old baby girl greedily watched her mother downing a scale of 'bantu beer' . . . the baby began to cry. The mother got the message and passed the scale somewhat grudgingly, saying: 'Here, drink'. *Drum* 22.9.73

scandal *vb. intrns. substandard.* Gossip, speak ill of someone: see *skinder.* [*prob. translit. Afk.* skinder gossip *fr. Du.* schenden to slander, revile]
You don't protect me you know. You let them scandal about me. Fugard *People Living There* 1969

scandal stories *n. phr. substandard.* Gossip: see *scandal* also *skinder.* [*translit.* skinderstories (q.v.)]
Scandal stories . . . tells financier
. . . rejected recently published attacks alleging that the . . . chairman had not levelled constructive criticism but had indulged in 'scandal stories.' *Cape Times* 5.2.77

scare *vb. trns. colloq.* Usu. among campers or caravaners sig. to cook meat partially to prevent its becoming tainted. [*poss. fr. Afk.* skrikmaak, to scare (*metaphorical to undercook*)]
It is inadvisable to 'scare' the meat in an attempt to keep it fresh. Article on Camping. Date and paper unknown.

schans [skans] *n. pl.* -en, -es. Also *skans* (q.v.) and *schanz :* barricade or fortification of stones, earth and even bushes used as cover from which to fire upon the enemy, esp. in the *Anglo-Boer War* (q.v.). [*Afk.* skans *fr. Du.* schans breastwork]
There are still . . . some signs of the schanzes, earthworks, hastily thrown up when General Smuts and his commandos approached the district. *Grocott's Mail* 28.3.69
In combination *wind* ~ a windbreak planted for the protection of vines and other crops. *cf. Austral.* breakwind.
The benefits of shelter to young vines have also been shown by planting narrow strips of common rye, as wind-schanzen, about 25 yards apart, at right angles to the line of the prevailing winds. Wallace *Farming Industries* 1896

schelling [ˈʃelɪŋ, -ɪŋ] *n. pl.* -s. *hist.* Also *skilling* (q.v.) a unit of currency worth six *stuiwers* (q.v.) or one eighth of a *rixdollar* (q.v.): see quot. at *rixdollar.* [*Du.* schelling cogn. shilling]
The only money in general circulation, is

small printed and counter-signed pieces of paper, bearing value from the triilng sum of one schelling, or sixpence currency, upwards to five hundred rix-dollars each. *Burchell Travels I* 1822

schelm See *skelm*.

schepel ['skepəl, skɪə-] *n. pl.* -s. An old Dutch dry measure by volume consisting of a box 368.3 x 368.3 x 215.5 mm (14½ x 14½ x 8½ inches) used for grain, equal to ¼ *muid* (q.v.) [*Du. cogn. Eng. skep*, coarse round basket, *skepful* amount held by a skep: *also Old Norse skeppa* bushel]

> CORN MEASURE, 4 Schepels, equal to 1 muid. 10 Muids equal to 1 Load. The Muid of Wheat weighs on an average about 180 lbs. Dutch, being somewhat over 196 lbs. English. *Greig's Almanac* 1833

scherm. See *skerm*.

schimmel. See *skimmel*.

schlenter ['ʃlentə(r), slen-] *n. pl.* -s. **1.** *Sect. Mining.* A fake diamond sold as genuine, or used either for detection of (2) *IDBs* (q.v.) or for salting claims: see also *snyde* (*diamonds*). [*Du. slenter* a trick, *sch- poss. fr. Ger. Schleicher* smuggler *or fr. influence of Yiddish*]

> *Schlenters*, or *slenters*, are fake diamonds. The best *Schlenters* in South West are made from the marbles in the necks of the lemonade or mineral-water bottles that can be found in dozens at the old German diggings. White *Land God Made* 1969

2. *colloq.* Later used as *adj.* of anything spurious, counterfeit or imitation, including gold coin; also sig. illicit, see quot. [*fr.* (1) ~]

> . . . confidence men found customers in plenty for schlenter gold bricks and amalgam. Cohen *Remin. Johannesburg* – 1924

3. also *vb. slang.* To trick, 'wangle'. *cf. Austral.*, *schlanter*, *slanter* a trick, (also underworld *slang slinter*, *slanter*).

schlep [ʃlep] *n. and vb. slang* Also *U.S. usu. figur.* A drag, bore, effort, also to drag; see quot. at *uitpak: cf. sleep²* [*prob. fr. US use fr. Yiddish schlepp* haul, drag *cogn. Du slepen* to drag, tow etc.]

> . . . my mother-in-law . . . is outclassed by her sister-in-law who cannot resist asking anyone going to Israel to mitschlep some litchis for her daughter . . . no wonder that the El Al flight is described as an airborne delicatessen. Jane Mullins cit. *Fair Lady* 2.3.77

> Once again it's a night-time directionless, woeful schlep, where no T.V. aerialed lounges invite us in for a cuppa comfort. *Oppidan* Vol

III No I Feb. 1978

schof(t). See *skof(t)*.

schoolbag *n. pl.* -s. A satchel for school books; see also *bookbag*. [*trans. Afk. skoolsak* satchel]

school leaving certificate *n. pl.* -s. A certificate of having completed the minimum requirements of *Standard X* (q.v.) without qualifying for university entrance: also given in the case of a combination of subjects, not conforming to the requirements of full *Matriculation exemption* (q.v.) see also *conditional exemption cf.* Brit. *school certificate.*

schoon [skʊən] *adj.* Beautiful: see *skooi* [*Du. schoon* beautiful]

schuur [sky:r] *n. usu. suffix.* Barn [*D schuur, Afk. skuur* barn] **1.** Found names of houses and buildings e.g. Kleinschuur, Groote Schuur.

2. *rare* barn, store: see quot.

> They dared the torrent to report that the homes of some settlers of Cannon Island had been swept away, but that the wheat stored in the main *schuur* was safe. Birkby *Thirstland Treks* 1936

scoff *n.* (*and vb. intrns.*) Food: see *skof, skaf.* [*fr. Du. schaften* to eat, *schaftijd* dinner time]

scoffle See *skoffel, skoffler* and *skoffler oxen.*

Scotch cart *n. pl.* -s. A small stout two-wheeled tip cart, horse- or ox-drawn, used for carting rubbish, manure etc.; also formerly used for taking *blue ground* (q.v.) for washing or sorting on the Diamond Fields. *cf. Canad. dump car. [fr. Afk. skotskar fr. High Ger. Schuttekarren* tip cart *fr. Ger. schutten* to spill, pour out; *see also etym. at cocopan*]

> As the stuff is landed on the staging, it is quickly thrown down, and in Scotch carts or barrows is carried away to the gravel mounds where the sorters sit. Schreiner *Diamond Fields* [1882–4]

In combination ~ *oxen.*

> 1 pair Scotch Cart Oxen. 1 Afrikander Bull, 4 years. *Daily Dispatch Advt.* 6.3.71

scriba ['skribə, -ba] *n. pl.* -s. **1.** Secretary or keeper of records for the synod of the *Dutch Reformed Church* (q.v.) [*Lat. scriba* scribe]

> Synod of the Dutch Reformed Church . . . Rev. A. Faure, *Scriba, Actuarius, Archivarius*

and *Quaestor Synodi. Cape of G.H. Almanac* 1845

2. Parish Secretary to the clergy and congregation of a Dutch Reformed Church. [*as above*]

It is becoming more a custom to appoint women as scribas ... Mrs ... has been scriba of the Dutch Reformed Church, George South, for the past three years. *E.P. Herald* 13.10.70

sea cat *n. pl.* -s. Any of several species of *Octopus* esp. *O. vulgaris :* see quot. at ollycrock. [*prob. translit. Afk. seekat fr. Du. zeekat* ink fish]

We began our dinner and supper with white wine with aloes for bitters, this day we eat sea-cat soup, and Sasaartjies, which is mutton roast on skewers, mighty tasty, and a tart with custard in it. Fitzroy *Dark Bright Land* 1955

sea cow *n. pl.* -s *obs.* The hippopotamus or 'river horse' *Hippopotamus amphibius,* was so called in the early days of the settlement: see quot. at *heer.* [*translit. Du. zeekoe* hippopotamus]

More inland are lions, tygers, buffaloes, elephants and in the river hippopotami, called by the Dutch sea-cows. *Afr. Court Calendar* 1807

In combination ~ *hole;* also *seekoegat* (q.v.) a deep pool. *cf. kaaimansgat,* see *kaaiman.*

... a deep part of the river, called, by the natives, a sea-cow hole, where several of these huge animals then were. Rose *Four Yrs in S.A.* 1829

Also found in place names e.g. Seacow Lake, Seacow Valley, also in form *seekoe* (q.v.) and *zeekoe.*

I have ... mention'd the *Sea-Cow* Valley ... This valley has its name from an amphibious creature, vulgarly call'd a Sea-Cow, and by the learned, *Hippopotamus.* Kolb (trans. Medley) *Descr. of Cape II* 1731

Sechuana [ˌseˈtʃwɑna] *n. prop.* Earlier name of Tswana (*Setswana*) the Bantu language of the *Tswana* (q.v.) people and of Botswana, formerly known as Bechuanaland. [*Bantu prefix sig.* language + *Chuana* (*Tswana*)]

The Gospel is regularly preached within these Districts by the Wesleyan Missionaries in four different languages; viz the English, Dutch, Kaffir, and Sechuana. Shaw *My Mission* 1860

Second New Year *n. usu. Tweede Nuwejaar* (q.v.) the second day of January [*Eng. trans. Afk. as above*]

SECOND NEW YEAR
The offices of this Society will be closed to the public on Tuesday the 2nd January 1979. *Printed Notice, Building Society, George,* Dec. 1978

Secretarius [ˌsekrəˈtɑrɪəs] *n. prop. hist.* Dutch term for the Government Secretary during the term of British rule at the Cape. [*Lat. and Du.* secretary]

A number of boors also, who were beginning to get reconciled to the English government, came to wait on the 'Secretarius' and the Landrost, partly from curiosity, partly from policy. Barnard *Letters & Journals* 1797–1801

secretary bird *n. pl.* -s. A long-legged S.Afr. bird, *Sagittarius serpentarius,* which lives on mice, snakes and small vermin. ⫿The name is derived from its crest or tuft of black feathers thought to resemble quill pens.

... a pair of tall secretary-birds, with the appearance, from which their name had fancifully been given, of pens behind their ears, and knee-breeches, stalked in the distance. Alexander *Western Africa 1* 1837

sections, make, See *make sections.*

Secunde [seˈkʊnde, ˌŭ,-ə] *n. hist.* The deputy governor or second-in-command of the Cape Colony during the regime of the *Dutch East India Company* (q.v.): see also quot. at *rondloper : Anglo-Ind. Naib,* deputy governor. [*Du. fr. Lat. secundus* second]

The honourable the 'Secunde Persoon,' Samuel Elzevier (the colonial secretary of the day) ... bore his treasure in triumph to the Governor's residence. *Cape Monthly Magazine II* 9 Sept. 1857

... the Secunde was a most important man at the Cape. Not of the same consequence as the Governor, to be sure, but still most important. Rousseau *Van Hunks, trans.* 1966

seekoegat [ˈsiˌkŭˌxat, ˈsɪə-] *n.* See *sea cow.*

sell-out *n. pl.* -s. *Afr. E.* Mode of address or reference to a collaborator with whites or with the Government.

Last year while addressing a gathering in Soweto ... was heckled by a crowd which kept shouting 'sell-out!' *Drum* 8.10.73

'Opportunist,' 'Sell-Out.' 'Uncle Tom,' ... detractors hissed because of the stand he took on Bantu Education. *Ibid.* 8.4.74

selonsroos [səˌlɔnsˈrʊəs] *n. pl.* -rose, ∅. Also *Ceylon rose:* a highly toxic flowering shrub, *Nerium odorum,* very similar to the Oleander (*N. oleander*), with which it is freq. confused in S.A.: see first quot. [*Afk. selon(s)* Ceylon, *roos cogn.* rose]

The oleander – Selon's roos it is called in Marico ... is hardy and stands up well to drought conditions. H. C. Bosman *A Cask of Jerepigo* 1964

I can smell a whole row of assegais . . . the stabbing assegai has got more of a selons-rose sort of smell about it than the throwing spear. The selons-rose that you come across in graveyards. *Ibid. Unto Dust* 1965

Selous Scouts [səˈluː] *pl. n. phr Rhodesian.* A reconnaissance force of black and white Rhodesians which penetrates behind terrorist lines and feeds back information. ℙ ~ do not fight unless forced to, and live off the land when on reconnaissance since airdrops of supplies would betray their presence in the area. [*fr. surname* of explorer, Frederick Selous]

Perhaps the most famous Rhodesian military unit is the Selous Scouts . . . a secret, mixed-race tracker group of about 300 men . . . renowned for their ability to survive in the bush: if water is not available they will slake their thirst by sucking moisture from the stomach of a slaughtered kudu . . . Black members of the Scouts have masqueraded as guerillas. Consequently whenever . . . innocent civilians have been tortured or murdered by guerillas the nationalists . . . answer that any alleged atrocities were committed by 'Selousies' in disguise. *Time* 13.6.77

Rhodesia's elite Selous Scouts . . . the awe-inspiring Selous Scouts . . . formed five years ago. The Scouts' training course culminates in a gruelling 160 km survival march through bush in the hostile north territory . . . the Scouts are one of the toughest fighting units in the world. *Sunday Times* 11.3.79

You know the Selous Scouts are the absolute elite among Rhodesian troops – now the over 45s called out on Bush duty reckon they've pipped them – they call themselves the 'Salusa* Scouts'. *O.I. Serviceman* 14.4.79 *A patent medicine for rejuvenating the middle-aged.

semels [ˈsɪəməls, ˈsi-, -z] *pl. n.* Bran, added to bread flour or *boermeal* (q.v.) in making brown or health bread. [*Afr. fr. Du. zemelen* bran]

Storage chest or meelkis in Swellendam Museum with compartments for fynmeel, growwemeel and semels. Baraitser & Obholzer *Cape Country Furniture* 1971

Sendingkerk [ˈsendɪŋˌke(r)k] *n. prop.* The 'sister' mission church to the *Ned. Geref. Kerk, Dutch Reformed Church* the work of which lies among coloured and black peoples. [*Afk. sending* mission + *kerk* church]

SENDINGKERK SLAMS COP SPIES
The Moderator of the N.G. Sendingkerk, ds . . . accused the Security Police this week of using Broederbond-tactics . . . to recruit clergymen as informers. . . . It was disclosed by several delegates at the Sendingkerk's synod . . . A clergyman was asked to act as an informant on

. . . black delegates. *Sunday Times* 1.10.78

sensitive area *n. phr. pl.* -s. *Rhodesian* A danger zone.

The village is in what is known as a 'sensitive area'; the Zambesi Valley. *Fair Lady* 16.3.77

separate development *n.* The policy of 'grand' *apartheid* (q.v.) particularly in the development of the African *homelands* (q.v.) or *Bantustans* (q.v.) in which social, geographical and political separation is envisaged as well as a measure of self-rule: see quot. at *Transkei*.

The policy of separate development – a term the South African government prefers to 'apartheid' – has been relentlessly pursued since 1948, when the National Party came to power. Its goal is the eventual creation of national states for South Africa's tribes through a staged process bringing more autonomy at each step. *National Geographic Mag.* Dec. 1971

sering(boom) [ˈsɪərɪŋ(ˌbuəm)] *n. pl.* -s. Either of two species: *Burkea africana,* (*red/rooi sering(a)*) cr *Melia azedarach,* the fruits of which are toxic to animals and people, also called Chinese Umbrella Tree: see quot. at *boekenhout.* [*fr. Du. sering cogn.* syringa, lilac]

In West-street an effort has been made to run a brick footway or pavement on either side, and as seringa trees have been planted . . . one can manage to trudge along with a little more comfort than when wading through the other streets. A Lady *Life in Natal* 1864–5

[sesbania] *n.* See *rooikeurtjie* and quot. at *proclaimed weed*

Pull out your sesbania now (Headline) *Evening Post* 28.4.79

Sesotho [seˈsutŭ] *n. prop.* Also *Sesuto*, the language of the Basotho (Basuto) people of *Lesotho* (q.v.) also *Sotho* in combinations *North* ~, *South* ~. [*Bantu prefix sig.* language + *Sotho*]

'Greetings Morena Dumela!' he exclaimed in Sesuto. 'But the Moreno speaks Sesuto. . . . Why did the Morena not tell me the other day he speaks our language . . .?' Krige *Dream & Desert* 1953

I am a Mosotho; I love Lesotho; I love my language Sesotho. Lanham *Blanket Eoy's Moon* 1953

sestiger [ˈsesˌtɪxə(r)] *n. usu. pl.* -s and *n. modifier.* One of a group of *avant garde* Afrikaans writers of the 1960s. [*Afk. sestig* sixty + *personif. suffix -er*]

At a conference in Pretoria, where prominent Afrikaans cultural leaders made slashing attacks on the 'sestigers' . . . *Daily News* 16.10.70

as *modifier* ~ group, ~ writing, ~ author etc.

Sestiger author and poet André P. Brink is one of the speakers invited. *E.P. Herald* 5.6.73

Sesuto See *Sesotho*.

set(t)laar [ˈsetₗlɑ:(r)] *n. pl.* -s. *obs*. Dutch term for a British *settler* (q.v.) of 1820, usu. derogatory. ⎮See, however, Shaw quot. at *kerk*. [*fr. Eng.* settle(r) + *Du. agent. suffix -aar*]

In ringing his brass plate at the commencement of a sale his stereotyped condition was – 'drie monts crediet for de Christemens – no krediet for de Settlaar.' The latter term was then and for some years afterwards, one of opprobrium. Meurant *Sixty Yrs Ago* 1885

settler *n. pl.* -s. *hist*. A British settler of 1820: sometimes, then, a derogatory term: see quot. at *settlaar*. *cf. Canad. settler* esp. one who settles on a Frontier.

8.1.1826. Evening a prayer meeting Mr. Kemp, two Boers & a Settler present. Barker *Diary* 1815–28

seven single *n. phr. pl.* -s. *Sect. Army* An unworn recruit's beret so called from its size and the fact that it is flat and disc-like. [*fr.* familiar name for 7 inch long-playing gramophone record]

7-single . . . a recruit's beret before being shaped onto head. *Informant* H. C. Davies *Letter* 2.2.79

seventy-four *n. pl.* ∅, -s. *Polysteganus undulosus*, an edible marine fish found from the Cape to Mozambique: see quots.

I there for the first time tasted a most delicate fish, called the 'seventy four' . . . which undoubtedly was as good as the best cod or turbot out of England. A Lady *Life at Cape* 1870

The *Seventy four* is characterised by several very distinct bright blue bands running along the body, not unlike the rows of guns of an ancient man-of-war, one carrying seventy-four guns being considered a well-equipped vessel in those days. Gilchrist *Local Names of Cape Fishes* 1902 *cit.* Pettman

se voet [ₗseˈfũt] *interj. slang*. An exclamation of derision, disbelief, equiv. of 'my foot', also (vulgar) *se gat* (q.v.). [*Afk. possessive suffix se* + *voet, poss. fr. Eng.* 'my foot!']

'Shame Lorna . . . where's yore stiff upper lip man?' 'Stiff upper lip se voet!' she snarls. *Darling* 14.4.76

sewentiger [ˈsɪəvənₗtɪxə(r)] *n. pl.* -s. An author of the nineteen-seventies: by *analg. sestiger* (q.v.) and *dertiger* (q.v.) [*Afk. sewentig* seventy + *-er personif. suffix*]

In the . . . book Professor Du Rand claimed that leading sestiger and sewentiger authors

and others . . . worked to end the 'white minority Government' and even destroy it with violence if necessary. *Sunday Times* 30.4.78

sewejaartjie [ₗsɪəvəˈjɑ:(r)kĭ, -cĭ] *n. usu. pl.* Flower of any of numerous species of *Helichrysum* and *Helipterum;* see *everlasting(s)* [*Afk. fr. Du. zeven* seven + *jaar cogn.* year + *dimin. suffix -tjie (pl. -s)*]

sgomfaan [ˈsgɔmₗfɑn] *n. Afr.E.* A highly intoxicating illicit home-brew, *lit.* that which makes you bent, 'doubles you up'. [*Zu. isigomfane – causative form of gomfa (vb.),* be bent, crouch]

Mrs . . . who travels all the way from . . . Meadowlands for her regular fruit can of 'sgomfaan'. *Post* 16.2.69

Shaka's Spear *n. prop.* See *Mkonto ka Shaka,* also quot. at *B.O.S.S.* [*trans. Zu.*]

The truth is that Shaka's Spear is fully supported by BOSS. The constitution, Press releases, the money and everything comes from BOSS. *Drum* 8.8.74

shame *interj*. Exclamation, not equiv. of *Brit*. ~ sig. 'how disgraceful' etc. but either of sympathy or of warmth towards something endearing or moving, attractive, or small: see quots. at (2) *skinder, stadig* and *poppie*. [*unknown: poss. equiv. Afk. foeitog,* fie ('shame' in *Afk.* is *skande* (q.v.))]

Shame, he's got a bad head cold. Says his ears are giving him hell and he feels dizzy. McClure *Caterpillar Cop* 1972

. . . when somebody said of the bride. Isn't she beautiful? the circle of women chorused: 'Shame!' The exclamation seems to be an all-weather word used not merely in the usual fashion but to express, for example, admiration and approval. *Cape Times* 19.6.73

'Oh, look, look! . . . those foals. Oh, shame, aren't they sweet . . . that tiny, little one . . . can't be more than a day old. Look at it's mother nuzzling it.' *Sunday Times* 14.11.76

shambok [ˈʃæmₗbɒk, -ₗbɔk] *n. pl.* -s, also *vb trns*. See *sjambok.* [*form of sjambok*]

9 Apr 1811 The *shambok,* here mentioned, is a strip, three feet or more in length, of the hide of either a hippopotamus or of a rhinoceros, rounded to the thickness of a man's finger, and tapering to the top . . . universally used in the colony for a horsewhip. Burchell *Travels I* 1822

Shangaan [ˈʃaŋₗgɑ:n] *n. pl.* -s. **1.** A member of a tribe of Zulu origin living in Rhodesia, Mozambique and S.A., a name also given to the Livingstone antelope: see also *Tsonga.* [*poss. fr.* name of *Zu. chief Soshangane, leader of the*

Northern revolt from Shaka, or Zu-
-shangane a wanderer, roving person]
He lived on fish, turtles, berries, honey and the game he ensnared, and never sowed or reaped; whenever he felt the need, he got cereals from the Shangaans or other tribes in exchange for dried meat. Pohl *Dawn and After* 1964

2. The language of the ~ people.
Besides Afrikaans and English he speaks all the African languages in the Republic except Shangaan and Venda. *Rand Daily Mail* 28.7.71

Sharpeville *n. prop.* Place name: the scene of a demonstration in 1960 at which there were a number of fatal casualties: still commemorated, esp. in phr. *another* ~ : see also quot. at *SASO.*

There are conflicting versions of what . . . happened at Derry, but the fact remains that 13 people were shot . . . As Miss Bernadette Devlin said, this was Ireland's Sharpeville . . . As at Sharpeville, the forces of order felt under threat and the situation got out of hand. *Daily Dispatch* 2.2.72

Professor Demps van der Merwe of the theological school said . . . that if the whites commemorated Blood River, the Blacks should be entitled to commemorate Sharpeville. *Sunday Times* 17.10.76

shebeen [ʃəˈbiːn, ʃɪ-] *n. pl.* -s and *n. modifier.* **1.** Also Irish: an establishment in which illicit home-brewed liquor is sold, or other liquor sold illicitly without a licence, usu. by African women known as ~ *queens*, but also occ. by whites: see second quot. also (2) *queen :* see quots. at *beerhall, patha patha* and *call. cf. U.S. speakeasy, Austral. shanty, sly grog shop; Canad. shebang, tolerance.* [*Irish Gaelic sibin* bad ale]

I doubt very much whether all shebeens can be completely wiped out, beause we have no place to go in the evenings except home or to our favourite shebeen. *Drum* 27.8.67

. . . Durban's flamboyant dockside personality . . . Mrs. . . . was convicted several times of running a shebeen . . . at the sound of an approaching police side-car the pet monkeys . . . on her Point Road balcony would rattle their chains and give her time to hide any illicit liquor. *E.P. Herald* 24.6.76

Shebeen raids in Ciskei
Ciskei police have declared war on shebeens . . . There is a shebeen in nearly every street in Mdantsane. These are known as nophepha bags* who sell small quantities daily of usually cheap wine. *Indaba* 27.4.79　*(q.v.)

In combinations ~ *queen*, see also (3) *auntie* and quots. at (2) *queen;* ~ *king,* ~ *trade,* ~ *party* etc.

Shebeens want to start trade union Soweto's top shebeen queens and kings plan to meet . . . today to decide what to do about a very serious problem – the new liquor prices. *Post* 4.4.71

The shebeen kings and queens in the Black townships of Nyanga, Guguletu and Langa are angry because of raids on their shebeens by students and have threatened to retaliate. *E.P. Herald* 14.10.76

2. *vbl n. prob. obs.* ~*ing*, running or living upon the proceeds of a ~. [*fr.* (1) ~]

Tom . . . a kafir, living in Zwartkops Street, was brought before the magistrate on Monday morning charged with shebeening. *E.P. Herald* 17.2.1897

sheepskin (dance) *n. pl.* -s. An informal dance held in a barn or shearing shed, also known as *opskud/opskop* (q.v.) see quot. at (2) *opskud,* and *vel(d)skoen.*

Weddings are occasions for the capers known as 'sheepskin dances' because they are held in the shearing-shed. Green *Giant in Hiding* 1970

sheep's tail fat *n.* The rendered down fat from the tail of the Cape or *Africander* (q.v.) sheep, also *ronderib* (q.v.) ⫽Esteemed for all culinary purposes from the earliest days and regularly remarked upon by travellers and visitors, from the C18 onwards: see *fat tailed sheep,* also *kaiings.*

All the game of this country is so dry that it requires plenty of sheep's tail fat with it in the dressing. The Dutch eat everything swimming in it . . . Sheep's tail fat is useful for many purposes – lamp oil, frying fish; and salted and dried it eats like the best bacon. The Dutchmen praise it most highly and certainly most deservedly. The tail weighs in general from four to six pounds. Philipps *Albany & Caffer-land* 1827

Sheilas' Day *n. colloq. Urban Afr. E.* Thursday, the day of most *Manyano* (q.v.) meetings, on which most women in domestic service are off duty: see *off* and allusion in quot. at *bunny chow.*

It's 'Sheila's Day' and Babsy gets a special dispensation from her lawyer boss to share the privilege of a day-off with the Jo'burg domestics. Melamu *cit. New Classic* Vol. 3 1976

Shellhole see *M.O.T.H.*

shift boss *n. pl.* -es. A miner in charge of an underground shift, usu. a white man: *cf. Canad. shifter.*

. . . the young shiftboss . . . who courageously rescued three Africans . . . nearly two kilometres down, while his own life was in grave danger. *Rand Daily Mail* 28.8.71

He started as a miner, was promoted to shift boss, studied in his spare time . . . to obtain a mine overseer's certificate. *Panorama* May 1972

shimiyana ['ʃīmĭ₁jɑna] *n.* Also *shimiyaan:* a home brewed liquor of water and treacle fermented in the sun, from which the illicit *govini,* see second quot., is distilled. *cf. Canad. hootch(inoo)* a fierce home-brew. [*Zu. isiShimeyana* intoxicating drink made from treacle]

For many years much trouble was caused by the manufacture . . . of *Shimyaan* . . . This beverage was maddening in its effects and the parent of much crime. Sir J. Robinson *A Lifetime in S.A.* 1900 cit. Pettman

Govini – a prostituted word, coined to mean 'the government is against it.' Distilled from shimiyana. Extremely potent. Edmondstone *Thorny Harvest* (date unknown)

Shoeshoe ['ʃwe₁ʃwe] *n. prop.* See *Qwaqwa.*

Shona [₁ʃɔna] *n. pl.* (a)ma-, -s. The language or one of the group of closely related languages of the *Mashona* (q.v.) peoples of Rhodesia: the peoples themselves, also a single member of one of them.

My mother language is Shona and I can speak English and Ndebele. *Drum* 18.1.73

shongololo *Natal: Zu.* form of *songololo* (q.v.)

shot out *vb. partic.* Denuded of game through indiscriminate shooting or *hunting* (q.v.) *cf.* phrs. *eaten out, tramped out, beaten out, hunted out.* [*trans. Afk. uitgeskiet* lit. shot out]

Southern Angola has never been an elephant sanctuary, and now it has been almost shot out. Green *Last Frontier* 1952

shottist ['ʃɒtɪst] *n. pl.* -s. Shot: equiv. of *Brit.* usage 'He is an excellent shot.' [*Eng.* shot (personal noun) + *redundant personif. suffix -ist: unknown cf. Afk. skoot* shot (single discharge), *skroot* shot (lead), *skut* shot (person)]

Top shottists to compete Leading South African shottists will take part in the Eastern Province smallbore championships . . . At least 40 shottists are expected to take part. *E.P. Herald* 21.5.73

shushu ['ʃu₁ʃu] *adj. colloq. poss. reg.* E. Cape. Hot: used of the weather etc. In combination (*obs.*) ~ *broekies:* the fashion of shorts formerly called 'hotpants'. [*Xh. shushu* hot *also Afk. sjoebroekie* hot pants]

It has been shu-shu today and we're sure to have rain after this heat. *O.I.* 1970

Hot-Pants or Shu-shu broekies Call them what you may but see our selection now. *Grocott's Mail Advt.* 26.11.71

side-bar *n.* See quots. In S.A. 'the term is used to connote the general body of attorneys, just as the bar is used to describe the general body of advocates' *Informant* Mr Sydney Kentridge S.C.: see also *advocate.*

Lawyers have continued to fall into one of two groups, the advocates (making up the Bar) and the attorneys (composing the Side-bar), approximating to the barristers and solictors of England. Hahlo & Ellison Kahn *Union of S. Africa* 1960

-siekte ['sĭktə] *n. suffix.* Disease, sickness: suffixed esp. to the names of diseases of domestic animals e.g. *dun* ~, *dom* ~, *gal* ~, *jaag* ~, *lam* ~, *slap* ~, *snot* ~, *styf* ~, *spons* ~ etc. all (q.v.) [*Afk. fr. Du. ziekte* sickness]

sies [sĭs] See (1) *sis* and quot. at *little.*

siestog ['sĭs₁tɔx] *interj. colloq.* An expression usu. of sympathy or dismay for the person addressed, see also *foei tog* and *shame;* also equiv. of 'how sweet' or 'how touching': *cf. sis* (q.v.), *sies* (q.v.). [*Afk.* what a pity, *shame* (q.v.), poor thing *etc.*]

The other day my mother arrived from the farm . . . Siestog Dainty, says my ma . . . city life must be taking it out of you . . . Are you getting enough to eat? *Fair Lady Advt.* 10.21.75

sieur [sĭœ(r)] *n.* Respectful mode of address or reference to the master or other superior, poss. archaic. [*Afk. poss. abbr. Fr. monsieur,* or *fr. seigneur,* lord]

'But Nooi,' protested the Hottentot cook . . . 'the Sieur will kill us.' . . . her nooi would have none of it . . . 'tell me when you see the men coming and I will myself speak with the Sieur.' McMagh *Dinner of Herbs* 1968

siffie ['sɪfĭ] *n. pl.* -s. *Haliotis spadicea,* Venus ear: see *klipkous* and quot. at *ollycrock.* [*Afk. siffie* strainer, small sieve]

-sig [sɪx] *n. suffix.* View: found in S. Afr. place names e.g. Bergsig, Vooruitsig, Bothasig: more freq. in form -*gesig* (q.v.) -*gezicht,* also -*uitsig* (q.v.) [*Afk. sig fr. Du. zicht* cogn. sight]

'silver' *n. pl.* Ø. *Sparodon durbanensis,* the *musselcracker, musselcrusher* (q.v.), also known as silver steenbras. [*fr. colour*]

. . . along the Kleinemonde coast . . . in two outings caught eight janbruin, two small 'silvers' and two galjoen. *Grocott's Mail* 13.10.72

silverfish *n. pl.* Ø. *Argyrozona argyrozona:* see (3) *doppie,* (2) *Kaapenaar;* see also quot. at *roman.* ⫿Also used of the silver moth or fish moth *Lepisma saccharina* (non-S.A.E.).

Hottentot fish, a small fish of the shape of the perch, covered with scales of a dirty brown colour, silver fish, something of the size and shape of the former, but of a bright silver colour. Ewart *Journal* 1811–14

silver tree *n. pl.* -s. Also *witteboom* (q.v.), *silwerboom Leucadendron argenteum*, a tree of the Proteaceae preferring almost exclusively as habitat the slopes of Table Mountain, Cape: see quot.

The most striking object . . . is the Silver Tree (Protea Argentea) which is seen exuberantly clothing the least rugged part of the Table Mountain (to which it is indigenous), making a pleasant break in its dark and gloomy sides, with its downy and silver color'd foliage. Ewart *Journal* 1811–14

. . . a vast grove of silver trees similar to those which are to be seen on the slopes of Table Mountain at Cape Town. I had never before met with them in all my wanderings, except at the Cape, and their appearance here astonished me greatly. Haggard *Solomon's Mines* 1886

silvie See *flatty.*

Sindebele [ˌsɪndeˈbeːle] *n.* The language of the *Ndebele* (q.v.). [*Bantu prefix sig.* language + *Ndebele*]

-singel [ˈsɪŋəl] *n. suffix.* Crescent: found in street names e.g. Buitensingel. [*Afk. fr. Du. singel* boulevard, promenade]

singular *n. pl.* -s. A white person: facetiously used as an opposite of *plural* (q.v.) hence ~*stan(ian)* etc. and numerous now (1979) obsolescent puns: see second quot. at *plural.*

Two Pluralstanians . . . are invited to a conference in West Germany. They would like to attend because this conference is vital to Pluralstanians and Singularstanians here. But they can't. *Drum* Aug. 1978

sink [zɪŋk] *n. and n. modifier substandard.* See (1) *zinc.* ‖A kitchen sink by similar confusion is often pronounced 'zinc'. [*fr. Afk. sink(plaat)* galvanized iron]

Everything one needed could be found there . . . nails, spades, sink baths and buckets, groceries, saddles, bridles, harnesses. *Panorama* Apr. 1974

sinkhole *n. pl.* -s. Ground subsidence sometimes engulfing large areas in mining districts: not equiv. *Canad. sinkhole.*

Frightened families are preparing to evacuate eight houses near the . . . gold mine this week because of the enormous sinkhole which appeared nearby three weeks ago . . . The new sinkhole, measuring 100 ft. by 70 ft. and about 70 ft. deep . . . has been cordoned off, and is

continuing to cave in. *Rand Daily Mail* 16.11.70

sinkings [ˈsɪŋkɪŋs] *n.* Neuralgia. [*Afk. fr. Du. zinkings* fluxion, rheum]

ZINKINGS or FACE-ACHE Calf's Antifebrile Lotion is the speediest alleviator 1s. 6d. per bottle. *Cape Town Directory* 1866 *Advt.*

In combination ~ *bossie*, any of several aromatic species used in the treatment of neuralgia including *Myrothamnus flabellifolius* (bergboegoe), *Chenopodium ambrosioides* (galsiektebossie) and *Pelargonium ramosissimum* (dassie buchu); also ~ *wortel* (*root*) *Clematis brachiata* (traveller's joy).

sis¹ [sɪs, sɨs] *interj.* Also *sies :* an exclamation of disgust, see also *ga :* often in combination *ag* (q.v.) ~ : see quots. at *little* and *kaal.* [*anglicization Afk. sies poss. fr. Hott. si or tsi*]

. . . you have been fighting again. Sis, a man of your age and with a grown daughter still fights like a small boy. Dikobe *Marabi Dance* 1970

. . . amid calls of 'sies' and 'skande' he pointed out that White and non-Whites were forced to make use of the same changing rooms, showers and toilets. *Argus* 12.10.72

. . . a number of Koinon words which passed directly into Afrikaans . . . Eina (to express pain), sies (disgust), soe (heat) and arrie (astonishment) *Post* 28.7.74

In combination ~ *on you*, a type of imprecation esp. among children *cf. Irish* 'bad cess to/on you' and *Austral.* 'good on you'.

sis² [sɪs] *n.* Printed cotton material usu. in combination *Duitse* ~ (see *German print*) *voersis* (q.v.). [*Afk. fr. Du. sits* chintz]

Paris fashion plates had not reached the 'Backveld', dress materials being limited to the 'Dutch' and 'German' Blaudruck print, checked Ginghams, Kapje Sis (a light cotton dress material always with a very small pattern). Jackson *Trader on Veld* 1958

sisi [ˈsɨˌsɨ] *n. pl.* -s. *Afr.E.* Sister, a mode of address or reference to an African woman by Africans, men or women, unless the woman is old: see *mama.* ‖Also adopted by many whites esp., children and students; see *buti;* written erron. as *sissie, sissy.* [*Bantu sisi* sister]

I am appealing to you to stop publishing such disgusting poses . . . My six-year-old boy . . . asked me 'Why does this sisi sit like this when she sits for a photograph?' *Drum* 22.9.73

. . . of the . . . black staff a Grahamstown psychologist says the sissies have the roughest deal . . . no maternity leave and they badly need the money they earn . . . two daily free meals provided for sissies. *Rhodeo* 29.4.76

. . . a big woman . . . who wore a long dress and a German print apron . . . and she gave Poppie advice. 'You must get papers, Sisi . . . if you don't have papers, the white people do just what they want with you.' Joubert *Poppie Nongena (trans.* Smuts *Fair Lady* 11.4.79)

sister *n. pl.* -s. *Afr.E.* Also *brother :* a mode of reference to first cousins among Africans: see also *father, small/big* and *mother, small/big.*

sitatunga [ˌsɪtaˈtũŋgə, -a] *n. pl.* ∅. Also *situtunga : Tragelaphus spekei* an aquatic antelope of the swamps of E. and Central Africa ; usu, known in S.A. as *water-skaap* or *waterkudu* [*Swahili name*]

Only once did we see the sitatunga, a rare aquatic antelope which spends its whole life in water . . . The sitatunga . . . browse happily in areas where other antelope would sink. *Personality* 10.9.71

sit-en-kyk [ˈsɪtn̩ˌkeɪk] *n. lit.* 'sit and look': an open pattern railway coach not divided into compartments: called *(m)bombela* (q.v.) by Africans.

. . . what goes on in the Third Class sit-en-kyk . . . after six o'clock on a Friday night. O.I. Judge 28.9.74 Grahamstown

sit up *vb phr.* Also as *vbl n. sitting-up :* see quot. at *opsit.* [*trans. Afk. opsit* court, woo *fr.* Du. *opzitten* to sit up]

The Boer youth's idea of love-making is expressed in the term 'sitting-up' . . . Should a young man wish to make advances to a young woman, he asks her to sit up with him, and if she wishes to encourage him, she consents. Brinkman *Breath of Karroo* 1914

sjambok [ˈʃæmˌbɒk, ˈʃamˌbɔk] *n. pl.* -s. **1.** Also *shambok* and *sambok* both (q.v.). A stout rhinoceros or hippopotamus hide whip: see quot. at *shambok;* now used erron. of riding crop or other whip. [*Du. fr. Malay tjambok fr. Persian and Urdu chabuk* a horsewhip]

Riding sjambok – 55c cash *Farmer's Weekly Advt.* 12.5.71

. . . plastic Sjamboks 65c, longer 85c. Cash *Ibid.* 30.5.73

2. *vb. trns.* To horsewhip, beat with a ∼. [*fr.* (1) ∼]

Mrs . . . who left home with her children after her husband sjambokked their pregnant daughter, returned to the . . . farm yesterday. *Rand Daily Mail* 23.9.71

skaal *n. pl.* -e, (-s). See *scale* and quot. below.

Classy establishments like The Stetson in the township of Molapo, Sportsman in Meadow-lands, Falling Leaves in Dube and Kwa Dutch

serve anything from imported beer to Scotch whisky. But don't ask for a skaal of Bantu beer. Venter *Soweto* 1977

skaam [skɑːm] *adj. colloq.* Embarrassed, shy. [*Afk. skaam,* shy, ashamed *cogn.* shame]

'. . . cut it out, man . . .' Frik mutters, getting all skaam at the way she's performing . . . 'Don't give us a hard time now, doll.' *Darling* 14.4.76

skaamblom [ˈskɑːmˌblɔm] *n. Serruria florida,* see *blushing bride,* also called *Pride of Franschhoek.*

skaap [skɑp] *n. pl.* -s. *slang.* **1.** Derogatory: *lit.* 'sheep', a fool: see also (3) *dikkop, houtkop, pampoenkop.* [*Afk. skaap* sheep, simpleton]

He was never noisy . . . there was no shouting at or to his men. When irritated he found the use of the word 'skaap' most effective. 'Tweede in Bevel' *Piet Kolonel* 1944

2. skaap- [skɑp] *n.* Sheep: prefixed to numerous plant names sig. either that they are relished by sheep or have some characteristic associated with sheep, e.g. ∼ *boegoe, Agathosma 'ovina'* ; ∼ *doring (thorn)* etc. Also combination *water*∼ see *sitatunga.*

skaapplaas [ˈskɑpˌ(p)las] *n.* A grass: either *Cenchrus ciliaris* or *Eragrostis denudata :* see also third quot. at *kweek.* [*Afk. skaapplaas lit.* sheep farm]

SKAAPPLAAS – The most sought-after grass, usually in short supply, now available again. *Farmer's Weekly* 3.1.68

A Transvaal company specialising in turf uses skaapplaas, one of the finest of lawn grasses. *S. Afr. Garden & Home* Jan. 1973

skaapsteker [ˈskɑpˌstɪəkə(r)] *n. pl.* -s. Any of several back-fanged snakes of the genus *Psammophyllax :* poisonous but generally harmless and not capable of killing sheep as the name implies. [*Afk. fr.* Du. *schaap cogn.* sheep + *steker fr. steken* to sting, pierce + *agent. suffix -er*]

The farmer, or his herdsman, comes along, finds the dying sheep, and seeing *Schaapstekers* about, immediately concludes they are the guilty parties, hence the name *Schaapsteker,* which means 'sheepsticker'. Fitzsimons *Snakes of S.A.* 1912 *cit.* Pettman

skaf [skaf] *n. Afr.E.* Food: (skoff, scoff) in combination ∼ *tin* also ∼ *box,* a sandwich or lunch tin: see quot.: see also *varkpan.* [*Zu. isiKhafu* food *prob. via S.A.E. fr.* Du. *schaffen, schaften* to eat, dine]

I left next morning by boat, after having

arranged the loads that I had with me, a deck-chair, camp-table, a couple of shovels and a skoff-box. Hemans *Log* 1935

skande ['skandə] *interj.* (*n*). Exclamation equiv. *Brit.* shame! disgraceful! etc.: see quot. at *sis*[1]. [*Afk. fr. Du. schande* disgrace, scandal]

skans [skɑns] *n.* See *schans; cf. Anglo-Ind. sungar,* a breastwork of stone. [*Afk. fr. Du. schans* breastwork, redoubt]
. . . the British thought the Boers would be hiding behind these skanses or heaps of stone. *Radio & TV Magazine* 6.5.74

skat [skat] *n.* Treasure: an endearment usu. *my ~, ou ~* or *~ tebol,* freq. to a baby: see quot. at *skop-die-blik.* [*Afk. fr. Du. schat* treasure]

skebengu, skebenga [ske'beŋgŭ, -a] *n. pl.* -s. A gangster, bandit, robber. [*Zu. isigebengu fr. vb. -gebenga* plunder, act the bandit, assault]
These practices must not be confounded with those of the ordinary skebenga . . . a low-class thief and foot-pad who would rob his mother's hen-roost. Campbell *Old Dusty* 1964

skeef [skɪəf] *adj./adv. slang.* Crooked, off-beat. [*Afk. fr. Du. scheef* askew, awry, *cogn.* skew]
. . . we're a minority group and these cats tune us skeef. *Rhodeo* 23.3.72
Why are you looking at me so skeef? Fugard *Boesman & Lena* 1969

skei[1] [skeɪ] *n. pl.* -s. A yoke pin (skey), or later skittle-shaped modification of one used as the missile in playing *jukskei* (q.v.) [*abbr. jukskei* yoke pin/skey]
The monarch was King George VI and the picture of him lobbing a skei, watched by the two Jannies – Smuts and Hofmeyr – is on view at the South African Archives. *Argus* 14.5.71

skei[2] [skeɪ] *interj.* (*imp. vb.*). *colloq.* (oop. among children). Equiv. *Cave!* sig. watch out, look out etc. [*prob. Afk. fr. Du. scheiden* to divide, separate]
Skei, Ma there's a car coming up on your tail. O.I. Child 13, 1971

skel [skel] *vb. intrns. and trns. slang.* To scold, abuse, 'carry on' at someone; see *pick out. cf. Austral. give curry to, go crook at.* [*Afk. fr. Du. schelden* to berate, abuse, revile *cogn.* scold]
. . . just wait till ouma sees you then she'll skel you out . . . *Quarry '77 New S.Afr. Writing* 1977

skelm ['skeləm, -m] *n. pl.* -s, also *adj. and adv.* **1.** *colloq.* A rascal, villain used usu. of a man but occ. of an animal also: see quot. *cf. Brit. and Scottish skellum*

and *skillum* (archaic). *E. Afr. shenzi, n.* bad character; *Anglo.-Ind. nut-cut* rogue. [*Afk. fr. Du. schelm* villain, rogue]
. . . but both the lion and saddle had disappeared, and nothing could be found but the horse's clean picked bones. Lucas said he could excuse the schelm for killing the horse . . . but the felonious abstraction of the saddle . . . raised his spleen mightily. Thompson *Travels II* 1827
War correspondent, uncle? That is one of those skelms who writes all the lies about us in the newspapers. Brett Young *City of Gold* 1945
2. *adj. colloq.* Sly, cunning, wicked; also of man and beast. [*Afk. skelm* sly, knavish]
Diederik . . . determined on shooting it, declaring that no schelm beast should kill his horse. Philipps *Albany & Caffer-land* 1827
One foreman – he refused to give his name for fear of being fired – told me 'You've got to be "skelm" (sly) when you're working on a project like this.' *Sunday Tribune* 25.6.72
3. *adv. m. rare* Cunningly, slily. [*Afk. skelm* slily]
He says the Jan Bruin were there but he could not hook them. They were biting 'skelm', probably wanting a different type of bait or needing to be coaxed. *Daily Dispatch* 7.12.71

skepsel ['skepsəl] *n. pl.* -s. Creature: usu. derogatory, esp. offensive when used of an African. [*Afk. fr. Du. schepsel* creature *fr. scheppen* to create *cogn.* shape]
This 'skepsel' (creature) has ravaged a white man's wife and daughter. Klein *Stagecoach Dust* 1937
The Native to many . . . was a *schepsel* ('creature') not wholly human; standing intermediate, as it were, between the game and the livestock, on the one side, and White humanity, on the other. Hoernlé *S. Afr. Native Pol.* 1939

skerm [skerm, -rəm] *n. pl.* -s. A primitive screen-type shelter made usu. of branches and brushwood either serving as protection in camps of hunters, woodcutters etc., or more freq. as a dwelling for nomadic peoples esp. in desert areas. *cf. Austral.* (Aboriginal) *wurlie/ey, Canad. lean-to, commoosie, siwash camp. Anglo.-Ind. pandal.* [*Afk. fr. Du. scherm* screen *rel. Afk. vb. beskerm* to protect]
. . . we went to work to build a 'scherm' near one of the pools . . . This is done by cutting a quantity of thorn bushes and piling them in the shape of a circular hedge. Then the space enclosed is smoothed, and dry tambouki grass, if obtainable, is made into a bed in the centre, and a fire or fires lighted. Haggard *Solomon's Mines* 1886
They live in the open air with only a bush

skerm as shelter, and they like it. Green *Karoo* 1955

... in a primitive skerm of branches at night with a bubbling ... pot over a permanent fire, they would chip away until the great tree was down. Hjalmar Thesen *cit E.P. Herald* 18.5.73

skey[1] [skeɪ] *n. pl.* -s. Also *skei* (*q.v.*) and *yokeskey:* a notched wooden peg or yoke pin. See quot.: ~ *yoke,* a yoke of the type used with ~*s :* see also *jukskei.* [*Afk. skei* yoke pin *poss. rel. Du. scheiden* to separate]

Through these holes went the skeys – pieces of hard wood notched to take the leather strops or straps that passed under the ox's throat and were adjusted in the notches according to the size of the animal. Cloete *Rags of Glory* 1963 ... Skeys R3,60 doz; Skey Yokes R4,50; *Farmer's Weekly Advt.* 30.5.73

skey(ear)[2] [skeɪ] *n. and n. modifier.* A V-shaped notch cut or punched as an identification mark in the ear of an ox or other domestic animal: ~ *ear,* an ear so marked. [*Afk. skei poss. fr. vb. skei* divide, split *fr. Du. scheiden*]

One Red Ox with brown markings, swallow tail and skey right ear, hole in left ear. *Grocott's Mail* 25.5.71

skiet en(and) donder [ˈskɪt ənˈdɔn(d)ə(r)] *n. phr. colloq.* Also *skop* ~ : equiv. of 'blood and thunder', 'cloak and dagger' etc. ; pertaining to films, plays or literature. *cf. snot en trane.* [*Afk.* (*skop* kick) *skiet* shoot + *en* and + *donder* thrash]

It's the organizer's task to bring the script to life. International intrigues, bloody *skiet en donder,* and dreamy romance are staged, here in Cape Town, all as quickly, cheaply and smoothly as possible. *Cape Times* 19.6.73

... It's a cowboy fillum. Skop, skiet en donner addict, that's my boet. *Darling* 26.11.75

skietgat [ˈskɪtˌxat] *n. pl.* -te. *Sect. Army* The pit of the shooting range from which targets are raised and lowered acc. instructions from the *skietpunt.* [*Afk. skiet* shoot(ing) + *gat* hole]

'When we go shooting *ou swaer* ... you will have to spend many hours in skietgat and afterwards you must pick up all the doppies ...' Picard in *Eng. Usage in SA* Vol. 6 No. 1 May 1975

skiets [skɪts] *n. slang.* Vulgar term for diarrhoea esp. but not exclusively among children. *cf. Austral. the trots :* see *apricot sickness.* [*Afk. skyt fr. Du. schijten* to defecate *presum. rel. Afk. skittery* diarrhoea *cogn. Brit. slang squitters*]

The whole school gets skiets every time we have cottage pie. Boy 11 1972

skilling [ˈskɪlɪŋ] *n. pl.* Ø, -s. *hist.* See *schelling* and quot. at *rixdollar.* [*cogn. shilling*]

One and a Half Skilling per diem for herding and grazing each Head of Cattle, or Pigs, and Two skillings for each Dozen of Sheep or Goats ... shall be levied by the Pound-master. (17th June 1825) *Greig's Almanac* 1833

skilpad [ˈskɪlˌpat] *n.* Tortoise. [*Afk. fr. Du. schild* (*cogn.* shield) *pad,* toad, *see padda*] **1.** *prefix* sig. either made of tortoise (see quot. at *pastei*) or being like a tortoise in action or appearance as in ~ *trek,* ~ *loop* (pull, go) a tug-of-war for two persons played on all fours.

We held shooting matches, we went out after buck and we wrestled and played skilpadloop and jukskei. Green *When Journey's Over* 1972 **2.** *usu. prefix* to plant names sig. a likeness to a tortoise or affording food or shelter to tortoises, ~*blom* (flower) *Hyobanche sanguinea;* ~*bossie* (bush) any of several species of *Crassula;* also ~*kos,* various *vygies* (q.v.) ; ~*knol, Dioscorea elephantipes* or *olifantsvoet* (*elephant's foot*), see *Hottentot bread.*

skilpadbessie [ˈskɪlˌpatˌbesï] *n. pl.* -s. See (2) *skilpad Mundia spinosa,* also *duinebessie* the juicy fruits of which are edible. ᵽ. . . . eaten by ostriches, natives, children and tortoises. Smith *S. Afr. Plants* 1966. [*Afk. bessie,* berry]

skilpadkos [ˈskɪlˌpatˌkɔs] *n.* Any of several species of *Dorotheanthus,* see *vygie* and quot. at *gifboom;* also (2) *skilpad* [*Afk. kos fr. Du. kost* food]

skilpadtrek [ˈskɪlˌpatˌ(t)rek] *n.* See (1) *skilpad.*

skimmel [ˈskɪml̩, -əl] *n. and n. modifier.* Roan or dappled horse, used either sig. the animal itself as does *Eng.* 'roan', or pertaining to the colour as in combination *blou* ~ or ~*blou(blue),* blue roan or dapple gray; *rooi* ~ (red). [*Afk. fr. Du. schimmel* grey horse, mottled *fr. schimmelen* to mildew, grow mouldy *prob. cogn.* shimmer]

... the Kaffir led round the horse ... a valuable schimmelblouw grey that had been salted by recovering from horse-sickness. Brett Young *City of Gold* 1940

Stud comprises ... tip-top mares ... Vonk the sire (Skimmel) ... a former South African Boerperd Champion. *Farmer's Weekly* 12.3.71

In combination *figur*. ~ *day/dag* – early dawn presum. fr. dappled sky.

skinder [skɪnə(r)] *vb. intrns. also n. and n. modifier. colloq.* Gossip. [*Afk. prob. fr. Du. schenden* to slander, revile, *cogn.* scandal (q.v.) + *bek* (q.v.) mouth *cogn.* beak] **1.** *vb.* To gossip about, slander; see also *scandal*.

The women may discuss Lena and 'skinder' about her. But the men are silent. They keep their own thoughts. Meiring *Candle in Wind* 1959

2. *n.* Gossip, evil talk freq. in combination ~*bek*, *slang*, a gossip monger ~*stories* (q.v.) *cf. Anglo-Ind. gup*, gossip.

. . . perhaps that will teach you not to believe everything that old skinderbek says! *Ibid.* 1959
. . . the boys . . . setting up the braai fire, and us chicks . . . catching ouma up with the latest skinner. Shame, she . . . craves to know what's going on. *Darling* 15.10.75

skinderstories [ˈskɪnə(r)ˌstuərɪs] *pl. n. also sing.* Gossip, scandalmongering tales: see *skinder*, and *scandal stories*.

. . . 's men remain scornful of the clinging stain of the . . . debacle, calling it 'skinderstories'. *Sunday Times* 1.10.78
His press secretary said that the . . . Minister did not want to react to every bit of gossip (skinderstorie). *Evening Post* 9.3.79

skip, skep *n. Sect. Mining*. The cage or lift in use on the mines: In combination ~*man*, the signalman or operator of a ~.

. . . yearly the underground man must pass the mine doctor to keep his red ticket. Without it he may no longer drop in clanging skip down miledeep shaft nor clump iron-shod with his men through labyrinths of tunnels to the gold-veined rock face. Lighton *Out of the Strong* 1957

skipjack, *n. pl.* ∅, -s. Also *skipper:* the ~ *tuna Katsuwonus pelamis* of the Scombridae fam. ▮The name ~ is also loosely applied to the *elf(t)* (q.v.), also to the 'tenpounder' *Elops machnata*, see (2) *Cape Salmon*, sometimes called 'river salmon' from its habit of penetrating far up rivers to spawn.

One was a skipjack (Elops saurus . . . also called Cape Salmon, springer, or wildevis), weighing 3.600 kg, caught in the surf from a high rock. *Grocott's Mail* 28.1.72

skipper[1] See *skipjack*.

skipper[2] *n. pl.* -s. *Afr.E.* Usu. a long sleeved tee-shirt in knitted cotton, also used of other knitted garments. *cf. Brit. jumper.* [*presum. analg. jumper*]

Men's knitted skippers, long sleeves, three buttons in front. *Post* 27.6.71
. . . members of Youth for Christ . . . were

wearing skippers with a finger pointed upwards to a small cross with the legend: One Way. *Drum* 8.5.73

skit [skɪt] *vb. slang* Pinch, steal. [*prob. fr. Afk. skut* (q.v.)]

If a ol' lady ankles across . . . do yous . . . Offer the old bat a ride on your pillion . . . Beat her up and skit her Pension-Book? *Bike S.A.* Oct. 1976

skof[1] [skɔf] *n. pl.* -te. Also *skoft:* a lap, stage or leg of a journey, sometimes the distance or period of travel between (2) *outspans* (q.v.): *cf. Canad. march*, leg of a journey: ▮Also (rare) a shift or period of work. [*Afk. skof* stage (of a journey) *fr. Du. schoft* 3 hours of work *cogn.* shift, *schoften* to rest, leave off working for a time]

That last skof was hard. Against the wind . . . Heavier and heavier. Every step. Fugard *Boesman & Lena* 1969
The normal day on the road was made up of two or three stages known as *skofte*. Green *When Journey's Over* 1972 *figur.*
The year 1977 then sees the world 'op die laaste skof' of its quickening journey towards the birth of a new century. *Argus* 1.1.77

skof(f)[2] [skɒf, skɔf] *n.* [*and vb.*] Also *scoff, skaf* (q.v.) and combination ~*box/tin* see quot. at *skaf*. Food, a meal also, *rare*, as *vb.* see first quot. *cf. Canad. chuck*. [*S.A.E. (not Afk.) fr. Du. schaffen, schaften* to eat, dish up: *schaftijd* (also *schofttijd*) dinner time: *poss. via Afr. E. fr. Zu. isiKhafu* food]

. . . not one of them would scoff or eat – they shook their heads with a look of horror – 'nee, nee' – and I found that, owing to the ham having been put in the same basket . . . everything was contaminated with them. Barnard *Letters & Journals* 1797–1801
. . . after a great deal of sputtering on both sides I recognized the words, 'Scoff, myneer,' – Food, sir. Buck Adams *Narrative* 1884
When Christmas came I arranged with the Police to join their celebrations – the agreement being that I should provide the drink and they the skoff. Jackson *Trader on Veld* 1958

skoffel [ˈskɔfəl, -ļ] *vb trns. and intrns. and n.* To cultivate or hoe the ground to clear it of weeds: also as *vbl n.* ~*ing*. [*Afk. fr. Du. schoffelen* to hoe *cogn.* shuffle]

For the rest is the woman's sphere – the planting, the *skoffeling* or hoeing, the weeding and the harvesting. Walker *Kaffirs are Lively* 1948
She . . . hates working in the vegetable patch. 'I can't take a hoe and skoffel' . . . 'After five minutes I'm dead tired.' *Darling* 3.3.76

as *n.*, a hoe.

Those who are interested should ask themselves if they are prepared to take a skoffel – a hoe – and work with the labourers. Kavanagh *Merry Peasants* 1963

skoffler [ˈskɔflə(r)] *n. pl.* -s. Also *scoffler*: a cultivator, hand-operated or ox-drawn hence ~*oxen*. [*fr. vb. skoffel* (q.v.) + *agent. suffix* -*er*]

2 Separators 13 Skofflers; Ox Ploughs and various Farm Tools. *Daily Dispatch Advt.* 12.5.73

8 Scoffler Oxen *Ibid* 17.5.72 *Advt.*

skoft [skɔft] *n. pl.* -en See *skof*[1] [*fr. Du. schoft, see at skof*[1]]

skokiaan [ˈskɔkĭˌɑn] *n.* A powerful, illicit home-brew made with yeast. *cf. Canad. hootch(inoo):* see quots. at *Ma* and *Barberton.* [*etym. dub. poss Ngu.*]

When he wanted skokiaan – brewed with yeast and water – he went to Cape Location where Coloured people lived . . . Skokiaan being much stronger than malt beer. Mpahlele *Down 2nd Avenue* 1959

A 37-year-old White mother of nine submitted in the Bloemfontein Magistrate's Court yesterday that she had brewed 50 litres of sifted skokiaan (an illicit brew) in an attempt to get her alcoholic husband 'off other liquor'. *Friend* 25.6.75

In combination ~ *queen* see also *shebeen queen* and *queen*: a woman dealing in illicit liquor.

Often, a well-known 'Skokiaan Queen' was sent to prison without the option of a fine. In such cases the Stokveld helped with the home and children till the member came out of jail. Abrahams *Mine Boy* 1946

skolly, skollie [ˈskɒlĭ, skɔ-] *n. pl.* -ies A Coloured street hoodlum, usu. a criminal or potential criminal and member of a gang. *cf. Austral. bodgie,* a teddy boy or '*larrikin*': see also at *tsotsi* and ~*broek.* [*Afk. prob. fr. Du. schoelje* rogue, rascal, *also thought rel. Afk. skorriemorrie* rifraff *fr. Hebr. or Yiddish* (*Afk. skorrie* a teddy boy)]

. . . tsotsis and skollies who lurk in street corners, waiting to assault and rob innocent people? *Post* 14.3.71

skolliebroek [ˈskɔlĭˌbrŭk] *pl. n.* Cut off 'half mast' trousers: at one time almost invariable *skollie* (q.v.) dress: now *cf. tsotsi* (q.v.) *suit, trousers* etc. [*Afk. skollie* (q.v.) + *broek* (q.v.) trousers]

Take those halfmast skolliebroek home. That's not a skolliebroek – they're Bermuda shorts. O.I. *Child* 13 1971

skolollie [ˌskɔˈlɔlĭ] *n. poss. reg.* E. Cape.

A ball game played by two teams in which members crossing a line are bowled out: one succeeding in crossing twelve times shouts ~! which frees the whole team. [*unknown*]

skoon [skʊən] *adj. prefix.* Also *schoon* beautiful: found in S. Afr. place names Skoonberg, Schoongezicht, Skoonuitsig. [*Afk. fr. Du. schoon* fair, beautiful, pure etc.]

skop [skɔp] *n. and vb. trns. and intrns. slang.* Kick: as in 'I gave him a ~' or have a good time as in 'We're going to give it a ~ that night.' O.I. E. Cape. [*Afk. fr. Du. schoppen* to kick, swing]

No sooner does he chuck me down . . . then I throw him with a skop from my right boot . . . One more skop, this time from behind, and he's out for the count. *Darling* 12.3.75

as *vb.* to kick.

Now here we make friends . . . Skop man, Skop, Skop. *Star* 9.6.73

In combination *op*~ (q.v.), ~ *die blik* (q.v.), ~ *skiet en donder* (q.v.) and ~ *lawaai*, to kick up a row.

We were not out to skop lawaai . . . we wore expensive clothes and moved around with pretty dolls. *Drum* 8.4.72

skop-die-blik [ˈskɔp dĭ ˈblĭk] *vb. phr. as n.* A children's game of hide and seek or 'block' for any number in which if one player can kick the tin which is the block and shout ~! all those already blocked are free: also occ. *kick the tin* (q.v.) and *blikskop.* Also *U.S.* children *kick the can.* [*Afk. sig.* kick the tin]

Of course skat. Doctor says as soon as your leg is strong enough you'll be able to do everything. Even play Skop die Blik. *New S. Afr. Writing* 3 (no date)

skop skiet en donder [ˈskɔp ˌskĭt ən ˈdɔnə(r)] See *skiet en donder.*

skottel [ˈskɔtl̩, -əl] *n. pl.* -s. Dish, platter or bowl: used loosely of many container-utensils. In combination *blik*~ a term of abuse equiv. blighter, blackguard etc., *lit.* 'tin dish'. [*Afk. fr. Du. schotel* prob. cogn. chattel, dish or bowl *fr. Lat. scutella dimin. of scuta* shield (+ *blik* tin)]

Gore became increasingly commonplace . . . and I delighted in the predictable paling of my fellow first aiders . . . 'If you find someone with his entrails beside him, place them in a skottel or even your skirt and transport them together. This is important.' *Fair Lady* 19.7.78

skraal [skrɑːl] *adj. colloq.* Thin, scrawny.

[*Afk. skraal* thin, gaunt, bleak etc.]
She looks . . . And he's hang of a skraal and His hair's only short, hey; . . . Smith *Contrast* 46 Vol. 12 No. 2 Dec. 1978
also figur. cf. 'slender resources', 'slim purse' etc.
She's growing tomatoes, carrots and the whole *tutti* in her window box. She says her pension is too skraal. *S.A.B.C.* (*English Service*) 2.4.78

skrik [skrɪk] *n.* and *vb.* **1.** *colloq.* A fright: see quot. at *papnat.* [*Afk. fr. Du. schrik* fright, alarm]
'Who, me?' . . . meantime feeling the skriks running up and down my spine. *Darling* 12.2.75
Also *figur. Sect. Army:* A particularly ugly person. *cf. Brit. to look a fright.*
Some of the troepies with particularly dirty habits are adequately censured by their mates when they receive the title *vuilgat* whilst a very ugly specimen becomes a *skrik.* Picard in *Eng. Usage in S.A.* Vol. 6 No. 1 May 1975
2. To frighten or be frightened. [*Afk. fr. Du. schrikken* to start, be frightened]
The Britstown Era comes out with . . . 'THE DREAD MESSENGER' . . . The Era should be more careful about 'schrikking' folks. *Grahamstown Journ.* 10.9.1892

skuif(ie) [ˈskœif(i)] *n. pl.* -s, -we. *prison slang.* A cigarette or *zol(l)* (q.v.), also *rokers'* (q.v.) and schoolboys' term for a 'puff': see *skyfie².* [*Afk. skuif* puff of smoke]
The roker did not touch liquor . . . What did he get out of smoking dagga? 'It kept me calm . . . I had to have it everyday – I couldn't go to sleep at night unless I had a good few skuifs.' *Drum* 27.8.67
The shaven heads were thrust forward. 'Skyf' they said, 'Smoke!' The dagga zol was a three-out zol and he drew it quick and long into his body. Muller *Whitey* 1977

skurk [skœrk] *n. pl.* -s, -e. Form of abuse equiv. blackguard, scoundrel etc. [*Afk. fr. Du. schurk* a villain, scoundrel]
Mr . . . told me that he had no comment to make, but said that Mr . . . was a 'skurk'. *Sunday Times* 8.4.79

skut [skœt, skət] *n. pl.* -s. **1.** A (usu.) municipal pound in which straying animals are confined. [*Afk. fr. Du. schut* screen, fence, partition, *vb. schutten* to impound, fence]
Stock was often sent by angry farmers to the skut (pound), and only released on payment of a fee of so much per head. Jackson *Trader on the Veld* 1958
In combination ~*meester* see *poundmaster;* ~*vee,* impounded stock; ~ *kraal.*

. . . on calling for the horses I had engaged, found they had been put in the schut-kraal or pound. Thompson *Travels II* 1827
~ *geld* (money) pound fees.
As his neighbours made their pocket money from '*schutgeld*' (pound fees) by catching his stock trespassing on their farms, Rooi was advised to sell one thousand of his horses to provide the where-withal for the cost of fencing. Jackson *Trader on Veld* 1958
2. *skut* [skət] *vb. trns. slang* among children, and motor cyclists: see *skit.*

skyfie¹ [ˈskeifi] *n. pl.* -s. **1.** *colloq.* A section of fruit, usu. citrus. *cf. Brit. pig* (of orange.) [*Afk. fr. Du. schijf* slice + *dimin. suffix -ie*]
'Garlic . . . two cloves like it says. Two of those little onion things.' '*Two!* Sis man Bloss don't you know a "clove" means a skyfie, not the whole thing?' *Darling* 24.12.75
2. *colloq.* Potato chips (*Brit.* '*crisps*') *usu. pl.:* see also at *slap.* [*fr.* (1) ~]

skyfie² [ˈskeifi] *n. pl.* -s. *slang.* Also *skuifie,* (q.v.), a cigarette, *lit.* a 'puff' or 'draw' *cf. Brit. slang* 'drag'. [*Afk. skuif* puff (of smoke) + *dimin. suffix -ie*]

slaams [slɑːms] See *slamaaier.*

slag [slax] *vb.* and *n.* **1.** To slaughter: also in combination ~ *skaap,* sheep ready for slaughter, and ~ *tery,* see *butchery.* [*Afk. fr. Du. slachten* to kill *cogn.* slay, slaughter]
8 July 1850 Day lovely . . . Slachting sheep at Scott's & William's. J. D. Lewins *MS. Diary*
We will have to slag a sheep next week because we are having visitors. O.I. Schoolgirl 1970
figur. equiv. colloq. 'slay' as in 'Ma'll slag me if I'm late today.'
2. *n. poss. obs.* see quot. also *slagter.* [*fr.* (1) ~]
10th October 1836 Pato said to Colonel Somerset, he had heard . . . that Colonel Smith's last great meeting was to be the time of the great 'slag', and that all the big wigs and friendly chiefs were to be knocked on the head. J. M. Bowker *MS. Diary*

slagter [ˈslaxtə(r)] *n. pl.* -s. *Sect. Army.* Medical Officer. *lit.* Butcher, killer: in place name Slagtersnek: see also *butchery.* [*Afk. fr. Du. slachter* butcher, slayer]

slamaaier [ˌslaˈmaɪə(r)] *n. pl.* -s. Also *slaams:* a *Malay* (q.v.) equivalent of a *herbalist* or *witchdoctor* dealing in the supernatural: see *goëlery* and *toor.* [*prob. fr. Islam*]
The slamaaier whom the 'ghost family'

consulted and who gave them several 'jumats', tiny parcels wrapped in silver paper and brown cotton, has now asked them to send him a black fowl to be slaughtered . . . his jumats did not work. Fire broke out . . . where his jumats were stored. Then a jumat taped across the lintel of a door . . . exploded in flame. *Het Suid Western* 30.7.75

slang [slaŋ] *n. Snake* **1.** Suffixed to the names of several snakes e.g. *boom* ~ (q.v.), *spuug* ~ (q.v.), *bakkop*(~) (q.v.). [*Afk. fr. Du. slang* snake]
2. *prefix* in numerous names often of poisonous plants usu. suggesting some association with snakes, and 'not infrequently involving some popular fallacy' (Smith *S.Afr. Plants* 1966) e.g. ~ *appel Solanum burchellii*, or *S. tomentosum* see *snake apple;* ~ *bessie* (*boom/bos*) species of *Lycium* (also *Solanum burchellii*); ~ *blom Monsonia speciosa;* ~ *bos* (q.v.); ~ *brood*, see *slangkos;* ~ *kop* (head) any of several species including *Lycoperdon hyemalel* see *devil's snuffbox* and other species, some named for the toxic principle contained in them, others from a likeness to a snake's head.
3. Found in place names Slangriver, Slangkop and in combinations ~ *meester*, snake charmer and ~ *steen*(*tjie*) [stɪəŋkĭ, -cĭ] see *snake stone*. [*Afk. fr. Du. slang* snake + *bosch cogn.* bush]
. . . among the Bushmen are found individuals called *slang-meesters* (serpent-masters) who possess the power of charming the fiercest serpents, and of readily curing their bite. Thompson *Travels I* 1827
. . . – the *slangsteentjies* hundreds of years old which came originally from the Dutch East Indies and have been preserved as heirlooms. Seldom can the owner of a snake stone be persuaded to sell it. Green *Karoo* 1955

slangbos [ˈslaŋˌbɔ̃s] *n.* Either *Lycium krausii* (*slangbessie*) or *Elytropappus scaber*. [*Afk. fr. Du. slang* snake + *bosch cogn.* bush]
slangbos . . . This is the plant mentioned by Thunberg (1773) as 'Seriphium' as being a good remedy in the form of decoctions against intestinal worms and also used as an antidote for snake bite. Smith *S. Afr. Plants* 1966

slangkos [ˈslaŋˌkɔ̃s] *n. Amanita phalloides*, the deadly poisonous 'death cup' or 'death cap': also called *slangbrood* (*bread*) and *duivels* (*devil's*) *brood* ⫿The name is from the old belief that snakes

feeding on this fungus obtained their venom from it. [*Afk. fr. Du. slang* snake + *kost* food, victuals]
slap [slap] *adj. colloq.* Feeble, limp, without energy. [*Afk. slap* weak, flaccid]
I'm well over the operation but I still feel a bit slap. O.I. 1971
deriv. ~ *heid n.* Slackness, limpness. [*n.* forming *suffix -heid -ness*]
She shambles across the room . . . and throws herself into a chair arms dangling. There is a familiar *slapheid* about her actions which doesn't quite level with the new look . . . *Darling* 20.12.78
In combination *slang*, (*vulgar*) ~ *gat:* [ˈslapˌxat] slovenly, slack, sloppy.
Mrs . . . said in class this morning that we were the most slapgat school she'd ever been in. Child 13 O.I. 1974
~ *chips, colloq.* Fried potatoes as served with fish or steak not (2) *skyfies*[1] (potato crisps) ; ~ *siekte* dourine: a disease of horses and donkeys similar to *nagana* (q.v.) also caused by a trypanosome. ⫿It can be transmitted by stallion to mare in mating.

slaphakskeentjies [ˈslapˌ(h)akˈskɪənkĭs] *pl. n.* A cooked onion salad *lit.* 'weak little heels'. [*Afk. lit. slap* flabby, *hakskeen* heel + -*tjie dimin. suffix* + -s *pl.*]
Salads, beetroot, tomato . . . and 'slaphak-skeentjies' which for those who don't know, is a salad of small onions in a sweet-sour sauce. Van Biljon *cit. Sunday Times* 24.12.78

slat [slat] *n.* and *vb. slang:* see quot. at *gooi.* Beat, hit or strike: as *n.* esp. among children, a blow, clout etc. as in 'He cheeked him today and so he gave him a helluva slat'. *cf. klap, Jam. Eng. clate, Austral. stoush.* [*Afk. slaat fr. Du. slaan* to hit]

slave *n.* Prefixed to several terms, relics of the slave-owning days of the colony: as in ~ *lodge* (q.v.), ~ *quarters*, usu. crypts under C18 Dutch houses ; ~ *hole*, sleeping place indoors for a female slave ; ~ *bell* a bell set usu. in a tall, whitewashed arch for summoning slaves esp. on Cape farms ; ~ *chair*, see quot.
The utility chairs made of stinkwood or yellowwood with solid square legs but attractively styled wooden backs are usually referred to as Slave chairs denoting that many of these were made by slaves, particularly on farms. Fehr *Treasures at Castle* 1963
Tokai slave bell stays silent Legend says that if the slave bell is tolled on New Year's Eve the ghost will ride again. But nobody dares to

ring it. *E.P. Herald* 4.12.75

Slave Lodge *n. prop.* The building in which the slaves of the Dutch East India Company and later those of the British Government were housed in Cape Town. ⁇Later the old Supreme Court Building, now the Cultural Museum.

Department of the Slave Lodge. . . . a large building in which the government slaves, to the number of 330, are lodged. Afr. Court Calendar 1807

In that year [1811] under British rule, the building was altered by Thibault from the Slave Lodge to the Supreme Court. *Argus* 30.8.73

sleep¹ [sli:p] *vb. intrns. Afr. E.* To lie or to lie down, not necessarily to be asleep: *substandard* but regularly heard as a substitute for 'lie (down)' in the language of children influenced by Africans. see *lala kahle*. [*fr. Ngu. ukulala* to sleep, lie down]

'I was sleeping on the ground in the middle of the scrum and someone stood on me.' Boy 17 O.I. 1974

sleep² [slɪəp] *vb. trns. colloq.* Drag, pull along: *cf. U.S. fr. Yiddish schlepp* haul, drag [*Afk. fr. Du. slepen* drag, tow]

. . . thear apeard to be serfishent blood for a Ox but they Kaffers had sleped him into a bead of rushes. Goldswain *Chronicle I* 1819–36

It's me, that thing you *sleep* along the road. Fugard *Boesman & Lena* 1969

deriv. sleper, a type of bush cutter or eradicator dragged over veld for clearance.

. . . where the veld is fairly level, farmers in this area use, today, an implement known locally as a 'sleper' which, by being dragged through a patch, uproots the bushes. *Farming in S.A.* Dec. 1931 cit. Swart *Africanderisms: Supp.* 1934

and in combination ~ *mist* low lying ground mist or fog.

A policeman described the heavy mist at Luderitz yesterday as 'sleep mist' (rolling fog), which brought visibility down to a minimum. *E.P. Herald* 4.4.74

slenter See *schlenter* also *snyde diamonds, snide diamonds.*

slim [slɪm] *adj. colloq.* Clever, usu. with connotations of cunning, crafty or wily freq. in combination ~ *kêrel*, or with first name ~ *Jannie*, ~ *Piet* etc. See quot. at *kêrel*. [*Afk. fr. Du. slim* sly, cunning, clever]

He is a *slim kêrel*, that . . . to think of turning a hoot into a cough which needed a glass of water. Fairbridge *Which Hath Been* 1913

. . . he was known among his Boer comrades-

at-arms as 'Slim Jannie'. . . . 'Slim' as they used it, simply meant 'clever' and not, as some of his detractors often suggested, 'cunning' or 'crafty'. *Daily News* 19.5.70

Also *inflect. form* ~ *mer.*

The Dutch about here are a slim lot, and the Kaffirs are slimmer. Trust no man, that's my motto. Buchan *Prester John* 1910

slip-slops *pl. n.* Backless sandals: see *beach thongs* and *champals: cf. Brit. flip-flops.* [*fr. sound and appearance*]

Boys going to and from the Swimming Bath, after changing in their Houses, are to wear dressing gowns or overcoats. Permitted footwear: tackies, slipslops or leather shoes. Headmaster's school rules, Grahamstown 1975

sloot [slʊət, slu:t] *n. pl.* -s. Also freq. sp. *sluit* (q.v.). [*Afk. fr. Du. sloot* ditch] **1.** A man-made *furrow* (q.v.) for irrigation purposes or other water supply.

There is no water but what runs down the streets in the sloot, a paved channel, which brings the water from the mountain and supplies the houses and gardens. Duff Gordon *Letters* 1861–2

2. A ditch or other small watercourse usu. worn by rain: see also *donga.* [*fr.* (1) ~]

Water is already running quite strongly in the sloots and veld pitfalls are completely disguised by the snow. *Star* 18.9.71

slopie [ˈslʊəpɪ] *n. pl.* -s. *Rhodesian* Derogatory term for a member of the *S.A.P.* (q.v.) [allegedly *fr.* sloping forehead]

sluit [slu:t] *n. pl.* -s. *Freq.* sp. form in S.A.E. *erron.* for *sloot* (q.v.) poss. *fr. analg.* with sp. 'fruit', or *fr.* confusion with *spruit* (q.v.), also a watercourse. [*pron. sp. Afk. sloot* (q.v.)]

. . . charged and brimming with flood waters from the Drakensberg peaks . . . and from a hundred rushing sluits. (see (2) *sloot*) Dirkby *Thirstland Treks* 1936

Soon he had turned into Dorpstraat, the long oak-shadowed street leading out of the town. On either side the water sang in the sluits. (see (1) *sloot*) Krige *Dream & Desert* 1953

sluk [slœk, slɔk] *vb‾ trns. and n. pl.* -s. *slang.* (To) swallow, gulp. [*Afk fr. Du. slikken* to swallow]

Here – have a drink. Hey, hey, don't sluk the lot! *The Loser* (Radio S.A.) 23.5.72

. . . a tiekie draai coming out the loud speaker and the crowds shifting and muttering and taking sluks out the half jacks they got in they back pockets. *Darling* 26.2.75

. . . hot water melon . . . no shade to keep it cool isn't a blerry joke man. Let alone trying to catch a sluk of yore cane and Coke . . . *Ibid.* 12.2.75

Also *figur.* esp. among children, e.g. 'Make up some story. Ma'll sluk anything.'

slyt [sleɪt] *modifier.* Of or pertaining to sheep usu. ewes with worn-down teeth, usu. fr. age but also fr. hard or sparse (2) *veld* (q.v.). *cf. Austral. gummy*, an old and toothless sheep. [*Afk. fr. Du. slijten* to wear away, wear out]

15 Registered Slyt Ewes *Evening Post Advt.* 7.7.73

1 159 MERINO EWES full mouth – slyt *Farmer's Weekly Advt.* 18.4.73

smaak [smɑk] *vb. trns. slang.* To fancy a girl, clothes etc., to relish food or drink: generally applied sign. 'approve of' esp. among children as in 'I smaak his new jeans, . . . this ice cream, . . . playing rugby when it's cool etc.' [*Afk. fr. Du. smaken* to relish, savour, *etc.*]

She's not so bad, but I smaak the ones with their padding a little lower,' returned the other. *Forum* Vol. 6 No. 2 1970

'The okes will . . . smaak the bit where he saws that blonde in half.' *Darling* 1.10.75

small father, small mother *n. pl.* -s. *Afr.E.* See quot. at *father.*

small little *adj. phr.* See *little.*

He was a small little fellow – no more than 15cm high and 45cm long. *Rand Daily Mail* 8.6.71

small stock unit *n. phr.* See *stock unit.*

smallworking *n. pl.* -s. A privately owned gold mine worked by owners, usu. in Rhodesia.

smear *vb. trns.* To spread or treat floors with *mis* (q.v.) varying mixtures containing cowdung: *cf. Anglo-Ind. leep*, to wash with cowdung and water: *~ing*, a method of floor dressing in country districts: see quots. at (2) *mis* (also *misvloer* and *dung floor*). [*fr. Afk. smeer* to spread, rub]

. . . when the mud floor . . . was freshly smeared with *mist* it smelt of bullock's blood and cowdung as well. Smith *Little Karoo* 1925

'But I can't put them on the floor . . . Wet cow-dung.' The post-office floor had just recently been smeared. Bosman *Jurie Steyn's P.O.* 1971

smeerlap [ˈsmɪə(r)ˌlap] *n. pl.* -s, -pe. *slang.* An abusive mode of address or reference, equiv. blackguard, cad, bastard. [*Afk. fr. Du. smeerlap* lit. 'grease clout', blackguard]

'Ja, I know,' said Van Ryn in Dutch, 'but when an Afrikander's turned khaki like that – King George and the Union Jack stuck all over

him – he's not a brother-in-law, he's a smeerlap.' Black *Dorp* 1920

smeerperskes [ˈsmɪə(r)ˌpɛə(r)skəs,-ˌpeːr-] See *perskesmeer*, also *peach leather.*

smell out *vb. trns.* Also as *vbl n. smelling out*, the process of detection of a witch or other evildoer by a *witchdoctor* (q.v.) by means of medicines and incantations, freq. assisted by corroboration or the reverse by his hearers: as *vb.* to detect a witch by these means. [*trans. Bantu/ Ngu. ukuNuka* to divine, smell out, smell]

. . . there will be a smelling out, but a smelling out of a new sort, for he and you shall be the witch-finders, and at that smelling out he will give to death all those whom he fears. Haggard *Nada* 1895

. . . The remedy was for the 'victim' to go to a witchdoctor, who would smell the witch out, sometimes using information gleaned from the unwitting patient. *Grocott's Mail* 9.5.72

Angry villagers believed that the little girl had been cursed by witches . . . and called in a sangoma, or witchdoctor from Komatipoort to 'smell them out'. *Sunday Times* 9.1.77

smelter [ˈsmeltə(r)] *n. pl.* -s. Also *same-smelter*, fusionist, a member or supporter of the coalition of the *Nat* (National) and *Sap* (q.v.) (S. Afr.) Parties which formed the *United Party* (q.v.) in 1933, of which the present Nationalist Party (*Nats*, (1) *HNP*) was a splinter group. [*Afk. smelt* to coalesce, merge + *agent. suffix -er fr. Du. smelten* to melt (*same* together)]

There was a big photograph of General Smuts on the wall. Mr. . . . described himself as a 'smelter', a sort of coalition Sap. *Cape Times* 8.11.72

smokkel [ˈsmɔkəl] *vb. slang prob. reg. Cape.* To deal in illicit liquor. [*Afk. fr. Du. smokkelen, cogn.* smuggle]

'. . . We know that you smokkel here!' the cop snapped . . . 'We know that these skolly bastards gather here to buy wine . . . one of these days I'll catch you at it. Muller *Whitey* 1977

deriv. ~ing, vbl. n. also *modifier: cf. shebeening*

. . . Willy came through a door which led to the kitchen – probably the smokkeling room used at night. . . . I had three little children and I had to feed them, so I turned to smokkeling. *Ibid.*

in combination ~huis a *shebeen* (q.v.)

The madam of the 'smokkelhuis' . . . that international mother figure of the underworld . . . is well drawn. *E.P. Herald* 4.5.77

smoor [smʊə(r)] *vb. trns.* Braise, stew.

. . . fry (or 'smoor' as they say in Dutch) in boiling fat or butter. Mix with the stock. *Duckitt's Recipes* 1966 *ed.* Kuttel
In combinations ~ *snoek* (q.v.) ~ *vis* also partic. ~ *ed fish, snoek, chicken*, mixed dishes of fish braised with onion, potatoes, spices and sometimes tomatoes, usu. served with rice: also occ. made with chicken: see quot. at *stockfish*. [*Afk. fr. Du. smoren* to stew *cogn.* smother]
His smoorvis was magnificent and this added the aroma of onions and chillies to the air. Green *When Journey's Over* 1972

smous [sməʊs] *n. pl.* -es. Formerly an itinerant pedlar, often Jewish, who made a living hawking various goods from farm to farm: *cf. Jam. Eng. higgler.* [*Afk. smous* hawker, pedlar *fr. Du. smous* (abusive name for a Jew, *fr. mausche rel.* Moses)]
. . . but brandy (the only luxury besides tobacco in which the poorer boors indulge) is purchased from *smouses*, or hawkers, who traverse the remotest skirts of the Colony with waggon-loads of this detestable beverage. Thompson *Travels II* 1827
In combination *vrugte* ~ (fruit), *lappie* (q.v.) ~, a soft goods pedlar dealing in cloth and other materials; *gold* ~ a ~ dealing in jewellery.
Craftiest of all these traders was the gold smous . . . When he entered a new district he selected a man of standing . . . and sold him a first-class gold watch at far below cost price. That was all the advertisement necessary. Green *Land of Afternoon* 1949
fish ~, an itinerant vendor of fish. See also *fish cart/fish horn.*
. . . a fish smous of Boekenhout Street . . . *Het Suid Western* 26.5.76

smous(e) [sməʊs, smaʊz] *vb. usu. intrns.* To peddle goods in outlying districts: also *partic.* or *vbl n.* ~ *ing*: also, loosely, to ~ *around*: see *fossick.* [*fr. smous* (q.v.)]
Having given up Kurveying – I thought a Smousing trip might pay, I got a waggon load of goods from W.R. Thompson at 6 months Credit, and started. Stubbs *Remin. I* 1876

snaaks [snɑks] *adj. slang.* Peculiar, strange etc., funny (amusing). [*Afk. fr. Du. snaaks* comical, droll]
Yesterday, when it was nearly dark, I knew there was something snaaks about the bushes; I was trying to remember. Philip *Caravan Caravel* 1973

snake apple *n. pl.* -s. *Solanum burchellii*, dwarf spiny species of the Solanaceae with poisonous fruit: also *slangappel* (q.v.) *bitterappel* and *devil's apple*: see also *satansbos.* [*poss. trans. slangappel poss. vice versa*]
A family which covers a very great variety of plants is the Solanaceae family . . . Included is the plant known as 'devil's apple' or 'snake apple', the fruit of which is very poisonous. *Evening Post* 30.6.73

snake stone *n. pl.* -s. Also *slang(e)steentjie* (q.v.): a porous stone or bone formerly believed to extract and absorb venom from a snake bite. Also *Brit.* and *U.S.* snakestone, adderstone, serpent stone.
The Hottentots are acquainted with several vegetable antidotes against the poison of serpents; but the most approved remedy amongst the Dutch is the *slange-steen* or *snake-stone*, which they hold to be infallible. Barrow *Travels I* 1801 *cit.* Pettman

snaphaan [ˈsnapˌhɑn] *n. pl.* -s. Early name for a muzzle-loading flintlock musket, also called *baviaanboud* (baboon's buttock). [*Afk. fr. Du. snaphaan* (*early Eng. snaphance*) early type of flint lock *fr. snap(pen)* + *haan* hammer of a gun: *also sig.* cock]
The gun known to the trekker as a snaphaan was of the flintlock pattern and it fired round, lead bullets. Botha *Our S.A.* 1938
Flintlocks were the guns mostly used by the Voortrekkers, since percussion caps reached them only between 1834 and 1846. These expert hunters did not speak of a roer but used the term snaphaan (flintlock). *Daily Dispatch* 22.7.72

sneeu- [ˈsniu] *n.* Snow: *prefix* found in place names Sneeuberg, Sneeukraal and in flower name ~ *blom Protea cryophila* which is densely white and grows above the snow line. [*Afk. fr. Du. sneeuw cogn.* snow]

sneezewood *n.* The large forest tree *Ptaeroxylon obliquum* (*P. utile*) and its timber which is exceptionally hard and durable, used for fencing posts, railway sleepers and some furniture. ¶Its smell causes sneezing and hay fever. [*trans. Afk. nies* sneeze + *hout* wood]
. . . it was soon found that, apart from stinkwood, sneezewood was the only really durable material, and this proved extremely difficult to shape. Lewcock *C19 Architecture* 1963
Sneezewood is too hard for nails, so the steps . . . are kept in place by pegs. *Grocott's Mail* 27.7.71
Sneezewood, where I have seen it polished, has an almost glassy gleam. *E.P. Herald* 30.7.73

snelskrif [ˈsnelˌskrɪf] *n. prop.* A system of shorthand devised for the Afrikaans language. [*Afk. fr. Du. snel* swift, rapid + *skrif cogn.* script]

PART-TIME CLASSES IN Shorthand Snelskrif Typewriting Bookkeeping *Cape Times Advt.* 3.5.72

snik [snɪk] *vb. and interj. colloq.* Sob: as verb or as interpolation in conversation sig. semi-jocular, self-pity or sympathy, *cf. shame.* [*Afk. fr. Du. snikken* to sob]

... she starts off on the snot and trane again. You should jis hear the carry-on ... And she starts off snikking much better. *Darling* 20.8.75

snide [snaɪd] See *snyde.*

snoek [snŭk] *n. pl.* ∅. *Thyrsites atun,* the edible, small-scaled marine fish of the Gempylidae fam. common in Cape waters. It is frequently salted and dried especially at the Cape where it is used for the traditional dishes *smoor* (q.v.) ~ and ~ *pekelaar* (q.v.). Also combinations *China* ~, a smaller variety appearing later in the season, and ~*kierie,* the club with which this fierce fish is dispatched when caught or landed, ~(*ing*) *season,* ~ *boats,* ~ *fishing* etc. *cf. U.S. snook, Jam. Eng. lanternjaw snook* etc. ⚲ ~ is loosely applied in Natal to the 'queen mackerel' *Scomberomorus lineolatus* (occ. confused with the 'king mackerel' or *katonkel* (q.v.) the Natal barracuda). In Transkei ~ is used of various spp. of the true barracuda (Sphyraenidae fam.). See quot. at (2) *duiker.* [*Afk. fr. Du. (zee) snoek* (sea) pike]

In naming the snoek 'zee snoek' the pioneers obviously realized that they had found something different from the Dutch fresh-water pike. Small snoek, up to twenty-four inches, are often referred to as *China snoek.* Fishermen declare that China snoek are caught after the ordinary snoeking season is over; they have thicker bodies and shorter heads than the large snoek. Green *Grow Lovely* 1951

The men had caught a lot of snoek. Lena had never seen snoek before. She was glad she had never had to deal with those hideous jaws with the wicked, needle-like teeth. Meiring *Candle in Wind* 1959

snoep [snŭp] *adj. colloq.* Stingy, miserly, in freq. use usu. sig. mean with money, food or belongings. *cf. Austral. mangy.* [*Afk. snoep* greedy, grasping *fr. Du. snoepen* to eat sweets]

... try and pick up some free samples ... only the make up folks is getting hang of a snoep like everyone else these days. *Darling* 9.7.75

snot-en-trane [ˌsnɔtənˈtrɑːnə] *n. phr. usu. modifier. slang. lit.* 'Snot and tears':

a state of maudlin misery, of or pertaining to such a state, or to literature or other entertainment inducing it: see also quot. at *snik.* Brit. and *U.S. tearjerker.* [*Afk. fr. Du. snotteren* to snivel + *en* and + *traan* tear + *pl. suffix -e*]

Now is the time, surely, for Afrikaners to get away from what someone has called the 'snot and trane' approach to drama. *Grocott's Mail* 10.10.72

He suggests that up to now Afrikaans dramas have been of the 'snot en trane' variety. *Ibid* 13.10.72

snotsiekte [ˈsnɔtˌsiˑktə] *n.* Any of several animal diseases manifested in copious mucous discharge from the nose: malignant catarrhal fever of stock, generally thought to be transmitted by *wildebeest* (q.v.) or veld on which wildebeest have grazed. [*Afk. fr. Du. snot* nasal mucus + *siekte fr. Du. ziekte* disease]

Many years ago when game was plentiful, it was observed by farmers in the Transvaal and elsewhere in South Africa that a peculiar disease which they called snotsiekte ... broke out among their cattle in contact with wildebeeste. *E.P. Herald* 24.7.1925

snuff box, devil's, old maid's *n. Lycoperdon hyemale :* see *devil's snuff box.*

sny¹ [sneɪ] *vb. trns.* To castrate: see *cut.* [*Afk. fr. Du. snijden* to cut, castrate, geld]

sny² [sneɪ] *vb. trns. freq. pass. colloq.* To oust or 'cut out' someone with a boy or girl friend: hence *to be* ~*d,* to be superseded by someone else in the affections of the beloved. [*poss. fr. Eng.* cut out]

He's always so madly jealous of her because he's terrified of being sny-d. Girl 15, O.I. 1974

snyde diamond(s) [snaɪd] *n. phr.* Imitation diamonds used in the early days of the Diamond Fields for swindling or other illicit purposes, or for trapping: see also *schlenter (slenter):* also applied to counterfeit gold coins or other spurious articles, see second quot. [*prob. fr. Du. snijden* to overcharge or fleece, *or poss. fr. Yiddish*]

A brief examination satisfied the disgusted inspector that the astute Yankee had once more turned the laugh against him. The things were 'schlenters' or *snyde diamonds. cit.* Pettman *Africanderisms* 1913 (no source given)

... men who travelled in Cape Carts laden with 'snide' jewellery and merchandise were the curse of the country. Cohen *Remin. of Kimberley* 1911

snysels ['sneɪsəls, -z] *pl. n.* Noodles: see quot., also *melk* ~ (q.v.) [*Afk. snysels* noodles *fr. Du. snijdsels* clippings, cuttings]
. . . how fond the children had been of her snysels! . . . flour mixed with well-beaten egg till it was stiff dough; Then you had to roll the paste, lifting it as you rolled . . . like a carpet and then you cut it lengthwise into strips. Cloete *Watch for Dawn* 1939

so *substandard.* **1.** Used either in deviant word order e.g. 'he gets so on her nerves' or redundantly in S.A.E.: in phr. ~ *a little.* e.g. 'It was my birthday: we danced so a little and had a few drinks.' Coloured informant, Cape Town 1971. [*trans. Afk. so 'n bietjie* just a little, a bit]
'Can you read and write?' 'Ja meneer, so a little.' Marchand *Dirk, A South African* 1913 cit. Swart 1934 *Africanderisms Supp.*
2. Also in phr. *or* ~, as in 'let's have a cup of coffee or so.' [*trans. and abbr. Afk. of so(iets)* or something of the sort, like that, *etc.*]

sobo sobo ['sɔbɔˌsɔbɔ] *n.* Also *msobosobo*: *Solanum nigrum*: also *nas(ter)gal*, the berries of which are used for a jam similar to blackberry: see quot. at *nas(ter)gal.* [*Ngu.* (*u*)*msobosobo plant name for this species*]
. . . the plant is a species of Solanum . . . some of which are poisonous, and some of which are edible. This complex of species includes the European species Solanum nigrum, Black or Garden Nightshade. . . . The name is usually given as msobo or umsobo, sometimes as msobo-sobo in the Transkei. *Farmer's Weekly* 21.4.72

soet- [sŭt] *adj.* Sweet: *prefix* sig. usu. sweet-tasting as in ~ *koekies* (q.v.) ~ *riet* (q.v.) etc., as in the names of plants, sweet smelling or 'sweet' sig. opposite of 'bitter' or 'sour' and therefore suitable for stock e.g. ~ *gras* (q.v.); ~ *veld* see (2) *veld, sweet;* ~ *spekboom* (see *spekboom*) etc. Also found in place names e.g. Soetmelksrivier, Soetdorings. [*Afk. fr. Du. zoet* sweet]

soetes See *sweetwine.*
. . . do just one last little thing for me – if you'll just get me a bottle of soetes, and then lock me in here. Muller *Whitey* 1977

soetgras ['sŭtˌxras] *n.* Numerous species of palatable grazing grasses including all varieties of *Eragrostis*: also *sweet grass* (q.v.) (*veld*) (q.v.). [*Afk. fr. Du. zoet* sweet + *gras cogn.* grass]

soetkoekie ['sŭtˌkŭkĭ] *n. pl.* -s. Traditional spiced sweet biscuits. [*Afk. fr. Du. zoet* sweet + *koek* cake + *dimin. suffix -ie*]
Competitions involve . . . six soetkoekies made from a traditional South African recipe and an arrangement of dahlias. *E.P. Herald* 21.1.71

soetriet ['sŭtˌrĭt] *n. Sorghum dochna*, also *imfe* (q.v.). [*Afk. fr. Du. zoet* sweet + *riet cogn.* reed]
Soet riet, the indigenous type of sugar-cane, was to us what ice cream is to the modern child. Pohl *Dawn and After* 1964

solder ['sɔ̃ldə(r)] *n. pl.* -s. Loft or attic under the roof of a house: see also *dakkamer* and *brandsolder.* [*Afk. fr. Du. zolder* loft]
There are no ceilings; the floor of the zolder is made of yellow wood, and, resting on beams, forms the ceiling of my room, and the thatch alone covers that. Duff Gordon *Letters* 1861–2

soldier *n. pl.* Ø, -s. **1.** *Kniphofia uvaria* and related species: see *red hot poker.*
2. *Cheimerius nufar*, an edible marine fish also known as 'santer': in combination *black king* ~. [*presum. fr. military appearance*]
It is seldom that a soldier is caught from the surf. The fish . . . was a nice specimen of 2 lb 15oz when gutted and gilled. *Albany Mercury* 19.1.70
. . . yesterday boated some unusual and rare big 'black' king soldiers averaging 2.7kg to 3.6kg. These are excellent table fish. *Daily Dispatch* 26.2.71

soldoedie [ˌsɔ̃ldˈdŭdĭ] *n. pl.* -s. *colloq.* George District: A member of the Women's Army from the S. Afr. Army Women's College at George: see *Botha('s) Babes.* ['*portmanteau word' Afk. soldaat* soldier + *doedie* (q.v.) popsy]
Girls will be arriving . . . from all over the country today when this year's 152 soldoedies start their 10-month training at the Civil Defence College . . . Already 587 soldoedies have completed their training. *Het Suid Western* 30.1.75

so long *adv. colloq.* Used in S.A.E. as *adv. of duration*, equiv. of 'in the meanwhile,' 'for the time being'. ⫽Not equiv. of conditional ' ~ *as*' or of *colloq. so long* as form of farewell. [*translit. Afk. solank* meanwhile, for the time being]
'You go and sit down so long,' Vincent said. *Evening Post* 10.6.72

something else *n. phr. colloq.* Good, excellent, exceptional: common term of approbation usu. among younger speakers. ⫽A similar usage was noted by Lady

Anne Barnard 1797–1802. 'Ah! Lord Macartney, that is a different thing!' – is the sort of praise given by those whose stock of English does not afford a more particular list of the qualitys they allude to. [*prob. trans. Afk. iets anders*, something else, something different]

Suddenly he's 'something else': . . . South African theatre's 'child of fortune' is the hottest pop star in the country – and he hates it . . . the idol of thousands of teenybopper girls. *T.V. Times* 6.6.76

sommer ['sɔmə(r)] *adv. colloq.* Various meanings; usu. equiv. of 'just', 'simply', 'merely', also used redundantly in phr. *just* ~: see Pettman quot. and quot. at *mos*[1]. [*fr. Afk. sommer(so)* simply, just, *fr. Du. zo maar zo* so-so]

The Dutch word *somar* . . . is also a word to which I think I could challenge the most learned schoolmaster in the Colony to attach any definite meaning. It is used by both Boers and Hottentots in almost every sentence; it is an answer to every question; and its meanings are endless. Gordon Cumming *Adventures I* 1850 *cit.* Pettman

. . . a lot of the goodies . . . were found in abandoned henhouses . . . or sommer standing around in the rain. Van Biljon *cit. Star* 16.6.73

somoosa See *samoosa*.

songololo [ˌsɔngəˈlɔlɔ] *n. pl.* -s. *Jurus terestris*: any of several of the so-called 'pill' millipedes with hard shiny exterior armour, which roll into a flat, spring-like coil if touched or alarmed. ⟨Their coming indoors is taken by many to be a sign of imminent rain. [*Ngu. songololo fr. vb. ukusonga* to roll up]

So we decide on shongalolos . . . we train them to crawl up and down . . . instead of curling up like swiss rolls when I touch them. *Darling* 24.12.75

A songololo . . . is also a plague in Tamboerskloof where after the first rains they have marched in their serried ranks down the mountain into . . . crevices, baths, curtains, up walls, down paintings, into cupboards. Van Biljon *cit. Sunday Times* 1.5.77

Xhosa women call the coiled intra-uterine contraceptive device songololo. *Fair Lady* 17.7.77

soojee See *Indian terms*.

sopie ['suəpĭ] *n. pl.* -s. *colloq.* A drink or tot, usu. of spirits, of variable quantity: 'one man's soopje would be another man's overthrow' (Pettman 1913): see also quot. at *Cape Brandy*. cf. *Scottish dram*, *Canad. smash, snort, U.S. slug, Austral. nobbler*. [*Afk. fr. Du. zoopje* a little glass *fr. zuipen* to tipple or

270

drink *cogn. Eng.* sup, sip]

He pressed me to take a soopie with him, to which I willingly agreed, as the night was very chilly, but asked for water to mix with the brandy. Thompson *Travels I* 1827

sorghum beer *n.* Traditional African beer brewed with malted sorghum millet also known as *mabela* (q.v.) and *kaffircorn* (q.v.) or '*kc*' ⟨The generic name *sorghum* is now being adopted: see quot. below and at *mqomboti*.

The sale of sorghum beer had not been prohibited in Transkei and as it was included in definitions of the Liquor Act it could be sold by the holders of bottle store and hotel licences. *Indaba* 6.4.79

sosatie(s) [səˈsatĭs, -z] *pl. n.* See quot. and description at *sasatie*: similar to kabob or kebab: curried meat on skewers. ⟨~ can be bought ready prepared for cooking from some butchers: see also quot. at *sea-cat*. [*Afk. fr. Du. prob. fr. Malay saté*, spiced meat on skewers, pork, beef, mutton or chicken. ⟨Said by various authorities to be *fr. Malay sisateh*, minced or chopped meat]

Less frequent items on the braaivleis menu are sosaties, the kabobs of Europe and America. . . . At a braaivleis the sosaties are eaten straight from the grill. *Farmer's Weekly* 25.4.73

In combination: ~ *Western* cf. '*Spaghetti Western*' A *skiet en donder* (q.v.) S. Afr. film: see also at *boerewors*.

He can't be real. He looks gorgeous and talks like a bad movie: Robert Redford in a Sosatie-Western. *Fair Lady* 21.7.76

so size *adj. phr. colloq.* See also *tamaai*. Used esp. but not exclusively among children equiv. of 'as big as this' or 'so big' also 'so high' usu. accompanied by gesture: see quot. at (1) *pikkie*. [*poss. fr. Afk. so groot* as big as, so large, *so hoog*, so high, *etc. See etym. tamaai*]

Sotho ['sutŭ] *n.* **1.** Also *Sesotho, Sesuto,* the language of the *Basotho* people of *Lesotho* (q.v.): also in combination *North* ~, *South* ~ for the principle dialects of ~.

I managed . . . to get a neatly typewritten letter on how he was converted to this religion. It is in Sotho. *Drum* Sept. 1968

2. In combination *North* ~ and *South* ~, used of the main peoples of Lesotho or a member of either: usu. *sing.* form *Mosotho*: see quot. at *Sesotho*.

I am 32, a Southern Sotho. *Drum* 8.3.73

. . . he had had consultations with the territorial authorities of the . . . South Sotho, the

Tswanas, the Venda, the North Sotho and the Machangana. *Daily Dispatch* 4.2.71

sour *adj.* In S.A. used of or pertaining to grass (2) *veld* (q.v.) or grazing of a particular type: also occ. of the soil which supports it: see (2) *veld, sour* (also *sweet*) and quot. at *swaer*. [*trans. Afk. suur fr. Du. zuur*]

The best grazing districts, more especially for sheep, are at a distance from the sea; the grass near the coast, being constantly fed by the heavy dews which prevail in that region, becomes long, coarse, and 'sour'. Shaw *My Mission* 1860

Local variations in . . . annual rainfall lead to certain differences in the type of grazing veld . . . commonly indicated by the terms 'sour' and 'sweet'. . . . Where the rainfall is high 'sour' conditions prevail while 'sweet' conditions are found where the rainfall is comparatively low. *H'book for Farmers* 1937

sour fig *n. pl.* -s. Also *suurvy, ghokum* (Hott. *gaukum*), *Hottentot fig.* The sour fruit of either of two species of *Carpobrotus, C. acinaciformis* or *C. muirii*, used for jams and preserves, see *ghokum;* the plant itself often planted in dune lands as a sand binder. [*trans. Afk. fr. Du. zuur + vijg* fig]

In the Duine, . . . of the Southern Cape coast . . . where almost nothing grows except thatch reed, farmers are turning more and more to the cultivation of sour figs. *Argus* 13.4.73

Preserve, Hottentot Fig or Sour Fig. The Hottentot fig is the fruit of a kind of mesembreanthemum which grows wild at the Cape. *Duckitt's Recipes* 1966 ed. Kuttel

sour porridge *n.* Also *suurpap* (q.v.) Mealie meal porridge, soured usu. by natural fermentation, made and eaten by Africans: see also *mahewu*.

sous-[1] [səus] *n.* Sauce: prefixed to certain items prepared in a sauce e.g. ~*boontjies* (q.v.) ~*kluitjies* (dumplings) served in a cinnamon and butter sauce. [*Afk. fr. Du. saus cogn.* sauce]

sous[2] [səus] n. Sect. *Army.* Petrol. [*Afk. lit.* gravy *cogn.* sauce]

'You stuck?'
'We've run out of sous, dammit.' *O.I. Ex-Serviceman* 14.2.79

sousboontjies ['səusˌbŭĭŋkĭs] *pl. n.* Brown haricot beans in a sweet-sour sauce, usu. served cold as a side dish, occ. hot as a vegetable: commercially available in jars. [*Afk. fr. Du. saus cogn.* sauce + *boon* bean + *dimin. suffix* -tjie + *pl.* -s]

Sousboontjies
2 cups dried sugar beans (brown flecked dried beans)
½ cup sugar
¼ cup vinegar
Salt
. . . Serve with venison or roast mutton. *Evening Post* 17.10.70

sout [səut] *n.* Salt [*Afk. fr. Du. zout cogn.* salt] **1.** *prefix* sig. 'prepared with salt' as in ~*ribbetjie* (q.v.); ~*appelkoos* apricots pickled with salt, *cf. mebos;* ~*happies*, cocktail snacks, see *hap;* 'containing salt' as in ~*vaatjie;* ~*balie*, a vessel for the storage of salt; or, as in the case of plant names, 'flourishing in brak or saline soils' as in ~*bos(sie)* (q.v.); ~*ganna, Salsola strobilifera*, see *ganna;* and ~*vygie, Sceletium varians,* see *vygie:* also ~*pan* (q.v.) and ~*piel* see *soutie.*

2. *n. prefix* found in place names Soutkuil, Soutpan (q.v.) and Soutpansberg.

soutbos ['səutˌbŏs] *n. Atriplex capensis:* applied also as a general term for other species including *Chenopodium* and *Exomis* which grow in *brak* (q.v.) saline soils: see also *brakbos* and *ganna* [*Afk. fr. Du. zout cogn.* salt + *bosch cogn.* bush]

The sun blisters the driedoorn* and the soutbos that battle to keep alive in the desert. Birkby *Thirstland Treks* 1936
*see *driedoring*

south easter *n. pl.* -s. The prevailing wind at the Cape during the summer months also known as the *Cape Doctor* (q.v.): in combinations *black* ~, *blind* ~, *tablecloth* (q.v.) ~, see quot. at *fish cart. cf. Austral. (southerly) buster.* [*fr. S.E.* wind]

It was the season of the 'south-easter'; and the Cape of Storms soon proved that its original name was not undeserved. Alexander *Western Africa I* 1837

Though the long afternoons are hot, there is a movement of cloud and at last an end to the roaring South-Easter that so bedevils the Cape. Thompson *Richard's Way* 1965

South West *n. prop. abbr.* South West Africa: see quot. at *schlenter.*

. . . he's just made a fortune in South-West. He sold his civil engineering group, the largest in South-West . . . in one of the biggest takeovers in the territory's history. *Daily News* 10.6.70

deriv. ~*er*, an inhabitant of ~: see *Herero, Damara, Baster, Namaqua,* also *Namibia.*

It was decided to have a pensions scheme . . .

for everyone, better housing for various groups and a uniform identity document for all South Westers. *Panorama* Aug. 1976

South West Africa Specialist Unit *n. prop.* See '*Specs*'.

The SWA Specialist Unit, as it is designated was 'declassified' this week when military correspondents were flown to its training base in the vicinity of the Etosha Pan . . . The unit contains one of the oldest forms of warfare and hunting . . . dogs and horses – with the latest in powerful Scrambler motor cycles to track down terrorist elements. *Daily Dispatch* 14.6.79

soutie [ˈsəutĭ] *n. pl.* -s. *slang.* An Englishman: *abbr. soutpiel* (vulgar). Common in Army usage: euphemisms *soutkop*, *kopsout* and *soutriem* (q.v.) are also found. [see second quot.]

In Mafeking they are more used to the English. There was a time when the only people in Mafeking were the souties and their servants, and this has made the area different from the Marico. *Star* 29.8.78

. . . soutie for an Englishman – one foot in England one in South Africa, appendage in the sea. *Informant Serviceman: Letter* 4.2.79

soutpan [ˈsəutˌpan] *n.* Salt pan. [*Afk. fr. Du. zoutpan* salt lake/pan] **1.** See *pan, salt.*

2. Found in place names Soutpan and Soutpansberg.

soutribbetjie [ˈsəutˌrĭbĭkĭ, -cĭ] *n. pl.* -s. Loin chops, see also *ribbetjie*, pickled with salt and dried after which they are grilled or *braai'd* (q.v.) over wood-fire coals. (q.v.) [*Afk. fr. Du. zout cogn.* salt + *rib* + *dimin. suffix* -(*b*)*etjie*]

Soutribbetjie: Take ribs of mutton and salt them well . . . Let the ribs lie in a covered basin for two to three days . . . then take them out of the brine and hang them up in a sheltered place to dry slowly. Gerber *Cape Cookery* 1950

soutriem [ˈsəutˌrĭm] *n. pl.* -e. Euphemism for *soutpiel*, an Englishman: see *soutie*. [*Afk. sout cogn.* salt + *riem* thong]

Without asking the price the Englishman had offered a hundred and fifty pounds . . . and bought oxen worth ten pounds for sixteen. The Dutchman chuckled, for he had the 'Salt reim's' money in the box under the bed. Schreiner *Story of an Afr. Farm* 1883

so waar [ˌsŭˈvɑːr] *interrogative and interj.* Interrogative: an expression of amazed incredulity: see also *wragtig?* and *is it?*: and an *interj.* to emphasize the truth of the speaker's statement, occ. in form *so wragtie waar!* cf. *true as God*, see quot. at *Meraai*. [*Afk. so waar* really and truly, *prob. abbr.* ~ *as ek lewe*, or

~ *as God*]

Don't you worry . . . And, Oom Karel silently added, if you call me 'my dear fellow' once more, sowaar I might just do something irresponsible. *Farmer's Weekly* 4.12.74

Soweto [səˈweːtŭ, ˌsɔ̃-] *n. prop.* Place name: the scene of the first of the 1976 student riots *cf. U.S.* use of *Watergate*. [*acronym* South Western Townships (q.v.), Johannesburg.]

And now that Soweto has happened, the solidarity between the two generations is genuinely impassioned. *New Statesman* 10.9.76

. . . three possibilities . . . a mixture of the two . . . sporadic Sowetos on a wider scale. *cit. Fair Lady* 27.10.76

spaanspek [ˈspănˌspek] More usu. *spanspek* (q.v.)

span¹ [spæn] *n. pl.* -s. **1.** A team of draught animals usu. oxen though also used occ. of horses, also *Canad.* see first quot. at *tulp.* [*Afk. fr. Du. span* team]

. . . it requires from twenty to thirty oxen, divided into two teams or spans, for a weight of two thousand pounds. Thompson *Travels II* 1827

2. A team or gang of workers: see also quot. at (2) *agterryer* [*fr.* (1) ~]

The group with which each boy paraded was called his *span*, which is the Afrikaans word for 'team' or 'gang'. Paton *Kontakion* 1969

I have six labourers . . . I try to lead by example, to instill the will to work . . . I can do this with a small span of six, but when there are more their negative attitudes are too strong for me. *Darling* 3.3.76

span² [spæn] *colloq.* 'A lot', 'very much' or indefinite measure esp. in expression 'Thanks a ~' or *pl.* as in 'She's got spans of friends, money, clothes etc.'. [*Afk. span poss. fr. Du. span* span of the hand, rough linear measure]

. . . platform all set up with ropes and floodlights and a span more folks packed around on folding chairs . . . a span of titters: then the bell rings and the fight's on. *Darling* 26.2.75

span³ [spæn] *vb. rare.* To harness or yoke draught animals usu. in combination ~ *in* (see *inspan*) or ~ *out* (see *outspan*). [*fr. Du. spannen* to put (horses) to]

The ox wagons brought us oranges . . . Every year at this time they come . . . The wagons all span out at the outspan place. The drivers chase the oxen to the dam to drink water then let them eat. Vaughan *Diary circa* 1902

spanspek [ˈspanˌspek] *n. pl.* -s. The musk melon or canteloupe *Cucumis melo* of the *Cucurbitae* is so called in S.A.: see quot. at *pampelmoes¹*: ¶Said to be called 'Spanish bacon' from Sir Harry Smith's Spanish wife's preference for it at breakfast: prob. apocryphal; see first quot. [*Afk. fr. early Du. spaenspek musk melon*]

. . . a melon called 'span-spek' (Span-ham) a name which Ouma . . . maintained was a contraction of Spanish-ham, being eaten as an hors-d'oeuvres with ham. Van der Post *Story Like Wind* 1972.

Some farmers who have irrigation water have started growing 'spanspek' (sweet melon) which is exported particularly to France. Within 70 hours of being picked at Prince Albert they are on the market in Paris. *Panorama* Apr. 1974

Spear, the *n. prop. abbr. Shaka's Spear* (q.v.) see also *Mkonto ka Shaka.*

The Spear was perpetrated by Whites to work against . . . promised the Spear that . . . would be toppled. *Drum* 8.8.74

spears *to wash hist. prob. obs* See *wash spears.*

Special Branch, See *S.B.*

speck See *spek.*

Specs [speks] *pl. n.* The *South West African Specialist Unit,* a tracker unit in use since 1977 operating on foot, with dogs, on horseback and on scrambler motor cycles: *cf. Selous Scouts* (q.v.). (*Afk.* form *Swaspes*).

. . . trained 'Specs' as they like to call themselves, contribute increasingly to anti-insurgency. The unit consists of men trained as expert trackers on foot . . . with a tracker dog and a mounted section with men who can track from horseback. *Daily Dispatch* 14.6.79

speel [spɪəl] See *spiel.*

spek [spek] *n. and n. modifier. Also anglicization speck:* fat salt pork not usu. smoked, used for larding venison or other game, available under the name ~ from butchers: in combination ~ *strips,* prepared lardoons: see also quot. at *zeekoe.* [*Afk. fr. Du. spek* fat pork, bacon]

Lard the meat with speck strips. Mix all the ingredients and press on top of the meat. Wrap meat . . . and bake . . . *Sunday Times* 6.2.72

spekboom ['spek,buəm] *n. Portulacaria afra,* also *olifantskos* or *elephants' food:* a small leaved succulent shrub with pink to lilac flowers: a valuable fodder plant in times of drought. ⁋It is planted in certain areas to assist in *veld reclamation* (q.v.), and can also be planted to form a garden hedge in the E. Cape. [*Afk. fr. Du. spek* fat pork, bacon +

boom cogn. Ger. Baum tree]

One of the most valuable shrubs . . . is the spek-boom (portulacaria afra). It is found in great abundance on the stony ridges and affords excellent food for those large flocks of sheep, and especially of goats . . . In severe droughts this bush is truly invaluable. *C. of G.H. Almanac* 1843

Trials conducted by the Dohne Research Station on bushveld farms have shown . . . that elephant's food or spekboom contributes the most to the diet of goats . . . Elephant's food recovers slowly after being grazed. *Grocott's Mail* 31.12.74

spider[1] *n. pl.* -s. A light carriage with disproportionately large and slender wheels: in combination *American* ~, ~ *cart, wagon;* also *German* ~ more heavily built, and *four wheeled* ~; see second quot. and quot. at *splinter new.* [*fr. wheels*]

On Monday a pair of horses belonging to Mr. Robt. Warren bolted with a spider from the Episcopal Church at Kei Road and made a complete smash of the vehicle. *E.L. Dispatch and Frontier Advertiser* 15.1.1881

Horsemen leading the procession of carriages and four-wheeled spiders in which the Prime Minister and other dignitaries arrived for the celebrations. *Panorama* June 1974

spider[2] *n. pl.* -s *slang.* See *rock spider.*

The spiders all crawl out of their holes and from under their stones and put on their suits on voting day. Letter Schoolboy 1972

spiel [spɪəl, ʃpiːl] *n. pl.* -s *colloq.* [Not exclusively S.A.E.] Usu. sig. a lot of talk, etc. as in 'He spun me a long ~ about . . .' ⁋Freq. used in S.A.E. poss. because of likeness to *Afk. speel,* to play, *cogn. Ger. Spiel* game. *cf. Austral. spieler* swindler, cardsharper. [*Eng. fr. Ger. spielen* to speak in a voluble, extravagant manner]

Here is a party which understand that White domination cannot endure forever; that nothing, absolutely nothing, can bring about a separation of the races . . . In our kind of political climate this is not exactly a vote-catching spiel. *Daily Dispatch* 4.5.70

Us chicks in 'Durbs' waits with bated breath for Bloss's latest spiel . . . that's wot we buy yore mag for – see. *Darling* 12.3.75

spinnekop *n. pl.* -pe, -s. A minesweeping device towed by any of several types of vehicle, consisting of pairs of wheels placed at varying distances apart, for the purpose of covering areas of differing width.

splinter new, *adj. phr. colloq.* Brand new. [*translit. Afk. splinternuut, splinternuwe*

brand new]
   Those are his 'bles' (white-faced) horses; but where has he come by a *splinter new* spider like that? Watkin *From Farm to Forum* 1906 cit. Pettman

spog [spɔx] *vb. intrns. slang.* Boast, show off with new clothes, horse, motor car etc. cf. *Austral. blow, skite,* boast; *Anglo-Ind. buck,* bragging talk. [*Afk. fr. Du. pochen* to boast]
   . . . the young man would arrive from a neighbouring farm mounted on a prancing horse, on which he would proceed to 'spog', or show off in front of the abode of the girl. Jackson *Trader on Veld* 1958

spoilers *n. usu. pl.* Formerly *n. prop,* name of a specific gang, see first quot., later extended and applied to African men of the *tsotsi* type generally: ⫰ poss. obsolescent: a term of the riots of the 1960s. [*fr. n. prop and presum. destructive ways*]
   The *townees* or *tsotsis* are also called '*location boys*', *ooclever, bright boys* and *spoilers* after a gang which terrorized Alexandra township in Johannesburg. Wilson & Mafeje *Langa* 1963
   . . . the wife of a Xhosa policeman . . . told of the way spoilers had dragged out their furniture and burnt everything, even her husband's uniform. Louw *20 Days* 1963
   '. . . why do they call them "spoilers"? Everyone is using that word . . .' Philemon was puzzled at her ignorance . . . 'They spoil your pass, Madam. They tear it, or burn it or throw it away.' Gordon *Four People* 1964

sponssiekte [ˈspɔnsˌ(s)ĭktə] *n.* A stock disease known elsewhere and in S.A. as *blackleg, blackquarter, quarter evil:* an infection transmitted by a bacillus, the symptoms of which are high fever and spongy swellings containing gas in the muscles of one or more quarters. ⫰Immunization is possible: see quot. at *quarter-evil.* [*Afk. fr. Du. spons cogn.* sponge *and ziekte* disease]
   Black Quarter, Quarter Evil or Sponsziekte. These names and several others such as quarter ill and black leg, are used for another disease attacking cattle, and occasionally sheep and goats. *Farmer's Annual* 1914
   Recently immunised against heartwater, red water, gallsickness, anthrax, sponsiekte, botulism. *Farmer's Weekly* 3.1.68

spook[1] [spuək] *vb. trns or intrns. colloq.* To haunt someone, or to walk as a ghost. ⫰Not exclusively S.A.E., but from earliest record, *O.E.D.* (see quot.) seems to have originated in S.A. [*Afk. fr. Du. spoken* to haunt]
   . . . but three nights ago she heard a rustling and a grunting behind the pantry door, and knew it was your father coming to 'spook' her. Schreiner *Afr. Farm* 1883

spook[2] [spuək] *n. pl.* -s. *Also Brit. and U.S.* Ghost, spectre: cf. *Jam. Eng. duppy, jumbie.* [*Afk. fr. Du. spook* haunting spirit, ghost.]
   'Spooks, mynheer. Plenty spooks in cellar.' . . . 'I am going to look for the spook, Dantje. Bring a lantern.' Fairbridge *Which Hath Been* 1913
   As modifier *substandard.*
   . . . although I'm not a superstitious man I could not shake off the idea that it was a spook thing that had happened. Bosman *Mafeking Road* 1947
   figur. 'Spectre', pervasive ill-feeling esp. in political sense. cf. (2) *gogga.*
   If this spook is to be the basis of political debate – 'You hate me, therefore I now hate you' – what is to become of South Africa? *Sunday Times* 21.5.72
   In combination ~ *voël* (q.v.).

spook[3] *n.* and *vb. Sect. Army.* A mine proof vehicle: as *vb.* to sweep for landmines, as in 'Has that road been ~ *ed*?': see also *minesweeper.*
   I tell you they kla you on if you ride on a road that hasn't been spooked. *O.I. Serviceman* Mar. 1979

spook loop [ˈspuəkˌluəp] *vb.* and *n. Sect. Army* To stalk in a particular manner without sound, lifting the feet high: as *n.* this manner of stalking: see also *leopard crawl* and *bobbejaankruip, loop.*
   We had to spook loop round the outside of the whole camp without getting seen in the first few weeks of basics. *O.I. Serviceman* 9.3.79

spookvoël [ˈspuəkˌfuəl] *n. pl.* -s. The grey-headed bush shrike *Malaconotus hypopyrrhus,* the largest of S. Afr. shrikes, named for its usual invisibility and strange call. [*Afk. fr. Du. spook* ghost + *vogel cogn. fowl* bird]
   Spookvoël . . . Keeps mostly to thick cover . . . Has an uncanny call, a frequently repeated, prolonged mournful whistle preceded by a cluck. Gill *Guide to S.Afr. Birds* 1959
   . . . those she could rescue . . . Tortoises with broken shells, . . . a hadedah with an injured wing, a rare spookvoël . . . found sanctuary and healing at her hands. *Grocott's Mail* 23.9.75

spoor[1] [spuə(r)] *n. pl.* -s, ∅. The track, trail or footprint of man or beast, including snakes, now also extended to the track of a wagon or motor vehicle: *Anglo-Ind. pug.* [*Afk. fr. Du. spoor* trace, track, footprint, road]
   . . . we only saw the *spoor,* or foot-marks, of some Kaffirs on the road. Alexander *Western*

Africa I 1837
... shortly afterwards the Fingoe called out that he had found the 'spoor' of the wounded deer. Buck Adams *Narrative* 1884
He has spent years roving the desert in a jeep, examining ... the tracks of simple people who foolishly imagined that their movements were entirely unsuspected ... Byleveld can tell the age of a motor-spoor at a glance. Green *Last Frontier* 1952
Hong Kong ... its history ... has been short and rarely glorious; its progress marked not by splendid old buildings, but by the bulldozer's barren spoor. Robt. Elegant *Hong Kong* 1977

spoor² [spʊə(r)] *vb. trans. and intrns.* To track, to hunt by following ~. [*Afk. spoor* to track]
His Excellency [Sir H. Smith] then said: how was it that you cannot spoor them? Goldswain *Chronicle II* 1838-58
derivatively *vbl. n.* ~*ing* equiv. tracking.
Hunters call the grey loerie the Go Way bird because just after you creep up to your quarry after perhaps hours of spooring, this bird calls out from a tree-top in a coffee-grinder voice 'Go Way!' *Argus* 10.6.71
and *agent. n.* ~*er* a tracker. *cf. Anglo-Ind. puggee.*
Ventvogel ... was one of the most perfect 'spoorers', that is, game trackers, I ever had to do with. Haggard *Solomon's Mines* 1886

spoorie [ˈspʊərĭ] *n. pl.* -s. *slang.* A member of the lower grade artisan class typified by the railway worker; see etym.: see also *padkamper*. [*fr. Afk. spoor(weg)* railway + *personif. suffix -ie*]
I remember when he was responsible for the railway medical practice and his waiting room was packed with spoories morning, noon and night. O.I. 1975

sport, play See *play sport*.

spotted grunter *n. pl.* Ø. *Pomadasys commersonni* known in the E. Cape as *tiger*: see *grunter*.

spreeu [spriu, spru:] *n. pl.* -s. Any of several species of the Sturnidae fam., (starlings): in combination *red-winged* ~, *rooi vlerk* ~, *blue-eared* ~, *blou oor* ~, *swartpensglans* ~, *black-bellied glossy* ~: see quot. [*Afk. fr. Du. spreeu* thrush]
I particularly remarked two sprews of a dark though glossy green, that, when they met the sun's rays were of exquisite beauty. Philipps *Albany & Caffer-land* 1827

springbok, springbuck [ˈsprɪŋbɔ̆k, ˈsprɪŋbʌk] *n. pl.* Ø, -s. **1.** *Antidorcas marsupialis*, an antelope peculiar to S.A.; a swift gazelle well known for its habit of jumping considerable distances, whether escaping from pursuers or engaging in display, see *pronk:* also formerly known as *pronkbok.* ¶As a national emblem it is seen on aircraft, airways uniforms etc.: also on sports blazers: see (2) *springbok* [*Afk. fr. Du. springen* to jump + *bok cogn.* buck]
An elegant springbuck or two appeared near the road; and, as with fairy-bound they cleared obstacles, showed large patches of white among their light brown skins. Alexander *Western Africa I* 1837
as modifier: ~*biltong* (q.v.), ~ *herds* (see *trekbokke(n)*), ~ *migration.*
The first white uitlander intrekkers – as the Afrikaners called the foreign pioneers – then saw in Bushmanland the amazing *springbok* migrations that to-day sound like travellers' tales. Birkby *Thirstland Treks* 1936
2. *n. pl.* -s, and *n. modifier.* A South African sportsman or woman representing the country in international matches or contests, freq. in *abbr.* '*Bok(s)* (q.v.) see also *Leopards* (2)*Proteas:* as modifier: ~ *blazer*, ~ *captain*, ~ *colours* ~ *team* etc. [*fr.* ~ *as national emblem*]
The 1906 Rugby team which toured Great Britain gave the name its birth, the springbok being an antelope peculiar to S.A. Swart *Africanderisms Supp.* 1934
The Minister ... said ... no South African non-White chosen to represent South Africa even as part of a multi-racial team can ever be called a Springbok. According to ... the term 'Springbok' is reserved for Whites only. *Daily Dispatch* 17.5.71
3. *n. prop.* The commercial channel of Radio S.A. [*as above*]
It is very much the mixture as before in 'Sarie 73', radio's top talent contest of the year which opened on Springbok this week ... *Evening Post* 14.7.73
4. *n.* Springbok, springbuck: found in place names Springbok, Springbokvlakte, Springbokkraal, Springbokpoortjie.
5. *n. Prefix* to several plant names usu. sig. that they were formerly browsed upon by (1) ~ e.g. ~*bos*, *Hertia pallens*, not liked by stock and thought to be poisonous: several other species are so named, ~ *ganna*, *Salsola humifusa*, see *ganna;* and ~*karoo*, *Nestlera conferta*, see *kar(r)oo bush*: both species said to have been favoured by (1) ~

Springboks *pl. n.* World War II term for S. Afr. troops.
The Springboks ... men fitted for bush warfare and already in the wilds. The Governor read them the king's message ... The Spring-

boks whipped off their sunhelmets and gave three rousing cheers for the king . . . Kenya took them at once to its heart. The Nairobi daily newspaper announced their arrival under a six column banner headline saying 'The Springboks are Here!' Birkby *Springbok Victory* 1941

springer *n. pl.* Ø. Any of several fish, known for their habit of jumping out of the water, including *Elops machnata,* see *skipjack* and (2) *Cape Salmon:* also used of several species of mullet, Mugilidae fam. including the *harder* (q.v.) ‖This has been favoured as a table fish from the early days: see quots. [*fr. habit*]

> The springer also . . . boiled in large slices – admirable! It is a fish that would make the fortune of anyone who could convey it by spawn to England. Barnard *Letters & Journals* 1797-1801
> . . . the springer, a flat fish, of a heavy, fat, luscious quality, particularly well adapted for the palate of a Dutchman. Percival *Account of Cape of G.H.* 1804
> The fish caught here are principally of the mullet species . . . The best is the springer so called from his frequently springing many yards out of the water. Philipps *Albany & Caffer-Land* 1827

springhaas, springhare [ˈsprɪŋˌhɑs] *n. pl.* Ø. Also *springhare: Pedetes capensis (P. cafer)* an animal of nocturnal habits hunted by farmers after dark. ‖*Jerboa capensis (berghaas)* is also known as ~. [*Afk. fr. Du. springen* to jump + *haas cogn.* hare]

> One of the party shot a spring-haas, or jumping hare, formed like the kangaroo, with very short fore-legs and long hind ones. Alexander *Western Africa I* 1837
> Springhares . . . break into the garden and lands at night and cause ruinous havoc . . . The springhare thing is a kangaroo-like creature, and although some people consider it as edible I'd as soon think of feeding on a cooked baboon. Stormberg *Mrs P. de Bruyn* 1920

sprinkaan [ˈsprɪŋˌkɑn] *n. pl.* -e, -s. *Sect. Army.* An extra-long anti-land mine vehicle on which the driver sits at the extreme rear, and which has an extra pair of wheels to drop down should the front ones be damaged. *Descr. O.I.s Serviceman* Feb. Mar. 1979. [*Afk. sprinkaan* grasshopper, locust, *prsum. fr.* appearance of folded-up supports to extra wheels]

sprinkaanvoël [ˈsprɪŋˌkɑnˌfʊəl] *n. pl.* -s. Also *locust bird* (q.v.) any of several species of stork (Ciconiidae fam.)

which feed upon and destroy locusts, or the *klein ~ Glareola nordmanni*. [*Afk. fr. Du. sprinkhaan* grasshopper, locust + *vogel cogn.* fowl, bird]

> Farmers know that the sprinkaanvoël can be relied upon to clear the veld of locusts and other unwelcome insects. Green *These Wonders* 1959

spruit [sprœɪt] *n. pl.* -s. **1.** A natural watercourse, often dry, usu. a tributary or other offshoot feeding a larger stream or river. *cf. Austral.* and *U.S. creek:* see (2) *sloot.* [*Afk. spruit* offshoot, tributary brook]

> From its rocky and precipitous sides issue the various streams called here *spruits,* which, uniting lower down form the Koonap river. *Cape of G.H. Almanac* 1843
> While all the spruits on other farms were muddy torrents, the main spruit on Hillside which marks its northern boundary, was running almost as clear as crystal with very little silt. *Farmer's Weekly* 21.4.72

2. *n. usu. suffix* found in place names e.g. Sterkspruit, Nelspruit, Tweespruit, Bronkhorstspruit; as *prefix* Spruitdrif. [*as above*]

> By the evening we reached a tributary rivulet known as the Sand Spruit – a term usually applied to small watercourses running to river beds. But 'spruit' as it was, we found about six feet of water in the drift. Du Val *With a Show I* 1882

spuugslang [ˈspyxˌslaŋ] *n. pl.* -s, -e. *lit.* 'Spit(ting) snake': see *rinkhals.* [*Afk. spuug* to spit *cogn.* spew + *slang* snake]

square-face, *n. prob. obs.* Also *~ gin:* in the early days gin was so called on account of the square-sided bottles in which it was sold: see also quots. at *winkel* and *canteen.*

> . . . having drunk a 'tot' of squareface and smoked his pipe, he went to bed beneath the after-tent of his larger wagon. . . . fortified with gin, or squareface, as it is called locally. Haggard *Nada* 1895
> 'God only knows what sort of head mine will be by the time I've finished,' he added lugubriously; 'you don't know the way these beggars shift square-face gin.' Brett Young *City of Gold* 1940

squash *n. pl.* -es. Usu. suffixed in certain species of *Cucurbitae, gem~* (q.v.) *Hubbard ~,* also *U.S.* [*Eng. fr. Narragansett Indian askuta* squash]

> There are an abundance of plantains, guavoes, pumpkins, melons, squashes, or water-melons. Percival *Account of Cape of G.H.* 1804

SSB [ˈesˌesˈbiː] *n. prop. Sect. Army* acronym Special Services Battalion, the

Armoured Corps: also called *koolkoppe* (cabbage heads) *fr.* triple protea on flashes and beret: see quot. at *eland*[2].

S.S.R.C. [ˈesˈesˌɑ(r)ˈsi] *acronym. Afr. E.* Soweto Students' Representative Council: see *student.*

At the funeral service itself, the new SSRC president, Mr . . . told the crowd that the student march . . . had been a victory for the students. *World* 26.6.77

SSU [ˈesˌesˈju] *n. pl.* ∅. Small stock unit: *stocking-rate* (q.v.) of grazing land is calculated as so many ~ per hectare or vice versa: see quot. at (2) *stock* and *stock reduction scheme.* [*acronym*]

. . . carrying capacity . . . 2,4 hectares a small stock unit (S.S.U.) . . . Many farmers . . . believe that 1,7ha a SSU is a very safe stocking rate. *E.P. Herald* 17.6.75

staaldak [ˈstɑːlˌdak] *n. pl.* -ke. *Sect. Army.* A military steel helmet: see also *mosdop-(pie)* and *doibie* and quot. at *balsak, cf. Austral. battle bowler, panic hat.* [*Afk. staal cogn.* steel + *dak* roof]

Troepies with *min* or *baie dae* or those who have already hung up their staaldakke have been invited to contribute towards a new dictionary of South African military slang. *Cape Times* 17.1.79

stable door *n. pl.* -s. Also *bo-en-onderdeur Afk. lit.* 'top and bottom door': see quot. *cf. U.S. Dutch door.* [*fr. similarity*]

. . . main door . . . was of sturdier, simple construction, with two horizontal halves in 'stable-door' fashion. The 'stable-door' entrance was especially suitable for Cape climatic conditions for it allowed the upper half to be opened for cool fresh air while the lower half remained shut against the farm animals. Gordon-Brown *S.Afr. Heritage* 1965

stad [stat] *n.* City: suffixed in place names e.g. Ventersstad, Wolmaranostad, Manthestad, and as *trans.* 'town' in Kaapstad (Cape Town), Grahamstad, Simonstad. See also *stat. cf. Brit. suffix* -ton. [*Afk. fr. Du. stad* city *cogn. O.E. stede, Eng. suffix* -stead]

Peace is declared, and I return
To 'Ackneystadt, but not the same . . .
Kipling *Five Nations* 1903

stadhuis, stadhouse [ˈstatˌ(h)œis] *n. prop. hist.* Old Town House: formerly the administrative building of the *Burgher Senate* (q.v.) at the Cape. [*Afk. fr. Du. stadhuis* town hall]

The Stadhuis, or Burgher Senate-house, is a large, handsome building, appropriated to the transacting of public business of a civic nature. It stands in the middle of the town, on one side

of the square called Groente Plein, in which a daily market for vegetables is held. Burchell *Travels I* 1822

stadig [ˈstɑdɪx] *adv. or interj.* Slowly: on street signs, also *ry(drive)* ~ : or *colloq.* as warning to a driver or child also ~aan! (on) equiv. of 'steady on!'. [*Afk. stadig* slowly *cogn.* steady]

. . . then they started playing the wedding march and we started a hundred-mile-an-hour sprint down the aisle and I kept on saying to my dad 'stadig, stadig.' And shame he cried when he gave me away and I cried also. *Darling* 20.12.78

staffrider *n. pl.* -s. *Afr. E.* One who rides illegally on the outside of suburban trains without paying. ⊩ acc. some this is pure bravado, acc. others to avoid the risks of theft or assault inside the carriages esp. on Fridays. [*unknown*]

The charred remains dangled from a tree . . . His widow identified the jacket . . . this was not an isolated case. The mortality rate of so-called staff riders is high. Venter *Soweto* 1977

Being a staffrider is dangerous. Riding illegally . . . is fraught with danger . . . on a fast-moving train. It becomes even more dangerous when you dance on the roof of a train as it hurtles into the city. *Daily Dispatch* 10.2.79

deriv. vb. to ~

For many it is a way of life – every morning and every evening – to staffride between Soweto and Johannesburg. For many the bravado ends in death – and yet the habit persists. Venter *Soweto* 1977

stamp *n. pl.* -s *and n. modifier.* Also *stamper:* a device for crushing auriferous ore on the goldfields; not exclusively S.A.E. [*pressum. fr. Eng. vb.* stamp]

. . . tube mills which crushed faster and more finely than stamps made possible a still more complete extraction of gold. De Kiewiet *Hist. of S.A.* 1941

In combination ~ *battery.*

. . . the 10-stamp battery, used by the Langlaagte Estates Goldmining Company when crushing operations began in 1886 . . . *Sunday Times* 23.4.72

stampblock *n.* A hollowed out wooden mortar sometimes made of the wood of the *marula tree* (q.v.) in which grain, usu. maize is crushed with a wooden pestle: see (2) *stamper.* [*prob. fr. Afk. stampblok* pounding block, pounder]

Father managed to lay hands on a 'stampblock', the genuine egg-cup-shaped hollowed-out tree-trunk mortar with the wooden pestle used by the natives to pound their grain. McMagh *Dinner of Herbs* 1968

stamper *n. pl.* -s. **1.** See *stamp.*

2. Wooden pestle used by Africans in a *stampblock* (q.v.). [*vb.* stamp + *agent. suffix* -er]

stampmealies *n.* Also *samp* (q.v.) *stamped mealies, stampkoring(corn)*: maize (*mealie*) kernels coarsely crushed: formerly in a *stamp-block* (q.v.) but now usu. by machine unless homegrown in remote districts. *cf. U.S. hominy, Jam. Eng. corn-corn.* [*fr. Afk. vb. stamp sig.* crush + *mealies* maize]

One night Oom Fanie and his family had just asked a blessing on their supper of pumpkin and stamped mealies, when a tap came at the door. Prance *Tante Rebella's Saga* 1937

stamvrug [ˈstamˌfrœx] *n. pl.* te. *Chrysophyllum* or *Bequaertiodendron magalismontanum* the so called wild plum of tropical and subtropical Africa which bears its edible slightly acid red fruit on the stems. [*Afk. stam cogn.* stem + *vrug* fruit]

There was certainly no beer in Mr . . .'s rockery . . . but there WAS refreshment. The gnarled old silver-green and copper leafed stamvrug trees – Bequaertiodendron magalismontanum – which twisted up out of cracks in nature's much sculpted rocks had cherry-red fruit on the stems. *Panorama* April 1977

-stan [stɑn] *n. suffix. freq.* facetiously or ironically used after the coming of *Bantustan* (q.v.) sig. 'special area set aside for . . .' In combinations: see quot. and quots. at *Griqua* and *mlungu: Coloured~, Griqua~, Homo~, Multi~, Verkramp~, White~*, prob. nonce coinages. [*prob. orig. fr. Bantustan analg. Hindustan, Urdu -stan sig.* 'country of']

. . . the Government have created two Xhosastans, a Zulustan, Vendastan etc. If this is good for the Blacks why is it not good for the Whites? Why do the Government not have an Afrikanerstan, and an Englishstan, a Frenchstan, Jewishstan, Germanstan etc.? *Sunday Times* 2.9.72

stand *n. pl.* -s. A building plot, formerly an urban lot in a new town or (2) *township* (q.v.) now used of a seaside, suburban or industrial building site. See quot. at *erfpacht. cf. U.S. stand*, building site. [*prob. fr. Afk. standplaas fr. Du. standplaats* stand 'standing place', *poss. U.S.*]

The plantation has been marked out as a township . . . it is 200 acres within three-quarters of a mile of the centre of Johannesburg, which could be divided into 1000 stands. Raymond *Barnato Memoir* 1897

Make your future seaside home at Cannon Rocks . . . All stands within easy walking distance of the sea. *E.P. Herald* 28.12.71

standard *n. pl.* -s. Usu. followed by numeral e.g. ~ eight, 8, VIII: term in S.A. roughly equiv. *Brit. form, U.S. grade:* ten numbered ~*s* form the latter, major part of a child's schooling: ~ *ten:* the final year in government schools at the end of which the Senior Certificate examination for *Matriculation* (q.v.) is written: see also *Substandards, Sub. A, Sub. B.*

Until the end of this year she will stay in Pretoria while her four children complete their standards at school. *E.P. Herald* 26.8.74

If the child starts in a higher standard there are bound to be problems . . . because their initial education was so bad. Some . . . people who teach them have even failed JC (Standard 8). *Sunday Times (Extra)* 8.7.79

standpoint *n. pl.* -s. Point of view: usu. in a political context. [*prob. translit. Afk. standpunt* point of view]

It has always been the standpoint of the church that appeals should not be referred to courts but to a panel or appeal board which should consist of more people than only jurists. *Evening Post* 4.3.74

stat [stat, stæt] *n.* Also *stad* a rural African village; see also (4) *kraal. cf Canad. rancherie*, an Indian village. [*presum. fr. stad* (q.v.)]

. . . during the rest of the time that he remained the head of the tribe, he would not allow a white man to enter his stat again. Bosman *Unto Dust* 1963

In place name *Dingaan* ~.

Today a huge cross – the largest in southern Africa – soars above Dingaanstat, a mission station and church close by, symbol of the victory of Christianity over heathen superstition and fear. *Panorama* Mar. 1972

State President *n.* Title of the highest ranking personage in the Republic of S.A. equiv. of *President* elsewhere: see also quot. at *Order of Good Hope.* [*fr. Afk. Staatspresident*]

Former State President C. R. Swart is at present working on a book of his political memoirs. It will largely deal with prominent leaders that he knew over a long career. *News/ Check* 15.5.70

States, the *Sect. Army.* Border usage for the Republic of South Africa, 'home'.

Hell, I can't wait to get back to the States. Had a letter from the States today. *Servicemen: Ex-Border Informants* Mar. 1977

stay *vb. intrns. substandard.* Live: freq. used as equiv. 'reside' implying permanent residence as in *Scottish stay* (live): see quot. at (2) *location* [*unknown poss. Scottish.* ⁋ *Afk. bly* to stay *and woon* to live, *do not appear to be used interchangeably*]

The date for the funeral has been provisionally set for Saturday this week at Buntingville, where they stayed. *Daily Dispatch* 30.9.69

stay well *imp. vb. phr.* A farewell greeting to those remaining behind: see also *go well, hamba kahle, mooi loop :* see quots. at *go well* and *umnumzana.* [*trans. Ngu. greeting sala(ni) kahle* (q.v.)]

steeg [stɪəx] *n. and n. suffix.* Lane, alley: see quot. still found in Cape Town street names. [*Du. steeg* lane, alley]

Adamse widow, Mossel-steeg
Adriaanse J. fisherman, Krabbe-steeg
Agom widow, bonnetmaker, Hilleger-steeg...
Greig *Cape Town Directory* 1833
In the old town the houses usually had lanes between them four feet in width . . . Such a lane was called a 'steeg' and the 'steeg' names were often picturesque. Off Waterkant . . . one found Dopper Steeg, Crabbe Steeg, Mossel Steeg, Klipfish Lane and Lelie Steeg. Green *Grow Lovely* 1951

steek [stɪək] *vb. trns. colloq.* among children: To put up a certain number of marbles at which opponent throws his *ghoen* (q.v.) e.g. ' ~ you a six, ~ you a four' etc. [*unknown, poss. fr. Du. steken* to put, *poss. cogn. Eng.* stake]

steekappel [ˈstɪəkˌapəl, -l] *n. pl.* -s. Also thorn-apple, mad apple, Apple of Peru, the prickly fruit of *Datura stramonium :* see *stinkblaar,* also *malpitte.* [*Afk. fr. Du. steken* to prick, stab + *appel* cogn. apple ; thorny fruit]

steekgras [ˈstɪəkˌxras] *n* Also *stickgrass :* any of numerous species of *Aristida,* except for those known as *Bushman* or *toa(twa)-grass* (q.v.) and of several species of *Andropogon,* with sharp wiry stems and foliage and spiky fruits or awns, which cling to fleeces and even penetrate the skin. *cf. Canad. spear grass, Austral. porcupine grass (spinifex)* [*Afk. fr. Du. steken* to stab, prick, cut + *gras* cogn. grass]

'Steek-grass' is the colonial name applied to a number of species of the natural order Gramineae, having long sharp awns attached to their seeds, by which they adhere to the wool of sheep. Wallace *Farming Industries* 1896
The rooigras had disappeared almost completely . . . to be replaced by steekgras. The

main reason why steekgras has spread to such an extent . . . is probably because the veld had been grazed too heavily . . . and had never been rested in order to enable the rooigras to recover and seed. *H'book for Farmers* 1937

steekhaar [ˈstɪəkˌhɑːr] *n. and n. modifier.* The coarse-haired strain of the *Afrikander* (q.v.) *ronderib* (q.v.) sheep: those which are not *blink(shiny)haar* (q.v.). [*Afk. fr. Du. steken* to prick, stab + *haar cogn.* hair]

. . . the occurrence of coarse, opaque, brittle, kempy fibres . . . in contrast to the soft, silky hair of the greater majority. This is . . . the origin of a subdivision in Ronderib sheep, namely the 'Blinkhaar' and the 'Steekhaar'. *H'book for Farmers* 1937

steeks [stɪəks] *adj. colloq.* Balky, obstinate, inclined to jib, usu. of or pertaining to horses: *figur.* recalcitrant (of persons), *cf. figur.* remskoen. [*Afk. fr. Du. steegs* unwilling to budge]

There we stand at the bottom of a steep hill, struggling with our horses, who have taken it into their heads not to move an inch further – they have become *steeks,* as the Boers say. Mackinnon *S.Afr. Traits* 1887 *cit.* Pettman

steen-¹ [stɪən] *n.* Stone: *prefix* in names of flora and fauna usu. sig. either the choice of a rocky habitat e.g. ~ *bok* (q.v.), ~ *bras* (q.v.) ; or a likeness to a stone e.g. ~ *klawer* (cogn. clover), *Melilotus indica (M. parviflorus)* ; ~ *tjieskweek* (q.v.), *Cyperus rotundus.* [*Afk. fr. Du. steen* cogn. stone]

Steen² [stɪən] *n. modifier.* Steen: name of both the grape and the wine made from that variety, now *usu.* known as *Chenin Blanc.* ⁋ ~ is regularly confused with *Stein* (q.v.) even by experts, see second and third quots. below.

. . . by the time he's [the grysbok] finished we've actually lost what would have been ten tons of Steen or Riesling grapes. *Sunday Times* 8.6.75

Steen (or Stein) South Africa's commonest white grape, said to be a clone of the CHENIN BLANC. It gives strong, tasty and lively wine, sweet or dry, usually better than S.African RIESLING. *Hugh Johnson's Pocket Wine Book* 1977

Formerly Chenin Blanc was known in S. Africa as Steen. It is possible that in the early days the word was loosely interchangeable with stein, the German for stone. Many local wineries give the name Stein to their wines made from the Chenin Blanc grape, and some estate wineries keep the old name of Steen to avoid confusion with the German product. The

Chenin Blanc wine differs greatly from German Stein wine which is made mostly from Riesling and from Sylvaner grapes. Bolsman *S.Afr. Wine Dictionary* 1977

There are two or three kinds of sweet wine made, but too heavy to drink after meals. The Steen wine has a sparkling quality and tartish taste, something like Vin de Grave, but much inferior in flavour. Percival *Account of Cape of G.H.* 1804

steenbok, steenbuck [ˈstɪənˌbɔk, -bʌk] *n. pl.* ∅. Any of several small antelopes of the genus *Raphicerus* (*R. campestris*) usu. preferring a rocky habitat. [*Afk. fr. Du. steen cogn.* stone + *bok cogn.* buck]

7th July 1811. . . . brought home a Steenbok (Stone-buck) . . . This is a small antelope, of nearly the same size of the Duyker, but of a lighter and reddish color, having the under part of the body white . . . but there is not, as in the Duyker, any tuft of hair between the ears. Burchell *Travels I* 1822

This year steenbuck were declared ordinary game but the council . . . decided . . . to ask the Department of Nature Conservation to have them declared protected game in 1974. *E.P. Herald* 26.9.73

steenbras [ˈstɪənˌbras] *n. pl.* ∅. Used of numerous species esp. of the Sparidae fam.: *silver* ~ or musselcracker *Sparodon durbanensis; black* ~ or poenskop *Cymatoceps nasutus; white* ~ *Lithognathus lithognathus; sand* ~ or bontrok *Lithognathus mormyrus; red* ~ (also called *yellow* ~) *Petrus rupestris:* ⟪Also *bank* ~ *Chirodactylus grandis* (Cheilodactylidae fam.) See quot.: also quot. at *allewêreld.* [*Afk. fr. Du. steen cogn.* stone + *brasem* bream]

Other common estuarine fish – the white stumpnose (Rhabdosargus globiceps) and the white steenbras (Lithognathus lithognathus) are carnivores and feed mainly on the little crustaceans in the mud banks. Grindley *Riches of Sea* 1969

The word 'steenbras' is Dutch for 'stone bream', a description which does not seem to fit our white steenbras very well as his favourite feeding grounds are sandy banks and estuarine mud banks. *Grocott's Mail* 16.8.72

Stein 1. Said to be grape variety peculiar to S.A. (prob. erron. but see quot. below and second and third quots. at *Steen²*.) ⟪ a *Stein* wine does not bear the mark signifying a particular cultivar on the label on the neck of the bottle: ~ is thus not *certified* (q.v.) as a cultivar: see *wine-of-origin* and *W.O. Seal*

South Africa has types of grape which are peculiar to this country. Pinotage is one. This represents a cross between the French variety, Pinot Noir, and Hermitage . . . In the field of white wines, Stein is a mutation, possibly, of a Sauvignon Blanc, and, like Pinotage, grows only in South Africa. It makes an outstanding wine, fuller than the delicate Riesling. It has no connection with the German Stein wines. *Panorama* Mar. 1975

2. Any of numerous semi-sweet S. Afr. wines.

I. . . said to Mr B, 'O Fye why do you give us this today, it is some of our fine hock' a certain Lieut Col . . . on this filled his glass, 'Lord Bless me what fine wine this is, said he, I have not tasted a glass such as this since I came here' – I found on asking that it was a Stejne wine which Mr B. had not liked and ordered for common use in the family – in a moment the Col found out 50 faults with it – Barnard *Letter* 29.11.1797

. . . a 'Stein' wine is a semi-sweet white wine. A 'Steen' wine is in actual fact a wine that has been made from the Chenin Blanc grape. This is a cultivar grape and when marketed it will have the 'cultivar' band on the bottle. It would be correct . . . that the name 'Steen' was an earlier name for *Chenin Blanc. Informant* F. J. Malan, Simonsig, *Letter* 19.2.79.

stellasie¹ [ˌsteˈlɑsǐ] *n. pl.* -s. A framework, scaffolding or series of racks for various purposes, drying skins, biltong, fruit or fish, growing mushrooms etc. *cf. Canad.* (*fish*) *flake*, a drying device. [*Afk. fr. Du. stellage* scaffolding *fr. vb. stellen* to place]

Almost everyone in South Africa has tasted ostrich biltong . . . Mile after mile of this biltong may be seen drying on the frames and wires called *stellasies* near Oudtshoorn. Green *Karoo* 1955

stellasie² *Sect. Army lit.* scaffolding: a raised observation point for a prison-camp guard. [*Afk. stellasie* structure, scaffolding]

Our experience, guarding 'Ities' at Zonderwater . . .
But the SWAIS, yes the SWAIS
Guarding prisoners fell to their lot,
Though they yearned to shoot Nazis,
They'd stood on 'stellasies',
And not one wretched Iti was shot.
S.W African Infantry Battalion Song cit. 'Twede in Bevel' 1944

Stellenbosch [ˈstelənˌbɒʃ] *vb. trns.* and *n. prop.* **1.** *vb.* Anglo-Boer War term: to relegate an incompetent or foolhardy officer to a post where he is unable to do harm, usu. as *pass: to be* ~ *ed figur.* to put a difficult or controversial person in a position on the shelf. *cf. Fr. Limogef* to relegate to Limoges. [*fr. n. prop. or*

town, then a military post]
An' it all went into the laundry,
But it never came out in the wash.
We were sugared about by the old men
(Panicky, perishin' old men)
That 'amper an' 'inder and scold men
For fear o' Stellenbosch!
Kipling *Five Nations* 1903
While some poor generals were being Stellenbosched banks were burning in the Free State . . . [Caption, Photograph 1902] *E.P. Herald* 17.4.79
2. *n. prop.* The site of a prehistoric Stone Age culture. [*fr. place name*]
The so-called 'Stellenbosch' culture range of artifacts to be found in such large numbers in the Boland and Western Cape must have been smashed from the huge pebbles of sandstone by homonids of large and powerful proportions. *Sunday Times* 14.11.71

Stem, Die [ˌdi ˈstem] *n. prop.* The S. Afr. national anthem: a poem by C. J. Langenhoven set to music. [*abbr. Die Stem van Suid Afrika* The Voice of S.A.]
. . . as he appeared in the gangway the band struck up 'Die Stem' and the assembled crowd of ambassadors, Queen's representatives, Foreign Office and Service dignitaries stood to attention. *Daily News* 23.5.70
An enthusiastic crowd . . . cheered as Transkei's ochre, white and green flag was hoisted . . . a large choir sang the Transkei national anthem, Nkosi Sikelele Afrika. Earlier the South African flag had been lowered for the last time and Die Stem sung by the same choir. *E.P. Herald* 26.10.76

stemvee [ˈstemˌfiə, -ˌfiː] *pl. n. lit.* 'Voting cattle': simple minded and gullible persons 'herded' to vote by often unscrupulous politicians: also used of parliamentary yes-men: see quot. [*Afk. fr. Du. stemmen* to vote + *vee cogn. O.E. feoh* cattle]
Stemvee . . . back-benchers whose duties in parliament are confined to voting on, and not discussing, the measures brought forward. Swart *Africanderisms Supp.* 1934

step-outs *pl. n. phr.* Military dress uniform also *mooi-moois* (q.v.). [*trans. Afk. uitstap(klere)*]

ster [steːr, stɛə(r)] *n.* Star: found in place names Morgenster and Aandster. [*Afk. fr. Du. ster cogn.* star]

ster- [steːr, stɛə(r)] *n.* Star: as *prefix* in plant names usu. sig. star-like shape, leaf arrangement etc., e.g. ~*boom, Cliffortia arborea;* ~*dissel, Centaurea calcitrapa* (*star thistle*); ~*gras Cynodon plectostachyus* star grass or *Ficinia radiata.* [*Afk*

sterk [sterk, stɛə(r)k] *adj. prefix.* Strong, powerful: found in place names e.g. Sterkstroom, Sterkspruit, Sterkwater. [*Afk. fr. Du. sterk* strong]

sterkte [ˈste(r)ktə] *interj. colloq.* 'Courage' 'more power to your arm', *etc.* an encouraging *interj. cf. hou moed.* [*Afk. sterkte* strength]
Hoezit with you these days, ou pel? I loved your Christmas day letter . . . but it sounded as though things were getting . . . rather . . . Sterkte, ou maat. All things must pass. Jenny Hobbs *Letter* 16.1.78

sterloop [ˈsterˌlʊəp, ˈstɛə(r)-] *n. pl.* -s. Gun: a smooth bore flintlock in use in the mid-C19. [*Afk. fr. Du. ster* star + *loop* barrel]
By the middle of the 19th century the greatest damage to the game herds of the plains was being done by the famous sterloop – a smooth-bore flintlock of about 1842 manufactured in Birmingham with the much prized star on the barrel. *Daily Dispatch* 22.7.72

stertriem [ˈstertˌrim, ˈstɛə(r)t-] *n.* A loincloth of hide worn by male members of certain tribes after attaining manhood. [*Afk. fr. Du. staart* tail + *riem* strap, *Du. staartriem* crupper]
Proud in their wearing of the stertreim – sign of manhood – these young boys, clad in their tribal clothes of well-trimmed, round-cut goatskins. Lanham *Blanket Boy's Moon* 1953
. . . he pretended that he was just laughing at . . . the thought of himself wearing a Bushman's wild-cat-skin loin cloth . . . 'Isn't that a scream . . . me wearing a stert-riem in the hiernamaals*?' Bosman *Bekkersdal Marathon* 1971
*hereafter

stick *n. pl.* -s. *Sect. Army.* not exclusively SAE, but in S.A. used of a group of trainee *Parabats* (q.v.) in the charge of a single instructor, who jumps with them. [*analg.* 'a stick of bombs' dropped so as to fall in a straight line]
. . . at this stage trainee paratroopers have been divided into small groups of twelve known as sticks . . . to enable the troops to receive personal attention from instructors. Marks *Our S. Afr. Army Today* 1977
The number one in the stick is now ready for his jump. *Ibid.*

stick *vb. trns. colloq.* To stab. [*translit. Afk. steek* to stab]
You've got to watch out they don't stick you with a knife. O.I.

stick away *vb. phr. substandard.* To hide: as in 'I stuck it away somewhere and now I can't find it.' [*trans(lit). Afk.*

wegsteek to hide away]

stick fast *vb. phr.* To become stuck usu. in mud etc. when travelling. [*translit. Afk. vassteek* to stick, or get stuck]

Once indeed we did stick fast where water and mud lay thick on the track, and we came five hours late to Upington. Birkby *Thirstland Treks* 1936

stickgrass See *steekgras* [*translit. Afk. steekgras*]

stiffsickness, *n.* See *styfsiekte* and quot. below. [*fr. Afk. styfsiekte* (q.v.)]

The term stiffsickness is frequently used in S.A. to describe an affection of the locomotor system . . . Popularly it included various widely different conditions such as Three days sickness, Lamsiekte, Gallamsiekte, Gallsickness etc. . . . to avoid confusion the term should be confined to only two conditions (a) stiff-sickness caused by the stiffsickness bush . . . and (b) stiffsickness due to phosphorous deficiency (aphosphorosis). *H'book for Farmers* 1937

stil(le) [stɪl(ə)] *adj.* Quiet, calm: found in place names in *attrib.* inflect. form Stille-water and in *uninflect.* form Stilbaai-oos, Stilbaai-wes, Stilfontein. [*Afk. stil* quiet]

still *adv.* Used spatially in S.A.E. sig. further, more, as well, in space or distance, or redundantly, see second quot. [*trans. Afk. nog* still, yet, further]

There's a little garden then there's still a stoep between us and the street. O.I. 1973

. . . turned to . . . saying 'Look how that bus is smoking – it is still going to explode today' as they passed the vehicle. *Daily Dispatch* 24.11.71

stinkblaar [ˈstɪŋkˌblɑː(r)] *n. Datura stramonium* (also *D. metel* and *D. ferox*): an annual weed of rank growth and fetid smell with trumpet-shaped mauve flowers and thorny fruit (see *steekappel*) containing *malpitte* (q.v.). ⎧The plant source of the drug Stramonium is used for various medicinal purposes: see quot. below, but is poisonous to stock. See quots. at *malpitte*. U.S. *Jimson weed, stink weed, Apple of Peru.* [*Afk. fr. Du. stinken* to stink + *blaar* leaf]

Stinkblaar . . . esteemed as a remedy for asthma and for this purpose the dried leaves are smoked in a pipe. The freshly warmed leaves or the vapours from an infusion are used as a sedative in cases of neuralgia, rheumatic or other pains, and in the form of poultice are also applied to cancerous ulcers and rolled up into little pellets, are employed . . . to ease earache. Smith *S. Afr. Plants* 1966

stinkbug, stinkbeetle *n. pl.* -s. Any of several insects which eject evil-smelling

fluid as a form of defence when touched or disturbed: see also (*h)oogpister.*

Effective and economic control of grain stinkbugs is obtained by spraying . . . emulsifiable concentrate. *Farmer's Weekly Advt.* 12.5.71

stinkfish Also *bamboofish, bamvoosie;* see quot. at *mooinooientjie.*

stinkhout [ˈstɪŋkˌ(h)əut] *n.* See *stinkwood: wit* ~ (white), *rooi* ~ (red), *swart* ~ (black). [*Afk. stink* + *hout* wood, timber]

stinkwood *n.* The wood of three indigenous hardwood trees, or any of the trees themselves, which have in common the fact that their timber smells unpleasant when cut. [*trans. Afk. stinkhout* (q.v.)]

. . . a great cabinet of stinkwood – that beautiful ill-named wood of South Africa, odorous only in its freshly cut state. Whitney *Blue Fire* 1973

A record price of R1 360 a cubic meter was realised for prime quality black stinkwood at the four-day annual sale of timber from the indigenous forests of the Southern Cape . . . Last year's top price for stinkwood was R720 a cubic meter. *Het Suid-Western* 26.9.74

(*black*) ~ (*swart stinkhout*) the hard, finely grained and exceptionally heavy timber of *Ocotea bullata*, much prized for furniture, souvenirs etc.: also called Cape laurel, Cape Mahogany, Cape Oak, African Oak. *white* ~ or *camdebo* ~ (wit stinkhout) *Celtis africana*: also corrupted to *cannibal* ~.

Its texture is such that it does not work easily and it bends when wet. White stinkwood wa-kiste are fairly commonly found. Baraitser & Obholzer *Cape Country Furniture* 1971

red ~ (*rooi stinkhout*) *Pygeum africanum* or bitter almond, the timber of which is red, borer-proof and valued for wagon work: all parts of the tree smell of prussic acid when cut or bruised.

stiver [ˈstaɪvə] See *stuiwer.*

stock *n. and n. modifier.* 1. Livestock, farm animals occ. extended to game. [*abbr. livestock* or *trans. Afk. vee* (q.v.) *cogn. O.E. feoh,* cattle]

The pasturage is also diversified and is found to be suitable to all descriptions of stock. *Cape of G.H. Almanac* 1856

In combinations ~ *fair,* a sale of ~; ~ *theft;* ~ *unit* (q.v.); ~ *reduction scheme* (q.v.); ~ *farmer,* etc. see also *S.S.U.*

2. Also as *vb.* esp. *vbl n.* ~*ing* in phr.

~ing rate [fr. (1) ~]
Thickly grassed sour veld – stocking rate 2 small units per morgen. Natural bush provides ample shelter for stock. *Daily Dispatch* 25.3.72

stockfel *n. pl.* -s. See *stokfel*.

stockfish *n. pl.* ∅. Also *stokvis:* the commonest S. Afr. edible fish: either of two species of the Merlucciidae fam., *M. capensis*, the shallow-water hake and *M. paradoxus*, the deep-water hake. ⫡*M. capensis* (thought by some to be the same fish as *M. merluccius*, the European hake) is described as 'commercially the most important single species in S.A. being taken in great quantity by trawl . . .' Smith *Sea Fishes of S.A.* 1961 ed. [*fr. Afk. fr. Du. stokvis* hake]
Stockfish, or hake . . . is most obliging. You can do almost anything with it – fry, boil, bake or make into fish cakes or fish pies. It is too soft for pickling or for the making of smoorvis. *Farmer's Weekly* 18.4.73

stock reduction scheme *n.* A government scheme operative in certain areas, by which a farmer who reduces stock by one third of the carrying load (see *stocking rate*) of his (2) *veld* (q.v.) is paid compensation for that number of *stock units* (q.v.) in return for which he guarantees to rest 25% of his grazing (veld) for the period of a year. ⫡The penalty for not complying with this condition is forfeiture of the compensation plus interest.
. . . has gone in for the stock reduction scheme and keeps a flock of stud Dormers in a 'house camp' or kraal where they are fed. *Farmer's Weekly* 4.7.73

stock unit *n. pl.* -s. Also *animal unit.*
'. . . the equivalent of one ox, horse, mule or donkey (calves and foals are counted as full-grown as soon as they are weaned) two calves, two foals or four sheep or goats. Pigs, lambs and kids are disregarded in computing animal units.' *Tomlinson Commission* 1955. See quot. at *stocking rate:* also *S.S.U.* (small stock unit) [*fr.* 'animal unit' and *stock* (q.v.)]

stoelriempies ['stŭl₁rĭmpĭs] *n. usu. pl.* Fine strips of softened hide sold by weight, see quot. at *riempie*, for making *riempie* (q.v.) seats and backs for chairs, stools, *rusbanks* (q.v.). [*Afk. fr. Du. stoel, chair cogn.* stool + *riem* (q.v.) + *dimin.*

suffix -*pie*]

stoep [stup, stŭp] *n. pl.* -s. A raised platform or verandah in front of or all round a house, orig. in town houses raising them above street level: now loosely used of any covered veranda or open porch: also *U.S.* and *Canad. stoop*. [*Afk. stoep* veranda *fr. Du. stoep* steps or paved elevation in front of a house *cogn.* step]
In the evening, the family parties of the respectable classes enjoyed themselves walking slowly about the raised stoep, or terrace, in front of their houses, or sat on chairs watching the passers by. Alexander *Western Africa I* 1837
This platform extended in front of the house to form a promenade or 'stoep' flanked at each end by a low brick seat, which was reached from street level by one or two flights of steps. Lewcock *C19 Architecture* 1963
In combination ~ *bank* (bench) described above, or occ. a *rusbank* (q.v.) used on a ~ ; ~ *chair*, any chair suitable for outdoor use ; ~ *farmer*, see ~ *sitter;* ~ *kamer* portion of a ~ built-in as an extra room often seen, esp. in the country.
. . . it was apparently intended that the line of the roof of the stoepkamer should be acknowledged on the wall of the new fortified tower. Lewcock *C19 Architecture* 1963
~ *plant* ornamental pot plant ; ~ *talk* general conversation, odds and ends, title of certain newspaper columns ; ~ *sitter* (or ~ *farmer*) an idler who prefers to sit on his ~ drinking coffee and giving instructions to participating actively in any work himself: occ. known as a 'cheque book farmer'. *cf. Brit. armchair traveller, philosopher* etc.
One of the devices in the play is a chorus of three old men, 'stoepsitters' who represent the traditional small-town community and comment on the main action. *Evening Post* 9.9.72
THIS WEEK'S STOEP TALK *Het Suid Western* 3.7.74

stofie ['stŭəfĭ] *n. pl.* -s. Also *voetstofie* and (2) *konfoor* usu. a small footwarmer consisting of a lined wooden box with a perforated top, containing a pan or *tessie* (q.v.) of burning coals: see also quot. at (1) *Cape Dutch.* [*Afk. fr. Du. stoofje, stoof* footstove or warmer + *dimin. suffix* -*je* or -*ie*]
. . . the *stofie*, a small hollow wooden stool, in which charcoal was burnt to warm the feet. Mills *First Ladies* 1952
It had been the custom for elderly folk to have a couple of slaves precede them to church, one carrying a 'stoofie', a padded metal-lined

footstool containing glowing charcoal. McMagh *Dinner of Herbs* 1968

stokkies-draai [ˈstɔkĭsˌdraɪ] *vb phr.* or *n.* To play truant: *cf. Canad. play hookey.* [*Afk.* to play truant]

. . . were involved in a hammer-and-tongs shouting match that disgraced the council chamber. Over what? Why . . . hadn't attended a meeting. Was he really sick or was it stokkies-draai? That's what . . . wanted to know – as if it matters a row of beans. *Het Suid Western* 8.8.74

stokvel [ˈstɒkˌfel, ˈstɔk-] *n. and n. modifier.* African syndicate or club of closed membership . . . operating mainly on food and liquor sales at parties held at members' homes in rotation: these may be small-time or *tiekie-line* ~, with an entrance fee to members and their guests as low as 50c, or *big-time* ~, see second quot. The ~ also operates as a mutual aid society for members in prison, in debt or other trouble: see quot. at *skokiaan queen.* Also *vb phr.* to *play* ~ and *slang abbr.* **stokkie;** see quot. at *non-voter. cf. Canad. potlatch* sig. 'donation feast'. *Hong K hui,* mutual savings club.

The origin of the word 'Stockfel' is unknown, but this form of organization occurs . . . under a great many different names. Of course they did not call this arrangement a Stockfel, because the name Stockfel has become associated with things 'primitive' and 'backward' and 'drink and all that'. They proudly spoke of their 'Society'. This '*Manyano*-Society' even had a written Constitution. Brandel-Syrier *Black Woman* 1962

In contrast to the 'Tiekie-Line' parties which attract predominantly female 'followers', Big-Time *Stokvel* has a special lure for males. Stakes are high. The 'table money' is never less than R10 . . . the amount spent by members at any one party must be reciprocated by the host as the *Stokvel* gatherings rotate. *The 1820* June 1973

As modifier ~ *party,* ~ *gathering,* ~ *members,* ~ *rules,* etc.

He used to take me to 'stokvel' parties every weekend and sometimes he would get so drunk that I would have to carry him home . . . One day at a 'stokvel' party he got very drunk again and was very rowdy. *Drum* Aug. 1971

stokvis [ˈstɔkˌfɪs] See *stockfish.*

stomp [stɔmp] *adj. and adj. prefix.* Blunt, cropped. [*Afk. fr. Du. stomp* blunt, cropped off *cogn.* stump] **1.** *colloq.* Used of a knife etc. squared off, cropped: see quot. at *winkelhaak.*

I black Ox, right ear stomp and square, left ear square behind. *Grahamstown Journ.* 11.4. 1833

2. Blunt, stumpy: as *prefix* in names of fauna and flora e.g. ~ *neus* (q.v.) also *stumpnose;* ~ *stertjie* (little tail), a small bird *Sylvietta rufescens;* ~ *doring, Gardenia spatulifolia;* ~ *stertbobbejaantjie, Babiana erectifolia,* see *bobbejaantjie* etc.

stompi [ˈstɔmpĭ] *n. pl.* -s. A small stove designed with sloping sides for the heating of sad-irons: now also used as room heaters. [*prob. fr. size*]

Order stompies for your servants' rooms: they're excellent and work up a terrific heat. *O.I.* 1966

stompie [ˈstɔmpĭ] *n. pl.* -s. Stump(y). **1.** *colloq.* A cigarette butt. *cf. Austral. bumper.* [*fr. Afk. stomp cogn.* stump + *dimin. suffix -ie*]

So be . . . careful with lighted cigarettes, matches, stompies, glass bottles – because they can start fires, remember – not only on Signal Hill either. *E.P. Herald* 11.1.73

figur. in phr. *picking up* ~s, breaking into a conversation by picking up the tail-end of a previous remark.

2. Small fish used for bait, see also *flatty.*

K. was fishing for flatties (stompies) for bait for his father. *Ibid.* 2.5.74

3. See *stompi.*

4. Certain dwarf plants are so named.

stompneus [ˈstɔmpˌnīœs] *n. pl.* ∅. See *stumpnose* usu. in combination *rooi*(red) ~ or *wit*(white) ~: see quots. at *baardman, steenbras.* [*Afk. fr. Du. stompneus* snub nose]

stone(s) *n. usu. pl.* **1.** *Sect. Mining.* Diamond trade term sig. a gemstone of particular size: see quot.

'Stones' are usually over one carat (a carat being 200 milligrams) Anything smaller falls in the 'melee' category. *Panorama* Dec. 1972

2. *colloq.* diamonds: see also *klip, blinkklip(pie).*

The cooling mud has closed round the stones taking the impress of every angle and facet. Glanville *Fossicker* 1891 cit. Pettman 1913

stone pine *n. pl.* -s. *poss. reg.* Cape: the Mediterranean *Pinus pinea,* the cones (see *dennebol*) of which contain pine-nuts, see *dennepit(s),* also the 'umbrella pine'. Used erron. for the Mediterranean cluster pine *P. pinaster.*

. . . two are abundant at Kirstenbosch, the stone pine (*Pinus pinea*) and the Cluster or Maritime pine (*Pinus pinaster*) . . . The stone pine and the cluster pine can be distinguished . . . the former is characteristically umbrella shaped with a flat or rounded crown while the latter is pointed or pyramidal. The cones of *P. pinea** are large round and blunt ended, while those of *P. pinaster* are smaller, conical and pointed. *Journ. Botanical Soc. of S.A.* Part XIII, 1927 *See *dennebol*

It looks as though the avenue of huge stone pines at the Oudtshoorn entrance to George . . . is doomed. The Town Clerk . . . reported at this week's council meeting that most of the stone pines were rotten and in danger of being blown over. *Het Suid Western* 5.12.74

stops [stɔps] *n. slang* (vulgar). Constipation: see also *skiets*. [*unknown, poss. fr. Afk. stopsel* plug]

storing [ˈstʋerɪŋ] *n. pl.* -s. *Sect. Army.* Trouble generally, more esp. in sense of jamming or blockage of a firearm: see also quot. at *mammajoor*. [*Afk. fr. Du. storing* disturbance, interruption trouble, failure, also *vb. storen* to jam *etc.*]

. . . had storing with my rifle at shooting practice last Wednesday. I set the gas pressure too low and it jammed and refused to go off. *O.I. Serviceman* 10.3.79

stormjaer [ˈstɔrmˌjɑə(r)] *n. pl.* -s. **1.** A member of the militant core of the *O.B.* (q.v.). [*Afk. fr. Du. storm + jager* hunter]

. . . ex-Minister of Defence and the Ossewa Brandwag – Ox-wagon Sentinel – movement and its inner core of stormjaers – fighters. Walker *Kaffirs are Lively* 1948

2. Also *stormja(g)er :* a flour dumpling fried in fat or cooked in hot embers. *cf. vetkoek;* also called *maagbom(men)* (q.v.) *cf. Austral. damper, buck jumper, Jam.-Eng. stick-in-the-middle.* [*see quot.*]

Moolman taught the boys how to cook the flour they drew in boiling fat. These delicacies were known as *stormjagers* or *maagbommen*, that is to say, storm hunters because they were rapidly cooked, or stomach bombs, owing to their effect on the digestion. Cloete *Rags of Glory* 1963

stormwater *n.* Water caused to run rapidly in streets etc. by sudden heavy downpours of rain: see also quot. at *dwarswal:* usu. in combinations ~drain, ~culvert, ~gutter, ~grid etc. ⫫ Presum. so-called to distinguish *fr.* floodwaters caused by flash floods, see *come down*, or overflowing rivers: usu. an urban term.

. . . the Council has also energetically tackled the provision of basic services – gravel roads, stormwater drains and night soil removals in the slum parts. *Cape Times* 3.7.71

. . . a horse trapped in a stormwater trench . . . examination revealed a terrified struggling

bay gelding trapped in a stormwater drainage trench . . . beneath a disused railway line. *Grocott's Mail* 12.1.79

storosh(a) [ˈstɔˌrɔʃ(a)] *n.* The grass-thatched hut, usu. of a farm labourer, built on the traditional circular pattern: see also *struis: figur.* nickname for a boy with hair too long or untidy. *cf. Brit. 'thatch'* as nickname. [*fr. Xh. isithorosha* 'hut of one in employment.' (*Xhosa informant*) *prob. rel. stroois or struis* (q.v.)]

'Storosh played a good game for Firsts last week.' Girl, 16 Grahamstown 1974

stove *n. pl.* -s. In S.A. usu. an electric, gas, coal, paraffin (kerosene) or wood cooking stove or range, as opposed to a room heater: *equiv. Brit. cooker, U.S. range.*

-straat [strɑt] *n.* Street: found in Afk. or translated street names e.g. Kerkstraat, Burgstraat, Adderleystraat and Yorkstraat. [*Afk. fr. Du. straat cogn.* street]

straight *n. pl.* -s. *Afr.E. colloq.* A whole 750ml bottle (formerly 25 fl oz): see also *half-jack (ha-ja)*, and *nipinyane. cf. square-face.* [*unknown*]

. . . African women who buy large quantities of liquor . . . measure out tots for 2s 6d, nips for 5s, half-jacks for 10s, or a straight (a full bottle) for £1 or 30s. Longmore *Dispossessed* 1959

He takes out a wad of notes and instead of buying this cherie a nipinyana he collects a whole straight of mahog. *Drum* 8.1.74

strand [strant, strænd] *n.* Beach, strand. **1.** Found in place names e.g. Strandfontein, Paradysstrand. [*Afk. strand* beach *cogn.* strand]

2. *prefix* to certain plant names sig. a preference for a sandy or coastal habitat e.g. ~boegoe, see *buchu;* ~dagga, *Leonotis hirtiflora;* ~gousblom, *Gazania uniflora;* ~vy, *Carpobrotus acinaciformis,* see *sour fig.*

strandlopers [ˈstrantˌlʋəpə(r)s] (*usu.*) *pl. n.* **1.** A prehistoric coastal race of S.A., poss. forerunners of both Bushman and Hottentot: see quot. [*Afk. fr. Du. strand* beach + *loop* go + *agent. suffix* -er]

. . . the camp sites (kitchen middens) of the Strandlopers, a Stone Age people, who inhabited the Cape coast up to the time of the arrival of the White man . . . Most authorities describe the Strandlopers as being of a later Stone Age culture, but Desmond Clark suggests in his descriptions of their tools that they were middle Stone Age. *E.P. Herald* 18.12.72

2. A bushman race living on the coast near S.W. Africa; also *sandlopers*. [*prob. fr.* (1) ~]

> By far the most mysterious people . . . are members of a group known to the Hottentots as 'Sandlopers'. It is possible that these are the last of the old Strandlopers, a Bushman race inhabiting the beaches of South Africa centuries ago, but thought to have become extinct. Green *Last Frontier* 1952

3. *n. pl.* -s *figur.* a beachcomber, vagrant (freq. to a child) as in 'Put on your shoes you look a proper ~.'

4. *n. pl.* -s. Also ~ *tjie*, any of several coastal plovers of the Charadriidae which frequent the seashore. [*fr.* (1) *strandloper*]

Strandveld ['strant‚felt] *n. prop.* The coastal area S. of Bredasdorp and Riversdale. [*Afk. fr. Du. strand* beach + *veld* country]

strandwolf ['strant‚vɔlf] *n. pl.* -we, Ø. *Hyaena brunnea*, lit. 'beach wolf' which scavenges along the sea coast: see *wolf*. [*Afk. fr. Du. strand* beach + *wolf Afk.* hyena]

> . . . then there is the hyena (wolf of the colonists), of which there are different species, such as the aarde-wolf, the tiger-wolf, striped hyena, and the strand-wolf or maned jackal. McKay *Last Kaffir War* 1871

streep- [striəp] *n.* Stripe(d), striated, banded: prefixed to names of fauna and flora e.g. ~ *dassie* (q.v.), ~ *ha(a)rder* (q.v.), ~ *vis* (q.v.), ~ *muis* striped field mouse, ~ *muishond* (q.v.), banded mongoose polecat, ~ *aalwyn Aloe striata*, ~ *vygie* (q.v.), *Aloinopsis rubrolineata*. [*Afk. fr. Du. streep cogn.* stripe]

streepdassie ['striəp‚dasĭ] *n. pl.* Ø. *Diplodus cervinus*, a marine fish also called *zebra*, *bontrok*, *wildeperd*. [*Afk. fr. Du. streep* stripe + *das* badger + *dimin. suffix -je*]

streepha(a)rder ['striəp‚ha(r)də(r), -‚hɑd-] *n. pl.* -s. *Liza tricuspidens*, one of the mullet also called *springer* (q.v.). [*Afk. streep cogn.* stripe(d) + *harder* (q.v.) mullet]

stre(e)pie ['stripĭ, 'striəpĭ] *n. pl.* -s. *Sarpa salpa:* small fish variously named commonly used for bait: see *karanteen* [*Afk. streep cogn.* stripe + *dimin. suffix -ie*]

> I hear that some *geelbek* up to 8kg have been caught in the deep water . . . these live strepies should tempt them to feed. *Farmer's Weekly* 4.12.74

streepvis ['striəp‚fĭs] *n. pl.* -se. See *seventy four*. [*Afk. streep cogn.* stripe + *vis fr. Du. visch cogn.* fish]

strik See definition at *wip*. [*Afk. strik* noose, loop, bow]

-stroom [struəm] *n. suffix.* Stream: found in place names e.g. Nylstroom, Sterkstroom, Wakkerstroom. [*Afk. fr. Du. stroom cogn.* stream]

strop [strɔp] *n. pl.* -s. The leather throat strap holding an ox between the yoke-skeys: see quot. at *yoke-skey*, *skey* and *nek-strop*. [*Afk. fr. Du. strop* halter *cogn.* strap]

> Boermaak riems: 10 ft. R5,00 . . . Strops R3,00; Agterslags. *Farmer's Weekly Advt.* 13.6.73

struis [stroeĭs] *n. pl.* -e(s). Also *strooĭs:* see *storosh(a)*, thatched roofed hut of an African labourer usu. on a farm, built in the traditional circular style. [*Afk. contraction of strooi* straw + *huis cogn.* house]

stryddag ['streĭt‚dax] *n. pl.* -dae. An Afrikaner political party rally: lit. 'day of struggle'. [*Afk. fr. Du. strijd* conflict, struggle + *dag cogn.* day]

> The day of the big stryddag is over. Those large crowds will never again drive across the veld to hear emotional appeals to the blood. *Daily Dispatch* 14.4.72

strykgeld ['streĭk‚xelt] *n. hist.* The premium or bonus paid to the *opslag* (upward) bidder: a term in the Dutch method of auctioneering.: see quot. at *opslag*, also at *afslag*. [*Afk. fr. Du. strijkgeld* premium, (*geld* money)]

> The usual procedure was reversed. It was first sold by opslag (advance bidding) and then put up again by afslag (downward bidding). The bidder in the first instance had no intention of buying, but was out to increase the final amount. For his service he received a bonus or premium, or, as it was called, strykgeld. Botha *Our S.A.* 1938

student *n. pl.* -s. *Uṣu. Afr. E.* A black school child, now in regular use in the press.

stuiwer ['stoeĭvə(r)] *n. pl.* -s. *hist.* Also *stiver* (*Eng.*) and *stuiver:* the smallest monetary unit in use at the Cape until the mid *C19:* the sixth part of a *schelling* (q.v.) or *skilling* (q.v.) and the forty-eighth part of a *rixdollar* (q.v.), see quot. at *rixdollar*. [*Du. stuiver* stiver, halfpenny]

> Six stivers are equal to one schelling, and

eight schellings to one rix-dollar or four shillings currency; but the value of this currency is excessively reduced by the rate of exchange, which, in 1810, was 33 per cent, in favour of England. Burchell *Travels I* 1822

stukvat [ˈstœkˌfat] *n. pl.* -s. *prob. obs.* A wine or spirit vat: see quot. at (2) *yellowwood*. [*Du.* (*S.A.*) *stukvat* large wine vat]

. . . has continually on sale Stuckvat Staves and Teakwood do., large and small Iron Hoops. *Greig's Almanac Advt.* 1831

stump See *stomp*.

stumpnose *n. pl.* ∅. An edible marine angling fish: any of several sea bream of the Sparidae fam.: esp. in combination *red* ~ (*rooi stompneus*) *Chrysoblephus gibbiceps* variously known as *miss lucy, dageraad, stump, magistraat, michael, miggiel* etc.; also *white* ~ (*wit stompneus*) *Rhabdosargus globiceps*. [*translit. Afk. fr. Du. stompneus* snub nose]

Red and white stumpnose have much more in common than red and white steenbras. *Farmer's Weekly* 18.4.73

styfsiekte [ˈsteɪfˌsɪktə] *n. lit.* '*stiffsickness*' (q.v.) a term used for several diseases of domestic livestock including *three-days-sickness* (q.v.): these diseases affect the locomotor system and are characterized by stiffness of the limbs and even paralysis: see also *lamsiekte* (q.v.). ⁋One form which is caused by plant poisoning (through eating the ~ *bossie, Crotalaria burkeana*) is marked by an abnormal growth and elongation of the hoofs. [*Afk. fr. Du. stijf* (*attrib. stywe*) *cogn.* stiff + *ziekte* disease]

Two kinds of stywesiekte are recognised, namely the so-called aphosphorosis . . . and the stywesiekte caused by ingestion of the plant *(rotalaria burkeana* (Klappers) *H'book for Farmers* 1937

stywe pap [ˈsteɪvəˌpap] *n.* Firm *mealie meal* (q.v.) porridge used either with meat and gravy esp. by Africans, or with *braaivleis* (q.v.): see quots. at *mopane worm*, and *pap*. [*Afk. stywe* (*attrib. form*) *fr. Du. stijf cogn.* stiff + *pap* porridge]

Mieliepap for every taste. Have you tried it with meat and gravy? A real old-time South African favourite, this – perfect at braais. Make the firm 'Stywe pap' or the dry and crumbly 'Krummelpap'. *Fair Lady* 30.10.68

stywesiekte [ˈsteɪvəˌsɪktə] See *styfsiekte* and quot. at *stiffsickness*.

Sub A, Sub B See *substandard*.

substandard *n. pl.* -s. *Sub. A* and *Sub. B:* the first two years of primary schooling before the child enters *standard* (q.v.) one.

I sent my child there from Sub A and . . . cannot see any problem. At that age children will mix irrespective of race. *Sunday Times (Extra)* 8.7.79

sugar baron *n. pl.* -s. Usu. slightly derisive term in Natal for any person who has made a fortune in the sugar industry. *cf. Randlord* (q.v.); *Anglo-Ind. nabob; Brit. merchant prince; Hong K. taipan; Canad.* grain king, lumber king, sawdust nobility, timber baron etc.

A Natal South Coast sugar baron . . . who died . . . last year aged 69, has left an estate of R1,104,606. *E.P. Herald* 2.4.74

sugar beans *pl. n.* Dried beans of mottled colouring: see quot. at *sousboontjies*, freq. cooked with *samp* (q.v.) by Africans.

It was hard to swallow, and sugar beans, cooked with meat and served over rice, tasted as flat as her spirits. Whitney *Blue Fire* 1973

sugar bird *n. pl.* -s. Any of several species of the Promeropidae, long-tailed birds which feed on nectar and insects. See quot. at (1) *suikerbos*. [*prob. trans. Afk. suikervoël* sugar bird]

The little sugar-birds are extremely elegant, and are allied to the humming-birds. Webster *Voyage I* 1834

sugarbush *n. pl.* -es. Any of several proteas but most freq. *P. mellifera :* see *suikerbos*. ⁋Also *Canad.* sig. a grove of sugar maples. [*trans. Afk. suikerbos, suiker cogn.* sugar + *bos cogn.* bush]

. . . The Protea of Linn–, . . . called by the Colonists the sugar bush, from the quantity of sweet juices the large and beautiful flowers contain, which they often extract and use as a substitute for that article. Ewart *Journal* 1811–14

-suid- [sœɪt] *n. modifier prefix and suffix.* South: found in place names freq. in translation e.g. Suid-Afrika, Suidwes-Afrika etc. Also *Du. zuider* in Zuider Paarl. [*Afk. fr. Du. zuid* south, *zuider* southern]

suikerbekkie [ˈsœɪkə(r)ˌbekĭ] *n. pl.* -s. Afk. name for any of the Nectariniidae: see *sunbird*. [*Afk. fr. Du. zuiker cogn.* sugar + *bek cogn.* beak + *dimin. suffix* -(*k*)*ie*]

suikerbos(sie) [ˈsœɪkə(r)ˌbɔ̃s(ĭ)] *n. pl.* -s.
1. Also *sugarbush* general name for many varieties of flowering *proteas* (q.v.)

esp. *P. mellifera* (honey bearing) because of its exceptionally sweet nectar. Also found in place name Suikerbosrand. [*Afk. fr. Du. zuiker cogn.* sugar + *bosch cogn.* bush]

The delicate Humming-birds . . . here called by the Dutch colonists Suiker-vogels (sugar birds), from having been observed, . . . to feed . . . on the honey of the flowers of the Suikerbosch (sugar-bush). [Protea mellifera] Burchell *Travels I* 1822

2. Title of folk song. [*see quot.*]

One of the best-known of all South African folk songs, 'Suikerbossie ek wil jou hê' (How I want you, sugar bush) was written about it, and interestingly enough, it came to be known as 'suikerbos' because of the nectar in its calyx. *Panorama* Feb. 1972

suikerbrood [ˈsœïkə(r)ˌbrʊət] *n.* A traditional Cape sponge cake. [*Afk. suiker, cogn.* sugar + *brood* loaf *cogn.* bread]

. . . that triumph of Cape cookery, the suikerbrood, of whose glories the 'sponge cake' can give but a poor reflection. Marchand *Dirk, A S. African* 1913 cit. Swart *Africander- isms Supp.* 1934

Spongecake (suikerbrood) seems to have been by far the most usual sort of sweet cake made in the 18th Century Cape. Cook *Cape Kitchen* 1973

suikervoël [ˈsœïkə(r)ˌfʊəl] See *sugar bird* and quot. at (1) *suikerbos.*

suiwer [ˈsœïvə(r)] *adj.* Pure, unsullied, usu. in phr. ~ *Afrikaans.* [*Afk. suiwer* pure]

We hope in all sincerity that . . . will go down in history as the man who . . . fulfilled the promise he made moments after he was elected Prime Minister – to give us 'suiwer administra- sie'. Can he do it? *Het Suid Western* 6.12.78

suka [ˈsuga, -ka] *interj.* Go away: some- times in combination *hamba* (q.v.) ~ or *figur. Afr.E.* an expression of dis- belief equiv. of 'get along with you ' 'Go on !' etc. [*Ngu. ukuSuka* to leave, get away]

. . . he called out to the Caffer suka, an expression of great contempt, as if driving away a dog; *Grahamstown Journ.* 16.5.1833

sukkel [ˈsœkəl, ˈsək-] *vb intrns. colloq.* Struggle, have difficulties, battle, etc. [*Afk. fr. Du. sukkelen* to live poorly, plod, drudge: *sukkelaar* a poor fellow, a drudge]

Every night you lie there and sukkel with things you'd like to forget. Mclennan *Winter Vacation* 1970

~ *aar* one who struggles against odds, a plodder. *cf. Austral. battler :* also in the political sense ; see quot. at *remskoen.*

sunbird *n. pl.* -s. Any of numerous Nec- tariniidae: also known as *suikerbekkie* (q.v.) including the *malachite* ~ or *Jan groentjie* (q.v.) and other brilliant- coloured species which live on the nectar of flowers: see quots. at *sakabula, white eye* and *painted lady.*

sundowner *n. pl.* -s and *modifier.* A drink at the end of a day's work ; in com- bination ~ *party,* pre-dinner drink party, 'cocktail' party. ▐Not *equiv. Austral. sundowner,* a *swagman* arriving at sunset. [*fr. time*]

. . . it was a usual place for them to gather at that hour, where over their sun-downers they compared their 'finds' of the day. Klein *Stagecoach Dust* 1937

. . . bearing in mind some rather tart remarks made at last night's sundowner party, he made the forthright statement . . . *Argus* 24.2.73

suur- [sy:r] *adj. prefix.* Sour. **1.** Prefixed to numerous items e.g. ~ *veld* (q.v.) ; ~ *anys* (q.v.) ; ~ *deeg* (q.v.) ; ~ *pap* (q.v.) also *sour porridge;* ~ *vy,* see *sour fig.* [*Afk. fr. Du. zuur cogn.* sour]

2. Found in place names Suurbron, Suurbraak, *Suuranys* (q.v.).

3. Prefixed to numerous plant names usu. sig. acidity of the sap, and in the names of certain grasses sig. unpalata- ble, as opposite of *soet,* sweet e.g. ~ *gras cf. soetgras,* numerous grass species, see ~ *veld,* ~ *karree,* see *karree;* ~ *spekboom* (q.v.) *Portulacaria afra;* ~ *turksvy* (q.v.) *Opuntia opuntia* (prickly pear), see *turks- vy,* etc.

suuranys [ˈsy:raˌneɪs] *n.* Also ~ *wortel* (q.v.) *Annesorrhiza hirsuta,* the roots of which are anise-flavoured and were formerly used with the 'vinkelbol' *Chamarea capensis* (see also *vinkel*), roasted or stewed, as a vegetable by Hottentots and travellers alike. [*fr. Afk. suur cogn.* sour + *anys fr. Du. anijs cogn.* anise]

suurdeeg [ˈsy:rˌdɪəx] *n. lit.* 'Sour-dough' used as equiv. of 'yeast' of various types: see also *moer².* [*Afk. fr. Du. zuurdeeg* leaven, yeast] **1.** Leaven consisting of a piece of dough retained from the pre- vious baking: *cf. Canad. sourdough, salt-rising.*

2. Home-made leaven made by souring flour and water by fermentation.

3. Commercially prepared '∼' occ. obtainable. Used in the country but now usu. supplanted by granular or compressed yeast. [*Afk. gis*]

Both Vetkoek and Roosterkoek were made either with unleavened dough or with a dough leavened with suurdeeg. Gerber *Cape Cookery* 1950

suurpap ['sy:r₁pap] *n.* See *sour porridge.* [*Afk. fr. Du. zuur cogn.* sour + *pap* porridge]

He could not swear that the beer produced was made of kaffir corn and mealies. He knew that it was kaffir beer from experience and the smell. Witness had no experience of 'Zuurpap'. *E.P. Herald* 27.8.1925

suurveld ['sy:r₁felt] *n.* Sourveld. [*Afk. fr. Du. Zuurveld*] **1.** *n. prop.* The name of the area now called Albany.

The south-western division . . . designated 'Albany proper' is exclusively that portion formerly called the Zuurveldt, which sufficiently indicates its character and is a description of pasturage against which old colonists are strongly prejudiced. *Greig's Almanac* 1831 **2.** *n. veld* of the type designated *sour* (q.v.): grazing usu. found in a high rainfall area, see *sourveld, sourgrass.*

suurvy ['sy:r₁feɪ] *n.* Any of several species of *Carpobrotus*: see *sour fig.* Hottentot *fig. ghokum.* [*Afk. fr. Du. zuur cogn.* sour + *vijg cogn.* fig]

swaer [swaə(r)] *n. pl.* -s. Brother-in-law, also *colloq.* friend, pal, etc. see below. [*Afk. fr. Du. zwager* brother-in-law]

. . . I went with Jan and his swaer (brother-in-law) to the top of the hill. Farini *Through the Kalahari* 1886

. . . David, dearly loved son and swaer – so brave and courageous, missed and deeply mourned. *E.P. Herald* 20.12.73

freq. in combination *as* ∼, mode of address usu. between men equiv. 'old chap.'

Listen . . . swaer, I promise you, ou swaer, my veld's so sour that when the tortoises eat it their eyes water. O.I. E.Cape Farmer 1971

swak [swak] *adj. colloq.* Weak, feeble. [*Afk. fr. Du. zwak* weak, feeble]

How are you? Still feeling a bit swak-ish? O.I. Doctor 1970

Swakara ['swa₁kara] *n. prop. and n. modifier.* Trade name for Persian lamb (karakul) produced in S.W. Africa; as modifier ∼ *pelts,* ∼ *coat* etc. [*acronym* South *West* African *Kara*kul]

So 1970 will see a great deal of karakul or swakara as the South West African product is more often called. But mink is still the best

seller. *News/Check* 15.5.70

swaelstert ['swaɔl₁stert, -₁stɛət] *n.* See *swallow ear.* [*Afk. fr. Du. zwaluw cogn.* swallow + *staart* tail]

swallow ear *n.* An identification mark clipped in the ear of a domestic animal: also *swaelstert* (q.v.) see quot. at *skey ear.* [*trans. Afk.* swaelstert *oor*]

Swapo ['swapəʊ] *n. prop. usu. capitalized.* South *West* African *People's* Organization: a banned political organization of South West Africa, also used attributively of its members or activities: See also quot. at *Namibia.* [*acronym*]

Vorster still refused to deal directly with the South West African People's Organization (SWAPO), Namibia's main liberation (and guerilla) movement. *Time* 20.9.76.

swart- [swart, (swɑt)] *adj. prefix.* Black. **1.** Prefixed to numerous items sig. usu. black or dark colouring e.g. *Swartland* (q.v.), ∼ *gevaar* (q.v.), ∼ *witpens* (q.v.). [*Afk. fr. Du. zwart* black *cogn.* swarthy] **2.** Prefixed to numerous plant species: e.g. ∼ *aasblom, Stapelia gemmiflora,* see *aasblom;* ∼ *haak* (q.v.) ∼ *doring* (q.v.) ; ∼ *hout,* see *blackwood;* ∼ *stinkhout, Ocotea bullata,* see *stinkwood.* **3.** Found in place names e.g. *Swartland* (q.v.), *Swartkrans* (q.v.), Swartberg, Swartkops etc.

swartdoring ['swart₁dʊərɪŋ] *n. pl.* -s. *Acacia karroo,* also called *soetdoring* and *sweet thorn;* a thorny tree with fragrant yellow flowers. ¶It yields gum and is used medicinally in the form of a decoction made from the bark. [*Afk. fr. Du. zwart* black + *doorn cogn.* thorn]

Swartdoring . . . is the 'Doringboom', 'Mimosa' or 'Mimosa illotica' of the early colonists, travellers and writers. Smith *S. Afr. Plants* 1966

swartgevaar ['swart₁xə'fɑ:r] *n. lit.* 'Black peril,' used by politicians in reference to the possibility of being swamped by the black majority. *cf. Brit. and U.S.* yellow peril. Also derivatively *rooigevaar* (communism), *khakigevaar* see *khaki, Roomsegevaar* (q.v.), and *swarthaak gevaar* (quot. at *swarthaak(doring)*) as modifier ∼ *election,* ∼ *tactics.* [*Afk. fr. Du. zwart* black + *gevaar* danger]

. . . these rights were scrapped by white South Africa in the light of what they call 'the black peril' or 'swart gevaar' or simply their fear of being swamped by hordes of blacks. *Daily Dispatch* 1.3.72

swarthaak(doring) ['swɑ(r)t‚hɑk(‚dʊərɪŋ)] *n. pl.* -s. *Acacia mellifera (detinens)*, a tree with paired, hooked thorns which yields very hard red timber, resistant to water and fire: it is troublesome in some areas to farmers; see quot. [*Afk. fr. Du. zwart* black + *haak cogn.* hook]

The greatest threat is the Swarthaak thorn bush, commonly known as the Swarthaak gevaar. The Swarthaak is a sort of galloping consumption of the veld. It spreads so quickly the local goats have to watch it in case it creeps up behind them and strangles them. *E.P. Herald* 27.8.73

swarthout ['swart‚hɑʊt] *n.* Blackwood.
1. *Afk.* for *blackwood* (q.v.) *Acacia melanoxylon*.
2. *Rothmannia capensis*, also called *duine* ~.
3. *Maytenus peduncularis*, a tall forest tree yielding hard tough timber, of which the heartwood of older trees is nearly black. [*Afk. fr. Du. zwart* black + *hout* wood, timber]

Swartkrans, Swartkranz (ape) man, ['swɑ(r)t‚krăns] *n. Paranthropus* or *Australopithecus crassidens;* skulls, teeth and other remains of which were found at Swartkrans near Johannesburg. ⟪ ~ is comparable with Java man in brain capacity and man-like jaw formation. *cf. Kromdraai (ape) man;* see also *Boskop man.* [*fr.* place name *Swartkrans*]

Swartland ['swart‚lant] *n. prop.* The wheat-growing area of the Cape Province is so called from the dark appearance of the (1) *veld* there: see quot. [*Afk. fr. Du. zwart* black + *land* country]

. . . I can see one by-way that leads into the heart of the Swartland . . . it was not the soil that gave the district its name . . . It was the dark, notorious and aggressive shrub, the renosterbos. Green *Land of Afternoon* 1949

swartwitpens [‚swart'vɪt‚pe:ns] *n. pl.* Ø. The sable antelope, also known as the *Harris buck.* [*Afk. fr. Du. zwart* black + *wit* white + *pens* belly *cogn.* paunch]

Then he groped in his shirt and brought out what I thought was a Boer tobacco pouch of the skin of the Swart-vet-pens or sable antelope. It was fastened with a little strip of hide, what we call a rimpi. Haggard *Solomon's Mines* 1886

Swazi ['swɑzĭ] *n. pl.* -s. **1.** A member of the Bantu race of Swaziland or the language spoken there, allied to both Zulu and Xhosa with which it forms the Nguni group of languages.

ZULUS RAIDED BY ARMED SWAZIS A large Swazi impi, armed with guns and assegais, crossed the Swaziland border into Zululand and opened fire on the Zulus in the Ingwavuma district on Monday, a witness claimed yesterday. *E.P. Herald* 4.10.74
2. The *homeland* (q.v.) of the (1) *Swazi* consisting of territory in the E. Transvaal.

There are three regional authorities in the Swazi areas, but as yet no central territorial body. Horrell *Afr. Homelands of S.A.* 1973

Swazi print *n. pl.* -s. Colourful African cotton print material. [*see quot.*]

. . . top and skirt are made up of Swazi print – a new material designed in Swaziland and doing nicely in the colour fashion stakes. *Drum* Mar. 1971

sweat-leaf *n.* Also *sweetkruie* (sweat herbs): *Centella glabrata*, used to induce perspiration. [*fr. Afk. fr. Du. zweet cogn.* sweat + *kruie* herbs]

But it might be the will of the Lord that a pack of dried 'sweat-leaf' sprinkled with brandy might yet save her. Yes, she would – spread the leaves out on a blanket, sprinkle them with 'dop' and wrap the child tightly into this pack to sweat. Smith *Beadle* 1926

sweep [swɪəp] *n. pl.* -s. Whip: also ~*stok* found in form 'whipstick'. [*Afk. fr. Du. zweep* whip]

He used a sweep to chase the cows. O.I. 1970

sweet *adj. substandard.* Well behaved, good: usu. of a child, sometimes by mothers e.g. 'My second one was really a sweet child, but the first was a devil' etc. [*mistrans. Afk. soet fr. Du. zoet* good, well behaved, *opposite of stout,* naughty, bold]

I'll be a sweet boy for a change and copy out the words for you. O.I. Schoolboy 16 1973

sweet cane, sweet reed *n. Sorghum saccharatum* or *S. dochna:* see *imfe*, also *soet riet.* [*prob. trans. Afk. soetriet*]

Before we left one of the men, said he should like to give me some sweet cane, to chew. Shaw *Diary* 26.2.1828

sweethearts *pl. n.* The hooked seeds of *Bidens pilosa:* see (1) *blackjack.* [*see quot.*]

Sweethearts . . . The vernacular name is more properly applied to the fruits, a poetic conceit, probably in allusion to the clinging of these to the clothing etc. Smith *S. Afr. Plants* 1966

sweetkruie [swɪət‚krœɪə] See *sweat-leaf.*

sweetgrass *n. pl.* -es. Any of numerous species of palatable pasture grasses. See also *sour*. [*trans. Afk. soetgras* (q.v.)]

There are 6 Veld Camps (sweetgrass), and the 11 Land Camps total about 160 morgen. *Farmer's Weekly* 3.1.68

sweetveld *n. pl.* -s. A (2) *veld* type in which *sweet* grasses predominate. *cf. sourveld, suurveld.* [*prob. trans. Afk. soetveld*]

The types of grass found in the sweetveld maintain their feed value after maturity and to a large extent even after they have been frosted. *H'book for Farmers* 1937

sweetwine [ˈswitˌwaɪn] *n. colloq. prob. reg Cape* Fortified muscatel, port, *jerepigo* (q.v.) or sherry-type wines, also known as *'soetes'* (q.v.) (*slang*). [*translit. Afk. soetwyn*]

I was fast asleep . . . I was having a lekker dream about a full can of sweetwine. Muller *Whitey* 1977

sysie [ˈseɪsǐ] *n. pl.* -s. Any of numerous species of Fringillidae (siskin or canary) usu. in combination *berg(mountain)* ~ ; *dikbek* ~ see (3) *dikbek; geel(yellow)* ~ ; *streepkop* (*stripy-headed*) ~ etc. [*Afk. fr. Du sijsje* linnet *cogn.* siskin]

They had left Stellenbosch at an hour when even the sparrows and seisjes were still sleeping, long before the false dawn had whitened the eastern sky. Fairbridge *Which Hath Been* 1913

T

taai [taɪ] *adj.* Sticky, tough. **1.** *colloq.* Resilient, also difficult to remove, of a tyre etc. [*Afk. fr. Du taai* tough, tenacious]

That lady is a taai old political campaigner. O.I. 1970

2. *prefix* to numerous plant names: ~ *boom, Rhus pyroides;* ~*bos* (q.v.); ~ *bosdoring, Rhus spinescens;* ~*doring, Acacia nebrownii;* ~ *gras, Eragrostis gummiflua;* ~ *heide,* ~ *heath,* any of several species of *Erica* with sticky flowers ; ~*pit* (q.v.).

taaibos [ˈtaɪˌbɔs] *n.* Any of several species some of which are 'invader' plants (see *plant migration*): mainly applied to numerous species of *Rhus* on account of their flexible and resilient branches, formerly used by Bushmen and Hottentots for making bows and *kieries* (q.v.), and for tanning, see quot.: also in place name Taaiboschkraal. ❡Applied to certain species with tough bark, *esp.*

Passerina filiformis. [*Afk. fr. Du. taai* tough, resilient + *bos cogn.* bush]

One of the Cape sumach (Taaybosch) has been recommended for culture; where this is done the branches should be cut down annually . . . consequently contain a greater proportion of *tannin.* Greig's *Almanac* 1831

taaipit [ˈtaɪˌpɪt] *n. pl.* -s *and n. modifier.* Clingstone peach varieties are so named ; as modifier ~ *peaches.* [*Afk. taaipit* clingstone *fr. taai* tough + *pit* stone, pip]

. . . making peach jam, . . . a big box of 'taaipits' . . . so I've been peeling & boiling & cutting. *Letter* 17.2.64

taal [tɑːl] *n. and n. prop.* Language: usu. in the form *the Taal* sig. *Afrikaans* (q.v.) see quots. at *Dutch* and *Taalbeweging.* [*Afk. fr. Du. taal* language, tongue, speech]

TAAL vs DUTCH – At the same time he . . . disagreed with the present movement for the study of the Taal in preference to the Dutch language proper with its fine literature. *E.P. Herald* 9.12.1911

colloq. praat the ~, *praat die* ~: to speak Afrikaans, to be *bilingual* (q.v.).

It appears to be shameful and unpatriotic if the Engelse cannot 'praat die taal.' *Sunday Times* 25.2.73

In combinations ~*beweging* (q.v.), ~*bond* (q.v.), ~ *toets* (q.v.), ~*fees* (q.v.), ~*stryd* (q.v.).

Taalbond [ˈtɑːlˌbɔnt] *n. prop. abbr. Suid Afrikaanse* ~, a society formed in 1890 to strengthen and encourage the use of Dutch, later of Afrikaans: it offers national examinations in Afk. known as the *Higher* ~, *Lower* ~ and *Voorbereidende* (*Preparatory*) ~. [*abbr. Suid Afrikaanse Taalbond* (Language League) *eksamen*(*s*) examination(s)]

In October 1890 the Taalbond was inaugurated, to propagate the Dutch language and culture in schools. Davenport *Afrikaner Eond* 1966

Taalbeweging [ˈtɑːlbəˌvɪəxɪŋ] *n. pl.* -s. Either of two language movements firstly to recognize Afrikaans as a written language (1875–1900) and later to have Afrikaans made an official language of S.A. (1905–1925). [*Afk. fr. Du. taal* language + *beweging* movement]

On its hundredth birthday Afrikaans finds its future uncertain . . . As it stands on the threshold of a new era it is in the throes of a Derde Taalbeweging facing challenges as great as those which confronted the Genootskap van Regte Afrikaners. *Sunday Times* 17.8.75

Taalfees ['tɑːlˌfɪəs] *n. prop.* The language festival celebrating a hundred years of the Afrikaans language, 14th August 1975. [*Afk. fees* festival *cogn.* feast]

It must learn this language of Africa . . . to speak to Africa in words that it will accept. If it does . . . this week's Taalfees will mark not the end of one chapter, but the beginning of another. *Sunday Times* 17.8.75

taalstryd ['tɑːlˌstreɪt] *n. lit.* 'Language conflict', here sig. the struggle to establish and maintain the Afrikaans language. [*Afk. strydfr. Du strijd* struggle, battle]

And yet in a year which marks its triumphs Afrikaans is racked by a taalstryd. Once again academics are prophesying its demise, once again it is under stress. *Sunday Times* 17.8.75.

taaltoets [tɑːlˌtŭts] *n. colloq.* Test of proficiency in both official languages, obligatory for all hotel employees in S.A. [*Afk. fr. Du. taal* language + *toets* test]

. . . recently became the first hotel employee in Knysna to pass her 'taal toets'. Now she has a certificate to prove that she can say 'boerewors' in both official languages. *Het Suid Western* 14.12.72

table butter See *butter.*

Tablecloth, the *n. usu. capitalized.* The cloud which covers the top and sides of Table Mountain at Cape Town, a phenomenon remarked upon by travellers from the earliest times.

But there is a remarkable circumstance connected with the south-east wind at Cape Town, viz. the dense mantle of vapour which rests upon Table Mountain, and rushes over its precipitous sides like a cataract of foam or vapour. This . . . is called by the inhabitants the Table-cloth . . By the evening, about nine, the table-cloth is gone, and with it the wind, and a beautiful, calm and serene night ensues. Webster *Voyage I* 1834

In combination ~ *south easter.*

South easters . . . are of three kinds (1) Tablecloth South-easters (2) Blind south-easters (3) Black south-easters. *Addresses* British and S. Afr Association 1905 *cit.* Pettman 1913

tackie, tacky ['tækĭ] *n. usu. pl. -ies.* A rubber-soled, laced canvas shoe usu. white occ. black: ⫽Sometimes, also used of track shoes; see quot. at *slip slops; cf. U.S. sneakers; Brit. plimsolls; Austral. sandshoes.* [*unknown*]

We all have to wear blue pleeted [*sic*] skirts and blouses and white tackies on the feet and drill with wood dumb bells and broom sticks. Vaughan *Diary circa* 1902

For R370, you can buy a tackie you can sleep

in . . . Made with supple canvas and giant laces, the so-called 'sneaker bed' comes with foam mattress. Sneaker is the American term for tackie. *E.P. Herald* 6.10.76

figur. motor tyres, esp. in combination *fat* ~ (q.v.) radial ply tyres.

. . . Super Premium radial is the safest tyre . . . ever built. Squat and square, it hangs onto corners. And sticks in the wet like a big fat sticky tacky! *Daily Dispatch* 3.10.72

In combination ~ *lips* (q.v.) *tough* ~ *interj.* Hard luck ! *cf. Austral. hard kack.*

tackie-lips *pl. n. colloq.* A large, slack or protruding mouth. ⫽Not a pout ; see *dikbek, langbek.* [*unknown; poss. sig. rubbery fr. tackies*]

. . . has such tackie lips he can't keep his mouth closed shut. Child 14, 1974

tafel ['tɑfəl] *n.* Table (sig. flat-topped): found in place names as *trans.* table e.g. Tafelberg (Table Mountain), Tafelbaai (Table Bay): see quot. at *kopje.* [*Afk. fr. Du. cogn.* table]

tagati [taˈgatĭ] *usu. modifier.* Loosely used in S.A.E. sig. bewitched, of person or thing (not by African speakers: see etym. also quot. at *umthagati*). [*Ngu. umthagathi* wizard *fr. ukuthagato* to work magic, bewitch]

But they had not allowed for the expansion caused by the fizz in the wine, and, feeling themselves swelling, rolled about in the bottom of the boat, calling out that the good liquor was 'tagati' – that is, bewitched. Haggard *Solomon's Mines* 1886.

. . . and then turning to me with his usual positiveness, 'Rooiland is mad. Umtagati! Bewitched!' FitzPatrick *Jock of Bushveld* 1907

. . . he had never seen a bottle behind the books, and would not have dared to open it if he did, in case it might be 'tagati' which is what you people call 'bewitched'. Prance *Tante Rebella's Saga* 1937

tail truck See *fat-tailed sheep,* and quots.

tailings *pl. n.* Also *debris:* the residue or leavings of crushed ore from which gold has been removed, sometimes a profitable source of further gold: also *U.S.* [*presum. equiv.* remains, remnants]

. . . a considerable quantity of pyrites with the gold, which prevented the amalgamation process to such an extent that stuff which by assay gave twenty dwts. of gold to the ton of reef left only five dwts. on the plates, the rest passing away with the tailings. Raymond *Barnato Memoir* 1897

The new McArthur-Forrest method of extracting almost all the residue of gold from the 'Tailings' (sand or slime) by dissolving it in cyanide solutions was not developed sufficiently for general use until the 'nineties'. *Cambridge Hist. VIII* edit. Walker.

takes me up to . . . *vb. phr. colloq. among children*, equiv. of 'reaches my (neck, shoulder, waist)' etc. e.g. 'The water takes me up to my neck.' [*poss. fr. Zu. ifika lapa ngi mi* reaches here against me *or equiv. phr.*]

take me up to here transl. of a Ba. expression used to indicate how tall a person is, the hand held close to the body; the hand is usually cupped, palm upwards, as superstition holds that a child will cease to grow if the palm is turned down. Beeton *Eng. Usage in S.A.* Vol. 3 No. 1 June 1972

. . . she's so tall her pundus [buttocks] take me up to my chest. O.I. (English speaking *Student*) 9.7.74

take the gap, to *vb. phr. Rhodesian* To leave the country. [see quot.]

. . . White phrase for leave the country derived from rugby maneuver [sic] of breaking past other players. The emigration route once known as the Chicken Run, is today widely referred to as the Owl Run, because it is more wise than cowardly to take the gap. *Time* 30.4.79

takhaar ['tak₁(h)ɑ:(r)] *n. pl.* -s, -e. A crude rustic usu. with connotations of uncouth manner and unkempt appearance: *see backvelder* and quot. at *gawie*. [*Afk. tak* branch + *haar cogn.* hair]

His lanky dust-coloured hair, fading with age instead of turning grey, and worn long like a Tak-Haar Boer's from the Transvaal, gave him a wild and unkempt look. Smith *Little Karoo* 1925

Oom Schalk did not always show that sardonic knowledge of human weakness for which he is beloved, and one sometimes wondered whether to respect the old takhaar, or to indulge him. *Daily Dispatch* 8.9.71

figur. a mess (*cf. strandloper figur.*)

You look a proper takhaar – didn't you bring a comb in your luggage? O.I. 1971

takkies ['takĭs, -z] *pl. n.* Dry sticks and twigs: brushwood for kindling; cf. *Jam-Eng.* brush-brush. [*Afk. fr. Du. tak* branch + *dimin. suffix -(k)ie + pl. -s*]

Haven't we got some takkies to get this fire going? O.I. 1971

tamaai [₁ta'maɪ] *adj. colloq.* See also *so size :* vast, as in 'We found a ∼ dagga bush growing in the veld.' O.I. 1970. [*Afk. tamaai* very large *fr. Malay tamaju fr. Port. tamanho, Lat. tam magnum* so big]

They say they want independence and yet suitable personnel is a tamaai obstacle and . . . they too have decided to have an army . . . *Bolt* July No 12 1975

tamatiebredie [tə'matĭ'briədĭ] *n.* A traditional Cape dish of stewed mutton with tomatoes: see also *bredie.*

And where was the fragrant *tamatie bredie* made with Karoo lamb? Why must we have a so-called Spaghetti Bolognese in the heart of our country? *Darling* 16.3.77

Tambookie [₁tæm'bukĭ, tam-] *n. pl.* -s. *hist.* name for the *Tembu* (q.v.) people, AmaTembu, originally one of the three divisions of the *Kaffir* (*hist.*) nation: see quot. at (1) *Kaffir: obs.* except as prefix in plant names ∼ *grass* (q.v.); ∼ *doring* see *wag 'n bietjie;* ∼ *twak* (*tobacco*), *Peucedanum caffrum* etc.

. . . it may be considered as sufficiently established, that the tribes commonly called Caffers, or Koosas (Amakosae), the Tambookies (Amatymbae). Thompson *Travels I* 1827

tambookie, tamboekie gras(s) [₁tæm'bukĭ, ₁tam'bŭkĭ] *n.* Also *tambuki-,* and erron. *tamboti :* any of several species of *Cymbopogon, Hyparrhenia, Miscanthidium,* tall grasses used for thatching African huts: see also first quot. at *skerm.* [*fr. name of tribe*]

. . . a small company of troopers could be seen moving cautiously and watchfully through the tall tambookie grass and the tangled, matted vegetation. MacDonald *Romance of Rand* 1933

tambutiboom, tambuti wood [₁tam'butĭ, ₁tɔm-] *n.* Usu. *tomboti boom, tambuti wood* (q.v.), see quot. at *boekenhout.* [*form of tomboti(e)*]

TAMBUTI WOOD. – Furniture of high quality . . . is being produced from the best known, renowned indigenous Tambuti Wood. *Farmer's Weekly Advt.* 21.4.72

Tambotie is a protected species . . . rarer than stinkwood because . . . it is not propagated. Tamboti wood was first used by the early Transvaal pioneers for furniture and other articles. *Panorama* Sept. 1975

tammeletjie [₁tamə'lekĭ, -cĭ] *n. pl.* -s. Taffy or 'stick jaw': a hard toffee: sometimes with pine nuts added. Also *occ.* a slab or roll of sugared, dried fruit ∼ *rol/roll :* see *perskesmeer, peach leather.* [*Afk. fr. Du. tabletje* tablet, *poss.* pastille]

. . . dennebol pips brought back memories of the walks . . . where we searched among the pine needles for pips and collected cones . . . tameletjies, a delicious sweet made with butter, brown sugar, syrup, vinegar and water, and the addition of the pips, which my elder sister made for us. *Cape Times* 12.1.74

tampan ['tam₁pan] *n. pl.* -s. Any of various ticks of the Argasidae esp. the fowl or chicken tick which has a painful or irritating bite: see quot. at *kameeldoring*. [*prob. fr. Tswana tampane*]

... the driver said '... Baas dit is maar tampans.' They were not proper bugs ... They only ate on fowls ... the buzz had been kept beside a fowl hock. Vaughan *Diary circa* 1902

tankjokkie ['tæŋk͵jɔ̃ki, 'teŋk-] *n. pl.* -s. *Sect. Army.* Also *tenkjokkie pron.sp.* see second quot., a member of the Armoured Corps: see also *S.S.B.*

A member of the Armoured Corps is a *tankjokkie*. An officer giving a particularly poor salute *waai vliee* whilst a useless soldier becomes a *vuiluil*. Picard in *Eng. Usage in S.A.* Vol. 6 No. 1 May 1975

It doesn't matter if you are a *kanondonkie* or a *tenkjokkie* or even an ordinary bokkie with a kettie (infantryman with a rifle) you must have a host of expressions to pass on. *Cape Times* 17.1.79

tannie ['tani] *n. pl.* -s. **1.** Auntie: used as friendly but respectful mode of address by children or young people to older women, whether they are related to them or not: see also *tante* and *auntie*. [*colloq. form Afk. tante* (q.v.), aunt]

I am finding that most Afrikaans children call one 'Aunty', or 'Tannie', and are most charmingly co-operative. Vaughan *Last of Sunlit Years* 1969

... the way patients and visitors were addressed by hospital staff. 'Their complaint was specifically against the use of Afrikaans terms – presumably to Afrikaans patients – such as "tannie", "mevroutjie", "vroutjie", "oumatjie" and "oumoedertjie" to women, and "oomie", "meneertjie" and "oupatjie" to men.' *E.P. Herald* 3.12.76

2. Somewhat deprecatory usage for a simple, strait-laced or narrow-minded small-town woman. [*fr.* (1) ~]

... a fine performance as the strong-charactered Hanna Verster, essence of every stiffbacked, gloved and hatted tannie who ever sat in judgement. *Grocott's Mail* 15.9.72

In combination: *ticket* ~ a meter maid.

If our traffic cops would do more about controlling the traffic and less patrolling for expired parking meters like so many tickettannies we'd be better off. O.I. Motorist 1970

Joeys, where the ticket tannies come in pink and the jollers come in black leather and the mining-house smoothies come in grey pinstripes. *Darling* 4.8.76

tante ['tantə] *n. pl.* -s. Aunt: usu. a formal and respectful mode of address or reference with or without an appended Christian name, to older women by children or young people. ⫐Some-

times shortened before a following vowel as in Tant' Elisabeth: see quot. at *neef*. [*Afk. fr. Du. tante*, aunt *cogn. Fr. tante*]

The word 'tante', meaning 'aunt', and 'oom' meaning 'uncle', are still used by the young Dutch as a mark of respect when addressing their elders ... To the whole countryside she was known as 'Tante Let'. No one would ever have dreamed of calling her 'Mrs. Uijs', and had anybody done so, she would have put him down as 'uppish'. Brinkman *Breath of Karroo* 1915

tarentaal [͵tarən'tɑ:l] *n. pl.* -s. A guinea fowl: any of several species of the Numididae fam. [*Afk. fr. Lat. Terra Natal*]

... the dogs drew on some game ... and soon sprung ten guinea fowl, exactly like those now bred in Europe, called here, by the Hottentots tarentalls. Philipps *Albany & Caffer-land* 1827

tasselfish, *n. pl.* ∅. See *baardman*.

At Lumley reef on Sunday afternoon he speared a ... yellow tail kingfish and 7kg. tasselfish. *Daily News* 14.5.70

Tassies ['tæzĭz, 'tæsĭz] *n. prop. colloq. abbr.* Tassenberg, a popular red wine.

Boeuf Bourguinonne ... ½ bottle good red wine (Tassies will do). *Darling* 19.7.78

Booze consumption is a good barometer of changing times. They've taken Tassies away from us and the shelves bulge with strange-looking jugs. Jane Mullins *cit. Fair Lady* 13.9.78

tata ['tata] *n.* Father: a mode usu. of address, also reference, by Africans to an older African man, also by children to their fathers: see *father, small, big*, also *buti, sisi* ⫐*Mama* (q.v.) is used in the same way to older African women. *cf. Jam. Eng. taata* or *tata*, father and grandfather. [*Ngu. utata* father: *poss. fr. Du. tate*]

'Tata I'm a man now ... Nama, tata if you want to be a sissy, say so' ... 'Kwedini, you are not going to do as you like, not while we live ...' 'Tata, what have I done?' Dike *First S.African* 1979

Taung [tauŋ] *n. prop.* and *modifier.* A place in the N. Cape Province which has given its name to an australopithecine homonid, the skull of which was discovered there in 1924 by Professor Raymond Dart: usu. in combination ~*skull*, ~*child.* See also *Boskop, Kromdraai, Stellenbosch, Swartkrans*.

... 50th anniversary of the Taung skull ... evidence of an early 'relative' of man who lived in South Africa a long time ago. ... In Dart's opinion the position of the opening on the base of the cranium . . . indicated that the australo-pithecine youth must have walked upright. *Panorama* Nov. 1974

taxis *pl. n. Sect. Army.* See at *area.*

You've got to keep your 'taxis' – mine are two pieces of an old blanket – in a bush outside the bungalow so you can get to your own area without messing up the other ous'. *O.I. Serviceman* 4.2.79

I keep my taxis under my trommel so it'll run without grating in inspection – no bushes where we are. *O.I. Serviceman* 10.3.79

tea *n.* See *tearoom, tea-water* and *tee-:* also *bush* ~, *rooibos* ~.

tearoom *n. pl.* -s. The shop of a merchant usu. dealing in perishables, soft drinks, confectionery and cigarettes, with groceries as a side-line, open on Sundays, holidays and in the evenings, in addition to normal trading hours: sometimes also known as a *Greek shop* (q.v.). ¶A ~ unless at some resort or beauty-spot, and trading as a tea-garden-cum-restaurant, is usu. without facilities for serving refreshments. In Natal ~ is freq. used of a regular grocer's shop. See quot. at *café.*

It's a nice place . . . they've got ablution blocks now, and lots of taps and bins and braai places scattered about, plus a tearoom for odd stores. Letter 9.10.63

I told the girl in our local tearoom I wanted 40 cigarettes of a brand I named and she said: 'Sorry, only 20s, they don't make them in 40s.' *Personality* 22–29.2.76

In combination ~ *keeper's licence:* ¶This empowers the licensee, unlike the holder of a general dealer's licence, to remain open for the sale of perishables, cigarettes and sweets on Sundays, holidays etc. though usu. no groceries may be sold after 6 p.m.: not equiv. of a restaurant-keeper's licence.

tearoom cinema *n. pl.* -s. Usu. formerly known as a *bio-café* (q.v.) or *café-bio.*

A schoolboy went into a coma from a suspected drug overdose in a city tearoom cinema yesterday . . . Mrs . . . who attends the refreshment counter . . . saw . . . rocking back and forth in his chair. *Cape Times* 21.7.78

tears, queen's. See *queen's tears.*

tea-water, *n. obs.* An infusion of tea as opposed to tea in leaf form: see also *treksel:* found in *Afk.* form in place name Theewaterskloof. [*prob. trans. Afk. teewater,* the beverage itself]

I learned that he was a man of singularly abstemious habits, seldom or never being induced to take anything but tea-water.

Philipps *Albany & Caffer-land* 1827

. . . sits the mistress of the house, with a tea-urn and chafing-dish before her, dealing out every now and then tea-water, or coffee, and elevating her sharp shrill voice occasionally to keep the dilatory slaves and Hottentots at their duty. Thompson *Travels I* 1827

In combination ~ *konfyt,* ˌpreserves served with tea: see quot. at *konfyt.*

tecoma [təˈkəʊmə] *n. Tecomaria capensis,* also *Cape honeysuckle* (q.v.) and *Kaffir honeysuckle* usu. in combination ~ *hedge.* [*fr. botanical name*]

TDC [ˈtiˌdiˈsiː] *n. prop.* Transkei Development Corporation: see also *XDC.* [*acronym*]

Blacks would be able to serve on the board of Tembalethu, the Transkei Development Corporation's giant wholesale division the TDC announced recently. The corporation will also give Blacks in Transkei the opportunity to acquire R5 million worth of shares. *E.P. Herald* 23.8.76

-tee [tiː, tɪə] *n. suffix.* An infusion or decoction of some leaf, bark, root, flower or other herb: usu. taken for medicinal purposes: see also *treksel, -water.* [*see quot.*]

The young leaves and flowering tops of many herbs and shrubs are used for making a tea . . . The vernacular name is often generally employed to any infusion which is used medicinally, but most plants used as the source of a 'tee' bear special names and several of the non-medicinal kinds e.g. *bossietee* and *rooibostee* have been exploited commercially. Smith *S. Afr. Plants* 1966

teegoedbalie [ˈtɪəˌxŭtˌbalɪ] *n. pl.* -s. A basin reserved exclusively for the washing and storage of tea-making utensils: a feature of the early S.Afr. kitchen: see quot. [*Afk. teegoed* tea things + *balie* vat, vessel]

. . . the teegoedbalies in which cups and saucers were not only washed, but also kept lying under clean water, being taken out and dried only when needed . . . the teegoedbalie under its white cloth standing ready to hand. Cook *Cape Kitchen* 1973

teff (grass) [tef] *n. Eragrostis tef (E. abyssinica),* a quick growing grass excellent for hay, a native of N. Africa naturalized in parts of the Transvaal highveld: also in combination ~ *(grass) hay.* [*fr. botanical name*]

Best quality racehorse quality teff, up to R1,20; first grade teff R1,10; medium teff 80c; and inferior teff, 60c, sweet grass and eragrostis are both fetching up to 50c and veld grass up to 40c. *Farmer's Weekly* 12.5.71

Tekwini, Tegwin(i) see *Thekwini.*

telling you, I'm, *vb. phr. colloq.* Used for emphasis to increase the impact of the statement made or the event related, e.g. 'He drives like hell, I'm telling you!' See also *ek sê.* [*trans. Afk. ek sê* (q.v.) *vir jou*]

. . . the smoke and boerewors fumes of 500 campfires. It works on yore nerves, I'm telling you. Not to mention bringing on Ouma's asthma. *Darling* 12.2.75

Tembu [ˈtembŭ] *n. pl.* ama, -s. One of the three *Xhosa* (q.v.) speaking nations; formerly known as *Tambookie, Tambuki* (q.v.): also a single member of this people.

Paramount Chief of the powerful Tembu tribe Chief Sabata Dalindyebo, has turned down the offer of the presidency of Transkei . . . he has been offered the post of the first president of the Republic of Transkei when it attains independence later this year. *World* 22–24.7.76

10/1 [ˈten₁wʌn] *n. Urban Afr. E.* An African, classified under the Native Administration Act 10(1)B.

Move out of Langa, but why, I'm a 10/1 and so are my parents. I qualify here. My skin! I thought my skin had nothing to do with this. If I was classified as a 10/1 in the beginning, I don't see why that should change now. You and tata fall under 10/1, I don't understand . . . Dike *First S. African* 1979

-tent- *n. pl.* -s. Usu. *suffix* as in *wagon~* sig. the tilt of a covered waggon or *prefix* as in *~cart.* [*Afk.* tent hood of a vehicle, tilt]

Country people can be supplied with . . . 3–inch Canvas for wagontents and Horse-cribs . . . Chain Trektouws etc. *Cape of G.H. Almanac Advt.* 1845

It's a nice kind of wagon. It has 4 horses to pull us all. It has had lots of bugs. When it got hot the bugs ran out the tent. Vaughan *Diary circa* 1902

. . . five ox-wagons, full of people who had been to the Zeerust Nagmaal, were trekking . . . back to the Groot Marico. Inside the wagontents sat the women and children, listening to the rain pelting against the canvas. Bosman *Mafeking Road* 1947

terr [tɜ] *n. pl.* -s. Terrorist: Rhodesian 'code' term. See quots at *Uz(z)i, zap* and *auxiliaries.*

'We'll shoot too, if we have to,' says Sally . . . 'and the Terrs know it.' *Radio & T.V.* 22–29.8.76

'Nothing to worry about . . . The Terrs [terrorists] don't like to take on convoys . . . but we can't take chances. We have to show the Terrs who's boss.' *Time* 22.11.76

. . . they play soldiers and terrs rather than cowboys and Indians. *Fair Lady* 16.3.77

Terrs. Short for terrorists, the term whites use

when referring to Patriotic Front guerrillas. Also: **gooks, floppies, oxygen wasters.** *Time* 30.4.79

In combination tiny ~ s and *deriv. termites* see quot.

Tiny terrs. Children used by guerrillas as spotters. Also: **termites.** *Ibid.*

terro *n. pl.* -s. See *terr, terry.*

TERROS FIRE ON BORDER TOWN *Citizen* 9.4.79

terry [ˈterĭ] *n. pl.* -ies. Terrorist: S. Afr. and S.W. Afr. use: *pl.* -ies a boy's game; updated version of 'Cowboys and Indians': see quot. above. [*fr.* terrorist]

. . . or Terry-bashing up in lekker ol' Rhodeesha. *Bike S.A.* Oct. 1976

No terrys here! Three pretty Rhodesian holiday makers relax at Wilderness. *Herald Phoenix* 5.1.79

Territorial Authority *n. and n. modifier,* pl. -ies. The legislative body of an African *homeland* (q.v.) see quots. also *separate development.* [*official term*]

. . . the Chief Executive officer of the newly created Zululand Territorial Authority, Chief Gatsha Buthelezi. *Daily Dispatch* 12.6.70

Discussing the clause of the Bill which provides for the granting of advancement of the political status of a Territorial Authority to that of a full Legislative Assembly, Mr. – said it could be done without consulting Parliament. *Grocott's Mail* 16.2.71

as modifier ~.

The Bantu Homelands Citizenship Bill . . . had to be postponed until 1970. It provided that there shall be a citizenship in respect of every territorial authority area. *S.A.I.R.R. Survey* 1969

tessie [ˈtesĭ] *n. pl.* -s. A firepan of iron or copper to contain coals for placing in another container: see quots. *konfoor, (voet)stofie;* also an ornamental table brazier often of silver used by pipe smokers. [*Afk. fr. Du.* test chafing dish + *dimin. suffix* -je]

The tessie or kolebakkie is the small dish into which the glowing coals or embers are actually placed . . . the tessie is then put into a voetstofie or konfoor . . . most voetstofies and konfore found today are without their tessies.

. . . The tessie used on the table . . . was often made of silver . . . mounted on a wooden base, and always had a shallow copper lining in the bottom to protect the silver from the heat. *Letter* 6.7.72 Curator, Stellenbosch Museum

Old Voetstoofie complete with Original Tessie. *Grocott's* 20.5.75

Thaba- [ˈtaba] *n. prefix* Also (n)*taba*: Mountain found in place names Thaba Nchu, Thabazimbi, Thaba Bosiu, Thaba Chitja, Entabeni, Ntabamhlope(white), Ntabebomvu(red). [*Bantu*]

... chief of the Thaba Bosiu district ... gave the institute the use of a piece of land ... The site ... is overlooked by a small mountain called Thaba Khupa. *Farmer's Weekly* 8.1.75

thali [ˌtɑlĭ] *n. pl.* -s. A cord worn round the neck by every Hindu married woman symbolically similar to a wedding ring. [*presum. Hindi*]

Mrs N. ... Mr N's first wife – removed her thali (the sacred cord which symbolizes that a Hindu woman is married) and attempted to place it round Miss G's neck. *Sunday Times* 23.3.75

thank you *vb. phr.* Equiv. of 'no thank you' e.g. 'Have a cigarette?' 'Thank you, I'll have one of mine.' ⫿This expression has been a stumbling block to non-S. Africans since early C19: see first quot. [*trans. Afk. fr. Du. bedanken* to decline, refuse]

One thing more, however, is to be recommended; if he value his meals, – and that after a long day's ride no doubt he will, – never when invited to eat, reply with a genteel thank ye, (dank u) as that piece of politeness is understood throughout the colony as a negative, the disagreeable consequences of which the writer of this has more than once found to his cost. *Greig's Almanac* 1833

When anyone asked him to have a meal, he would say: 'Thank you I have eaten.' But his friends knew that he was starving. *Drum* 8.10.73

thebe [ˈteːbe] *n.* See (2) *pula*

(e)Thekwini [eˌtekˈwinĭ] *n. prop. colloq.* Durban. *cf. eGoli.* [*locative form of itheku* bay, harbour]

'So I drink ... tell him many stories. He not been to Tegwen (Durban) ... never see the sea ... the big ships and the horses that ... run races.' McMagh *Dinner of Herbs* 1968

there- *adv. prefix.* The prefix ∼ as in 'therein', 'thereafter' etc. is freq. found as a substitute for the *prn.* 'it', *esp.* in the writing of persons influenced by *Afk.* usage. ⫿By this process of substitution 'in it', 'for it' *etc.* are not rendered in Afk. as *in dit* and *vir dit* but as *daarin* and *daarvoor*. [*fr. Afk. prefix daar*]

The planting was also encouraged and the government offered prizes therefor in the territories and also in Basutoland. Edgar Brookes *Hist of Native Pol.* 1924

The Malays, ... are not only partial to music, but display a marked aptitude therein.* Du Plessis and Lückoff *Malay Quarter* 1953 *see in

... the snake will strike with a distorted open mouth displaying the ... interior colouring thereof ... Although superficially resembling

the night adder it can be easily distinguished therefrom. *E.P. Herald* 28.11.78

thick *adj. colloq.* Stupid, dense, uncomprehending, *cf. thick head* etc.: see also *dik.* [*prob. trans(lit). Afk. dik* dense]

Don't be so thick, man. O.I. 1975

thing, my *n. phr.* As mode of address, an endearment *cf.* '*old thing*', e.g. How are you my thing? [*translit. Afk. my ding, an endearment*]

third person form of address Avoidance of direct address 'you' in the second person leads to substitution of forms such as *oom, oupa, ouma, uncle, tannie, the baas, (your) goodself,* all (q.v.) and third person verb forms; also the use of a redundant article: usu. to elders or superiors esp. among children, e.g. 'When does Tannie want me to bring it?' 'Madam, is the madam ready for tea?' 'Will Doctor lend me two rand?' 'Would Ouma like Ouma's shawl?': see second quot. at *morena* and third quot. at *auntie.** [*fr. Afk. custom of oblique form address, pos. rel. Malay: see below*] *and at (1) *meneer*

Please, baas, I will dig for the baas. Jacobson *Dance In Sun* 1957

He had moreover a habit ... of addressing his hearers in the third person, in preference to the more direct 'you'. This is a linguistic peculiarity in which many people at the Cape indulge. Fairbridge *The Torch Bearer* 1915

⫿ Note: It is possible that this usage is related to that of Malay in which a system of pseudo-pronouns* is adopted for modes of address. 'If the Malay verb were inflected to show 'person', the verbs that follow these* would show third person endings, not second person endings. Malays avoid the use of direct second person address. Beside these pseudo-pronouns the following types of words are used as substitutes for second person pronouns.' Lewis *Teach Yourself Malay* 1947

Thirstland *n. prop.* More usu. Afk. *Dorsland* (q.v.) The large desert area of the Kalahari crossed by the *Thirstland/Dorsland trekkers* (q.v.) in a series of treks from the Transvaal into Angola from 1874 to 1905: see also quot. at *wragtig.* [*trans. Afk. Dorsland* (q.v.)]

'Fever is raging amongst us, and a great part have died. We stand here in the *Thirstland* by some wells. Our cattle and sheep are almost all dead ... but I hope that our God will save us from this wilderness of hunger, care, and sorrow.' Trek Boer's letter, 1875, *cit.* Pettman 1913

A four-cent postage stamp to commemorate

the Thirstland (Dorsland) Trek, which began 100 years ago from the Transvaal, through the Kalahari Desert, to Angola, will be issued in South West Africa on November 13. *E.P. Herald* 5.11.74

thorn *n. prefix* and *suffix*. A regular rendering of Afk. *doring* in names of plants, e.g. *camel~* (kameeldoring); *sweet* ~ (soetdoring); *wait-a-bit* ~ (wag-'n-bietjie); ~ *tree* (doringboom). [*trans. Afk. doring fr. Du. doorn cogn.* thorn]

I plucked from the great thorn-trees some of their prickles, of which I send you a few; they exactly resemble the horns of the cattle. I hear the plant has found its way to Kew Gardens, and is there called the cuckold-tree. Barnard *Letters & Journals* May 1798

Also in term ~ *veld* sign. an area in which varieties of thorn (doring) (Acacia) are the predominant vegetation: cf. Austral. *mulga country* (Acacia).

The thornveld was like a park, with flat-topped trees scattered over the grassy plain as far as the mountain slopes. Rooke *Lover for Estelle* 1961

three day's sickness *n.* A short-lived, not usu. fatal disease of cattle, commonly known as *styfsiekte* or *stiffsickness* (q.v.) [*fr. duration of symptoms*]

The dreaded heart-water increased, gall sickness caused many deaths, three day sickness, which as a rule animals recovered from if left alone, meant abandoning beasts since there was no time to wait for them to recover. Cloete *Turning Wheels* 1937

through *prep. substandard* Usu. in phr. '*through the face*' equiv. of 'across' or 'in'. [*trans. Afk. deur* through, across]

. . . testified he was slapped through the face; hit on the body . . . bruised on his left shoulder. *S.W. Herald* 22.10.76

throw- *vb. substandard* Used in S.A.E. in various combinations, ~ *with* (a stone or other missile), see third quot.; ~ *wet* to sprinkle with water; ~ *dead* to stone to death, or cause death by some other missile: see also quots at (2) *with, doodgooi.* [*trans. Afk. met 'n klip gooi* (with a stone), *natgooi* (wet), *doodgooi* (to kill by means of missiles, *lit.* 'throw dead']

. . . farmer will say of wheels the spokes of which have become loosened by the dryness of the atmosphere. 'I must throw them wet,' the wetting process causing them to swell and so to become tight. The expression is a literal rendering of the Dutch *Nat gooien*. Pettman *Africanderisms* 1913

He retreated a little way and picked up stones,

saying he would throw her dead with stones. *Grahamstown Journ.* 16 (month missing) 1892 . . . But Hamlet . . . switches foils and rams Laertes. (Ramming with a foil is the Shakespearean equivalent of our own South African habit of throwing him with a stone.) *Cape Times* 5.1.75

throw (the) bones, to *vb. phr.* To divine the source of trouble or to foretell the future, a method used by *witchdoctors*, and *sangomas cf. Austral. point the bone*: see quots. at *bones* and *sangoma.* [*trans. Ngu. phr. ukuposa amathambo*, to throw bones]

Millionaire inyanga Sethuntsa Khotso, who last year threw the bones and successfully tipped Naval Escort, says that Golden Jewel will win the Durban July next Saturday. *Post* 28.6.70

thula, thwala, thwasa (Thixo) See under *t.*

tick, bont See *bont* and quot. at *heartwater*

tick bird *n. pl.* -s. Also *bosluisvoël*: the cattle egret *Ardeola ibis*, which keeps close to, often perching upon, grazing cattle, and feeds on ticks and other insects. [*prob. trans. Afk. bosluisvoël* tick bird]

The beak of a tick-bird is long and strong – capable of gripping and disposing of a mouse. *Panorama* May 1973

ticket tannie See at *tannie.*

tickets *pl. n. slang. equiv.* Finish, end: as in 'That last car smash was nearly tickets for him.' [*unknown poss. non-S.A.E.*]

It was luck
but it was bad luck, maaster.
I am Hotknife
of Capricorn
and she was in de Crab sir. It was tiekets.
Clouts *Hotknife fr. One Life* 1966

tickey ['tɪkĭ] *n. pl.* -ies and *n. modifier.*
1. The obsolete S. Afr. threepenny piece now superseded by the five cent coin. [*etym. dub. Afk. fr. Malay tiga* three *poss. rel. Austral. trizzie, trezzie*, threepence *or Brit. dial. tizzy*, sixpence]

Schiel demanded complete subjection and tributary taxes. Mujaji grudgingly sent him an ox horn filled with coppers and tickeys. Bulpin *Lost Trails* 1951

As modifier ~ *beer*, ~ *phone*, ~ *shop*, cf. U.S. a *five-and-ten* (cent store); ~ *snatcher*, ~ *snatching* one who is exceptionally close-fisted or out for small, quick profits.

Martienssen's brewery . . . lasted a long time, but it is now merely a memory of the happy

days of 'tickey beer'. Green *Grow Lovely* 1951
'. . . there's a tickey phone. I want to phone
Gran' . . . 'Dad, you're out of the Ark . . . There
hasn't been a tickey phone for years. It costs
double. It's five cents.' 'There aren't even
tickeys any more,' added Peter. Philip *Caravan
Caravel* 1973
   . . . you cannot get him out of a bazaar, or
what we in our day used to call a 'tickey-shop'.
Spilhaus *Under Bright Sky* 1959
   . . . came bounding on 'Change . . . and asked
how the – I dared sell my own shares, and that
if I wanted three hundred threepenny bits why
didn't I ask him and not go tickey-snatching
like a blasted . . . Cohen *Remin. Johannesburg*
1924
   . . . I bought shares, to sell again on the first
rise: tickey-snatching, Father called it. Rooke
Margaretha 1974
In combination *tickie/ ~ aand/evening*
(q.v.), *tickie/ ~ draai* (q.v.), *~ drive*
(q.v.), *tickie/ ~ line* (q.v.); also *long
~* and *sticky ~* : see *tickey, long.*
2. Something very small: a common
nickname for a small person, *cf. titch.*
Also *n. prop.* a famous circus dwarf:
see second quot. [*fr.* (1) *tickey*]
   . . . full credit must go to former Durban
jockey 'Tickey' Carr who got the other . . .
favourite Idol Worship up for a spectacular
neck win . . . at Milnerton on Saturday. *Argus*
10.5.71
   . . . the story of Eric Hoyland, a dwarf . . .
who overcame hardships and his diminutive
stature to becoming a South African circus
institution as 'Tickey the clown'. *E.P. Herald*
2.11.76
In combination *two bricks and a ~ high,*
see second quot. at *boep*: also *half a
brick and a ~ high,* very small, usu. of
a child. *cf. U.S. knee high to a grasshopper.*
   . . . in the days when he was two bricks and a
tickey high . . . a tickey was his reward for every
furry digger caught. *Cape Times* 30.5.70
   . . . inside there's this tiny little ou half a
brick and a tickey high, with a smashed nose
and curled-over ears. *Darling* 26.2.75
tickey-draai *n. pl.* -s. Also *tiekiedraai*:
a fast dance in which couples turn round
and round on one spot, also the music
for such a dance: see quot. at *sluk*. [*Afk.
fr. Du. draaien,* to turn, twist, twirl]
   There were vigorous tiekiedraais, gay
moments when a man swung his partner round
and round on one spot, turning, as it were,
upon our then smallest coin, the silvery threepenny piece we know as a 'tickey'. Henshilwood *Cape Childhood* 1972
tickey drive *n. pl.* -s. A function similar
to a *tickey aand, tickey evening* (q.v.)

cf. beetle-drive. [*prob. analg.* beetle-
drive]
tickey evening *n. pl.* -s. Also *tiekieaand,*
usu. a small-scale fund raising venture
for Church or charity in which entrance
for an event or game was formerly a
tickey (threepence), now five cents
though the name is still in use. [*trans.
Afk. tiekieaand* (evening)]
   Members of the Dutch Reformed Church . . .
held a five cent function (Tiekieaand) in the
church hall . . . Various games which were much
enjoyed by young and old were played. *E.P.
Herald* 13.10.71
   . . . thanked the local businessmen who had
generously donated goods for the recent
'Tickey Games Evening'. *Grocott's Mail*
11.5.76
tickey line[1] *n.* Also *tiekie line:* used of
small time women's *stokfel* (q.v.) (second quot.) also of a girl or woman out
of the 'big-time' African social swing:
see quot.
   These are people who have a very high
choice in women and to find a woman who
would suit their choice, it would be very hard.
Drum: You mean you can't plant a 'tickey
line' in their ranks? *Drum* Nov. 1964
tickey-line[2] *n. pl.* -s. *Afr. E.* A cheap
prostitute: also *penny-line.* ℙ *cf. ~* in
quot. at *stokfel.*
   . . . a Tickey-line who worked as a kitchen-
girl . . . his only Johannesburg contact outside
the compound . . . Tickey lines were, as the
nickname implies, cheap prostitutes who knew
how it felt to be the underdog. Venter *Soweto*
1977
tickey, long *n.* Also *sticky tickey:* both
devices used in public telephone boxes:
the *long tickey* being usu. a coin on a
thread or some other means of with-
drawing it, the *sticky tickey* a metal
disc.
   . . . found guilty . . . for telephoning with a
'long tickey' from a public telephone booth.
He . . . pleaded guilty to using . . . a 'long
tickey'. The 'long tickey' exhibited in court
was a ten cent coin suspended from a cotton
thread which was attached with a piece of
cellotape. *Het Suid Western* 13.3.75
tiekie ['tīkī] See *tickey* and combinations.
[*Afk. form*]
tier [ti:r] *n. Leopard(s)* prefix in certain
plant names *~bek(mouth) vygie* se-
veral species of *Mesembryanthemum;
~hoek* (*place name*) heath; *~ hout, Lox-
ostylis alata*
   . . . came across the biggest tierhout tree . . .
almost certainly a survivor from Settler days.
Tierhout is so called because leopards ap-

parently prefer its trunk to the trunks of other trees for sharpening their claws. *E.P. Herald* 27.9.78

Also found in place names Tiervlei, Tierfontein, Tierhoek: see also *tiger* and *tyger*[*Afk. fr. Middle Du. tiger* tiger, leopard]

tiffie, tiffy [ˈtɪfi] *n. pl.* -ies. *Sect. Army.* A member of the Technical Services Corps: see also quot. at *push beat.* [*acc. Partridge abbr.* artificer, 'nautical 1890s']

Other stories they have lined up for the series [Paratus] concern the tiffies (army slang' for mechanics believed to be derived from the British army term 'artificers'). *Sunday Times TV* 7.8.77

In the workshops some of the technicians or *tiffies* may feel *sterk** about their girlfriends at home . . . Picard in *Eng. Usage in S.A.* Vol. 6 No. 1 May 1975 *strongly

Sgt. Major . . . who was well known as the 'boss of the tiffies' joined the Permanent Force in 1954 *Grocotts* Mar. 19.6.79

In numerous combinations *pot*~ an army cook, also a *'fitter and turner'* (q.v.) *cf. Austral.* rissole king; *aspro* ~, *pill* ~, *sick bay* ~ (Navy) etc. a 'medic' *cf. Austral.* iodine (also, civilian, a *Zambuk*); *soul* ~ or *sieltiffie* an army chaplain *cf. Austral.* sinshifter.

tiger *n. pl.* -s. Leopard [*fr. Middle Du. tiger,* tiger, leopard] **1.** Also *tyger (obs)*: *Panthera pardus,* the leopard: ⫿Tigers do not occur in S.A., see quots. at *under* and *palmiet.*

. . . the tyger of South Africa, a large and fierce species of the leopard. Ewart *Journal* 1811-14

It was very large and the skin was beautifully marked. The Dutch call them tigers, but we were informed that there is no tiger in the colony, and this certainly was a leopard. Philipps *Albany & Caffer-land* 1827

2. The *spotted grunter* (q.v.)

3. As *prefix* sig. spotted (not striped) e.g. ~ *wolf,* ~ *cat :* see quot. at *rooikat.*

tiger's eye *n. pl.* -s. Also *cat's eye :* a semi-precious form of the mineral crocidolite, yellowish brown in colour. [*fr. appearance*]

. . . the inhabitants . . . will tell you about the exporter who wanted to market Tiger's Eyes overseas and received an enquiry as to the cost per carat. The reply was: 'We sell them by the ton.' *Panorama* Jan 1974

tiger's milk *n. colloq.* Strong liquor, spirits: *cf.* E. Cape *kudu milk;* also *Canad.* moose milk, wolf juice, *Austral.* snake juice (also ~) [*trans. Afk.* tiermelk booze, liquor]

His wagons were loaded with lead and powder, with bolts of material, copper and brass rings, with muskets, tobacco, with brandy-wine and tiger's milk in casks. Cloete *Watch for Dawn* 1939

tiger-wolf *n. pl.* -ves. *Crocuta crocuta,* the spotted hyena: see also *wolf* and *strandwolf.* [*fr. Afk. fr. Middle Du. tiger* tiger, leopard + *wolf* hyena]

We instantly seized our guns and ran to the rescue of the remaining horse, and found him beset in a corner of the thicket by a ferocious tiger-wolf (*hyæna crocuta*) who was attempting to break in upon him. Thompson *Travels II* 1827

tikoloshe [ˌtĭkɔˈlɔ́ʃe] *n.* See *tokoloshe.*

time, in . . . *adv. phr. substandard* sig. 'Within the period of . . .' as in 'I completed these displays in three years' time' (Museum Curator 1974). [*prob. trans. Afk. in* . . . *se tyd,* '(with)in a period of . . .' *which lacks the Eng. future sense of* 'in two year's time']

. . . it is an effect that doesn't stay very long. In about ten minutes' time the sky is streaked with crimson and the magic of the grey light is gone. Bosman *Cask of Jerepigo* 1972

ting-tinkie [ˌtɪŋˈtɪŋkĭ] *n. pl.* -s. Any of numerous warblers of the Sylviidae: the name is *onomat.,* as are the African language names for these birds. [*Afk. tingtinkie onomat.*]

The variety is overwhelming – from the tiniest tintinkie, honey bird or kingfisher to the awesome albatross and the ungainly ostrich. *Evening Post* 28.4.73

tit *adj. slang.* General term of approval equiv. 'terrific', 'super', 'great' etc. a ~ party, a ~ car, ~ hey? etc. [*unknown*]

But I quit the back chat double-quick . . . This is one tit-looking ou . . . all curly blond hair. Hobbs *Blossom* 1978

titihoya [ˌtĭtĭˈhɔja] *n. pl.* -s, -∅. Zulu name for the crowned plover or *dikkop* (q.v.) or the black-winged *Vanellus melanopterus.* [*Zu. titihoye* stone plover, *onomat. fr. call*]

. . . the forlorn crying of the titihoya, one of the birds of the veld. Paton *Cry, Beloved Country* 1948

Tixo [ˈtĭkɔ +] *n. prop., interj.* God: usu. as *interj.* or mode of prayer among Africans; see quots. at *Unkulunkulu* and *Uvuko.* ⫿ Some Xhosa-speaking congregations strongly influenced by tradition are today wishing to adopt the

term *Qamata* (q.v.) on the grounds that *Tixo* is a God of the white man's creating who answers only white men's prayers. *Informant* African Anglican Priest Nov. 1978.
[*Xh. uThixo* God, the Almighty, *also Zu. fr. Xh.: prob. fr. Hott. Tixwa or Tiqwa*]

UTIKXO, the word adopted for God by the early missionaries among the Kxosa or Frontier Kafirs, was not a word known to the natives of these parts, but was introduced by missionaries and others. And it it generally supposed that the word does not properly belong to the Kxosa or any other of the alliterative dialects spoken in Southern Africa, but has been derived from the Hottentots. Callaway *Religious System of the Amazulu* 1870 (1884)

Tixo, watch over me, he says to himself. Tixo, watch over me. Paton *Cry, Beloved Country* 1948

tjeers [tʃɪə(r)z] *interj. slang* Goodbye, see you, etc.: see *cheers*.

Hoezit ous!!?
... trying to proof yous not, even if you isn't. Tjeers!!!
Eike S.A. Oct. 1976

-tjie [kǐ, cǐ] Also *-djie* see (1) *-ie*, and second quot. at (1) *tannie*.

tjap [tʃap] *n.* (and *vb*.). *colloq.* Also *'rubber tjap'*: a rubber ink stamp: as *vb. Afr. E* see quot. [*orig. fr. Chinese chop* a seal, stamp. *prob. via Malay*]

He would not be satisfied. 'He will not i'tjap me?' he asked fearfully, having learnt that once the papers had been stamped by a White official all would be well. Louw *20 Days* 1963

tjokka [ˈtʃɔka, tʃ▯-] See *chokka*, (also sp. form *tshokka*)

tjommie [ˈtʃɔmǐ] *n. pl.* -s. *colloq.* Pal: see *chommie*.

tjor(rie) [ˈtʃɔr(ǐ)] *n. pl.* -s. *colloq.* Car: see *chorrie*, *knortjor* a go-kart. *cf. Austral. bomb.*

Any tjorrie tied together with faith, hope and bits of string can get through the safety checks on this road. *Sunday Times* 21.9.75

tjor-tjor [ˈtʃɔr₁tʃɔr] *Chor-chor*: see *grunter*.

toa grass [twa-] *n.* Bushman grass: see *twa(a)grass*.

toby, tobie(fish) *n. pl.* See (1) *blaasop* [*unknown*]

A tobyfish is also known as a puffy or blaasop. *Daily Dispatch* 25.7.72

toe [tǔ] *adj. colloq. lit.* 'Closed up': impenetrably stupid. [*Afk. toe* closed]

This girl is so toe, man, that she never gives me any messages. I don't even let her answer the telephone. O.I. 1971

Who me? You must be toe ou pel. Not me! It's not my job. *Het Suid Western* 18.1.78

toe maar [ˈtǔ₁maː(r)] *interj. colloq.* A consolatory expression equiv. of 'never mind', 'there there' etc. [*Afk. toe maar* never mind]

'Toe-maar, ou Maggie!' she said, 'don't cry; don't worry.' . . . 'Ag, toe-maar,' she added, as Lena glared at her, 'don't be cross now man. Alles sal regkom.' Meiring *Candle in Wind* 1959

toenadering [ˈtǔ₁nadərɪŋ] *n.* Rapprochement between two political parties or other opposing forces or nations: see quot. [*Afk. vbl. n. fr. toenader* to approach, meet half way]

As the session drew to a close, however, everyone was fiercely denying 'toenadering', or should I rather use the new 'in' term 'consensus'. *E.P. Herald* 16.6.73

. . . it was the Daily Dispatch that pioneered South African toenadering with Malawi and pointed to that country as an obvious market for South African export. *Daily Dispatch* 23.8.71

toering [ˈtuːrɪŋ] *n. pl.* -s. The broad brimmed straw hat formerly worn by Cape Malays usu. over a small crimson turban or handkerchief: now supplanted by the fez or tarboosh: see also quot. at *kapparrang* and fourth quot. at *fishcart fishhorn*. [*Afk. fr. Du. fr. Malay toudang*, covering, lid]

The toering is still worn by Malay coachmen when driving the wedding group. Du Plessis *Cape Malays* 1944

Very small figures appear in some of the drawings . . . They are usually Malays in which the 'toering', or conical straw hat worn by the men, is prominent. Gordon-Brown *Artist at Cape of G.H.* 1965

tog [tɔx] *intensifier or modal adv.* As an interpolation roughly equiv. 'really'; see quot. at *niggie*, or as equiv. of 'do' or 'please' adding emphasis to *vb*; see quot.; also combinations *foei* ~ and *sies* ~ both (q.v.). [*Afk. yet nevertheless, all the same, do etc.*]

'Ma, Ma! wake up, toch!' . . . Tante Jacoba at last opened her little, faded blue eyes. 'Wake up, toch, Ma,' repeated her daughter, impatiently. Bancroft *Veldt Dwellers* 1912

togt[1] [tɔxt] *n. modifier.* Of or pertaining to casual or day-labour, usu. in combinations: ~ *labourer*, ~ *boy*, casual labourer not contracted to an employer. *cf. Canad. trip man*, lumber worker employed for a single trip. [*fr. Afk. togarbeider fr. Du. togt* march, *cogn.* tour]

CASUAL LABOUR, in terms of Native Pass Laws is synonymous with the term togt labourer. Sisson *Legal Dict* 1960

. . . male Native following the occupation of togt or casual labourer. Grahamstown Municipal Receipt form 15.12.69

togt² *n. and n. modifier hist.* A trading venture by wagon usu. in phr. *go on* ~ or as modifier in combination ~ *ganger*, a *transport rider* (q.v.) or an itinerant trader: see also *smous*. [*Du. togt* expedition *cogn.* tour, *Afk. togganger* itinerant trader]

He has made a fortune by 'going on togt', as thus: He charters two waggons, twelve oxen each, and two Hottentots to each waggon, leader and driver. The waggons he fills with cotton, hardware, etc. etc. – an ambulatory village 'shop' – and goes about fifteen miles a day, on and on, into the far interior, swapping baftas (calico), punjums (loose trousers), and voerschitz (cotton gown-pieces) . . . against oxen and sheep. Duff Gordon *Letters* 1861–2

tok, toc [tɔk, tɒk] *n. pl.* -s. *abbr. Tokkelok* (q.v.)

. . . this could be construed as a plot to take over the presidency – for what it's worth) and after all, why not? . . . they've had Tocs as President before. *Rhodeo* (Student Paper) 13.5.71

tokkelok [ˈtɔkəˌlɔk] *n. pl.* -s. A theological student is so called at an Afrikaans university: at English ones *toc* or *tok* (q.v.) is used. [*etym. dub. poss. rel.* Lat. *and Du.* theologica. ⁋Swart 1934 *suggests* Gk. *tokos*, birth + *logos*, discourse, *see quot.*]

A term humorously applied to theological students at Stellenbosch who are reputed to contemplate and discuss matrimony when the end of their course is in sight. Swart *Africanderisms: Supp.* 1934

tokoloshe [ˌtɔkɔˈlɔʃe, ˌtĭ-] *n.* An evil spirit widely believed in by both urban and rural Africans: it is invoked in witchcraft and offered as an extenuating circumstance in criminal cases. ⁋There are various beliefs concerning it: to look it in the face is death; it lives near water; it is fond of children; it is sexually insatiable; it can make itself invisible or take various forms though its usual form is that of a hairy dwarfish man: see also quots. at *worry*, *sangoma* and *mamlambo*. [*Ngu. Zu. utokoloshe, Xh. uthikoloshe*]

Tokoloshe haunted me says the mad Hillbrow killer. I had been attacked by the tokoloshe and I was hopelessly drunk . . . The first I knew of what I had done was when the police arrested me at my work on Monday. *Post* 18.1.70

Banquo's ghost becomes the Tokoloshe (a Zulu evil spirit), the witches are three bone-throwing Sangomas (witch-doctors). *Sunday Times* 5.4.71

toktokkie [ˌtɔkˈtɔkĭ, ˌtɒkˈtɒkĭ] *n. pl.* -s.
1. A large, blackish beetle of the genus *Psammodes*: ⁋Its mating call, tapping its abdomen on the ground or floor, is often mistaken for a knock at the door. [*Afk. fr. Du. tokken* to knock softly, tap]

I threatened to get the priest in the morning as the place was clearly haunted – then we found the ghost, a toktokkie beetle banging his way up the passage. O.I. 1964

. . . my wife returned with . . . a 'toktokkie' in her hand, which she had just heard calling for its mate outside her bedroom door. *Grocott's Mail* 26.1.73

2. A (children's) game: knocking on doors and running away, formerly declared a 'nuisance' in Cape Town: see quot. [*fr.* (1) *tok-tokkie sound*]

⁋ Municipal Regulations for Abating Nuisances 28. No person shall be allowed wilfully and wantonly to disturb any occupier by pulling or ringing any door bell or knocking at any door without lawful cause. *Cape of G.H. Almanac* 1843

3. A home-made toy consisting of a wooden cotton reel, matchstick, soap or candle for lubrication, and a twisted rubber band which causes it to move: *Brit.* 'tank'. [*fr.* (1) *toktokkie fr. slow movement and ungainly appearance*]

The toktokkie craze is on at the moment and . . . is in and out of my workbasket looking for cotton reels but the things are all plastic now. O.I. 1970

tol [tɔl] *n. pl.* -s. Top [*Afk. fr. Du. tol* top, *also Afk.* reel] **1.** A child's spinning top: see also *pik*.

If you don't get that car out of my way I'll pik it in half with my tol. Child 7 ex Pearston, Karroo 1968

2. Reel:

Thatching twine R– per tol. *Farmer's Weekly Advt.* Date unknown

3. In combination ~ *bos*, *Diospyros dichrophylla*: see also *tolletjie*, ~ *bossie*, *Salsola kali*, and *rolbos*.

tolletjie [ˈtɔlĭkĭ, -cĭ] *n. pl.* -s. The dried fruit or seed pods of several plants including those of the eucalyptus (*Austral. gum-nuts*) used by children as pipe-bowls: also cones of several species of *Leucadendron* and the fruits of *Dio-*

sporos whyteana. [*Afk. fr. Du. tol* top + *dimin. suffix* -(*le*)*tjie*]

> *tolletjie*(*s*) . . . a name applied to the cones of several species . . . from the manner of spinning like little tops. Smith *S. Afr. Plants* 1966

tollie ['tɔlĭ] *n. pl.* -s. **1.** A castrated bull calf. [*Afk. fr. Ngu. iThole,* a calf]

> 20 good Heifers and Tollies. 40 Good Tollies including a span of choice Afrikaners, 2 years old. *Grocott's Mail* 1.4.1932
> 5 Beef Shorthorn Heifers
> 5 Beef Shorthorn Tollies 15 months
> *Evening Post* 19.4.72

In *pl.* form *amathole,* found in place name Amatola Mountains.
2. ¶Also poss S.A.E. a large marble or *ghoen* (q.v.) ¶*Dict. of New Eng.* gives *tolley* (1970) a marble used to shoot at marbles [*of unknown origin*]

> Four of these allies are worth a tollie. O.I. 1970

tolofiya [ˌtɔlɔ'fija] *n.* The fruit of the prickly pear, *Opuntia* spp. [*Xh. itolofia, fr. Afk. turksvy* (q.v.)]

> . . . succulent little green globes that are the cheapest fruit in town . . . Mounds of prickly pears nature's gift to those with fingers hardened to their barbs . . . the unfriendly little itolofiya . . . Katie . . . knows a good itolofiya when she sees one. (*Caption*) *E.P. Herald* 11.2.76

tom [tɔm] *n. slang.* Money: as in 'to be in the ~'. [*unknown poss. fr. dial. tommy shop* – where payments, usu. in kind were made to merchants *or fr. tommy,* daily allowance, barter goods, rations]

> Plenty of muscle on the right places, . . . plenty of tom to chuck around on the girls . . . own personal Volksie, Vernon's only suave man. *Darling* 12.2.75

tomboti(e)boom, tamboti(e)-hout, tamboti(e) wood, [ˌtɔm'bʊatĭ bʊəm, həʊt, -'butĭ-] *n.* Also *tambuti* (q.v.) *Spirostachys africana,* a deciduous tree common in bushveld valleys. ¶The heavy, durable wood is permanently scented and used by Africans for necklaces and charms: it is also used for furniture: see quot. at *tambuti;* found in place name Tambotiepan. [*fr. Ngu. umThomboti* poison tree]

> *tombotieboom* (*-hout*) . . . the wood is heavy, close, but cross grained, very durable and untouched by borers or termites and is said to be indestructible . . . the sap is caustic and care should be taken it does not come into contact with the eyes. Smith *S. Afr. Plants* 1966

tonne *n. pl.* -s. A metric ton: see *metrication.*

> In the half tonne league the 1 275cm³ engined . . . takes very good care of your money. The panel van has the most space you ever saw in a half tonner. *Financial Mail* 22.10.76

tontel- ['tɔntəl] *n.* Tinder: *prefix* to several names of plants with woolly or hairy leaves or floss in the fruits which was formerly used in tinder boxes e.g. ~*blom, Hermas gigantea;* ~*doos* (*box*) *bossie, Haplocarpha scaposa* and ~*bos* any of several plants which furnished material for tinder: Also in place names Tontelbosch and Tontelboschkolk, Tonteldoos. [*Afk. fr. Du. tontel* tinder]

too good *adj. phr. Afr.E.* Equiv. of 'very good' used by many African speakers also *too much,* very much *adv.* also *Jam. Eng. too good.*

> Mrs. M. Gwizi . . . says . . . 'We get a big chicken, R1 beef . . . The children like it too much.' *E. P. Herald* 5.7.75

toor [tʊə(r)] *vb. usu. trns.* Also *tover:* to bewitch. [*Afk. fr. Du. toveren* to enchant, bewitch]

> Just so long as you 'toor' me, I do not care. . . . toor me now. The witchdoctor studied his client with interest. 'And how must I toor you?' . . . 'Toor me so that the police cannot catch me again.' . . . the little toor-doctor only wanted to make sure there was money in the bag. Only money could buy . . . magic. Meiring *Candle in Wind* 1959

In combination ~*doctor,* a witchdoctor dealing in spells and magic (see quot. above): ~*goed,* substance or material believed to be a source of, or containing, magic.

> . . . they had something in their possession which had bewitched her . . . They beat them with switches, demanding all the time that they should produce the toorgoed. *Grahamstown Journ.* 24.9.1892

topi, topee See ¶ at *kufija,* also quot. at *amatopi.*

Topiya *pron. sp.* of *Ethiopia:* see *Ethiopia, Order of,* also *Ethiopian* and *Zionist.*

> . . . the number of mission stations has greatly increased, and . . . the separatist churches (e.g. Church of Zion, Topiya, Ethiopian) are active, though with a smaller following. Hammond-Tooke *Bhaca Society* 1962

toppie[1] ['tɔpĭ, 'tɔpĭ] *n. pl.* -s. The Cape bulbul *Pycnonotus capensis* (Pycnonotidae fam.), which lives on fruit and berries.

> Toppie or Kaapse tiptol . . . lively manner and habit of whistling from which it gets the name of Tiptol, or 'Pietmajol' or 'Piet-pietpatata' in Afrikaans. Roberts *Birds of S.A.* 1944

toppie² ['tɔpǐ] *n. pl.* -s. *slang* Mode of reference to one of the older generation. man or woman; [see *amatopi*]

'Members only' says a certain toppie with grey hair. *Drum* 30.1.66
Toppie: common, widely used term for old man. *S. Afr. Gaol Argot: Eng. Usage in S.A.* Vol. 5, No. 1 May 1974.
'Hysterical'... says my mom. 'Maybe she's pregnant...' And she starts to count off the months in the way ou toppies do, for ever expecting the worst. *Darling* 14.4.76
Nina... whose ol' toppie's ol' toppie was once a corporal in the Grenadier Guards. *Scope* 10.1.75

tot siens [ˌtɔtˈsǐns] *interj.* Equiv. *Auf wiedersehen, Au revoir:* an informal farewell *cf. See you;* see also quot. at *alles van die beste.* [*Afk. tot* until + *siens* see(ing)]

'Tot siens', they shouted. 'Good-bye. Till we see you again'. Cloete *Rags of Glory* 1963

tot system, *n.* A system of paying coloured labourers particularly on wine farms, part of their wages in 'tots' (usu. mugs) of wine: see quot. at *vaaljapie.* [*fr. Eng.* tot, measure of liquor]

The Coloured Representative Council has called for the abolition of the tot system on farms... He said farm labourers received a mug of wine at regular intervals of the day as part of their wages... tot system made farm workers non-persons... *Argus* 10.8.72
Most farmers who still apply the tot system argue they will lose their labour if they stop it. *Ibid.* 14.4.73

tottie, his, your, my etc. name is *idiom. Sect. Army.* Also 'he has made his name tottie': a saying meaning that the person in question is in trouble, out of favour etc.: see also quot. at *dear Johnny*

If the... corporal tells you 'your name is tottie' you know you're in the dirt. *O.I. Ex-Serviceman* 4.4.79

totty, *n. pl.* -ies and *n. modifier or adj.* A mode of reference to a *Hottentot* (q.v.) still heard occ. as '*tot*', now regarded as offensive: see fourth quot. at (1) *Hottentot.*

I got a Hottentot to accompany me, as I preferred a Tottie to a Fingoe at any time. If accompanied by the former, he has always got some wild, improbable or laughable tale to tell; whereas the latter is sullen, morose. Buck Adams *Narrative* 1884

as modifier ~ *pink*, also *coolie pink.*

'totty pink –... puce or magenta. Later rendered fashionable by Schiaparelli as "shocking pink".' *O.I.* 1970

tough tackie, tough tacky, *interj.* Hard luck, tough luck: see *tackie.*

touleier ['tɔuˌleɪə(r)] *n. pl.* -s. Also *voorloper* (q.v.) or *leader* (q.v.) usu. a young coloured boy employed to lead a team of oxen for wagon travel: his means of leading them being the *voortouw* (q.v.) twisted round the horns of the first pair. [*Afk. fr. Du. touw* rope + *leiden* to lead + *agent. suffix. -er*].

By the time he was twenty-one he had a wife and three children, two coloured shepherds and a Bushman touleier to lead the oxen and find the way from one water-hole or vlei to the next. Green *Karoo* 1955

toutjies ['tɔukǐs, -cǐs] *pl. n.* Also ~ *vleis:* thin strips of salted, dried meat: see *biltong,* also *occ.* fish, see *bokkems.* [*Afk. fr. Du. touw* twine, string + *dimin. suffix -tjie*]

Town House¹ *n. usu. cap.* See *Stadhuis,* [*trans. Stadhuis* Town Hall]

Worth seeing is the... Museum, temporarily housed in the old Town House, where the visitor could spend hours. *Panorama* Mar. 1974

town house² *n. pl.* -es. A *nagmaalhuis* (q.v.) or *kerkhuis:* also occasionally called *Sondaghuis* 'Sunday house'.

The sheep farmers who make any pretension at all have their little 'town house'... which is occupied once a quarter or so, when the nachtmaal, or sacrament, draws all good followers of Zwinglius to church. Boyle *To the Cape for Diamonds* 1873

town house³ *n. pl.* -s. One of a row of duplex maisonettes usu. each with its own garden in front and yard behind ℙ *acc.* Barnhart *A Dictionary of New. English:* 'U.S. an attached one family house: row house. The equivalent British term is *terrace house*'.

When her husband died suddenly... she moved into a townhouse... Used to a large house she felt hemmed in. *E.P. Herald* 2.2.79
MAISONETTES FOR SALE
Most attractively designed town house with lovely garden R23,000. *Ibid.* 19.2.79
It's like a sort of town-house complex except that they're separate little houses. *O.I. (Johannesburg)* 8.4.79

township *n. pl.* -s. and *n. modifier.* **1.** An area set aside for Non-White occupation: Coloured ~, African ~ etc.

Even in Soweto, Johannesburg's vast, sprawling Bantu township of over three-quarters of a million people, the fifth-largest city in Africa, there are some who have more and who are seen to have more. *News/Check* 24.7.70

In addition the council has completed the planning of the new Coloured township on the site of the existing African township. *Daily Dispatch 27.5.72*
As *modifier* usu. referring to African ~s; ~ *council;* ~ *jazz* see *mbaqanga;* ~ *life* etc.
    The play is a departure from . . . most musicals which treat township life as an unending song, one bout of drinking; one long knife fight. *Drum 22.1.73*
In *abbr. Soweto* (q.v.)
    Soweto is merely a convenient abbreviation for the South-Western Townships of Johannesburg. *Personality 23.4.70*
2. *township, n. pl.* -s. Usu. in phr. *proclaim a* ~, a new area being developed by land or building speculators who, if they *proclaim a* ~, are responsible for roads, sewerage, water supplies etc. before *stands* (q.v.) can be sold. *cf. Austral. township.* (*hist.*) tract of surveyed land ; town site (now) small town or settlement.
    Mr. Justice . . . said according to . . . the purchase price was to be R50,000, payable in cash, enabling him to establish a township *Daily Dispatch 22.5.71*

tradesman *n. pl.* -men. One who has served his apprenticeship and is qualified in his trade, e.g. *mason* (q.v.), plumber, carpenter etc. ❡Not as in *Brit.*, a shopkeeper.
    . . . is a first class tradesman – served his apprenticeship in my firm but now he's started drinking there's no hope for him. O.I. Coloured Master Builder 1968
    Built to the highest standards by our European tradesmen and backed by 18 years of poolbuilding experience. *Star Advt. 20.7.74*

traffic circle *n. pl.* -s. *equiv. Brit.* roundabout: a usu. round island built in an open street space to facilitate the routing of town traffic.

tramp *vb. trns. substandard.* To run over: freq. pass. *to get* ~ed. [*unknown: poss. fr. Afk.* doodtrap to crush (to death) *fr. Du.* trappen to tread *or* trappelen to trample]
    I've shut the cat in the girls' bedroom because I don't want him to get tramped by the cars. O.I. 1969
    . . . pa's never been the same since . . . Uncle Max gets tramped by that car outside Carsten's place. *Darling 3.9.75*

tramp out *vb. trns.* as modifier. Usu. ~ed out : Of or pertaining to (2) *veld* which has been damaged or destroyed by overstocking or overgrazing: see

also *beaten out, eaten out.* [*poss. fr. Afk. uittrap fr. Du. uittrappen* to crush underfoot]
    Another result of the old system of farming was overstocking and tramping out the veld, erosion of the soil and a general tendency towards degeneration of the most valuable species of karoo bush, while 'opslag' vegetation increased. *H'book for farmers 1937*

Transkei [ˈtrănsˌkaɪ, ˈtransˌkaɪ, -ˈkeɪ] *n. prop.* The *homeland* (q.v.) of the *Xhosa* (q.v.) peoples: see also *Ciskei* and *T.D.C.* and quot. at *Nkosi Sikelel' iAfrika* : also *abbr.* '*Kei* see quots. below.
❡The inhabitants of Transkei belong to the South Nguni group who are said to have left the Lake Victoria area centuries ago, moved south, reached the Kei River by 1700, and settled in Transkei. *Panorama July 1976.* [*Lat. trans* across + *Kei* name of boundary river]
    . . . in 1962 . . . the Prime Minister, Dr. Verwoerd, announced plans to grant limited selfrule to the Transkei as a 'first step on the road to independence'. *Rand Daily Mail 31.5.71*
    At midnight on 25 October, 1976, Transkei will become independent as the Republic of Transkei. For South Africa this will be a proud moment and, for the young Republic, the realisation of a long-cherished ideal. B. J. Vorster in *Panorama July 1976*
    South Africa's policy of separate development passes a milestone on Tuesday when the first Homeland, Transkei (it dropped the definite article some months ago), gets its independence. *Financial Mail 22.10.76*
    Five Flee 'Kei S.B. Cops *World 21–23.10.76*
    Kei consul's R65,000 White house *Sunday Times 31.10.76*
Derivatively ~*an, n. pl.* -s. A citizen of ~ .
    More and more Transkeians are now filling key posts. *Panorama July 1976*

transport driver See *transport rider.*

transport rider *n. pl.* -s. Formerly a carrier of goods by oxwagon, also known as *transport driver* : see also *kurveyor.* [*translit. Afk.* transportryer transport rider ; *ry* convey, cart]
    . . . all roads led to Kimberley in the early eighteen-seventies, when transport riders made good money transporting machinery and other goods to the bustling diamond town before the coming of the railway. McMagh *Dinner of Herbs 1968*

transport riding *vbl n.* or as *vb phr.* In former times the trade of a carrier, conveying goods by wagon usu. fr. the coast to the interior: see *transport rider* also ride. [*fr. Afk. transport ry* to convey goods for a fee; *ry* convey, *see ride*]

. . . money could be had for the asking by any man who possessed a bullock-wagon and cared to go transport-riding, dragging fuel and farm-produce and goods through the sandbelt to Kimberley. Brett Young *City of Gold* 1940

Transvaal Teak *n. Pterocarpus angolensis :* see *kiaat* or *P. erinaceus :* see *bloodwood.*

Transvaler [ˌtrăns'falə(r)] *n. pl.* -s. An inhabitant of the Transvaal whether by birth or inclination. ❡ ~ s and *Kaapenaars* (q.v.) are traditional rivals: see quots. at *Kaapenaar* (2), *Vaalpens* (1), and *Capey, Capie* (2)

WHY WE ARE DIFFERENT FROM THE TRANSVALERS
It is not that we dislike Transvalers. It's just well . . . that they are *different* . . . Some of my best friends are Transvalers. It's just that . . . we don't want to associate with them . . . our drinking habits are different. Transvalers drink 'cane' . . . Kapenaars drink wine . . . Capeys and Transvalers even laugh differently. Lorraine *cit. Capetonian* May 1979

trapvloer ['trapˌfluːr, -ˌfluə(r)] *n.* A threshing floor. [*Afk. trap* to thresh *fr. Du. trappen* to tread + *vloer cogn.* floor]

23 Nov 1842 26 bro–t our two first loads of Corn to the Trap floor to day. Collett MS. *Diary II*

trassie ['trasĭ] *n. pl.* -s. *slang.* A hermaphrodite, homosexual or 'queer': abusive as a mode of address or reference: see *moffie.* ❡Also used by farmers, see quot. In combination ~*bos* or *terassiebos, Acacia hebeclada.* [*prob. fr. Nama taras,* woman, *rel. to tarase/i; see Hottentot apron : poss fr. transie fr.* transvestite]

The vernacular prefix 'terassi' appears to be a corruption of the original Hottentot word 'taras' – a woman and meaning hermaphrodite; further corruptions 'teransie', 'transie' or 'trassie' are in common use, especially among sheep farmers, for a hermaphrodite animal and applied to the plants in a figurative sense. Smith *S. Afr. Plants* 1966

tree aan [ˌtriə'aːn] *vb. intrns.* See *aantree.*

trek *Note :* This term has several meanings and uses in S.A.E. These appear below as: (1) 'travel, migrate', etc., a verb with various shades of meaning; (2) the corresponding noun, again with several meanings which are for convenience marked (a), (b) and (c); (3) a subsidiary use as a trade name; (4) a prefix with widely differing meanings which are marked (a), (b), (c), (d) and (e). The compounds listed at (4) *trek* follow the

main entry in alphabetical order. ❡Some settlers interpreted this word as *track* and substituted that spelling form. See quot. at *veeplaas.* [*Afk. trek,* to journey, travel, migrate *fr. Du. trekken,* to migrate]

trek [trek] *vb. usu. intrns. occ. trns.* **1.** To make a difficult and arduous overland journey: formerly involving making a permanent move after abandoning home and property: see quots. below and at *yokeskey.*

Because he was lazy he was a very bad farmer. In times of drought he would trek with his sheep to find pasture. Krige *Dream & Desert* 1953

figur. to make a journey.

I will arise an' get 'ence;-
I will trek South and make sure.
Kipling *Five Nations* 1903

Often enough, when the holiday seasons are really booming, they trek from one hotel to another. No joy. *Rand Daily Mail* 29.6.71

Derivatively ~ *ing, vbl n.* travelling usu. by wagon.

Mama refused flat to come, she declares she has had by far too much of trekking as the Boors call travelling by wagon, but indeed I don't know why she should dislike it, tis the pleasantest way imaginable of passing the day. Fitzroy *Dark Bright Land* 1955

In combination *pony* ~ [non S.A.E.] usu. Brit. form of holiday on horseback, also *mini* ~, presum. by mini-car. ~ (*k*)*er, n. pl.* -s. A member of the *Great Trek,* see *Voortrekker* or of the *Thirstland Trek* (q.v.) (see also *Dorsland*): see (2) *trek* (a): or anyone toilsomely or unwillingly on the move.

Many Coloureds, with despair in their hearts, trekked to the wasteland of the Cape Flats, but unlike other trekkers before them there was no promised land. *Sunday Times* 9.12.73

2. *trek, n. pl.* -s. **a.** *hist.* A journey by ox wagon esp. an organized migration of people overland as in the *Great Trek,* see *Voortrekker,* or the *Thirstland* (q.v.) (*Dorsland* (q.v.)) *Trek;* also in phr. *on* ~.

On trek, the women and children placed their mattresses on the katel, a wooden frame with leather thongs, and slept inside the wagon. Green *Land of Afternoon* 1949

Thus a great trek was decided upon, and northwards moved the whole community in a body – wagons, furniture, cattle protected by the men with their muzzle-loaders, northwards into the unmapped territory of South West Africa. Green *Last Frontier* 1952

Eastward Ho! The Wagons The long, sometimes treacherous 19th century Westward trek of covered wagons is deeply embedded in American legend, folklore and sense of national identity. Thousands of pioneers . . . braved the perilous Oregon Trail. *Time* 5.7.76

b. Any journey or migration as in *springbok* ~ ; see also *trekbok*.

. . . houses forsaken by commuters who could stick the daily trek no longer and have taken off for the suburbs. *Sunday Times* 17.10.71

I recall, as a small boy, hearing first-hand reports of the 'springbok treks'. *Farmer's Weekly* 1.3.68

c. A stage of a journey, see *skof(t)*, between one outspan and the next.

It was one day's trek from Cape Town by ox-wagon at that period, a hard trek along the sandy ou Kaapse wapad. Green *Land of Afternoon* 1949

d. Goods and chattels, belongings, a baggage train.

. . . speed was impossible – 'stick to the trek (or line of pack oxen) and the trek will stick to you' – so it was necessary to move slowly, to bring on all our cattle and stores. Alexander *Expedition II* 1838

Whenever we engaged a new labourer we had to make a trip to collect his trek, lock stock and barrel. *O.I.* 5.4.79

3. Trek used as n. prop. in trade name: . . . *Trek(shoes)* (Durban), *Trek – a thrilling Game for All Ages* (Toyshop, Kowloon).

. . .'s Treks. The shoe that made America relax. Treks are more than just sturdy, comfortable, long-lasting practical shoes. They're a lifestyle. *Seventeen Advt.* Aug. 1974

4. *trek.* n. *modifier or vbl prefix.* Equiv. of (a) 'migratory' in ~*boer,* ~*bok,* ~ *sheep,* ~ *swarm;* (b) 'of or pertaining to equipment for a journey' in ~ *bed,* ~ *gear,* ~ *goed,* ~*stoel,* ~ *wagon;* (c) 'something pulled or hauled' in ~ *chain,* ~ *net,* ~ *saw,* ~*touw;* (d) 'of or pertaining to movement or travel' in ~ *fever,* ~*gees,* ~ *pass,* ~*pad,* ~ *path;* (e) 'draught' in ~ *ox,* all (q.v.).

trek bed ['trek₁bet, -bed] *n. pl.* -dens, -s. A light bedstead suitable for travelling; see also (1) *katel,* cf. *Anglo-Ind. cot* [*Afk. fr. Du. trekken* to migrate, *hence* travel + bed]

. . . a smaller riempie 'trekbed' – the forerunner of the stretcher. *Panorama* Jan 1975

trekboer ['trek₁bu:r, -buə] *n. pl.* -s. Also *trek-farmer:* see quot. at *trek sheep* formerly a nomadic grazier moving with his flocks and travelling by ox-wagon (see also first quot. at (1) *trek :* also used to

refer to the *Voortrekkers* (q.v.) [*Afk. fr. Du. trekken* to migrate + *boer* farmer]

When the Trekboers entered it with their flocks and tented wagons, they left the current of European life . . . Though in strict legality the land tenure was not in freehold and was recoverable at the will of the administration, actually the Trekboers availed themselves of the land with the utmost freedom. De Kiewiet *Hist. of S.A.* 1941

It was the happy hunting ground of the Trek Boer, moving from place to place with all his livestock, wherever the grazing was good. Jackson *Trader on Veld* 1958

trekbok ['trek₁bɔk] *n. usu. pl.* -ke, -ken. Migratory buck, mostly *springbok* (q.v.) moving in herds away from droughtstricken areas: see quots. [*Afk. fr. Du. trekken* to migrate + *bok cogn.* buck]

It is from these tracts that the destructive trek-bokken or migratory springboks, pressed by the long droughts, occasionally inundate the northern parts of the Colony. Thompson *Travels II* 1827

As far as the eye could see they covered the country . . . grazing off everything eatable before them, drinking up the water in the street furrows, fountains and dams . . . It took about three days before the whole of the trekbokken had passed, and it left the country looking as if a fire had passed over it. This mass movement of millions of antelope, the trekbokke as the Afrikaner farmers called these springboks, was one of the most . . . inexplicable phenomena of the wildlife of Africa. *Daily Dispatch* 22.7.72

trek chain *n. pl.* -s. A *trektou(w)* (q.v.) made of chain instead of rawhide *riems* (q.v.) cf. *Austral.* trace-chain. [*Afk. fr. Du. trekken* to pull, haul]

. . . Chain traces 55c each; Trek chains R1.55. *Farmer's Weekly Advt.* 21.4.72

trek fever *n.* Restlessness, wanderlust, a longing for open spaces and open-air life: see also *trekgees.* [*fr.* (1) *and* (2) *trek*]

The unbearable 'trek fever'; of restless, sleepless, longing for the old life; of 'homesickness' for the veld, the freedom, the roaming, the nights by the fire, and the days in the bush. FitzPatrick *Jock of Bushveld* 1907

trek fishing *vbl. n.* Fishing with a *trek net* (q.v.); also *trek fisherman;* see quot. at *treknetter.* [*Afk.* trek to haul]

. . . another crew member of the trawler, had gone along with the six crew men to Strandfontein to assist them with the trek fishing. They were on the beach when the boat capsized. *Argus* 24.12.70

trek gear *n.* Equipment for wagons, carts etc., harness, yokes, *skeys* (q.v.), *strops* (q.v.), *trektous* (q.v.), also *trekgoed* (q.v.)

[*prob. trans. Afk. trekgoed* (q.v.) *fr. Du. goed* goods, things]

　　Trek Gear etc. Harness Double Cape Cart Horse Harness . . . Bridles R6; Halters R2.35 also R3.50; Chain traces R3 pair; Wheeler Reins R8.60; Leader R11.50; neckbar and Straps R10. *Farmer's Weekly Advt.* 13.6.73

trekgees(t) ['trek₁xɪəs(t)] *n.* Restlessness, wanderlust: see also *trek fever.* [*Afk. trek fr. Du. trekken* to migrate + *Afk. fr. Du. geest* spirit]

　　It very clearly stimulated the old trekgeest, the spirit of the open veld, of greener grass beyond the horizon, of a land where only those should be free who were free by God's design. De Kiewiet *Hist. of S.A.* 1941

trekgoed ['trek₁xŭt] *n.* See *trek gear.* [*Afk. trek* to haul *fr. Du. trekken* + *Afk. fr. Du. goed* goods, materials]

　　Spades; Picks; 2-wheel 3-ton Trailer; Dexter Tractor 3 000; . . . Trekgoed Sundries etc. *Grocott's Mail Advt.* 17.11.72

trekker¹ ['trekə(r)] *n. pl.* -s. See at *trek* (1).

Trekker² *n. pl.* -s. *non. S.A.E.* A fan of 'Star Trek' a U.S. television show, cancelled in 1969 occasionally still seen in re-runs. In combination ~ *slang* the special vocabulary of these fans.

　　STAR TREK LIVES: TREKKER SLANG
　　Trekkers clip words and then use the clipped forms in compounds . . . The new fan who calls himself a *Trekkie* signals his ignorance of in-group vocabulary. *American Speech, Spring,* 1978 Vol. 53

Trekkie See *Trekker²*

　　Trekkies take note . . . the ultracool, ever rational hemihuman of the Star Trek TV series is as accident prone as any non-Vulcan. *Time* 8.1.79

trek net *n. pl.* -s. A beach seine. ⟨A type of fishing net which hangs vertically in the water having floats on one edge and sinkers on the other, and which can be hauled ashore or aboard enclosing the fish. [*Afk. trek* to haul + net]

　　From the early days of the Dutch East India Company to the beginning of the present century most fishing was carried out from small rowing boats with hook and line and by beach trek-nets . . . Beach seines or trek-nets are to-day of little importance although they were the main method of fishing in the early days. Grindley *Riches of Sea* 1969

deriv. ~ *er* a *trek fisherman,* one who fishes with a ~ or seine.

　　A shark, two metres in length, was caught by trek netters operating from the beach at Fish Hoek yesterday. The heavily toothed shark . . . was killed with knives by the trek fishermen. *Argus* 30.11.72

trek-ox *n. pl.* -en. Draught ox. **1.** An ox usu. one of a *span* (q.v.) trained to the yoke as a draught animal. [*trans. Afk. trekos* draught ox *fr. trek* to pull, haul]

　　15 Young Heifers, 14 Tollies, 1 Span Red Trek Oxen. *Grocott's Mail* 13.1.1932

2. *figur.* Tough beef: also *attrib:* see quot.

　　They were near starving, living on a ration of half a biscuit and a chunk of rotting trek ox a day. Cloete *Rags of Glory* 1963

　　. . . our grading system . . . conceived in the good old trek-ox days when animals were seldom slaughtered before they had reached a ripe old age. *Farmer's Weekly* 3.1.68

trekpad ['trek₁pat] See *trek path.*

trek-pass *n. pl.* -s. A *pass* (q.v.) for moving within a limited area, also a dismissal pass: see quot. [*Afk. trekpas in phr. die trekpas kry* to 'get the sack,' be dismissed]

　　. . . those who were unwilling to renew their contracts on the new terms, were offered only a 'trek-pass,' entitling them to seek a position on another farm in the district, instead of the accommodation in the new Reserves to which they thought themselves entitled. Hoernlé *S. Afr. Native Pol.* 1939

trek path, trekpad *n.* A right of way across another's land: a rural servitude. [*Afk.* trail, trek-road, wagon road]

　　But on some farms the title-deeds say that the owner need allow no trek-path, but merely the ordinary right of way along the road a forty-feet track across his ground. Birkby *Thirstland Trek* 1936

　　Trek path, trek pad; the right of driving cattle, including large flocks of sheep across the land of another, and also of allowing the stock to graze as they travel. Wille *Principles S. Afr. Law* 1945 ed.

treksaw, *n. pl.* -s. A two handed 'cross cut' saw operated by two men pulling and pushing alternately. [*trans. Afk. treksaag* lumberman's saw, cross cut saw]

　　. . . large hole in the ground under the log and a double-handled treksaw pushed and pulled by two men, one down in the hole and the other above. *E.P. Herald* 28.5.73

treksel ['treksəl] *n. pl.* -s. *obs.* A brew or infusion; see also *-tea, -tee*; also enough tea or coffee for a single brew; see quot. [*Afk. fr. Du. trekse* infusion, brew]

　　I have been often asked first for a 'soepje' or dram by Griquas whose 'places' I was passing; and when my driver whispered that I was a missionary, nothing daunted, the beggar would then substitute the request for a '*treksel*' or a single infusion of tea or coffee. Mackenzie

1871 *cit.* M. D. W. Jeffreys

trek sheep, *n.* Sheep travelling long distances for pasturage in times of drought. [*trans. Afk. trekskape* trek sheep (*pl.*)]

That the Government be requested to take into immediate consideration the desirability of amending the railway tariff for *treksheep* to enable trek farmers to avail themselves of the railway when moving stock to winter pasture. *E. London Dispatch* 1.5.1912 *cit.* Pettman

trekstoel [ˈtrekˌstül] *n. pl.* -e. A folding chair suitable for wagon travel. [*Afk. fr. Du. trekken* to migrate, *hence* travel + *stoel* chair *cogn.* stool]

. . . featured a number of 'trekstoele' . . . it was the custom to take your own chair when you went visiting – then these ingenious folding chairs with 'riempie' (thong) seats were useful. *Panorama* Jan. 1975

trek swarm *n. pl.* -s. S.Afr. beekeeper's term for an absconding or migrating swarm of honey bees. [*fr. Afk. trek* to migrate + swarm].

trektou(w) [ˈtrekˌtəʊ] *n. pl.* -s. A heavy rope usu. of twisted oxhide *riems* (q.v.) attached to the *disselboom* (q.v.) of the wagon and upon which the oxen pull; see quot.: a chain is sometimes substituted: see quots at *tent* and *trek-chain.* [*Afk. trek* to haul + *Afk. fr. Du. touw* rope]

The trektouw (draw rope or trace) is a long rope made of twisted thongs of raw hide, made fast by a hook at the staple at the end of the pole, and having iron rings attached to it at proper distances, into which rings the yokes are hooked. Burchell *Travels I* 1822

One great advantage of the raw hide *trektou* is that it does not attract lightning. When metal is used, a whole span of oxen may be struck dead in an instant. Green *Karroo* 1955

trek wagon *n. pl.* -s. A covered wagon of the type used by *Voortrekkers* (q.v.), *trekboers* (q.v.) and similar to that of pioneers in the U.S. going out west known as a 'prairie schooner'. [*Afk. trek fr. Du. trekken* to migrate + wagon]

All their possessions had to be transported by boer trek wagons. Lewcock *C19 Architecture* 1963

tribesman *n. pl.* -men. A rural African living under tribal conditions.

The entire population of one kraal attacked by an impi of 250 tribesmen on Sunday have fled and it may be several days before . . . life at the kraal returns to normal. *Daily Dispatch* 3.8.72

trippel, tripple [ˈtrɪpəl] *n.* One of the gaits of a *Boerperd* (q.v.) or other five-gaited horse in which it moves both near and both off legs alternately. [*Afk. n. fr. Du. trippelen* to trip along; ~*er fr. Eng. vb.* tripple *or Afk. fr. Du. n. trippelaar* one that trips along]

He schooled his pony to pace the trippel – easy gait, a smooth-flowing lope – which tired neither of them; but when necessary, he could ride rough and fire his flintlock while giving the pony its head. Mockford *Here are S. Africans* 1944

Derivatively ~*er*, a horse with this gait.

. . . good ones, real Boer ponies. Two tripplers. All about fifteen hands. Cloete *Rags of Glory* 1963

troepie [ˈtrüpĭˌtru-] *n. pl.* -s. *colloq. Sect. Army.* One holding the lowest rank of National Serviceman.See quots. at *Ride-Safe.* [*Fr. Eng.* trooper, or trans. *Afk. kavalleris* trooper, cavalryman.]

Some of the troepies with particularly dirty habits . . . receive the title *vuilgat* . . . Of course some 'troepies' are impatient for their period of service to end. Picard *Rooities and Oumanne* in *Eng. Usage in S.A.* Vol. 6 No. 1 May 1972

The toll . . . 181 guerillas, 20 Rhodesian 'troopies', twelve white and 88 black civilians.

. . . Helmeted 'troopie' with machine gun aboard Rhodesian escort truck. *Time* 22.11.76 (caption)

What on earth is kav.?

It's short for *kavalier** – Afrikaans for troepie, trooper I mean. *O.I. Grahamstown Station* 4.4.79 *erron.

trommel [ˈtrɔməl] *n. pl.* -s. *colloq.* Regulation army tin trunk: See also quots. at *taxis* and *klaar out.*

We have got all our kit, even our R1 rifles big thrill hey . . . I have a 'trommel' and kitbag as well. *Letter Serviceman* 20.1.79

tronk [trɔŋk] *n. pl.* -s. *colloq.* Prison, gaol, lock-up: see also quot. at (2) *pass. cf.* Canad. *skookum house, box;* (see *kas*), Austral. *log(s)* (now *obs.*), Anglo-Ind. *cho(o)ky.* [*prob. fr. Malay trungku* to imprison *or Port. tronco cogn.* trunk, box]

The landdrost came himself next day and carried off ten men to the tronck, as they call the prison, a great to do. Fitzroy *Dark Bright Land* 1955

. . . we have had therefore to interfere, and have captured nineteen Bushmen, through means of the Kafir police, and they are now in the 'Tronk' awaiting their trial. Gray *Journal I* 1851

In combination ~*volk* (*people*), gaol birds.

Missionary boys perhaps, but *tronk-volk*, jail birds for certain. Brett Young *Seek a Country* 1937

troopie See *troepie*.

tropsluiter ['trɔpˌslœïtə(r)] *n. pl.* -s. *obs.* One forming part of the tail-end of a procession: see also quot. at *huilebalk*. [*Afk. trop* crowd, multitude, *vb. sluit* to close + *agent. suffix -er*]

Funerals were accompanied . . . by professional mourners known as 'huilebalken' . . . 'Tropsluiters', were paid merely to walk in pairs to lengthen the procession . . . they frequently changed places owing to the superstition that the last person in the procession would be the first to die, and they wanted to spread the risk! Gordon-Brown *S. Afr. Heritage* 1965

true as God, as *interj. adj. phr. interj.* Used to emphasize the genuineness of an assertion: also in form (*'s*) *true as bob.* [*trans. Afk. so waar as God*]

And she threw her bible at me 'strue's God – she did. Maclennan *Winter Vacation* 1969

Jislaaik, what a fight. What's going on in the ring's nothing to what's going on in the front row, 's true's bob. *Darling* 26.2.75

try for white *vb. phr. modifier or vbl. n.* Also *play white* (q.v.). To cross the colour line: see also *Kimberley Train,* cf. *U.S. crossing the line.*

In the United States they call it 'crossing the line'. The phrase in Cape Town is 'trying for white'. . . . 'trying for white' – a desperate, pathetic struggle it must be, for even success can give no lasting security or immunity from insult. Success, moreover, carries its own penalties. It means, often enough, the loss of family and friends. All the dark-skinned ones must be left behind, passed in the street without a word of greeting. Green *Grow Lovely* 1951

So she became his common law wife, had five children by him and lived with him in a White area – in a tragic 'try-for-White' existence. *Sunday Times* 1.9.74

tsam(m)a melon ['tsama] *n. pl.* -s. *Citrullus lanatus :* also *kaffir* (*water*) *melon, karkoer, makataan,* all (q.v.). [*fr. Hott. t'sama*]

Some patrols . . . the desert men must make are quite waterless. Then it is that the *tsamma* melon, which grows in the Kalahari after rain, serves as a thirst-quencher and life-preserver. Birkby *Thirstland Treks* 1936

tsessebe ['tsesebĭ, -be] *n. pl.* s, Ø. The antelope *Damaliscus lunatus,* similar to the *hartebeest;* see *sassaby, sassabi.* [*Tswana tshêsêbê*]

The old hunter spoke of . . . the tsesebe's rich purple coat. Birkby *Thirstland Treks* 1936

tsetse fly ['tsetsĭ] *n. pl.* -ies. *Glossina*

morsitans, the insect which by its bite transmits sleeping sickness to man and *nagana* (q.v.) to domestic animals: see also quot. at *nagana* and *fly, the.* [*Tswana tsê tsê*]

The tsetse fly, *Glossina morsitans* . . . is the fatal pest which destroys the horses of big-game hunters, and the oxen of up-country transport riders, when they unwittingly come within the fly-infested country. The bite of the fly, though innocuous to every species of game, is fatal to all domestic animals, including the donkey, the dog and the goat, which were once thought to be exceptions. Wallace *Farming Industries* 1896

tshwala ['tʃwala, -alə] *n. Kaffir beer* (q.v.) brewed with malted grain usu. *mabela* (q.v.) (*kaffir corn* (q.v.)) or maize: see *uitloop* and *mtombo.* [*Zu. utshwala* liquor, beer, *Xh. utywala*]

'Tshwala,' the beer brewed by the Bantu, is more than a mere social drink. The traditional village brew is extremely nutritious and is considered a national foodstuff. *Panorama* Nov. 1972

Tsonga ['tsɔ̃nga] *n.* An African people of Mozambique and the E. Transvaal: also one of the official languages of *Gazankulu* (q.v.): see quot. also *Shangaan.*

The Bantu languages in which the complete Bible is already available in South Africa are North and South Sotho, Tswana, Central Tswana, Venda, Xhosa, Zulu, Tsonga and the South West African Ndonga. *Panorama* Dec. 1971

tsotsi ['tsɔ̃tsĭ] *n. and n. modifier pl.* -s. A usu. flashily dressed African street thug freq. a member of a gang, armed with a knife or other weapon: see also *skolly* and *spoilers* cf. *Austral. bodgie,* a teddy boy, *larrikin, lair,* an overdressed lout. [*etym. dub. acc. Fr. Trevor Huddlestone corruption of zoot-suit, prob. fr. tsotsile perfective form of vb. ukutsotsa,* to dress in narrow trousers or clothing of exaggerated cut: *urban African usage: see ~ taal, ~ language below*]

. . . tsotsis and skollies who lurk on street corners. *Post* 14.3.71

Thugs, known as *tsotsis,* prowl the streets, particularly on payday. With murders running 1,000 a year, the all-black Soweto urban council has called for vigilante patrols. *Time* 28.6.76

Originally, the word tsotsi referred to a type of clothing. In the forties . . . men's fashions saw the birth of a new, sleeker pair of trousers. The legs were tapered like stovepipes, tight in the crotch and even tighter around the ankles. In black townships and shanty towns the male

youths immediately accepted the new fashion, and called it tsotsi trousers. Venter *Soweto* 1977
In combinations ~ *suit*, ~ *trousers*, ~ *language*, ~ *taal*:

'They wear Tsotsi trousers, sixteen inches round the bottom, that is how you will recognise them' . . . Monare noticed . . . a young African well dressed in a tsotsi suit, wearing a wide-brimmed American hat. Lanham *Elanket Boy's Moon* 1953

Afr.E. ~ *taal*, ~ *language*, urban argot containing numerous terms unknown to rural Africans or those new to the townships, e.g. *govini* (q.v.), *laani* (q.v.). See also *flaaitaal*, *fly-taal*.

. . . entered the room and picked a quarrel . . . M. spoke in a 'Tsotsi-taal' which . . . said he could not follow. *World* 24.9.76

Their tsotsi jargon was bred in the townships and is as urban as the people who live there. It reflects all the linguistic influences that the urban African has been exposed to; from a base of Bantu tongues, the tsotsi language has grown with Afrikaans, English and even Yiddish limbs. It is ironic that Afrikaans, the language most despised by urban youths, has become the jargon's very backbone. Venter *Soweto* 1977

deriv. ~ *sm.*

Tsotsism is spreading in a community which is socially ripe for it, where juvenile delinquency has become inevitable. It may be claimed that the tsotsi gang has come to stay. Longmore *Dispossessed* 1959

Tswana ['tswana] *n.* One of the three *Sotho* peoples, also their language, Setswana: see also *Sechuana* and quot. at *abba*. [*Bantu n. prop.*]

He is equally at home in Tswana, Afrikaans or English . . . The people of Bophuthatswana form the historical nucleus of the three Sotho peoples – the Tswana, the South Sotho and the North Sotho . . . the domain of the Tswanas of South Africa consists of a number of areas in the north western part of the country. *Panorama* Sept. 1975

Tuan ['tuan] *n. pl.* -s. A Muslim saint or holy man: see also *kramat*. [*Malay tuan* lord]

Few Capetonians know that their city is protected by no fewer than seven saints . . . buried in a circle around the Peninsula. The saints are 'Tuans' (holy men) most of them of royal blood and high rank. *Capetonian* May 1979.

tub chair *n. pl.* -s. Also *baliestoel* (q.v.). [*trans. Afk. baliestoel*]

The Cape Tub-Chair. Among the many local designs the tub chair can probably claim to be the most typically Cape in origin, nor does this pleasant and comfortable chair appear to have been in general use elsewhere. Fehr *Treasures at Castle* 1963

tuj See *Indian terms*.

Tukkie [tʌkǐ] *n. pl.* -s. A student or alumnus of the University of Pretoria: *Tuks*, the University itself. [*acronym Transvaalse Universiteits Kollege + personif. suffix -ie*]

What do Afrikaans students think? I put this to six political science students of the University of Pretoria. In the relaxed atmosphere of the Tukkies' recreational centre they got their discussion going. *Star* 14.6.73

CONNIE (a final year BA student at Tuks): I'm not much of a patriot. *Ibid.* 8.6.73

Tuks [tʌks] *n. prop.* See at *Tukkie*.

tula ['tula] *interj. imp. vb. colloq.* Exclamation or order equiv. 'shut up', 'be quiet' etc. [*Ngu. ukuThula* to be quiet, stop speaking]

'Thula!' shouted Gus, turning round quickly and then as if to show the attendant – in case he did not understand the Xhosa word – where his own sympathies lay, followed it up with the English equivalent, 'Shut up!' Gordon *Four People* 1964

Tulbagh chair ['tœl,bax] *n. pl.* -s. A chair similar to the Restoration Chair of the latter half of the C17 with a back consisting of a central splat or lozenge, often caned, between turned, twisted uprights. ⫦The origin of the name is obscure. See quot.: also at *jonkmanskas*. [*unknown: see quot.*]

. . . a few 'Tulbagh' chairs were doubtless made, but probably not many, for it is a fairly elaborate type. Why, people ask, is it called a Tulbagh chair? The name seems to be recent and not traditional. Certainly it has nothing to do with the place of that name. On the other hand, if it has been called after Governor Ryk Tulbagh (1715–1771), then it looks like an attempt to date it. But such a date is obviously too late. Baraitser & Obholzer *Cape Country Furniture* 1971

tulp [tœlp] *n. and n. modifier.* Any of several species of *Moraea* or *Homeria* of the Iridaceae, with graceful showy flowers. ⫦Highly toxic to stock, it is most dangerous after dry periods when it grows up together with new grass: cattle used to it know it and avoid it. [*Afk. fr. Du. cogn. tulip*]

Only twelve oxen remained to us out of the beautiful span of twenty . . . three had died from eating the poisonous herb called 'tulip'. Five more sickened from this cause, but we managed to cure them with . . . an infusion made by boiling down the tulip leaves. Haggard *Solomon's Mines* 1886

An abundance of the highly poisonous tulp plant (Moraea edulis) in the Grahamstown district could mean stock losses for unsuspecting farmers, and has already cost one more than R1 000. 'There has been a great abundance of tulp this year, probably because of the good summer rains,' he said. *Grocott's Mail* 19.6.73 as *modifier* and in combination ~ *poisoning, blou(blue)* ~, *geel(yellow)* ~, *rooi(red)* ~ etc.

There is a *blaauw tulp* or pale blue *moraea*, which grows there . . . and this when other vegetation is scanty the cattle devour with fatal effect to themselves. Alexander *Western Africa II* 1837

tulpbrand ['tœlp₁brant] *n.* Flagsmut: a disease which attacks grain, esp. wheat in S.A.

Plants robust with an untidy leaning straw . . . susceptible to rust and vaalblaar and tulpbrand. *H'book for Farmers* 1937

turf [tɜf, tœrf] *n. and n. modifier.* In S.A. a soil type: sticky black noritic soil, heavy, clayey and fertile: particularly favourable to the cultivation of tobacco. [*etym. dub. prob. rel. Du. turf* peat]

Black noritic turf soils . . . It is exceptionally high in clay content but produces a fine textured leaf. *H'book for Farmers* 1937

Wheat is cultivated on black clayey soil (turf), heavy red loam and also on sandy soils. *Ibid.* Also in combinations *black* ~, ~ *soil*.

He had seen Johnny Coen, . . . busy scraping some of the worst turf soil off his veldskoens. . . . taking all the trouble to get the turf soil off his veldskoens *and* to get the turf soil off his face. 'If he was coming here to see us, well he wouldn't care how much black turf there was on his face.' Bosman *Jurie Steyn's P.O.* 1971

turksvy [₁tœrks'feɪ, ₁tɜks-] *n. pl.* -e. The fruit of any of several species of *Opuntia* (prickly pear), also *tolofiya* (q.v.). [*Afk. fr. Du. Turks* Turkish + *vijg cogn.* fig]

. . . next to it a prickly pear, broad-leafed and steadfast: Turkish fig was the name the Boers gave to it. Bosman *Willemsdorp* 1977

Turlington *n. prop.* One of the *Dutch Medicines* see quot. at *Old Dutch Medicines*.

Turnhalle ['turn₁halə, 'tɜn-] *n. prop.* The *drill hall* (q.v.) in Windhoek, scene of the S.W. African constitutional talks begun 1 Sept. 1975, now used to refer to the talks or body of delegates themselves. [*Ger. turnen* to perform gymnastics + *Halle cogn.* hall]

The leaders and representatives of the country's eleven different population groups, with a history of tension behind them, have gathered around a conference table in the Turnhalle at Windhoek, capital of the territory. *Panorama* Aug. 1976

The Turnhalle's latest attempt to stave off collapse, with Friday's announcement of a renewed commitment to the speedy establishment of an all-race interim government, has been rejected outright by Swapo's internal wing. *E.P. Herald* 26.10.76

Support for a multi-racial Turnhalle party is on the increase. *Sunday Times* 31.10.76

. . . even if a new and exemplary South West Africa does emerge from the crucible of Turnhalle. *Argus* 1.1.77

turnoff ['tɜn₁ɔf] *n. pl.* -s. A fork or branch road leading off usu. a bigger road, freq. with *n. prop.* as in the *X* ~, being the road leading to or coming from X. [*fr. presum. vb.* to turn off (out of) a main road]

. . . a collision between a bakkie and a car on a national road outside Mossel Bay at the Vleesbaai turnoff. *Het Suid Western* 2.9.71

twa(a)gras(s) ['twa:xras] *n.* Also *toagrass, dwa-grass, Bushman grass:* any of several feathery species of *Aristida*, important fodder grasses in the Kalahari. [*fr. umTwa* the self-styled name of the *Bushmen*]

A species of Aristida, 'twa-gras', is the most abundant grass, so far as is known, in the Kalahari region and in the upper region of the Karoo. Wallace *Farming Industries* 1896

twak¹ [twak] *n. interj. slang. interj.* Equiv. rot, nonsense, rubbish. [*Afk. twak* nonsense]

'Don't forget, a third of that bat's mine!' 'Twak!' Pieter snorted, 'Just try and bluff us with that one . . .' Krige *Dream & Desert* 1953 Also as *n.*

Honestly I know I wrote a helluva lot of twak – I knew I'd fail. O.I. (Student) 1972

twak² *n. colloq.* Tobacco. [*contraction of tabak* tobacco]

If I have the convicts to work for me I always buy them fat cracklings, coffee and twak – but the twak's what they always ask for and what they like best. O.I. (Farmer) 1963

When the only rain is what the weather forecasters . . . call *plek plek** and the first rains this year turn out to be a hailstorm that flattens the *twak* then it is time . . . to discuss some of the big changes that are happening in the Marico. *Star* 17.1.79 *here and there, localized.

twala ['twala] *vb. and n.* An African custom practised among rural usu. heathen tribes, by which a bride is carried off by force by a prospective husband. [*Ngu. ukuThwala* to carry off, abduct]

TWALA. As a defence to a charge of abduction. Sisson *S. Afr. Legal Dict.* 1960

The accused was never consulted before arrangements for her 'twala' were made. . . . she was very unhappy with her husband. Pleading mitigation her councel . . . asked the court to be lenient because Noranga was forced by her parents to marry a man she did not love. She was 'twalaed' (abducted) he pointed out. *Daily Dispatch* 13.2.21

twasa [ˈtwasa] *n. pl.* ama-. A student *sangoma* (q.v.) or witchdoctor in the initiatory stages of training during which powers of clairvoyance become apparent. [*Ngu. thwasa*, witchdoctor during apprenticeship *vb. ukutwasa*, 'emerge for the first time, become possessed by a spirit, as of divination.' (*Doke & Vilakazi Zulu Dict.* 1948)]

Not everyone can become a diviner. Initiation is prefaced by a definite 'call' from the ancestral shades, sent in the form of a sickness called *ukuthwasa*. Hammond-Tooke *Bhaca Society* 1962

Sangoma training involves . . . continual instruction of the Baba (teacher) as she watches her twasa (pupil) . . . noting her particular spirit manifestation. *Panorama* Nov. 1975

Tweede Nuwejaar [ˈtwɪədəˌnyvəˌjɑː(r)] *n. prop.* The second of January: a public holiday in the Cape. *Second New Year* (q.v.). and usu. second day of the *Coon Carnival* (q.v.). [*Afk. tweede* second, *nuwe cogn.* new, *jaar cogn.* year]

The Tweede nuwejaar saw the troupes with shields and trophies cavorting through the streets followed by enthusiastic supporters. *Argus* 5.6.71

. . . this is likely to be the last year that shops in the Cape are forced to close on Easter Saturday and Tweede Nuwe Jaar . . . in terms of the new Shop Hours Ordinance. *Het Suid Western* 6.4.77

Normal editions of The Argus will be published on Tuesday January 2, Tweede Nuwejaar. *Argus* 29.12.78

tweetalig [ˌtwɪəˈtɑlɪx, twiː-] *adj.* See *bilingual.* [*Afk. twee cogn.* two + *taal fr. Du. taal* language, tongue + *adj. suffix -ig cogn.* -y + *n. forming suffix -heid equiv.* -ness *cogn.* -hood *fr. O.E. -had* state of being]

This wall chart, attractive . . . and tweetalig nogal, is to be distributed to all hospitals. *Cape Times* 21.6.73

deriv. as *n.* ~ *heid,* use of or proficiency in both official languages.

Insist on 'tweetaligheid' in the hotels if this makes you feel more at home when spending the day away from Clocolan. *Star* 26.10.72

twist [twɪs(t)] *n. and vb.* Quarrel: found in place names Twistniet (*as imp. vb.*) and Lover's Twist as *n.* Also occ. in political writing: see quot. also *broeder*

~ . [*Afk.* twis, *Du. twist* quarrel]

The . . . carried the . . .'s official denial . . . It says . . . our story was a load of lies. So it confirms the worst – as official denials so often do. Yes there is a 'twis'. *Het Suid Western* 10.11.76

tyger *n. and n. prefix.* Leopard: found in place names Tygerberg: see also (2) *tier* and *tiger.* [*fr. Du. tijger* leopard]

U

U.B.C. [ˈjuˌbiˈsi] *n.* Urban Bantu Council: an advisory board of elected councillors for an African *township* (q.v.): see also quot. at *beerhall.* [*acronym*]

. . . the U.B.C. has no more power than the old advisory boards – only a grand new building and the money would have been better spent on schools. *Post* 28.6.70

uclever [ŭˈklevə] *n. pl.* oo-. *Urban Afr. E. slang* See quot. at *spoiler* also *clever.* [*fr. Eng.* clever]

This clever man (uclever) – is a person who is town-rooted – he can either be a tsotsi himself or may not be. He knows the liquor outlets and the dagga smugglers. Because of his knowledge . . . he does not associate with migrant labourers (amaGoduka). *Letter* Mr Sabelo Sillie 10.2.79

U.D.F. [ˈjuˌdiˈef] *n. prop.* Union Defence Force, former designation of the Permanent Force, S.A.'s standing army: see *P.F.* [*acronym*]

UDI [ˈjuˌdiˈaɪ] *acronym.* Unilateral Declaration of Independence: now freq. metaphorical.: See quot. at Capey, Capie [*fr.* Rhodesia's UDI of 1965]

. . . they all decided to stay on but declared an editorial UDI saying that the editorial department . . . alone would dictate editorial policy. *Drum* Feb. 1979

The Minister . . . denied any knowledge of a move in South West Africa to declare UDI. *Citizen* 9.4.79

ufufunyana [ŭˈfŭfŭnˌjana] See *mafufunyane, mafufunyana.*

uhlaka [ŭˈɬaka, -ga] *n.* (*pl.* izinhlaka). An *inyanga's* (q.v.) apprentice or a travelling herbalist. [*Ngu. uhlaka*, travelling herbalist who carries his medicines with him]

UHLAKA . . . an inyanga's, native medicine man's apprentice. He requires a licence if he practices for gain. *Bell's Legal Dict.* 1951 ed.

Uhlanga [uˈɬaŋga] *n. prop* and *n.* **1.** The reed clump or primal source of being from which *Unkulunkulu* (q.v.) first sprang or broke off and continued to break off others in the act of creation of mankind. ℙ This is only one interpretation of many of the stories of the Creation from ~: others say a creator *Umvelinqangi* (q.v.) made the earth and everything in it, including the reed from which Unkulunkulu and his wife both sprang and afterwards begat primitive man. [*Zu.* (*u*)*hlanga* reed, original stem, stock, ancestry]

Uthlanga is a reed, strictly speaking, one which is capable of 'stooling', throwing out offsets. It thus becomes metaphorically to mean a source of being. A father is thus the *uthlanga* of his children from whom they broke off . . . it may be concluded that originally it was not intended to teach by it that men sprang from a reed. It cannot be doubted that the word alone has come down to the people whilst the meaning has been lost. Callaway *Religious System of the Amazulu* 1884

2. *metaph.* The father of a family: see quot at 1. ~.

'Are there any called Uthlanga now?'
'Yes. It is I myself who am an uthlanga'
'Because you have become the father of children?'
'Yes . . .' As he said this he tapped himself on his breast. *Ibid.*

Uhuru [ŭˈhŭrŭ] *n.* Freedom: independence. See also *Mayibuye i Afrika, Usutu.* and quot. at *plural.* [*Swahili*]

Basutoland opposition may appeal to the UN – UHURU PACT OPPOSED. *Argus* 18.6.66
For us in this country it is not yet Uhuru. *Star* 13.4.73

uintjie [ˈœɪnkĭ, -cĭ] *n. pl.* -s. *abbr.* **1.** Common name for a troublesome weed *uintjiekweek*, or yellow nut-grass, *Cyperus rotundus*, which produces strings of very small bulbs: see also at *kweek*. [*fr. Du. ajuintje* little onion]

I see that a preparation is being advertised for eradicating 'Uintjies' in lawns. I have not yet heard of any success with it. *Daily Dispatch* 18.5.74

2. *n. pl.* -s. Any of numerous edible bulbs esp. the corms of certain Iridaceae including species of *Moraea* formerly much used by colonists as well as the indigenous tribes: commented upon by early writers: see also *wateruintjie*, and quot. at *veldkos*.

Many varieties of *uintjies* (edible bulbs) are found in the Karoo. Some are eaten raw, others stewed with meat. . . . Certain *uintjies* taste like chestnuts and make a strong soup. Green *Karoo* 1955

uitkoop [ˈœɪtˌkʊəp] *n.* A right of 'making over' exercised in favour of a surviving spouse under the provisions of the Administration of Estates Act No. 66 of 1965 at the discretion of the Master of the Supreme Court. Informant Adv. Geo. Randell, Grahamstown. [*Afk. lit.* 'buy out' *fr. Du. uit cogn.* out + *kopen* to buy]

'– taking over by surviving spouse instead of realising the property of the deceased, to make it over* by the surviving spouse at a valuation to be made by a sworn appraiser – or by any such person as the Master may approve.' Wille *Principles S. Afr. Law* 1945 *ed.* *This is called *uitkoop.*

uitkyk [ˈœɪtˌkeɪk] *n. and vb.* Look-out post or place: also place name Uitkyk. [*Afk. fr. Du. uit cogn.* out + *kijken* to look]

Monday, May 28th 1798 – The day being a fine one, we proposed going to the Out Keek, or look-out post, about four miles distant to see the bay and adjacent country from the highest ground. Barnard *Letters & Journals* 1797–1802

uitlander [ˈœɪtˌlandə(r)] *n. pl.* -s. *lit.* 'Outlander', an alien or foreigner, a term current before the *Anglo-Boer War* (q.v.) applied esp. to English people resident in S.A.: see quot. at *make a plan.* [*Afk. fr. Du. uitlander* alien]

Kruger . . . was fully determined to grant no privileges to the Uitlanders who, he was certain, had it in their minds to filch his country . . . 'The Uitlanders,' he said, 'were never invited to settle in the Transvaal, and are not wanted there.' Cohen *Remin. Johannesburg* 1924
Obviously the major concern should be the urbanised African, who is being treated far worse than Paul Kruger ever treated the Uitlanders. *Daily Dispatch* 19.6.72

uitloop [ˈœɪtˌlʊəp] *n.* Sprouted grain, malt: usu *kaffircorn* (q.v.) also called *mtombo* (q.v.), used for brewing African beer; see *kaffir beer, K.B., tshwala, maiza, mqomboti.* [*Afk. fr. Du. uitlopen* to sprout]

1½ cups kaffircorn that has to grow and is dry (known as 'uitloop') . . . Ask your maid to get the 'uitloop' as many of them use it for their own beer. It is obtainable in shops in the African townships. *Evening Post* 8.1.72

uitpak inspeksie, inspection [ˈœɪtˌpak ɪnˈspeksĭ] *Sect. Army.* See *packing out parade* at *pack out* and quot. below unpack. [*Afk. uitpak*]

I am writing on my bed in the middle of what is called 'uitpak inspeksie'. It is a bit of a schlep.

Letter Serviceman 13.1.79

uitsig ['œɪt‚sɪx] *n*. View, prospect: found in place names Mooiuitsig, Blyvooruitsig: see also *sig, gesig, gezicht*. [*Afk. fr. Du. uitzicht* outlook, view]

ukuthwasa See *twasa*.

umac [ŭ'mæk] *n. pl*. oo-. *Urban Afr. E. slang*. A young African man-about-town: see also *uscuse-me* and *uclever*. [see quot.: deriv. ~*azi* abbr. *Ngu. umfazi* wife, woman]

Among the half educated two main divisions based on age are recognized *ikhaba* and *ooMac* . . . a little older than the *ikhaba*, from about 25 to 35 is the set called *ooMac* from the Scottish *Mac* which is a popular nickname among young men in town. They are expected to be more reasonable and responsible than the *ikhaba* boys . . . Their wives, *ooMackazi* being young married women are expected to settle down as housewives. Wilson & Mafeje *Langa* 1963

. . . somebody who 'understands' town life unlike the migrant labourer who is slow . . . or is not capable of doing what ooMac can do . . . their way of talk, their easy way of earning money. *Letter* Mr Sabelo Sillie 10.2.79

umfaan ['ʊm‚faːn] *n. pl*. -s. anglicization. A young African boy employed in a house or garden esp. in Natal: see quot. at *nkosazana cf. Anglo-Ind. cho(o)kra*. [*Zu. umFana* a young boy]

. . . complains about African youths being called umfaan which is, in a way, correct. *Drum* 22.11.72

. . . young umfaans dancing au naturelle at the roadside in the hopes of attracting sweets or coppers from passing motorists are in fact the young herd boys sent out to mind the family goats and cattle. *The 1820* Sept. 1973

umfazi [(ŭ)m'fazĭ] *n. pl*. aba-, -s. A mode of reference, also address, to a wife or married woman, *cf. ntombi*; see also *abafazi, fazi*. [*Ngu. umFazi* woman, wife]

Just as my men were in the act of firing, they called Umfasi, we then saw they were Women. Stubbs *Remin*. II 1874–6

We did not worry people in the streets. We were out looking for fun with our bafazi's and here come these Vikings. *Drums* 8.4.72

umfundisi [(ŭ)m'fŭndĭsĭ, -‚fŭn'dizĭ] *n. pl*. -s. A mode of address or reference to a priest occ. to a learned man or teacher, *cf. Rabbi*: see also *withalskraai*. Anglicized form *colloq. fundi/s* (q.v.), see also quot. at (3) *Xhosa*. [*Ngu. umFundisi* teacher, minister]

No Minister of religion doubted his profes-

sion or minimised his ability. 'Mfundizi – Moruti – Minister,' they reverently addressed him. Dikobe *Marabi Dance* 1970

Job's comforters are the umfundisi and a cheerful scavenger from the municipal rubbish dump. *E.P. Herald* 6.8.73

Umkhonto we Sizwe [‚ŭm'kɔntɔ̃ we'sizwe] *n. prop. lit*. 'The Spear of the Nation': see quot. below and *ANC²*. *cf. Umkhonto ka Shaka* (q.v.) [*Ngu.*]

He was arrested by Frelimo soldiers on suspicion of being a . . . spy and kept in prison until he agreed to join Umkhonto we Sizwe, the military wing of the ANC. *Daily Dispatch* 21.2.78

umlungu [ŭm'lungŭ] *n. pl*. -s, abe-. A white man: see *mlungu*.

umnumzana [(ŭ)m'nŭmzan(a)] *n. lit*. 'Chief' 'headman' used as equiv. Sir, Master: see *mnumzane, mnumzana*, and quot. at *imbongi cf. nkosi*. [*Ngu. umNumzane, umNumzana* chief, headman]

. . . there is the old African family retainer who went to the Boer War with Grandpa and now goes round saying: 'Go well, child,' and 'stay well, *umnumzana*'. Mackenzie *Dragon to Kill* 1963

umphakati [ŭm‚pa'gatĭ] *pl*. ama- *n*. See *amaphakati*.

umtagathi [ŭm‚ta'gatĭ] *n. and modifier*. (*pl*. ama-) Wizard: see *tagati*.

. . . the blow went crashing through shield and spear, through head-dress, hair, and skull, till at last none would come near the great white 'umtagati', the wizard, who killed and failed not. Haggard *Solomon's Mines* 1886

Umvelinqangi [ŭm‚velɪn'kaŋgĭ+] *n. prop*. See ⸆ at *Uhlanga*. ⸆ *acc*. some ~ was the creator of the earth before *Unkulunkulu* (q.v.) *acc* others ~ is one of the names of *Unkulunkulu*.

umzimbete, **-beet,** **-biet** [ŭm‚zɪm'bete, -'bit, -bĭt] *n*. Ironwood: the flowering tree *Millettia grandis* and its exceptionally hard timber, also called 'kaffir ironwood', a tree protected in Natal: ⸆See also *mopane* known by some as 'ironwood' [*Ngu. umSimbithi* iron wood (tree)]

Many South Africans know the wood of Umzimbeet (which word is a corruption of the Xhosa and Zulu umSimbithi, meaning ironwood). Until its use was protectively stopped, it was the favourite wood for the tribal craft of making walking-sticks and knobkerries. *E.P. Herald* 12.1.72

uncle *n. pl*. -s. A mode of address or reference to an older man, also *oom* (q.v.): or as a mode of reference sig.

'man', 'gentleman.': see also *tannie*, *tante* and *auntie*. [*usu. trans. Afk. oom* uncle ; *see etym. at oom*]

> We met a dingy old farmer going to his work on Bultfontein. 'Good morning, uncle!' said Mr. Fry. 'Good morning, brother!' returned the boer. Boyle *Cape for Diamonds* 1873
> . . . the child said he recognised the man: 'I know him . . . That uncle was at our house that night . . .' Mr Justice . . . asked him one more question . . . 'Uncle, I remember nothing. She lay in the yard and she died.' *E.P. Herald* 18.11.76

under *prep. substandard.* Used as equiv. of 'among'. [*translit. Afk. onder* among]

> 1839 Dec 23 frightful sickness under my Flocks about 150 affected & 10 dead the last three days.
> 1842 25 March Tiger got under the wethers & killed 8 or 9. Collett MS. *Diary II*
> The cataloguing is shocking: under the plays you find a volume of poetry. O.I. 1974

underground *n. Afr.E. colloq.* African term for any of several illicit home-brewed liquors which are, or were formerly buried or stored underground for fear of detection.

unimproved *partic. modifier.* Of or pertaining to land without *improvements* (q.v.): term used in advts. of land sales.

Union, the *n. prop.* Previous designation of the *Republic* (q.v.) before the referendum of 1960, *cf. the United States :* see also quot. at *S.A.R.* [*abbr. Union of South Africa*]

> . . . so far as our essential food requirements are concerned, the Union is today practically independent of imports. With the exception of luxury articles such as coffee, tea and whisky, rice is practically the only article . . . still being imported. *H'book for Farmers* 1937
> 'The Union' means the territorial limits of the Union of South Africa. *Union Statutes* 16.5.57

Unisa [ˌjuˈnisə, ˌjuˈnaisə] *n. prop. University of South Africa, Pretoria.* [*acronym*]

> *New posts at Unisa* A record number of new academic appointments has been made by the University of South Africa. *Rand Daily Mail* 29.1.71

United Party *n. prop.* Also *U.P.* The party formed in 1934 by the coalition of '*Nat en Sap*' both (q.v.) of which the present National Party (1) *H.N.P.* was orig. a splinter group. Now dissolved.

> DEBATE ON SWING TO U.P. A new emergence of 'a moral sense of difference between what is right and what is wrong,' had led the swing away from the Nationalist Party, according to . . . the former independent MPC

for Umkomaas, who recently joined the United Party. *Daily News* 3.6.70

Unkulunkulu [(ŭ)ŋˈkulŭŋˈkulŭ] *n. prop.* God, the Almighty: previously the 'first man': see first quot.: now the term of address in prayer or reference adopted by the Zulu for the Christian deity. See at *Tixo, Uhlanga* and *Umvelinqangi. cf. Tixo.* [*Zu. unKulunkulu lit.* Great-great, *hence* God]

> The Unkulunkulu *par excellence*, the first man, is no where worshipped. No isibongo* of his is known. The worship, therefore, of him according to native worship is no longer possible. Callaway *Religious System of Amazulu* 1884 **isibongo* (q.v.) clan name
> In the eastern Cape missionaries had, by common consent, used 'Thixo' for 'God', a word which had had very few prior associations. Colenso determined to use the Zulu 'Unkulunkulu', the name already used for the Creator. . . . There was considerable opposition but Colenso's view triumphed and the word has now presumably acquired Christian associations. Hinchliff *Anglican Church in S.A.* 1963

unsalted See *salted.*

unwisseld [ˌʌnˈvisəld] *neg. partic. modifier.* Of or pertaining to a lamb which has not shed its milk teeth: see *wissel.* [*Eng. neg. prefix.* un- + *past partic. formed with Afk. vb wissel* to shed teeth *fr. Du. wisselen*]

U.P. [ˌjuˈpiː] *n. prop. modifier.* See *United Party.*

up country *adv. pl. n. and modifier.* This phr. has several uses in S.A.E. [*prob. fr. Afk. use binneland n.* interior, and *binnelands adj. and adv.*] 1. As *adv.p.* sig. 'in the interior,' 'away from the coast' etc. Also *Austral.* ⫐Occ. sig. remote from cities *cf. backveld.*

> June 2 1864 In point of fact, putting coloured people to one side, you only meet Europeans here. There are scarcely any Dutch families in Durban, although they abound 'up country'. A Lady *Life at Natal* 1864–5

2. As *n.* sig. the interior, far-away parts etc.

> The Dutch around Cape Town (I don't know anything of 'up country') are sulky and dispirited. Duff Gordon *Letters* 1861–2

3. As modifier sig. 'from the interior,' 'platteland', see also quot. at *tsetse fly.*

> The highlight . . . will be the Southern Cape New Year regatta which is widely advertised. . . . throughout the country and attracts a large number of up-country yachtsmen. *E.P. Herald* 2.10.73

U.P.E. [ˈjuˌpiˈi] *n. prop., modifier. University of Port Elizabeth:* see also

Uppie. [*acronym*]

Record for U.P.E. sprinter. *Evening Post* 20.3.71

up north *adv.p.* World War II term: *going or being sent* ~ for S.Afr. soldiers sig. posting to the N.African (Egyptian) theatre of war: see *red oath, red tabs, red flashes.*

The Royal Natal Carbineers were called up and seven out of the eight male clerks at Electricity House went Up North with them. *Personality* 12.7.74

Uppie [ˈʌpĭ] *n. prop. pl.* -s. A student of the University of Port Elizabeth: see *U.P.E.* [*fr. acronym U.P.E.*]

Rhodes kicked off with Uppies facing a weak sun. *Evening Post* 19.5.73

upsaddle *vb. intrns.* To saddle a horse, 'saddle up': see also *off saddle* [*translit. Afk. opsaal fr. Du. opzadelen* to saddle up]

Leaping out of bed Christiaan roared out: 'Upsaddle, everybody!' and within an hour the whole camp had been shifted miles across the veld. Rosenthal *De Wet* 1946

uscuse-me [ŭˈskjuzˌmi] *n. pl.* oo-. *Urban Afr. E. slang.* An educated African of the professional or white-collar middle class: usu. a term of contempt or derision [see second quot.]

The urbanized whose homes are in town . . . some of whom form an educated middle-class – the *ooscuse-me* – . . . The educated people are referred to by others somewhat derogatorily as *ooscuse-me*, and accused of being aloof and conceited . . . ooscuse-me include those in profesional jobs – teachers, lawyers, doctors, ministers of religion, nurses, secretaries – as well as university students and others. Wilson and Mafeje *Langa* 1963

Ooscuse-me used to include professional men   teachers, clerks, lawyers, nurses etc. but it is no longer like that because some people with a low standard of education who lead decent lives, dress respectably and are gentle and polite in their manners are also called scuse-me. *Letter* Mr Sabelo Sillie 10.2.79

usutu [ŭˈsutŭ] *interj.* A Zulu battle cry: cf. *E.Afr. uhuru* (q.v.); now also in S.A.E. Traditionally the name of the ruling faction of the Zulu people under Cetshwayo and later Dinizulu. [*Zu. uSuthu*, the ruling section of the Zulu; Zulu Royal House]

Cetshwayo's personal following were traditionally known as the Usutu. Bulpin *White Whirlwind* 1961

. . . an Usutu war-party . . . shouted out their familiar, deep-noted war-cry of 'Usutu, Usutu'

as soon as they saw him. *Ibid.*

utywala [ŭˈcwala] *n.* See *tshwala. Xh.* general term for alcoholic drink, also for African beer: see *mqomboti, kaffir beer.*

Uvuko [ŭˈvugɔ̃] *n.* Traditional Xhosa term for the re-awakening of the dead: see quot. at (*u*)*Qamata* [*Xh.* awakening]

Religion is close to my people. We were never heathens – we have always believed in uQamata who is the one and only God and in Uvuko, the reawakening. Chief Lennox Sebe *cit. Bona* Oct. 1978

Uz(z)i [ˈuzĭ] *n. pl.* -s. A lightweight Israeli-made sub-machine gun. ⫿Manufacturer's or trade name now (1977) a household word in Rhodesia.

Introducing a BBC . . . programme about White Rhodesian farmers, producer David Dimbleby . . . went on to explain to British viewers that *terrs* is short for 'terrorists', *Uzzi* is the name of an Israeli sub-machine gun . . . One of the interviewed . . . 'I have my revolver in my handbag . . . even in town, whenever I go. But my Uzzi is what I carry when I'm travelling on the road.' *The Listener cit.* Rhodeo 4.3.77

It is a rare farmer who does not carry in his car an automatic rifle or even a 'rhogun', the local adaptation of Israel's Uzi submachine gun. *Time* 28.3.77

V

vaal- [fɑːl] *adj.* Grey, dun, fawn [*Afk. fr. Du. vaal* ash-coloured, tawny] **1.** *prefix* to numerous plant names e.g. ~*bos* (q.v.); ~*boom, Terminalia sericea;* ~ *brak,* see *brak*(*bos*); ~*doring, Acacia hæmatoxylon,* see *doring;* ~ *sewejaartjie, Helichrysum argyrophyllum,* etc. **2.** *prefix* to names of animals, birds and fish: ~*hartebees,* see *hartebees*(*t*); ~*ribbok,* see *ribbok;* ~*valk*(*ie*) see *valk;* ~ *haai* (*shark*); ~*korhaan, Otis* see *korhaan;* ~*spreeu,* see *spreeu,* etc.

3. Found in place names Vaalhoek, Vaalplaas, Vaalwater, Vaalriv(i)er, Transvaal.

4. *prefix* to other items ~*blaar* (q.v.), ~*japie* (q.v.), ~*pens* (q.v.).

Straight to that vaal koppje (grey hill), and there off-saddle for a bit. Farini *Through the Kalahari* 1886

vaalblaar [ˈfɑːlˌblɑːr] *n.* A disease to which certain varieties of wheat are subject. [*Afk. fr. Du. vaal* ash-coloured, tawny + *blaar* leaf]

In the past few years many of the well-known old wheat varieties have disappeared . . . This has been due chiefly to the invasions of rust . . . and of vaalblaar (Septoria spp.). *H'book for Farmers* 1937

vaalbos ['fɑːlˌbɔs] *n. pl.* -se. *Tarchonanthus camphoratus*, a shrub or small tree extremely drought resistant and therefore an important source of fodder in dry parts of the country such as Botswana and the Kalahari: such parts are known as ~*veld.* [*Afk. fr. Du. vaal* ash-coloured, tawny + *bosch cogn.* bush]

People say they enjoy life in Sishen, where not so long ago the bare ground sustained only vaalbos. These days, too, water is no longer such a problem. *Panorama* Feb. 1971

vaaljapie¹ ['fɑːlˌjɑpĭ] *n.* Raw, young wine: see quot., also at *kop-en-pootjies. cf. Austral. red ned*, bulk claret; *Canad. pinkie*, cheap wine. [*Afk. vaaljapie* rough new wine. *lit.* 'tawny Jim']

. . . Vaaljapie . . . the favourite everyday drink of the Cape coloured farm labourer; and in some districts it is difficult to get the heavy work of harvesting or shearing done without regular tots of Vaaljapie. Green *Land of Afternoon* 1949

. . . content with *vaaljapie*, our *vin ordinaire*, with the added sparkle of soda water bubbles. Kavanagh *Merry Peasants* 1963

Vaaljapie² *n. pl.* -s. usu. capitalized. A small greyish-coloured tractor. [*fr. colour*]

Tractor (This Vaal Japie). Excellent condition with many spares. *Rand Daily Mail Advt.* 5.3.71

as *modifier*:

VAALJAPIE belt pulley to drive hammermill, like new. *Farmer's Weekly Advt.* 30.5.73

Vaalpens ['fɑːlˌpeːns] *n. pl.* -e. **1.** A *Transvaler* (q.v.) one from the Transvaal. [*Afk. fr. Du. vaal* ash-coloured, tawny + *pens* belly *cogn.* paunch]

For the past 50 years and more Free Staters have been known among Dutch-speaking South Africans as Blikore (tin ears), Transvalers as Vaalpens and Cape Colonials as Woltone. *Star* 1.5.34 cit. Swart 1934

We'd hate you to miss our Bash-a-Vaalpens week . . . scheduled to start the day Parliament adjourns. Steenkamp *cit. Capetonian* May, 1979

2. A member of the Ba-Kalahari tribe of Bushmen. [*as above*]

Early on the morning of July 31st, the Vaalpens reported that one of our oxen had been mauled . . . We saddled up and with three Vaalpens soon found where the lion had caught the ox. *E.P. Herald* 28.9.1916

va(a)tjie ['fɑɪˌkĭ, -cĭ] *n. pl.* -s. A small keg, cask or firkin: also in *water* ~ a soldier's carrying canteen: other combinations *wyn/wine* ~, *sout/salt* ~. [*Afk. fr. Du. vaat* keg + *dimin. suffix -jie*]

No doubt many a shooting party assembled on the beach with a sheep for their braaivleis, a vaatjie of wine and a sack of sweet potatoes for their embers. Green *Grow Lovely* 1951

If you put raw witblits into a little wooden vaatjie and keep it there it'll be brandy at the end of three years. Wine Merchant, George, Cape 1972

vabond ['fɑˌbɔnt] *n. pl.* -s. Rascal, good-for-nothing. *cf. Austral. vag.* [*Afk. fr. Du. vagebond cogn.* vagabond]

The policeman laughed. 'All right, you old va'bond! You can have the whole five pounds. Meiring *Candle in Wind* 1959

Vaderland ['fɑdə(r)ˌlant] *n. (prop). n. modifier* Fatherland. [*Afk. fr. Du. vader cogn.* father + *land* country] **1.** As *n. prop. Die* ~ *:* an Afrikaans newspaper.

2. *n. modifier:* hist. ~*volk*, ~*cattle*, see *Fatherland cattle.*

They style themselves Africaners, and distinguish all those who come from even any part of Europe as Vaderland Volk, or Fatherland people. Philipps *Albany & Caffer-land* 1827

3. *prefix* to plant names: ~(s)*riet*, (reed) *Phragmites communis*; ~(s)*rooihout, Kiggelaria africana*; ~(s)*wilg-* (*erboom*) willow tree, any of several species of *Combretum* having a likeness to a willow (*Salix*) etc.

Vaderlands . . . given by the colonists to several plants they observed in their migrations which recalled or suggested similar forms in their homeland. It is also used in the sense of 'indigenous'. Smith *S. Afr. Plants* 1966

vadoek ['fɑˌdŭk, 'fɑ̌ˌdŭk] *n. pl.* -s. A linen tea or glass-cloth or more freq. any cloth for wiping up spills, cleaning surfaces etc. [*Afk. fr. Du. vaatdoek, vadoek* dish-cloth (*vaat* crockery, hollowware)]

Vaatdoek . . . A common clout used for the thousand and one things that a damp cloth is needed for in the kitchen. Pettman *Africanderisms* 1913

To call the Union Jack a 'rooi spinnekop' was definitely cleverer, but was it more, or less insulting than 'die rooi vadoek?' The Sappe found no metaphors as vivid as these. Butler *Karoo Morning* 1977

valk(ie) ['falk(ĭ)] *n.* Hawk, falcon, kestrel: suffixed to the names of several species of the Falconidae e.g. *blou* ~ (blue) *Elanus caeruleus; dwerg* ~ (dwarf) *Poliohierax* *semitorquatus; edel* ~ (noble) *Falco biarmicus; rooi* ~ (red) *Falco tinnunculus* etc. [*Afk. cogn.* falcon]

Van der Hum [ˌvændə'hʌm, ˌfandə(r)'hœm] *n. prop.* A S.Afr. liqueur flavoured with *na(a)rtjie* (q.v.) peel. [*see second quot.*]

That was the time when every farmer made three liqueurs for household use – Jan Groentjie (peppermint), aniseed and Van der Hum. They were known as 'the green, the white and the brown', and the guest had to taste all three before he departed. Green *Land of Afternoon* 1949

The most famous South African liqueur is undoubtedly Van der Hum with its tangerine flavour, and it has remained popular since the early days of settlement at the Cape. According to tradition the name is derived from a sea captain, Van der Hum, who had a special liking for this liqueur. Opperman *Spirit of Vine* 1968

Van der Merwe [ˌfandə(r)'mervə, '-mɜvə] *n. prop.* A comic Afrikaans folk figure *Koos* ~ from *Pofadder* (q.v.) the 'hero' of numerous ~ *jokes* or ~ *stories*

Being able to speak Afrikaans is suddenly 'in' and telling van der Merwe jokes with an English accent is really 'in'. (Even if the punchline seems faintly obscure.) *Star* 10.10.73

The South African social scene has not been the same since Van der Merwe first lumbered on to it a decade ago . . . Anecdotes about him are swopped from boardroom to boudoir. *E.P. Herald* 2.7.75

varklelie, varkblom, varkoor ['fa(r)kˌliəlĭ, -ˌblɔ̃m] *n. pl.* -s, -me. *Zantedeschia (calla) aethiopica*, usu. *pig-lily* (q.v.) or Arum lily fr. the similarity to the Arum of Europe; formerly used medicinally. [*Afk. fr. Du. varken* pig + *blo(e)m cogn.* bloom, *lelie cogn.* lily]

, , , the medicine chest of the Cape Flats is not without virtue . . . For rheumatism there is varkblom, which must be heated before being applied to the skin. Green *Tavern of Seas* 1947

varkpan ['fa(r)kˌpan] *n. pl.* -ne, -s. *Sect. Army. National Servicemen's* term: a compartmented metal lunch tin: see also *mindae, vasbyt, blouie, blougat, ouman, roofie.* [*Afk. vark* pig + *pan* dish]

OK, so you've finished complaining about the chow . . . polishing your area, washing your varkpan, . . . counting your baie dae – and you're slacking it up with nobody to bug you. *Scope* 10.1.75

vasbyt ['fasˌbeɪt] *vb. usu. imp., interj. Sect. Army.* Term among *National Servicemen*, equiv. of 'hang on', 'grin and bear it', 'bite on the bullet' etc.: see quot. at

ouman, cf. bite one's teeth. [*Afk. fr. Du. vas* firmly *cogn.* fast + *bijten* to bite]

. . . all we can say is 'Keep your hair on boys – Vasbyt Mindae!' *Grocott's Mail* 19.11.74

⫷ Used with varying meanings in the forces see quot. ~ as *prefix* as in ~ *mog*, ~ *Bedford* etc.: meaning is either an older, more worn model or applied to a vehicle stripped for Border action.

One thing I can tell you about army language – no two ous mean the same thing – when he says 'vasbyt' he means one of his Bedfords – when I say 'vasbyt' I mean Coke – gives me me vasbyt – can't do without it – bottle of vasbyt . . . *Ex-Border Serviceman* 6.4.79 (reputedly *'bossies'* (q.v.))

Also the slogan of the Parachute Battalion.

. . . if you live another month – Vasbyt! Vasbyt! – you may achieve the ecstasy of floating down the silent sky under a silken parachute. Nobody lets you forget that the Parachute Battalion's slogan is 'Vasbyt!' *Sunday Times* 12.3.72

vastrap ['fasˌtrap] *n. and n. modifier, also vb.* **1.** *lit.* 'Firm tread', a fast dance danced to *boeremusiek* (q.v.) often by country people: see also *tickey-draai, opskud.* [*Afk. fr. Du. vasttrappen* to stamp, tread down]

Lena's still got a vastrap in her old legs. You want to dance Boesman. Not too late to learn. Fugard *Boesman & Lena* 1969

2. A dance or 'hop' in the country, also called a *sheepskin, opskud* or (3) *velskoen*, all (q.v.). [*fr. above*]

A country dance is often referred to as a vastrap or a velskoen. Most of the early writers at the Cape had something to say about the fondness of the people for dancing . . . Many tunes heard at a vastrap . . . are composed on the farms by the players themselves. Green *Land of Afternoon* 1949

As modifier ~ *music.*

Concertina and guitar music goes on until dawn with few intervals, but with many rounds of *vaaljapie* and brandy. You remember the candle light, the *vastrap* music, the supper of *kop-en-pootjies.* Green *Giant in Hiding* 1970

As *vb.;* see quot. at *poffertjie.*

vat en sit [ˌfatən'sɪt] *vb. phr. or modifier colloq. Afr.E.* See quots.

A girl who is foolish enough to agree to live with a man, as man and wife, known as *'vat en sit'* 'just take and sit', or 'keep', hoping that eventually he will decide to marry her, is entertaining false hopes. There inevitably comes the day when he fancies someone else and tries to get rid of his *vat en sit* woman. Longmore *Dispossessed* 1959.

They call it 'living in sin', 'trial marriage'

elsewhere ... In South Africa these loose unions age-old as they are, are laconically referred to as 'vat en sit'. *Voice* 25-31.3.79

As *modifier:* equiv. of 'common law', see above ~ *woman,* also ~ *marriage.*

Somewhat of a philosopher he will tell anybody who is prepared to listen that vat en sit marriages have been blessed by the gods seeing as the Little Woman will never nag or threaten to institute divorce proceedings. *Drum* 8.3.74

vatterig ['fatə₁rɪx] *adj. slang.* Of or pertaining to one who cannot 'keep his hands to himself'. [*Afk. fr. Du. vatten* to take, catch + *adj. forming suffix -ig cogn -y*]

It's always a bit worrying when these adolescent boys start getting vatterig with each other. O.I. Mother 1974

vee [fi:, fɪə] *collective n.* Domestic livestock: usu. cattle, sheep, goats: see *stock.* [*Afk. fr. Du. vee* domestic livestock *cogn. O.E. feoh* cattle]

December 6th set off this Morning up the Country to Purchase Vee for Slaughter.

13 returned home to day having purchased 56 Sheep & 9 Beast. Collett *Diary I* 1836

In combination ~ *plaas,* ~ *place* (q.v.), ~ *boer* (q.v.), ~ *kraal,* a *kraal* (q.v.) in which to pen domestic animals; *skut* ~, *schut* ~ impounded animals; see *skut.*

veeboer ['fɪə₁bu:r, 'fi:-] *n. pl.* -s, -e. A stock farmer or grazier, sometimes nomadic: see *trekboer.* [*Afk. fr. Du. vee* (q.v.) livestock + *boer* farmer]

Here the free burghers, the vee-Boers, rapidly acquired qualities unknown to the more sedate residents of Table Valley ... They learned woodcraft, veldcraft. They became backwoodsmen, men of the veld ... The vee-Boers were grazing their cattle northward and eastwards in a world alive with a bewildering tangle of Hottentot and Bushman clans. Mockford *Here are S. Africans* 1944

veeplaas ['fɪə₁plɑs, 'fi:-] *n. pl.* -e, (-en). Also *veeplace;* a stock farm used for grazing not agriculture. also a place name E. Cape Veeplaas. [*Afk. fr. Du. vee* (q.v.) stock + *plaats cogn.* place *usu. sig.* farm]

The farms, 'few and far between', are mere *vee-plaatzen,* or cattle places, without in general the comfort of a garden, or the means of cultivating a single blade of corn. Thompson *Travels II* 1827

27 August Tracked to day with my Bucks to the Vee Place. Collett *Diary I* 1838

vee place See *veeplaas.*

veg- [fex] *n. prefix* Battle, combat: found in place name Vegkop. equiv. 'Combat' as prefix in army rank ~ *generaal* etc.

[*Afk. veg* to fight]

'We will appoint a burgher who knows all about fighting to act under you,' Kruger said, 'and call him the Veggeneraal.' Brett Young *City of Gold* 1940

veld(t) [felt] ⫐Note: This term in S.A.E. has several basic meanings or uses (see quot. below): viz. (1) The S. Afr. landscape, flat or hilly, or non-urban area; (2) Grazing or farming land; (3) An element in S. Afr. place names; (4) An administrative or military prefix. The material is therefore presented under four headings with the *relevant* compounds following each alphabetically and numbered accordingly, e.g. (1) *veld craft,* (2) *veld management,* (4) *veld cornet.* Also *obs.* form *field* (q.v.). [*Afk. fr. Du. veld, cogn.* field, *used as general equiv. of* 'country']

The word veld, as used by South Africans, can be puzzling. Many a stranger has the idea that it refers to some definite locality. ... But you soon learn that there are all kinds of veld. There is sweet-veld and sour-veld, high-veld and low-veld, back-veld and sand-veld, cold-veld and warmveld and bushveld ... the word is used ... to describe grazing country anywhere, and also to describe what kind of grazing it is. So a South African could sigh for the veld in general but if he were buying a farm, he would want to know what particular kind of veld it was. Morton *In Search of S.A.* 1948

1. veld *n.* The S. Afr. countryside, landscape etc. *cf. Austral. bush, U.S. & Canad. prairie.*

On Saturdays we generally went for a Hunt, and on coming Home all game that had been shot was equally divided, I never had happier days than with them out in the Veldt. Stubbs *Remin.* 1876

1. veld *attrib.* and in combination [see also *attrib.* and combination forms at (2) *veld,* (3) *veld,* (4) *veld*]. ~ and vlei (q.v.); ~ *broek,* ~ *craft;* ~ *fever* (q.v.); ~ *fire;* ~ *flower(s);* ~ *kos* (q.v.); ~ *ponde* (q.v.); ~ *rat;* ~ *remedy* (q.v.); ~ *skoen* see *velskoen;* ~ *sore* (q.v.); ~ *tent.* Also *back* ~ (q.v.); *bont(e)* (q.v.); ~ *bush* ~ (q.v.); *eland* ~ (q.v.); *Jonas* ~ (dolomite); *kalk* ~ (chalk, lime); *Kar(r)oo* ~ (q.v.); *sand* ~ (q.v.); *strand* ~ (q.v.); *thorn* ~ (q.v.); *vlakte* ~ (q.v.); *vlei* ~ (q.v.).

1. Veld-and-Vlei (school) *n. prop.* A camp, sometimes known as an 'Adventure School' founded in 1958 by F. Spencer Chapman for developing self-reliance and toughness in boys of school age. *cf. Brit. 'Outward Bound'* schools. [*fr.* (1) *veld* (q.v.) + *vlei* (q.v.), lake, swamp *etc.*]

. . . they all come with different ideas about what to expect from Veld and Vlei, South Africa's adventure school for boys. It takes just a few painful days for them to learn that Veld and Vlei has more than deserved its reputation for being one of the toughest schools of its kind. *Panorama* Aug. 1970

Although the Veld and Vlei Adventure Courses are not linked with the British Outward Bound courses, they are modelled on them. *Het Suid Western* 12.5.76

1. vel(d)broek ['fel(t)ₗbrŭk] See *crackers.*

1. veldcraft ['feltₗkrɑft] *n.* Knowledge of the (1) *veld* (q.v.) and of its phenomena. *cf. Austral.* bushcraft, *Canad.* bushlore, bushcraft : see also quot. at *veeboer.*

Nosikaas, that magnificent specimen . . . whose unrivalled veldcraft and wonderful constitution enables him to live happily and comfortably under conditions that no white man could have endured. Pohl *Dawn and After* 1964

1. veld fever *n.* A nostalgia for the veld and open spaces ⌐A common theme in much S. Afr. poetry, Afrikaans and English. *cf. trekgees* (q.v.).

Veld fever is a malady, a longing indescribable, which comes over many South Africans, who have lived much on the veld. Mrs. Lionel Phillips *S. Afr. Recollections cit.* Pettman

1. veldkos ['feltₗkɔs] *n.* Food such as the (1) *veld* (q.v.) will furnish, fungi and other plants, occ. rodents or other small animals. see quot. at *monkey orange. cf. Austral.* bush tucker. [*Afk. fr. Du. veld* field (country) + *kost* food, victuals]

The hind flap, like the men's, was provided with a pocket, for what the Dutch call 'veld kost,' country food, as bulbs, the fruits of the mesembryanthemum, & c. Alexander *Expedition II* 1838

. . . they are lazy folk, and when game is scarce they are content to loaf while their women gather *veldkos* or dig up *uintjie* bulbs to be baked in hot ashes. Birkby *Thirstland Treks* 1936

1. veldpond(e) ['feltₗpɔnd(ə)] *n. usu. pl.* Gold Kruger coinage struck during the *Anglo Boer War* (q.v.) on an improvised mint in the veld at Pilgrim's Rest. *cf. Kruger rand* at (4) *rand.* [*Afk. fr. Du. veld sig. either* (1) *veld* or (battle)field *as in* (4) *veld* + *pond cogn.* pound + *pl. suffix -e*]

On the site is the machinery of the 'mint in the field' with which the Boers made their coins, the 'veld ponde', during the Anglo-Boer War. *Panorama* Dec. 1973

1. veld remedy *n. pl.* -ies. Remedy consisting of plant or other materials which the (1) *veld* (q.v.) provides, e.g. cobwebs to stop bleeding, numerous medicinally used plants: see also *boereraat.*

As a medicine it is interesting, that's all. And like other veld remedies you have people who swear by them . . . and others who swear at them. *Cape Times* 1.6.73

1. veldskoen ['feltₗskŭn] See *velskoen.*

1. veld sore *n. pl.* -s. A skin eruption thought to be caused either by weather conditions or dietary deficiency. *cf. Desert sore, Austral. Barcoo rot :* see at *patat.*

My 'ands are spotty with veldt-sores, my shirt is a button an' frill,
An' the things I've used my bay'nit for would make a tinker ill!
Kipling *Five Nations* 1903

2. veld *n. and. n. modifier.* Grazing land, or the grazing, indigenous or planted, supported by it: see quots. at *swaer, veld camp,* also *field. cf. Canad.* rangeland. [*from* (1) *veld*]

As the Veldt was completely done, they asked me to help them to remove to another farm. Stubbs *Remin.* 1876

Around Stutterheim the sweet-veld was not as good as the sour-veld and the protein content of the veld started falling. *Daily Dispatch* 9.3.72

2. veld *attrib. and in combination.* [See also attrib. and combination at (1) *veld;* (3) *veld;* (4) *veld*] ~ *burning* (q.v.); ~ *camp* (q.v.); ~ *cattle* (q.v.); ~ *damage* (q.v.); ~ *deterioration;* ~ *grass* (q.v.); ~ *hay* (q.v.); ~ *management* (q.v.); ~, *off the* (q.v.); ~ *reared* (q.v.); ~ *reinforcement* (q.v.); ~ *replacement* (q.v.), ~ *resting* (q.v.), ~ *sickness* (q.v.), ~ *type* (q.v.); also *berg* ~ (q.v.); (ge)broken ~ (q.v.), see ~, broken; bush ~ (q.v.); grass ~ see *veld, grass; Kar(r)oo* (q.v.) ~; *mixed* ~ see *veld, mixed;* poor ~ ; *ranching* ~ see (2) *veld, ranching; redgrass* ~, *rooigras* (q.v.) ~ ; *sand* (q.v.) ~ ; *sour* (q.v.) ~ ; *sweet* (q.v.) ~ ; *sweetgrass* (q.v.) ~ ; *thorn* (q.v.) ~ ; *vlakte* (q.v.) ~ ; *vlei* (q.v.) ~ .

2. veld, broken *n. phr.* Also *gebroken veld* (q.v.) mixed veld: grazing of more than one type together usu. bush with grass.

Look there's the broken veld ahead – mixed bush and grass veld. O.I. Farmer 1972

2. veld burning *vbl. n. phr.* Common

farming practice: burning to clear the (2) *veld* of the previous season's grass, thus making way for the new: considered by some to be detrimental to the soil, see quot. *cf. Austral. to burn off.*

In the sourveld in the North East Cape there is still considerable difference of opinion over the desirability of veld burning. Some farmers consider it necessary while others think it is unnecessary and harmful ... the greatest problem with veld burning was the heavy and continued grazing of burnt veld as soon as it started to sprout. This impoverished the grazing and increased the danger of erosion. *E.P. Herald* 4.12.74

2. veld camp *n. pl.* -s. An enclosed grazing *camp* (q.v.) as opposed to a *land camp* (q.v.) enclosed for purposes of cultivation.

There are 65 morgen ... and nine veld camps, all with permanent spring water. The veld consists of mountain veld, mixed grass veld and vleis. *Grocott's Mail* 3.3.72

2. veld cattle *pl. n.* (2) *veld-reared* (q.v.) cattle: see also quot. at *veld, off the.*

... worm remedy has been found to be completely safe for dosing calves, pregnant cows, veld cattle, feedlot cattle, show cattle. *Farmer's Weekly* 18.4.73

2. veld damage *n.* Veld deterioration (q.v.) caused by over-stocking and over-grazing: see *beaten out, eaten out, tramped out.*

2. veld deterioration *n.* Loss of quality of the veld: see (2) *veld damage*, (2) *veld-reinforcement* and (2) *veld replacement;* also *stock reduction scheme.*

Having relatively little arable land and a rapidly increasing population, South Africa cannot afford to be complacent about veld deterioration but must make every effort to halt the process. *E.P. Herald* 14.6.73

2. veld grass *n.* Any of several indigenous S.Afr. pasture grasses used commercially as baled forage: see quot. at *oulandsgras.*

2. veld, grass *n.* (2) *veld* consisting largely of grasses without bush.

2. veld hay *n.* Hay consisting of dried indigenous (2) *veld grasses:* obtainable commercially for fodder.

According to the latest statistics the department's 20 prison farms ... last year produced the following: 640 tons of bean hay, ... 2 124 tons lucerne hay, 294 tons veld hay. *Farmer's Weekly* 20.3.74

2. veld management *n.* The control of (2) *veld* by the best use of varying (2) *veld types* (q.v.): rotation and *resting* (q.v.)

of *camps* (q.v.) so as to obtain maximum grazing for stock with minimum damage to or deterioration of the (2) *veld:* see *stock reduction scheme*, (2) *veld resting*, (2) *veld reinforcement*, (2) *veld replacement*, and quot. at (2) *veld deterioration.*

... Pasture Research Officer of the Dohne Research Station ... spoke about veld management which he defined as the manipulation of grazing animals and rest conditions most beneficial to the natural vegetation for high productivity. *E.P. Herald* 3.12.74

2. veld, mixed *n.* Also *broken* or *gebroken* (2) *veld:* see quot.

My veld is all mixed – karoo grassveld (it's couch grass with karroo bush) but it's all sweetveld. O.I. Farmer Carlisle Bridge 28.5.72

2. veld, off the *adv. or predic. modifier.* Of or pertaining to *veld-reared* (q.v.) slaughter cattle sold ~ without stall feeding or other fattening. [*trans. Afk. van die veld af* not stall fed *etc.*]

The animals are off the veld and have not been pampered in any way. The bulls are hardy veld ranch bulls and have NOT been fed. *Farmer's Weekly* 21.4.72

Super grades have been obtained off the veld. *Ibid.* 30.5.73

2. veld, ranching *n.* (2) *veld* or farming land suitable for cattle ranching or occ. *game ranching* (q.v.): see also *stock farm* and quot. at *veld, off the.*

2. veld-reared *modifier.* Of or pertaining to cattle reared solely on grazing (2) *veld* without stall-feeding, feedlot fattening or other aids to rapid weight-gain: see *veld cattle, off the veld* used also of a horse not accustomed to being stabled, see quot.

Their horses, having survived the trek, were as tough as their owners, true Boer horses, of Arab and Basuto strains, small, veld-reared beasts, which could forage for themselves. Cloete *Turning Wheels* 1937

2. veld reinforcement *n. phr.* Methods of conservation of thin, damaged or depleted veld consisting in planting usu. indigenous pasture grasses, *spekboom* (q.v.) or other nutritious bush between existing grazing: see (2) *veld deterioration, management, replacement* etc.

2. veld replacement *n. phr.* Replanting of *veld* which has been *eaten out, tramped out, beaten out* all (q.v.), or eroded, with soil-binding grass or bush: see (2) *veld reinforcement.*

2. veld resting *vbl. n. phr.* Various methods

of (2) *veld* conservation usu. by means of rotational grazing leaving certain *camps* (q.v.) unused for periods of up to one year so that the (2) *veld* may recover: see also *rest, stock reduction scheme,* and quot. at (2) *veld management.*

It was shown that about 50 per cent of farms had fewer than 10 camps, that only 17 per cent adhered to a controlled rotational grazing system, and that more than 30 per cent of the farmers applied no form of veld-resting. *Farmer's Weekly* 21.4.72

2. veld sickness, veldsiekte (ˈfeltˌsĭktə] *n.* A disease of sheep and cattle resulting in serious loss of condition and even death (from 'scours' or purging) when animals are moved from one type of veld to another. [*Afk. siekte fr. Du. ziekte* disease]

Animals brought from sweet veld suffer from what is called veld-sickness, which results from insufficient nutrition and the hard and irritating nature of the food consumed. Wallace *Farming Industries* 1896 *cit.* Pettman

2. veld type *n. pl.* -s. The variety of grazing borne by a particular soil type. ⫫There are numerous ~s esp. *sweetveld, sourveld, broken veld, bushveld, grass-veld, hardveld, kalkveld, rooigrasveld* etc. etc.

Where 2 distinct types of veld occur on the same farm, paddocking should be arranged in such a way as to keep them separate, and the sour veld grazed during spring and summer, while the sweet veld is kept until the seed has fallen. *H'book for Farmers* 1937

3. veld *n.* Region, country: found in S. Afr. place names usu. of districts e.g. Veld-drif(town), Bokkeveld, Highveld, Kaokoveld (S.W. Africa); Lowveld, Middleveld, Roggeveld, *Sandveld* (q.v.) *Strandveld* (q.v.) *Zuurveld* (q.v.) etc.

... Lichtenstein's travels to the north – that gentleman and his party having come by a route across the Bokkeveld and the great Karroo, and returned in the same track. Thompson *Travels I* 1827

4. veld- *n. prefix.* Field-, also district-: *prefix* denoting administrative and/or military rank as in ~*cornet* (q.v.), ~*corporal* (q.v.), ~*commandant,* ~*kommandant* (q.v.), ~*marshal,* ~*wagt-meester* (q.v.). See also (1) ~*ponde.*

4. veld-commandant, veld-kommandant [ˈfeltˌkɔmanˈdant] *n. pl.* -s. *hist.* Officer in charge of a commando.

20 Mar. 1812: They (the commando) wear no uniform, but are divided into squadrons under the command of a veld-commandant, who is also

a boor, nominated by the Government and who at all times retains that title, and with it a rank superior to that of *veld cornet.* Burchell *Travels II* 1824

4. veld-cornet, veld-kornet [ˈfeltˌkɔ(r)ˈnet, ˈkɔnɪt] *n. pl.* -s. *hist.* see also *field cornet :* an administrative official, similar to Brit. District Officer, answerable to the *land-drost* (q.v.) and with certain military duties.

The Veld-Cornet is a sort of petty magistrate, empowered to settle little disputes within a circuit of fifteen or twenty farms, to punish slaves and Hottentots, and to call out the burghers, over whom he presides in the public service, and act as their officer on Commandoes. Thompson *Travels II* 1827

4. veld corporal *n. pl.* -s. *hist.* Pseudo-military rank: see quot.

... the *land-drost* has appointed one of the farmers, with the title of *veld-corporal,* to command in these wars, and as occasion may require, to order out the country people ... for the purpose of defending the country against its original inhabitants. Sparrman *Voyages II* 1786

4. veld wachtmeester [ˌfeltˈvaxtˌmɪəstə(r)] *n. pl.* -s. *hist.* The official 'pacer' of land before surveying instruments were in use: see quot. at *baken.* [*etym. dub. Du. veldwachter* village policeman, *veld* field + *wachten* to wait, guard + *meester cogn.* master, *Afk. veldwagter* ranger]

There were no surveyors nor surveyor's implements in the country, and the "official pacer" or veld wagtmeester could take an extra long step when measuring for a friend or favourite. Macdonald (no title) 1890 *cit.* Pettman

... the surveyor in Potchefstroom who was mobbed because he measured land with some other sort of instrument, instead of pacing it out like the old veld-valktmaster? Brett Young *City of Gold* 1940

velkombers [ˈfelkɔmˌbers, -ˌbɛə(r)s, -ˌbɑ:(r)s] *n. pl.* -e, -en. A blanket made of skins: see also *ingubu* and (2) *kaross.* [*Afk. fr. Du. vel* skin, hide + *kombaars* (ship's) blanket]

I shall sleep in mine, and dream of African hill-sides wrapt in a 'Velkombaars'. Duff Gordon *Letters* 1861–2

Colin Fraser, ... was asleep one night under his *velkombers* (sheepskin blanket). Green *Karoo* 1955

velskoen [ˈfelˌskŭn] *n. pl.* -s, -e. *lit.* 'Hide shoe'. [*Afk. fr. Du. vel* hide + *schoen cogn.* shoe *cf. schoon*] **1.** Formerly a handmade rough shoe of untanned hide sewn without nails, thought to be first made by the Hottentots before the arrival of the white man: see second quot. cf. *Canad. moccasin, greenhide(s)*.

His unaccustomed feet are stuffed into stiff, shiny-leather boots, instead of his dear, old, easy-going 'Veldtschoons' of home manufacture. Roche *On Trek in Tvl.* 1875

. . . their unstockinged feet protected by handsewn veldschoens, the equivalent of Canadian moccasins. Mockford *Here are S. Africans* 1944

Another feature . . . is that the 'velskoene', or shoes made of raw hide, worn by his Hottentot or Malay figures always show conspicuous stitches round welts. Gordon-Brown *Artist at Cape of G.H.* 1965

In combination ~*s maak vb phr.* or *making* ~*s prob. obs.* The spending of a period of enforced inactivity by a countryman in a town awaiting and during the confinement of his wife.

Because the father, waiting for the happy event, would while away his time by making velschoens, this particular excursion was called 'making velschoens'. Jackson *Trader on Veld* 1958

2. Now any type of rough suede ankleboot or shoe usu. with a light rubber sole, similar to *Anglo-Ind. chukka boot* etc.

. . . welted Veldtschoens for Golf. Hong Kong Sports Shop Victoria, Oct. 1972

3. A 'hop' or dance in the country: see *sheepskin* (2) *opskud* and quot. at *vastrap* cf. *Canad. moccasin-dance.* [*presum. fr. country footwear*]

4. *prefix* to certain plant names sig. a likeness to soles, tongues, etc. of ~*s;* ~*blare*, leaves of *Haemanthus coccineus;* ~*klappe* (flaps, tongues) leaves of *Massonia candida;* ~*sole* (soles) any of several species of both *Haemanthus* and *Massonia*.

Venda ['venda, -ə] *n. pl.* Ba, -s. **1.** A Bantu people of the N. Transvaal, and their language: see quot. at *malombo :* also a single member of this people.

There was also laughter when he said Vendas who wanted a change to be made quickly in the development administration of their territories should leave Vendaland. *Rand Daily Mail* 28.7.71

2. The *homeland* (q.v.), also called ~ *land* of the ~ in the N. Transvaal.

Venda received self-governing status on the same day as did Gazankulu in terms of Proclamation 12 of 26 January 1973. Its seat of government is Sibasa, and Venda was recognized as an official language. Horrell *Afr. Homelands of S.A.* 1973

vendue ['vendju, 'vɔ̃ndy] *n. hist.* Used as equiv. of *vendusie* (q.v.) esp. by Eng.-speaking colonists. [*Du. vendu cogn. Fr. vendue fr. Lat. venditio* sale]

16 February held a Vendue to day at Elephant Fountain of my Buck Flock . . . things sold well upon the whole Vendue Roll amt. to about 540£ 17–9. Collett *Diary II* 1842

In combinations ~ *master*, ~ *meester*, an auctioneer ; ~ *clerk*, ~ *roll* (see quot. above), ~ *list*, ~ *accounts*, ~ *note* etc. cf. *Jam. Eng. vendue room*, a sale room for slave auctions.

. . . the vendue clerks . . . shall not be allowed to bid . . . That also the vendue masters . . . shall not themselves come forward at the sale. *Cape Statutes* 1862

Vendue Notes, or Vendue Accounts, including all other documents relative to public sales. . . . Vendue lists of immovable and movable property . . . Licences for the Vendue Masters. *Ibid.* 1862

vendusie [ˌfen'dysĭ, ˌven-] *n. pl.* -s. A public auction sale of goods or stock. [*Afk. fr. Du. vendutie*, auction sale *fr. Lat. venditio* sale]

In South Africa the sale of goods by auction or 'vendusie' has prevailed from early times. Botha *Our S.A.* 1938

ver- [fe:r, fɜə(r)] *adv.* and *adj.* Far: found in place names as *adv. prefix* in e.g. Vergeleë, Vergelegen, Verkykerskop (telescope), Vergenoeg (enough): or as *adj.* equiv. 'distant' in Verberg: see also *far.* [*Afk. fr. Du. ver cogn.* far]

verdiend, verdient [fə(r)'dĭnt] *vb. partic.* Deserved: found in place names e.g. Welverdient, Goedverdient, Nietverdient. [*Afk. fr. Du. verdiend partic. of verdienen* to earn, deserve, merit]

verdomde [fe(r)'dɔmdə] *partic. modifier. slang.* Damned, accursed: formerly in freq. combinations ~ *Engelsman*, Englishman, ~ *uitlander* (q.v.), ~ *rooinek* (q.v.). [*Afk. fr. Du. verdoemd* damned *cogn. O.E. dōm* doom, judgement]

I answered the usual interrogations as from whence I came and whither I was going, but was somewhat surprised when, after enquiring my name he asked whether I was a verdomde Engelsman. Mockford *Here are S. Africans* 1944

-verdriet [fə(r)'drĭt] *n.* Sorrow, grief: found in place name Langverdriet, also in *slang phr.* dronk ~, 'alcoholic remorse' ; see *babbala(a)s.* [*Afk. fr. Du. verdriet* sorrow, distress]

verdwaal- [fə(r)ˈdwɑːl] *vb.* Stray, be lost: in place name Verdwaalkloof: see also *dwaal*. [*Afk. fr. Du. verdwalen* to lose one's way]

verkramp [fə(r)ˈkramp] *adj.* Bigoted, ultra-conservative, orig. in a political sense, now also more widely applied. [*coinage, prob. rel. Afk. bekrompe* narrow minded *fr. Du. vb. bekrimpen,* to shrink, restrict, *partic. bekrompen + attrib suffix. -te and n. forming suffix -heid sig.* -hood *fr. O.E. -had* state of being]

Verkramptes are very much in the news these days, especially in Oudtshoorn where all three candidates, in the by-election caused by the resignation of . . . are said to be 'Verkramp' so I have designed a . . . quiz . . . to find out how 'Verkramp' YOU are. *Het Suid Western* 16.3.72
Also as *adj.* in *attrib. form* ~ *te,* of or pertaining to people, policies, ideas etc.
. . . the Afrikaner-dominated National Party . . . faces continuing tension between its moderate verligte (enlightened) and archconservative verkrampte (narrow-minded) wings. *Time* 1.5.72
Derivatively as *n.* with *n. forming suffix* also *-te;* see *verkrampte,* and as *n. abstr.* ~ *theid,* the state of being ~ ; see also *verlig(te).*
He said this was a definite choice in favour of verkramptheid versus verligtheid. *E.P. Herald* 30.1.74

verkrampte [fə(r)ˈkramptə] *n. pl.* -s. An ultra-conservative person: formerly applied to politicians, now extended to any one holding narrow-minded or bigoted social or religious views. *cf. Canad. mossback, Austral. wowser.* [*Afk. coinage verkramp + n. forming suffix -te*]

LONDON. – 'The Times' yesterday used the Afrikaans word 'verkramptes' to describe the Protestant Unionist extremists in Northern Ireland. It was the first notable use of the word in English to describe people in a situation outside South Africa. *Rand Daily Mail* 19.9.69
In combination *super* ~, *arch* ~ etc.
. . . campaign being waged by a band of super-verkramptes against certain Afrikaans financial institutions in general. *Sunday Times* 12.3.73.
His image underwent a subtle change from that of arch-verkrampte some years ago to one of pragmatic conservatism today. *Argus* 16.9.72

verlaten- [fə(r)ˈlatən-] *partic. modifier.* Deserted, isolated, forsaken: found in S. Afr. place names Verlatenvlei, Verlatenheid (desolation), Verlatenkraal: see also *verloren*. [*Afk. fr. Du. vb. verlaten* to leave, abandon *etc.*]

verlep [fə(r)ˈlep] *adj. colloq.* Wilted, of or pertaining to flowers and vegetables:

figur. of persons, flabby, under the weather etc.: see also (2) *pap, oes.* [*Afk. fr. Du. verleppen* to wither]
The lettuces don't look edible – they're pretty verlep. *O.I.* 1973

verlig [fə(r)ˈlɪx] *adj.* Enlightened, broadminded: formerly in a political sense only, now more widely applied: also in *attrib. form* ~ *te :* see also *verkramp(te).* [*Afk. fr. Du. verlichten* to light, enlighten *+ attrib. suffix -te + -heid abstr. n. forming suffix cogn.* -hood *fr. O.E. -had* state of being]
Professor de Klerk is the creator of the now famous words, verlig and verkramp. He is editor of the independent verligte Calvinistic monthly Woord en Daad. *Sunday Times* 25.2.73
deriv. as *n.* see verligte and as *n. abstr.* ~ *theid* the state of being ~ : see quot. at *verkramp.*

verligte [fə(r)ˈlɪxtə] *n. pl.* -s. One who is enlightened or broadminded: see *verlig.* [*Afk. fr. Du. verlichten* to light, enlighten *+ n. forming suffix -te*]
Although candidates were standing as individuals and there was no official line up, students were aware of the split between the verligtes and verkramptes. *E.P. Herald* 25.10.71

verloor [fə(r)ˈlʊə(r)] *vb.* Lose: in place name Moedverloor (despondency). [*Afk. fr. Du. moed* courage, heart, spirit *cogn. O.E. mōd* (heart etc.) *+ Afk. vb. verloor* lose *fr. Du. partic. verloren* to be lost]

-verloren [fə(r)ˈlʊərən] *vb. partic. suff.* Lost: found in place name Allesverloren: see also *verlaten-, -verloor.* [*Du. partic. verloren* lost *cogn.* forlorn]

verneuk [fə(r)ˈnɪœk] *vb. trns. colloq.* To cheat, swindle. [*Afk. fr. Du. verneuken* to cheat, swindle, deceive]
'Ach', she sighed, 'to think that Gert Kleinhouse, whose nose I used to wipe with a lappie, has verneuked me out of a lovely little pig.' Cloete *Turning Wheels* 1973
deriv. as *n.* ~ *ery* [~əˌreɪ]. Cheating or dirty dealing: also ~ *er* [~ə(r)]. One who swindles or deceives usu. in money matters, hence *boer* ~ *er* (q.v.) a dishonest dealer trading on the ignorance of simple country people, often exchanging worthless goods by barter for stock or produce: see also quot. at *boer.*
. . . When the richness of the Kimberley mine was an undisputed fact . . . thousands of eager ones left useful occupations to try their

luck . . . trekkers from the Transvaal, Dutchmen from the Free State, planters from Natal, Boer verneukers from Cape Town. Cohen *Remin. of Kimberley* 1911
and . . . a German share-dealer . . . waxed exceeding wroth; he had lost a sovereign. 'What is this *verneukery?*' he exclaimed. Cohen *Remin. Johannesburg* 1924

versterkdruppels [fə(r)ˈsterkˌdrœpəls, -ˈstɛə(r)k-] *pl. n.* One of the *Dutch medicines*, a tonic: see also quot. at *Old Dutch Medicines*. [*Afk. fr. Du. versterken* to strengthen, fortify + *druppel* drop + *pl. suffix -s*]
It will be a transformed countryside indeed when the last bottle of 'Versterk Druppels' is sold. Green *Land of Afternoon* 1949
'Calm yourself, my old husband,' said old Susanna, handing him a small glass of wine and versterkdruppels. He drank it, settled himself in bed again and closed his eyes sulkily. Louw *20 Days* 1963

verwacht See *verwag*.

verwag [fə(r)ˈvax(t)] *vb. partic.* Expected, awaited: found in place names freq. of farms usu. preceded by an *adv.* e.g. Langverwacht, Goedverwag, Nooitverwacht, or *neg.* Onverwacht, Onverwagslaagte. [*Afk. fr. Du. verwachten* to expect, look forward to]

vetkoek [ˈfetˌkŭk] *n. pl.* usu. ∅ or -ies. A cake usu. of yeast dough, deep-fried and similar to a doughnut though freq. not sweetened ; see also quot. at *roosterkoek*. [*Afk. fr. Du. vet cogn.* fat + *koek cogn.* cake (+ *dimin. suffix -ie*)]
Everything you make with . . . [yeast] – bread, rolls, cakes, vetkoek, and ginger beer – is good for you too. *Fair Lady* 30.10.69
Also *dimin.* ~ *ie*.
I also recall . . . 'mosbolletjies' . . . 'melk tert' . . . 'vet koekies' (a type of doughnut), and many other specialities of the Old South African kitchen. Jackson *Trader on Veld* 1958

vetsak[1] [ˈfetˌsak] *n. colloq.* A fatty, 'fatso': a mode of address or reference to a fat person, often offensive: see also (1) *mafuta*. [*Afk.* fatty, 'fat-guts', *vet cogn.* fat + *sak cogn.* sack]
That healthy baby is a real little vetsak. O.I. 1970

Vetsak[2] *n. prop.* Trade name of co-op selling farming equipment, fertilizers, feeds etc. [*acronym* Vrystaat en Transvaalse Sentrale Aankoops Koöperasie Beperk (Ltd)]

Vierkleur [ˈfiːrˌklïœ(r), ˈfɪə-] *n. prop.* The four-coloured flag of the old

Transvaal Republic (*Zuid-Afrikaansche Republiek* (q.v.) with three horizontal stripes of red, white and blue with a vertical one of green on the left side. [*Afk. fr. Du. vier* four + *kleur cogn.* colour]
. . . on April the twelfth, eighteen hundred and seventy-seven, the Proclamation of Annexation was read, amid cheering crowds, at Pretoria, while the Vierkleur was hauled down and the Queen's flag hoisted . . . folded in his saddle bag, the Vierkleur flag – green, white, red and blue – of Burger's design. Brett Young *City of Gold* 1940

vies [fis] *adj. slang.* Angry, disgusted. [*Afk. fr. Du. vies* nasty, loathsome]
Ja! See what I mean. This time I'm laughing, and you . . . ! Vies! You don't like it when somebody else laughs. Fugard *Boesman & Lena* 1969

vine stalk *n. pl.* -s. Vine: a (rooted) grapevine cutting for planting. [*prob. translit. Afk. wingerd-stok*]
. . . he started a model *erf* at Kakamas where the people could study the best methods and buy the right fruit trees, vine stalks and seeds. Green *Glorious Morning* 1968

vingerpol [ˈfɪŋə(r)ˌpɔl] *n.* Any of several species of *Euphorbia* usu. *E. caputmedusae*, also *E. gorgonis*, *E. pugniformis*, *E. clavaroides*, which have a bunch of strange finger-like growths of striking appearance. [*Afk. fr. Du. vinger cogn.* finger + *pol* tuft, tussock]

vingertrek [ˈfɪŋə(r)ˌtrek] *n.* The game of 'finger hooks' consisting of a tug-of-war between the linked forefingers of two players. [*Afk. fr. Du. vinger cogn.* finger + *trekken* to pull, haul]
The children of South Africa . . . have always played all manner of singing, dancing and marching games, leap-frog, hide and seek, *vinger trek*, and prisoner's base, for which no equipment is needed. De Kock *Fun They Had* 1955

vink [fɪŋk] *n. pl.* -s. Also *fink* (q.v.): any of numerous weaverbirds (Ploceidae fam.) including the *kaffir* ~ (q.v.), the *red bishop bird, rooi* ~ etc., but esp. the *geel* ~, yellow weaver bird the pendent nests of which are a common sight esp. over water. [*Afk. fr. Du. vink cogn.* finch]
. . . at this drift the willows are very big . . . The water is very deep here. Vinks nests hang all over the river. They look like little baskets. Vaughan *Diary circa* 1902

vinkel(bos) [ˈfɪŋkəl(ˌbɔs)] *n. Foeniculum vulgare,* the feathery-leaved fennel, the aromatic seeds of which can be used as a flavouring similar to aniseed, or chewed whole ; in Indian cookery known as *somf* or *saunf*, various sp. forms. [*Afk. fr. Du. venkel cogn.* fennel]

Vinkel, another famous shrub, is guaranteed to keep fleas away if you place a branch under your bed. Green *Tavern of Seas* 1947

vis [fɪs] *n.* Fish, common *suffix* to names of fish e.g. *stok~*, *leer~*, *wilde~* etc., also as *prefix* in place names Visrivier, Vishoek etc. [*Afk. fr. Du. vis(ch) cogn.* fish]

vlakte [ˈflaktə] *n. pl.* -s. **1.** Open plains, flat land. *cf. U.S. prairie.* [*Afk. fr. Du. vlakte* plain, level, flat]

Two years later, Captain Cornwallis Harris, an army engineer who was also a hunter and artist, crossed the same vlaktes, the prairies of South Africa. *Daily Dispatch* 29.7.72

Plain: in place names Sandvlakte, Kaapse Vlakte etc.

2. *n. prefix.* Plain: in several plant names, usu. sig. preference for this habitat e.g. *~ alwyn*, see aloe; *~ anyswortel Annesorrhiza capensis*, see *suuranys*; *~ suring Oxalis lawsoni; ~ vygie Nananthus aloides* etc.

vlam [flam] *n.* Methylated spirit, term used among down-and-outs: see *outies* and quots. at *arm* and *horries* [*Afk. vlam cogn.* flame]

. . . their own liquor: methylated spirits variously termed juice, mix or *vlam*. Of all the outies I met only one . . . did not drink *vlam*. It's cheaper than wine and more potent than any spirits you can buy . . . One bottle might kill a man who isn't used to it. One bottle keeps an outie going for a few hours . . . My outie friends hastily advised me not to touch *vlam* if I wasn't on it. *Family Radio & TV* 23-30.1.77

In combination *~ drinker cf. Austral. metho.*

. . . methyl alcohol, which is poison. Vlam drinkers invite blindness as well as the severe liver and intestinal damage that afflicts alcoholics. *Ibid.*

vlakteveld [ˈflaktəˌfelt] *n.* Flat land, plain. [*Afk. fr. Du. vlakte* plain, level, flat + *veld* field, country]

vlei [fleɪ] *n. pl.* -s. Lake, swamp etc. variously used. [*Afk. fr. Du. vallei cogn.* valley] **1.** A large shallow lake or swampy piece of ground (see *~ veld*, *~ land*, *~ ground*) a depression where rain collects in the rainy season etc.: also *U.S. vly* and *vlaie* (swamp): see also *pan*, and quot. at *ruigte*.

Having come to a small pond or vley, the water in it was so thick that the men had to keep their teeth closed to act as strainers. McKay *Last Kaffir War* 1871

In combination *~ grass, ~ ground, ~ lands, ~ lily/lelie, ~veld* all (q.v.) also *talking ~s.*

Along the Kalahari edge . . . are so-called 'talking vleis', little ponds left after the rare rains. Some gurgle and rumble, the strength of the sound ranging from a whisper or a moan to a shriek . . . the natives have reason to dread the 'talking vleis', for there are some where quicksands form during the rainy season. Green *These Wonders* 1969

2. Found in names of sheets of water, even quite large lakes e.g. Swartvlei, Langvlei, Zeekoevlei, Groenvlei, Ruigtevlei etc. *cf. Canad. marais* in place names sig. swamp, marsh.

3. Prefix to numerous plant names sig. a preference for this habitat e.g. *~ aandblom Gladiolus concolor*, see *aandblom; ~ biesie*, any of several reeds or sedges ; *~ gras* (q.v.) ; *~ kos*, see *wateruintjie; ~ lelie/lily* (q.v.) ; *~ pypie*, see *-pypie; ~ sewejaartjie, Helichrysum foetidum*, see *sewejaartjie; ~ tulp, Moraea glauca (geeltulp)*, see *tulp.*

vleigras(s) [ˈfleɪˌxras] *n.* Any of numerous grasses growing in *vleis* (q.v.) esp. *Echinochloa holubii.*

Excellent red and vlei grass grazing with carrying capacity 1 500-2 000 sheep and 200 head of cattle. *Farmer's Weekly* 21.4.72

vlei ground [ˈfleɪ ˌgraʊnd] *n.* Also *vleigrond:* swampy ground where waterloving plants flourish: see also *vleiveld* and *vleiland. cf. Austral. mickery/ie*, marshy ground, *Canad. muskeg soil/ country* bog, swamp.

. . . only a little way off we came into dry vlei ground where there were few trees and the grass stood about waist high. FitzPatrick *Jock of Bushveld* 1907

vleiland [ˈfleɪˌlænd, -ˌlant] *n.* Cultivated swamp *lands* (q.v.) usu. highly productive on account of extra moisture.

Approximately 140 morgen of lands, 70 morgen under established lucerne – all good vlei lands with abundant stock water from permanent fountains and borehole. *Daily Dispatch Advt.* 11.3.72

vlei lily, vlei lelie [ˈfleɪ ˌlɪəlĭ] *n. pl.* -ies. Any of several species: *Galtonia princeps; Crinum campanulatum; Nerine frithii* and *N. laticoma*, the *prefix vlei-* sig. preference for a damp or marshy habitat.

Vlei Lily . . . This handsome lily-like plant

lives in damp places (vlei in Afrikaans, dambo in Nyanja) and has an enormous bulb which is easily transplanted. Hoyle *Flowers of Bush* 1953

vleiloerie ['fleɪˌlurĭ] See *loerie*, also *rain-bird*.

vleiveld(t) ['fleɪˌfelt] *n.* Land, semi-water-logged in the rainy season, usu. highly productive of excellent fodder for stock: see *vlei ground, vleiland.*

This farm is well known for its high carrying capacity and vleiveldt. There is also exceedingly good grazing for sheep. *Farmer's Weekly Advt.* 3.1.68

vleis [fleɪs] *n. and n. prefix and suffix.* Meat: usu. in combinations ~ *pastei* (see pastei), *braai* ~ (q.v.), *pap-en-* ~. [*Afk. fr. Du. vlees* dial. *vleis cogn.* flesh, *Ger. Fleisch*]

He flashed that infectious grin. 'The food was quite different from my daily diet of vleis and mieliepap.' *Sunday Times* 15.4.73

. . . 800 . . . delegates and guests forsook the cuisine of the upper floors for good old 'pap en vleis' braaied in the hotel's parking basement. *E.P. Herald* 16.5.74

vlek [flek] *vb. trans.* To 'gut' or 'clean' a fish or a carcass: also *fleck* (*out*): [*Afk. vlek* to gut, flay *prob. fr. Middle Ned. vlekken* to cleave, split open]

All night a party remained by it to cut and 'vlek' the meat, for carrying off a quantity of it. Alexander *Expedition II* 1838

The fish has a firm but delicate flesh which rapidly goes soft . . . unless it is 'flecked' i.e. cut down the back and backbone and entrails removed. *Farmer's Weekly* 18.4.73

. . . hunched over the . . . carcass, skinning it out using the flecked-out hide to keep the raw meat clean of sand. Stander *Flight from Hunter* 1977

vlossie ['flɔsĭ] *n. pl.* -s. *Sect. Army.* A Hercules transport aircraft, usu. *flossie.*

The men on the border have a language of their own: an aircraft is a 'vlossie', a landrover a 'gerry' and a helicopter a 'lawnmower'. *Panorama* Jan. 1978

V.O.C. ['viˌəʊˈsi:] *n. prop.* **1.** The Dutch East India Company (see *etym.*) which was founded in 1602, went bankrupt in 1794 and disappeared in 1795 as a result of the revolution: The *Company* (q.v.) was responsible for the original settlement at the Cape: see quot. [*acronym* Verenigde Oostindische Compagnie]

On April 6th, 1652, Commander Jan van Riebeeck, a ship's doctor employed by the V.O.C. (Dutch East India Company), and his retinue first set foot on the shores of Table Bay to establish a victualling station at the Cape for ships en route to the East. *Panorama* May 1971

2. The monogram of the initials of the *Company* (q.v.). *V.O.C.* found in various forms, sometimes entwined, on silver, porcelain and glass, formerly the property of the Company, and on coins issued during its regime: as modifier ~ *monogram/mark,* ~ *glass,* ~ *Arita ware,* ~ *platter* etc. etc. [*see* (1) *V.O.C.*]

Dutch Silver Beakers in the Groote Kerk, Cape Town, probably late 17th century. One bears the V.O.C. monogram of the Dutch East India Company. Gordon-Brown *S. Afr. Heritage* 1965

V.O.C. is the monogram of the Dutch East India Company (Vereenigde Oost Indische Compagnie). This symbol of the Dutch East India Company's authority . . . appears on the gateway to that most important historic building in South Africa, the Castle of Good Hope in Cape Town. The VOC mark was decided upon by the Here XVII, directors of the mighty trading organisation as far back as 1603 when they insisted that these initials appear on all armaments purchased for their ships. *E.P. Herald* 5.9.74

voëi ['fʊəl] *n. usu. suffix.* Bird: suffixed to numerous usu. *Afk.* names of birds e.g. *bosluis* ~ (see *tickbird*); *brom* ~ ; *sprinkaan* ~ see also *locust bird; kou* ~ ; *spook* ~ etc. all (q.v.). [*Afk. fr. Du. vogel cogn.* fowl]

voëlvry ['fʊəlˌvreɪ] *adj.* Outlawed. [*Afk. fr. Du. vogelvrij verklaren* to outlaw, declare outlawed]

During the Boer War he was declared voëlvry by proclamation, which means he had to be shot on sight. He was actually an Englishman who fought with the Boers. O.I. 1972

voer [fu:r] *n.* Forage, fodder, animal feed. [*fr. Afk. vb. voer* to feed, *n. voer cogn.* fodder]

Cape Early rye is grown either alone for grain or together with oats to form 'voer' or grain feed normally fed to mules in the grain areas. *H'book for Farmers* 1937

voerchitz, voerchits ['fu:rˌʃits, -ˌsɪts] *n. hist.* Former sp. forms of *voersis :* see also *sis, Duitse sis,* cotton material, printed or plain, usu. sold in lengths − 'cotton gown pieces': see quot. at *bafta.* [*prob. fr. Du. voeren* to line (a garment) + *chits(z) fr. sits* chintz]

. . . Baizes, Flannels, Fancy Cambric Voerchitz Prints and Black Voerchitz, sheetings, Skirtings white and brown Punjums, white and brown Baftas. *Cape of G.H. Almanac Advt.* 1856

voersis ['fu:rsɪs] See *voerchitz, voerchits.*

voertsek [ˈfŭrtˌsek] See *voetsek, voetsak.*
voet, se See *se voet.*
voetganger [ˈfŭtˌxaŋə(r)] *n. pl.* -s. A
pedestrian. **1.** A locust at the 'hopper'
stage: see also (2) *rooibaadjie.* [*Afk. fr.
Du. voet cogn.* foot + *gang* passage,
gait *cogn. gaan* go + *agent. suffix -er*]
   . . . the said occupier shall define as nearly
as may be the locality on his land where flying
locusts have appeared or are depositing or have
deposited eggs or where voetgangers have
appeared. *Union Statutes* 10 June 1957
2. *colloq.* A pedestrian (usu. sig. vagrant)
or one without motor transport.
   In some ways the voetganger of a century
ago fared better than the modern tramp who
rides in limousines and covers a thousand miles
a week. Green *Land of Afternoon* 1949
voet-in-die-hoek [ˈfŭt ɪn (d)ĭ ˈhŭk] *adv.
phr. Sect. Army. lit.* 'Foot in the corner',
sig. with accelerator flat on the floor –
usu. *to drive* ~, to go flat out, or as an
imp. interj. of encouragement to a
driver. *O.I.s S.S.B. or Transport Service-
ment* 1979. [*Afk. voet cogn.* foot + *in
die cogn.* the + *hoek* corner]
voetjie-voetjie [ˈfŭĭkĭˌfŭĭkĭ, -cĭ] *n.* Usu. in
phr. *to play* ~ : to make non-verbal
contact (with the feet) either by courting
couples, by two persons who wish to
share a private joke or by those who
wish, for some reason, to signal un-
obtrusively to each other. [*Afk. fr. Du.
voet cogn.* foot + *dimin. suffix -je*]
   . . . a boy and a girl had each kicked off one
shoe and they were playing the delightful South
African dinner-table game known as 'voetjie-
voetjie'. *Het Suid Western* 1.4.71
voetpad [ˈfŭtˌpat] *n.* A rural servitude,
a right of way for persons on foot: term
in Roman Dutch Law: see also *trekpad,
vuispun⁰.* [*Afk. fr. Du. voet cogn.*
foot + *pad cogn.* path]
   Rights of Way. These include the following
varieties: Iter, footpath, voet pad; the right of
walking across the land of another. Wille
Principles S. Afr. Law 1945 ed.
voetsek, voetsak [ˈfŭtˌsek, -ˌsæk, -ˌsak]
interj. vb. usu. imp. (*n.*) **1.** A rough
command to be off, go away usu. to a
dog: offensive applied to a person:
numerous sp. forms including anglicized form *footsack* (q.v.). [*Afk. voertsek, voertsik fr. Du. voort* forward,
away, *seg ik* say I]
   Dogs attacked us as we approached; but on
the cry of 'voortzuk!' from the master, followed

by a stone, they left us. Alexander *Western
Africa I* 1837
   '. . . and I told him my name was Naude, and
asked him how he was,' At Naude said. 'He
told me to voetsek.' . . . Old Lemare was telling
the Indian to voetsek, Jurie Steyn said. Bosman
Jurie Steyn's P.O. 1971
2. *n. prop.* See quot. [*fr.* (1) ~]
   For all South African springs seem originally
to have been called Muddy Fountain just as
according to the humorist Leonard Flemming
all South African dogs are called 'Voetsak'!
Birkby *Thirstland Treks* 1936
3. *colloq.* Used as an indefinite numeral
usu. in phr. *in the year* ~, 18 ~, 19 ~ etc.
[*fr.* (1) *voetsek*]
   At least twice a week I get phone calls asking
who won the world racing championship in
19-voetsak. *Daily Dispatch* 16.6.73
   Whatever agreement Mr – made with some
nineteen-voetsek Council of the past . . .
Het Suid Western 15.12.76
voetstoots [ˈfŭtˌstʊəts] *adv.m. or modifier.*
As it stands: Roman Dutch term in law
of sale sig. with 'all defects latent or
patent' relieving the seller of liability
for faults in the thing sold. *cf. colloq.
Brit.* 'as is'. [*Afk. fr. Du. phr. met de
voet te stoten* to push with the foot]
   When the article is sold Voet Stoets the seller
on his side cannot complain if the article turns
out better than either party thought. Belcher
Law of Sale in S.A. 1961
   The trailer will be sold voetstoots without
registration documents. *Farmers' Weekly* 30.5.73
   As modifier ~ *sale,* ~ *clause* etc. or
predic. sale is ~ .
   Voetstoots Sales. A sale is said to be voetstoots when . . . Wille *Principles S. Afr. Law*
1945 ed.
   A so-called 'Voetstoets' clause does not
protect the seller against fraud or the non-
disclosure of a material defect of this nature.
Argus (Property) 5.6.71
voetstoof(stofie) [ˈfŭtˌɒtʊof(ĭ)] *n. pl.* ɒ. Soo
konfoor (2), also *stofie* and quots. at
tessie [*Afk. fr. Du. voetstoofje* foot-
warmer]
volbek [ˈfɔlˌbek] *modifier.* Also in form
full mouth : of or pertaining to a sheep
which has cut its full complement of
teeth and therefore usu. between 2½ and
3 years old: see quots. at *wissel, unwisselled.* [*Afk. fr. Du. vol* full + *bek*
(q.v.) mouth *cogn.* beak]
Volk [fɔlk] *n.* freq. capitalized. **1.** The
people usu. *Die* ~ sig. the Afrikaner
people, freq. only Afrikaner nationalists: see quot. [*Afk. fr. Du. volk* people

cogn. folk, O.E. *folc*]
The Afrikaner Volk proceeded to find itself along cultural and then along political lines. The Old Colony led the way. Walker *Hist. of S.A.* 1928
For instance, *die volk* is a paternalism when used in phrases such as *die volk werk op die land*; yet in Party slogans such as *Vir Volk en Vaderland!* it becomes the highest expression of Afrikaner Nationalism. *Cape Times* 19.5.73
In combination ~*seie*, ~*sleier*, ~*slied*, ~*swil*, ~*spele*, ~*sraad*, ~*svreemde*, all (q.v.). *Boere* ~ see quot. at (3) *nation.*
2. Labourers, usu. farmhands: see quot. above: also *volkies* (q.v.) in combination ~*shuisie*. [*Afk. fr. Du. volk* people *cogn.* folk + *huisie fr. Du. huisje* small house, cottage]
No old fashioned Cape farmer would have thought this . . . picturesque enough for his labourers. The traditional volkshuisie at least had a peaked roof, a couple of simple gables and the sort of commodious chimney piece. *Cape Times* 5.10.73

volkies ['fɔlkĭs] *pl. n. usu. Coloured* (q.v.) farm labourers. see also (2) *volk.* [*Afk. fr. Du. volk* labourers + *dimin. suffix*]
These old volkies . . . work until sundown and still have the energy to play their guitars and sing. Green *Land of Afternoon* 1949
He always said the volkies were drunker on Good Friday than any other day of the year. *O.I.* 1974

volkseie ['fɔlks‚eɪə] *n.* National identity: see quot. [*Afk. fr. Du. volk* people + *eigen* own]
A valuable half-truth is included in the doctrine of the *volkseie*, the national ethos – that each 'national' group has something of its own to contribute to humanity. Edgar Brookes *False Gods cit.* Spottiswoode *S.A. Road Ahead* 1960

Volksie ['fɔlksi] *n. pl.* -s. *colloq. abbr. Volkswagen* used to refer to the V.W. 'Beetle' only: see also quot. at *tom.*
. . . if there's one thing turns me off a ou, it's greasy hair . . . Hang man, you could keep a Volksie going for three weeks on he's hair, no kidding. *Darling* 9.10.74

volksleier ['fɔlks‚leɪə(r)] *n. pl.* -s. A political leader. [*Afk. fr. Du. volksleider* national leader, *leiden* to lead]
Perhaps the inner significance of the 'bohaai' is that there is a rising sense of panic abroad in South Africa today among 'volksleiers' who have clearly and irremediably lost their grip on the minds and hearts of young people. *Het Suid Western* 17.2.71

volkslied ['fɔlks‚lĭt] *n. prop. and n.* National anthem or hymn. [*Afk. fr. Du. volkslied* national song]
The singing of the Volkslied and a vote of

thanks to General Hertzog closed the meeting. *E.P. Herald* 18.1.1921

volkspele ['fɔlk‚spɪələ] *pl. n.* Afrikaans folk dances usu. performed in *Voortrekker* (q.v.) costume: see also quot. at *boereorkes.* [*Afk. fr. Du. volk* people + *spele* games *fr.* spelen to play]
. . . dancing that night will be confined to volkspele and other typically South African dances. *Evening Post* 5.2.72
deriv. volkspeler, a dancer of ~.
Teacher Mrs. Kruger helps two young 'Volkspelers' from Britain to strike the right pose. *Ibid.* Dec. 1972

Volksraad ['fɔlks‚rɑt] *n. prop.* Formerly used of the Legislative assemblies of the Transvaal (*Zuid Afrikaansche Republiek*) and Orange Free State Republics. ⸢Now *Afk.* for the House of Assembly equiv. of *Brit.* House of Commons. [*Afk. fr. Du. volks* people's + *raad* council]
British authority again having been withdrawn from the whole of this country [Orange Free State] on the 23rd February 1854 . . . its government is now in the hands of a President, freely elected by the inhabitants, assisted by an Executive Council and Landdrosts, and Heemraden in the several districts while the Volksraad exercises legislative functions. *Cape of G.H. Almanac* 1856
The diggers, who hated and derided the Free State, appealed to the law just passed by its Volksraad, though it had not yet been published. Boyle *Cape for Diamonds* 1873

volksvreemd(e) ['fɔlks‚frɪəmt, -də] *modifier.* Alien to the people, see (1) *volk*, or to their interests. [*Afk. fr. Du. volk* people + *vreemd* strange, alien + *attrib. suffix -e*]
From Broederbond documents I have seen over the past few years there emerges a startling story of the Broederbond fight against 'volksvreemde' organisations. *Sunday Times* 5.11.7 2

volkswil ['fɔlks‚vɪl] *n.* The will of the people, the National will. [*Afk. fr. Du. volk(s)* people's + *wil* wish, will]
. . . a Party which has claimed, until now, that its leaders are the heaven-sent interpreters of the volkswil – the people's will . . . it is difficult now for the Government to claim that they represent the 'volkswil'. *cit.* Rogers *The Black Sash* 1956

volume ⸢ Cubic capacity under the S. Afr. *metrication* (q.v.): system is calculated in *litres* (q.v.) and *millilitres* (not usu. ccs): 1 cubic foot = 28.3 litres: see also *mass* and *cumec.*

voor- [fʊə(r)-] *adj. and adv. prefix.* Fore, front, ahead: In place name Voorbaai, and numerous combinations ~*bok*, ~*huis*, ~*kamer*, ~*kis*, ~*laaier*, (1) ~*loop*, (2) ~*loop*, ~*skot*, ~*slag*, ~*span*, ~*touw*, (1) ~*trekker*, (2) ~ *trekker(s)*, all (q.v.) [*Afk. fr. Du. voor* before, in front *cogn.* fore]

voorbok ['fʊə(r)ˌbɔk] *n.* A goat usu. a *kapater* (q.v.) which leads sheep. *cf. Brit. bell-wether* [*Afk. voorbok* bell-wether, leader goat]

No herd of sheep was complete without one or more 'Voorbokkies' (leader goats). Sheep will neither go into nor out of a gate without a voorbok; without him they mill around, almost crushing each other to death. Jackson *Trader on Veld* 1958

figur. Leader: latterly equiv. of *opstoker* (q.v.), see second and third quots.

We are . . . very much the agteros rather than the voorbok in this field of research. *U.C.T.* Dec. 1971

Woman 'riots voorbok' is jailed for 18 months. *Het Suid Western* 17.11.76

A 25-year-old woman was sent to jail for eighteen months . . . The woman . . . was described in court as one of the 'voorbokke' of the riots who incited children to throw stones at White people's cars. *Evening Post* 20.11.76

voorchitz, voorchits See *voerchitz, voerchits voersis*

voorhuis ['fʊə(r)ˌ(h)œɪs] *n.* The entrance hall of a Dutch house, usu. large enough to be a general sitting room: see also *voorkamer*. [*Afk. fr. Du. voorhuis* fore part, hall of a house]

At about half an hour after nine, all retired to rest; some to a mat on the floor in the voorhuis (entrance-room, or hall) which is a large room used for general purposes, and occupying the middle and principal part of the ground-floor; the master to the bedroom at one end of the voorhuis, and the guests to a small chamber at the other. Burchell *Travels I* 1822

voorkamer ['fʊə(r)ˌkɑmə(r)] *n. pl.* -s. The 'front room' or *voorhuis* (q.v.) of a Dutch house as opposed to the *agterkamer*, the dining- or living room. [*Afk. fr. Du. voorkamer, voor cogn.* fore + *kamer cogn.* chamber, *Lat. camera*]

Papa and Leonora . . . are at Whisk in mijnheer's fore-kamer as they style it, a handsome Lofty chamber with floor of plum-coloured tiles high polisht daily by a Slave. Sophia Pigot *Diary circa* 1820 *cit.* Fitzroy 1955

A central front door, often under an ornamental gable, gave access to a 'voorkamer' or front room. Lewcock *C19 Architecture* 1963

voorkis ['fʊə(r)ˌkɪs] *n. pl.* -te. The front *wakis* (q.v.) or *wagon chest* (q.v.), serving both as storage space and as the driver's seat on wagon journeys. [*Afk. fr. Du. voor* fore, front + *kist cogn.* chest]

In his voorkis (which was also the driver's seat) he stored everyday items such as coffee beans and tea, canisters of rice and sugar. Green *When Journey's Over* 1972

voorlaaier ['fʊə(r)ˌlaɪə(r)] *n. pl.* -s. A muzzle-loader: see also *agterlaaier*. [*Afk. fr. Du. voor cogn.* fore + *laden* to load (*Afk. laai*) + *agent. suffix* -er]

. . . the percussion cap came in, to take the place of the flintlock that the old Boers used in their voorlaaiers. Bosman *Jurie Steyn's P.O.* 1971

Another call for Cape independence

Most Transvalers will reach for their voorlaaiers when they read a special report by three Capetonians. *Cape Times* 26.4.79

voorloop¹ ['fʊə(r)ˌlʊəp] *vb. intrns. rare* To lead a team of oxen. [*Afk. fr. Du. voor* in front + *lopen* to go]

Then he called Tom and handed the leading riem over to him, for it had been decided that Tom would voorloop for a while. Goldie *River of Gold* 1969

voorloop² *n. rare: poss. obs.*: see quot. [*Du. voorloop* 'heads', first runnings]

In distilling brandy the first to make its appearance is known as the voorloop: Pettman *Africanderisms* 1913.

Acc. some *witblits* (q.v.) consists of the ~ of the still.

voorloper ['fʊə(r)ˌlʊəpə(r)] *n. pl.* -s. The leader of a *span* (q.v) of oxen, usu. a young Hottentot or coloured boy: see also (1) *leader* and *touleier*. [*Afk. fr. Du. voor* in front + *lopen* to walk, go (*Afk. loop*) + *agent. suffix* -er]

Then a long wagon would pass . . . drawn by a span of ten or fourteen oxen under the guidance of a voorloper, a brown boy, holding occasionally a small rope attached to the horns of the leading bullocks. Alexander *Western Africa I* 1837

voorskot ['fʊə(r)ˌskɔt] *n.* Advance payment made to a farmer in respect of a crop, wool clip etc.: see also *agterskot*, *middelskot*. [*Afk. fr. Du. voorschieten* to lend, to advance (money)]

. . . many wool farmers . . . feel that they are losing out to others through the 'voorskotagterskot' system of interim payments for their clips. *Farmer's Weekly* 30.5.73

Voorskot for wool fixed The Voorskot for the 1974/75 wool season has been fixed at an average of R1,50 a kilogram for clean wool, the same as the voorskot for the first half of the previous season *E.P. Herald* 9.9.74

voorslag [ˈfʊə(r)ˌslax] *n. pl.* -s. A whip-point or lash: see quot. *cf. Austral. cracker* (on a whip). [*Afk. fr. Du. voorslag* first stroke]

The driver wielded a gigantic whip – a long bamboo, tapering like a fishing rod from thick base to slender tip, to which was laced a still longer rawhide thong ending in a spliced or *voorslag* or lash. Mockford *Here are S.Africans* 1944

. . . Strops R1.25; Agterslags 80c; Voorslags 45c; Skeys R2.35 all dozens. *Farmer's Weekly Advt.* 7.7.71

voorspan [ˈfʊə(r)ˌspan] *n. prob. obs.* An extra team or *span* (q.v.) usu. of oxen or horses. [*Afk. fr. Du. voor cogn.* fore + *span* team]

26 June 1811 This gentleman anticipated my wants, by proposing, as the passage of the Hex-river Kloof, and the ascent of the Rogge-veld mountain, would greatly exhaust the strength of my own oxen, that he might issue orders for a voor-span (relay of oxen) to meet me at those places. Burchell *Travels I* 1822

voortou(w) [ˈfʊə(r)ˌtəʊ] *n. pl.* -s. The *riem* (q.v.) or rope held by the (1) *leader, voorloper* or *touleier*, all (q.v.) to guide a *span* (q.v.) of draught oxen. [*Afk.fr. Du. voor cogn.* fore + *touw* rope, cord]

Her father had himself taken the 'voortouw' – the thong looped around the horns of the leading oxen, and led the span to keep it upon its course. McMagh *Dinner of Herbs* 1968

Voortrekker [ˈfʊə(r)ˌtrekə(r)] *n. pl.* -s.
1. A Boer pioneer *usu.* a member of the *Great Trek* (q.v.) from the Cape Colony to the Transvaal in 1834 and 1837 of those dissatisfied with British rule and with the abolition of slavery: see also quot. at *outspan*. [*Afk. fr. Du. voor* advance *cogn.* fore + *trekken* to travel, migrate, march + *agent. suffix -er*]

The first Dutch settlers, a party of the 'Voor-trekkers', arrived in Pietermaritzburg in 1838. Buchanan *Pioneer Days* 1934

As *modifier* ~ *costume,* (see at *volkspele* and *boereorkes*), ~ *dress,* ~ *monument.* ~ *victory,* ~ *wagon* see *trek wagon.*

It was the week before the foundation-laying of the Voortrekker Monument and at every station small groups of boys and girls in Voortrekker dress got on and off the train. Then the boys, with their coloured corduroy suits, and the girls, with their wide *kappies* and long gingham dresses made a bright show. Krige *Dream & Desert* 1953

They feel that the monument, commemorating the Voortrekker victory against the Zulus at Blood River in 1838, is 'a waste of money'. *Sunday Times* 19.12.71

2. An Afrikaner youth movement similar to the Boy Scouts, Girl Guides etc. [*fr.* (1) *Voortrekker*]

. . . the Voortrekkers hived off from the Boy Scouts, and the Noodhulpliga hived off from the Red Cross, and latterly the Rapportryers are going to break away from Rotary. E. G. Malherbe *Training for Leadership in Africa cit.* Spottiswoode *S.A. Road Ahead* 1960

Eleven George Voortrekkers have raised R2 000 for their commando . . . The money will be used to build their own commando hall. The picture shows the Voortrekkers with their commandant – on the extreme right. *E.P. Herald* 7.4.75

vos [fɔs] *adj.* Bay. [*Afk. fr. Du. vos (paard)* bay]

1 dark-foss Mare, wolf bite on right thigh. *Grahamstown Journ.* 3.1.1833

. . . the traveller would come past there, with two vos horses in front of his Cape-cart, and he would get off from the cart and shake hands . . . Bosman *Jurie Steyn's P.O.* 1971

vrede [ˈfriədə] *n. abstr.* Peace. Found in place names e.g. Vredefort, Vrededorp, Rust-en-Vrede. [*Afk. fr. Du. vrede* peace]

vrek [frek] *vb. intrns. colloq.* Die: usu. of an animal. [*Afk. fr. Du. verrekken* to disjoint, strain *cogn.* Ger. *verrecken* to die (*vulgar*)]

'. . . our stock, particularly in mountainous areas where it is difficult to round up every four days for foot bathing, are "vrekking" like flies', Mr – told congress. *Farmer's Weekly* 11.7.73

figur. as in 'I just about ~*ked* of the heat' or *slang* in '*gaan vrek*' *equiv.* 'drop dead'.

vroetel [ˈfrŭtl̩, -əl] *vb. slang.* Usu. in phr. ~ *with,* ~ *about with/in* etc. sig. 'mess about', 'root around' etc. [*Afk. fr. Du. wroeten* to rootle about (as of a pig) burrow]

I haven't got time to go vroeteling with hymns – let's do the anthem. O.I. Choirmaster 1971

vrot [frɔt] *adj. slang.* Rotten (*lit.*) *figur.* equiv. of 'lousy', 'no good' etc.: see combination ~*eier.* [*Afk. fr. Du. verrotten* to rot]

And what are the charges against me? That I broke a stick over Booy's head. It was a rotten stick, vrot, or it would not have broken. Cloete *Watch for Dawn* 1939

There are girls in the school long-jump and swimming teams. 'Only because the guys are vrot this year,' one of the boys said. *Fair Lady* 24.12.75

vroteier [ˈfrɔtˌeɪə(r)] *n. pl.* -s. *slang. lit.*

'Rotten egg'. among children one who is 'out' in a game, or a poor sport, poor specimen etc.: see *papsnoek, papbroek.* [*Afk. vrot* rotten + *eier* egg]

vrotpootjie [ˈfrɔtˌpŭĭkĭ, -cĭ] *n.* A virulent fungus disease of wheat and other cereal crops also known as 'white heads'. ℙ ~ (*Afk.*) is used of foot-rot in sheep, and of blackleg, eelworm and root gallworm in other crops. [*Afk. vrot* rotten + *poot* foot, leg + *dimin. suffix -tjie*]

RELENTLESS MARCH OF VROTPOOTJIE
In certain . . . major producing areas, vrotpootjie or 'take all' has become the greatest single menace to wheat. It has assumed far larger proportions than the wheat grower's old enemy, rust. . . . (Captions) . . . vrotpootjie symptoms showing the characteristic blackening of the stem bases . . . close-up showing white heads among green heads. *Farmer's Weekly* 21.3.79

vrou[1] [frəʊ] *n. pl.* -s. Woman [*Afk. fr. Du. vrouw* woman, wife, mistress etc.]
1. Also anglicized *fro(u)w*: a married woman usu. Dutch: mistress of a household. see quot. at (3) *man.*

. . . good looking cheerful Frouw, two neat daughters and the mynheer a very capital specimen of the best class of Dutch Boer. '*The Traveller's' Journal* 1832-3

2. equiv. 'Wife': a mode of address or reference *cf. juffrou* ℙAlso formerly occ. titular with surname as Ger. *Frau Schmidt.* see also second quot. at *tannie.*

Distance to them is no consideration, the boor puts his vrouw and kinders into the wagon, lights his pipe and sets off to travel five hundred miles with as much ease as we should ten in England. Philipps *Albany & Caffer-land* 1827

. . . the mystified man said, 'But, vrou, where I ask you, where do you come by all this money . . .?' His wife laughed. McMagh *Dinner of Herbs* 1968

In combination *boer(e)* ~, sig. either an *Afrikaans* (q.v.) woman or a country-woman.

'My husband, of course, is English', Lisbet explained. 'But I am pure Dutch – a regular Boer *vrou.*' Brett Young *City of Gold* 1940

vrou[2] [frəʊ] *Sect. Army. lit.* 'wife',. 'woman': slang in certain camps for a serviceman's rifle which, esp. on the *Border,* (q.v.) must accompany him everywhere. [*Afk. vrou* wife woman *cogn. Ger. Frau*]

Your vrou had to go with you – even to the pub – everywhere except the shower and if you

dropped it you had to kiss it better – you're wedded to the – thing. *O.I. Ex-Serviceman* 15.2.79

vry [freɪ] *vb. intrns. colloq.* To court, woo: or to indulge in kissing and caresses: derivatively as vbl. n. ~ *ing. cf. Austral. mash,* to court or woo, as *n.* a lover. [*Afk. fr. Du. vrijen* to court or woo]

I'm furious with that . . . chaplain. He said I did too much vrying. *O.I.* Schoolboy 18, 1974

~ *er,* lover; ~ *ery,* courting; ~ *hoek(ie),* a convenient corner for a courting couple.

Tant Sannie was well satisfied when told of the betrothal. She herself contemplated marriage with one or other of her numerous 'vrijers' and she suggested that the weddings might take place together. Schreiner *Afr. Farm* 1883

Vryheidsoorlog [ˈfreɪˌheɪdsˈʊə(r)ˌlɔx] *n.* The Afrikaans mode of reference to the S.Afr. Wars: see *Anglo-Boer War;* usu. as *eerste* (first) ~ or *tweede* (second) ~. [*Afk. fr. Du. vrijheidsoorlog* war of independence]

'We in South Africa have another "vryheidsoorlog" to fight – not against the cruel might of Britain, but against the selfishness which dominates our hearts and our laws, and which is contrary to the principles of the Kingdom of Heaven,' he said. *E.P. Herald* 31.5.73

Vrystaat [ˈfreɪˌstɑt] *n. prop. and interj.* Free State [*Afk. fr. Du. vrij* free + *staat cogn.* state] **1.** The Orange Free State: derivately ~ *er,* one born, bred or living in the 'Free State': see also *Blikoor.*

He tells me he is an old 'Vrystaater', and his people before him, . . . 'when I was a boy the Vrystaat was not like this.' Vaughan *Last of Sunlit Years* 1969

2. Exclamation equiv. of hurrah, whoopee etc.

. . . was in . . . the police station when he turned and saw a naked White man turn around at the door of the charge office . . . and run out shouting 'Vrystaat.' *E.P. Herald* 19.4.74

. . . was awarded the trophy and he was loudly cheered with the appropriate cry of 'Vrystaat . . .!' *Panorama* Feb. 1975

vuilgat [ˈfœɪlˌxat] *n. pl.* -te. *Sect. Army.* (vulgar) An extremely dirty person, insulting: see quots. at *troepie* and *skrik* [*Afk. vuil cogn.* foul + *gat* vent, anus]

vuiluil [ˈfœɪlˌœɪl] *n. pl.* -s. *Sect. Army. lit.* 'Foul-owl', a useless soldier: see first quot. at *tankjokkie.* [*Afk. vuil* dirty + *uil cogn.* owl]

vula ['vula] *vb. Afr. E. usu imper.* Open.
℗ Also in *advt.* for black consumers
'Buy . . . in our new litre bottle with the
vula-vala top'. [*Ngu. -vula* to open
(*-vala* close)]

'Vula, open up, or we break in!' a furiously
impatient voice with a Xhosa accent shouted . . .
followed by another nerve-wrecking [sic]
knock. *Staffrider* Vol. 1 No. 2 May/June 1978

vuma See **woema**.

vundu ['vŭndŭ] *n. pl.* -s, Ø. *Heterobranchus
longifilis*, a large air-breathing fresh-
water catfish of the Clariidae fam. fr.
the Zambezi. [*local name*]
• Dr Rex Jubb holding a small specimen of the
vundu, . . . of 40 lb. The split dorsal fin separates
it from the common Barbel. *Piscator No 90
Autumn 1974*

vygie ['feɪxĭ] *n. pl.* -s. *lit.* 'Little fig': name
for almost all species of *Mesembryan-
themum* and allied genera: often those
species of *Carpobrotus* with edible fruits
(see *sour fig, hotnotsvy, suurvy, ghokum*
(gaukum)) and numerous species with
showy flowers opening only during
hours of sunlight: see also *skilpadkos.*
[*Afk. fr. Du. vijg ʋogn.* fig + *dimin.
suffix -ie*]
. . . it's all in the new flower book this year,
'The Genera of the Mesembryanthemaceae.'
Better known as 'vygies', these bright little
plants owe a lot to the creators of the book who
spent a lifetime studying them. *Fair Lady*
10.11.71
In numerous combinations e.g. *bees (cattle)* ~ , *klip(stone)* ~ , *berg(mountain)* ~ ,
brakveld ~ , *bokbaai* or *sand* ~ , *Dorotheanthus* growing most prolifically at
Bokbaai (Buck Bay).
It was September when we visited Buck Bay
and what a joy it was to walk through fields of
. . . bokbaai vygies, yellow and gold sour-figs.
Farmer's Weekly
Dorotheanthus or bokbaai vygies are best in
late winter or early spring. *Evening Post* 10.10.70

W

wa [vɑ:] *n.* Wagon: suffixed in wagon
types *bok* ~ , see *buckwagon; osse* ~ , (see
O.B.); *kakebeen* ~ (q.v.): prefix in plant
name ~*boom* (q.v.) *Protea arborea*
(*grandiflora*), sig. that the wood was
used in wagon building (see also *wagon-
wood*) ~*pad* (q.v.); and pl. ~*en* in
waenhuis (q.v.). [*Afk. fr. Du. wagen*]

waai [vaɪ] *vb. intrns. lit.* Blow. **1.** In *phr. I*

must/Ek moet ~ , *equiv.* 'I must fly, dash
off' etc. *cf. Canad. hit the trail.* [*Afk. fr.
Du. waaien* to blow]
2. *prefix* in plant name sig. a tendency
to blow away etc. ~*bos, Tarchonanthus
camphoratus;* ~*bossie, Salsola kali* and *S.
tragus* 'tumbleweed' see *rolbos;* ~*gras,
Panicum atrosanguineum.*
The slightest breeze will send the Karoo
waaibosse tumbling along the weed-infested
platform. The door of the signal-box hangs on
one hinge. *Radio & TV.* 21-28.11.76
3. Blow(y): found in place names
Waaiplaats and Waaihoek.

waaier- ['vaɪə(r)] *n.* Fan [*Afk. fr. Du.
waaien,* to blow, fan + *agent. suffix -er*]
1. *prefix* In names of plants with fan-
shaped leaves or flowers: ~*boom, Cus-
sonica spicata,* see *kiepersol;* ~*lelie, Cur-
tonus paniculatus;* ~ *palm, Hyphaene
crinita,* fan palm: [also ~*stert meerkat*
(bushy-tailed) see *meerkat*]
2. *prefix:* see also *waai* sig. 'blown
about' e.g. ~*bossie Triumfetta sonderi*
and dimin. form ~*tjie, Witsenia maura.*

waai vlieë ['vaɪ 'fliə] *vb. fr. Sect. Army. lit.*
'Brush off flies' to give a dilatory or
careless salute: see first quot. at *tankjokkie* [*Afk. waai* fan, blow, + *vlieg
cogn.* fly, *pl.* -e]

waar [vɑ:(r)] *adj.* See *ware* and *so* ~ .

waatlemoen ['vɑtlə‚mŭn] *n.* Watermelon:
kaffer ~ see *kaffir melon, makataan,
karkoer;* ~*konfyt,* see *konfyt.* [*Afk.
corruption of Du. watermeloen*]

waboom ['vɑ:bʋəm] *n. pl.* -s. *Protea
arborea* or *P. grandiflora,* a tree-protea
of up to 5m in height: also called *suikerbos* (q.v.). [*Afk. wa* fr. *Du. wagen
cogn.* wagon + *boom,* tree *cogn. Ger.
Baum*]
14 April 1811 . . . We passed some large trees
of Wagenboom (Protea grandiflora), so called
by the colonists because the wood of it had
been found suitable for making the fellies of
waggon-wheels. It is reddish, and has a very
pretty, reticulated grain. Burchell *Travels I* 1822
The waboom is one of the tallest of the genus
Protea . . . often with a trunk a foot or more
in diameter. It is handsome, very blue foliage,
oval leaves, and large pale yellow flowers . . .
rather like a stiff, round shaving brush. Palmer
& Pitman *Trees of S.A.* 1961
Also found in place names e.g. Wa-
boomskraal, Waboomsrivier.

waenhuis ['vɑən‚hœɪs] *n. pl.* -e. Coach-
house or wagon-shed: a feature of old

farms, *drostdys, drostdies* (q.v.) or other old houses preserved or restored, esp. in the country. [*Afk. waen fr. Du. wagen cogn.* wagon + *huis cogn.* house]

The latest addition to the museum complex is a **waenhuis**, a traditional shelter for riding horses and farm implements, so-called because its dimensions had to be large enough for a span of oxen and a wagon to turn in. The new waen-huis is a whitewashed thatched building with a dung floor, the laying of which . . . 'was very messy'. *Fair Lady* 6.6.79

wagonchest, wagonbox, wagonkist *n. pl.* -s, -es, -te. A lidded chest used on wagon journeys both for storage and as a seat for the driver (see *voorkis:* the *agterkis* being the ~ under the *tent* (q.v.) both still prized as articles of domestic fur-niture: see *wakis.* [*trans. Afk. wakis fr. Du. wagen cogn.* wagon + *kist cogn.* chest]

The above, with a large gun, an axe, adze, and hammer, a couple of waggon-chests, a churn, a large iron pot for boiling soap, and one or two smaller ones for cooking, are all that is absolutely requisite. Thompson *Tra-vels II* 1827

. . . Four wagon kists. *E.P. Herald Advt.* 1.6.73

wagon wood *n.* The timber of *Protea ar-borea (grandiflora):* see quots. at *wa-boom* and at (2) *yellowwood.*

P. W. Keytel. Has always on hand a great variety of English and Swedish Iron, Wagon wood, Stinkwood, Deals and Boards. *Cape of G.H. Almanac Advt.* 1856.

wag-'n-bietjie [ˈvaxnˌbikĭ, -cĭ] *n. pl.* -s, *interj. lit.* 'Wait a little' **1.** Any of se-veral species of thorny Acacia *cf. Austral. wait-a-while* (Acacia): in various combi-nations e.g. *kaffer* ~ *;* see *katdoring.* [*Afk. fr. Du. wachten* to wait + *een* a + *bietje* little bit]

. . . a thorny shrub . . . well known in the Colony by the name of wagt een bitje (wait a bit), the prickles of which being shaped like hooks, there is no getting loose from them when they catch hold of one's clothes, except by tearing out the part entangled. Their grappling properties I soon experienced . . . for I was nearly pulled off my horse several times. Thomp-son *Travels I* 1827

2. *colloq. interj.:* equiv. of 'Wait a moment', 'Hang on' etc.

Bilingualism in hotels: Wag 'n bietjie. The Minister . . . has extended for two months the period within which receptionists and tele-phonists of licensed hotels are required to show proof that they are bilingual. *Cape Times* 7.12.72

wait-a-bit *modifier. prob. obs.* Travellers'

term for *wag 'n bietjie* (q.v.). [*trans. Afk. wag 'n bietjie*]

Torn and scratched by 'wait-a-bit' thorns, tripped and half strangled by parasitical mon-key-rope, the company tore a way through the treacherous bush. McKay *Last Kaffir War* 1871

wait on *vb. phr. substandard.* Equiv. of archaic 'wait upon' sig. 'wait for'. [*trans. Afk. vb. phr. wag op formal* 'wait upon' *usu. in biblical contexts*]

We're out of stock but we're waiting on supplies now. Cape Town Shopkeeper 6.7.71

The Langa police station looked like the out-patient department of a busy hospital as scores of people with serious wounds waited on am-bulances. *E.P. Herald* 26.3.73

wakis [ˈvɑˌkɪs] *n. pl.* -te. A chest or *kist* (q.v.) made specifically for wagon travel: the front or *voorkis* (q.v.) serving as the driver's seat and the rear or *agterkis* (q.v.) used under the *tent* (q.v.) for storage purposes: now much sought after collectors' pieces: see also at *wagon chest,* and quots. at *koskas* and *jonkmanskas.* [*Afk. fr. Du. wagen cogn.* wagon + *kist cogn.* chest]

A dealer in indigenous antiques warned . . . that certain shady dealers were buying up wakiste (wagon boxes) and used the wood of one to make four fake wakiste . . . Wakiste made of stinkwood and yellowwood that would have sold for R45 two years ago are now fetch-ing R250 *E.P. Herald* 24.5.74

want to, doesn't/don't *vb. phr.* Used with inanimate subject as equi... Eng. 'won't', without sense of negative volition. [*prob. mistrans. Afk. wil nie* doesn't/don't want to]

This door doesn't want to open. O.I. 1973 . . . ask them to help push the Kombie which does not seem to want to start. *Drum* 22.9.75

wapad [ˈvɑːˌpat] *n.* Wagon road: also in expression *as old as the Kaapse* ~, and in place name Wapadsberg: see quot. at (2) *trek(c).* [*Afk. fr. Du. wagen cogn.* wagon + *pad,* road, track *cogn.* path]

wapenschouw, wapenskou [ˈvɑpənˌskɔʊ] *n.* A military review at the Cape, invol-ving a display of marksmanship, shoot-ing at a target in the shape of a parrot known as a *papegaai* (q.v.) ℗ This is commemorated in the place name On-der-Papegaaiberg near Stellenbosch Cape. [*Afk. fr. Du. wapen cogn.* weapon + *skou* display *cogn.* show]

The 'wapenschouw' was an event of some importance, and the competition of shooting at the popinjay was keenly contested. The target was in the shape of a parrot placed upon a pole in the centre of a circle with a radius of five roods. Botha *Social Life in the Cape* 1927

wardmaster *n. pl.* -s. *hist. obs.* An unpaid civic official in charge of a ward in the early days of Cape Town: see also quot. at (1) *eating house.*

. . . the office of Wardmaster is absolutely necessary to the internal well-being of the town and the tranquility of the inhabitants, so that it cannot be permitted that anyone legally called upon shall refuse serving it . . . Should anyone refuse to fill the said office, he must state his reasons for so doing, in writing to the Burgher Senate . . . Should the grounds of refusal be deemed inadmissable . . . and the appointed Person continue to refuse, he shall be deprived of all his privileges as Burgher, and the Government approving, be moreover considered as an unwilling and refractory Burgher and be sent out of the Colony. *Afr. Court Calendar* 1819

ware [ˈvɑːrə] *adj. colloq.* Genuine, true: freq. in phr. ~ *Afrikaner* and ~ *boer cf. Austral. dinkum, Anglo-Ind. pukka.* [*attrib. form of Afk. waar* true *fr. Du. waar cogn. Ger. wahr*]

While pointing out that Smuts was probably the only political figure in our history with world stature, the author adds that he was not a *ware* Afrikaner. *Evening Post* 4.11.72

I'm a *ware* boeremeisie, I love porridge. *Cape Times* 3.7.73

warm *adj. substandard. equiv.* 'Hot' in *phr.* ~ *bath,* ~ *food,* ~ *drink* etc. [*translit. Afk. warm* hot]

. . .'s warm pies sold here. (Notice in Grahamstown shop) 1970

wasbessie [ˈvasˌbesĭ] *n. pl.* -s. *Myrica cordifolia:* see *wax berry* and *berry wax.*

. . . subtropical growth of taaibos, melkhout, kreupelhout, wasbessie and natural bush strung together with creepers. Louw *20 Days* 1963

wash spears, to *vb. phr. hist.* See quots. [(*Ngu.*) *Zu. ukugeza imikhonto* to wash spears, to carry out a ceremonial cleansing]

He restored the old military system, with regiments composed of young braves who could not marry till they had 'washed their spears' in blood. Keppel Jones. *Hist. of S.A.* 1948

On the very day that Maqubu's grandfather was washing his spear with English blood at Isandhlwana, Ian's grandfather was killing Zulus at the Inyezane River. *National Geographic Mag.* Vol. 140, No. 6 Dec. 1971

-water *n. suffix. substandard.* Used as equiv. of or sig. 'a mixture of X- and -water', e.g. sugar ~, dettol (disinfectant) ~, vinegar ~: or sig. a solution or in-

fusion: see *tea water:* also *Jam. Eng. sugar water.* [*prob. transference fr. Afk.* use] *cf. Ind.E. butter-bread.*

The powder increases itself with each brew, like the vinegar plant. It adds a splendid flavour to ordinary sugar-water, and as yeast for bread making simply cannot be equalled. *E.London Dispatch* 20.12.1911 *cit.* Pettman

waterblommetjie [ˈvɑtə(r)ˌblɔmĭkĭ, cĭ] *n. pl.* -s. *lit.* 'Little water flower' loosely applied to several species preferring a damp habitat or growing in water, esp. to *Aponogeton distachyos,* now being commercially cultivated in some dams: see *wateruintjie:* in combination ~ *bredie.* [*Afk. water* + *blom* flower *cogn.* bloom + *dimin. suffix -etjie*]

Wateruintjie (Waterblommetjie) Bredie: 2 soup plates of waterblommetjies (Aponogeton), plucked from their stalks. Gerber *Cape Cookery* 1950

waterduisendblaar[ˈvɑtə(r)ˈdœĭsəntˌblɑː(r)] *n.* A plant pest in watercourses: one of the *Myriophyllum* species.

. . . commonly known as the water duisendblaar already widely distributed along the major waterways of South Africa . . . directly attributable to man through his careless and irresponsible habit of dumping his unwanted rubbish anywhere. *E.P. Herald* 23.8.76

watereendjie [ˈvɑtə(r)ˌɪənkĭ, -cĭ] *n. pl.* -s. *prob.* corruption of *wateruintjie* (q.v.).

waterfiskaal, water fiscal [ˈvɑtə(r)ˌfɪsˈkɑːl] *n. pl.* -s. **1.** The boubou or bush shrike *Laniarius ferrugineus.*

About Cape Town this bird is known as the Waterfiskaal, and is common in thick bush but far more often heard than seen. Gill *Guide to S.Afr. Birds* 1959

2. Water bailiff, the official in control of the water *furrows* (q.v.) of a town: see also (water)-*erf* and *fiskaal.*

water, ride on See *ride (on water).*

waterskaap, waterkudu [ˈvɑtə(r)ˌskɑp, -ˌkŭdŭ] *n. pl.* Ø. The aquatic antelope *Tragelaphus spekei* see *sitatunga.* [*Afk. water cogn.* water + *skaap cogn.* sheep (*kudu*) (q.v.)]

wateruintjie [ˈvɑtə(r)œĭŋkĭ, -cĭ] *n. pl.* -s. *Aponogeton distachyos;* the bulbs and flowers of which are edible: see also *waterblommetjie;* also known as *watereendjies,* 'little ducks', *poss.* because of the floating white flowers, or more *prob.* corrupted form of ~. [*Afk. uintjie fr. Du. ajuintje* little onion]

31 Jan 1811 . . . a plant called Water-uyen-

tjies, the root of which, when roasted, is much eaten by the slaves and Hottentots. The heads of flowers, boiled, make a dish which may, in taste and appearance, be compared to spinach. Burchell *Travels I* 1822
In combination ~ *bredie, a bredie* (q.v.) made with the flowers of ~ *s.*
A recipe which probably dates back to Hottentot times is wateruintjiebredie. Early Springtime at the Cape is marked by the appearance in sluggish streams and shallow vleis of the sweet-scented white flowers of the wateruintjie ... Only the flowers are used and these must be young and fresh. *Farmer's Weekly* 25.4.73

wattle *n. pl.* -s. Also *basboom : Acacia decurrens* and *A. mollissima* both of which have been introduced from Australia and are cultivated in S.A. for the bark which yields tannin. In combination *black* ~ *Peltophorum africanum,* so called from its likeness to *A. mollissima; silver* ~ *Acacia dealbata,* (also *orig.* a native of Australia) both of which species can be troublesome to farmers on account of their tendency to encroach upon (2) *veld* (q.v.) ; and ~ *looper,* a moth, which in its caterpillar stage defoliates ~ .
The thousands of dark-coloured, medium sized moths which have suddenly appeared ... are ... commonly known as the 'Wattle Looper' ... a species occurring throughout the greater part of Africa. .... Crop plants which are defoliated by this species include black wattle in Natal, castor oil plant in the Transvaal and citrus in the Eastern Cape. *Grocott's Mail* 22.3.74

wax berry *n. Myrica cordifolia :* see *berry wax* and quot. at *mevrou.* [*trans. Afk. fr. Du. was* cogn. wax + *besje* berry]

weaver bird *n. pl.* -s. See *vink/fink.*

webbing *n. Sect. Army.* A skeleton kit harness holding ammunition, dixies (rectangular dishes) water bottle, rucksack and other equipment, worn by servicemen: apparently standard usage in S. Afr. Army where the name of the material of the harness is used for the entire kit (*O.I.s*) see also quots. below.
⫟ Given by Partridge et al *Dict. of Forces' Slang* 1948 as 'Web equipment: Army coll.: C 20.' [*fr. webbing,* the material which replaced leather in the British Infantry during World War I]

Another method of removing an injured soldier is shown: here he is being dragged by his webbing. Note how he holds his rifle... without

it the infantryman considers himself naked. Marks *Our S. Afr. Army Today* 1977
... kicking him ... while pulling him by the webbing. Men in detention barracks had to wear their webbing with their rucksacks and two smaller sacks while doing physical exercise ... On that morning ... filled with gravel. *E.P. Herald* 18.5.79

Wederdoper [ˈvɪədə(r)ˌdʊəpə(r)] *n. pl.* -s.
⫟ This word which is of considerable historical interest is still current in parts of S.A.
1. *hist.* A member of one of the Anabaptist sects. ⫟ A branch of the original sect, founded Zurich 1523, is said still to be extant in the W. Cape. (unconfirmed) [*Afk. wederdoper* anabaptist *fr. Du. wederdopen* to rebaptize + *agent. suffix* -er, cogn. Ger. Wiedertäufer (*Gk. prefix ana-* Eng. re- *Du. weder*)]
We had with us on the Cambrian a passenger, Swiss by birth, who is a leading glory of the Veder-doopers, as this Wellington sect is called ; ... There was ... much talk amongst Cape people of a certain action ... in a court of law some years back in which the most damnable facts and deeds were sworn against the Veder-doopers, but I never could gain a sight of the report, nor were the accusations retailed to me other than the vaguest. Boyle *To the Cape for Diamonds* 1873
Andrew Murray the younger was at Wellington at the time Boyle wrote and seems to be recognised as one of the earliest proponents of Pentecostal ideas in South Africa though he never left the D.R.C. – possibly Boyle is referring to some local group influenced by Murray or more remotely to Wellington members of Ds. D. P. Faure's Unitarian Church founded in the late 1860s. *Informant* Mr Michael Berning, Cory Librarian *Letter* 9.4.79
2. ~ is sometimes applied in a derogatory sense by Afrikaners to anyone who joins a church or sect which re-baptises by immersion any new member who has already been baptised in another church. ⫟ acc. one Baptist informant one Afrikaans-speaking prospective member of the Baptist Church gave up because of the humiliation of being called a ~ by her family.
3. ~ in certain parts of the W. Cape, Stellenbosch also poss. Wellington area (see Boyle quot.) ~ is applied by local people to the Assemblies of God, which re-baptize new members. *Informant* Revd. E. Goodyer, Stellenbosch. ⫟ The term may well be used of other Pentecostal sects but this is unconfirmed.

Weeds Act See *proclaimed weed*.

Wees- [vɪəs] *n. prefix* Orphan-: found in *hist.* terms ~*kamer* (*cf. Boedelkamer*) see quot. and ~*huis* the first Orphanage founded at the Cape [*Afk. fr. Du. wees* orphan]

The Weeskaamer or Orphan Chamber . . . like our Court of Chancery, regulates the affairs of orphan children and takes charge of property which may be disputable, in order to have justice done. Ewart *Journal* 1811-14

The Weeshuis, or Orphanage in Long Street was founded in 1815 by a public spirited widow called Mother . . . In 1799 she had submitted a plan to the Governor, who granted a plot of ground for her purpose . . . to provide succour for old women; but in 1808 she decided to build a place of residence for Orphans as well. Laidler *Tavern of Ocean* 1926

On 1st October, 1829, the South African College opened in rooms in the Weeshuis. *Ibid.*

-weg [vex] *n. suffix.* Road, way: found usu. in street names, Alexandraweg, Milnerweg, Hoofweg in contrast to *straat* (q.v.): also in route name Ou Kaapse Weg. [*Afk. fr. Du. weg*, road *cogn.* way]

wel- [vel] *adv. prefix.* Well: found in place names e.g. Welgevonde, Welgelegen, Welverdient. [*Afk. fr. Du. wel cogn.* well]

werf [verf] *n. pl.* -ven, -we, -s. An area, often enclosed, including homestead, barns and other outbuildings: roughly equiv. of farmstead/yard: also in place name Driewerwe (Three ~). [*Afk.* farmyard *fr. Du werf* yard *cogn.* wharf]

The homestead of the corn farmer was also a substantial and commodious structure with many outbuildings in the 'werf', or yard, including workshops and a smithy. Botha, *Our S.A.* 1938

The nearest English equivalent of a werf is a homestead, and reference to the previous Gold Laws shows that two kinds of werven are contemplated. Sisson *S. Afr. Judicial Dict.* 1960

-wes- [¹ves] *n. or adj.* West: found in place names either as trans. as in Somerset-Wes or in Afk. names e.g. Wes-Driefontein, Stilbaai-wes. [*Afk. fr. Du. west*]

what *prn. substandard.* Used redundantly after 'than' in comparative structures. [*trans. Afk. comparative structure e.g. ouer as wat ek is, lit.* 'older than *what* I am.']

'It's that sort of thing that gives us Marico farmers a bad name. . . . And we didn't want

any worse name than what we already had' Chris Welman reckoned. Bosman *Bekkersdal Marathon* 1971

'How do you expect me to say it any clearer than what I've just said it?' *Ibid.*

In the words of . . . Sandton Tvl: 'I feel fitter than what I have ever felt and very happy.' *Sunday Times Advt.* 27.10.74

White *n. pl.* -s, also *adj.* Also *European* (q.v.) sig. a white-skinned person: see (*u*)*mlungu*, as opposed to one of colour, freq. as adj. in phr. *a member of the ~ group*, and combination *non* ~ sig. any person of colour whether of African, Asian or mixed ancestry: see also *Coloured*, *non-European* and *Black*, and quot. at *reclassify*. [*fr. Afk. blanke pl.* -s (q.v.) *cogn. Fr. blanc* white]

. . . the first black to manage a chain store in Umtata . . . has a staff of seven, including a white. *Daily Dispatch* 3.5.72

Three Whites have been arrested, and will appear in the Magistrate's Court this morning to be charged with murder. *E.P. Herald* 10.4.73

white *adj.* Cheeky, insubordinate, disobliging: see *wit*.

Hell she's white that one –won't do a thing she's asked. *O.I. Settlers' Hospital* 14.10.78

'Don't get white with me' ('You are forgetting yourself') typical utterance, corporal to roofie. *O.I. Serviceman* March 1979

white-by-night *modifier.* Of or pertaining to an urban area which does not permit or make provision for living-in domestic servants, or to the policy dictating this.

The new White suburbs will not allow servants to live-in . . .to conform with the White-by-Night policy so beloved by verkramptes. *Daily News* 17.4.71

The women sing about their children, who have turned moles in West Deep Levels and Owls in white-by-night Johannesburg. *cit. Staffrider* Vol. 1 No. 2 May/June 1978

figur. see quot.

The dramatic 'white-by-night' life of a coloured part-time student ended this week when he was expelled . . . he had to study another subject . . . but no black college offered the course . . . 'I was determined to complete my course even if I had to cross the colour line and join a white College.' *Sunday Times* 8.10.78

white eye *n. pl.* -s. Any of several species of the Zosteropidae fam., usu *Z. capensis*, the Cape ~, *glasogie* or *kersogie* (*kersie*); small gregarious fruit and insect-eating birds recognizable by a white ring round the eye. *cf. Austral. wax-eye* (*Zosterops spp.*).

Birds had made themselves at home in the garden – brilliant sunbirds of several kinds, chattering white-eyes, finches, robins and thrushes. *Personality* 2.4.71

white flag *n. pl.* -s. *Sect. Army:* A signal to 'call a truce' with the weather, *viz.* that it is too hot to work, (Border usage.)
    . . . muggy, sandy, the sun scorching. Sometimes a white flag goes up during the day as a signal that it's too hot to work! *Darling* 7.2.79

white sore throat *n. phr.* Diphtheria. [*trans. Afk. witseerkeel* diphtheria]
    . . . it was found that white sore throat had set in, and she became decidedly worse. *Cape Town Directory* for 1866 (Obituary)

white tie *n. pl.* -s. Ambiguous in S.A.E.: more freq. the ~ worn by an *ouderling*, *dominee* or *predikant* all (q.v.) of the *Dutch Reformed Church* (q.v.); than the white bow tie worn with formal evening dress or 'tails': see also *manel*. [*trans. Afk. wit das*]
    . . . the minister presided . . . settled his white tie in place, and coughed portentously. Fairbridge *The Torch Bearer* 1915
    All the leading farmers were ouderlings (elders), who, on the Sabbath, clad in their 'swart manel' (black frock coats), 'swart keil' (high black top hat), 'wit boortjie en das' (white collar and tie), took their duties very seriously. Jackson *Trader on Veld* 1958

White Train *n. prop.* The train used officially by the head of State, formerly the Governor General, now the State President. ⓟ The ~ was renovated or rebuilt at the time of the Royal Visit in 1947. It was used for the confrontation between black and white African leaders on the Victoria Falls Bridge, Livingstone in 1975. [*fr. colour*]
    . . . the White train coasted down the hill . . . and even I gave it a wave which was returned with the best of royal humour by a steward in the dining car. *Rhodeo* 28 no. 1. 5.9.74

whitewash See *kaffir beer*.
    This is the working man's rendezvous . . . where 'white wash' or traditional type sorghum beer is plentiful and inexpensive. Peter Becker cit. *The 1820* July 1973

who-all *prn.* Used as a plural interrogative or quasi-relative pronoun as in ' ~ were there?' or 'I've no idea ~ are coming.' *cf. Southern U.S. you-all* (you, *pl.*). [*trans. Afk. wie-almal, lit.* 'who-everyone']

widow-bird *n. pl.* -s. Any of several species of the *Viduinae* (Ploceidae fam.) also *Euplectes progne*, usu. called *sakabula* (q.v.) esp. in Natal: see quots. at *sakabula* and *kaffir finch*. [*fr. similarity*

of plumage to black 'widow's weeds']
    By the time the widow-birds put on their funereal tails in November the whole crop had withered. Brett Young *City of Gold* 1940

wilby *n. pl.* -ies *Reg. Albany District:* An African, abbr. William Wilberforce: name coined in about 1965 *acc. Informant* Mr R. Forward, Bathurst, 'to get away from the 'native' vs 'Bantu' vs 'African' question' of terminology. ⓟ Now also used pejoratively by some e.g. 'A Wilby will be a Wilby and always will be,' etc. *O.I.* ⓟInterpreted by anthropologist R. Palmer as being from *'will be'* i.e. as holding the future: *prob.* idiosyncratic.
    'Wilbies don't think like we do' Bathurst Farmer's wife: Statement obtained during interviews for M.Soc.Sci. thesis 'Domestic Workers in the E.Cape – a socio-historical investigation'. *Informant* J. Cock 18.5.79

wild dog *n. pl.* -s. *Lycaon pictus:* a predatory dog, hunting in packs, destroying both flocks and game: *cf. Austral. dingo.*
    . . . where wild dogs and wolves devour the flocks . . . Webster *Voyage I* 1834
    The wild dogs, or 'wilde honden' as they are termed by the Dutch boers, are still abundant in the precincts of the Cape Colony . . . This interesting though destructive animal seems to form the connecting link between the wolf and the hyena. McKay *Last Kaffir War* 1871

wildebees(t) ['vɪldəˌbiəs(t)] *n. pl.* Ø, -s. Also *gnu* (q.v.) either of two S. Afr. antelopes with many of the characteristics of an ox: *Connochaetes taurinus* the *blue/blou* ~, or *C. gnou* the *black/swart* ~ : see also quot. at *snotsiekte*. [*Afk. fr. Du. wild(e)* wild + *beest cogn.* beast (q.v.)]
    That remarkable animal, the gnu or wildebeest . . . Forming the link which connects the ox tribe with the antelope, it partakes in some degree of the character of both . . . Great numbers are annually killed, and their flesh cut into strips, dried, and converted into excellent biltongue. *Cape of G.H. Almanac* 1843
    WILDEBEEST, is a gnu, and therefore an antelope. Sisson *S.Afr. Judicial Dict.* 1960

wildeperd ['vɪldəˌpert, -ˌpɛə(r)t] *n. pl.* Ø. Wild horse. [*Afk. fr. Du. wild(e)* + *paard,* horse *fr. Lat. equus*] **1.** *obs.* Name for the zebra (*Equus zebra*)
    Equus Zebra. The Zebra. *Wilde Paard* of the Cape Colonists. Harris *Wild Sports* 1839

2. *Diplodus cervinus,* also called *zebra* and *streepdassie* (q.v.). [*fr. stripes*]

wildevis [ˈvɪldəˌfɪs] *n. pl.* -s. *Elops machnata :* see *springer.* [*Afk. fr. Du. wild* + *attrib. suffix -e* + *vis, visch cogn.* fish]

One was a skipjack (Elops Saurus, . . . also called Cape salmon, springer or wildevis), . . . caught in the surf from a high rock. *Grocott's Mail* 18.1.72

wildevyeboom [ˌvɪldəˈfeɪəˌbʊəm] *n. pl.* bome. Also *wild fig. Ficus* (various spp.) see *Wonderboom.* [*Afk. fr. Du. wild cogn.* wild + *attrib. suffix -e* + *vijg cogn.* fig + *boom* tree *cogn.* Ger. *Baum*]

wilds [vɪlts] *modifier.* Game (*Afk.*)*:* see quot. at *withond.*

wind bird *prob.* erron. See *impundulu.*

The matter had embarrassed the . . . family because other squatters were saying Mrs . . . had been wearing mourning clothes for a 'wind bird' (impundulu). *Indaba* 29.9.78

wind(gat) [ˈvɪntˌxat] *n. pl.* -e. (-s). *Afk.* term for a braggart, 'blowhard' etc. *abbr.* to *adj.* form by servicemen to *wind* meaning having a high opinion of himself. [*Afk.* boaster, windbag etc.]

'Hell but he's wind' is used by someone else talking about a windgat. *O.I. Serviceman* 4.2.79

wine-of-origin *n. phr.* The S.Afr. equiv. of *Fr. Appellation Controlée:* introduced to bring S.Afr. wines into line with the European system so that S.A. could continue to export to the European Economic Community (E.E.C.): see quot. below and quot. at *certified.*

South Africa's wine-of-origin legislation came into operation on 28 September 1973, with the introduction of the WO seal. The seal . . . guarantees that what is on the label will also be found in the bottle insofar as it refers to geographical origin, variety and vintage year . . . Wine-of-origin is indicated by a blue band on the WO seal, accompanied by the designation of the area on the label . . . A red band guarantees the vintage year . . . a green band signifies the contents . . . are derived from the vine cultivar reflected on the label. Bolsmann *S. Afr. Wine Dictionary* 1977

winkel, winkle [ˈvɪŋkəl, wɪŋkl̩] *n. pl.* -s. A shop, usu. a countrified *general dealer's* (q.v.) or *algemene handelaar* (q.v.): in combination *Boer* ~, a country store: also in place names Winklespruit and Winkelpos. [*Afk. fr. Du. winkel* shop]

. . . the little roadside winkle – a composite shop, where you could buy moist black sugar, tinned butter, imported; tinned milk, also imported; cotton prints, boots, 'square face', tobacco, dates, nails, gunpowder, cans, ribbons, tallow candles, and the Family Herald. Glanville *Tales from Veld* 1897

When I went into a dorp winkel . . did I throw hysterics because the woman in the shop could not speak a word of English? *Sunday Times* 14.10.73

deriv. ~ *ler prob. obs.* a shopkeeper ; and combination *Boer* ~ *ler;* see quot. at (4) *kraal.*

. . . grabbing men called winklers were charging per shilling per pound for meat, flour, sugar, salt and similar necessaries. McKay *Last Kaffir War* 1871

He hurried back to the Boer winkler in a rage and said '. . In your blessed barrel there are pebbles at the bottom and coffee on the top.' Cohen *Remin. of Kimberley* 1911

winkelhaak [ˈvɪŋkəlˌhɑk] *n.* An ear-mark like a three-cornered tear for identification of sheep, cattle and other livestock. [*Afk. fr. Du. winkelhaak scheur* three cornered, right angled, rent, tear: *winkelhaak* a set square]

1 White Nanny Goat, L/Ear stump, R/Ear w/haak in front. *Daily Dispatch* 29.6.71

wip [vɪp] *n. pl.* -pe. A loop snare (*strik*), or any other of several kinds of trap set for game by poachers in S.A. [*Afk. wip* trap, gin]

'So long' said Amos, going off to . . . examine a tiger-trap he had placed in the kloof, and a *wep* set in a game-path . . . the *wep* was sprung, the loop tightly drawn, though the prey had escaped. Glanville *A Fair Colonist* 1894

wish you, to *vb. phr. substandard.* To wish someone well, congratulate on a birthday etc. [*prob. fr. Afk. om jou te kom wens,* to come to wish you (good luck, happiness, joy)]

'I'll be sure to come and wish you on the right day.' O.I. (English speaking) 1969

wissel [ˈvɪsəl] *vb. intrns.* To shed teeth: used of sheep usu. as *neg. partic. un ~ ed.* [*Afk. fr. Du. wisselen* to shed teeth]

1640 Merino Hamels – unwisselled to 6 tooth – well grown and in good condition. *Grocott's Mail* 28.3.69

In combination ~ *hamel* (q.v.) a castrated ram which has shed its milk teeth: see also *volbek.*

. . .771 6 tooth to full mouth Hamels . . . 100 'Wissel' Hamels. *Farmer's Weekly* 3.1.68

wit- [vɪt] *adj.* White. [*Afk. fr. Du. wit cogn.* white] **1.** Found in place names e.g. Witbank, Witkop, Witnek, Witpoort etc.

2. Prefixed to numerous plant and animal names sig. white colouration.

wit [vɪt] *adj. Sect. Army lit.* 'white' (q.v.)

used of a junior too forthcoming, or unmindful of his place: also in general use as both *white* and ~ with an extension of meaning, of anyone insubordinate, disobliging or impertinent. [*Afk. wit cogn.* white]

> In the mornings they must *aantree* (form up) and any recalcitrant behaviour evinces the exclamation 'ek sê he's wit'. Picard in *Eng. Usage in S.A.* Vol. 6 No. 1 May 1975

witbaas [ˈvɪtˌbɑs] *n. colloq.* Mode of address or reference to someone bossy or domineering over non-White people. [*Afk. wit* white + *baas* master *cogn.* boss]

> ... always likes to play witbaas here. O.I. 1972

witblits [ˈvɪtˌblɪts] *n.* Home-distilled 'brandy' or raw spirit, acc. some the *voorloop* (q.v.) *cf. U.S. white lightning, moonshine, Canad. whisky blanc, white whisky, Irish potheen, 'mountain dew': see also quot. at vaatjie. [*Afk. fr. Du. wit cogn.* white + *blits* lightning-flash]

> Witblits, of course, is home-distilled dop brandy with a high alcoholic content, . . . A slow fire then gives a pure, strong *witblits* – white lightning because it has none of the colour imparted to more respectable brandies by their casks. . . . Farmers are allowed to distil small quantities of witblits for their own use. Green *Karoo* 1955
>
> One remembers as recently as 1946 someone saying 'Whatever you do, don't get into the habit of drinking brandy.' A folk memory, I suppose, of witblitz and Cape Smoke. *E.P. Herald* 3.4.74

witchdoctor *n. pl.* -s. Also *sangoma, iqira isanusi, twasa,* all (q.v.). An African practitioner of magic, medicine and witchcraft both in tribal and urban societies, often a smeller-out of witches or other evildoers, usu. with a stock-in-trade of herbal and other remedies, incantations and spells. ⟦Some ~s are consulted by Whites also: see *throw the bones,* (2) *dolos* and *smell out,* and second quot. at *ritual murder.* [cf. *shaman, Jam. E. obeah man/woman, samfie man/woman; U.S. and Canad. (Indian) medicine man*]

> Belief in the power of witchcraft, and the ability of the witchdoctor to divine who practised it, is as strongly rooted in the minds of the people as ever. Whiteside & Ayliff *Hist. of Abambo* 1912
>
> Does the urban African believe in the divining powers of witchdoctors? . . . – the only doctor who has two 'degrees' in witchcraft, (sangoma, herbalist and witchdoctor) . . . can speak to a

snake, can turn a snake into wood with his herbs. *Bona* Mar. 1974

> . . . unashamedly admits that his agency pays a retainer to a witchdoctor to 'vet' ads specifically aimed at blacks. 'He tells us he knows the black man's psyche. We make use of him in the same way as we use psychologists and sociologists for whites. *Financial Mail* 16.6.78

witchweed *n.* Also *mieliegif* and *rooiblom* (q.v.) any of several species of *Striga.*

> Witchweed-resistant varieties of kaffir corn are being produced at the Potchefstroom School of Agriculture. *H'book for Farmers* 1937

witels [ˌvɪtˈels] *n. Platylophus trifoliatus,* a forest tree of up to 18m in height preferring watercourses or other damp habitat: also the timber used for furniture and the keels of boats: see *rooi-els* and quot. at *pear.* Found in place names Witelsbos, Witelsriver. [*Afk. fr. Du. wit cogn.* white + *els* alder]

> The other woods most in request, and found in Albany, are – Red and White Milk, Red and White Else. *Greig's Almanac* 1831

as *modifier;*

> One White Els Chest. *E.P. Herald Advt.* 1.6.73

witgatboom [ˈvɪtˌxatˌbuəm] *n. pl.* -bome. *Boscia albitrunca* an evergreen tree with a pure white trunk 'as if they had been whitewashed' (Burchell *Travels II* 1824) and edible roots which can be pounded into meal or roasted and ground as a substitute for coffee: also known as *wonderboom* (q.v.). [*Afk. fr. Du. wit cogn.* white + *gat* hole, opening + *boom* tree *cogn. Ger. Baum*]

> Where the witgatboom grows . . . a coffee shortage causes no trouble. This tree, also known as the shepherd's tree . . . grows by itself, to a height of twenty feet, sometimes offering the only shade for miles. Shepherds love the *witgatboom* for other reasons as well. It is an evergreen, the berries can be eaten by men and animals and sheep thrive on the leaves. Green *Karoo* 1955
>
> 'Labuschagne drank the rotten witgat coffee – not that he minded – and repeated his words.' Lighton *Out of the Strong* 1957

withalskraai [ˈvɪt(h)alsˌkraɪ] *n. pl.* -e, -s. *Corvus albicollis,* also *ringhalskraai,* the white-necked crow or Cape Raven, a predator troublesome to farmers, taking fowls and even lambs. ⟦Known among Africans as *umfundisi* (q.v.) from the white collars worn by missionaries in former times. See also quot. at *korhaan.* [*trans. Eng.* white-necked crow. *Afk. hals* neck + *kraai cogn.* crow]

with *prep. and adv. substandard.* Used in various ways in S.A.E. **1.** *adv.* redundantly in phr. *come/go/take* ~ equiv. of 'along'. [*mistrans. Afk. saam, adv.* along]

Let's . . . split to some other beach . . . Take our surfies with . . . *Darling* 12.2.75

That is not too bad since I shall stay only for a day, but she has insisted that Tom Jones comes with. *Friend* 9.7.75

Also *dial U.S.:*

In his dialect such sentences as (53a) are possible, and they are synonymous with sentences like (53b).
(53) a. Sid is coming with.
      b. Sid is coming with me.
John Ross *On Declarative Sentences* (ed 1970)

2. erron. in phr. *throw* ~ : see also quot. at *throw*. [*trans. Afk. met,* with, by, *in phr. gooi met, lit.* 'throw with']

The living are throwing me with things. I know . . . but it is dangerous, they will kill me. *Drum* Nov. 1964

withond [ˈvɪt(h)ɔnt] *n. poss. reg. Graaff-Reinet.* A distilled liquor said to be the E. Cape version of *mampoer* (q.v.): see quot. below. [*Afk. wit cogn.* white, *hond cogn.* hound]

There are lovely local delicacies in Graaff-Reinet by the way – prickly pear syrup, wilds biltong, lucern honey, pampelmoes konfyt . . . withond so I was led to believe, was distilled from prickly pears and has a kick equal to nuclear fall out . . . Your local told me all was a blatant lie. True withond is made from the purest of hanepoot grapes, not a drop of anything else and it's 90 per cent proof . . . it was about all they didn't offer me to drink. Von Biljon *cit. Sunday Times* 6.11.77

witkoppie [ˈvɪtˌkɔpi] *n. pl.* -s. *colloq. lit.* 'Little white head': a blonde child. [*Afk. fr. Du. wit cogn.* white + *kop* head *cogn. Ger. Kopf* + *dimin. suffix* -(*p*)*ie*]

With such a pack of witkoppies coming out of Prep. I can't tell which is my own child. O.I. 1968

witogie [ˈvɪtˌʊəxi] *n. pl.* -s. Known as *glasogie, trans. white eye* (q.v.), any of several of the Zosteropidae, also *kersogie, kraal* (q.v.) *ogie.* [*Afk. fr. Du. wit cogn.* white + *oog* eye + *dimin. suffix* -*ie*]

. . . the tiny witogies have for their own use a fruit laden pomegranate tree in the lush garden. Kavanagh *Merry Peasants* 1963

Wits [vɪts, wɪts] *n. prop.* The University of the Witwatersrand, ~ *ie*, a student or alumnus of ~ university. [*abbr. Witwatersrand* (+ *personif. suffix* -*ie* (q.v.)]

Then there is Wits. I was a student at the Witwatersrand University in the early days, when there was still the smell of wet paint and drying concrete about the buildings at Milner Park. Bosman *Cask of Jerepigo* 1972

witteboom [ˈvɪtəˌbʊəm] *n. pl.* -bome. *Leucadendron argenteum* (Proteaceae): see *silver tree,* also found in place names Witteboom, Wittebome. [*Du. wit(te) cogn.* white + *boom* tree *cogn. Ger. Baum*]

14 Feb. 1811. This place is called Witteboom, a name which, with great propriety, it has received on account of numerous plantations of large Witteboom, or Silver trees, which grow about it. The native station of this handsome tree, is the sloping ground at the foot of the eastern side of Table Mountain. Burchell *Travels I* 1822

woel [vul] *vb. intrns. slang.* 'Rush around', 'hare about' etc. [*Afk. fr. Du. woelen* to bustle, toss about]

Low water early. We'll have to woel if we want prawns. Fugard *Boesman & Lena* 1969

They're always woeling round town and their mother can't bear it. O.I. Schoolgirl 1973

woema [ˈvuma] *n. colloq.* Energy, power as in 'He's always full of ~ ', or 'I never have much ~ in the hot weather' etc.: see also *krag.* [*poss. fr. Zu vb.* -*vuma* thrive, grow well, used of persons or plants]

WOEMA

Let your car come alive . . . It'll save you money. And at the same time give you something really new. Woema! *Rand Daily Mail Advt.* 6.3.71

woer-woer [ˈvurˌvur] *n. pl.* -s. Child's toy, believed to have originated with the Bushmen, consisting of a flat object, often a button, threaded on a loop of string, which twisted and held taut between the hands is made to hum by moving them closer together and further apart alternately. [*onomat.*]

Get your magic woer-woer 5c each at this theatre. Cinema Cape Town Jan. 1973

woes [vus] *adj. slang.* Furious: ill-tempered esp. among children. [*Afk. fr. Du. woest* wild, fierce, unruly]

. . . one big happy jol. Not a insult to be heard or a fist shook. The only woes faces is those what's hung over. *Darling* 28.5.75

'Who says you can go snooping in . . . my wardrobe, hey Ma?' I screech. Hang but I'm only woes too. *Ibid.* 4.2.76

wolf [vʊlf, vɔlf] *n. pl.* -ves, -we. Any of various species of *Hyaena* inc. *Crocuta crocuta,* originally called ~ by the colonists from a real or fancied likeness to the wolf of Europe: see quot. at *wild dog*. [*Afk. fr. Du. wolf*]

Aug. 11th 1825. I hereby declare that I have purchased 5 ounces of arsenick of J. Hancock for the alone purpose of destroying wolves which infest my place – Witness my hand. Hezh. Sephton. Hancock *Notebook*

In combination: *berg* ~, *gestreepte* ~, *maned* ~, *strand* ~ (q.v.), *tiger* ~ (q.v.); also in place name Wolfhuis.

. . . the intelligence that the dogs tracked a maned wolf. . . . This animal is a species of hyena, and exceedingly destructive to the sheep, killing more than it devours. Philipps *Albany & Caffer-land* 1827

attrib.

Gillmer and Martin Have for Sale At their Stores No. 26 Grave-street Ironmongery in Great Variety . . . Mouse, Rat and Wolf traps. *Cape of G.H. Almanac Advt.* 1841

Woltone [ˈvɔlˌtʊənə] *pl. n.* Cape Colonials: see quot. *cf. Kaapenaar;* see also *Blikoor,* (1) *Vaalpens.* [*Afk. wol cogn.* wool + *toon* toe + *pl. -e*]

Woltone – Cape Colonials are so called because the main branch of farming in early days was with sheep, and wool used to be baled by two barefooted men who stood inside the suspended wool-bag and 'tramped' down the wool as it was being thrown into the bag. Swart *Africanderisms Supp.* 1934

wolwe- [ˈvɔlvə] *pl. n.* Hyenas: found in place names e.g. Wolwekraal, Wolwəhoek, Wolwedans. [*pl. form of wolf* (q.v.)]

wolwegift [ˈvɔlvəˌxɪft] *n.* Also *wolweboontjie, Hyaenanche globosa,* the fruits or seeds of which are highly toxic, and in powdered form rubbed into carcasses for destroying hyenas: see *wolf.* [*Afk. fr. Du. pl. wolven* wolves + *gift* poison]

The root, commonly known by the name of 'Wolve-gift' (a medical drug) grows here in abundance, and has become an article of commerce. *Greig's Almanac* 1833

Wonderboom [ˈvɔndə(r)ˌbʊəm] *n. usu. capitalized. Ficus pretoriae,* formerly *Ficus cordata,* the *wildevyeboom :* properly applied to a remarkable clump of these at the foot of the Magaliesberg outside Pretoria: also place name Wonderboom. ⟨A term used to refer to the *witgatboom* (q.v.) on account of its extraordinary usefulness. [*Afk. fr. Du. wonder* marvel + *boom* tree *cogn. Ger. Baum*]

Ficus Pretoriae This species is a spreading, usually evergreen tree, growing up to 70 feet in height and found in Natal, in various parts of the Transvaal, and northwards into tropical Africa. To this species belongs the famous Wonderboom, the unique groups of trees growing on the outskirts of Pretoria. Palmer & Pitman *Trees of S.A.* 1961

woodcutter *n. pl.* -s. One of a number of self-employed workers of timber in the Knysna forests from the establishment in 1776 of a Dutch East India Company woodcutting station in the area until their removal by Act of Parliament in 1939. ⟨On account of their primitive way of life and poverty they were known also as *poor whites* (q.v.); see quot. at *poor white* (Informant Hjalmar Thesen); and *erron.* as *bosbouers* (q.v.).

These woodcutters are the poorest class of white people in the colony: earning a livelihood with severe labour by conveying timber to the Knysna or to Cape Town, in wagons . . . which, they complain, affords them but a meagre subsistence. Thompson *Travels I* 1827

wooden orange *n. pl.* -s. See *klapper², kaffir orange.*

work *n. substandard* In S.A.E. sig. not the job but the 'place of work': see quots. at *tokoloshe* and *dwaal.*

works on my/your etc. nerves *vb. phr. substandard* equiv. of Gets on my/your etc. nerves: see quot. at *telling you, I'm. cf. nerves, on my.* [*trans. Afk. dit werk op my/ jou etc. senuwees* lit. 'it works on my/your etc. nerves']

worry *vb. trns. Afr.E.* Used by African speakers sig. to importune sexually or for money: see quot. at *umfazi.* [*prob. trans. Ngu. ukukhataza* to pester, bother]

. . . complained of a tikolosh which worried her at night in the servant's quarters. . . . 'The tikolosh which is planted here worries girls who sleep in the room. It keeps them awake and has sex with them.' *Drum* 22.2.72

Every Saturday afternoon go to the park and kick those dogs, booze, swear at them and worry their mrezans. *Ibid* 8.4.72

wors [vɔ(r)s] *n.* Sausage, usu. *suffix* or *abbr.* of *boerewors* (q.v.) Also in combinations such as *rook~* (smoked), *garlic* ~, *dried* ~. [*Afk. fr. Du. worst cogn. Ger. Wurst* sausage]

By the time we rock back footsore and weary the wors is burnt to a crisp, the pap's cold and ouma's kipping. *Darling* 15.10.75

Wors is just sausage until . . . you add a little of what's in this distinctive well known bottle. *Fair Lady Advt.* 20.12.78

W.O. seal. See *wine-of-origin,* also *certified.*

wragtig [ˈvraxtɪx] *interj. colloq.* Exclamation of surprise, incredulity or in form *so* ~ as emphasis to statement: also *wragtie* and *so wragtie waar:* see *so waar*. [*Afk. fr. Du. waarachtig* truly, indeed]

'Wragtig! We live in wonderful times!' ejaculated the astonished settlers of Thirstland. The running of the rivers was indeed a diversion after the dreadful drought. Birkby *Thirstland Treks* 1936

wyn [veɪn] *n.* Wine: *prefix* in place name Wynberg, Cape, a grape-growing area, and in plant names usu. sig. having fruits with a vinous flavour: ~ *bessie, Dovyalis rhamnoides fr.* which 'brandy' and vinegar were formerly made; ~*klapper, Strychnos cocculoides fr.* the vinous taste of the pulp: also ~*blommetjie, Geissorrhiza rochensis fr.* gobletshaped crimson-based flowers. [*Afk. fr. Du. wijn cogn.* wine]

wys [veɪs] *vb. slang lit.* To 'show' *cf. spog* used by children and young people *sig.* 'show off', 'make a display' etc. e.g. He's got to ~ with his belongings all the time; He's not happy unless he's ~ing about something; Nothing but a ... ~er; Always must ~ with his money etc. [*Afk. fr. Du. wijzen* to show]

X

X.D.C. [ˈeksˌdiˈsi:] *n. prop.* A government scheme for financial development and assistance in the *Xhosa* (q.v.) *homelands* (q.v.): see also *T.D.C.* [*acronym X*hosa *D*evelopment *C*orporation]

The XDC aims at planning and promoting the development of all sectors of South Africa's two Xhosa nations, namely of the Transkei and the Ciskei. *Panorama* May 1973

Historic Farm Goes Black. The sale of ... to the XDC follows four years of negotiations between the ... family and the Government. The XDC will take over ... in September. *E.P. Herald* 5.7.74

Xhosa [ˈkɔ̆sa +] *n. pl.* ama-, -s, also *modifier*. **1.** The African people of *Transkei* (q.v.) and the *Ciskei* (q.v.) formerly now *obs.* known as (1) *Kaffirs* (*hist.*) see quot. at (1) *Kaffir*: consisting of the AmaXhosa, Ama*Pondo* (q.v.) and Ama*Tembu* (q.v.): see quot. at *XDC*.

... the people residing on the border of the Cape Colony call themselves Amaxosa; but while all who belong to this nation call themselves by the national name, yet everyone belongs to some tribe, of which there are several. Shaw *My Mission* 1860

The newspapers in South Africa were still fighting the frontier wars and trying to divide the Xhosa nation . . . Chief Minister of the Ciskei, said at the weekend. *E.P. Herald* 16.12.74

2. no *pl.* The language of the Xhosa people, related to Zulu and to Swazi with which it combines to form the *Nguni* (q.v.) group of languages.

The Xhosa-speaking people, according to historian Basil Holt, welcomed the first Europeans with offers of sincoa – the Portuguese interpretation, he thought, of the Xhosa word for bread – 'sonka'. *Sunday Times* 31.10.76

3. A member of the ~ people.

I do not understand, he said.
– You are a Xosa, then, umfundisi?
– A Zulu, he said.
Paton *Cry, Beloved Country* 1948

But for the other million-odd Xhosas who live and work in 'White' South Africa, life under apartheid continues as usual. *Evening Post* 30.10.76

Xosa [ˈkɔ̆sa +] *n. pl.* ama, -s. Earlier sp. form of *Xhosa* (q.v.).

Y

yakkie [ˈjækĭ] *n. pl.* -s. *prob. reg.* E. Cape: An earthenware marble: *Austral. common* o clay marble, see also *al(l)ie, ghoen, ironie, queen,* (2) *tollie.* [*unknown*]

As a marble a yakkie is lowest of the low – a dull-coloured earthenware, not even glass. If you had a good big ironie as your ghoen you could pulverise your enemies' yakkies. Guy Butler 1971

yebo [ˈjeːbɔ̆] *interj. colloq.* Yes: see also *ja.* [*Zu. yebo* yes]

yellow-belly *n. pl.* ∅, -ies. *Epinephelus guaza:* the yellow-belly rock cod, also known as garupa (grouper). *Slang* Second World War a coloured girl, *cf. Anglo-Ind.* ~ a half-caste: see also *Vaalpens.*

yellowfish *n. pl.* ∅. Any of several S. Afr. river fish of the Cyprinidae, genus *Barbus*: in combinations Clanwilliam ~ *B. capensis; large/small mouth* ~ *B. kimberleyensis* and *B. holubi; large/small scale* ~ *B. marequensis* and *B. polylepis* etc.: also *geelvis.*

These hand lines which were about twenty feet in length were, as well as the hooks attached to them, very strong, and barbel or yellow fish rarely broke away from them. Pohl *Dawn and After* 1964

Carp dominated the contest, with a smattering of yellowfish, mudfish and barbel also in the weigh-in nets. *Sunday Times* 27.10.74

yellow ground *n.* The upper level in a *kimberlite* (q.v.) pipe, probably consisting of *blue ground* (q.v.) weathered by sun, which overlies the blue ground proper. [Orig. thought to be the only diamondiferous earth, see quot. [*fr. yellowish-green colour : prob. trans. Afk. geelgrond*]

Hadn't people said it was finished at the time when the yellow-ground ended and the blue-ground, actually incomparably richer, began? Brett Young *City of Gold* 1940

yellow rice *n.* Traditional dish: rice coloured yellow with *borrie* (q.v.) and usu. containing raisins: see also *geelrys*, *begrafnisrys*, and quot at *borrie*. [*trans. Afk. geelrys fr. Du. geel* yellow + *rijs* cogn. rice]

... roasted mutton, an' roasted potatoes an' ... sweet-potatoes an' yellow rice with raisins in it. Smith *Platkops Children* 1935

yellow route *n. equiv. Chicken run* (q.v.)

More than 1000 are leaving every month, taking the 'yellow route' and the 'chicken run' south to South Africa, east to Australia, even north to the much-despised United Kingdom. (*London*) *Sunday Times Magazine* 9.10.77

yellowtail *n. pl.* Ø. Any of the species of the marine fish *Seriola* of the Carangidae fam. esp. *S. lalandi* the Cape yellowtail or albacore: see *alfkoord. cf. U.S. amberjack, Jam. Eng. yellow tail.* [*prob. trans. Afk. geelstert fr. Du. geel* yellow + *staart* tail]

That fine fighting fish, the yellowtail, has started running already. *Daily Dispatch* 23.11.71

The struggling yellowtail attracts others of its shoal. ... yellowtail are inquisitive fish and when one is hooked others rush in to see what the fuss is all about. Deep sea anglers have observed this habit. *E.P. Herald* 1.8.74

yellowwood *n. pl.* -s. **1.** Either of two species of *Podocarpus; P. falcatus, Outeniqua* ~ see *kalander*[2]; or *P. latifolius*, '*Real*' ~, formerly known as *upright* ~, *mistrans.* or *translit.* of *opregte geelhout* sig. genuine ~. [*trans. Afk. fr. Du. geel*, yellow + *hout* timber]

We have forest trees too, tall as any elm, the one I admire is called yellow-wood tho- why I cannot tell, tis more grey and dark green than any thing. Kate Pigot *Diary* Aug. 1820 *cit.* Fitzroy 1955

The unbalanced exploitation of yore, which

brought the precious stinkwood to near-extinction and decimated the yellowwoods, was stopped long ago. *E.P. Herald* 18.4.73

2. *n. and n. modifier;* The wood of the above species, clear golden yellow in colour but not very hard, used for building purposes as well as for furniture: see quots. at *wakis, jonkmanskas.* [*as above*]

J. C. Truter, General Dealer ... Has always on hand ... Wagon and Cart Wood, Yellowwood and Stinkwood Planks. *George Advertiser* 14.4.1870

... the long white wine-house with its rows of massive oak stukvats bound with iron, and its ceiling of bamboos placed close together over heavy cross-beams of yellow-wood. Fairbridge *Which Hath Been* 1913

as *modifier* ~ beams, floors, furniture, panelling, table, *wakis* (q.v.) etc.

yes-no *interj. colloq.* An emphatic affirmative usu. in answer to a qn.: see also quot. at *ja-nee.* [*trans. Afk. ja-nee* sure, that's a fact *etc.*]

The lady enriched my vocabulary by a glorious word, not in the phrase-book, which may express affirmation, negation, approval, credulity, and incredulity at will. viz. 'ja-nee', which means yes-no. Morton *In Search of S.A.* 1948

yessus ['jəsəs] *interj. slang.* Form of 'Jesus' as exclamation. *cf. U.S. Gee, Jeez, Jeepers* etc.: see quot at (1) *only.* [*fr. Afk. Jesus* ['jɪəsəs]]

The world was open this morning. It was big! All the roads ... new ways, new places. Yessus! It made me drunk. Fugard *Boesman & Lena* 1969

yesterday, today and tomorrow *n. Brunfelsia*, a shrub with flowers which open purple, turn light lavender-mauve and finally white so that all three colours occur at the same time. *cf. Jam Eng. today-tomorrow mango* which ripens first one side then the other.[*fr. sequence of colouring*]

The common Brunfelsia which we know as 'Yesterday, today and tomorrow' is one of the most sweetly scented of all flowering shrubs. *Daily Dispatch* 16.6.73

yet *adv.t. substandard* Used in past tense sentences e.g. Didn't you get it yet? [*prob. fr. trans. Afk. nog* still, yet, as yet (*without restriction of tense in use*)]

In 1935 the river wasn't canalised yet. O.I. 1973

... came to pick me up to take me to dinner but I wasn't back from the ... yet. Letter Schoolgirl 16, 9.7.74

Also redundantly:

> By Monday morning she had not yet returned and had not yet been found by the two animal societies, by the police, or by the Settlers Park rangers. *E.P. Herald* 7.11.74

yl [eɪl] *adj. colloq.* Sparse, scanty, thin. usu. of or pertaining to growth of plants or hair. [*Afk. fr. Du. ijl* thin, sparse]

> His beard's too yl for him ever to grow a full one. O.I. 1969

yirra See *Here.*

yo [jɔ̃] *interj.* Also *yu*: African exclamation of surprise, *cf. hau* or *Eng.* whew !

> ... African housewives' shopping baskets ... 'Yo, food costs too much,' the women said, shaking their heads. *E.P. Herald* 5.7.75

yokeskey [ˈjəʊkˌskeɪ] *n. pl.* -s. A yoke-pin or *skei* (q.v.), see also quot. at *skei*. [*translit. Afk. jukskei*]

> But often it happens while trekking that something goes wrong with the gear – a yokeskey or a nekstrop breaks. FitzPatrick *Jock of the Bushveld* 1907

Young Turk *n. prop.* also *modifier pl.* -s. Formerly in S.A.E. a member of the reformist wing of the *United Party* (q.v.) later the *Reform Party*: see *Progressive Reform Party.*

> Young Turks spurn Prog link. The firm 'no' to a Progressive-United Party axis came from ... top members of the 'Young Turk' movement in the United Party. *E.P. Herald* 14.6.74

youth, the *collective n. substandard* Used in S.A. sig. young people in general. [*prob. translit. Afk. die jeug* young people]

> The magazine was a popular one and was available to all – including the youth. *Daily Dispatch* 27.11.71
> The almost universal Afrikanerism that translates 'Die jeug' into 'The youth' – which is certainly not English, except when used in an individual sense. 'The youth (name withheld) was found guilty ...' *Sunday Times* 9.12.73

yster [ˈeɪstə(r)] *n. prefix suffix.* Iron: usu. sig. extreme hardness or strength e.g. ~hout (q.v.); ~ (*boom*), *Olea capensis;* ~*bos, Zygophyllum incrustatum;* ~ *gras, Aristida diffusa:* sig. also 'made of iron' as in *strijkijzer* (*obs.*), colonial name for sad-iron; or 'containing iron' as ~*klip* (q.v.) etc.: see also ~*vark.* In place name Ysterfontein. [*Afk. fr. Du. ijzer* iron]

> William Farmer Tools of all descriptions ... Coffee and Pepper Mills, Strykyzers, Brushes, Kettles and Komfores. *Cape of G.H. Almanac Advt.* 1841

ysterhout [ˈeɪstə(r)ˌhəʊt] *n.* See *ironwood*

and *umzimbiet/bete.* So-called acc. Smith (*S. Afr. Plants*, 1966) from Van Riebeeck's time when it was regarded as too hard for anything but fuel. [*Afk. fr. Du. ijzer* iron + *hout* timber]

> As a rule these came from Knysna and other villages in that neighbourhood, where timber was plentiful, especially stinkwood, yellow-wood, and ysterhout (ironwood) enormously tough, heavy, and suitable for ox-wagon shafts. Jackson *Trader on Veld* 1958

ysterklip [ˈeɪstə(r)ˌklɪp] *n. or n. modifier.* Ironstone, dolerite: see also second quot. at (1) *rand* [*Afk. fr. Du. ijzer* iron + *klip* rock, crag, *Afk.* stone]

> ... there is no doubt that the ysterklip (ironstone) country in the Northern Cape attracts and distributes lightning. Some of those ysterklip koppies have been struck hundreds of times. Green *Land of Afternoon* 1949

ystermannetjie [ˌeɪstə(r)ˈmanɪkɪ̈, -kɪ̈] *n. lit.* 'Iron-mannikin' the children's game occ. called 'punching statues' in which the aim is to keep still. *cf. donkermannetjie.* [*Afk. fr. Du. ijzer* iron + *man* + *dimin. suffix* -(*n*)*etjie*]

ystervark [ˈeɪstə(r)ˌfark, -fɑk] *n. Hystrix africaeaustralis:* the porcupine. Found in place name Ystervarkfontein. [*Afk. fr. Du. ijzer* iron + *varken* pig]

> The *hystrix cristata* of Linnaeus, called by the colonists here *yzter-varken* (or *iron hog*), is the same animal as the Germans carry about for a show in our country by the name of *porcupine.* Sparrman *Voyages I* 1785 cit. Pettman
> ... we never had much respect for the school teacher ... all he had was book-learning and didn't know, for instance, ... that an ystervark won't roll himself up when he's tame. Bosman *Bekkersdal Marathon* 1971

Porcupine (also *poss.* sig. hedgehog): as *prefix* in plant names referring either to a bristly or prickly habit of growth as in ~ *bos*(*sie*) *Microloma burchelii, Aspalanthus spinosa* and other species; ~ (*pol*), *Euphorbia eustacei* or *E. multifolia;* or to their being eaten by porcupines e.g. ~ *kos* (food) *Pachypodium succulentum;* ~ *melkpol* (milk shrub) *Euphorbia inermis;* ~ *patat* (sweet potato), various species of *Kedrostis;* ~ *wortel Zantedeschia aethiopica,* see *varklelie, varkblom.*

yu [jŭ] *interj.* See *yo.*

Z

zakaat [zə'kɑt] *n.* Obligatory payment made for charity by all Muslims in terms of Islamic law: see quot. at *halim*. ¶In Durban *esp.* part of this is a handout to all beggars coming to any Muslim shop owner on a Thursday. [*Arab.*]

The purpose of this leaflet is to warn the Muslim public that it should be on guard when paying Zakaat. If Zakaat is handed to such persons who are not cognizant with the Islamic rules applicable to Zakaat-spending, . . . Remember that your Zakaat will not be spent according to the Shariah. And . . . your obligation of Zakaat will not be regarded as being discharged. In terms of the law of Islam, you are again liable for Zakaat payment. *Leader* 21.2.75

ZANU ['zɑnŭ] *n. prop.* Zimbabwe *A*frican *N*ational *U*nion. [*acronym*]

. . . the rift in the nationalist movement in Rhodesia in 1963. The movement divided into two groups: the Zimbabwe African Peoples Union (ZAPU) led by Mr. Joshua Nkomo, and the Zimbabwe African National Union (ZANU) led by Rev. Ndabaningi Sithole. The following years saw attempts by the OAU to reconcile the two parties. *African Review* Vol. 5 No. 1. 1975

zap [zæp] *vb. colloq.* Rhodesian: to shoot, 'get': *S.Afr.* and *U.S.* also.

I'm a good shot . . . if the terrorists come I'll zap them. I have a child . . . He knows all the slang . . . he knows when his Dad's off zapping terrs. *Fair Lady* 16.3.77

ZAPU *n. prop.* See *ZANU*.

Z.A.R. *n. prop.* The *Z*uid *A*frikaansche *R*epubliek (q.v.). [*acronym*]

On April 12, 1877, the erstwhile Zuid Afrikaansche Republiek (ZAR) was annexed by Sir Theophilus Shepstone as British territory. *Panorama* Jan. 1974

Many of our supporters have pointed out that the old ZAR anthem 'Kent Gij Dat Volk' has been virtually unused for 80 years. Steenkamp *cit. Capetonian* May, 1979

Zarp [zɑp] *n. pl.* -s. A member of the *Z*uid *A*frikaansche *R*epubliek *P*olitie: acc. some *Z. A. R*ijdende (mounted) *P*olitie (police). [*acronym*]

On one occasion he was being escorted to Pretoria by two ancient-looking burgher policemen, commonly known as 'Zarps' (Zuid Afrikaanse Republiek Polisie). Klein *Stagecoach Dust* 1937

zebra *n. pl.* -s or Ø. **1.** *Equus zebra*, formerly *wildeperd* (*Du. puard* (q.v.)) 'wild horse'.

2. *Diplodus cervinus*, marine fish also called *wildeperd, streepdassie* or *bontrok*.

Zebu *n. prop.* Non-S.A.E. Humped cattle, *Bos indicus* : in S.A. usu. *Africander* (q.v.) also presum. Brahman. [*Zebu, Bos indicus*, widely spread species of humped cattle found throughout Asia and Africa]

Supreme Zebu bull and Champion Africander Bull, Berlin Buffel. *Farmer's Weekly* 7.6.72

zeekoe ['zikŭ] *n. lit. Sea cow* (q.v.): travellers' or colonists' term for *Hippopotamus amphibius*.

30 Oct. 1811 . . . had never seen a Zee-koe (Sea-Cow), as the colonists call the Hippopotamus. Burchell *Travels I* 1822

In place names Zeekoevlei, Zeekoegat (see above), Zeekoerivier. [*Du. zeekoe* walrus, river horse (*hippopotamus*), sea cow or 'manatee']

In combination *hist obs.* ~ *spek*, 'bacon' made of fat hippopotamus meat.

The first was an omelette of ostrich egg, and the latter the salted and smoked flesh of the hippopotamus, or as it is called here Zee Koe spek (sea-cow pork), and from good pork it cannot be distinguished. Chase *Cape of G.H. & Algoa Bay* 1843

Also ~ *gat*, a deep pool in a river-bed. *cf. Canad. buffalo wallow*, a mud hole.

. . . large pools, or as the colonists call them, *Zeekoe-gats*, deep enough to float a man-of-war. Thompson *Travels I* 1827

Zimbabwe [ˌzɪm'babwe] *n. prop.* Rhodesia: see quot. and quot. at *Azania*: also *ZANU* and *ZAPU*. [*fr. earlier place name*]

He told an ANC rally in Sinoia that the freedom of 'Zimbabwe' (the African Nationalist name for Rhodesia) was coming fast. *E.P. Herald* 19.8.74

deriv. ~ *an, n. pl.* -s. A Black Rhodesian.

He will meet President Kaunda and make a broadcast to Zimbabweans (the nationalist term for Black Rhodesians) over Zambia radio and television. *E.P. Herald* 17.3.75

Zimbabwe-Rhodesia *n. prop.* The country previously known as Rhodesia and now called *Zimbabwe* (q.v.): see also *Rhodzim*.

'*I do not want Zimbabwe ever to become another banana republic*' So declared Bishop Abel Muzorewa . . . soon to become the first black Prime Minister of Rhodesia, or Zimbabwe-Rhodesia as it is henceforth to be known. *Time* 30.4.79

zinc[1] *n.* Also *zink:* in S.A.E. equiv. of galvanised iron as in ~ *bath*, see also *bath*, or corrugated galvanised iron roofing as in ~ *roof* etc. *cf. Jam. Eng. zinc roof* corrugated iron. ¶Formerly genuine zinc was used: see bracketed quot. see also *sink*.

The red walls of the farm-house, the zinc roofs of the out buildings, the stone walls of the kraals all reflected the fierce sunlight. Schreiner *African Farm* 1883

The brazier, zink bath and cooking utensils, sack for the empties . . . were obtained from Coloureds in the area when filming [Boesman and Lena] started six weeks ago. *E.P. Herald* 23.1.73

Shrunken peaches cling to the branches of a parched tree. A zinc-bath hangs from a nail on the privy wall . . . 'I fetch the zinc-bath from its nail on the lavatory wall and half-fill it with water. Then I stagger with it into the kitchen . . . I love a bath in the morning.' Venter *Soweto* 1977

[Zinc was the roofing material eventually put on the Bathurst English church, and seems to have continued in popularity until the arrival of the cheaper corrugated iron in the fifties. Lewcock *C19 Architecture* 1963]

zinc,² **zink** *n. pl.* -s. *substandard.* Used erron. for (kitchen) *sink* (q.v.).

I'll just clear the zink drain first. O.I. Plumber

Zionist *n. pl.* -s. A member of the African separatist Zionist church: also used pertaining to the Church itself, see quot. also at *Ethiopian*. [*unknown poss. fr. Zion 'the New Jerusalem'*]

Sundkler discussing African separatist churches of the Zionist type points out that they are hostile to European techniques of education and medicine. They are syncretistic, in the sense that they carry over into their doctrines and worship values and rituals derived from traditional religion. Longmore *Dispossessed* 1959

zol(l)¹ [zɔl] *n. pl.* -s, *vb. trns., n. prop.* A hand-rolled cigarette. [*Afk. zol*, a hand-rolled cigarette: *Ngu. form i-zoli: presum. rel. Mexican-border argot, zol* a marijuana cigarette (*American Speech* May 1955)] **1.** A smoke, (*colloq.*) now *obs.*, also trade name of miniature cheroots. *cf. Canad. rollie, twisting, Austral. 'makings'* i.e. the materials used.

2. A cigarette of or containing *dagga* (q.v.); *cf. 'reefer'*.

The defence told the court . . . had smoked numerous dagga 'zolls' that afternoon and also had consumed a third of a bottle of brandy. *Daily Dispatch* 22.2.72

3. A measure of *dagga* (q.v.), by size or weight. *cf. U.S. joint, stick* etc.

The customers in turn sell 'zolls' – cigarette size – to their customers. *Drum* 27.8.67

. . . smuggling a small amount of dagga – in each case less than a 'zoll' – into the prison . . . *Het Zuid Western* 10.2.72

4. *Dagga, Cannabis sativa:*

We grow it in between the mealies and the zoll gets better attention than the mealies, because people prefer smoking zoll to chewing mealies . . . even the hungry people . . . so we make more money with the zoll. *Darling* 8.11.78

I started smoking zoll when I was thirteen. It was during my school holidays, and I made friends with a guy who . . . sold zoll. Through him I met other regular smokers. *Ibid.*

zoll² *vb.* and *n. Children's slang:* to pinch or steal, hence *~er* one who *~s*, e.g. 'Somebody's zolled my ballpoint': extended to sig. a dirty trick. [*unknown, poss. fr. zol(l)¹*]

'It was a zoll I'm telling you, he set stuff in the exam paper we've never done in class.' O.I. Child

Also as *n. prop.* a card game – usu. of children – similar to beggar-my-neighbour or 'cheat'.

zone *vb. trns.* To proclaim or set aside a certain area or district for occupation by a particular racial group; also to *rezone* to change either the purpose of an area e.g. from 'residential' to 'business' or the racial group for which it was formerly *~d:* see also *Group Area* and *release.* [*official term*]

In terms of a Group Areas proclamation . . . this area, and another mainly residential, adjacent area . . . were zoned for White occupation. . . . all buildings, land or premises in the newly zoned trading area shall . . . be occupied or used only for the purposes of business as defined in the proclamation. *Daily Dispatch* 27.3.70

zoning *vbl. n.* See quot. [*fr. zone*]

He and many people had the wrong impression about zoning. Zoning meant an area or portion set aside to be owned by a race group – Africans, for instance. *Daily Dispatch* 2.10.71

Zoo train *n. prop. prob. reg.* Cape Town. The train on which the members of Parliament and civil servants arrive from Pretoria for the parliamentary session from January to June each year.

Parliament isn't in session yet, but somehow I thought the Zoo Train had arrived weeks ago. O.I. 20.1.75

Zuid Afrikaansche Republiek [ˈzœit afriˈkɑnse ˌrepœˈblik] *n. prop. hist.* The old Transvaal 'Boer' Republic: see also *Vierkleur, Z.A.R., Zarp* and quot. at *furrow.* [*Du.* South African Republic]

The first prospectors began trickling into the Transvaal during the 1860's. At that time it was formed into the old Zuid Afrikaansche Republiek, and was in an impoverished state. *Panorama* Dec. 1973

zuider [ˈzœïdə(r)] *adj.* Southern: Dutch form in place name Zuider Paarl. [*Du. zuider,* southern]

Zulu [ˈzulu] *n. pl.* -s, ama-. **1.** The African people or nation concentrated in Zululand and Natal: see first quot. See also *Kwazulu, Territorial Authority* and quot. at *Swazi.*

16th May 1829 We this evening obtained some information respecting the country around Port Natal & the nation over whom Chaka ruled: the very names of the Chiefs, afford sufficient evidence of the pride of this people, . . . Chaka's grandfather was called Zulu, which signifies High, or the Heavens from him the Nation is now called Amazulu, or people of Heaven . . . Shaw *Diary*

2. The language of the ~ people which with *Xhosa* and *Swazi* makes up the *Nguni* (q.v.) group of Bantu languages: see also (2) *Ndebele.*

Natal: The natives are all of the Zoola nation and all speak the Kaffir language with slight variation in their dialects from the language spoken by the Kaffirs on the border of the Cape Colony. Shaw 1860 *cit* Sadler 1967
. . . who grew up in Natal and learned Zulu along with English. *Drum* 8.1.73

3. *n. pl.* -s. *Also modifier.* Small hardy indigenous cattle. [*fr.* (1) *Zulu prob. ex. Zululand*]

These Zulu cattle are small and light, not more than half the size of the Africander oxen, which are generally used for transport purposes; but they will live where the Africanders would starve. Haggard *Solomon's Mines* 1886
. . . the shafts of the Scotch-cart were re-

placed with the pole or disselboom, and a span of 16 ZULUS (the term for small oxen) . . . were harnessed to the cart. Tait *Durban Story* 1961

¶Other uses non-S.A.E. 'The fastest fishing vessel in the British Isles developed in 1878.' Goldsmith-Carter *Sailing Ships and Sailing Craft* 1969: also a *C19th U.S. and Canad.* railroad car or train. *Webster's Third International Dict.*

zut [zʌt] *prn. slang.* Esp. among children: equiv. of 'nothing', 'zero', 'blow all'. *cf. Brit.* 'blob'. [*unknown*]

. . . what with all this inflation and so forth . . . I've got zut left over at the end of the month . . . even living at home. *Darling* 4.2.76

zuur- [zy:r, sy:r] *adj. prefix.* Sour. [*Du. cogn. sour*] **1.** Found in S. Afr. place names now usu. *suur-* (q.v.), *hist.* ~*veld.*

. . . it may justly be said that the countries of *Auteniqualand* and the Zuurveld are extremely beautiful. Burchell *Travels I* 1822
. . . a farmhouse named Zuur-Plaatz. Thompson *Travels I* 1827

2. See *suur-*

zwager [ˈzwɑgə(r)] *n. prefix.* Brother-in-law, former sp. of *swaer* (q.v.) found in place name Zwagershoek. [*Du. zwager,* brother-in-law]

It is known by the name of Zwagershoek, or 'Brother-in-law's Corner.' Thompson *Travels I* 1827

zwart- [swart] *adj. prefix.* Black: formerly in plant, animal and place names, now *swart-* (q.v.). [*Du. zwart* black *cogn.* swart, swarthy]

EARLIEST RECORDED DATES OF SELECTED ITEMS

The following is a list of approximately 300 items, showing our earliest date for each one. They have been chosen either for their historical and general interest, or their continued usage in South African English, in many cases for more than a century. Among these in particular are the names of plants and animals, first used by the early travellers and naturalists, such as *dassie*, *buchu* and *korhaan*.

In compiling a *Dictionary of South African English on Historical Principles* it is our practice, for all printed sources, to use the first known date of publication in English as the source of our quotations. In the case of unpublished manuscript material, such as dated letters, journals and diaries etc., the stated date of composition has been accepted. We have used 1822 as the publication date of Burchell's *Travels in Southern Africa*, Volume I, based on his notes made between November 1810 and February 1812; likewise we have used 1824 as the publication date for Volume II, written on notes made between February and September 1812, so that

our earliest date for *kombaars*, for instance, is 1824 instead of 1st May 1812. We are thus more conservative than some other reference works. Our earliest date for Thunberg's *Travels* is 1795, when the first publication of the English translation appeared.

Many of these words have been absorbed into general English. The word *laager* was used figuratively in 1901 in the *Daily Telegraph;* in 1941 in the *Illustrated London News;* in 1958 in the *Times Literary Supplement*, and in 1960 in *The Economist*.

More such recent words are *kragdadig*, first used in a political sense in the South African Press in 1949, and in *Time* magazine in 1976. *Boerehaat* made its first appearance in the Press in March 1972; in April of that year it appeared in the London *Times*, and in *Time* magazine in May.

It is of interest to note that from this list 47% of the words originate from Dutch, 14% from Nguni, 4% from Khoisan and 3% from Malay.

M.Britz

aandblom	1822	borrie	1798	eat up	1827	inspan	1828
aardvark	1786	boslemmer	1821	enkosi	1835	inyanga	1836
aasvoël	1835	brak (adj.)	1731	erf	1812	isanusi	1886
abakwetha	1823	brakbos	1824	faction fight	1926	izibongo	1869
after-ox	1822	bredie	1815	fanagalo	1947	ja	1786
Africaner	1820	brei	1822	fat-tailed sheep	1857	jong	1812
Afrikaner	1850	buchu	1731	Fields, the	1871	kaffir beer	1837
aikona	1901	bush tea	1768	frikkadel	1870	kaffirboom	1827
alles sal reg kom	1822	bywoner	1886	fundi	1970	kaparrang	1867
amalaita	1925	Cape Corps	1820	ganna	1786	kapater	1833
amapakathi	1829	Cape doctor	1843	geilsiekte	1838	kappie	1834
amatongo	1884	Cape Smoke	1834	gesondheid	1875	karbonaatjie	1822
apartheid	1947	chinkerinchee	1795	gha(a)p	1819	karos(s)	1731
baasskap	1935	commandeer	1873	good-for	1821	karree	1795
babbalas	1959	compound	1886	gorah	1786	katel	1850
Baby	1886	concentration		guarri	1789	katjiepiering	1795
backveld	1902	camp	1901	hadedah	1786	kaya	1810
bandiet	1795	Constantia	1786	hamba	1827	kehla	1875
banket	1887	dagga	1670	hamba kahle	1838	kêrel	1837
Bay, the	1816	dagha	1878	hamel	1831	kerk	1821
bayete	1835	dassie	1786	hammerkop	1834	keurboom	1731
beerdrink	1895	dassievanger	1867	handsupper	1901	kgotla	1840
biltong	1815	dekriet	1822	hartebeest hut	1815	khalifa	1856
bittereinder	1906	deurmekaar	1901	heemraad	1795	kiaat	1801
blaasop	1853	disselboom	1822	hlonipa	1850	kierie	1731
blatjang	1902	District Six	1867	honey beer	1731	kiewietjie	1786
blueback	1866	doek	1798	Hottentot	1677	klaar	1852
blueground	1882	dominee	1846	Hottentot fig	1795	klipkous	1731
bobbejaan-		donga	1875	huisbesoek	1824	knobkier(r)ie	1832
spinnekop	1879	dop	1871	I.D.B.	1882	kloof	1731
bobotie	1870	Dopper	1815	impala	1824	kombaars	1824
boer	1776	dorp	1801	impi	1838	konfyt	1862
bokkems	1866	drift	1795	impundulu bird	1894	kopje-walloper	1886
boerehaat	1972	dubbeltjie	1795	indaba	1827	K.W.V.	1932
bokmakierie	1834	duiker	1731	induna	1835	korhaan	1731
bonsella	1908	Dutch medicines	1833	inkosi	1824	kraal	1731

350

kragdadig	1949	nagmaal	1883	silver tree	1731	treckers	1850
kramat	1833	off-load	1850	sis	1862	trek ox	1833
kreef	1863	off-saddle	1835	sjambok	1645	tripple	1880
kreupelboom	1731	oom	1822	skepsel	1844	tronk	1732
kukumakranka	1795	opgaaf	1800	skerm	1835	tsamma	1886
kwaai	1890	opsit	1883	skimmel	1832	tshwala	1826
kwedien	1912	opstal	1804	skoffel	1882	tula	1899
kweek	1904	ouderling	1818	skolly	1950	tulp	1835
kwela	1958	outspan (v)	1802	skrik	1887	uintjie	1786
laager	1835	padkos	1849	slim	1804	Uitlander	1884
lamsiekte	1790	pampoen	1798	smous	1796	umfaan	1852
leguaan	1790	pap	1858	sommer	1835	umfazi	1833
lekker	c.1840	pas op	1835	sopie	1790	Unkulunkulu	1857
lobola	1836	piet-my-vrou	1790	sosatie	1833	umlungu	1826
magtig	1891	pitso	1822	South-easter	1801	va(a)tjie	1838
malgas	1731	pondok	1821	spanspek	1863	vadoek	1880
mamba	1862	poor white	1896	spekboom	1688	Van der Hum	1870
mahem	1826	predikant	1821	spog	1870	velskoen	1731
marula	1857	ramkie	1806	spruit	1832	vendusie	1799
matjiesgoed	1795	ratel	1731	stand	1873	verdomde	1850
mealie-meal	1855	renosterbos	1731	stat	1897	verneuk	1871
mealie pap	1880	riem	1817	steekgras	1844	vlakte	1786
mebos	1795	rinderpest	1875	stinkblaar	1835	voerschitz	1831
meerkat	1801	roer	1824	stoep	1797	voetganger	1824
meester	1798	rondloper	1863	strandloper	1846	voetsak	1837
melee	1911	¹rooibaadjie	1848	strandwolf	1786	volksraad	1840
mis	1852	²rooibaadjie	1858	stuiver	1697	voorhuis	1816
mompara	1899	rooikat	1795	suikerbos	1795	voorloper	1837
mooi	1797	rooinek	1896	sukkel	1912	vry	1887
mopani	1875	roosterkoek	1852	tagati	1836	vryer	1883
mosbolletjie	1902	sakabula	1877	takhaar	1899	vygie	1795
mossie	1884	sambal	1870	tammeletjie	1838	wag-'n-bietjie	1786
mouse-bird	1822	schans	1846	tampan	1861	werf	1818
muchi	1836	schlenter	1891	tickey	1877	wildebeest	1824
muishond	1796	secretary bird	1786	Tixo	1731	winkel	1827
muti	1882	sheep's tail fat	1785	tokoloshe	1833	wragtig	1897
na(a)rtjie	1790	shimiyana	1870	tok-tokkie	c.1902	Zarp	1894

WORD SOURCES QUOTED

ABRAHAMS, PETER *Mine Boy*
Faber and Faber, London 1946, 1954 ed.
ADAMS, BUCK *The Narrative of Private Buck Adams*
7th (Princess Royal's) Dragoon Guards on the Eastern Frontier of the Cape of Good Hope 1843–1848
MS. dated July 4th 1884
ed. Gordon-Brown, A. The Van Riebeeck Society, V.R.S. Publication No. 22, Cape Town 1941
AGRICULTURE AND FORESTRY, Department of, *Handbook for Farmers in South Africa* ed. D. J. Seymore
Government Printer, Pretoria, 1937
ALEXANDER, SIR JAMES EDWARD
Narrative of a Voyage of Observation Among the Colonies of Western Africa in the Flag Ship Thalia and of a Campaign in Kaffir-land on the Staff of the Commander-in-Chief in 1835, in two volumes.
Henry Colburn Publisher, London, 1837
——*An Expedition of Discovery Into the Interior of Africa Through the hitherto undescribed Countries of the Great Namaquas, Boschmans, and Hill Damaras* in two volumes
Henry Colburn, Publisher, London, 1838.
Facsimile Reprint
C. Struik, Cape Town, 1967
——MS. annotated sketchbook dated 1878
Collection A. Gordon-Brown.

Almanacs, Cape
——*African Court Calendars* for 1807, 1809, 1815 1819.
Compiled and printed by Geo. Ross, Cape Town.
——*South African Almanac and Directory for 1827*
Government Printing Office, Cape Town.
——*South African Directory Advertiser for 1831*
——*South African Almanac and Directory for the year 1831*
George Greig, Cape Town.
——*South African Directory and Advertiser for 1833 and South African Almanac and Directory for 1833*
George Greig, Cape Town.
——*South African Directory and Advertiser for the Year 1834 and The South African Directory and Almanac for the year 1834*
George Greig, Cape Town
——*Cape of Good Hope Almanac and Annual Register for 1841*
B. J. van de Sandt de Villiers, Cape Town.
——*Cape of Good Hope Almanac and Annual Register for 1843*
B. J. van de Sandt de Villiers, Cape Town
——*Cape of Good Hope Almanac and Annual Register for 1845*
B. J. van de Sandt de Villiers, Cape Town.

——*Cape of Good Hope Almanac and Annual Register for 1853*
B. J. van de Sandt de Villiers, Cape Town
——*Cape of Good Hope Almanac and Annual Register for 1856*
B. J. van de Sandt de Villiers, Cape Town
——*Cape Town Directory, 1866*
Chas. Goode, Cyrus J. Martin, Cape Town.
ANDERSSON, C. J *Notes of Travel in South Africa 1875,* 1969 ed.
Hurst and Blacket, London.
Reprint C. Struik, Cape Town, 1969
ANDERSON, ANDREW A. *Twenty-Five Years in a Waggon in the Gold Regions of Africa*
Chapman and Hall Ltd., London, 1887
APPLEYARD, J. W. *The War of the Axe and the Xosa Bible*
The Journal of the Rev. J. W. Appleyard 1841–1859
ed. John Frye
C. Struik, Cape Town, 1971
ATMORE, M. G. *Cape Furniture*
Howard Timmins, Cape Town, 1965
AYLIFF, THE REV. JOHN see Whiteside, the Rev. Jos.
BAKER, SIDNEY J. *The Australian Language*
1966 ed. Currawong N.S.W. Publishing Co. Sydney
BANCROFT, F. *The Veldt Dwellers*
Hutchinson, London, 1912
A. GORDON BAGNALL *Wines of South Africa*
An account of their history, their production and their nature.
K.W.V. Paarl, Cape 1972
BARAITSER, M. and OBHOLZER, A. *Cape Country Furniture*
A. A. Balkema, Cape Town, 1971
BARKER, GEORGE *Diary Transcript*
1815–1828 Theopolis and Bethelsdorp
BARNARD, LADY ANNE, also BARNARD, ANDREW Extracts from *Lady Anne Barnard at the Cape 1797–1802*
Dorothea Fairbridge, Oxford, 1924
BARNARD, LADY ANNE *Journal* in *The Lives of the Lindsays* Vol. 3
John Murray, London, 1849
——*Letters and Journals* in *South Africa a Century Ago*
Maskew Miller, Cape Town, 1924
BARNHART, C. L. STEINMETZ, S. BARNHART, K. *The Barnhart Dictionary of New English 1963–1972*
Longman, London, 1973
BECK, HASTINGS *Meet the Cape Wines*
Purnell and Sons, Cape Town, 1955
BECKER, PETER *Sandy Tracks to the Kraals*
Dagbreek, Johannesburg, 1956
BEETON, D. R. ed. *Poetry 1974*
University of South Africa English Studies, Pretoria, 1974
BEETON, D. R. and DORNER, HELEN ed.

English Usage in Southern Africa Vols. 1–7, 1970–1976

BELCHER, C. I. ed. *Norman's Law of Sale and Purchase in South Africa* 3rd ed.
Hortors, Johannesburg, 1961

BELL, W. H. SOMERSET *Bygone Days* Being Reminiscences of Pioneer Life in the Cape Colony and the Transvaal with some account of the Jameson Raid and its Consequences
H. F. & G. Witherby, London, 1933

BENNETT, JACK *Mister Fisherman*
Michael Joseph, London 1964

BERGER, LUCY GOUGH *Where's the Madam?*
Howard Timmins, Cape Town, 1966

BERTHOUD, J. see VAN HEYNINGEN, C.

BIDEN. C. LEO. *Sea-Angling Fishes of the Cape (South Africa)*
Oxford University Press, London: Humphrey Milford, 1930

BIRD, W. WILBERFORCE *State of the Cape of Good Hope in 1822 by a Civil Servant in the Colony*
John Murray, London, 1823

BIRKBY, CAREL *Thirstland Treks*
Faber and Faber, London, 1936
—— *Springbok Victory*
Libertas Publications, Johannesburg 1941.

BISSET, C. B. *Sport and War in Africa*
John Murray, London, 1875

BLACK, STEPHEN *The Dorp* 2nd ed.
Andrew Melrose, London, 1920

BLORE, HAROLD *An Imperial Light Horseman*
C. Arthur Pearson, London, 1900

BOLSMANN, E. H. *The South African Wine Dictionary*
A. A. Balkema, Cape Town & Rotterdam, 1977

BOSMAN, HERMAN CHARLES *Mafeking Road* 1947 ed.
Human and Rousseau, Cape Town, Pretoria, 1969 ed.
—— *Unto Dust* ed. Lionel Abrahams
Human and Rousseau, Cape Town, Pretoria, 1963
—— *A Bekkersdal Marathon* ed. Lionel Abrahams
Human and Rousseau, Cape Town, Pretoria, 1971
—— *Jurie Steyn's Post Office* ed. Lionel Abrahams
Human and Rousseau, Cape Town, Pretoria, 1971
—— *Bosman at his Best* ed. Lionel Abrahams
Human and Rousseau, Cape Town, 1965
8th Impression, 1974
—— *Willemsdorp*
Human & Rousseau, Cape Town & Pretoria, 1977

BOTHA, C. GRAHAM *Social Life in the Cape Colony In the 18th Century*
Juta & Co. Ltd., Cape Town & Johannesburg, 1927

BOTHA, G. GRAHAM *Our South Africa: Past and Present*

Cape Times, Cape Town for the United Tobacco Company, 1938

BOWKER, T. H. *A Journal 1834–1835*
Cory Library MS. 1951

BOYCE, W. B. *Notes on South African Affairs* 1838 Facsimile Reprint ed. G. Mears
C. Struik, Cape Town, 1971

BOYLE, F. *To the Cape for Diamonds*
Chapman and Hall, London, 1873

BRADLOW, EDNA and FRANK *Here Comes the Alabama*
A. A. Balkema, Cape Town, 1958

BRANDEL-SYRIER, MIA *Black Woman in Search of God*
Lutterworth Press, London, 1962

BRETT YOUNG, FRANCIS *They Seek a Country*
Heineman, London, 1937
—— *City of Gold*
The Book Club, London, 1940

BRINK, ANDRÉ P. & HEWITT, A. H. *The Birds*
An adaptation and translation of *Die Hand vol Vere* André P. Brink based on the *Birds* of Aristophanes
Typescript lent by authors 1973

BRINKMAN, L. H. *Breath of the Karroo*
Herbert Jenkins, London, 1915

BRITS, J. P. see DU TOIT, P. J.

BROOKES, EDGAR *A History of Native Policy in South Africa*
Nasionale Pers, Cape Town, 1924

BROSTER, JOAN A. *Red Blanket Valley*
Hugh Keartland, London, 1967

BRUCE, M. C. *The Golden Vessel*
Juta, Cape Town and Johannesburg, 1919

BRUMMER, E. de S. *Problems and Tensions in South Africa*
Political Science Quarterly, New York, 1955

BUCHAN, JOHN *Prester John* (1910)
Penguin Books 1956, 1961 impression.

BUCHANAN, BARBARA I. *Pioneer Days in Natal*
Shuter and Shooter, Pietermaritzburg, 1934

BULPIN, T. V. *Lost Trails of the Low Veld*
Howard Timmins, Cape Town, 1951
—— *White Whirlwind*
Thos. Nelson & Sons, London, 1961

BURCHELL, WILLIAM *Travels in the Interior of Southern Africa*, Vol. I
Longman, Hurst, Rees, Orme, Brown and Green, London, 1822
—— *Travels in the Interior of Southern Africa*, Vol. II
Longman, Hurst, Rees, Orme, Brown and Green, London, 1824

BUTLER, GUY *Take Root or Die*
1966, published A. A. Balkema, Cape Town, 1970
—— *Cape Charade*
1967, published A. A. Balkema, Cape Town, 1968
—— *Karoo Morning*
An Autobiography (1918–1935)
David Philip, Cape Town, 1977

CALLAWAY, REV. CANON, M. D *The Religious System of the Amazulu* 1870
Publications of the Folk-Lore Society XV (1884)
CAMPBELL, GEORGE *Old Dusty of the Low Veld*
Howard Timmins, Cape Town, 1964
CAMPBELL, ROY *Collected Poems*
Bodley Head, London, 1955–1957
Cape Almanacs, see *Almanacs*.
Cape of Good Hope Literary Gazette, Vol. I
W. Bridekirk, Heerengracht, Cape Town, June 1830
CARLISLE, R. *et al.* ed. *Family of Man*, People of the World, How and Where They Live, Vol. 3, Part 42, 1975; Vol. 6, Part 80, 1976
Marshall Cavendish Great Britain
CARSTENS, W. P. *The Social Structure of a Cape Coloured Reserve*
Oxford University Press, Cape Town, 1966
CHARTON, NANCY *Afrikaners, Political Ring-masters*
Typescript lent by author. Paper delivered at *The Afrikaner Today*, Abe Bailey Centre for Intergroup Studies, March 1974
CHASE, JOHN CENTLIVRES *The Cape of Good Hope and her Eastern Province of Algoa Bay*
1843 J. S. Christophers, London
Facsimile, C. Struik, Cape Town, 1967
CLARK, PERCY M. *The Autobiography of an Old Drifter*
Harrap, London, 1936
CLOETE, STUART *Turning Wheels*
Collins, London, 1937
——*Watch for the Dawn*
Collins, London, 1939
——*The Hill of Doves*
Houghton Mifflin, Boston, 1942
——*Rags of Glory*
Collins, London, 1963
CLOUTS, SYDNEY *One Life*
New Coin Poetry
Purnell, Cape Town, 1966
COHEN, LOUIS *Reminiscences of Kimberley*
Bennett & Co., London, 1911
——*Reminiscences of Johannesburg and London*
Robert Holden, London, 1924
COLLETT, JAMES *MS. Accounts, Diary and Memoranda*
Lent by 1820 Settler Museum, Grahamstown
COLLINS, R. *The Impassioned Wind*
Jarrolds, London, 1958
COOK, M. A. *The Cape Kitchen* A description of its position, lay-out, fittings and utensils.
Printpak (Cape), Epping, Cape. 1973
COTTERILL, H. B. Homer's Odyssey a line-for-line translation in the metre of the original
Harrap & Co., London, 1911
COWIN, KAY *Bushveld, Bananas and Bounty*
Michael Joseph, London, 1954
DAVENPORT, T. R H. *The Afrikaner Bond*
Oxford University Press, Cape Town, 1966
DAVIDSON, BASIL *Old Africa Rediscovered* (1959) Seventh Impression
Victor Gollancz, 1970

DE JONGH, S. J. *Encyclopaedia of South African Wine*
McGraw-Hill Book Co. (South Africa) Pty. Ltd. Isando, Transvaal, 1976
DE KIEWIET, C. W. *A History of South Africa Social and Economic*
Oxford University Press, London, 1960
——*Fears and Pressures in the Union of South Africa*
1954, Virginia Quarterly Review, S.A. Pamphlets 89, Cory Library
DE KOCK, VICTOR *Those in Bondage*
Howard Timmins, Cape Town, 1950
——*The Fun They Had*
Howard Timmins, Cape Town, 1955
DE VILLIERS, A. R. W. ed., *English Speaking South Africa Today*
Oxford University Press (Southern Africa), Cape Town, 1976
DICKASON, G. B. *Cornish Immigrants to South Africa: The Cousin Jacks' contribution to the development of mining and commerce 1820–1920*
A. A. Balkema, Cape Town, 1978
DIKE, FATIMA *The First South African*
Ravan Press, Johannesburg 1979
DIKOBE, M. *The Marabi Dance*
Unpublished Typescript, 1970
DODD, A. D. ed. *An Anthology of Short Stories by South African Writers*
No date
DOKE, JOSEPH J. *The Secret City*
Hodder and Stoughton, London, 1913
DORNER, HELEN and BEETON, D. R. ed.
See Beeton, D. R.
DUCKITT, HILDAGONDA ed. M. KUTTEL *Book of Recipes*
A. A. Balkema, Cape Town, 1966
DUFF GORDON, LADY *Letters from the Cape 1861–1862* ed. Fairbridge, Dorothea
Juta, Cape Town, 1925
DUGMORE, H. H. *Reminiscences of an Albany Settler* 1870
Grahamstown, 1871
DUGMORE, ELIZA JANE *Diary 1871* (Covering a Journey to the Diamond Fields)
Typed transcript lent by Professor Guy Butler
DU PLESSIS, I. D. *The Cape Malays*
Maskew Miller, Cape Town, 1944
DU PLESSIS, I. D. and LÜCKHOFF, J. *The Malay Quarter and its People*
A. A. Balkema, Cape Town, 1953
DU TOIT, P. J. *Diary of a National Scout 1900–1902* ed. J. P. Brits
Human Sciences Research Council, Pretoria, 1974
DU VAL, CHARLES *With a Show Through Southern Africa*, 2 Vols.
Tinsley Bros., London, 1882
Eastern Province Directory for 1848
Godlonton and White, Grahamstown
EDMONDSTONE, F. J. *Thorny Harvest*
Central News Agency South Africa Ltd.
No date
ELIOVSON, SIMA *The Complete Gardening*

Book for S.A.
Howard Timmins, Cape Town, 1960
ELEGANT, ROBERT *Hong Kong*
Time-Life Books, U.S.A., 1977
EMSLIE, ETHEL *Diary 1901*
Pringle Collection MS.
EWART, JAMES, ed. GORDON-BROWN, A.
James Ewart's Journal covering his stay
at the Cape of Good Hope 1811–1814 and
his part in the expedition to Florida and
New Orleans ed. with an introduction by
A. Gordon-Brown.
C. Struik, Cape Town, 1970
FAIRBRIDGE, DOROTHEA *That Which Hath
Been*
Maskew Miller, Cape Town, 1913
——*The Torch Bearer*
Juta, Cape Town, 1915
——*Lady Anne Barnard at the Cape 1797–1802*
Clarendon Press, Oxford, 1924
——ed. See Duff Gordon, Lady
FARINI, G. A. *Through the Kalahari Desert. A*
narrative of a journey with gun, camera and
notebook to Lake Ngami and back.
Sampson, Low, Marston, Searle and Riming-
ton, London, 1886
Farmer's Annual and S.A. Farm Doctor Far-
mer's Weekly, Bloemfontein, 1914
FEHR, WILLIAM *Treasures at the Castle of
Good Hope* Board of Trustees, Castle Art
Collection with Howard Timmins, Cape
Town, 1963
FITZPATRICK, SIR PERCY *Jock of the
Bushveld*
1907, 1909 ed. Longman Green & Co. 1909
FITZROY, V. M. *Dark Bright Land*
Maskew Miller, Cape Town, 1955
Forum
1968 Vol. 4, No. 1
1970 Vol. 6, 1 and 2
FITZSIMONS, BERNARD, ed. *The Heraldry
and Regalia of War*
Phoebus, London 1973
FRYE, JOHN see APPLEYARD, J. W.
FUGARD, ATHOL *The Blood Knot*
Simondium Publishers, Johannesburg, 1963
——*People are Living There*
Buren, Cape Town, 1969
Oxford University Press, London, Paper-
back ed. 1970
——*Boesman and Lena*
Buren, Cape Town, 1969
——*Boesman and Lena* (1969)
Oxford University Press, London
Three Crowns Paperback ed. 1973
——with Ross Devenish *The Guest* Ad. Donker,
Johannesburg 1977
FULTON, ANTHONY *The Dark Side of Mercy*
Purnell, Cape Town, Johannesburg, 1968
——*I Swear to Apollo*
Purnell and Son, Cape Town, 1970
GERBER, HILDA *Cape Cookery Old and New*
Howard Timmins, Cape Town, 1950
GILL, ERIC *A First Guide to South African Birds*
5th ed., Maskew Miller, 1956
GIBBON, PERCEVAL *Stories South African* ed.

A. Lennox Short
APB Publishers, Johannesburg, 1969
GLANVILLE, E. *A Fair Colonist*
Chatto & Windus, London, 1894
——*Tales from the Veld*
Chatto and Windus, London, 1897
GOLDIE, FAY *River of Gold*
Oxford University Press, 1969
GOLDSWAIN, JEREMIAH *The Chronicle of
Jeremiah Goldswain 1819–1858*, 2 volumes,
ed. Una Long
Van Riebeeck Society, Cape Town, 1946,
1949 (V.R.S. numbers 27 and 29)
GOOLD-ADAMS, R. J. M. *South Africa Today
and Tomorrow*
John Murray, London, 1936
GORDON, GERALD *Four People*
McDonald, London, 1964
GORDON-BROWN, A. *South African Heritage*
from van Riebeeck to Nineteenth Century
times
Human and Rousseau, Cape Town and
Pretoria, 1965
——ed. see Adams, Buck; Ewart, James; Travel-
ler, the
——*Christopher Webb Smith, An Artist at the
Cape of Good Hope 1837–1839*
Howard Timmins, Cape Town, 1965
——*An Artist's Journey*
A. A. Balkema, Cape Town, 1972
GRAY, THE RIGHT REV. ROBERT D. D.
*A Journal of the Bishop's Visitation Tour
through the Cape Colony*
Part I, 1849; Part II 1851
Society for the Propagation of the Gospel,
London, 1849 and 1851
GRAY, STEPHEN ed. *On the Edge of the World*
Southern African Short Stories of the
Seventies
Ad Donker, Johannesburg, 1974
GREEN, LAWRENCE G. *Where Men Still
Dream*
Howard Timmins, Cape Town, 1945
——*So Few are Free*
Howard Timmins, Cape Town, 1946
——*Tavern of the Seas*
Howard Timmins, Cape Town, 1947
——*In the Land of Afternoon*
Howard Timmins, Cape Town, 1949
——*Grow Lovely, Growing Old*
Howard Timmins, Cape Town, 1951
——*Lords of the Last Frontier*
Howard Timmins, Cape Town, 1952
——*Under a Sky Like Flame*
Howard Timmins, Cape Town, 1954
——*Karoo*
Howard Timmins, Cape Town, 1955
——*There's A Secret Hid Away*
Howard Timmins, Cape Town, 1956
——*South African Beachcomber*
Howard Timmins, Cape Town, 1958
——*These Wonders to Behold*
Howard Timmins, Cape Town, 1959
——*Full Many a Glorious Morning*
Howard Timmins, Cape Town, 1968
——*A Giant in Hiding*

Howard Timmins, Cape Town, 1970
——*When the Journey's Over*
Howard Timmins, Cape Town, 1972
GREENE, L. PATRICK *The L. Patrick Greene Adventure Omnibus*
John Hamilton Ltd., London, 1928
GREIG, GEORGE *The South African Almanac and Directory for the year 1831*
Cape Town 1831
——*The South African Almanac and Directory for the year 1833*
Cape Town, 1833
GRIFFITHS, REGINALD *Man of the River*
Jarrolds Ltd., London, 1968
GRINDLEY, J. R. *Riches of the Sea*
National Commercial Printers, Cape Town, 1969
HAAGNER, A. and IVY, R. H. *Sketches of South African Bird Life*
Maskew Miller, Cape Town, 1923
HAGGARD, H. RIDER *Nada the Lily*
Longman, Green & Co., London, 1895
——*King Solomon's Mines* 1886
(Moscow 1972)
HAHLO, H. R. and KAHN, E. *The Union of South Africa;* the developments of its laws and constitution
Juta, Cape Town, 1960
HAMMOND-TOOKE, W. D. *Bhaca Society,* A people of the Transkeian Uplands South Africa.
(I.S.E.R. Rhodes University) Oxford University Press, Cape Town, 1962
HANCOCK, JAMES *MS. Notebook 1820–1837*
Photocopy lent by 1820 Settler Museum
HANCOCK, KEITH *Smuts* Vol. II
Cambridge University Press 1968
Handbook for Farmers in South Africa
see Agriculture
HARRIS, W. CORNWALLIS *Wild Sports of Southern Africa*
John Murray, London, 1839
HEADLAM, C. ed. see Milner, Sir Alfred
HELLMANN, ELLEN and ABRAHAMS, LEAH ed. *Handbook on Race Relations in South Africa*
Oxford University Press, Cape Town, 1949
HEMANS, A. N. *Log of a Native Commissioner*
Books of Rhodesia, Bulawayo, 1935
HENSHILWOOD, NORAH *A Cape Childhood*
David Philip, Cape Town, 1972
HERRMAN, L. *A Note on Cape English Idiom*
English Studies in Africa, Vol. 2, No. 2, September, 1959
HEWITT, A. G. *Cape Cookery*
Darter Bros. & Walton, Cape Town, 1890
HEWITT, A. H. and BRINK, ANDRÉ P. see Brink, André P.
HIGHAM, MARY *Household Cookery for South Africa,* (1st ed. 1916) 8th ed.
R. L. Esson, Johannesburg, 1939
HINCHLIFF, PETER *The Anglican Church in South Africa*
Darton, Longman & Todd, London, 1963
HOBBS, JENNY *Darling Blossom*
Don Nelson, Cape Town, 1978

HOCKING, ANTHONY *Pride of South Africa: Diamonds*
Purnell S.A., Cape Town, 1973
HOERNLÉ, R. F. S. *South African Native Policy and the Liberal Spirit*
Phelps-Stokes Fund of the University of Cape Town, 1939
HORRELL, MURIEL *The African Homelands of South Africa*
South African Institute of Race Relations, Johannesburg, June 1973
HOYLE, JOAN *Some Flowers of the Bush*
Longmans, Green & Co., London, 1953
HUDDLESTONE, TREVOR *Nought for Your Comfort*
Collins, London, 1956
HUDSON, M. B. *South African Frontier Life*
2 Vols. John Patterson, Port Elizabeth, 1852
I.C.I. Veterinary Products Handbook
I.C.I. (South Africa) Pharmaceuticals, no date
HUNT, P. M. A. *South African Criminal Law and Procedure*
Juta, Cape Town, 1970
JACKSON, ALBERT *Trader on the Veld*
A. A. Balkema, Cape Town, 1958
JACKSON, P. B. N. *Common and Scientific Names of Freshwater Fishes of Southern Africa* Piscator No. 90, Autumn 1974
JACOBS and ROSENBAUM *Readings in English Transformational Grammar*
Ginn, Lexington, Maine, U.S.A., 1970
JACOBSON, DAN *A Long Way from London* 1953, 1958 ed.
Weidenfeld and Nicolson, London, 1958
——*A Dance in the Sun*
Weidenfeld and Nicolson, London, 1956
JEFFREYS, M. D. W. *Afrikanderisms I, II, III*
I Africana Notes and News, Vol. 16, June 1964
II Africana Notes and News, Vol 17, No. 5, March 1967
III Africana Notes and News, Vol. 19, No. 1, March 1970
JENKINS, G. *A Bridge of Magpies*
Collins, Glasgow, 1974
JOHNSON, H. *Hugh Johnson's Pocket Book of Wine*
Mitchell Beazley, London, 1977
KAHN, E. see HAHLO, H. R.
KAVANAGH, M. *We Merry Peasants*
Howard Timmins, Cape Town, 1963
KEET, B. B. *Whither – South Africa?*
Trans. N. Marquard
University Publishers, Stellenbosch, 1956
KEPPEL-JONES, ARTHUR *A Short History of South Africa*
Hutchinson's University Library, London, 1948
KING, N. L. *Tree Planting in South Africa*
Reprinted from the Journal of the South African Forestry Association, No. 21, October, 1951
KIPLING, RUDYARD *The Five Nations*
Doubleday, Page & Co., New York, 1903
KLEIN, HARRY *Stagecoach Dust*
Pioneer Days in South Africa

Thos. Nelson & Sons, London, 1937
KOLB, PETER trans, MEDLEY *The Present State of the Cape of Good Hope*, or a particular account of the several nations of the Hottentots . . . together with a short account of the Dutch Settlement at the Cape. Done into English by Mr. Medley.
W. Innys, London, 1731
KRAUSS, FERDINAND ed. O. H. SPOHR *Travel Journal 1838–40*
A. A. Balkema, Cape Town, 1973
KRIGE, UYS *The Dream and the Desert*
Collins, London, 1953
KROPF, REV. A. *A Kaffir-English Dictionary*
Lovedale, 1899
KUTTEL, MARY ed. See Duckitt, Hildagonda
LADY, A. *Life at Natal a Hundred Years Ago 1864–1865*
C. Struik, Cape Town, 1972
——*Life at the Cape* (possibly by Louisa Grace Ross) 1870
C. Struik, Cape Town, 1963
LAIDLER, P. W. *A Tavern of the Ocean*
Maskew Miller, Cape Town, 1926
LANHAM, PETER and MOPELI-PAULUS, A. S. *Blanket Boy's Moon*
Collins, London, 1953
LARSON, KENNETH *The Talbots, Sweetnams and Wiggils*
Unpublished typescript, Cory Library, 1943
LATROBE, C. J. *Journal of a Visit to South Africa* in 1815 and 1816.
Seeley, London, 1818
LEIPOLDT, C. LOUIS *Three Hundred Years of Cape Wine*
Stewart, Cape Town, 1951
LE VAILLANT, M. *Travels into the Interior Part of Africa* 1780–85, 2 Vols. (trans. E. Helme)
G. G. J. and J. Robinson, London, 1790
LEWCOCK, RONALD *Early Nineteenth Century Architecture in South Africa, 1795–1837*
A. A. Balkema, Cape Town, 1963
LEWINS, JAMES DALRYMPLE *MS. Diary transcript* 1849–1850
lent by Dr and Mrs J. V. L. Rennie
LEYLAND, J. *Adventures in the Far Interior of South Africa*
George Routledge & Sons, Liverpool, 1866
LICHTENSTEIN, HEINRICH (trans. ANNE PLUMPTRE) *Travels in Southern Africa,* 1803–1806
Colburn, Henry, London, 1812
[V.R.S. No. 10, 1928, No. 11, 1930]
Van Riebeeck Society, Cape Town, 1928, 1930
LIGHTON, R. E. *Out of the Strong* School ed. (Revised)
Macmillan & Co. Ltd. London, 1958
LIVINGSTONE, DAVID *Missionary Travels and Researches in South Africa;* including a sketch of Sixteen Year's Residence in the Interior of Africa, and a Journey from the Cape of Good Hope to Loanda on the West Coast; thence across the Continent, down the river Zambesi, to the Eastern Ocean.

John Murray, London, 1857
LONG, UNA ed. See Goldswain, Jeremiah
LONGMORE, LAURA *The Dispossessed.* A study of the sex-life of Bantu women in urban areas in and around Johannesburg.
Jonathan Cape, London, 1959
LOUW, A. M. *20 Days That Autumn* 21st March – 9th April 1960
Tafelberg Uitgewers, Cape Town, 1963
LUCAS, THOS. J. *Camp Life and Sport in South Africa*
Chapman and Hall, London, 1878
LÜCKHOFF, J. See du Plessis, I. D.
MACDONALD, T. *Transvaal Story*
Howard Timmins, Cape Town, 1961
MACDONALD, W. *The Romance of the Golden Rand*
Cassel & Co., London, 1933
MACKENZIE, K. *A Dragon to Kill*
Eyre and Spottiswood, London, 1963
MACKRILL, JOSEPH *MS. Diary* 1806–1816
Cory Library
MACLENNAN, D. A. C. Unpublished typescripts lent by the author
——*A Winter Vacation*, 1961
——*(In) The Dawn Wind*, 1970. Also called *Great Wall of China*
——*The Wake*, 1971
1968
MAFEJE, A. See Wilson, M.
MARAIS, EUGENE *Road to Waterberg and Other Essays*
Human & Rousseau, Cape Town & Pretoria, 1972
MARKS, BERNARD *Our South African Army Today*
Purnell, Cape Town & Johannesburg, 1977
MASSON, F. *Botanical Travels*
Royal Society, London, 1776
MATSHIKIZA, TODD *Chocolates for my Wife*
Hodder and Stoughton, London, 1961
MAYAT, ZULEIKA ed. *Indian Delights*
Published by the Women's Cultural Group, John Ramsay, Durban, 1961
McCARTHY, JAMES REMINGTON *Fire in the Earth*
Robert Hale, London, 1946
McKAY, JAMES (late Sergeant in Her Majesty's 74th Highlanders) *Reminiscences of the Last Kaffir War*
1st ed. Grahamstown, 1871
C. Struik, Cape Town, 1970
McLURE, JAMES *The Caterpillar Cop* 1972
Penguin, England 1974
——*Rogue Eagle*
Macmillan, London, 1976
McMAGH, K. *A Dinner of Herbs*
Purnell, Cape Town and Johannesburg, 1982
MEARS, G. see BOYCE, W. B.
MEDLEY trans. See Kolb, Peter
MEER, FATIMA *Portrait of Indian South Africans*
Premier Press, Durban, 1969
MEINTJES, JOHANNES *Manor House*
Dial Press, New York 1964

MEIRING, JANE *Candle in the Wind*
Frederick Muller, London, 1959
MENTZEL, O. F. *Description of the Cape of Good Hope* translation 1944
Van Riebeeck Society, Cape Town, 1944 (V.R.S. No. 25)
METELERKAMP, SANNI *George Rex of Knysna*
Howard Timmins, Cape Town, 1955
METROWICH, F. C. *Frontier Flames*
Books of Africa (Gothic Press), Cape Town, 1968
MEURANT, L. H. *Sixty Years Ago* 1885,
Facsimile reprint
Africana Connoisseur's Press, Cape Town, 1963
MILLIN, SARAH GERTRUDE
Rhodes 1853–1902
Chatto & Windus, London, 1933
MILLS, GWEN M. *First Ladies of the Cape*
Maskew Miller, Cape Town, 1952
MILNER, SIR ALFRED later VISCOUNT
The Milner Papers 1897–1899
ed. C. Headlam
Cassel, London, 1931
MITFORD, B. *Aletta*
Spottiswoode & Co., London, 1900
MOCKFORD, JULIAN *Here are South Africans*
A. C. Black, London, 1944
'Money Maker' Manuals for Investors No. 2
A new dictionary of mining terms.
London, no date.
MOLLOY, BOB *South African CB Dictionary plus Codes and Regulations*
Don Nelson, Cape Town 1979
MOPELI-PAULUS, A. S. See Lanham, Peter
MORRIS, J. see WEST, M.
MORTON, H. V. *In Search of South Africa*
Methuen & Co., London, 1948
MPAHLELE, EZEKIEL *Down Second Avenue*
Faber and Faber, London, 1959
MULLER, D. *Whitey*
Ravan Press, Johannesburg, 1977
NAIPAUL, V. S. *The Suffrage of Elvira*
Penguin Books, Great Britain, 1969 (1958)
NEAME, L. E. *The History of Apartheid*
Pall Mall Press, London 1962
New South African Writing, Vols. 1, 2, 3, 4, 5
South African P.E.N. Centre, Purnell, no dates.
NICHOLLS, HEATON *Bayete*
George Allen & Unwin Ltd., London, 1923
NIDETCH, JEAN *Weight Watcher's Program Cookbook*
Hearthside Press Inc., New York 1973
OBHOLZER, A. and BARAITSER, M. See Baraitser, M.
O'CONNOR, J. K. *The Afrikander Rebellion*
Geo. Allen and Unwin, London, Maskew Miller, Cape Town, 1915
OPPENHEIMER, H. *The Fifth Stock Exchange Chairman's Lecture*
Supplement to *Optima One*, 1976
(Delivered at the Stock Exchange, London, 18 May 1976)
OPPERMAN, J. D. ed. *Spirit of the Vine*
Published on the 50th Anniversary of the

K.W.V.
Human and Rousseau, Cape Town and Johannesburg, 1968
PALMER, EVE and PITMAN, NORAH *Trees of South Africa*
A. A. Balkema, Cape Town, 1961
PARTRIDGE, E. *A Dictionary of Slang and Unconventional English* 6th ed.
Macmillan & Co., New York, 1967
PATON, ALAN *Cry, the Beloved Country*
Jonathan Cape, London, 1948
——*Too Late the Phalarope*
Jonathan Cape, London, 1963
——*Kontakion for you Departed*
Jonathan Cape, London, 1969
PERCIVAL, CAPTAIN ROBERT of His Majesty's 18th or Royal Irish Regiment *An Account of the Cape of Good Hope*
C. & R. Baldwin, London, 1804
PETTMAN, THE REV. CHARLES *Africanderisms, A Glossary of South African Words and Phrases and of Place and other Names.*
Longmans, Green and Co., London, 1913
PHILIP, MARIE *Caravan Caravel*
David Philip, Cape Town, 1973
PHILIPPS, THOMAS *Scenes and Occurrences in Albany and Caffer-Land*
William Marsh, London, 1827
PIGOT, ELIZA SOPHIA *Journal* 1819–1821
Typed transcript lent by Margaret Rainier, since published A. A. Balkema, Cape Town 1974
POHL, VICTOR *The Dawn and After*
Faber and Faber, London, 1964
POLSON, LIEUT. NICOLAS *A Subaltern's Sick Leave* or rough notes of a visit in search of health to China and the Cape of Good Hope.
G. H. Hultmann, Bengal Mily.
Orphan Press, Calcutta 1837
PRANCE, C. R. *Tante Rebella's Saga*
Witherby, London, 1937
PRINGLE, THOMAS *Ephemerides*
Smith, Elder & Co., London, 1828
——*African Sketches*
Edw. Moxon, London, 1834
RABONE, A. edit. *Record of a Pioneer Family* 1851–1890
C. Struik, Cape Town, 1966
RAYMOND, HARRY *B. I. Barnato, A Memoir*
Isbister and Co., London, 1897
REED, DOUGLAS *Somewhere South of Suez*
Jonathan Cape, London, 1950
REITZ, DENEYS *Commando;* A Boer Journal of the Boer War
Faber and Faber, London, 1929
Report of the Commissioners of Inquiry: *Inquiry on the Cape of Good Hope* ordered by the House of Commons to be published.
George Greig, Cape Town, 1827
ROBERTS, A. *The Birds of South Africa*
Witherby, London
Central News Agency, Johannesburg, 1948
ROBERTS, BRIAN *The Churchills in Africa*
Hamish Hamilton, London, 1970

ROBERTS, SHEILA *Outside Life's Feast*
Ad Donker, Johannesburg, 1975
ROBINSON, A. M. LEWIN ed. *The Letters of Lady Anne Barnard to Henry Dundas 1793–1803*
A. A. Balkema, Cape Town, 1973
ROCHE, HARRIET A. *On Trek in the Transvaal, or Over Berg and Veld in South Africa*
Sampson Low, Marston, Searle and Remington, London, 1878
ROGERS, MIRABEL *The Black Sash*
Rotonews, Johannesburg, 1956
ROOKE, DAPHNE *A Lover for Estelle*
Victor Gollancz, London, 1961
——*Margaretha De La Porte*,
Victor Gollancz, London 1974
ROSE, COWPER *Four Years in Southern Africa*
Henry Colburn and Richard Bentley, London, 1829
ROSENBAUM see JACOBS
ROSENBERG, V. *Sunflower to the Sun* The Life of Herman Charles Bosman
Human & Rousseau, Cape Town & Pretoria, 1976
ROSENTHAL, ERIC *General De Wet*
Dassie Publications, 1946
ROSS, LOUISA GRACE See Lady, A. *Life at the Cape*
ROUSSEAU, LEON trans. NANCY BAINES *Van Hunks and His Pipe*
Howard Timmins, Cape Town, 1966
SACHS, B. ed. *Herman Charles Bosman As I Knew Him*, with citations from S.A. Opinion and Trek Anthology
Dial Press, Johannesburg, 1971
SADLER, CELIA *Never a Young Man* Extracts from the letters and journals of The Rev. William Shaw
HAUM, Cape Town, 1967
SAMPSON, ANTHONY *Drum*
Collins, London, 1956
SCHOLEFIELD, ALAN *Wild Dog Running*
Mayflower Books Ltd., London, 1972
SCHREINER, OLIVE *Story of an African Farm*, 1833
Chapman & Hall, London, 1890
Hutchinson ed. no date.
——*Thoughts on South Africa*
T. Fisher Unwin Ltd., London, 1923
——*Diamond Fields* a fragment
ed. Richard Rive in *English in Africa*
Vol. 1, No. 1, March, 1974
SEAGRIEF, S. J. *Reading the Signs* Inaugural Lecture,
Rhodes University, delivered 24.5.76
SHAW, GEORGE BERNARD *Pygmalion*, 1914 published 1916
1969 reprint, Penguin edition 1969
SHAW, THE REV. WILLIAM *Diaries 1820–1822, 1826–1829*
Typed transcript lent by Professor David Hammond Tooke, since published A. A. Balkema, Cape Town, 1972
——*The Story of My Mission in South-Eastern Africa*
Hamilton, Adams & Co., London, 1860

SHONE, THOMAS *Diaries 1838–1867*
Typed transcript and MS.
Cory Library
SISSON, J. J. L. *The South African Judicial Dictionary*
Butterworth & Co. (Africa) Ltd., Durban, 1960
SLATER, F. C. ed. *Centenary Book of South African Verse* (1925 re-issued 1946)
Longmans, London, 1946
SMITH, ANDREW *The Diary of Dr. Andrew Smith*, 1835
ed. Percival R. Kirby
Vol. I V.R.S. 20, 1939
Vol. II V.R.S. 21, 1940
Van Riebeeck Society, Cape Town, 1939, 1940
SMITH, C. A. *Common Names of South African Plants*
eds. Phillips, E. Percy and van Hoepen, Estelle
Department of Agricultural and Technical Services, Government Printer, Pretoria, 1966
SMITH, J. L. B. *The Sea Fishes of Southern Africa*
1950 and 1961 ed.
Central News Agency, South Africa
SMITH, PAULINE *The Little Karoo*, 1925
1936 ed. Jonathan Cape, London, 1936
——*The Beadle*, 1926
1929 ed. Jonathan Cape, London, 1929
——*Platkops Children*
Jonathan Cape, London, 1935
SOUTH AFRICAN INSTITUTE OF RACE RELATIONS S.A.I.R.R. Report 1948–49
S.A.I.R.R. Survey, see also Race Relations 1969 S.A.I.R.R.
SOUTH AFRICAN NATIVE RACES COMMITTEE *The Natives of South Africa* Their Economic & Social Condition
John Murray, London, 1901
SPARRMAN, ANDERS *A Voyage to the Cape of Good Hope*
2 Vols. 1785 (translation G. Forster)
2nd ed. G. G. J. and J. Robinson, London, 1786
SPILHAUS, M. WHITING *Under a Bright Sky*
Howard Timmins, Cape Town 1959
——*Doorstep Baby*
Juta & Co., Ltd., Cape Town, Wynberg, Johannesburg 1969
SPOTTISWOODE, HILDA, ed. *South Africa – The Road Ahead*
Howard Timmins, Cape Town, 1960
STANDER, SIEGFRIED *The Horse*
Victor Gollancz, London, 1968
—— *Flight from the Hunter*
Gollancz, London 1977
Statute Law of the Cape of Good Hope Saul Solomon, Cape Town, 1862
STEEDMAN, A. *Wanderings and Adventures in the Interior of Southern Africa*
1835 in two volumes
Longman & Co., London 1835
Facsimile Reprint Struik, Cape Town 1966
STORMBERG, R. Y. *Mrs. Pieter de Bruyn*

Maskew Miller, Cape Town, 1920
STRUTT, DAPHNE *Fashion in South Africa 1652–1900*
A. A. Balkema, Cape Town, 1975
STUBBS, THOMAS *Reminiscences 1820–1877*
M.A. thesis Rhodes University, 1965
SWART, C. P. *Africanderisms:* A Supplement to the Rev. Charles Pettman's Glossary of South African Colloquial Words and Phrases and Other Names.
M.A. thesis University of South Africa, Pretoria, 1934
TAIT, BARBARA CAMPBELL *The Durban Story*
Knox Printing Co., Durban, 1961
TELFORD, A. A. *Yesterday's Dress:* A History of Costume in South Africa
Purnell, Cape Town, 1972
THOMPSON, GEORGE *Travels and Adventures in Southern Africa by George Thompson, Esq.*
Eight years a resident at the Cape, comprising a view of the present state of the Cape Colony with observations on the progress and prospects of the British Emigrants.
2 Vols. Henry Colburn, London, 1827
THOMPSON, KATE *Richard's Way*
George Harrap, London, 1965
THUNBERG, CHARLES PETER, M.D. *Travels in Europe, Africa and Asia Made Between the Years 1770 and 1779*
(translation published 1795)
2 Vols. F. & C. Rivington, London, 1795
TOMLINSON COMMISSION *Summary of the Report of the Commission for the Socio-Economic Development of the Bantu Areas within the Union of South Africa*
Government Printer, Pretoria, 1955
TRAVELLER, THE ed. A. GORDON-BROWN *Journal* in *An Artist's Journey*
Thought to be made by Lieut. Frederick Knyvett between 1832–33 at which time he was at the Cape from India.
A. A. Balkema, Cape Town, 1972
'TWEEDE IN BEVEL' *Piet Kolonel and His Men*
Knox Printing & Publishing Co., Durban, 1944
UYS, PIETER-DIRK *Paradise is Closing Down.*
Theatre One ed. Stephen Gray
Ad Donker, Johannesburg, 1978
VAN ALPHEN, J. G. *Jan Venter, S.A.P.*
Maskew Miller, Cape Town, 1929
VAN DER POST, LAURENS *A Story Like the Wind*
Hogarth Press, London, 1972
VAN DER SPUY, U. *Gardening in Southern Africa* (The Flower Garden) 1953
Juta, Cape Town 1959
VAN HEYNINGEN, C. & BERTHOUD, J. *Uys Krige*
Twayne Publishers Inc., New York, 1966
VAUGHAN, IRIS *The Diary of Iris Vaughan circa* 1902
Central News Agency, South Africa, 1958

Howard Timmins, Cape Town, 1969
——*Last of the Sunlit Years*
Howard Timmins, Cape Town, 1969
VENTER, P. C. *Soweto, Shadow City*
Perskor Publishers, Johannesburg, 1977
WALKER, ERIC *Lord de Villiers and his Times*
Constable, London, 1925
——*A History of South Africa*
Longmans, Green and Co., London, 1928
——ed. *Cambridge History of the British Empire*
Vol. VIII South Africa
Cambridge University Press, 1936
WALKER, OLIVER *Kaffirs are Lively*
Victor Gollancz, London, 1948
——*Wanton City*
Werner Laurie, London, 1949
——*Shapeless Flame*
Northumberland Press, for T. Werner Laurie 1951
WALLACE, R. *Farming Industries of the Cape Colony*
P. S. King & Son, London, 1896
WAR OFFICE PUBLICATION *The Native Tribes of the Transvaal*
H. M. Stationery Office, London, 1905
WEBSTER, W. H. B. *Narrative of a Voyage to the Southern Atlantic Ocean in the Years 1828, 29, 30 Performed in H.M. Sloop Chanticleer*
From the private journal of W. H. B. Webster – surgeon of the sloop, In two volumes
Richard Bentley, Publisher in Ordinary to His Majesty, London, 1834
(Vol. I, pp. 215–335)
WELCH, R. *The Brothers* Unpublished Story, 1978
WEST, M. & MORRIS, J. *Abantu* An Intoduction to the Black People of South Africa
C. Struik, Cape Town & Johannesburg, 1976
WESTWOOD, GWEN *Bright Wilderness*
Mills and Boon, London, 1970
——*Ross of Silver Ridge*
Mills and Boon, London, 1975
WHITE, JOHN MANCHIP *The Land God Made in Anger*
George, Allen and Unwin, London 1969
WHITESIDE, THE REV. J S. and AYLIFF, THE REV. JOHN *History of the Abambo*, generally known as Fingos.
Printed at the 'Gazette', Butterworth, Fingo-land 1912
Facsimile C. Struik, Cape Town, 1962
WHITNEY, PHYLLIS A. *Blue Fire*
David Bruce and Watson, London, 1973
WILLE, GEORGE *Principles of South African Law*
2nd ed. Juta, Cape Town, 1945
——*The Law of Mortgage and Pledge in South Africa*
Juta, Cape Town 1961
WILSON, M. and MAFEJE, A. *Langa*
A study of social groups in an African Township.
Oxford University Press, Cape Town, 1963

OTHER SOURCES

1. Oral Informants
2. Private letters
3. Radio programmes and bulletins
4. Songs by Pip Freedman, Jeremy Taylor, Des Lindberg and A. E. Voss
5. Philip's School Atlas
6. B.P. and Shell Road Maps of South Africa and Place Name Lists
7. Advertising or Information leaflets
8. National Geographic Magazine, December, 1971
9. Matriculation Handbook, 1973
10. Lennon's Dutch Medicines Handbook
11. Telephone Directories and Postal Code
12. Press Digest, 1948
13. Hansard
14. Goldsmith-Carter, George: *Sailing Ships and Sailing Craft.* Paul Hamlyn, London 1969

NEWSPAPERS AND MAGAZINES

African Review
Albany Mercury (E. Cape)
American Speech
Argus, The
Bolt
Eona (Black)
Cape Argus (Cape Town) now *The Argus*
Cape Herald (Cape Town)
Cape Monthly Magazine
Capetonian, the
Citizen, the
Commonwealth – Journal of the Royal Commonwealth Society
Contrast
Darling
Daily Dispatch (East London)
Daily News (Durban)
Drum (Black)
Eastern Province Herald (Port Elizabeth)
English Alive
Evening Post (Port Elizabeth) freq. Saturday ed. known as *Week End Post*, cited throughout as *Evening Post.*
Fair Lady
Farmer's Weekly
Fiat Lux (Indian)
Financial Mail
Forum
Friend, The (Bloemfontein)
George Advertiser (George Town, Cape of Good Hope)
Grahamstown Journal
Graphic, The (Indian)
Grocott's Mail (Grahamstown)
Herald Phoenix
Het Suid Western (S.W. Cape)
Ilanga (African)
Indaba, Supplement to the Eastern Province Herald & the Daily Dispatch
Journal of the Botanical Society of South Africa
Lantern
Leader, The (Indian)
Life
Listener, The
London Magazine

London Sunday Times Magazine
Muslim Digest
National Geographic Magazine
Natal Mercury
New Classic
News/Check
Newsweek
Optima
Pace
Panorama, South African
Paratus
Personality
Post also *East Cape Post*
Pretoria News
Progress
Quarry
Rand Daily Mail (Johannesburg)
Rhodeo (Student Paper, Rhodes University, Grahamstown)
S.A.B.C. Bulletin
S. African Journal of Science
South African Outlook
S.A. Radio and T.V.
Scope
Seventeen
South African Garden and Home
Speak
Staffrider
Star, The (Johannesburg)
Standpunte
Suid Westelike Herald (S.W. Cape)
Sunday Times (Johannesburg)
Sunday Tribune (Natal)
The 1820
The Reader
The World (African)
Time
To the Point
T.V. Times
U.C.T. (Alumni Magazine)
Voice, The

Note: This includes Newspapers and Magazines both current and obsolete.